Evolutionary Pathways and Enigmatic Algae: *Cyanidium caldarium* (Rhodophyta) and Related Cells

Developments in Hydrobiology 91

Series editor

H. J. Dumont

Evolutionary Pathways and Enigmatic Algae: *Cyanidium caldarium* (Rhodophyta) and Related Cells

Edited by

Joseph Seckbach

SPRINGER SCIENCE+BUSINESS MEDIA, B.V.

Library of Congress Cataloging-in-Publication Data

Evolutionary pathways and enigmatic algae : Cyanidium caldarium
(Rhodophyta) and related cells / edited by J. Seckbach.
p. cm. -- (Developments in hydrobiology ; 91)
Includes bibliographical references and index.
ISBN 978-94-010-4381-6 ISBN 978-94-011-0882-9 (eBook)
DOI 10.1007/978-94-011-0882-9
1. Cyanidium caldarium. 2. Algae--Cytology. 3. Algae--Evolution.
I. Seckbach, J. (Joseph) II. Series.
QK569.C956E96 1994
589.4'1--dc20 93-42980

ISBN 978-94-010-4381-6

Printed on acid-free paper

Cover illustration

Electron micrograph of *Cyanidium caldarium*, the predominant staring microalga of this treatise exhibits a dividing cell into two autospores, each cyanidiospore contains a dividing chloroplast (still connected with a Plastid Dividing Ring). We acknowledge Professor T. Kuroiwa (U. of Tokyo) for donating this photo.

Originally published by Kluwer Academic Publishers in 1994
Softcover reprint of the hardcover 1st edition 1994

This book is dedicated to:
Professor *Ikujiro Fukuda* (Science University of Tokyo), a professional colleague, supporter, host and catalyst

Contents

Foreword *by* L. Bogorad ix
Preface *by* J. Seckbach xi

I. Models for the eukaryotic cellular origins and evolutions

Chapter 1. Origin of Eukaryota from cyanobacterium: membrane evolution theory 3
by H. Nakamura
Chapter 2. Glaucocystophyta model for symbiogenous evolution of new eukaryotic species 19
by H.E.A. Schenk
Chapter 3. Alternative pathway (cyanobacteria to eukaryota) 53
by T.E. Jensen

II. Green enigmatic algae

Chapter 4. *Nanochlorum eucaryotum* – a green enigmatic alga: morphology, biochemistry and molecular biology 69
by R.K. Zahn
Chapter 5. Enigmatic chlorophycean algae forming symbiotic associations with ciliates 87
by W. Reisser

III. The paradoxical Cyanidiophyceae

The genus Cyanidium
Chapter 6. The natural history of *Cyanidium* (Geitler 1933: past and present perspectives 99
by J. Seckbach
Chapter 7. A review on the taxonomic position of the algal genus *Cyanidium* Geitler 1933 and its ecological cohorts *Galdieria* Merola in Merola *et al.* 1981 and *Cyanidioschyzon* De Luca, Taddei and Varano 1978 113
by F.D. Ott and J. Seckbach
Chapter 8. Systematic position and phylogenetic status of *Cyanidium* Geitler 1933 133
by J. Seckbach and F.D. Ott
Chapter 9. New classification for the genus *Cyanidium* Geitler 1933 145
by F.D. Ott and J. Seckbach
Chapter 10. *Cyanidium* investigations in Japan 153
by I. Fukuda

Chapter 11. Italy's contribution to the systematics of *Cyanidium caldarium 'sensu lato'* 157
by G. Pinto, P. Albertano and A. Pollio
Chapter 12. The study of Cyanidiophyceae in Russia 167
by O.Y. Sentsova
Chapter 13. *Cyanidium*-like algae from caves 175
by L. Hoffmann

Physiology, biochemical pathways and natural products
Chapter 14. The function of peroxisomes in the Cyanidiaceae 183
by W. Gross
Chapter 15. Nitrogen assimilation in a thermophilic acidophilic alga 193
by C. Rigano, A. Fuggi, V. Di Martino Rigano and V. Vona
Chapter 16. Natural products of the Cyanidiophyceae 201
by H. Nagashima

Fine structures of the Rhodophyta and Cyanidium
Chapter 17. Ultrastructure of unicellular red algae 215
by S.T. Broadwater and J.L. Scott
Chapter 18. Ultrastructure of cytoplasmic organelles in *Cyanidium caldarium* 231
by K. Ueda
Chapter 19. *Cyanidium caldarium* as a model cell for studying division of chloroplasts 239
by T. Kuroiwa, H. Kuroiwa, T. Mita and N. Ohta

Studies on the molecular level
Chapter 20. Molecular organization and expression of the plastid genome of *Galdieria sulphuraria* (= *Cyanidium caldarium*) and other unicellular red algae 255
by K. Zetsche
Chapter 21. The molecular aspects of pigments and photosynthesis in *Cyanidium caldarium* 263
by R.F. Troxler

Bibliographic list of the Cyanidiales and allied enigmatic cells
Chapter 22. List of relevant literature references and index for disciplinary fields 283
by F.D. Ott and A.J. Petrik-Ott

Appendix. Methods for algal investigation 333

General index 337

Foreword

I first met *Cyanidium caldarium* about 35 years ago on the pages of a paper by Mary Belle Allen (Arch. Mikrobiol. 32:207–277, 1959). I knew her as a teaching assistant in the summer microbiology course I had taken at the Hopkins Marine Station taught by C. van Niel. So, in line with the peculiar personal and emotional reactions we have as scientists, I felt a sort of personal connection with *C. caldarium* even before I met it in person. I knew that Mary Belle Allen was a wise and skillful biologist who would discern the truth before sitting down to write about an organism and its habits. Among other virtues, she and Roger Stanier seemed to be the two students who were at that time closest to the great, admirable and very wise van Niel.

C. caldarium seemed to be a wondrous and possibly wonderful organism for certain experiments that I had in mind. At that time I was mostly interested in the biosynthesis of tetrapyrroles, including chlorophylls and the then poorly characterized phycobilins. *C. caldarium* seemed to have plenty of both and besides it required light for their formation! The latter, together with its facultative heterotrophic nature it grows on 1% glucose in the dark, made it a good candidate for mutagenesis to isolate mutants with pigment deficiencies, to study action spectra for pigment production, etc. Superimposed on all this was the exotic nature of its habitat: acid hot springs no less! Not all that many organismus would live there (and therefore the contaminants in the laboratory would be likely to be few). Allen (1959) could grow it in 1N H_2SO_4 in the laboratory. Mary Belle Allen kindly provided a culture and Kenneth E. Nichols, then a graduate student in the laboratory, kindly agreed to work on the organism. I am indebted to *C. caldarium* for its being the source of interesting questions for many years and to the people who have found fascinating answers.

Some of the questions we had about *C. caldarium*, as we got to know the organism, are reflected in the chapters in this volume. Some questions are answered! Many additional questions, that we didn't know enough to ask or even think about then, have found their places in this volume together with answers in many cases. I am indeed pleased to see questions raised, addressed and answered. The study of *C. caldarium* has moved through and left a trail of pockets of active research in the fields of evolution, natural history and now on into hot areas of biochemistry and molecular biology. These days it is thought by some that an organism hardly exists as an acceptable research subject unless one of its genomes has been or is being sequenced in its entirety! Even this may not be totally out of sight for the chloroplast genome of *C. caldarium*.

Joseph Seckbach saw other people in the laboratory working on *C. caldarium* when he was a graduate student. He was appropriately fascinated by its strange growth habits as we all were. When Joseph Seckbach became an exo-biologist the scientific environment at UCLA seemed to be a strange enough one for even *C. caldarium* to fit right in. Thus, I have a second debt to *C. caldarium*. It has kept me in closer touch with Joseph Seckbach over the years then I might have been otherwise. My interests shifted although *C. caldarium* has come back into my scientific life from time-to-time but I knew that I could always find Joe Seckbach near 'our favorite organism'.

Simply by its existance *C. caldarium* has raised many questions about evolution, taxonomy, physiology and molecular biology. Its (sometimes) friendly nature has let it be used by all of us to answer some of these questions. Those of us who have been involved with *C. caldarium* know that others who have not met this alga are missing great opportunities and pleasures. I believe that the authors of all the chapters in this volume hope that the information will be helpful both for those who have been working with the organism and those who have not yet discovered it for themselves.

Harvard University
Cambridge, MA
January 21, 1994

Lawrence Bogorad

Preface

The origins and evolution of the eukaryotic cell and its organelles has become a major point of interest in modern biology. Currently, there are two principal theories concerning the eukaryogenesis: the classical autoendogenous compartmentalization and the more popular and widely accepted the (Serial) Endosymbiotic theory. Two traditional evolutionists present their view on the direct filiation of the bacterial and blue-green algal (prokaryotic) cell into the nucleated (eukaryotic) organism. The symbiotic speculation is stressed in the chapter of the Glaucocystophyta and is also scattered in the articles throughout this treatise.

The idea of organizing a treatise of a special group of algae and about the evolutionary pathways has been born about a dozen of years ago (see cover illustration of Ann. N.Y.Ac.Sc. **361**, 1981). During the recent years the interest in these organisms has increased enormously, hundreds of articles have been published in various journals or books, and a great amount of data has been accumulated and scattered all over the literature (see e.g., the bibliography chapter by Ott and Ott, in this book). However, there has not been a central reference source to present the background and present day picture of these interesting algae.

This treatise presents most of our current state of the art and present day knowledge of these organisms in one volume. It deals with enigmatic algae which indicate unusual features. They exhibit simple level of cellular organization but yet may serve as models for evolutionary pathways and are distinctive in their biochemical and molecular properties. The algal members assigned in this volume are special unicellular aquatic or hydrophilic plants which are important as primary food producers in the food chain, and as oxygen generators for their (cellular) environments. Their global distribution is ubiquitous, however, they are endemic to their characteristic habitats.

The leading algae presented in this text are acido thermophilic organisms, i.e., *Cyanidium caldarium* (Red alga) and its related algae within the Cyanidiaceae. The other unicellular algae staring in this volume, are some glaucophytes (e.g., *Cyanophora paradoxa*). They contain cyanobacterium-like **'cyanoplasts'** or cyanelles which function as photosynthetic organelles. Therefore, these distinctive algae have been recognized as the classical transitional protistan bridging the blue-green and the eukaryotic algae (see also extended chapter of Schenk in this volume). The green enigmatic micro-algae displayed in this manual are the newly discovered *Nanochlorum eucaryotum* as well as microscopic unicellular chlorophytes (*Chlorella* sp.) residing within other protistan single celled animals (ciliates) and surviving in symbiotic relationships with their host cells.

The recent usages of high power tools of molecular biology (for sequencing macromolecules) to detect phylogenetic relationships among various organisms) is also presented with these fascinated algae. In this composition several experts gathered to expose their studies and reviewed their long time investigations with these puzzling algae or on evolutionary speculations. Therefore this book can be considered an authoritative text. It is our hope and expectation that this book will undoubtedly stimulate more interest in the use of these algae as experimental organisms. Indeed these published reference work should be a lasting definitive scientific contribution to our knowledge of *C. caldarium* and its enigmatic allied.

This collection of enigmatic algal essays involved multi-disciplinary studies. We discuss and present fields of ultrastructure, cytology, physiology and biochemistry, molecular sequencing, as well as taxonomic and a new proposed classification of these organisms. It is our expectation that this treatise will assist graduate students, researchers or lecturers of natural and life sciences in their investigations.

Our present guidebook may well circulate in harmony with the recently new books dealing with *Algae and Symbioses* (1992; ed. Reisser, W. Biopress, Bristol, England), *Symbiosis in Cell Evolution* (1981, Margulis, L. Freeman & Comp. San Francisco), *Five Kingdoms* (1988, Margulis, L. and Schwartz, K.V. Freeman and Comp. N.Y.) or the proceedings of the *Endocytobiology* Colloquia (**I** and **II** eds. Schwemmler W. and Schenk, H., 1980 and by Schenk, H. and Schwemmler, W. 1983, both published by W. de Gruyter, Berlin; **III**, eds. Lee, J.J. and Fredrick, J.F. see **New York Acad. Sci. 503**, 1987; **IV**, eds. Nardon, P. *et al.* 1990, INRA, Paris; **V**, eds. Sato *et al.* 1993. Tübingen U. Press, Tübingen).

Finally, I would like to thank all people who carried with me the investigations of these fascinating cells and their evolutionary hypotheses. Special appreciations are due to Prof. Ikujiro Fukuda (Sci. U. of Tokyo) for initial encouragement and support, and to our contributors for their chapters and patience.

Hebrew University of Jerusalem
Mount Scopus, Jerusalem, Israel
April 15, 1994

Joseph Seckbach

Note: We use the 'short style' of references because the full bibliographic list of Ott and Ott (in this volume) contains a detailed information of most citations.

PART I

Models for the eukaryotic cellular origins and evolutions

Chapter 1

Origin of eukaryota from Cyanobacterium: membrane evolution theory

HAKOBU NAKAMURA
Biological institute, Faculty of Science, Konan University, Kobe 658, Japan

Outlines:

1. Introduction
2. Metabolic evolution
3. Membranous evolution
4. Symbiotic theory on the origin of eukaryotic cell
   4.1 Rebuttal against endosymbiosis
      4.1.1 Organellar morphology
      4.1.2 Internal genomic evolution
      4.1.3 DNA sequence, organellar products and plasmids
      4.1.4 The origin of introns
      4.1.5 Gene sequence of bacteria and organelles
      4.1.6 Ribosomal size and protein content
      4.1.7 Genomic content and intercellular genetic exchange
      4.1.8 Antibiotics in inhibition of protein synthesis
      4.1.9 tRNA species in mitochondria and bacteria
      4.1.10 The status of cyanelles, cyanobacteria and chloroplasts
      4.1.11 Origin and physiology of chloroplasts and mitochondria
5. Membrane evolution theory
6. Conclusion
7. References

1. Introduction

It has been currently estimated that the life generated about four billion years ago in the primitive sea as a result of the chemical evolution. There is no doubt that, at that time, the protoorganism was a prokaryotic monad, formed from a certain aggregate of primitive proteins, nucleic acids, and other macromolecules and wrapped with (phospho)lipid membrane. Some workers have proposed that the macromolecules involved in the formation of the life were adsorbed on surface of clay particles in the primeval soup (Cairns-Smith 1982). However, it seems very unlikely that such a simple aggregation of the macromolecules would lead to the formation of a vital cell. The reasons for such uncertainty which have been discussed (see Nakamura 1987c) are: (1) The extant cells, without exception, are wrapped by a single membrane consisting of phospholipid bilayer. (2) Life is expressed only in a closed, rather than open, system isolated from the environment by the plasma membrane. A closed system can make a life mechanism independent from the environment and thus further an unique evolution of the mechanism. Structural independence is one of the most important factors for life. Yet, the cell-free system that has usually been used by biochemists is without the self-government from the environment and thus has no life. However, it is of course clear that the vital cell can not be completely independent from its environment, and that exchanges of substances, energy, and others are necessary between them. (3) Phospholipid membrane can create an uneven distribution of ions, especially protons, between its two sides, namely a proton gradient. The latter generates a membrane potential which supplies energy for synthesis of ATP. All of the present-day living cells use such a mechanism in phosphorylation of ADP. Lipid membrane is one of the most important structure for persistence of the cell life. (4) Lipid membrane can give a selective permeability by which certain substances involved are more permeable than others. When proteins and ionophores were integrated into the lipid membrane, they made channels and worked as specific carriers of the organic substances and metallic ions. Active transport systems in the extant plasma membrane can transport particular substances under consumption of ATP even contrary to the concentration on one side. Direction of the transport is specific for the membrane protein, permease. (5). Plasma membranes, particularly of the prokaryotic cells, contain a number of enzymes of cellular metabolism, and thus their protein contents are much higher than those of the eukaryotic

J. Seckbach (ed.), Evolutionary Pathways and Enigmatic Algae: Cyanidium caldarium *(Rhodophyta) and Related Cells,* 3–18, 1994.

cells, as discussed later. The plasma membrane and its derivatives of the extant prokaryotes have been demonstrated to contain enzymes (and proteins) responsible for macromolecular synthesis, photosynthesis, respiration, permeability, and other functions. It may be not too much to say that the prokaryotic plasma membrane contains the most significant machineries for the life. (6) Lipid membrane plays a role as a kind of the shelf for the arrangement of macromolecules. DNA content of cell has increased with cellular evolution since the origin of life. The increase of nucleotide sequence in the genome does not necessarily mean a parallel accumulation of the genetic information because of the addition of untranslated sequences, such as introns, spacers, etc. However, there is a known fact that the more protein types involved, the higher the cellular evolution. On the other hand, size of the cell is not parallel to its DNA content. It seems, therefore that the cell has increased the membrane area with its evolution of life mechanism. In other words, the intracellular membrane system has a role in the development of the arrangement of enzymes newly added. (7) As suggested above, the intracellular membrane plays a role in isolating metabolism according to its specificity. The isolation of metabolism also is very useful to guarantee its high velocity and specificity.

On the basis of the cell fossils with large size of eukaryotic level, it is assumed that eukaryotic cells originated about 1.5 billion years ago. In addition to the plasma membrane, the eukaryotic cells developed various membrane systems absent in the prokaryotic cells. Nucleus, mitochondrion, chloroplast (in plant), endoplasmic reticulum, Golgi-apparatus and other organelles are constructed of the lipid membranes, in which the specific enzymes and proteins are positioned. The former three organelles contain DNA and are usually enveloped by double membranes, an inner and an outer layer (Nakamura 1987c). It should be pointed out here that the exclusive era of the prokaryotes (about 2.5 billion years) elapsed much longer ago than that of the eukaryotes.

According to the current knowledge of molecular evolution, mutations have been accumulated in the genome at nearly steady state, and the rate was dependent on function of the gene. The latter fact means that, although accumulation of the mutations might be equivalent in every gene, the physiological significance of each gene in the organism determines the rate of evolution, the more important the gene, the greater its stability. It is probable in the biological world that DNA content of a cell has increased with its evolution. For example, the cell of human contains DNA about a thousand times (as haploid) more than that of *E. coli*, and such an increment in DNA content appears to be caused by repeated duplications of the nucleotide sequence. Although the elongation of DNA sequence does not necessarily mean an increase of genes to be translated, as mentioned above, the quantitative and then qualitative evolution of DNA brought about the increase of enzyme species in the cell. In time the informative evolution did push forward elongation and the spread of branches of the metabolism (Nakamura 1987a).

It is said that the primitive atmosphere of the earth did not contain oxygen gas and thus the organisms at that period evolved metabolism which could anaerobically produce ATP. The power metabolism in absence of oxygen is designated as fermentation. The fermentative organisms were followed by the appearance of photo- and chemo-synthetic organisms. It is considered that heterotrophs, such as the fermentatives, were bred too long in the primitive soup leading to the exhaustion of their nutrients. Thus new types of the autotrophic mutants were generated. The photosynthetic organisms anaerobically fixed CO_2 (with light as energy source) and H_2S as reducing power source. Similarly, the chemosynthetic organisms also anaerobically fixed CO_2 to synthesize sugar, using energy and reducing power produced from inorganic chemical reactions. Then, photosynthesis evolved into using H_2O as reductive power source instead of H_2S and, as result, emitted O_2 gas into the atmosphere. The latter rapidly became oxidative with an ecological development of the O_2-emitted photosynthesizers and was compelled to select new types of energy metabolism, O_2-respiration and other oxidative reactions. That is, the O_2 molecule started to be used as an acceptor of electrons, which were produced from many reactions within the fermentation, though the electron acceptors in the metabolism were organic compounds. This was a natural result because much more energy was produced when O_2 molecule accepted the electron (Nakamura 1987a).

As it will be discussed later, the eukaryotic cells contain various paths of metabolism for anabolism and catabolism. However, it should be stressed here that these metabolisms were primarily accomplished in the first 2.5 billion years of the prokaryotic age (Nakamura 1987c).

The present communication deals with evolutional process in the genesis of eukaryotic cells. At present, there are two hypotheses on this, (endosymbiotic theory and membrane evolution theory). The former has been re-emphasized by Margulis in 1970 and 1981 and the latter discussed by Nakamura in 1975. We will explain and examine the contents of these theories in detail in the following sections. Please note that as this communication reviews evidence reported by many investigators, the experimental methods will not be described here in detail. For further elucidation on specific works, the readers should contact the author. Cf. the chapter by Jensen (on the direct filiation) and article by Schenk (endosymbiosis) in this volume.

2. Metabolic evolution

We will define the term 'metabolism' as a series of biochemical reactions, by which an end product is produced. It is natural to consider that a proto-cell which was generated through the chemical evolution in the primeval soup had no metabolism, although some metals, low molecular organic compounds, and primitive peptides had to possess weak ability as catalyst for natural reactions (Nakamura 1982).

Metabolism in the extant cells are long chained and branched. It is an interesting problem to ask how metabolism became elongated and branched. Horowitz (1945) proposed that the organisms acquired a new reaction step by step through mutations in order to obtain its product as nutrient. According to this theory, a final reaction in the present-day metabolic stream results from the starting one.

In other words, the metabolism has evolved in the reverse direction of the present flow. This theory also is reasonable in the consideration of molecular genetics, since it says that a gene responsible for synthesis of an enzyme duplicated, and then new type of the gene generated through repeated naturally occurring mutations and that natural selections selected for the best gene (Nakamura 1982).

Fermentation would be the first energy-acquiring metabolism in the proto-cells, as explained above. The present fermentative organisms produce various alcohols and organic acids as the end products. The metabolism can be divided into two parts: from glucose to pyruvate and from pyruvate to the end products via various branches. Today, several pathways of the former are known; (i) Embden Meyerhof (EM) pathway, in which glucose reaches pyruvate (Py) via glyceraldehyde 3-phosphate (GA3-P), (ii) Entner-Doudroff (ED) pathway, in which glucose goes to GA3-P via 6-phosphogluconate and then 2-keto 3-deoxy 6-phosphogluconate, (iii) phosphoketolase (PK) pathway, in which glucose arrives at GA3-P via 6-phosphogluconate, ribulose 5-phosphate and then xylulose 5-phosphate, (iv) pentose phosphate cycle, in which glucose is metabolized through many passes for production of pentoses for RNA and DNA and the GA3-P intermediate, and (v) others (de Witt 1977; Nakamura 1983, 1987a). In the biological world, EM pathway and pentose phosphate cycle are most widely distributed, and ED and PK pathways are used by restricted species of micro-organisms of prokaryote and eukaryote. It should be stressed that the pathway from GA3-P to Py is distributed in all kinds of the fermentative metabolism and that the pathway reasonably taken to be older than pentose phosphate cycle because the former is simpler than the latter (Battley 1987; Broda 1978; Holland *et al.* 1987). So, we will refer to the pathway from GA3-P to Py as primitive fermentation system. If Horowitz's theory is applied to the primitive fermentation system, then we can reach the interesting conclusion that the most primitive reaction for energy production is a change of phosphoenolpyruvate to pyruvate with a production of one mole ATP, which is catalyzed by enzyme pyruvatekinase (Nakamura 1982). In succession, after the fermentation emerged autotrophic metabolism, photosynthesis and chemosynthesis. Perhaps, this evolution occurred as a result of environmental shift which followed the organic nutrients termination from the primitive sea. Several types of photosynthetic prokaryotes have currently been classified; (i) green sulfur bacteria, (ii) purple sulfur bacteria, (iii) purple non-sulfur bacteria, (iv) *Erythrobacter*, and (v) cyanobacteria. Halophilic bacterium *Halobacterium* can synthesize ATP by absorbing light energy but has no ability to fix CO2 (Nakamura 1987b).

Green- and purple-sulfur bacteria are anaerobes and photosynthesize under an anaerobic condition alone. Their photosynthetic metabolism can divide into two phases, namely light reaction (photosystem) and dark reaction (CO_2 fixation). The purpose of the light reaction is to synthesize ATP and NADH which are energy and reducing power sources, respectively, for CO_2 fixation. The dark reaction is also called reduced pentose-phosphate cycle, in which CO_2 is incorporated into ribulose 5-phosphate finally to form glucose. In this dark reaction, six incorporated moles of CO_2 can form one mole of glucose. The reduced pentose-phosphate cycle is a back current of the pentose-phosphate cycle (de Witt 1977; Nakamura 1986, 1987a). This is very important fact in the metabolic evolution. Another example is the dark reaction in some photosynthetic bacteria. This particular reaction is called 'reduced carboxylic acid cycle' and is a back current of TCA cycle, in which four incorporated moles of CO_2 can form one mole of oxaloacetate (Nakamura 1986, 1987a).

Therefore, it is possible to say that the evolutionary origin of photosynthesis is a formation of the light reaction, rather than the dark reaction. The light reaction of primitive photosynthesis consists of two separated photosystems. These systems are, cyclic and non-cyclic electron transport systems for syntheses of ATP and NADH, respectively. In the former, pigment system contains bacteriochlophyll a that has a maximal wave-length absorbance at 980 nm, called P980, and its light-excited electrons are transferred along a series of several carriers including ubiquinone (Q), cytochrome b (b), and cytochrome c (c). Energy for ATP synthesis is supplied by big drops in the electron flow. Finally, the energy-lost electron returns to P980, and the circulation of electron is characteristic of this photosystem. In the latter system, on the other hand, light-excited electron in P980 is transferred among carriers consisting of ferredoxin, flavoprotein, and other electron carriers and, finally, the electron is delivered to NAD to form NADH. In this system, electrons do not circulate thus must always be supplied from some outsiders, H_2S and others, although an electron transfer system (H_2S)$\rightarrow$ Q ,$\rightarrow$ b $\rightarrow$ c $\rightarrow$ P980 (Nakamura 1987a, c).

Purple non-sulfur bacteria are evolutionarily interesting because they carry all histories of energy metabolisms, fermentation, anaerobic photosynthesis and O_2-respiration. Further, these metabolisms are able to be used adaptively in response to changes of the environmental conditions.

Cyanobacteria have acquired rather progressed photosynthesis as compared to purple- and green-sulfur bacteria as follows. (1) Cyanobacteria use H_2O as an electron source to reduce CO_2. As the amounts of H_2O and CO_2 were inexhaustible on the earth, the bacteria were able to distribute widely. In fact, we can now find their fossil records everywhere. (2) In cyanobacteria, photosystems to produce ATP and NADPH (not NADH as the old type in purple- and green-sulfur bacteria) are connected in a single chain. This photosystem has been demonstrated to be much higher in energy efficiency than the old type (de Witt 1977). These evolutions in the photosystem became a basis for subsequent evolution of the plant world because the cyanobacterial type of photosynthesis was inherited by higher plants with little modification. However, the emergence of a H_2O-reducing photosystem brought about a serious effect on the primitive ocean and atmosphere. This is because O_2 gas was poisonous to anaerobic organisms which had occupied the earth's surface. Therefore, the organisms including cyanobacteria at that time would have been compelled to adapt to the oxidative circumstance. It is a fact that O_2-sensitive cyanobacteria are living even now. The organisms have acquired a system for O_2-detoxication and all of the extant aerobic and O_2-tolerant organisms possess an enzyme superoxide dismutase. Superoxide (O_2^-), which is produced by one-electron reduction of O_2 molecule, is a powerful oxidant to cellular substances, and superoxide dismutase catalysis reaction $2O_2^- + 2H^+ \rightarrow H_2O_2 + O_2$. This H_2O_2, although also toxic, is decomposed to H_2O and O_2 by enzymes catalase and peroxidase. For further data on peroxisomal activity, see chapter by Gross, q.v. However, it is presumed that the early organisms were very adaptive and used O_2 gas as a final electron acceptor in energy production, that became depended upon O_2-respiration and aerobic chemosyntheses, since the O_2 molecule is at the lowest position of redox-potential among cellular substances (Nakamura 1987a).

The extant metabolism for O_2-respiration can separate into three parts, namely EM pathway, TCA cycle, and O_2-respiratory chain, each of which is different in the evolutional origin. EM pathway derived from fermentation. This is important fact in showing that the evolution of metabolism is an accumulation and then recombination of the previous system which was partially modified, rather than de novo. TCA cycle also was of a fermentative origin and derived by a combination of glutamate and propionate fermentations (Holland *et al.* 1987). The glutamate fermentation consists of reactions from EM pathway to glutamate via pyruvate, citrate, isocitrate, and α-ketoglutarate. Right half of TCA cycle is the pathway of pyruvate to α-ketoglutarate. On the other hand, the propionate fermentation consists of reactions from EM pathway to propionate via oxaloacetate, malate, fumarate, succinate, and succinyl-CoA. Left half of TCA cycle is the pathway of pyruvate to succinyl-CoA, although it is back flow (Tandlau de Marsac and Houmard, 1987; Carrn and Whitton, 1982). There is an incomplete TCA cycle in the biological world, called glyoxalate cycle, in which the chain of reactions is unconnected between α-ketoglutarate and succinyl-CoA, and thus some by-paths are made via glyoxalate. The incompleteness of TCA cycle is due to a deficiency of α-ketoglutarate dehydrogenase. Further on this issue in chapter by Gross, this volume.

Glyoxalate cycle is distributed in many species of prokaryotes including anaerobic bacteria and even aerobic cyanobacteria. Further, anaerobic cultivation of the aerobes also leads to formation of glyoxalate cycle. However, there is an evolutional fact that aerobic photosynthetic bacteria including purple non-sulfur bacterium *Rhodospirillum rubrum* and cyanobacterium *Chlorogloea fritshii* have a complete TCA cycle. Interestingly, even propionate-fermentative bacterium (anaerobic) *Propionibacterium acidi-propionici* has the complete one (Bryant 1980; Douce 1985). Therefore, TCA cycle has been completed by the evolutionary acquisition of α-ketoglutarate dehydrogenase which is a complex assembling a few subunit enzymes as in pyruvate dehydrogenase.

The respiratory chain is an electron-transfer system which makes the final step of the long O_2-respiratory mechanism to produce most ATPs. It is a series of electron-carriers including the set of Q → b → c, in which the energy-lost electrons are finally accepted by O_2 to become H_2O. It is strongly suggested from the electron-carrier arrangement that the respiratory chain has derived from an evolved photosystem (Nakamura 1987b). In fact, a high homologies in the amino acid sequences of cytochromes b and f (c) has been demonstrated between the photosystem of chloroplast and the respiratory chain of mitochondrion (Nakamura 1987c). Therefore, there is little doubt that the O_2-respiratory metabolism was evolutionarily constructed from the pathways of some parts in fermentation and photosynthesis which had previously been formed in the older prokaryotes. Again, we can see the biological economy is demonstrated in that evolution has made up the highest mechanism by spending the lowest energy. There is the additional interesting fact that purple non-sulfur bacteria commonly use the electron-transfer system Q → b → c for photosynthesis and respiration (Nakamura 1983, 1987b). Therefore, if the photosynthetic metabolism is blocked genetically and conditionally, the cells can live using the respiratory line. This problem again will be taken up later.

Another type of the autotroph is a chemotroph which uses energy from inorganic reaction for pro duction of ATP and NAD(P)H for CO_2 assimilation. For example, sulfate-reducing bacteria, one group of the anaerobic chemotrophs, use sulfate as electron acceptor and H_2 (and others) as electron donor and produce ATP and NADH through this redox-reaction. It should be pointed out here that the electron transfer system of the chemosynthesis is simpler than that of the photosynthesis. Therefore, it may be true that the latter evolved from the former, although some chemotrophs possess an arrangement of electron carriers similar to the photosystem (Nakamura 1987a,c).

We can summarize various lines of the metabolic evolution as in Figure 1 and can reach a conclusion that the cellular metabolic ways are of phylogenic products.

3. Membranous evolution

We can assume that membranous system of the proto-cells were plasma membrane only as in the

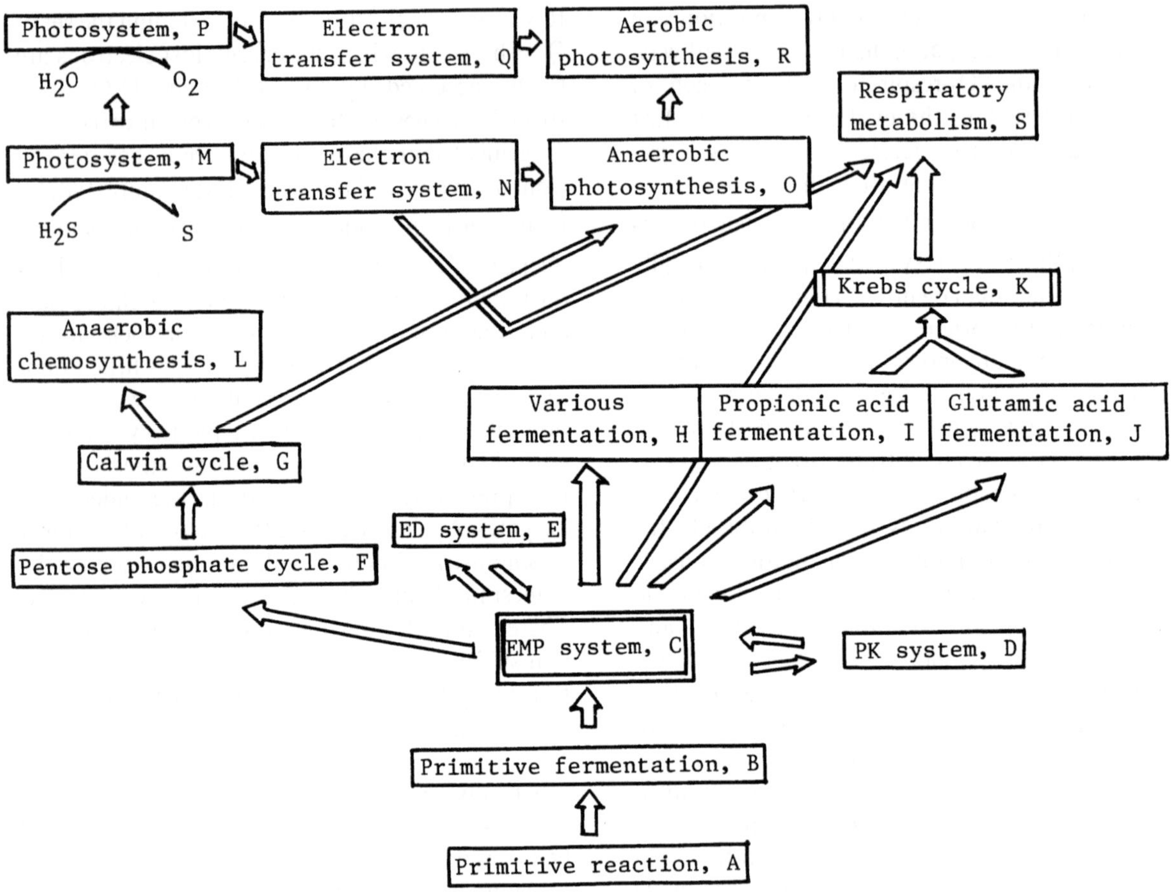

Fig. 1. Phylogenicity of the energy flow in metabolic pathways. Origin of each of the metabolic paths is ordered as fermentation, photosynthesis and then O_2-respiration. Chemosynthesis is deduced to be generated in the time similar with or earlier than the photosynthesis. The metabolisms have evolved through acquisition of new reactions and their recombination.

present-day Mycoplasma. Mycoplasma cell is in the smallest organism in size (125–250 nm in diameter) in the biological world and its DNA content is much lower on the average (4.4–12 x 10^8 in molecular weight) than that of bacterium (8–39 x 10^8). It is believed that the smallest cell unit of Mycoplasma is critical in its size to keep the life apparatus and the DNA amount (containing the lowest information) for survival. However, we have no evidence to show that they are the most primitive organisms living at present. The plasma membrane of Mycoplasma shows a typical unit membrane structure but its molecular species of phospholipids is quite simple (Nakamura 1975).

Archaebacterial cells are phylogenetically unique as compared to prokaryota and eukaryota (Woese *et al.* 1966) and their plasma membrane also is uniquely constructed by a monolayer of particular lipids.

In Gram-positive bacterium *Bacillus subtilis*, we can observe a membrane complex, called mesosome, which is derived from the plasma membrane through its elongation and folding. The mesosomal membrane shows an oxidation-reduction of 2,3,5-triphenyltetrazorium and trinitrobluetetrazorium which are redox indicators. Therefore, some investigators refer to this organelle as chondrid and believe it to be protomitochondrion (Nakamura 1987c, 1988). On the other hand, mesosome have been observed to connect with the nucleoid of the bacterial cell and hypothesized that the organelle contains machinery for DNA replication (replicon theory, Jacob *et al.* 1963).

Generally, *Escherichia coli* does not contain an intracellular membrane system except for the plasma membrane. However, we have found that, when the *E. coli* cells were treated with acriflavine (a kind of basic dyes) this drug can effectively eliminate plasmid from the cells. In an alkaline broth (pH 8.1), a formation of folded membrane system is induced, essentially at end(s) of the rod cell. We have isolated an acriflavine-sensitive (*acrA*) mutant from the wild type (acrA$^+$) strain of *E. coli*. The *acrA* gene is located at min 10.6, just near gene *dnaZ* for DNA synthesis, on the chromosome (Nakamura 1965). Plasmid becomes quite unstable in the mutant cell and is eliminated in the presence of membrane-attachable agents in addition to acriflavine and by heating at 42 °C. On the other hand, the plasmid is stable in the wild type cell under these conditions (Nakamura 1975; Nakamura *et al.* 1975). Further, induction of the membrane complex by acriflavine occurs more easily and more complexly in the *acrA* mutant cell than in the *acrA*$^+$ cell. Biochemical analysis showed that acriflavine permits an increased synthesis of phospholipids, particularly phosphatidyl ethanolaminc and cardiolipin. From these results, we can conclude that the formation of intracellular membrane is induced according to the physiological conditions (Nakamura 1983, 1988). In fact, as shown in the following, the occurrence of such membranous organelles has been known to be conditional in the prokaryotic and eukaryotic worlds.

The cell of purple non-sulfur bacterium *R. rubrum* is filled with a particular membranous system for photosynthesis, called chromatophore, and is deeply purple colored when photosynthesis is active, whereas the system disappears and the cell is non-colored when photosynthesis is not carried out. Our laboratory has found that when the wild type strain of *R. rubrum* was treated with drug acriflavine in the alkaline broth medium for 3 to 5 h at 30 °C, non-photosynthetic mutation is induced at a frequency of 10^{-1} to 10^{-5} orders, depending on condition applied (Nakamura and Yoshikawa 1986). It should be emphasized that the starting wild-type strain of *R. rubrum* used is genetically marked with a character of streptomycin resistance to avoid or examine the contaminations. Therefore, of course, the white mutant obtained also is streptomycin resistant. Kuhl *et al.* (1983) have isolated the same type mutant of *R. rubrum* by treatment with a mutagen, ethylmethane sulfonate. We demonstrated by molecular analyses that the wild type cell contains rather large circular plasmid of ca. 50 kbp.

Acetone and methanol (7:2) extraction of pigment from the wild type cell showed clear absorbance for carotenoid at 495 and 525 nm and for bacteriochlorophyll at 775 nm, but that of the white mutant cell had no absorbance for both the pigments. Further, electron micrographs of the wild type cell which was cultured under somewhat anaerobic condition showed that the chromatophore are loaded to full, whereas there is no or little chromatophore in the white mutant cell. Therefore, apparently, the plasmid contains genes which are responsible for syntheses of photosynthetic pigments and for formation of chromatophores in the cell.

On the other hand, we demonstrated that when the white mutant cell was transformed with the plasmid DNA, the wild type pigments and the membranous organelles could be regenerated. It is very interesting that the genes responsible for photosynthesis and morphogenesis in the photosynthetic bacterium are located on plasmid DNA. Currently, it is demonstrated that bacterial DNA, both of the main chromosome and plasmid, bind to its plasma membrane and/or membranous organelle, such as mesosome (Nakamura 1975, 1978; Kornberg 1980), and the replicon theory of Jacob (1963) has supposed that replication and division to both poles of DNA are controlled by a machinery which is integrated in the membrane system. Recently, our laboratory has found that, when the chromatophores of the wild type *R. rubrum* cultured under the photosynthetic conditions were isolated and purified by repeated density gradient centrifugations of osmotically disrupted spheroplasts, the organelles bound small DNA fragments. The following analyses by restriction enzymes and Southern hybridizations showed that the DNA fragments are 50 kb plasmid but little contaminated with the chromosome. Further, chromatophore fraction could be directly mixed and digested with restriction enzyme *Eco*R1 and, after washing well, hybridized with specific segments of the plasmid DNA as probe. Result showed that the chromatophore-binding DNA was a specific region of the plasmid (Nakamura 1990, 1993). Similar examples have been obtained to show that some specific genes may be involved in the DNA-membrane connection (e.g., in *B. subtilis* see Yoshikawa 1965) and each of chromosomes in the eukaryotic cell may be attached to specific region of its nuclear membrane (e.g., Berns 1983).

The combination of DNA and membrane also is well known about mitochondrion and chloroplast. We are accumulating evidence to show that the *acrA* mutation of *E. coli* cell leads to the defect of a specific protein (MW 60,000) in the plasma membrane and to instability of plasmids, such as sex (F) and resistance (R) factors (Nakamura 1974, 1976). Further, the *acrA* mutation decreases frequency of the genetic recombination to a few percent of the normal cross (Nakamura *et al.* 1975). It is reasonable from these results to conclude that the *acrA* gene product in the membrane is responsible for stability and gene recombination of the plasmid and the exogenote, respectively. We hope to analyze the membranous mechanisms of the binding and replication of DNA.

We are now studying cyanobacteria, which evolved from a line of the photosynthetic bacteria and thus acquired a new type of the energy metabolism and which, interestingly, possess various sizes of plasmids (Tandeau de Marsac and Haumard 1987).

Generally, eukaryotic cells develop various intracellular membranes and thus metabolic compartmentalization. The unicellular acidothermophilic algal class Cyanidiophyceae is of intriguing interest because it has been considered to be at a lowest evolutionary level among the eukaryota. Seckbach (1987) and Fredrick (1987) have hypothesized that these algae belong to Pre-Rhodophyta and bridge phylogenetically between cyanobacteria and Rhodophyta. These Cyanidian species are in a phylogenetic transitional position and link between cyanobacteria and eukaryotic algae as proposed on the basis of: cell morphology, manner of reproduction and some biochemical components such as storage glucans. The primitive species, for example *Cyanidioschyzon merolae* and *Cyanidium caldarium*, have almost no 'typical' membranous system other than the plasma membrane and the DNA organelles of nucleus, chloroplast and mitochondrion. Further, their chloroplasts seem to be encircled with a single membrane. For further information on the cyanidiophytes see Ueda, Kuroiwa *et al.* and Seckbach, in this volume.

Fukuda and Nagashima (1987) have determined the DNA contents of *Cyanidium* cells [using a method of 4,6-diamidino 2-phenylindol (DAPI) staining and fluorescence microscopy including the video intensified microscopy system] and have showed that their strain of *C. caldarium* is much lower in DNA amount per cell than cyanobacterium *Anabaena variabilis*.

Eukaryotes developed in the course of cellular evolution various non-DNA organelles, such as endoplasmic reticulum, Golgi-apparatus, dictyosome, peroxisome, lysosome and other intracellular structures. These membrane systems have become a completed network with a rise in the evolutionary level.

The number of mitochondria within a cell can dynamically change depending on physiological conditions, especially on phases of the cell cycle. In a certain phase, all of the mitochondria are inter connected to form a macro-complex. It makes polymorphic branches and resembles root system of a tree as revealed by the electron- microscopic figures from re-composed serial thin sections of a single cell. On the other hand, when the macro-complex of mitochondria was stained with DAPI, the DNA fiber contained therein appeared in a manner similar to a continuous light track of a firefly. However, in other phases of the cell cycle, the complex is divided into many units of individual mitochondria as seen in regular electron-micrographs of the eukaryotic cell. These mitochondrial figures can be most clearly observed in yeast cells. Although earlier investigators have emphasized that each mitochondrion is considered as a bacterial symbiont, this is an incorrect idea and the organelle can behave dynamically as the other cellular component.

Similar dynamic figures also have been observed in chloroplasts of *Euglena* (Ehara *et al.* 1985). The composed electronmicrographs of the serial thin sections of a cell show a polymorphic macro-complex of chloroplasts in a certain phase of the cell cycle which surrounds the nucleus. Early observers considered, on the basis of chloroplast ultrastructural morphology that the chloroplast is a cyanobacterial symbiont which is similar to mitochondrion previously invaded the host cell, but this is also an inaccurate concept. In mitochondrion and chloroplast, the electron transfer systems for respiratory and photosynthetic metabolism (resp.) are localized in the inner membranes and the carriers are arranged along with the reaction flows. The internal mitochondrial membranes are called cristae, while the chloroplast intra-membranes constitute the thylakoids. Both types are working in the same manner to produce energy for syntheses of ATP and NAD(P)H as observed in the purple nonsulfur bacteria and cyanobacteria (Carr and Whitton 1982). Development of these membrane systems is harmonized with the physiological requirement of the cell. For example, yeast cells which were cultured anaerobically, completely lose the mitochondrial structures and even their outer membranes are broken. However, when the cells were aerated, those structures are gradually constructed according to the activities of the respiratory enzymes. Chloroplasts also lose their interior structures in dark condition and convert into very small proplastids which possess little or no inner membranes. However, if the plant is illuminated, the proplastids become enlarged and the inner membranes are elongated to make lamellae or grana. We have to point out that the chloroplast thylakoids are singly arranged in the lower plants but are joined into two, three, or more bands to compose grana structures as occurred with the evolutionary development of the plant world. Therefore, the respiratory and photosynthetic organelles can dynamically change their sizes from 1–3 to 5 μm and larger, and their inner membranes go from absent or lower degree of membranes to complex systems, depending on evolutional, physiological, environmental and differential parameters. We must stress here that the organelle's morphologies are never static but dynamic as in metabolic activities. This is the same chain of development also with endoplasmic reticulum, Golgi-apparatus, and other membranous organelles.

There are interesting membrane movements specific for eukaryotic cells: endo- and exo-cytoses (Nakamura 1987c). The former is a phenomenon whereby plasma membrane invaginates to ingest a drop of liquid (called pinocytosis) or solid particle (called phagocytosis) by wrapping it with the plasma membrane and digesting the incorporated nutrient in the cytoplasm. Phagocytosis is a more complex movement than pinocytosis, and the vesicle (phagosome) taken in is conjugated with lysosome, which contains various kinds of hydrolytic enzymes, to digest the phagosomic content.

Exocytosis is a reverse phenomenon of the endocytosis, and the cytoplasmic vesicle is transferred to the plasma membrane and then eventually conjugated to excrete its content outside of the cell. Such membrane movement is actively and widely carried out in secretory cells to secrete hormones and exoenzymes, and in the excretory cells to excrete some stool and other wastes. Here is an interesting problem in the exocytosis; when the vesicles are continuously conjugated with the plasma membrane, size of the membrane, i.e., the cell, must be enlarged. However, in fact, the cell size does not steadily increase. This means that the exocytic membrane is degradated in harmony with the enlargement, and thus there is a dynamic turnover between the plasma membrane and, perhaps, Golgi-apparatus.

4. Symbiotic theory on the origin of eukaryotic cell

The idea of symbiotic origin of organelles, e.g., mitochondrion and chloroplast (in plant) within the extant eukaryotic cell was already presented toward the end of the 19th Century. It was derived from microscopic observation that shapes of the organelles are analogous to rod bacterium and unicellular cyanobacterium, respectively. Margulis (1970, 1981) re-emphasized this idea and also contended that flagellum also is a symbiotic derivative of spiral bacterium, spirochaeta. However, she has not scrutinized the origin of nucleus, although it is most important point of the problem.

Some supporters of the symbiotic theory have stressed that DNA-containing organelles, e.g., mitochondrion, chloroplast, and nucleus, have double (inner and outer) membranes. They believe, therefore, that the inner membranes of mitochondrion and chloroplast have derived from the corresponding prokaryotic plasma membranes (of the endosymbiont) while the outer membranes are from the plasma membrane of the host cell. In other words, the endosymbiotic prokaryote was wrapped by host's own membrane when invaded, and then cell wall of the invader disappeared. Of course, the inner and outer membranes are different in their compositions of phospholipids and enzymes and thus in metabolic functions. However, we have no evidence to explain as yet how the specific composition of inner and outer membranes in the organelle relates to their origins in the cellular evolution. On the other hand, there are many examples in the biological world which do not fit this idea: a single in *Cyanidium* and triple membranes in *Euglena* and Dinophyceae surrounding chloroplasts; quadruple membranes enclosing chloroplasts in Cryptophyceae, Phaeophyceae, Bacillariophyceae, Chrysophyceae, and Xanthophyceae; a single membrane of nucleus in *Noctiluca*.

The symbiotic theory explains quite simply processes of the generation of mitochondrion, chloroplast, or flagellum by mosaic designing of prokaryotes which have specific characters corresponding to the organelles. It is generally said that the simpler the theory, the better its acceptance. However, we have to keep in mind that the simplicity of theory is independent of the fact and, rather, such the simplicity leads to grater complexity when coming to explain details. This is true in the symbiotic theory, which complicates the explanation of the genetic differentiation from the four kinds of symbiotic prokaryotes to a single eukaryotic cell. In fact, it is impossible to explain, at least by the current knowledge of genetics, the genetic differentiation in the above theory. Some advocates of the symbiotic theory say that there was a big jumping step in the cellular evolution from prokaryote to eukaryote. However, we do not know of such an example of positive (progressive) jumps in the biological world although we do know of examples of negative (extinguished) jumps. The positive evolutions can be even considered to represent the repeated accumulations of minor mutations and selections. The macroevolution in morphology also is a result of accumulation of the microevolution in genetics.

4.1 Rebuttal against endosymbiosis

Some critical comments on the main problems within the symbiotic theory are given here.

4.1.1 Organellar morphology

The first problem is whether mitochondrion and chloroplast have bacterium- and cyanobacterium-like morphologies. The figures which the pioneers of cytology observed may appear on a single thin section of the cell. If they had observed many figures taken from serial thin sectioning and reconstituted them in a three dimensional structure, they would have seen that a real organelle's morphology appears to be much more complex than that of the corresponding prokaryote. Further, if they had observed the stearic structure through the progressive cell cycle, then the organelle would have indicated dynamic behaviors. Such behavior depends on physiology of the cell where the organelles assemble in one phase of the cycle but disjoin in the other, as discussed above.

4.1.2 Internal genomic evolution

All of the organelles, in spite of prokaryote and eukaryote, are products of differentiations along the cellular evolution. The increase in the number of certain genes by duplications may be followed by their expression in producing a variety of many enzymes and proteins through the accumulation of mutations and selections. The cells will exhibit an ever-present tendency to increase their metabolic complexity. The organelle plays an important role in compartmentalizing metabolism according to the functional unit. That is, the organelle guarantees specificity and velocity of metabolism. The cellular evolution from prokaryote to eukaryote follows the same line of metabolic compartmentalization through membranous and non-membranous structures.

Biological morphology is essentially discontinuous in any cells and organisms. However, we know at present that, in spite of such a discontinuity in morphology, the evolution of DNA sequence is quite a continuous process. Apparently, this results from the fact that a morphological change is derived by a minor alteration of the nucleotide sequence. For example, we can see a gorilla which is obviously quite different in the external morphology from human. However, there is little difference in the protein sequences between both of them, and it is even said that there might be no difference between them, except for the culture.

4.1.3 DNA sequence, organellar products and plasmids

Sequencing of whole DNA in mitochondrion and chloroplast performed by Anderson *et al.* (1981), Shinozaki *et al.* (1986) and Matsubayashi *et al.* (1987) have clarified the roles of the genes responsible for morphogenesis and the functions of these organelles. It shows first that almost all of the related content of genes within the nuclear DNA and their products are exported from cytoplasm to the corresponding organelle. Organelle DNA codes for its own tRNAs, rRNAs, small numbers of enzymes and proteins only, and syntheses of some subunits of enzymes including cytochrome oxidase, cytochrome b and ATPase in mitochondrion. The coding for the subunits of the photosynthetic enzyme ribulose-1,5-bisphosphate carboxylase (Rubisco) in the chloroplast is shared between the cytoplasm and the organelle itself. Second, it has been shown that there are homologous sequences among DNA of mitochondrion, chloroplast, and nucleus, and some of them are high in the homology and the others are low. Cytochromes *b*/*f* (c) in both mitochondrion and chloroplast are examples (Anderson *et al.* 1981: Matsubayashi *et al.* 1987). These facts suggest that some nucleotide sequences have duplicated repeatedly before the division of DNA among mitochondrion, chloroplast, and nucleus. Some believers of the symbiont theory may say that there occurred repeatedly large scale, but no minor, genetic re-distributions among the DNA organelles. However, no evidence has been demonstrated and none of the mechanisms hypothesized. In fact, we have no example to show such sequence transfer in large scale between DNA of the symbiont and the host and between the extant organelles. Although some unfunctional genes may be lost from the symbiont because of having no selective advantage, we have no evidence to show that the lost gene from the symbiont was picked up and integrated into the host DNA.

On the other hand, it has currently been shown that plasmids are located not only in prokaryotic cell but also in the organelle and cytoplasm of eukaryotic cell. However, generally, the plasmid is defined as a free genetic element which is not responsible for the basic life mechanism and thus can be eliminated conditionally without loss of the life of cells.

4.1.4 The origin of introns

Intron is a particular nucleotide sequence which is embedded in some eukaryotic genes and is trimmed off by a splicing enzyme before transcription. The intron has been found not only in nuclear genes but also in mitochondrial and chloroplast genes but, interestingly, not in prokaryotic genes. Some investigators explain this fact as follows. In the early times, the prokaryotic genes also contained the introns but lost them over the evolutionary duration. On the other hand, the eukaryotic genes continued to maintain them up to the present. However, such an explanation contains a serious contradictory concept, since all the introns are simultaneously eliminated from the entire prokaryotic world, a fact which is impossible to reason. The most conceivable explanation is that the introns were produced while the eukaryotic cell was generated from a prokaryotic line. That is because the eukaryotic DNA contains many duplicated sequences and the intron works to partition the primitive genes, the duplication of which led to the large sequence of the present gene. Therefore, it is quite difficult to consider DNA of mitochondrion and chloroplast as a direct-descendant product of the prokaryotic corresponding genes despite the positive approach to this by the symbiotic advocates.

4.1.5 Gene sequence of bacteria and organelles

The symbiotic proponents say that some gene sequences of mitochondrion and chloroplast are highly similar with those of the corresponding aerobic bacterium and cyanobacterium. However, there are stark facts to show that the gene sequences of non-photosynthetic bacteria are also similar with those of chloroplast and cyanobacterium. For example, sequence of 16S rRNA in maize's chloroplast is highly similar with that of *E. coli*, and sequence of 16S rRNA in Rhodophyceae's chloroplast is similar not only with cyanobacterium but also with aerobic bacteria, *E. coli* and *B. subtilis*. However, there also are many dissimilar sequences between genes of the prokaryote and the eukaryotic organelle. The second important fact is that the DNA prokaryotic sequence is also enclosed in many cases within the nucleus (Loomis 1988). A meaningful point in the molecular evolutional consideration is that homology in DNA sequence depends largely on the physiological significance of the sequence. This is true in the whole sequence of a gene and the partial sequence within a gene. The more important the gene or its higher level of activity (or its activity center), the more conservative is its sequence. The reason is that accumulated mutations in an important gene or a central part of a gene must lead to the disadvantage of the organism in the natural selection. Therefore, practically, it is difficult to deduce the phylogeny of the gene or organism from the averaged sequence (Nakamura 1990).

4.1.6 Ribosomal size and protein content

The endosymbiotic supporters say that mitochondrial and chloroplast ribosomes have a sedimentation coefficient of prokaryotic type 70S while cytoplasmic ribosomes are non-prokaryotic 80S. However, in fact, there are many varieties in the coefficient of ribosome; it ranges from 55S (mammals) to 80S (yeast) in mitochondrial ribosome and from 66S (spinach) to 78S (maize) in chloroplast ribosome. Cytoplasmic ribosome ranges from 77S (*Neurospora*) to 83S (mammals), depending on the organism. Further, the variety of the sedimentation coefficient is remarkable in each of small and large subunits of the ribosome among species of the organisms, and some ribosomes are deficient in 5 or 5.8S rRNA. The number of proteins constructing a single ribosome is distributed in a range of 50 to 107 according to the organism.

The cytoplasmic type of ribosome is generally larger than the prokaryotic type, but the former must be derived from the latter. Therefore, the evolutional interest is rather to analyze how the organelle evolved to make the larger one. After all, the ribosomal size can not be determinative for deducing the synthetic origin of mitochondrion and chloroplast.

4.1.7 Genomic content and intercellular genetic exchange

If the organellar DNA has descended indeed from a symbiotic origin, then the eukaryotic cell, must at first become a genetic mosaic of tetraploid consisting of symbionts aerobic bacterium (mitochondrion), cyanobacterium (chloroplast), spirochaeta (flagellum), and Mycoplasma (nucleus) through complex recombinations and deletions of the various genes. The symbiotic advocates have as yet not offered any mechanism addressing that problem. Further, it seems to us that such a scenario could not possibly have been occurred for following reasons: (a) The genetic cross is not observed beyond species, nothing between different kingdoms as eubacterium (Bacteriomycota), cyanobacterium (Cyanophyta) and Mycoplasma (Mycoplasmomycota, supposed as the host for the symbionts (Margulis 1970). (b) When a phylogenetically different line of DNA directly invades a host cell, restriction enzyme endonuclease of the latter must digest the invader to preserve his own phyletic line. All of the biological lines have been maintained by such a mechanism. There is no evidence to show that genetic exchange occurs between the symbiont and the host during their symbiosis. Cf. chapter by Schenk q.v.

4.1.8 Antibiotics inhibition of protein synthesis

According to the symbiotic theory, antibacterial inhibitors of protein synthesis are effective in mitochondrion and chloroplast but not in cytoplasm, whereas non-antibacterial inhibitors are effective in cytoplasm but not in the other two organelles. For example, chloramphenicol and mikamycin are antibacterial, and cycloheximide and anisomycin are non-antibacterial. However, inhibitive effects of emetine, erythromycin and lincomycin on protein synthesis of the organelle and cytoplasm depend on biological species and other conditions. In summary, the effect of the antibiotics on protein synthesis is not conclusive as evidence of the symbiotic origin of mitochondrion and chloroplast. Again, it should be remembered that the nucleus, namely cytoplasm, is also of the prokaryotic origin.

Streptomycin inhibits protein synthesis of mitochondrion and chloroplast as in bacterial cells. Streptomycin-resistant mutants of bacteria have ribosomes, a protein of which is resistant to the antibiotic. However, the protein is synthesized in the cytoplasm under nuclear control. Synthesis of rRNAs specific for the organelle are under the control of the organelle's DNA, whereas ribosomal-proteins of the organelle are imported from the cytoplasm.

4.1.9 tRNA Species in mitochondria and bacteria

The number of tRNA species involved in the mitochondrial protein synthesis is much lower than that in the bacterial one: mitochondrion contains usually 22–24 species of tRNA, while a bacterial cell possesses two to three times more species than the organelle. Further, there are large differences also in the codon usage between mitochondrion and bacterium. Therefore, it is difficult to consider that the mitochondrion is a direct symbiont of the bacterium, as in the symbiotic theory. See also Seckbach *et al.* 1990.

4.1.10 The status of cyanelles, cyanobacteria and chloroplasts

The protistan *Cyanophora paradoxa* is classified, on the one hand, into Protozoa (animal kingdom) and on the other into Cryptophyta or Glaucophycophyta, within the algal divisions. See chapter by Schenk, q.v.. This photosynthetic unicell contains cyanobacterial symbionts in its cytoplasm. The symbiont is distinctively called 'cyanelle' to mean that it is at an intermediate state from cyanobacterium to chloroplast. The main reason is that the cyanelle's DNA content is at a 10% level of the usual cyanobacteria and is comparable with that of chloroplast in the higher plant. However, the cyanelle has a cyanobacterial cell wall and possesses both types of ribulose 1,5-bisphosphate carboxylase, Rubisco, subunits (8 large and 8 small). Rubisco is a gate enzyme of CO_2 fixation in the reduced pentose phosphate (Calvin) cycle and of which synthesis is determined by specific genes on a single DNA molecule as in cyanobacterium. As shown in the foregoing section, in the higher plants and some algae, synthesis of the large subunits is determined by the chloroplast DNA and that of the small subunits is determined by the nuclear one. Further, it is said that isolated cyanelle can grow without the host. These facts clearly show that the cyanelle is undoubtedly the cyanobacterium *Cyanocyta korschikoffiana* which once invaded a host (Nakamura 1987c). The cyanobacterial world has adaptively radiated into quite wide phyletic lines and their DNA contents are also known to range from 1.2 x 10^{-9} dalton in an unicellular species to 10 x 10^{-9} dalton in a multi-cellular species. Therefore, it is reasonable to infer that cyanelle *C. korschikoffiana* is a self-growing unicellular species of Cyanophyta and is not an example in a modified way of chloroplast. It is natural that there are similar DNA sequences in the cyanelle to those in chloroplasts because the latter contains some DNA sequences homologous with cyanobacteria (Nakamura 1987c).

4.1.11 Origin and physiology of chloroplasts and mitochondria

The chloroplast of the higher plant cell is known to change largely in morphology and function depending on the environmental and physiological conditions. Chloroplasts are one type of the general plastids (also containing the chromoplasts) which (a) are surrounded by double membranes, (b) contain specific DNA and proteins, and (c) play various roles including in photosynthesis and the storage of starch and other substances. Dark-grown plant and non-illuminated cells within a light-grown plant have semi-autoreproducible proplastid — a quite small DNA organelle — but contain little or no thylakoid and photosynthetic pigment. When a dark-grown plant is illuminated, the proplastid develops into a matured chloroplast which can actively photosynthesize. Although the symbiotic theory has stressed that the chloroplast is a symbiotic derivative of cyanobacterium, the cyanobacterium does not possess the protoplastid phase within its life cycle.

Currently, there is a hypothesis that a mitochondrion can also develop from promitochondrion according to the cell physiology. If this is the case, then aerobacterium as the ancestor of mitochondrion has no special phase of promitochondrion in the life cycle.

5. Membrane evolution theory

We have inspected and analyzed thus far the content of the symbiotic theory on the origin of eukaryotic cell. As a result, we have concluded that the theory does not reasonably explain the process of the origin eukaryotic cell. So, we proposed a 'Membrane Evolution Theory' in 1975 instead of the symbiotic theory of Margulis (1970, 1981). We now deal with a discussion of our new theory.

It has been established that purple non-sulfur bacteria including *R.rubrum* contains both metabolisms of the (ancestral) photosynthesis and O_2-respiration (Nakamura 1983, 1987b). Similarly, cyanobacteria, for example *C. fritschii* which was derived from the photosynthetic bacteria and is believed to link phylogenetical between unicellular and simple filamentous forms (Fogg *et al.* 1973), provides more evolved photosynthesis which consists of a serial photosystem I and II, evolving O_2 gas. Furthermore, the species has a complete O_2 – respiration metabolism which is consists of EM pathway, TCA cycle, and respiratory chain (Ragan and Chapman 1978). In cyanobacteria, the DNA content of the cell has been demonstrated to increase duplicatively from unicellular to filamentous form (Herdman *et al.* 1979). As pointed out

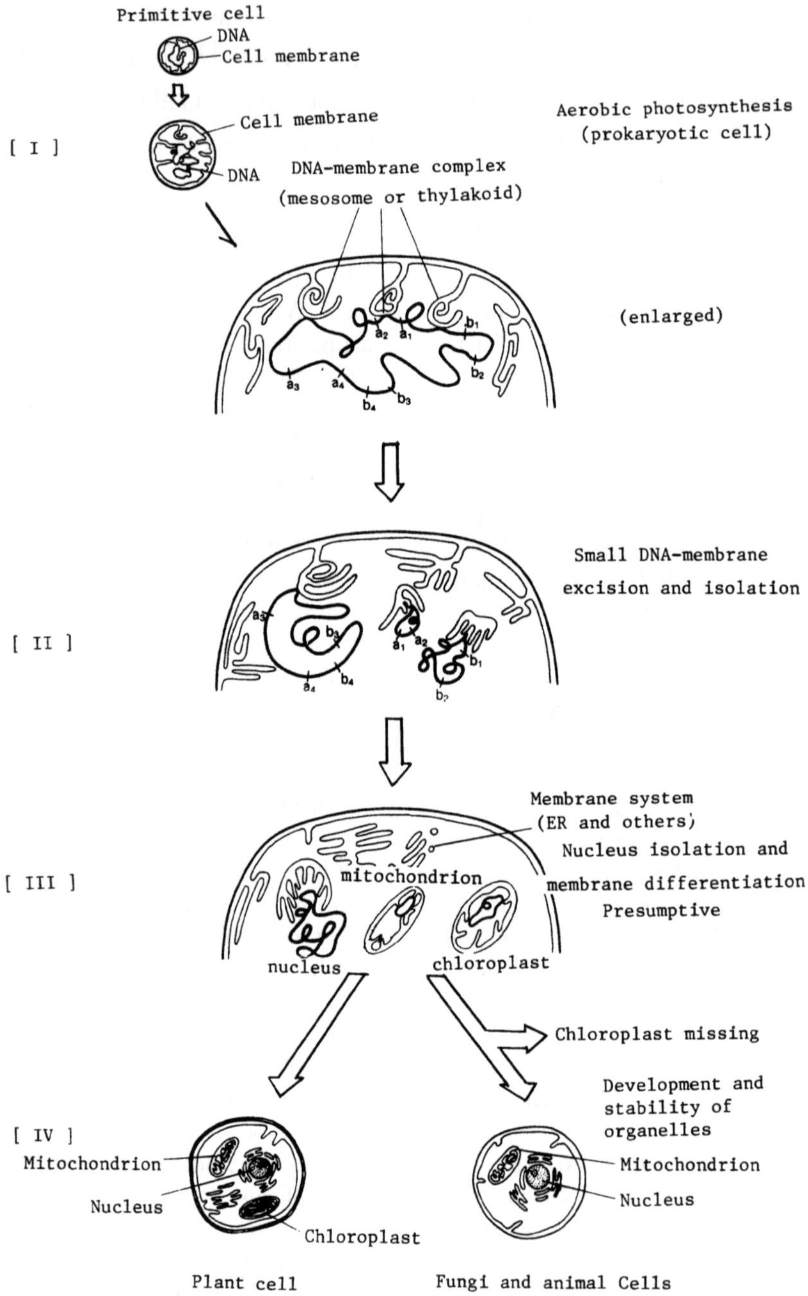

Fig. 2. Membrane evolution theory. The DNA and membranous system of an evolved cyanobacterium were once fragmented and differentiated toward formation of presumptive mitochondrion, chloroplast and nucleus. Non-photosynthetic animal and fungal cells were generated as a result for spontaneous elimination of the chloroplasts. Their regulatory mechanisms of genetic expression among the DNA-organelles have been acquired from the prokaryotic age.

above, the evolved species of cyanobacteria have already been endowed with genes coding for the eukaryotic photosynthesis and respiration (Loomis 1988; Peschek 1984, 1987). Therefore, it is quite reasonable to consider that the DNA-carrying organelles, such as mitochondrion, chloroplast and nucleus, were once generated as a result of genetic and membranous differentiation from a line of the evolved cyanobacteria at the same stage when there occurred differentiation of endoplasmic reticulum,

Golgi apparatus, and other membranous organelles in the pro-eukaryotic cell.

Here, it is unnecessary to suppose that the photosynthesis and respiration genes were derived from some separate organisms which once penetrated into an unknown prokaryotic host. Originally, the cell does not accept foreign DNA and does not allow recombination between DNAs of different species. Species-specific restriction enzymes are primarily so-called arms which defend the cell against the invader DNAs. Therefore, all biological species must have specific enzymes which digest the foreign DNAs. We must stress here that such a careful mechanism has preserved the phylogeny in the biological world. In fact, there occurs no natural hybridization without a single, or near, taxonomic species. Therefore, it is apparent that in the genetic recombination little occurs between organisms beyond the species, let alone the impossibility of combining organisms of the different Phyla, although the symbiotic theory has supposed it can take place. (On the other hand, symbiosis or parasitism occurs naturally and is popular beyond the taxonomy or even between animal and plant kingdoms).

Figure 2 illustrates the points of our argument stated above together with other biological phenomena. A long DNA fiber in the evolved cyanobacterium was broken into at three parts, two small fragments and one large one. Then, each of the DNA fragments was wrapped by the intracellular membrane, such as thylakoid and mesosome. A small DNA containing some genes relating to respiratory chain, t- and r-RNAs, and other components were organized as the mitochondrion. Another DNA fragment containing some genes relating to photosystem, t- and r-RNAs and other constituents was organized into a chloroplast. Further, the remainder of these DNAs is located in the nucleus. We have considered that t- and r-RNAs coded for by the mitochondrial and chloroplast DNAs are duplicative offsprings which have an ancestral nucleotide sequence common which nuclear t- and r-RNAs, but they became phylogenetically divergent with the cellular differentiation. When the differentiated cells containing the presumptive chloroplasts, lost their plastidal genes, they became non-photosynthetic as animal and fungus organisms. Spontaneous or induced, but not lethal, elimination of chloroplast from green algae has well been observed in the laboratory, for example in *Chlorella* and *Euglena.*

6. Conclusion

Since the proto-cell was first generated in the primeval soup about four billion years ago, its offspring continued to evolve the cellular metabolism and organelle and multi-cellular morphology throughout prokaryotic and eukaryotic worlds. Such a developmental course to raise the efficiency of the life was a natural result of repeated mutations and selections. The mutations always occur by spontaneous and induced mistakes of base pairing in DNA. The biological and non-biological conditions in nature also select for some organisms to adjust well in their habitat. In other words, the cells have always been compelled to adapt their function through the metabolic and morphological alterations. The generation of a new organelle and its development to enhance cellular function was an essential process for life and, if not done so, then the organism had to be selected against. Therefore, we may conclude that the eukaryotic organelles were not born as an accidental event of the symbiotic phenomenon but by the evolutional inevitability of the viving world.

The genesis of eukaryotic cell seems to have been largely advanced by a development of the membrane system. However, the latter has developed with the functional complexity along the evolution in the prokaryote. The membrane system plays a very important role in metabolic compartmentalization. Therefore, we can observe a fact that membranous organelles increase the morphological, namely functional, complexity and specialty with evolution even in the eukaryotic world. On the other hand, basic metabolisms for life are well preserved throughout prokaryotes and eukaryotes but their efficiency must be raised with morphological evolution.

It has been demonstrated that base sequence of chloroplast DNA coincides completely among liverwort (Bryophyta) in lower plants (Ohyama *et al.* 1986) and tobacco (Corolliforiidae) (Shinozaki *et al.* 1986) and rice (Glumiforiidae) (Hiratsuka *et al.* 1989) in higher plants. Together with other data, it is apparent that DNA of the organelle has specific conservativity throughout the plant kingdom. If symbiotic origin of the eukaryotic organelle was usual in the past, then we may be able to observe various types of mitochondria, chloroplasts, and other DNA-carrying eukaryotic organelles at present. Otherwise, the extant eukaryotes must be explained as derivatives of a single ancestor which was selected for among the organelle- heterogenic eukaryotes. Currently, evidence has been accumulating to show that amino acid sequence is present in enzymes which is common to all cells, from eubacteria to higher organisms. The origin of this sequence might have emerged in a chemical evolution which once occurred in the primitive soup (Loomis, 1988).

The organisms primarily have a strong compulsion to preserve, rather than mix, their phylogeny. Such a defense of the genetics has been accomplished by the biological mechanism and the chemical stability of DNA. However, there remain as yet many problems to be illuminated concerning our membrane evolution theory (Nakamura 1990).

References

Anderson, S. *et al.* (1981). *Nature* 290: 457.

Battley, E. W. (1987). *Energetics of Microbial Growth*, John Wiley & Sons, New York.

Berns, M. W. (1983). *Cells* (2nd Ed.) CBS Publishing, Los Angeles.

Broda, E. (1978). *The Evolution of Bioenergetic Process*, Pergamon, N.Y

Bryant, C. (1980). *The Biology of Respiration*, Edward Arnold, London.

Cairns-Smith, A. G. (1982). *Genetic Takeover*, Cambridge University Press, Cambridge.

Carrn, N. C. and Whitton, B. A. *The Biology of Cyanobacteria.* Univeristy of California Press, Berkeley.

Douce, R. (1985). *Mitochondria in Higher Plants*, Academic Press, New York.

Ehara, T. *et al.* (1985). *Plant and Cell Physiol.* 26: 1155.

Fukuda, I. and Nagashima, H. (1987). *Ann. New York Ac. Sci.* 503: 575.

Fredrick, J. F. (1987). *Ann. New York Ac. Sci.* 503: 438.

Fogg, G. E. (1973). *The Blue-Green Algae.* Academic Press, London.

Herdman, M. *et al.* ((1979). *J. Gen. Microbiol.* 111: 73.

Hiratsuka, J. *et al.* (1989). *Mol. Gen. Genet.* 217: 185.

Holland, K. T. *et al.* (1987). *Anaerobic Bacteria*, Blackie, Glasgow.

Horowitz, N. H. (1945). *Proc. Natl. Acad. Sci.* USA, 31: 153.

Jacob, F. *et al.* (1963). *Cold Spring Harbor Sym., Qunt. Biol.* 28: 329.

Kornberg, A. (1980). *DNA replication*, W. H. Freeman and Company, San Francisco.

Kuhl, S. A. *et al.* (1983). *J. Bacteriol.* 156: 737.

Loomis, W. F. (1988). *Four Billion Years*, Sinauer, Associated Inc., Sunderland, Mass.

Margulis, L. (1970). *Origin of Eukaryotic Cell*, Yale University Press, New Haven.

Margulis, L. (1981). *Symbiosis in Cell Evolution-Life and its Environment of the Early Earth.* W.H. Freeman and Comp, San Franscisco.

Matsubayashi, T. *et al.* (1987). *Mol. Gen. Gebet.* 210: 385.

Nakamura, H. (1965). *J. Bacteriol.* 90: 8.

Nakamura, H. (1974). *J. Gen. Microbiol.* 84: 85.

Nakamura, H. *et al.* (1975). *J. Gen. Microbiol.* 91: 45.

Nakamura, H. (1975). *Viva Origino* 4: 43.

Nakamura, H. *et al.* (1975). *J. Bacteriol.* 122: 874.

Nakamura, H. (1976). *Jap. J. Genet.* 51: 393.

Nakamura, H. (1978). In: *Origin of Life (ed. Noda, H.), Center for Academic Publication, Tokyo.* p. 515.

Nakamura, H. (1982). *Origin and Evolution of the Cell*, Baifukan Publisher, Tokyo.

Nakamura, H. (1983). *Mem. Konan Univ. Sci. Ser.* 29: 55.

Nakamura, H. and Yoshikawa, H.(1986). *Mem.Konan Univ. Sci. Ser.* 33: 1.

Nakamura, H., (1986a). In: *Origin of Life, 5th ISSOL and 8th Intl. Congr. Berkeley.* p. 215.

Nakamura, H. (1987a). *Biological Evolution on Aspect of Microorganisms, Baifukan Publish, Tokyo.*

Nakamura, H. (1987b). *Viva Origino* 15: 100.

Nakamura, H. (1987c) *Cellular Evolution*, Baifukan Publisher, Tokyo.

Nakamura, H. (1988). In: *Report of Special research Project on Evolution of Matter (Tsuba University)* 2: 185.

Nakamura, H. (1990). *Mem. Konan Univ. Sci. Ser.* 37: 87.

Nakamura, H. and Hase, K. (1990b). *Origin of Life and Evolution of the Biosphere.* 20: 499.

Nakamura, H. (1993). In: *Endocytobiology V. (eds. Sato et al.)* Tübingen University Press. Tübingen, FRG. pp. 337–344.

Ohyama, K. *et al.* (1986). *Nature* (London) 322: 527.

Peschek, G. A. (1984). *Subcell. Biochem.* 10: 85.

Peschek, G. A. (1987). In: *Cyanobacteria (eds. Fay, P. and van Baaken, C.) Elsevier, Amsterdam*, p. 119.

Ragan, M. A. and Chapman, D. J. (1978), *A Biochem. Phylogeny of the Protists.* Academic Press, New York.

Seckbach, J. (1987). *Ann. N.Y. Ac. Sci.* 503: 424.

Seckbach, J., Nagashima, H. and Fukuda, I. (1990). *Endocytobiology* IV. p. 575–578.

Sere, P. A. and Estabrook, R. W. (1978). *Microenvironments and Metabolic Compartmentation.* Acad. Press.

Shinozaki, K. *et al.* (1986). *EMBO J.* 5: 2043.

Tandeau de Marsac, N. and Houmard, J. (1987). In: *Cyanobacteria*, (eds. Fay, P. and van Baalen, C.) Elsvier, Amsterdam, p. 261.

de Witt, W. (1977). *Biology of the Cell – An Evolutionary Approach.* W. B Saunder Comp., Philadelphia.

Woese, C. *et al.* (1966). *Cold Spr. Harb. Symp. Quant. Biol.* 35: 723.

Yoshikawa, H. (1965). *Proc. Natl. Acad. Sci. USA* 53: 1476.

Chapter 2

Glaucocystophyta model for symbiogenous evolution of new eukaryotic species

H.E.A. SCHENK
University of Tübingen, Botanisches Institut, Auf der Morgenstelle 1, 72076 Tübingen, Germany

Dedicated to Prof. Dr. A. Hager on the occasion of his 65th birthday

Outlines:

1. Introduction
2. Symbiogenesis versus endogenous and other evolutionary mechanisms
   2.1 Endogenous (autogenous) hypotheses
   2.2 Exogenous (symbiogenous) hypotheses and theories
   2.3 Discussion of the different views
3. The Glaucocystophyta
   3.1 The eukaryotic cell
   3.2 The chromatophore and the cyanelle/cyanoplast problem
   3.3 *Glaucocystis nostochinearum* Itzigs.
      3.3.1 The eukaryote
      3.3.2 The cyanoplast
   3.4 *Cyanophora paradoxa* Korsch.
      3.4.1 The eukaryote
      3.4.2 The cyanoplast and subcellular concerted actions
         a) Envelope and pigments
         b) Photosynthesis and metabolite exchange
         c) Fatty acids and lipids
         d) Nitrogen assimilation and storage
         e) Proteins, biosynthesis and coding sites
4. Glaucocystophyta as model for eukaryotic cell evolution
   4.1 Endocytobiotic coevolution of symbiotic partners
   4.2 The end of intracellular symbiosis (exogenosome formation)
   4.3 Cyanoplast, model for proof of gene transfer
      4.3.1 Cyanoplast, an exogenosome
      4.3.2 Attempt to proof the gene transfer
      4.3.3 Glaucocystophyta, enigmatic algae
5. Glossary
References

Key words: *Cyanophora paradoxa*, *Glaucocystis nostochinearum*, Glaucocystophyta, metacyanomes, cyanelles, cyanoplasts, primary plastids, exogenosomes, evolutionary model, endogenous origin, autogenous origin, exogenous origin, Endosymbiosis Theory, symbiotic processes, antisymbiotic processes, gene transfer, unequal intertaxonic combination, pattern and biosynthesis of fatty acids, nitrogen storage, metabolism, ferredoxin-NADP$^+$-oxidoreductase, preprotein, transit sequence, phylogeny of cyanoplast.

1. Introduction

Since the discovery of chloroplasts and mitochondria as eukaryotic cell organelles in the last century until the late seventies of the twentieth century the evolution of eukaryotic cells was generally understood as an autogenous or *endogenous* process (Allsopp 1969; Kirk 1971; Raff and Mahler 1972; Bogorad *et al.* 1973; Parthier 1974, for review: Gray and Doolittle 1982), despite some daring hypotheses expressed early on (e.g. Mereschkowsky 1905, 1910; Margulis 1970). Considering the incredible, fantastic variability of capabilities in eukaryotic cell systems, this longstanding and undisputed acceptance of an endogenous origin by the biological community is not surprising. Over the past 30 years, however arguments for an *exogenous* origin of the named cell organelles gained considerable weight (e.g. Ris and Plaut 1962; Nass and Nass 1962; Sager and Ishida 1963; Sagan 1967; Margulis 1970; Schenk and Hofer 1972; Tomas and Cox 1973; Taylor 1974; Schwemmler 1979; Whatley and Whatley 1980; Gray 1989; Brandt 1991).

In view of the evolution of the different eukaryotic cell organelles I distinguish between organelles which evolved endogenously (endogenosomes) and those with an exogenous origin (exogenosomes, e.g. plastids and mitochondria; Schenk 1992b). Nevertheless I think it necessary to consider and discuss other evolutionary possibilities for plastids and mitochondria than their

J. Seckbach (ed.), Evolutionary Pathways and Enigmatic Algae: Cyanidium caldarium *(Rhodophyta) and Related Cells,* 19–52, 1994.

symbiogenous origin. I make this cautious remark for we *know* nothing about the first evolutionary steps from the prokaryotic to the eukaryotic cells. Cavalier-Smith (1991) assumes that it was an endogenous process (but see the Endocytobiological Cell Theory by Schwemmler, 1989). Consequently only mitochondria and plastids (perhaps two or three further organelles. e.g. the euflagellum or undulipodium, Margulis 1981) are thought to have an exogenous origin.

But what is really known up to now? Firstly, many intracellular symbioses were discovered in the last hundred years. Moreover, the spontaneous formation and firm establishment of an endocytobiotic system within a relative short time seems to be recently observed (Jeon and Jeon 1976, Rogerson *et al.* 1989). Secondly, plastids and mitochondria have many prokaryotic characteristics and a distinct genetic semiautonomy. Both features are explained easiest by endocytobiotic events having taken place early on the evolutionary time scale. But until now we lacked the symbiotic system which is demonstrating the gene transfer at a very early stage of symbiogenesis. Therefore it remained an open question whether photoautotrophic eukaryotic cells really have had enough time for the numerous postulated events of gene transfer. This question becomes even more pressing if one considers that plastome genes are not once but repeatedly present in the nuclear genome. If gene transfer is in fact a common and frequent process, why then is it so rarely observed, e.g. also not within different clones of species? Well, since the (Endo-) Symbiosis Theory (Mereschkowsky 1905; Margulis 1970), and by it the symbiogenesis (Mereschkowsky 1910) of distinct eukaryotic cell organelles (at least that for the origin of plastids, e.g. Gray 1989; Bayer and Schenk 1986; Palmer 1990; Schenk *et al.* 1992), seems to be largely accepted as an evolutionary macromechanism *sensu* Schwemmler (1989), two foremost questions need be answered: (a) What is symbiogenesis in reality, and (b) what steps are required in detail? In this chapter I shall discuss these questions with regard to the evolution of phyletic primary plastids (cyanoplasts, rhodoplasts and chloroplasts). With the Glaucocystophyta I present a model for the exogenous origin of plastids as well as for the variation of cell structures and functions following the processes of symbiogenesis and intertaxonic combination.

2. Symbiogenesis versus endogenous and other evolutionary mechanisms

2.1 Endogenous (autogenous) hypotheses

a) Elementary Hypothesis (Bell and Mühlethaler 1962). The plastids (and also the other cell organelles) evolved by a series of mutations of the nuclear genome which altered the eukaryotic cell membrane structures. Consequently, it should be still possible to observe *de novo* formation of plastids (up to now impressive, respective electron micrographs of germinating oocytes of the fern *Pteridinium aquilinum*, which were thought to demonstrate this process, were neither confirmed nor refuted).
b) Transformation Hypothesis (Allsopp 1969; Jensen 1991). The cell nucleus (and the other cell organelles) evolved simultaneously during transformation of a prokaryotic cell into a eukaryotic one. Cyanobacterial thylakoids evolved from the plasmalemma and changed later in distinct cyanobacterial strains partially on the one hand into the perinuclear cisterns and on the other in plastids, in this way forming DNA containing cell organelles. Accordingly, the envelope membranes of the nucleus and of the plastids must have the same origin.
c) Plasmid Hypothesis (Raff and Mahler 1972; Parthier 1974; Nakamura 1993). The mitochondria (and also the plastids) have evolved during the prokaryote/eukaryote transition by conservation of an easily detachable part of the still prokaryotic chromosome, a plasmid, over all coding for the proteins of the evolving new cell compartment. Inner and outer membrane have the same origin (the plasmalemma).
d) Cluster Clone Hypothesis (Bogorad *et al.* 1973) is an expansion of the Plasmid Hypothesis.
e) Endospore Hypothesis (Schenk 1973; The locked cell hypothesis, Keyhani 1983). An interruption during endospore formation is discussed as a possibility for evolution of mitochondria (Keyhani 1983) or for plastid evolution (corresponding endospore formation is found in many eubacteria but not in recent cyanobacteria). The spore mother cell evolved to the eukaryotic cell, while the intracellular embedded and at this stage fixed prespore cell

give rise to the plastid. Such a process could lead to or over an intermediate stage which would represent a 'homologous' endocytobiosis though no symbiosis (*per definitionem*, after de Bary, 'symbiosis' can be only 'heterologous').

f) Gene Duplication Segregation Hypothesis (Parthier *et al.* 1975). It is assumed that two different chromosomes developed by repeated gene duplication in a cyanobacteria-like cell. During and after this process the cell volume increased and the two chromosomes segregated. Subsequently the smaller chromosome was surrounded by two envelope membranes representing the very conservative plastome of the preplastid (Scheme 1).

2.2 Exogenous (symbiogenous) hypotheses and theories

a) Symbiosis Theory (Mereschkowsky 1905) and Theory of Symbiogenesis (Mereschkowsky 1910). Evolution of plastids by adaptation of a cyanobacterial like intracellularly living prokaryote within a prokaryotic host cell. Coevolution of both partners led to plastids and eukaryotic cells.
b) Endosymbiosis Theory (EST, Margulis 1970). Development of exogenosomes by adaptation of an endocytobiotic prokaryotic guest cell to a eukaryotic host cell under the regime of gene transfer from guest to host genome. In addition to plastids and mitochondria also the euflagellum (undulipodium) is discussed as a symbiogenetic eukaryotic cell organelle.
c) Serial Endosymbiosis Theory (SET, Taylor 1974; Margulis 1981). It is an enlargement of the EST and states that mitochondria and phyletic primary plastids (and the euflagella) evolved by a series of symbiotic events separated in time. Later evolution repeated itself by further symbiogeneses between eukaryotic cells (photoautotrophic eukaryotic cells within heterotrophic eukaryotic cells) leading to phyletic secondary (complex) plastids with three or four envelope membranes (see e.g. plastids of Heterokontophyta; Scheme 2).
d) Endocytobiological Cell Theory (Schwemmler 1989). Further development of the idea of symbiogenous evolution of the eucyte and understanding of the eucyte as a biological system, composed (in relation to many different cell characteristics) and gifted with the capability to accumulate genetic information of other biological systems.

At the beginning of the finally organelle forming symbiogenous processes stays — in the view of the SET — the formation of a stable endocytobiosis. A host cell (the macrobiont, regularly a eukaryotic cell) incorporates and retains a guest cell, the microbiont (symbiont or endocytobiont, often a prokaryote) and forms an extraordinary cell compartment containing one or more cells of the endocytobiont. The newly developed compartment is called symbiosome (Park and Jeon 1988). The outer envelope membrane originates from the host and is more than a pure 'vacuolic' membrane. This symbiosome membrane is also called periguest (perialgal, peribacteroid etc.) membrane, if the symbiosome contains only one endocytobiotic guest cell. The second (inner envelope) membrane is the plasmalemma of the guest. Already these first events of a symbiogenesis give rise for many questions. Does the formation will be possible for example by

a) an 'infection' of the macrobiont through the microbiont which is able (as xenobiont or as gnorimobiont) to break up the cellular defense of the host, inducing the formation of a symbiosome (a 'xenosome' *sensu stricto*, Soldo 1983, or a 'gnorimosome', if the infection is induced by recognition and accompanied with mutualistic processes of the host as in the case e.g. of *Rhizobium* and *Fabales*).
b) 'endocytosis' (incorporation without digestion)?
c) 'phagocytosis' (incorporation followed by partial digestion of the microbiont, e.g. plastid retention) or mycocytosis (Schnepf and Elbrächter 1988)?
d) another form ('Fusiogenesis Hypothesis', new proposal) of incorporation, e.g. by physically (e.g. dielectric breakdown of membranes by lightning) or chemically (by naturally arising weak detergents) induced, but until now unknown fusion processes (Zillig *et al.* 1989; Schwemmler 1991) between the microbiont and the macrobiont, perhaps directly combining the two cellular points of information (even possibly without forming an symbiotic consortium *sensu* de Bary)?

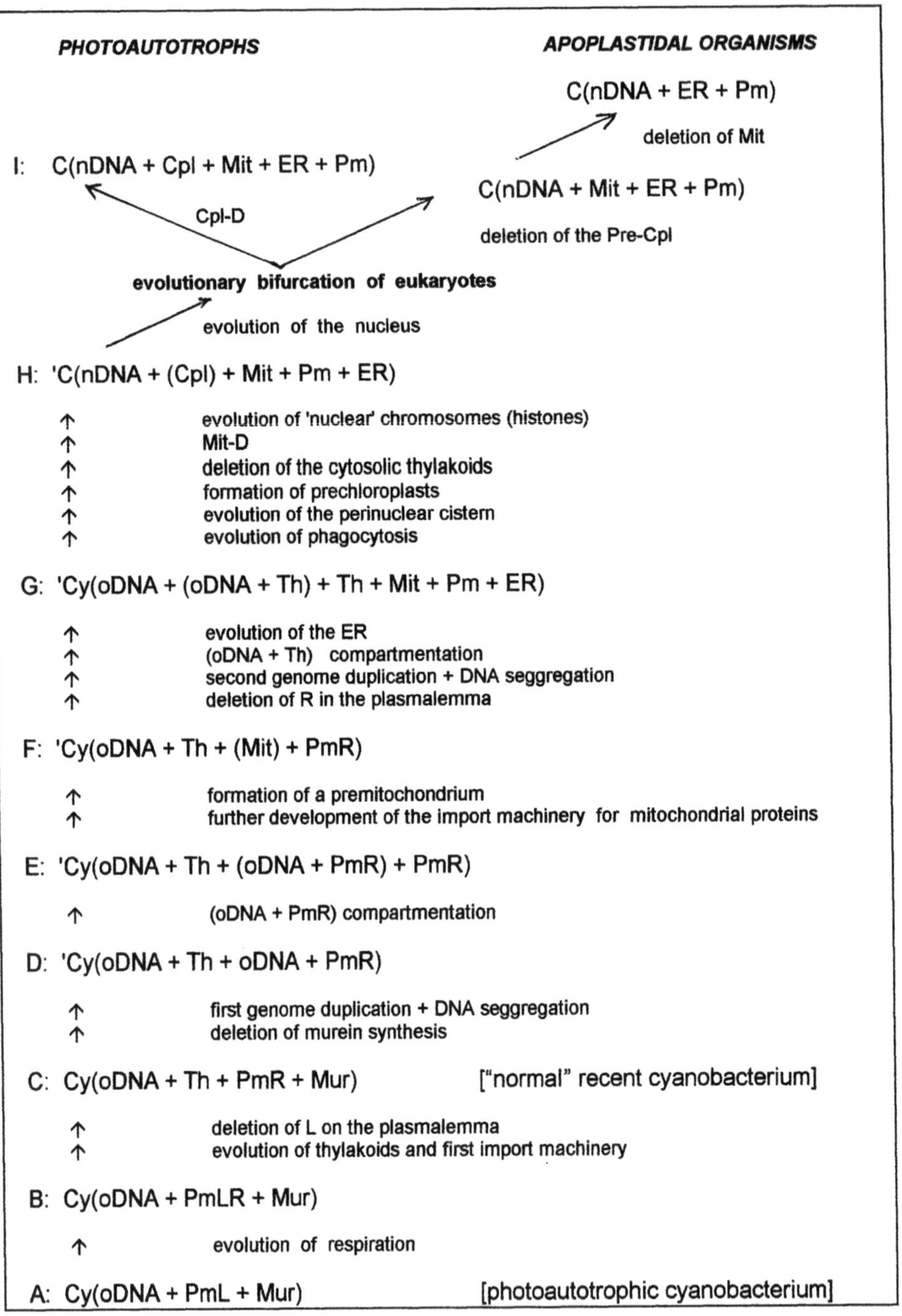

Scheme 1. Hypothetical ways regarding a possible endogenous evolution of phyletic primary plastids and mitochondria.
Abbreviations: A: original cyanobacterium (Cy) without thylakoids, photosynthetic e-transport (hydrophotoautotrophy) in the plasmalemma; B: evolved cyanobacterium with respiratory e-transport in the plasmalemma; C: higher evolved cyanobacterium with thylakoids; D-G: steps of the hypothetical cyanobacterium/eukaryote transformation ('Cy); H: steps of a primitive eukaryote ('C); I: forms of more recent eukaryotic cells (C); Cpl, chloroplast; (Cpl), prechloroplast (cell compartment still containing most of the oDNA information, evolution of Cpl-specific protein import); Cpl-D, development of chloroplasts (different deletions of photosynthetic and other proteins in the chloroplast genome and in the nucleus genome, development of carrier systems); oDNA, 'prokaryotic' DNA; nDNA, 'eukaryotic' nuclear DNA; ER, endoplasmatic reticulum; L, photosynthetic reaction complex (hydrophotoautotrophy); Mit, Mitochondrion; (Mit), premitochondrion (still containing most of the oDNA information, evolution of Mit-specific protein import);

Possibly, all these events are realized in some manner. For plastid evolution, I and other authors favor, however, symbiogenesis (see above and later). Therefore the start process should has been initiated by endocytosis or infection. In all cases, observed until now, the photoassimilating cell compartment is surrounded at least by two envelopes (possible exception see Seckbach 1992). Plastids, surrounded by two envelope membranes, are called 'phyletic primary plastids', assuming that they are direct descendants of cyanobacteria like progenitors.

2.3 Discussion of the different views

As mentioned in the introduction, today the SET is the generally accepted theory. What are the reasons which have given the SET such a reputation? There are old and new postulates and theorems of cell biology which now could also be summarized as typical theorems of endocytobiology:

a) The Continuity Theory (Schimper 1883) for plastids is supported by the following results:
   a1) Proof of the genetic continuity by cross experiments (plastome genome cooperation, Van Wisseling 1920).
   a2) Cytochemical indication of DNA (Chiba 1951) or of DNA and RNA (Metzner 1952) in different plastids.
   a3) Electron microscopic indication of the plastome DNA (Ris and Plaut 1962).
   a4) Isolation of chloroplast DNA (Sager and Ishida 1963). Similar experimental data are available for mitochondria.
b) The Biogenetic Rule (Haeckel). The statement, 'The ontogeny is often a short repetition of the phylogeny', could be used, on a trial basis, to win insight into the phylogeny of eukaryotic cell organelles. By strict use of this rule, the observed ontogenetic continuity of plastids would counter argue the explanation of an autogenous origin of these organelles.
c) The Compartmentation Rule (Schnepf 1966) says, that a cellular membrane (a unit membrane) is structured polarly and separates in main a watery Mischphase (mixed phase) from a plasmatic (DNA containing) Mischphase. That means a double enveloped organelle should possess its own DNA. That was confirmed until now. Admittedly, it gives also exceptions: The so called 'complex' plastids (Sitte 1987) found in most algal groups are enveloped by three (or four) membranes (Dodge 1973). That led to further speculations and finally to the postulate of 'phyletic secondary' plastids: It was proposed (Tomas and Cox 1973; Gibbs and Gillot 1980), that such complex plastids derive from ingested eukaryotic photoautotrophs, already containing plastids. During following evolution the eukaryotic endocytobiont should have been degenerated until up the remaining membranes, in any cases with some additional cytoplasmic inclusions, e.g. in the space between the outer membrane pair (periplastidial ER) and the inner one (the plastid membranes proper) of the cryptophytan plastids.
d) The Law of the unequal Intertaxonic Combination (Sitte 1990, see later and Glossary).

Also Molecular Biological and Phylogenetic Data (Brandt 1991) of eukaryotic and prokaryotic genomes, plastomes and chondriomes, or/and of model organisms such like Glaucocystophyta and Cryptophyta are supporting the SET (see later).

The main competing theories differ not only in respect to the 'exogenous' or 'endogenous' evolution of distinct cell organelles, but even more fundamentally in the question of how the eukaryotic cell evolved at all. Following the SET the ur-eukaryote was a heterotrophic, phagocytosing cell. Contrarily to the old Symbiosis Theory of Mereschkowsky or the new Endocytobiological Cell Theory of Schwemmler no information is given with regard to the first evolutionary steps of the eucyte. More speculative information is given by the 'endogenous' (autogenous) hypotheses. They explain the evolution of the eucyte as a continuous process from the prokaryotes to the eukaryotes (Scheme 1). Because the photosynthetic

---

Mit-D, development of mitochondria (different deletions of 'respiratory' and other proteins in the mitochondrion and in the pre-nucleus genome, development of necessary translocators); Mur, murein synthesis; Pm, plasmalemma; R, respiratory reaction complex; Th, thylakoids.

The scheme represents a part of the consequences following the hypotheses (e.g. Klein and Cronquist 1967) of the 'botanical myth' (Margulis 1981) without considering e.g. the evolution of the undulipodium.

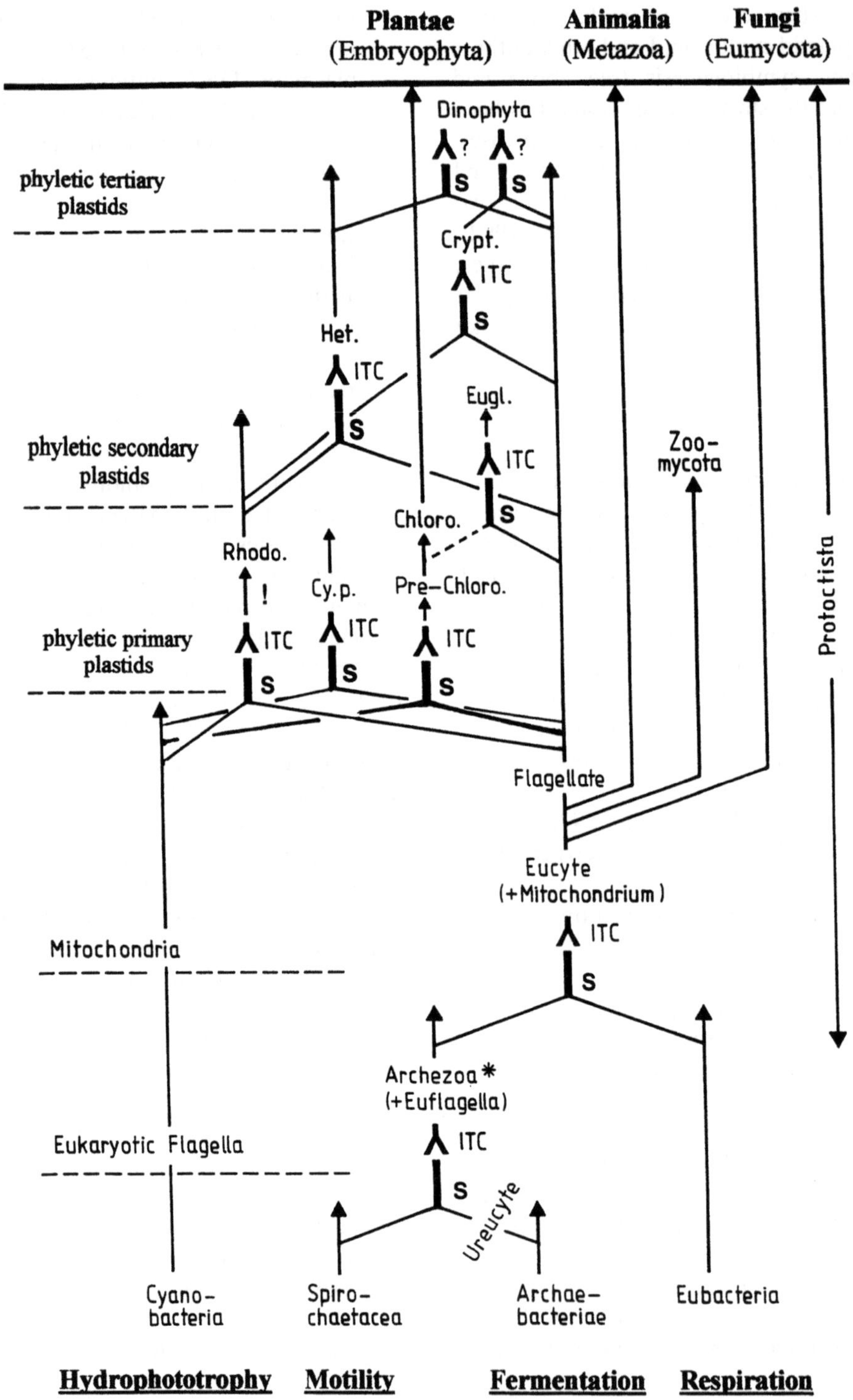

Scheme 2. Serial Endosymbiosis Theory (Taylor 1974; Dodge 1980; Margulis 1981, 1988; Whatley and Whatley 1980). Simplificated evolutionary scheme (without time scale) of different eukaryotic cell types, in particular in view of the acquisition of distinct postulated exogenosomes (eukaryotic flagella, mitochondrium, various types of plastids). Four eukaryotic kingdoms are recognizable: the animalia (metazoa), the plantae (embryophyta), fungi (Eumycota) and the protoctists (Margulis 1981; see also Cavalier-Smith 1990, 1991). Only the phyletic primary plastids descend directly from symbiotic associations between prokaryotes and eukaryotes (first sym-

apparatus of cyanobacteria and plastids is homologous and very complex, it must be excluded that oxygen evolving photosynthesis has evolved twice and convergently. Therefore, in view of the endogenous hypotheses, the ur-eukaryote must have been evolved directly from cyanobacteria or from cells very similar to cyanobacteria. That means that from the very beginning the ur-eukaryote must have been a photoautotrophic, a 'plant' cell, whereas the heterotrophic eukaryotic cells of fungi and animals must have been developed later from 'plant' cells. In view of the exogenous theories the ur-eukaryote was a heterotrophic, phagocytosing, protist like cell with a cytoskeleton and so, the 'plant' cell must be a later evolutionary product of eukaryotes. In endogenously evolved cells all different 'genomes' derive from one ur-genome. Exogenous evolution results in a cell, which is a complex product of combination or fusion due to the participating and interacting genomes. Qualitatively the product is more than the sum of the parts, that means not merely a simple product of addition (like a symbiosis). In a qualified sense and at least with regard to the DNA information one could compare this combination with a nuclear-chemical fusion process. In this view the product must be attested its own (non-symbiotic) evolutionary history.

In the following we will focus our interest only on the evolution of 'phyletic primary plastids' comparing it with the evolutionary stage of the Glaucocystophyta.

3. The Glaucocystophyta

3.1 The eukaryotic cell

The classification of the division Glaucocystophyta Kies et Kremer 1986b (Glaucophyta Skuja 1954), a group of eukaryotic, unicellular or colonial, monadoid, capsalean or coccoid, due to the chromatophores bluish green freshwater algae, with one class, the Glaucocystophyceae Schaffner 1922, and three orders (see Table 1), has been revised by Kies and Kremer (1986b) on the background of biochemical and ultrastructural data (see also Cavalier-Smith 1987).

The Glaucocystophyceae (for more detail see Kies 1979, especially for characterization, recognition and cultivation Kies and Kremer 1990) are characterized mainly by a system of peripheral lacunae, by an apical depression, if motile by two subapical inserted heterokont and heterodynamic flagella (Kies 1979; Heimann *et al.* 1989), both with non-tubular mastigonemes, representing dorso-ventrally shaped motile stages. The location of dictyosomes is parabasal (in *Glaucosphaera* perinuclear). Mitochondria have flattened cristae. Starch grains without envelope membrane lie free in the eukaryotic cytoplasm (Schnepf *et al.* 1966; Kies 1979). Cells reproduce by mitosis, sexuality is absent. During the mitosis an open spindle forms, centrioles and phycoplast were not found; the nuclear membrane fragments during the prophase (for *Glaucosphaera* unknown). Cytokinesis takes place by the infurrowing of the plasma membrane (for *Glaucosphaera* unknown).

3.2 The chromatophore and the cyanelle/cyanoplast problem

The bluish-green chromatophores of Glaucocystophyta, firstly termed 'cyanelles' by Pascher 1929 (= symbiotic cyanobacteria, see also Glossary), are responsible for the photoautotrophy of the whole algal cell. It is relatively easy to isolate the chromatophores, but it is impossible to cultivate them (Trench *et al.* 1978; Trench 1982; Marten *et al.* 1982; Kies 1984a), a result which has opened early the symbiont/organelle problem of the cyanoplasts (Herdman and Stanier 1977, Löf-

---

biosis developing plastids). The phyletic secondary plastids are relics of photoautotrophic eukaryotic endocytobionts (second plastid developing symbiosis) and may still consist of membranes and organelles of the eukaryotic endocytobiont including the real plastid. If such a eucyte with a secondary plastid will be newly endocytosed and accepted as endocytobiont by a third host, then a third mix (IITC-see Glossary) of DNA on the nuclear genome of the new genetic hypersystem (GHRS) could follow. Abbreviations: S: symbiogenesis; Chloro: Chloro- and Prasinophyceae; Crypto: Cryptophyceae; Cy.p.: *Cyanophora paradoxa* as representative of the Glaucocystophyceae; Eugl: Euglenophyceae; Het: Heterokontophyta (Bacillariophyceae, Chloromonadophyceae, Chrysophyceae, Phaeophyceae, Xanthophyceae); ITC, InterTaxonic Combination, see Glossary; Pre-Chlo: hypothetic Pre-Chlorophyceae; Rhod: Rhodophyceae; **Y**: 'origin cascade' of new species during the processes of IITC; ! : loss of flagellum.

Table 1. Glaucocystophyta Kies et Kremer 1986 (Glaucophyta Skuja 1954) Classification and availability of strains from Culture Collections of Algae, doubtful species are not considered, see (1), (2)

Metacyanomes	Cyanoplasts	Cult. Coll.	Org.	References
Glaucocystophyceae Schaffner 1922				
Glaucocystales Bessey 1907				
Glaucocystaceae West 1904				
Glaucocystis Itzigsohn 1868			cc	
G. bullosa	(mb) ?			(2)
G. duplex	(mb) ?			(2)
G. nostochinearum	mb ?(*1)	A	cc	(3)
G. oocystiforme	(mb) ?			(2)
Archaeopsis monococca	(mb) ?			(2)
Glaucocystopsis africana	(mb) ?			(2)
Cyanophorales Kies et Kremer 1986				
Cyanophoraceae Kies et Kremer 1986				
Cyanophora Korshikov 1924			m	
C. paradoxa	mb Y (*2)	B		(4)
C. tetracyanea	(mb) ?			(2)
Peliaina cyanea	(mb) ?			(2)
Strobilomonas cyaneus	(mb)(?)			(2)
Gloeochaetales Kies et Kremer 1986				
Gloeochaetaceae Skuja 1954				
Gloeochaete Lagerheim 1883				
G. wittrockiana	mb ? (*3)	C	m/c	(5)
Cyanoptyche Pascher 1929	?			
C. gloeocystis f. *dispersa*	mb ?(*4)	D	m/p	(6)
Glaucosphaeraceae Skuja 1954	(?)			
Glaucosphaera Korshikov 1930	(?)			
G. vacuolata	mf (?)	E	c	(7)
Chalarodora Pascher 1929				
C. azurea	(mf)(?)			(8)

The Glaucocystophyta will be shown here not as endocyanomes as Pascher thought (Kies and Kremer 1986a) but as 'metacyanomes' (Schenk 1992a), that means, as descendants of original endocyanomes (Pascher's definition 1929), and therefore as new species, originated by gene transfer from former endocytobiont to former 'host' nucleus. Another important group of metacyanomes seems to be the algal class of Cyanidiophyceae with the one family Cyanidiaceae including at least 3 genera (*Cyanidioschyzon merolae* (9), *Cyanidium caldarium* (9) and *Galdieria sulphuraria* (9), see chapters by Seckbach and by Ott, this volume).

Strain number and Algal Culture Collection: A: CCAP 229/1 = UTEX 64 = SAG B 229/1 (isolated by George, 1952), IABH 2344 (Kies, 1975), SAG 27.80 (Tassigny), SAG 28.80 (Brown), SAG B 229/2 (Pringsheim, 1955), SAG b 229/3 (Lewin), UTEX B 1929 (Lewin); B: CCAP 981/1 = UTEX LB 555 = SAG 29.80 (Pringsheim, 1943), IABH 1555 (Kies, 1967); C: IABH 2323 = CCAP 54/1 = SAG B 46.84 (Kies, 1973); D: IABH 2643 (Kies, 1989); E. UTEX LB 1662 = CCAP 130/1 (Starr). – CCAP: Culture Collection of Algae and Protozoa, Cambridge, CB3 ODT, England; IABH: Institut für Allgemeine Botanik der Universität, D-2000 Hamburg, Germany, (Collection of L. Kies); SAG: Sammlung für Algenkulturen, Pflanzenphysiologisches Institut der Universität, D-3400 Göttingen, Germany; UTEX: The Culture Collection of Algae, University of Texas at Austin, Texas 78712, U.S.A..

Abbreviations: mb: murein bearing cyanoplast, (*1), (*2): occurrence of a murein sacculus (peptidoglycan) has been proven (see text), (*3): preliminary biochemical indications for the presence of a murein sacculus were found (Kies 1988, Kies and Schenk unpublished data), (*4): murein sacculus probably present (Kies 1989); mf: murein free cyanoplast; Org.: Organization of the metacynome: a) vegetative cells: c = capsalean, cc = coccoid, m = monadoid, b) m/ = monadoid zoospore; Y: cyanelle-cyanoplast conversion (gene transfer) has been 'proven' (Schenk *et al.* 1992); ?: cyanelle-cyanoplast conversion is very probable; (?): uncertain systematic position of the algal species (including the open question, if the photosynthetic compartment is a cyanoplast, a rhodoplast, or a 'cryptoplast')

References: (1) Pringsheim 1958, (2) Kies and Kremer 1986b, 1990, (3) Griffiths 1915, Schnepf *et al.* 1966, Scott *et al.* 1984, (4) Schenk 1970, Herdman and Stanier 1977, Schenk *et al.* 1992, (5) Kies 1976, (6) Kies 1989, (7) Kies 1980, McCracken *et al.* 1980, (8) Pascher 1929, (9) Seckbach *et al.* 1983, Seckbach 1992.

felhardt *et al.* 1980; Marten and Brandt 1984; Schenk 1977). In gross morphology observed in the light microscope as well in fine structure according to electron microscopical investigations the chromatophores are indeed similar to certain free-living cyanobacteria. Therefore, Hall and Claus raised them into the rank of cyanobacterial species: *Cyanocyta korschikoffiana* (1963) in *Cyanophora paradoxa* and *Skujapelta nuda* (1967) in *Glaucocystis nostochinearum*. That was obviously wrong, because these cell compartments, with chloroplast like functions, are metabolically (Floener and Bothe 1983) as well as (up to 90 %) genetically dependent on nuclear (eukaryotic) DNA (Bayer and Schenk 1986). For that reasons the term 'cyanoplasts' (cell organelles as descendants of cyanelles) was proposed (Schenk and Hofer 1972; Schenk *et al.* 1987b; Schenk 1990, 1992a), and it was suggested to retain the given species names for the yet unknown cyanobacterial progenitors. The pigments of the cyanoplasts were chlorophyll a, β-carotene, zeaxanthin, C-phycocyanin and allophycocyanin (Chapman 1966) (in *Glaucosphaera vacuolata* additionally R-phycocyanin) and in less concentration β-cryptoxanthin (Schmidt *et al.* 1979, predicted by Schenk 1977). Typical cyanobacterial carotenoids are absent. Dinitrogen fixation (a property of many cyanobacteria) could not be found (Floener and Bothe 1982) and it remains still a question if these chromatophores never have had the property to fix dinitrogen or if they have it lost during evolution. The cyanoplasts of Glaucocystophyta can be of globular (coccoid) or more longish (rodlet) structure. They are surrounded, except the cyanoplasts of *Glaucosphaera*, by remnants of a cell wall consisting of a murein sacculus (peptidoglucan) demonstrated biochemically for *Cyanophora* (Schenk 1970, 1977; Heinz 1973; Aitken and Stanier 1979) and for *Glaucocystis* (Scott *et al.* 1984). This rudimentary wall around the cyanoplasts (of *Cyanophora*, *Cyanoptyche*, *Glaucocystis*, *Gloeochaete*, Kies 1984b) is one of the main proofs that cyanoplasts are phylogenetically derived from endocytobiotic cyanobacterial ancestors (= (endo)cyanelles). The cyanoplasts have regularly an outer chromatoplasm with concentrically arranged thylakoids and a lighter centroplasm. In the centroplasm a dense polyhedral body, a carboxysome (Blank 1985; Mangeney and Gibbs 1987) is either centrally located (e.g. *Cyanophora*, *Gloeochaete*) or asymmetrically at the smaller end of the cyanoplast (*Glaucocystis*). The carboxysomes of *Cyanoptyche* (Kies 1989) and *Gloeochaete* (Kies 1976) are surrounded by 'some sort of membrane which is not a unit-membrane'. (For deeper introduction, also with regard to the cyanelle/cyanoplast problem, see the reviews by Pascher 1929; Geitler 1959; Pringsheim 1958; Schenk 1977; Trench 1981; Wasman *et al.* 1987; Schenk *et al.* 1987b; Kies and Kremer 1990; Kies 1992; Schenk 1990; 1992a; Bohnert and Löffelhardt 1992). In *C. paradoxa*, the polyploid, circular DNA is located within the space between the chromatoplasm and around the carboxisome (Figure 2, and Bohnert *et al.* 1983).

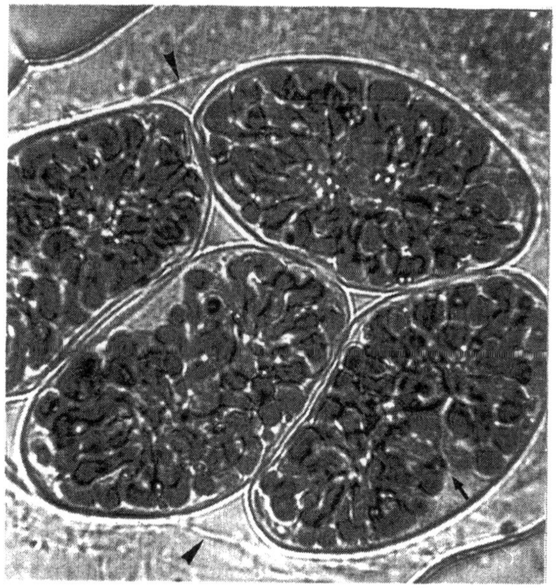

Fig. 1. Glaucocystis nostochinearum: four autospores (full of 'longish' cyanoplasts; position of nucleus, see arrow) within relics of the mother cell wall (triangles) (light microscopy, 1250:1; J. Hanf).

3.3 Glaucocystis nostochinearum *Itzigs. (Itzigson 1868)*

3.3.1 The eukaryote

Representatives of the genus *Glaucocystis* are distributed around the world (e.g. Prasad 1961; Fenwick 1966; Tell 1979). They show asexual reproduction by binary cell division. Cyanoplast division and cytokinesis are only loosely coupled and partially independent (Colwell and Wickstrom 1976).

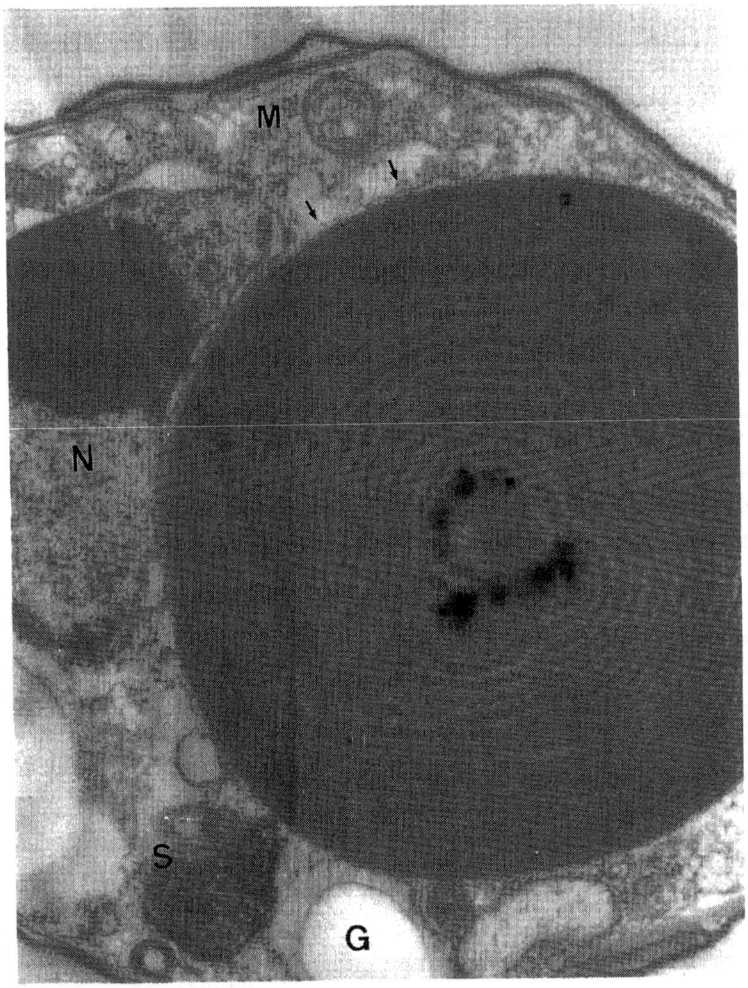

Fig. 2. Cyanophora paradoxa: part of a cell with nucleus (N), mitochondrion (M), slime droplet (S), starch grain (G), and under the pellicula/plasmalemma with scale like structures, partly enveloped by membranes (lacunae, Kies 1979); cyanoplast with centroplasm, with thylakoids and phycobilisomes within the chromatoplasm, and enveloped by an inner envelope membrane and a closely attached murein sacculus, and by an undulating (arrows) outer envelope membrane (40000:1; M. Amiressami).

Ultrastructural investigations about *G. nostochinearum* were published by Ueda 1961; Schnepf 1965; Lefort 1965; Echlin 1966; Schnepf 1966; Schnepf and Koch 1966; Schnepf *et al.* 1966; Hall and Claus 1967; Robinson and Preston 1971 a,b; Schnepf and Brown 1971; Willison and Brown 1978; Kies 1979; Scott *et al.* 1984. The coccoid alga (Figure 1) is characterized by a thick, multilayered cell wall consisting of cellulose fibrils. The wall can be maintained during several life cycles (until 3), in the manner, that daughter cells (until 8) are pairwise surrounded by the walls of the respective autospore mother cells. The peripheral lacunae remind of the earlier pellicle. Also the two rudimentary flagella under the cell wall are an indication for the earlier monadoid nature of the alga.

3.3.2 The cyanoplast

The longish, nearly hook-like and polarly organized cyanoplasts bear a pyrenoid, probably a carboxysome, on the smaller end. In autospores they are centripetally arranged around one, in older cells around two coordination centers of the cell with the smaller pyrenoid side directing to the coordination center. The two envelope membranes were investigated by electron microscopy (Lefort-Tran 1981). After two negative results (Holm-Hansen *et al.* 1965; Schenk 1971) the murein sacculus was detected in an excellent manner by Scott *et al.* (1984). Nothing is known about a nucleus related, genetic dependency of the chromatophores. Therefore it is not absolutely sure, if they are still cyanelles or already cyanoplasts. I call it in analogy to the relations in *C. paradoxa* cyanoplasts. Notwithstanding that is still to prove. For pigments see 3.2. The lipid pattern (digalactosyldiglyceride, phosphatidylglycerol and probably monogalactosyldiglyceride) was described preliminary and only in part (Scott 1987), and also the fatty acid pattern is not investigated systematically, until now. Preliminary data of GC-investigations (Schenk, not published) show a quantitatively different, but qualitatively similar pattern to *C. paradoxa* (presence of polyunsaturated eikosenoic acids, as arachidonic and eikosapentaenoic acid). Photosynthetic carbon-dioxide assimilation, first mentioned by Schenk and Hofer (1972) due to indications of the Calvin-Benson Cycle and a short light enhanced dark fixation period of carbon dioxide, was described more in detail by Kremer *et al.* (1979). The results of the last authors suggest that, from a taxonomic point of view, *Glaucocystis* may not be regarded as a rhodophycean representative as proposed by Schnepf and Brown (1971).

3.4 Cyanophora paradoxa *Korsch. (Korschikov 1924)*

3.4.1 The eukaryote

This monadoid alga (Fig. 2) represents the best investigated species within the Glaucocystophyceae (for a broad information the reader is directed to the reviews by Trench 1982; Wasman *et al.* 1987; Schenk *et al.* 1987b; Kies and Kremer 1990; Bohnert and Löffelhardt 1992; Schenk 1990). The two described *C. paradoxa* strains show remarkable differences in the size of cyanoplast DNA: the 'Pringsheim' strain with 127 kbp, the 'Kies' strain with 136 kbp (Löffelhardt *et al.* 1983). This cypDNA diversity between the two strains resembles plastid DNA diversity in order of different *Chlamydomonas* species. Multiplying (asexual reproduction) of the flagellate (Kies 1979) goes by binary fission, incompletely synchronized with cyanoplast division. The flagellar development cycle is described by Heimann *et al.* 1989. The flagellate has no phototactic orientation, but shows a weak photokinetic effect (photophobic response) which is dependent on photosynthetic pigments (carotenoids excluded) and the linear photosynthetic electron transport (inhibition, e.g. by DCMU, Häder 1985). It is to assume that this behavior is a result of coevolution of the original symbiotic partners. Indeed, *C. paradoxa* can now been considered as the classical example for a long-lasting coevolution of two originally symbiotic partners which led finally to a new species. Contrarily to *Euglena* the flagellate is today obligate photoautotroph (Provasoli and Pintner 1952) and it is not possible to cultivate the separated and isolated compartments: the isolated cyanoplasts (Trench 1981, Trench *et al.* 1978) and the cyanoplast-free eukaryote (Kies 1988). Nitrogen uptake goes via nitrate. Ammonia (1.3 mM) is a strong cell poison for *C. paradoxa*. The eukaryotic cell compartment is responsible for the nitrate assimilation and reduction to nitrite (up to 25 μmol/mg Chl*h). The following nitrite reduction takes place partially in the cytoplasm, but mainly in the cyanoplast compartment (Boettcher *et al.* 1982; Floener *et al.* 1982).

The cytoplasmic space contributes to the cyanoplast phosphatidylcholin and eikosenoic acids (Zook and Schenk 1986; Schenk 1990). The more or less known metabolic relationships between cyanoplast and cytoplasmic compartment are summarized in Figure 3.

3.4.2 The cyanoplast and subcellular concerted actions

a) Envelope and pigments. The inner envelope membrane of *C. paradoxa* cyanoplast resembles cyanobacterial plasma membranes (freeze-fracture technique) and is dissimilar to the plastid envelope membranes of red or green algae (Giddings *et al.* 1983). This membrane and the surrounding peptidoglucan (murein sacculus) are — unlike to chloroplasts — directly involved in the division mode of the cyanoplast (Hall and Claus 1963; Kies 1984a). The pigments (Chapman 1966; Trench and Ronzio 1978; Schmidt *et al.* 1979) are comparable with those of *Glaucocystis* (see above). The typical cyanobacterial carotenoid echinenone is absent. With regard to a quantitative estimation of carotenoids, chlorophyll a, phycocyanin and allophycocyanin a method for routine measurements by *in vivo* VIS (= visible light) spectroscopy was given by Schenk at al. (1983).

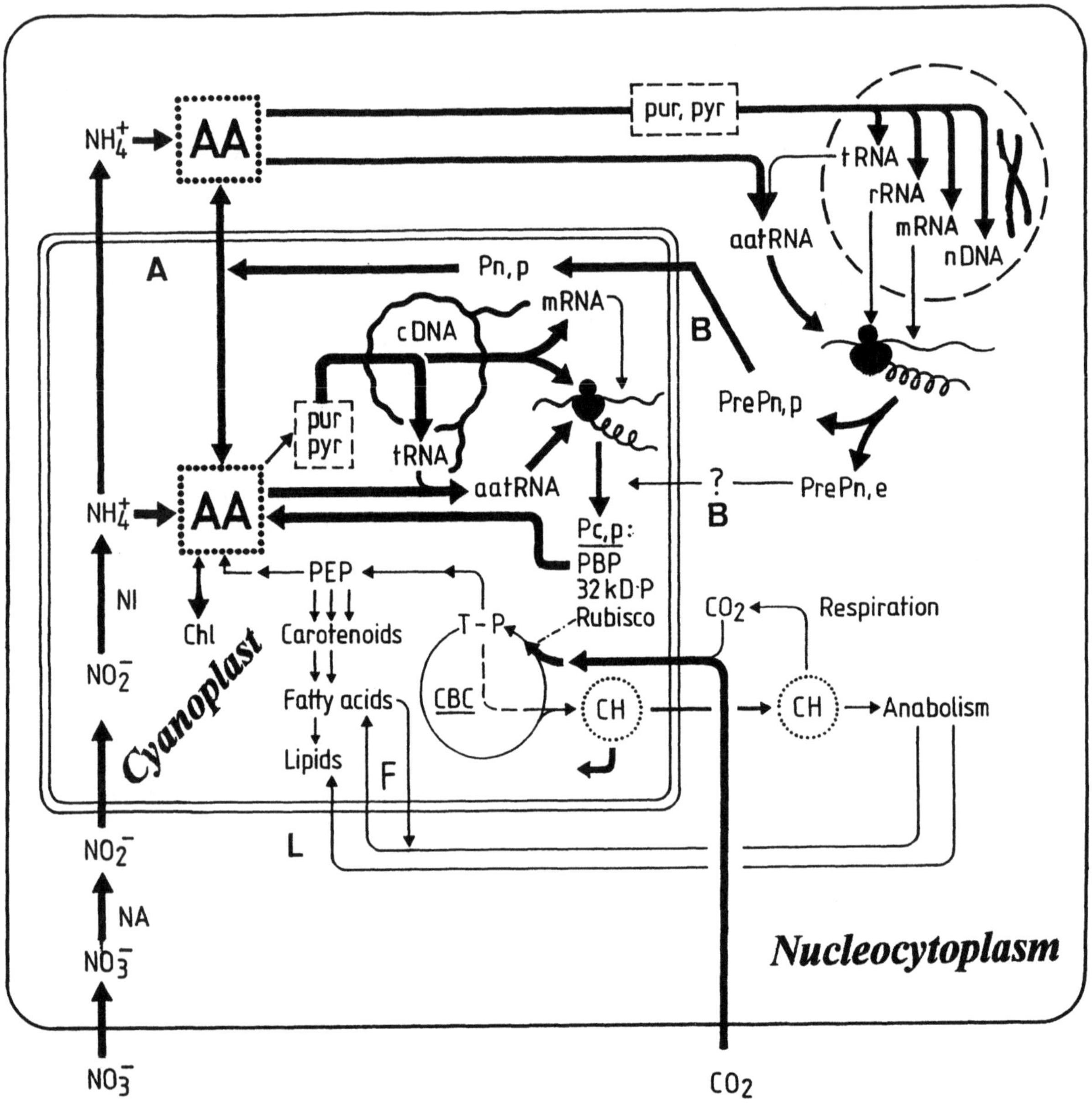

Fig. 3. Cyanophora paradoxa. Metabolic and genetic relationships between the nucleocytoplasm (outer rectangle) and the cyanoplast (inner rectangle) compartment.

Abbreviations: A: export of ammonia and amino acids from cyanoplast to cytoplasm; AA: presumed communicating amino acid pools in the cytoplasm and in the cyanoplast; B: import of nucleus encoded proteins; CBC: Calvin-Benson-Cycle; CH: carbohydrate pools; F: (regarding the cyanoplast) export of C18 fatty acids and import of C20 fatty acids; L: possible phospholipid import (overall phosphatidylcholine); NA: nitrate reductase; NI: nitrite reductase; PEP, phosphoenolpyruvate; Pn.p: mature cyanoplast protein (nucleus encoded, originally prokaryotic, like FNR = Ferredoxin-NADP-Oxidoreductase); PrePn,p: nucleus encoded cyanoplast preprotein, like preFNR; PrePn,e: hypothetical nucleus encoded preprotein (originally eukaryotic) for the cyanoplast compartment (proposed protein substitution or protein supplement, as yet not found).

Michalowski *et al.* (1990a) described a third phycobiliprotein, a second allophycocyanin (apcD gene, Table 3), also coded on cyanoplast DNA. Chlorophyll biosynthesis is inhibited by both chloramphenicol and cycloheximide (Neu-Müller and Schenk 1983).

b) Photosynthesis and metabolite exchange. Tracer studies showed that upon ^{14}C carbon-dioxide photoassimilation the radioactivity is found in the metabolites of the Calvin-Benson-Cycle, but there is also a rapid incorporation into dicarboxylic acids (Schenk and Hofer 1972; Trench *et al.* 1978; Kremer *et al.* 1979). The latter authors discuss the glucose as major carbohydrate translocated from cyanoplasts into the cytoplasm. Fixed carbon is stored mainly at two sites, in the phycobiliproteins of the cyanoplasts and in envelope-free starch grains in the cytoplasm. Catabolism is supported by an incomplete tricarboxylic acid cycle (Schenk and Hofer 1972; Kremer *et al.* 1979).

Exchange of metabolites between the cyanoplast compartment and cytoplasm was investigated by Schlichting *et al.* (1990) and by Schenk (1990, most unpublished). Both groups have found that sucrose and dicarboxylic acids penetrate the cyanoplast envelope with a very weak rate, unlike that of chloroplasts; a slightly higher penetration rate was measured for ATP. Basic amino acids (Schenk, unpublished) and inorganic phosphate (Schenk

Table 2. Fatty acids (as % of total fatty acid content) of *C. paradoxa* in comparison with other photosynthetic organisms

	%						
Fatty acid	Eustig. 1)	C. par. 2)	P. cru. 3)	P. aer. 4)	C. cal. 5)	Cyano. 6)	Crypto. 7)
10:0	n.d.	n.d.	–	–	–	–	–
12:0	n.d.	n.d.	–	–	–	–	–
14:0	6–7	t	2	8	t	+,++C	+,++C
15:0	n.d.	t	–	–	–	–	–
16:0	27–34	39	23	30	8	+,++B	++A
16:1	27–32	?	2	7	t	+	+A
18:0	1	1	2	11	7	+	+C
18:1	10–14	7	3	28	34	+	+,++C
18:2	1–2	6	16	8	11	+A,++B	+B
18:3	–	1	t	3	30	+A	+,++B
18:4	n.d.	–	–	–	–	+C	++A
20:0	n.d.	–	–	–	–	–	+C
20:1	n.d.	–	–	–	–	–	++B
20:2	n.d.	–	–	–	–	–	–
20:3	n.d.	t–2	2	–	–	–	–
20:4	2–5	15–22	36	–	–	–	+C
20:5	8–22	22–33	17	–	–	–	+,++B
22:0	n.d.	1–2	–	–	–	–	–
22:1	n.d.	–	–	–	–	–	–
22:2	n.d.	–	–	–	–	–	–
22:3	n.d.	?	–	–	–	–	–
22:4	n.d.	6	–	–	–	–	+B
22:5	n.d.	–	–	–	–	–	+B
22:6	n.d.	–	–	–	–	–	+B

References: 1) Eustigmatales (*Nanochloropsis* sp. strain 1982/NL-2 and strain 1982/NL-3 (Menzel and Wild 1989); 2) *C. paradoxa*, whole organism (Schenk *et al.* 1985, Kleinig *et al.* 1986, Zook *et al.* 1986); 3) *Porphyridium cruentum.* (Nichols and Appleby 1969); 4) *Porphyridium aeruginosum* (Wood 1974); 5) *Cyanidium caldarium*, (20°C) (Kleinschmidt and McMahon 1970), after Ikan and Seckbach (1972). *C. caldarium* has only traces of linolenic acid; 6) Cyanobacteria (Kenyon *et al.* 1972); 7) Cryptomonadales, 10 species (Wood 1974).

Abbreviations: n.d. = not determined; – = not found; ? = too little for estimation; t = less then 1%, + = 1–10%, ++ = 10%; A = found nearly in all investigated species, B = in the half or most, C = only in a few; fatty acids: 16:0, palmitic; 18:1, oleic; 18:2)9c, 12c), linoleic; 18:3, α- and γ-linolenic; 20:4, (5c, 8c, 11c, 14c, arachidonic) and (8c, 11c, 14c, 17c); 20:5 (5c, 8c, 11c, 14c, 17c), eikosapentaenoic acid.

1990) displayed the best penetration rates among all substrates investigated. The inhibition of inorganic phosphate influx by 0.1 mM SH-reagents (mersalic acid, p-hydroximercuribenzoate, N-ethylmaleimide, p-chloromercuriphenylsulfonate) is low, about 25%, but the competition with 5 mM 3-phosphoglycerate is high, about 80% (unpublished). Therefore it is to conclude that cyanoplasts have no dicarboxylic acid translocator and also the phosphate influx cannot be compared with the phosphate translocator of chloroplasts. A very interesting observation reported Schlichting *et al.* (1990). They reported a relatively high penetration rate for sorbitol, which is remarkable different to the poor uptake capabilities of chloroplasts. Sorbitol is often used as osmoticum in chloroplast isolation buffers. In conclusion it is obvious that the semipermeability of chloroplasts and cyanoplasts differ in decisive points.

c) Fatty acids and lipids. An unexpected observation was the fatty acid pattern of the alga (Table 2). In addition to palmitic acid (C16) polyunsaturated eicosenoic acids [C-20, e.g., arachidonic (C20:4) and eicosapentaenoic (C20:5) acids] were discovered (Schenk *et al.* 1985; Kleinig *et al.* 1986; Zook *et al.* 1986). The eicosenoic acids, as yet not found in cyanobacteria (Table 2), are the main fatty acids, also within the thylakoid membranes. There is evidence that they are synthesized by elongation and desaturation (of cyanoplast C-18 fatty acids) in the cytoplasmic space and then retransferred (Fig. 4, Schenk 1990). This pathway is so much more surprising because the cyanoplast is still able to synthesize at least oleic and linoleic acid. From my point of view incorporation of higher unsaturated eicosenoic acids into thylakoid lipids represents a typical case of 'metabolite substitution' (Schenk 1990, 1992b, Fig. 6).

Firstly Kremer (1982) has reported about lipids in *C. paradoxa*, namely monogalactosyl diglyceride (MGDG), phosphatidyl glycerol (PG) and phosphatidyl inositol (PI) or/and phosphatidyl serine (PS). Later (Zook and Schenk 1986) cyanoplasts were found to contain (in decreasing order) MGDG, DGDG, phosphatidyl choline

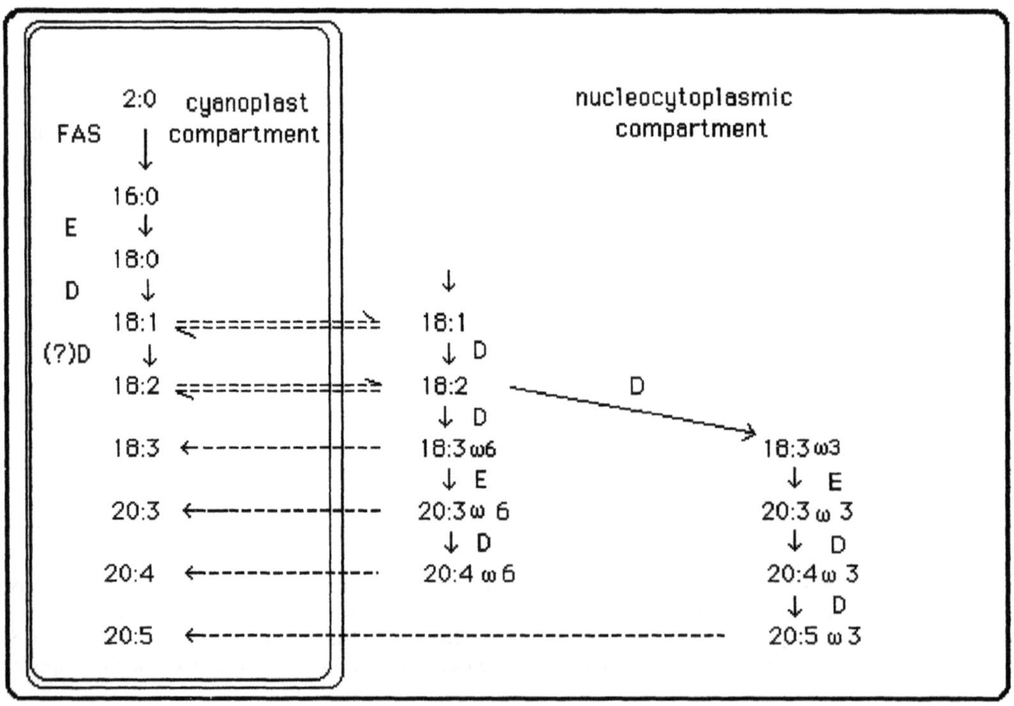

Fig. 4. Cyanophora paradoxa, compartmentation of fatty acid biosynthesis.
Abbreviations: D, desaturase; E, elongase; FAS, fatty acid synthetase complex (proposal for main pathways, the 'eukaryotic' FAS is not considered) (Schenk 1990, altered).

(PC), PG and sulfoquinovosyl diglyceride, whereas the flagellate membranes (including mitochondria) are composed mainly by PC and phosphatidyl ethanolamine. In whole cells, PI (but not PS) could be re-detected (Gottwald *et al.*, unpublished data). The main fatty acids (C16:0, C20:4, C20:5) are present in all listed lipids, but in different relation (q), also depending on the cell compartment in which they are located (Zook and Schenk 1986): e.g. in cyanoplast lipids is q (20:4/20:5) < 0.5 and in cytosolic lipids > 3. In PI and PE was q > 5 (whole cells) indicating their cytoplasmic origin and occurrence. In PI the C20:4 and C20:5 content was very low against the content of C16:0, C18:1, C18:2 and C20:3. In my view this is a further indication for the biosynthetic pathway of the eikosenoic acids. The q (20:4/20:5) of PC (whole cells) was found a mean value of 1.3 to 2.6 indicating the occurrence of PC in two pools: in cytoplasmic as well as in cyanoplast membranes.

d) Nitrogen assimilation and storage. Dinitrogen fixation could not be detected, nor the nitrogen storage within cyanophycin granules [Arg-poly (Asp) = multi-L-arginyl-poly(L-aspartic acid)], the typical cyanobacterial mode. Similar to chloroplasts of higher plant cells, the cyanoplasts reduce only nitrite, but not nitrate (Floener *et al.* 1982). There is a considerable controversy whether the reduced nitrogen is released into the cytoplasmic space as ammonia (Floener *et al.* 1982) or as amino acid (Schenk and Hofer 1972; Boettcher *et al.* 1982;

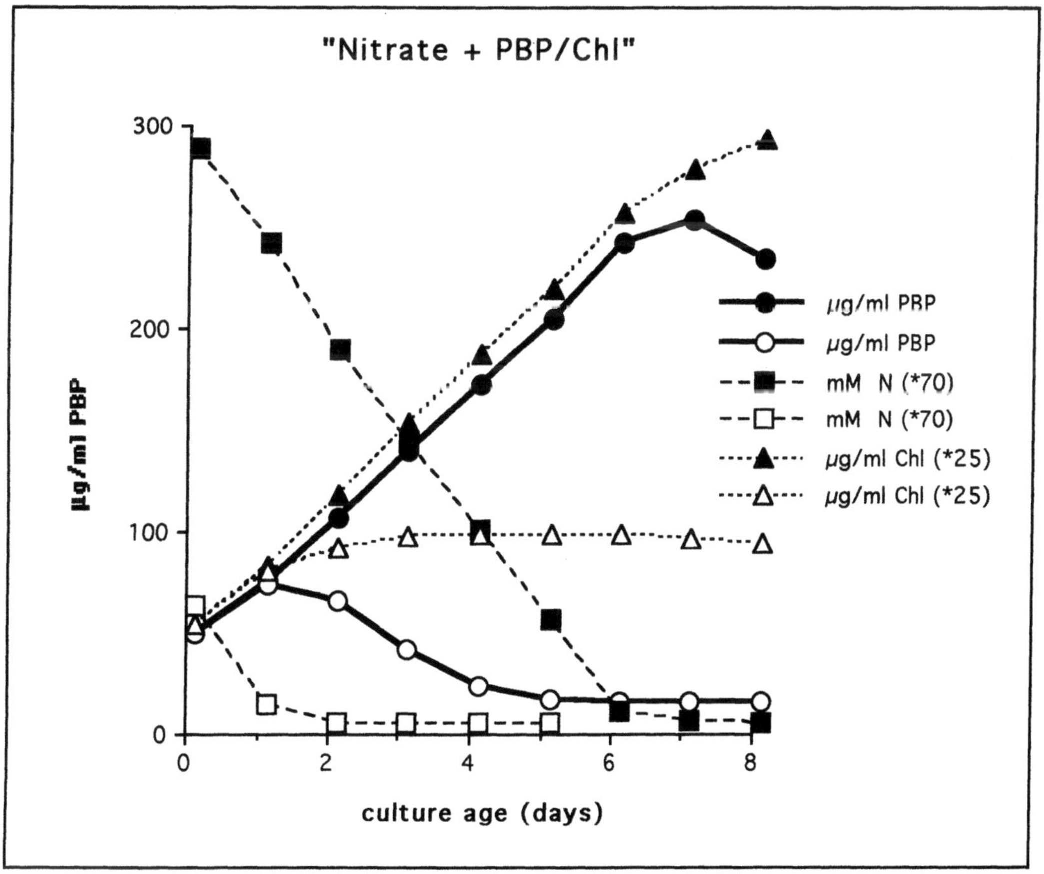

Fig. 5. Cyanophora paradoxa: Nitrogen storage function of phycobiliproteins. Nitrate assimilation (squares), biosynthesis/biodegradation of phycobiliproteins (circles) and biosynthesis of chlorophyll, depending on nitrate content of the medium. Two experimental series with different original nitrate concentrations in the medium. (empty signs: ca. 1.3 mM nitrate, filled signs: ca. 6.6 mM nitrate); equal algal concentrations (relative cell number: 4 million cells/ml); nitrate assimilation about 25 µmol/mg Chl*h (borrowed from Schenk *et al.* 1987).
Abbreviations: Chl, chlorophyll a; N,nitrate; PBP,phycobilliproeins (allophycocyanin and phycocyanin together).

Schenk unpublished data). It is noteworthy that the phycobiliproteins (PBP) in *Cyanophora paradoxa* function not only as accessory pigments but also as nitrogen (and carbon) storage proteins (Schenk *et al.* 1983, 1987a, Fig. 5; compare Marten *et al.* 1981).

The regulation of PBP content depends on the nucleus: The synthesis of PBP is inhibited by both chloramphenicol and cycloheximide, but the decomposition of PBP is inhibited only by cycloheximide (Schenk *et al.* 1983). This observation supports the idea that amino acid pools are present in both compartments and correspond through the cyanoplast envelope membranes by permeation. — In contrast to *C. paradoxa* the PBP in *Glaucocystis nostochinearum* do not seem to share in nitrogen storage functions (Schenk *et al.* 1987a).

e) Proteins, biosynthesis and coding sites. Jehn and Zetsche (1988) reported about *in vitro* synthesis of cyanoplast proteins by isolated cyanoplasts and *in vitro* translation by cyanoplast RNA. Gene localization of known cyanoplast proteins is compiled in Table 3 (in-depth information about the molecular biology of *C. paradoxa* cyanoplasts is found in the reviews by Löffelhardt *et al.* 1990 and by Bohnert and Löffelhardt 1992). Unlike the plastome of green algae and embryophyta the cyanoplast genome codes for the small subunit (SSU) of RUBISCO (Heinhorst and Shively 1983), which was also found for some chromophytic (Boczar *et al.* 1989; Valentin and Zetsche 1990a; Assali *et al.* 1991) and for some rhodophytan species (Valentin and Zetsche 1990b). On the cyanoplast DNA (cDNA) there are also the genes for ferredoxin (Bayer and Schenk 1989) and for eight ribosomal proteins (Löffelhardt *et al.* 1990; Löffelhardt *et al.* 1992), which are not present on chloroplast DNA. But notwithstanding, about 90% of the soluble and membrane bound cyanoplast proteins are encoded on the nuclear genome as demonstrated by crossed inhibition experiments for translation and transcription (Bayer and Schenk 1986; Schenk *et al.* 1987; Burnap and Trench 1989a).

Table 3. Cyanophora paradoxa. Gene localization of cyanoplast proteins on the cyanoplast or nuclear genome in comparison with the situation in chloroplasts of higher plants (from T.L. Maier, with permission)

Gene	Gene product	Gene location			References
		C. paradoxa		Chloroplasts	
		Cyanoplast	Nucleus		
Components of photosystem I					
psaA	Apoprotein P700a	+		+	1, 3, 37
psaB	Apoprotein P700b	+		+	1, 3, 37
petH	Ferredoxin-NADP-oxidoreductase	–	+	–	18
Components of photosystem II					
psbA	D-1 (herbicide binding)-protein	+		+	1, 3, 4, 37
psbB	Apoprotein P680 (51 kD)	+		+	1, 3, 37
psbC	Apoprotein (44 kD)	+		+	1, 3, 37
psbD	D-2 protein	+		+	1, 37
psbE	Apoprotein cytochrome b-559	+		+	1, 30
psbF	Apoprotein cytochrome b-559	+		+	1, 30
psbI	I-protein	+		+	30
psbJ	J-protein	+		+	5, 30
psbK	K-protein	+		+	6
psbL	L-protein	+		+	5
ATP-synthase complex					
atpA	α-subunit, CF1	+		+	1, 3, 4, 37
atpB	β-subunit, CF1	+		+	1, 3, 4, 37
	γ-subunit, CF1	+		+	19
atpE	ϵ-subunit, CF1	+		+	1, 4

Table 3. Continued

Gene	Gene product	Gene location: *C. paradoxa* Cyanoplast	Gene location: *C. paradoxa* Nucleus	Gene location: Chloroplasts	References
atpF	subunit I, CF1	+		+	17
atpH	subunit III, CF0	+		+	3, 44, 37
Cytochrome-b6/f complex					
petA	Apoptotein cytochrome f	+		+	1, 4
petB	Apoptotein cytochrome b6	+		+	3, 37
petD	subunit IV	+		+	3
petG	subunit V	+		+	7, 37
petFI	Ferredoxin I	+	–	–	8, 20, 22
Phycobiliproteids					
pcyA	α-subunit, C-phycocyanin	+		n.p.	1, 2, 13, 37
pcyB	β-subunit, C-phycocyanin	+		n.p.	1, 2, 13, 37
apcA	α-subunit, allophycocyanin	+		n.p.	1, 2, 14, 37
apcB	β-subunit, allophycocyanin	+		n.p.	1, 2, 14, 37
apcD	δ-subunit, allophycocyanin	+		n.p.	16
apcE	Linkerphycobiliprotein LCM100	+		n.p.	15
	Linker polypeptides (L 1-4)	–	+	n.p.	19
Ribulose-1,5-bisphosphate carboxylase/oxygenase					
rbcL	Large subunit	+		+	9, 1, 2, 3, 4, 24, 37
rbcS	Small subunit	+		–	9, 1, 2, 3, 37
Ribosomal proteins (small subunit)					
rps3	Protein S3	+		+	28, 29
rps4	Protein S4	–		+	17
rps5	Protein S5	+		–	28, 29
rps7	Protein S7	+		+	17, 26, 28
rps8	Protein 8	+		+	10, 28, 29
rps10	Protein 10	+		–	21, 22, 28
rps11	Protein 11	–		+	17
rps12	Protein 12	+		+	17, 26, 28, 29
rps17	Protein 17	+		–	28, 29
rps18	Protein 18	+		+	12
rps19	Protein 19	+		+	3, 12, 28, 37
Ribosomal proteins (large subunit)					
rpl2	Protein L2	+		+	12, 28
rpl3	Protein L3	+		–	11, 12, 28
rpl5	Protein L5	+		–	10, 28, 29
rpl6	Protein L6	+		–	10, 28, 29
rpl14	Protein L14	+		+	28, 29
rpl16	Protein L16	+		+	28, 29
rpl18	Protein L18	+		–	28, 29
rpl20	Protein L20	+		+	10
rpl22	Protein L22	+		+	12, 28, 29
rpl33	Protein L33	+		+	12
rpl35	Protein L35	+		–	10
rpl36	Protein L36	–		+	29
Translation					
infA	Initiation factor IF-1	–		+	17
rpoA	RNA-polymerase, α-subunit	–		+	17

Table 3. Continued

Gene	Gene product	Gene location: *C. paradoxa* Cyanoplast	Gene location: *C. paradoxa* Nucleus	Gene location: Chloroplasts	References
Components of photosystem I					
tufA	Elongation factor Tu	+		−	26
fus	Elongation factor G	−		−	26, 34
° RF65	Open reading frame 65	+		+	25
° RF102	Open reading frame 102	+		−	28
° RF492	Open reading frame 492	+		+	29
nadA	Quinolinate synthetase	+			27
crtE	Prephytoene PP dehydrogenase	+			23
16SrDNA	16SrRNA	+		+	35, 36, 37
23SrDNA	23SrRNA	+		+	35, 36, 37
5SrDNA	5SrRNA	+		+	33
trnA	Alanine-tRNA	+		+	35, 36
trnI	Isoleucine-tRNA	+		+	35, 36
trnN	Asparagine-tRNA	+		+	36
trnR	Arginine-tRNA	+		+	36
trnD	Aspartyl-tRNA	+		+	36
trnE	Glutamine-tRNA	+		+	32
trnG	Glutamyl-tRNA	+		+	31
trnH	Histidine-tRNA	+		+	36
trnL	Leucine-tRNA	+		+	32
trnK	Lysine-tRNA	+		+	36
trnM	Methionine-tRNA	+		+	36
trnF	Phenylalanine-tRNA	+		+	36
trnP	Proline-tRNA	+		+	36
trnS	Serine-tRNA	+		+	36
trnT	Threonine-tRNA	+		+	36
trnW	Tryptophane-tRNA	+		+	36
trnY	Tyrosine-tRNA	+		+	36
trnV	Valine-tRNA	+		+	36

Legends: (+), detected; (−), not detected; (n.p.), protein not present in chloroplasts of higher plants. References: 1, Lambert *et al.* 1985; 2, Starnes *et al.* 1985; 3, Bohnert *et al.* 1985; 4, Ko *et al.* 1985; 5, Cantrell and Bryant 1987; 6, Stirewalt and Bryant 1989a; 7, – 1989b; 8, Bayer and Schenk 1989; 9, Heinhorst and Shively 1983; 10, Bryant and Stirewalt 1990; 11, Evrard *et al.* 1990a; 12, – 1990b; 13, Lemaux and Grossman 1984; 14, Bryant *et al.* 1985; 15, Bryant 1988; 16, Michalowski *et al.* 1990a; 17, Wasmann *et al.* 1987; 18, Bayer *et al.* 1990; 19, Burnap and Trench 1989b; 20, Neumann-Spallart *et al.* 1990; 21, Neumann-Spallart *et al.* 1991; 22, Bryant *et al.* 1991; 23, Michalowski *et al.* 1991a; 24, Valentin and Zetsche 1990; 25, Evrard *et al.* 1990; 26, Kraus *et al.* 1990; 27, Michalowski *et al.* 1991b; 28, Löffelhardt *et al.* 1992; 29, Michalowski *et al.* 1990b; 30, Cantrell and Bryant 1988; 31, Kuntz *et al.* 1988; 32, Evrard *et al.* 1988; 33, Maxwell *et al.* 1986; 34, Löffelhardt *et al.* 1990; 35, Janssen *et al.* 1987; 36, Kuntz *et al.* 1984; 37, Breiteneder *et al.* 1988.

Burnap and Trench (1989b) gave the first electrophoretic indications to precursor proteins containing 'transit sequences' of 1.5 to 7 kDa. Recently it was possible to characterize the first specific protein, the photosynthetic enzyme ferredoxin-NADP$^+$-oxidoreductase (FNR), that is coded on nuclear DNA but destined for the cyanoplast (Bayer *et al.* 1990). After isolation and purification of mature FNR the N-terminal amino acid sequence was determined (Gebhart *et al.* 1992). After screening a cDNA expression library and cloning FNR-cDNA the nucleotide sequence of the full length protein including the transit peptide of 65 or 66 amino acids was analyzed (Jakowitsch *et al.* 1991; Jakowitsch *et al.* 1993). Together these results advance the knowledge twofold: (a) FNR is the first sequenced nuclear encoded preprotein of *C. paradoxa*, which is imported into cyanoplasts by a still unknown translocation mechanism, and (b) it is the first protein for which it is possible to demonstrate directly the gene transfer from the original symbiont (in phylogenetic terms after the developmental sequence: cyanobacterium → cyanelle → cyanoplast) to the corresponding host genome (see later).

4. Glaucocystophyta as model for eukaryotic cell evolution

Up to that point I have given a general description of some characteristics of the Glaucocystophyta. Now, I shall discuss what the Glaucocystophyta is making such fascinating model organisms with regard to the Endosymbiosis Theory. Firstly, I shall reflect upon different coevolutionary constraints within an endocytobiotic consortium, secondly I shall consider processes which change or determinate a symbiotic relationship. Then we show on the one side properties which indicate that cyanoplasts are descendants of cyanobacteria (Table 4), and on the other, we look at properties of cyanoplasts which are notwithstanding typical for plastids (Table 5). Furthermore, I shall demonstrate adaptations of the eukaryotic cell (of Glaucocystophyta) as a result of the former endocytobiosis (Table 6). The last point will be the demonstration of a possibility to prove gene transfer, a main demand of the Endosymbiosis Theory and a way leading to symbiogenous new species.

4.1 Endocytobiotic coevolution of symbiotic partners

In the introduction I focused on mechanisms which could have led to a composite biological system. In the main that were symbiogenous processes. This description requires some closer consideration: After the installation of the symbiotic relationships and in their following coevolutionary development we can discriminate (partially following the nomenclature of Schaede 1962) three basic endocytobiotic ways, three distinct positions of the 'symbiotic continuum' (Starr 1975), occurring also in dependency on the ontogenetic stage of the partners:

a) the dyssymbiotic (parasitic) way,
b) the parasymbiotic (neutral) way and
c) the eusymbiotic (mutual) way.

Table 4. Cyanobacterial characteristics of *C. paradoxa* cyanoplasts

Morphology:	coccoid shape like *Gloeocapsa* sp.	(1)
Anatomy:	inner envelope membrane	(2)
	chromatoplasma	(1)
	concentrically arranged thylakoids	(3)
	centroplasma	(1)
	division mode: infurrowing of inner envelope membrane and murein sacculus	(4)
Physiology:	incomplete tricarboxylic acid cycle	(5)
Biochemistry:	murein sacculus (peptidoglucan)	(6)
	hydrogenase properties	(7)
	in contrast to chloroplasts no inhibition of rRNA synthesis by cycloheximide	(8)
	presence of C-phycocyanin	(9)
Coding side:	ferredoxin	(10)
(in contrast	rRNA	(11)
to CPLs)	Rubisco SSU	(12)
Sequences:	amino acid-, nucleotide-, see text	

Abbreviation: CPLs, chloroplasts of embryophytes.
References: (1) Korschikov 1924; (2) Giddings *et al.* 1983; (3) Hall and Claus 1963; (4) Kies 1984a; (5) Schenk and Hofer 1972; Kremer *et al.* 1979; (6) Schenk 1970, 1977; Aitken and Stanier 1978; (7) Bothe and Floener 1978; (8) Trench and Siebens 1978 (9) Chapman 1966; Trench and Ronzio 1978; (10) Bayer and Schenk 1989; Neumann-Spallart *et al.* 1990; (11) Hori and Osawa 1987; (12) Heinhorst and Shively 1983; see also Table 3.

Table 5. Organellar characteristics of *C. paradoxa* cyanoplasts

Anatomy:	absence of cyanobacterial storage bodies (starch, polyphosphate, cyanophycin)	(1)
Physiology:	no respiration	(2)
	no nitrate reduction	(3)
	nitrogen storage regulation by nucleus	(4)
	polyunsaturated eikosenoic acids	(5)
Genome size:	ca. 127/137 kbp (like chloroplasts)	(6)
Polyploidy:	ca. 40 copies (like chloroplasts)	(7)
G+C content:	higher than in cyanobacteria	(8)
Genome structure:	inverted repeats	(9)
Coding capacity:	only about 10% of cyanoplast proteins (about 90% of proteins nucleus encodes)	(10)
	FNR nucleus encoded	(11)

References: (1) Schnepf and Brown 1971; (2) Floener and Bothe 1982; (3) Floener *et al.* 1982; (4) Schenk *et al.* 1987a; (5) Schenk *et al.* 1985; (6) Herdman and Stanier 1977; (7) Mucke *et al.* 1980; (8) Klein *et al.* 1981; (9) Wasmann *et al.* 1987; (10) Bayer and Schenk 1986; Burnap and Trench 1989a; (11) Bayer *et al.* 1990.

Table 6. C. paradoxa, synergism of cyanoplast and nucleocytoplasm, a result of coevolution and evolution

Photoautotrophy	(1)
Fatty acid biosynthesis	(2)
Membrane composition (fatty acids and lipids)	(3)
Nitrate assimilation	(4)
Nitrogene storage	(5)
Light perception and motility	(6)

References: (1) Pringsheim 1958; Trench *et al.* 1978; Trench 1982; (2) Zook *et al.* 1986; Kleinig *et al.* 1986; Schenk 1990; (3) Zook and Schenk 1986; Zook *et al.* 1986; (4) Floener *et al.* 1982; Böttcher *et al.* 1982; (5) Marten *et al.* 1981; Schenk *et al.* 1983, 1987a; (6) Häder 1985.

The dyssymbiotic way normally offers mechanistic possibilities of an infection by invasion and colonization of the macrobiont by the microbiont, or by ingestion of microsymbionts which thereby reach the intracellular host compartments. Virulent or aggressive reactions can originate from either the micro or macrosymbiont. If the one partner survives the first attacks of the other then dyssymbiosis consists of responsive rounds of attack and defense. Such mechanisms are also found in symbiotic relations which are classified under eusymbioses (e.g. endomycorrhizae or Rhizobiaceae-Fabales symbiosis).

The parasymbiotic way consists of mutual toleration of the partner (Toleration Phase) during or after which the microbiont is regularly transformed into a parasome by degeneration (Degeneration Phase). A case in point are the more or less indifferent 'particles' in ciliates. Thus the Degeneration Phase parallels or follows the Toleration Phase. The endocytobiont participates as commensalist in the metabolites of the host, but unlike predators does not use them for logarithmic growth (density independent strategy, Ahmadjian and Paracer 1986). That means that the endocytobiont does not exploit its own repertoire of metabolic pathways. The biological system as a whole accepts random deletions in some of the endocytobiont's (and host's) pathways which in turn induces the Degeneration Phase of the cytobiont. Here it should be recognized that such 'macrodeletions' change the endocytobiont into an entity comparable to a 'new' species, which in terms of metabolism is not able to live as autonomously as the original symbiont did. The processes of the Degeneration Phase, I think, justify the subcategory parasymbiotic or even antisymbiotic way.

Two parallel evolutionary trends are also observed for the eusymbiotic way, which often are not clear-cut enough in order to discriminate: (a) the optimalization of the symbiotic association (acceleration of fitness of the symbiotic association, Margulis 1981; Law and Lewis 1983) and (b) the subjection of the microsymbiont (in the sense of de Bary's symbiosis definition a kind of antisymbiotic processes). Adaptation and innovation or in the sense of Lewis (1985) integration of the partners are understood as 'symbiotic processes' or 'symbiosis advancing processes', e.g. after invasion or ingestion of guest: growth adaptation of the endocytobiont by K-strategy (Ahmadjian and Paracer 1986), growth regulation by developing growth inhibiting substances, guest retention, positioning, transmission or metabolic adaptation. Already most of these adaptation steps fall within the terms of colonization or domestication (Smith and Douglas 1987). Insofar, in part, these events are also standing under the regime of a 'Subjection Constraint', which can be seen in such processes as e.g. guest growth regulation by partial change of symbiosomes into phagosomes, to strong specialization or adaptation by degeneration of the endocytobiont. The 'Optimalization Constraint' can evolve symbioses with more or less equal autonomous partners or in combination with the 'Subjection Constraint' at least (by ending the symbiotic relation) a new cell type.

4.2 The end of intracellular symbiosis (exogenosome formation)

Whatever all these different constraints cause, the processes finally terminate in either of two outcomes after the long row of 'symbiotic' and 'antisymbiotic' adaptive intracellular and later also morphological (intra- and extracellular) adaptation steps: (a) a final deletion or loss of the symbiosome or (b) induced by a Gene Combination Constraint an unequal Intertaxonic Combination (IITC). IITC means a mainly one-directional gene transfer from endocytobiont to host in combination with further necessary accompanying processes (Sitte 1991), the genetic integration of the endocytobiont into the genetic system of the host, in reality a genomic fusion process. The genomic hypersystem (GHRS) integrates by IITC in time

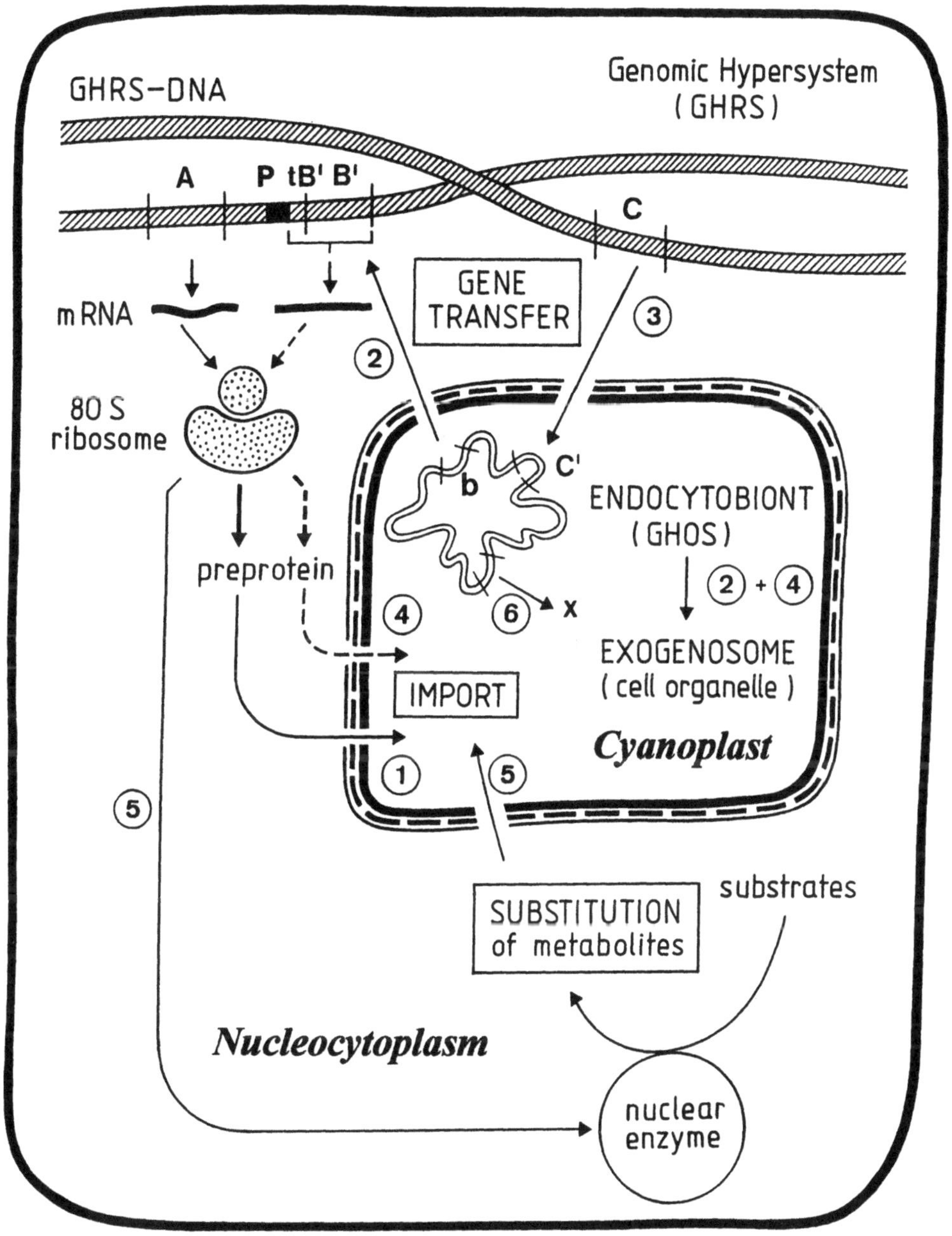

Fig. 6. Symbiosome/exogenosome transformation by Intertaxonic Combination. Transformation of a(n) (endo)cytobiont towards a cell organelle, some molecular adaptation steps (Schenk 1992b,altered).

1. Import of nucleus encoded 'host' proteins (protein substitution or protein supplement); 2. Up-gene-transfer, gene transfer from cytobiont to host nucleus (from GHOS to GHRS) for definitions see Glossary; 3. Down-gene-transfer, gene transfer from GHRS (nucleus) to the GHOS (genome of the endocytobiont or exogenosome); 4. Protein import (of an exogenous nucleus encoded protein into the exogenosome following a successful up-gene transfer); 5. Metabolite substitution in the epi/endocytobiont; 6. Gene deletion in the GHOS (epi/endocytobiont, exogenosome DNA).

Abbreviations: A, an original nuclear gene; b, an original endocytobiontic gene; B^1, endocytobiontic gene transferred and incorporated into nuclear DNA; P, promotor region of the new nuclear gene; tB^1, gene region of the transit sequence for the expression product of B^1.

more and more the genetic information of the genomic hyposystem (GHOS) in its own genome complex. Thereby the genomic hyposystem looses its genetic autonomy; the GHOS remains with incomplete DNA information and hence requires DNA information from the GHRS for its ontogenetic formation or completion. At that point the final transition of the (sym)biont into a cell organelle, symbiosome into an 'exogenosome' (*per definitionem* a cell organelle, Schenk 1992b) is reached. This interpretation is in agreement with the Cavalier-Smith Demand (Cavalier-Smith and Lee 1985) that the gene transfer from symbiont to host changes the symbiont into a cell organelle (Figure 6).

Besides reception, retention and continual intracellular 'adaptation' steps this course of events (IITC) needs two mechanisms: (a) the unequal transfer of genes between the genomic hyposystem and the genomic hypersystem and (b) during the more or less strong coevolution the installation of the post-translational protein import machinery with regard to the now GHRS nucleus encoded proteins of the former endocytobiont (GHOS). Obviously, there are complicated and complex processes involved, which are not really understood but just start to become amenable for partial description. For all the earlier improbability of these events to have happened it seems the more important to find model organisms, which offer to prove an unequal Intertaxonic Combination. Such model organisms should fulfill at least two properties: (a) the cell compartment of interest (symbiont or 'exogenosome') should be recognizable without doubt (according to our current knowledge) as descendant of an originally autonomous endocytobiont, and (b) the adaptation should be advanced so far that at least one gene transfer event (more precise an IITC event) has already happened between the GHR and GHO system. Clearly, model organisms with such properties counterargue the Cluster-Clone Hypothesis (Bogorad *et al.* 1973) or Gene Duplication Segregation Hypothesis (Parthier *et al.* 1975) and the Fusion Hypothesis.

4.3 Cyanoplast, model for proof of gene transfer

4.3.1 Cyanoplast, an exogenosome

Analyzing the cyanoplast of *C. paradoxa* as representative for the chromatophores of Glaucocystophyta it can be differentiated between properties resembling those of a free-living prokaryote (Table 4) and properties which share with organellar characteristics of other plastids (Table 5). There is no doubt that cyanoplasts have maintained properties, in particular the rudimentary cell wall (see 3.2) and some molecular biological features (see later data), which clearly point to their cyanobacterial ancestry (Table 4). On the other hand, properties listed in Table 5 clearly demonstrate that 'cyanelles' of *C. paradoxa* have lost the status of endosymbionts and reached the status of eukaryotic cell organelles (see definitions in Glossary). They can no longer be regarded as symbiotic cyanobacteria (= cyanelles), since cyanelles are *per definitionem* genetically autonomous symbionts like zoochlorellae, while cyanoplasts require nuclear encoded protein like chloroplasts or rhodoplasts.

The subcellular coordination of properties of the original guest with properties of the original host cell of *C. paradoxa* as presently known is summarized in Table 6. This overview makes clear that the coordination processes opened a new evolutionary alley for the ancestors of the present day cells of *C. paradoxa*, which neither partner could have pursued on its own.

At present it is not possible to decide at which developmental stage the adaptations exactly took place. Was there a coevolution of the endosymbiont (cyanelle) and the host, while still in the stage of symbiotic consortium? Or was there mainly an evolution of the formed metacyanome after a first successful intertaxonomic Combination (ITC)? In any case it is remarkable to which high degree the genomic hypersystem (here genome of the heterotrophic flagellate) was able to integrate properties of the genomic hyposystem (genome of the former cyanobacterium) and to use them for new evolutionary strategies.

What makes now Glaucocystophyta to representatives and to the analogical main model for the evolution of primary plastids? In Table 7, some answers and considerations are summarized. Up to the present period the metacyanomic Glaucocystophyta are the only known group of organisms which demonstrates with very high probability, that 'primary plastid' like cell organelles are descendants of forgoing endocytobiotic prokaryotes. Also that symbiotic con-

Table 7. What makes the Glaucocystophyta to symbiogenous metacyanomes, to representatives of new species?

Properties developed during or after symbiosis	Species of Glaucocystophyta			
	G.nost	*C.para*	*G.witt*	*C.gloe*
a) The new cell organelle	+ +	+ +	+ +	+ +
b) Loss of endocytosis	+ +	+ +	+ +	+ +
c) Loss of symbiosome/ phagosome conversion	+ +	+	(+ +)	(+ +)
d) Photoautotrophy	+ +	+ +	+ +	+ +
e) Degeneration of flagellae	+ +	−	+	+
f) Light dependent motility	−	+	n.d.	n.d.
g) New cell wall properties	+ +		+	+ +
h) Nuclear guest genes	(+ +)	+ +	(+ +)	(+ +)
i) Evolved new nuclear genes	+ +[1]	+[2]	n.d.	n.d.

Abbreviations and legend: *C.gloe: Cyanoptyche gloeocystis, C.para: C. paradoxa, G.nost: G. nostochinearum, G.witt: Gloeochaete wittrockiana;* + +: property evident; (+ +): property assumed; +: property observed, but weak expressed or not completely evolved; −: not found; n.d.: not determined; +[1]: e.g. cell wall; +[2]: e.g. light dependent motility.

sortia like cyanomes (exactly the as yet still not known original cyanomes) are able to change by IITC in new species (metacyanomes, e.g. the recent Glaucocystophyta). In theory the continual integration of different properties of the GHOS into the GHRS should result in a cascade of new species. Unfortunately no data exist on the phylogenetic relationships within the group. Hence it remains an open though challenging question for future research to find out whether the cyanoplasts of the Glaucocystophyta are of monophyletic or polyphyletic origin.

4.3.2 Attempt to proof the gene transfer

Hitherto, the model may be convincing with regard to an exogenous origin of the *C. paradoxa* chromatophore. Furthermore, the observation that numerous cyanoplast proteins are encoded on nuclear DNA makes gene transfer very probable, the proof though is not yet provided. With other words, only in case of the demonstration of a successful IITC it will be allowed to accept the assumption that the prerequisites for the Endosymbiosis Theory are fulfilled. At that point it needs to resort to the available molecular sequence data. On the basis of such data it is to show that the two compared originally symbiotic genetic systems in a recent cell (e.g. here those of the cyanoplasts and of the nucleocytoplasm) differ still fundamentally and that their past histories can be traced back to different simple organisms (demand of Gray and Doolittle 1982). This requirement holds true if the subcellular structures of *C. paradoxa* are compared as outlined in the forgoing sections. And this notion is also supported by sequence comparison of proteins and rRNA. For example, ferredoxin is a typical cyanobacterial protein regarding sequence comparison and phylogenetic trees have been constructed (Stevanovic *et al.* 1989; Neumann-Spallart *et al.* 1990; Lüttke 1991). Likewise cluster analyses with 5S rRNA (Van den Eynde *et al.* 1988) and 16S rRNA partial sequences (Giovannoni *et al.* 1988) showed a close relationship of *C. paradoxa* cyanoplasts with cyanobacteria. The cyanoplasts of *C. paradoxa* are found within the cyanobacteria-plastid cluster but rather contained within the cyanobacterial radiation. Following Wolters *et al.* (1990), the sequence of the cyanoplast 5S rRNA groups with most or all plastids, so far investigated, nearest to the cyanobacterial cluster. On the other hand, interestingly, the cyanoplast 16S rRNA forms the earliest branch in the plastid line rooting in the cyanobacterial cluster (Wolters *et al.* 1990). Whereas in a phylogenetic tree, the eukaryotic (nuclear) 5S rRNA of *C. paradoxa* is closest to the Euglenozoa (Hori and Osawa 1987) and far apart from its cyanoplast counterpart.

Let me recall that the compartmentation hypotheses state, that plastids evolved endogenously and more or less simultaneously with the eukaryotic cell as a whole by gene and/or genome duplication followed by DNA segregation and compartmentation via membrane enveloping. Subsequently the two formed genomes would drift further apart due to their different microenvironment and they need to develop new properties. With regard e.g. to the photosynthetic complex and to other important cellular properties the two segregated genomes would have been specialized, in main also by different types of deletions. Basically, such evolution processes should have been driven by *necessity* rather then by *chance*. In explanation of this type of evolution the assumption of gene transfer is not necessary. Therefore neither the small plastid genome nor the found import of proteins into the plastid-compartment can be used as proof for a gene transfer. Of course, in this moment of our consideration, these arguments hold true also for cyanoplasts.

In order to view gene transfer as support for the Endosymbiosis Theory, it must be found in a symbiotic association or extant organisms 'which without doubt, have evolved from a symbiotic consortium', e.g. metacyanomes. Though surveying the literature one deepens the impression, that the problem on the origin of plastids is settled and gene transfer proven. I think closer inspection makes this conviction arguable. Here I recall some examples discussed in relation to an indication of gene transfer: e.g. plastid-gene fragments of rRNAs, the gene of the plastome encoded large subunit (LSU) of the plastid-enzyme ribulosebisphosphate carboxylase in the chondriome (mitochondrial genome) of maize, or an active gene copy for the subunit 9 of ATPase in the nucleus and a silent one in the chondriome in *Neurospora crassa* (reviews: Schuster and Brennicke 1988; Behrens *et al.* 1990). A particular convincing case for an indication of a 'late' gene transfer is given by the location of the gene for the plastid-protein synthesis elongation factor, tufA. The gene is encoded by the plastome of cyanoplasts and plastids of chlorophytes and euglenophytes, but found on nuclear DNA in the embryophyte *Arabidopsis thaliana* (Baldauf and Palmer 1990). Regarding the homologous sequence in the cyanobacterium *Anacystis* and overall sequence comparison the authors rule out a gene substitution. Obviously these events were phylogenetically younger (on the evolutionary time scale later) than a 'symbiogenous' gene transfer. That means that these events happened at a time when the protein import machinery was already installed, while no such import machinery was present at the very first (original) gene transfer. For that reason we should take recourse to model organisms like the Glaucocystophyta, which 'under the veil of the eukaryote' still display the symbiotic partners (Table 4). This assertion is the fundamental statement of the following attempt of a direct indication of an 'old', 'primary' gene transfer (Schenk *et al.* 1992).

Already the following earlier results let speculate a gene transfer in *C. paradoxa* because: a) the cyanoplast genome comprises only 5% of that of recent cyanobacteria (Herdman and Stanier 1977), and b) cyanoplasts depend for about 90% of their proteins on nuclear encoded genes (Bayer and Schenk 1986). Though these discoveries let presume a gene transfer, neither of these observations is a direct proof for it. The detected protein import could also be the result of an evolutionary protein substitution (Schenk *et al.* 1987b, see also Figs. 6 and 7.IIb). Therefore it was proposed (Schenk 1992b) to investigate by sequence comparison the 'history' of cyanoplast proteins which are encoded on the nuclear genome and imported into the cyanoplast compartment.

With such protein at hand several possibilities exist to explain its origin (Fig. 7). In scenario a) (Fig. 7.I) the investigated protein has an endogenous (autogenous) origin by endogenosome development, in b) (Fig. 7.II) the protein could originate from the host which had it evolved endogenously together with a forgoing plastid ('protein substitution' of a cyanoplast protein by a 'host' protein, Schenk *et al.* 1987b, or by protein supplement from the 'host' genome), and in c) (Fig. 7.III) the protein originates really from the former endocytobiotic cyanobacterium (IITC). In the cases a) and b) the amino acid sequence should be significantly more similar to sequences of the recent plastid proteins than to cyanobacterial proteins, in the case c) more similar to the cyanobacterial proteins. Case a) can be excluded for the cyanoplast of *C. paradoxa* (see the upper fundamental statement). The possibility of case b) must still be excluded by argumentation. This way is indeed thinkable because e.g. some diatoms are known which harbor cyanelles (e.g. the marine *Rhizosolenia* with the cyanobacterial *Richelia intracellularis* or the freshwater alga *Rhopalodia gibba*, see Drum and Pankratz 1965; Floener and Bothe 1980; Lowe *et al.* 1984). Here, the prerequisite would be the replacement of the (endogenously or exogenously developed) original host plastid by the endocytobiotic cyanobacterium (cyanelle).

Ferredoxin-NADP oxidoreductase (FNR) was now not only the first purified FNR from a protist (Gebhart *et al.* 1992) but also the first reliably detected nucleus encoded cyanoplast protein of *C. paradoxa*. (Bayer *et al.* 1990). Amino acid sequencing (24 N-terminal residues, Gebhart *et al.* 1992) and nucleotide sequencing of the full length protein

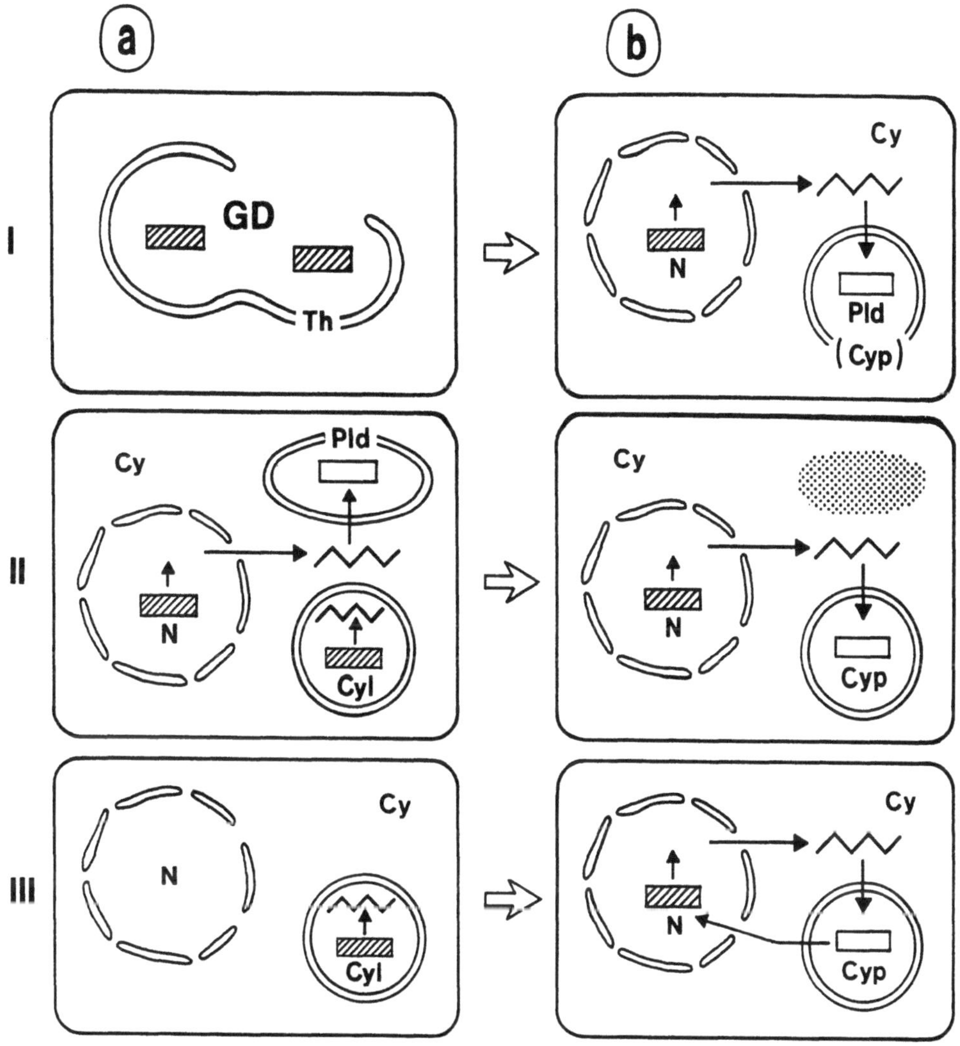

Fig. 7. Origin of a nucleus encoded cyanoplast protein. Illustration of possible evolutionary ways (Schenk *et al.* 1992, altered). I. Endogenosome protein development (endogenously by genome duplication/ segregation/ compartmentation); II. Exogenosome protein development (by gene product substitution: no proof for gene transfer and therefore no proof for exogenous origin of primary plastids); III. Exogenosome protein development (by unequal intertaxonic combination). Further explanations see text. Abbreviations: ▨ : gene of cyanoplast protein, active for expression; ▭ : gene of cyanoplast (or plastid) protein, disturbed regarding the expression (inactivated, deleted or transferred); ⋀⋀⋀ : gene product (protein) for the cyanoplast; Cy: cytoplasm, Cyl: cyanelle; Cyp: cyanoplast; GD: genome duplication; N: nucleus; Pld: plastid; Th: thylakoid.

including the transit sequence (Jakowitsch *et al.* 1993) allows a comparison with the known sequences from 3 higher plants and 3 cyanobacteria, for which the N-terminal portion is shown in Table 8 (Schenk *et al.* 1992). It is noteworthy that the *Synechococcus* sequence (Schluchter and Bryant 1992) contains an N-terminal stretch of about 90 amino acid residues with probable linker function, which is not present in the other cyanobacterial sequences and therefore makes the alignment difficult. For the alignment (used in Table 8) it was assumed that the gaps, found in the other cyanobacteria and *C. paradoxa*, also exist in the *Synechococcus* sequence. The similarity scores of the N-terminus of the *C. paradoxa* sequence are closer to the scores of the cyanobacteria than to the higher plants (Table 8). The cross comparison for the N-terminal part (gaps considered) shows two clearly separable clusters, one with 56–72% identity for the cyanobacterial sequences including the cyanoplast one, and the other with 64–77% identity for the embryophytes (Table 9, upper right

Table 8. Ferredoxin-NADP oxidoreductase (amino acid partial sequences)

Organism	Position 1	10		20		30		40		Similarity (%) a)	b)
A:	XQ-AVD	-----	---AK	-----	KKG--	DIPLN	LFRPA	NPYIG	K..	100	100
B:	------	-----	---AK	-----	--T--	DIPVN	IYKPK	NPYIG	K..	62	72
C:	-M--TQ	-----	---AK	-----	AKHA-	DVPVN	LYRPN	APFIG	K..	57	69
D:	xKI-VH	-----	---PK	-----	TTDT-	SVPVN	IYRPK	TPFLG	K	38	56
E:	XQIASD	VEAPP	PAPAK	VEKHS	KKMEE	GITVN	KFKPK	TPYVG	R..	43	33
F[c)]	XV-ASD	VEAPV	---AK	VEKHS	KKMEE	GVIVN	KYKPK	NPYTG	R..	38	38
G:	XQ--VT	TEAP-	---AK	VVKHS	KKQDE	NIVVN	KFKPK	EPYVG	R..	43	38

N-terminal amino acid partial sequence of mature *C. paradoxa* FNR and alignment with known N-terminal sequences of mature FNRs from other organisms: A: *Cyanophora paradoxa* (Gebhart *et al.* 1992; Jakowitsch *et al.* 1993) (it is not finally decided if the cleavage side is before or after Q (Jakowitsch *et al.* 1993), B: *Spirulina* sp. (Yao *et al.* 1984), C: *Anabaena* sp. PCC 7119 (Fillat *et al.* 1990), D: *Synechococcus* sp. PCC 7002 (Schluchter and Bryant 1992), E: *Spinacia oleracea* (Karplus *et al.* 1984; Jansen *et al.* 1988), F: *Mesembryanthemum crystallinum* (Michalowski *et al.* 1989), G: *Pisum sativum* (Newman and Gray 1988); x, proposed sequence continuation of *Synechococcus* FNR (Schluchter and Bryant 1992), alignment after Gebhart *et al.* 1992 considering the gaps; X, transit sequence known; –, amino acid omissions for maximum alignment with respect to the spinach enzyme. Similarity (%, Sneath and Sokal 1963) with respect to the partial sequence of *C. paradoxa* FNR: a), comparison of positions 21-41; b), maximum alignment as described within this table, comparisons (including all gaps) within positions 3 to 41; c), after Michalowski *et al.* (1989) begins the mature protein with three further amino acids: IRA(V-AS..).

Table 9. Ferredoxin-NADP-oxidoreductase: comparison of amino acid sequences. Mature protein of *C. pardoxa* (A) against the corresponding sequences of cyanobacteria (B to D) and eukaryotes (E to G) (references in Table 8)

		% identity/similarity						
Species		A	B	C	D	E	F	G
Cyanophora paradoxa	A:	–	72	69	56	33	38	38
Spirulina sp.	B:	62	–	69	67	28	38	41
Anabaena variabilis	C:	58	70	–	69	21	33	33
Synechococcus sp.	D:	59	72	64	–	21	31	33
Spinacia oleracea	E:	65	56	52	53	–	77	64
Mesembryanthemum crystallinum	F:	66	58	53	53	91	–	64
Pisum sativum	G:	65	57	55	55	89	87	–

Upper right triangle (similarity), N-terminal partial sequence (positions 3 to 41, including all gap positions, Table 8. b; Schenk *et al.* 1992, altered). Lower left triangle (identity), comparison of all 321 amino acid positions of the mature protein, but all 24 gap positions excluded (Jakowitsch *et al.* 1993).

triangle). Taking the full length sequences into account (gaps excluded), there are two clusters also, but the cyanoplast sequence does not definitely fall into one or the other (Table 9, lower left triangle). The unrooted phylogenetic tree (Jakowitsch *et al.* 1993), calculated with the neighbor joining method (Saitou and Nei 1987) using a distance matrix based on a log normal model, shows a similar picture (Fig. 8): two main branching sides (one for the eukaryotic, one for the cyanobacterial proteins) and a third branch in the middle between the two other groups, somewhat closer to the cyanobacteria.

Interestingly, the cyanobacterial FNR sequences and that of the *C. paradoxa* cyanoplast have long 'branches' (indicating an old phylogenetic age) in contrast to the eukaryotic sequences with shorter 'branches'. From these data it could be concluded that the nucleus encoded FNR in *C. paradoxa* descends from a cyanobacterial sequence, but it occupies a bridge position between the cyanobacterial sequences and that of em-

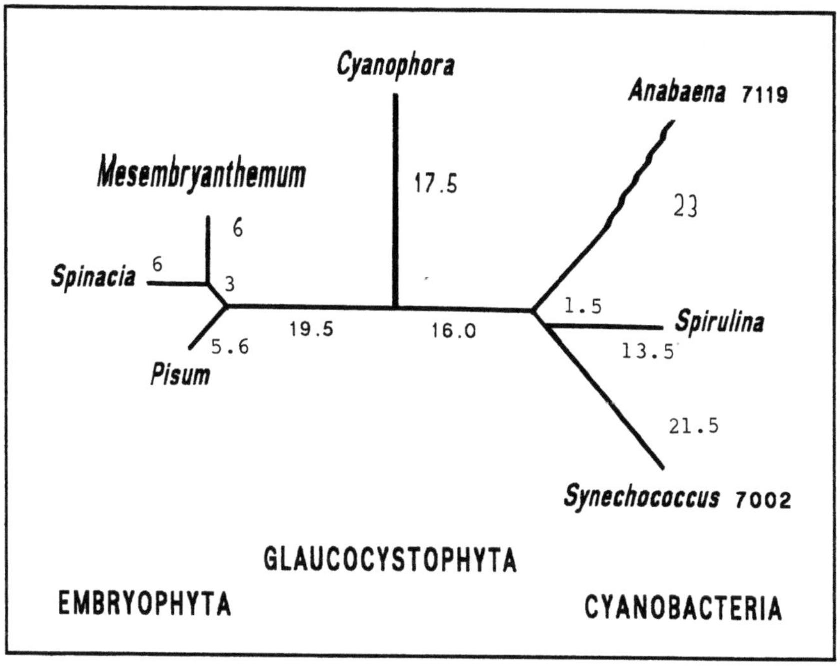

Fig. 8. Ferredoxin-NADP oxidoreductase (FNR): unrooted gene tree of mature FNR proteins from four eukaryotic and three cyanobacterial species (Table 8, calculated with a distance matrix method (Lüttke and Fuchs 1992) and represented by the 'neighbor joining' method, gaps not considered; Jakowitsch *et al.* 1993, altered). Numbers are distance values (see text).

bryophyta. In comparison with the other properties of the *C. paradoxa* cyanoplast, these observations can be taken as proof (or direct indication) of the FNR gene transfer from the former endocytobiont to the original host genome. Therefore this result is taken to be a fundamental experimental support of the Endosymbiosis Theory. The original but unknown cyanome, a symbiotic dibiont system (flagellate + endocytobiotic cyanobacterium) was combined genetically by unequal intertaxonic combination (IITC) to a monobiont (flagellate with a new cell organelle), a representative of a new cell type, a metacyanome, developing to the recent *C. paradoxa*. This statement supports also the classification of the Glaucocystophyta as a new algal class (Kies and Kremer 1986b).

4.3.3 Glaucocystophyta, enigmatic algae

With the simple transfer of a gene from one to the other genome the successful 'gene transfer' is still not established (Fig. 6). It needs furthermore an adaptation of the post-translational protein import machinery with respect to the new GHRS (see above) nucleus encoded proteins of the original endocytobiont. The unknown mode of gene transfer leads to the insertion of prokaryotic genes in an eukaryotic chromosome and to an enlargement of such a transferred nucleotide sequence, at least with a targeting sequence, forming the nucleotide sequence of a 'plastidic' precursor protein. Only this preprotein may be enabled to penetrate the envelope membranes of the newly formed exogenosome, supported by a protein import machinery. This means that at the same time as the gene transfer occurred, receptors recognizing the transit-sequence and proteins responsible for the protein import must have been evolved or adapted to the two envelope membranes of the endocytobiont. These 'very improbable' events must be happened, as shown with high probability. Comparison (Jakowitsch *et al.* 1993) of the transit sequence of the cyanoplast FNR preprotein with the transit sequences of higher plant chloroplasts shows no amino acid sequence homology, notwithstanding a significant overall structural similarity between these two sequence types. That may speak for a similar and evolutionary well conserved protein import mechanism. But, the extremely low sequence identity may be more in line with the assumption of a separate endocytobiotic event and a convergent evolution of the transit sequences in embryophytes and *C. paradoxa* (Jakowitsch *et al.* 1993).

Further sequence data support the close phylogenetic relationship between cyanoplasts and cyanobacteria. The sequence for ferredoxin, located near the middle of the large single copy region of the genome (Neumann-Spallart *et al.* 1990) and most similar to that of the cyanobacterium *Synechococcus* sp. PCC 7418 (= *Aphanothece halophytica*; Stevanovic *et al.* 1989, Neumann-Spallart *et al.* 1990) branches off from the cyanobacterial lineage in the unrooted gene tree (Lüttke 1991). Similarly by sequence comparison and tree construction the sequence of RUBISCO small subunit is more closely positioned to the cyanobacterial counterparts than to those of other eukaryotes (Valentin and Zetsche 1989, Lüttke 1991).

But the overall relations are more complicated and the phylogenetic position of *C. paradoxa* and its cyanoplasts far from settled. Indeed, the ambivalent situation with regard to the phylogenetic position of *C. paradoxa* is rather irritating, comparing the gene organization of the ATPase complex (Kowallik 1992) with four to three F10 genes (atpF to atpI) in cyanobacteria and/or plastids and three to five F11 genes (atpA to atpE) in eukaryotes or cyanobacteria. Overlapping of the genes atpD and atpF is a cyanobacterial property, also found in *C. paradoxa* and some chromophytic algae, but not in Embryophyta. On the other hand deletion of atpC, also found in *C. paradoxa*, is characteristic for all plastidic ATPases so far investigated. Gene tree construction from the sequences of rps 7, rps 12 and tufA (Kraus *et al.* 1990) corroborates the *C. paradoxa* cyanoplast position between the cyanobacterial and chloroplast branches, though slightly closer to the cyanobacterial one. In the comparison based on the nucleotide sequences of the tufA gene, however, cyanoplasts are placed within the chloroplast cluster (Kraus *et al.* 1990). The cyanoplast psbA gene codes for 360 amino acids. The amino acid sequence (Janssen *et al.* 1989) is cyanobacteria-like with regard to a seven amino acid-insertion close to the carboxyl terminus. This sequence is missing in all chlorophyll b type chloroplasts. In contrast, the overall sequence comparison shows slightly higher identity to the compared chloroplasts than to the cyanobacterial sequences. Maid *et al.* (1990) do see here a convergent development during the evolution and give the 7 amino acid insert a higher phyletic weight. That would support the idea of a polyphyletic origin of chloroplasts and rhodoplasts. From these data taken together it seems evident that different evolutionary clocks are working in the different genes of *C. paradoxa* cyanoplast proteins (for comparison see Palmer 1990).

It remains to tackle the question whether *C. paradoxa* cyanoplasts are an analogous model for the plastid evolution, which implies a polyphyletic origin of plastids (Raven 1970), or whether they represent a descendant of a missing link in the evolutionary series from an initial cyanome to the late embryophytes. Surveying the literature it is clear that this question is discussed controversially. Recently, a monophyletic evolution of all plastids was proposed (Kowallik 1992), based on one unique endocytobiotic event leading to an unknown symbiotic consortium and from this progenitor to all primary plastids, by branching to Rhodophyta on the one side, and to Glaucocystophyta, Prasinophyceae, Chlorophyta and land plants, on the other. All other plastids should then be phyletic secondary plastids, developing after different secondary endocytobiotic events. Accordingly, then the *C. paradoxa* cyanoplast would be a descendant of a missing link in the series of primary plastid evolution. To this end I shall follow the 5S rRNA phylogenetic tree of Hori and Osawa (1987) despite the caution expressed by Halanych (1991) and use their given relative time scale for a phylogenetic overview. According to that work, the Rhodophyta-Eukaryota diversification should have happened before about 1.5 billion years, that of Chlorophyta since ca. 1.1, that of *Euglena* and *Cyanophora* since ca. 0.5 billion years. This means that the endocytobiotic event leading to *C. paradoxa* should be much younger than that leading e.g. to the rhodophytes or chlorophytes. Consequently, *C. paradoxa* would be an analogous model organism, demonstrating the polyphyletic origin of primary plastids. At least in part, the found similarities and heterogeneities of phyletic relevancy on the molecular level should have evolved within the prokaryotic radiation before the symbiogenous evolution of plastids. The origin of cyanoplasts remains an enigma of Glaucocystophytes.

5. Glossary

antisymbiotic process: cellular processes which are appropriate to destroy the biological balance between the partners in a symbiosis (de Bary's definition 1879) in favor of one of them (the other partner will be deleted, degenerated or partially or totally integrated in the GHRS).

biont: Biological system with the ability of genetically independent reproduction (genetic autonomous biological system).

cell organelle: cell compartment with distinct function (normally surrounded by at least one envelope membrane) composed by (obligate) proteins of which at least some are encoded on the nuclear DNA.

chondriome: recent genome of a mitochondrion (=DNA within a mitochondrion).

cyanelle (Pascher 1929): genetically autonomous, symbiotic living cyanobacterium (like 'zoochlorelle' or 'zooxanthelle'). The term should not more be used for the photosynthetic organelles of Glaucocystophyta which are cyanoplasts. (Pascher 1929: 'Cyanelle bezeichnet eben eine endosymbiontisch lebende Blaualge gegenüber der freilebenden ... Cyanelle soll so wenig ein systematischer Begriff sein wie Zoochlorelle und Zooxanthelle').

cyanome: symbiotic consortium, consisting of host and cyanelle (Pascher 1929).

cyanoplast: eukaryotic, photo(hydro)trophic cell organelle, with regard to distinct cyanobacterial characteristics obviously originating from a cyanelle (Schenk 1990) by IITC (see unequal intertaxonic combination).

dibiont: symbiotic consortium of two bionts, e.g. lichens (myco- and phycobiont), or *Paramaecium bursarria* and *Chlorella* sp., or *Hydra viridis* and *Chlorella* sp..

dyssymbiosis: parasitic association (Schaede 1962).

endocyanosis: syncyanosis with intracellular cyanobacteria (Pascher 1929).

endocytobiology: science field dealing with the biology and phylogeny of (endo)cytobioses, exogenosomes and other biologically relevant cell compartments and constituents (with an exogenous origin).

endocytobiosis: intracellular association of two bionts. Endocytobiosis is called homologous if both cell types derive from the same taxon or same biont (like e.g. synergid cells in the macrogametophyte of seed plants), or heterologous if partner cells come from different taxa (see symbiosis). This special symbiotic association has the facultative ability to an evolutionary change (by IITC) of the dibiont-system into a new monobiont-system bearing a new cell organelle, the exogenosome.

endogenous (or *autogenous*): endogenous cell properties are (phylogenetically) developed in dependence on the own capabilities of the cell.

endogenosome: (phylogenetic term) cell organelle with an endogenous origin (Schenk 1992b) like ER or vacuoles.

endosymbiosis: symbiotic association between a host and a guest within the host. It can be intracellular (endocytobiosis) or extracellular (endosomabiosis, Schenk 1992a).

eucyte: eukaryotic cell.

eukaryogenesis: evolution of eukaryotic cells.

eusymbiosis: mutual association (Schaede 1962).

exogenous: term used for distinct properties (on organellar/molecular and/or gene level) of a biont which originate from another biont of the same or of a different taxon.

exogenosome: (phylogenetic term) *per definitionem* a cell organelle (Schenk 1992b), but descending of a former cytobiont, which has been come finally (by IITC) under the cybernetic control of the genomic hypersystem (GHRS).

GHOS: (phylogenetic term), the 'genomic hyposystem' looses genes under the influence of another genome during phylogenesis: a) by deletion, b) by gene transfer towards a partner genome (GHRS).

GHRS: (phylogenetic term), the 'genomic hypersystem' is in case of gene transfer the main receiver genome.

gnorimobiont: symbiont which is recognized during the infection of the host (or during endocytosis by the host) and received by the host accompanied by mutual processes (e.g. *Rhizobium* and Fabales).

gnorimosome: symbiosome (consisting of gnorimobiont and symbiosome membrane) which can be ontogenetically always form again.

IITC: see: unequal (*Inaequalis*) InterTaxonic Combination.

intertaxonic combination (ITC, Sitte 1991): intracellular processes (including gene transfer) changing a symbiotic consortium towards a new species by irreversible combination (mixing) of the original symbiotic genomes. In addition to mutation and recombination the ITC is the third evolutionary process which can change the genotype (Sitte 1991). It is a macromechanism.

ITC: see: intertaxonic combination.

macromechanism, evolutionary (Schwemmler 1989): evolutionary reaction complex (e.g. ITC) leading to new species.

metacyanome: new species, descendant of a cyanome, originated by IITC (e.g. Glaucocystophyta are metacyanomes).

organelle: see cell organelle.

plastome: the recent genome of a plastid (the DNA inside a plastid).

symbiogenesis: a) evolution of a symbiotic consortium by two steps. Step I: Installation of a facultative symbiosis, e.g. endocytobiosis with formation of a symbiosome. Step II: 'Stabilization' of the facultative consortium (step I) towards an obligate symbiosis, e.g. with transmission of the endocytobiont from one to the following generation. b) Exogenosome formation (symbiogenesis step III): Origin of a new organism (species) by IITC of two symbiotic partners.

symbiosis: '... Zusammenleben ungleichnamiger Organismen ...' (de Bary 1879), the (permanent or long-lasting, Smith and Douglas 1987) association of differently named bionts. This broad definition includes both 'mutualistic' and (in contrast to the usual language) parasitic associations. The intracellular symbiosis is a special form of symbiosis (see endocytobiosis). In accordance with this definition 'symbiosis' can be only 'heterologous' (see endocytobiosis).

symbiosome (Park and Jeon 1988): double enveloped cell compartment, composed by the endocytobiont (symbiont) with its plasmalemma (as inner envelope) and the symbiosome membrane (as outer envelope), also called perisymbiotic (e.g. perialgal, peribacteroid) membrane. Symbiosomes can be newly formed, artificially or naturally, dependent on the ontogenetic stage of the partners, or can be transmitted from one to the next generations.

syncyanosis (Pascher 1914): symbiosis of cyanobacteria with other organisms as hosts.

unequal intertaxonic combination (IITC): the observed gene transfer (combined with other necessary processes) goes mainly from the GHOS (e.g. endocytobiont) towards the GHRS (e.g. host nucleus) (this 'endocytobiological rule' was firstly expressed by Sitte 1991). IITC is changing a dibiont towards a monobiont system.

xenobiont: infectious symbiont which is able to break up the cellular defense of the host forming a xenosome *sensu stricto* (see xenosome).

xenosome (Soldo 1983): *sensu strictu* a symbiosome (consisting of xenobiont and symbiosome membrane) which can be formed ontogenetically at any time again.

Acknowledgements

The work was supported by the Deutsche Forschungsgemeinschaft (Sche 98/10-6). I am grateful to Prof. Dr. W. Löffelhardt (Wien), Prof. Dr. K. Poralla (Tübingen), Prof. Dr. G. Jung (Tübingen), Dr.habil. H. Frank (Tübingen), Prof. Dr. E. Hartmann (Mainz/Berlin) and Dr. A. Lüttke (Köln) for their very collegial and constructive collaboration. Special thanks go to my co-workers Dr. M.G. Bayer, Dipl. Biochem. U.B. Gebhart and Dr. T.L. Maier. Without their enthusiastic experimental work these considerations could not be referred.

References

Ahmadjian, V. and Paracer, S. (1986). *Symbiosis. University Press, Hanover, London.*

Aitken, A. and Stanier, R.Y. (1979). *J. Gen. Microbiol.* 212: 218–223.

Allsopp, A. (1969). *New Phytol.* 68: 591–612.

Baldauf, S.L. and Palmer, J.D. (1990). *Nature* 344: 262–265.

Bary, A. de (1879). *Die Erscheinung der Symbiose. Verlag Karl J. Trubner, Strassburg.*

Bayer, M.G. and Schenk, H.E.A. (1986). *Endocytobiosis & Cell Res.* 3: 197–202.

Bayer, M.G. and Schenk, H.E.A. (1989). *Curr. Gen.* 16: 311–313.

Bayer, M.G., Maier, T.L., Gebhart, U.B. and Schenk, H.E.A. (1990). *Curr. Genet.* 17: 265–267.

Behrens, M., Lisowsky, T., Pratje, E., Riemen, G. and Michaelis, G. (1990). In: *Endocytobiology IV (eds. Nardon, P. et al.)* INRA, Paris, pp. 467–474.

Bell, P.R. and Mühlethaler, K. (1962). *J. Ultrastruct. Res.* 7: 452–466.

Blank, R.J. (1985). *Endocytobiosis & Cell Res.* 2: 113–117.

Boczar, B.A., Delaney, T.P. and Cattolico, R.A. (1989). *Proc. Natl. Acad. Sci. USA* 86: 4996–4999.

Boettcher, U., Brandt, P., Müller, B. and Tischner, R. (1982). *Z. Pflanzenphysiologie* 106: 167–172.

Bogorad, L., Mets, L.J., Mullinix, K.P., Smith, H.J. and Strain, G.C. (1973). *Biochem. Soc. Symp.* 38: 17–41.

Bohnert, H.J., and Löffelhardt, W. (1992). In: *Algae and Symbioses (ed. Reisser, W.) Biopress Limited, Bristol.* pp. 379–397.

Bohnert, H.J., Michalowsski, C., Koller, B., Delius, H., Mucke, H. and Löffelhardt, W. (1983). In: *Endocytobiology II: Intracellular space as oligogenetic ecosystem (eds. Schenk, H.E.A. and Schwemmler, W.) De Gruyter, Berlin, New York.* pp. 433–448.

Bohnert, H.J., Michalowski, C., Bevacqua, S., Mucke, H. and Löffelhardt, W. (1985). *Mol. Gen. Genet.* 201: 565–574.

Bothe, H. and Floener, L. (1978). *Z. Naturforsch.* 33c: 981–987.

Brandt, P. (1991). *Evolution der eukaryotischen Zelle. Thieme, Stuttgart/New York.*

Breiteneder, H., Seiser, C., Löffelhardt, W., Michalowski, C. and Bohnert, H.J. (1988). *Curr. Gen.* 13: 199–206.

Bryant, D.A. (1988). In: *Light-Energy Transduction in Photosynthesis: Higher Plants and Bacterial Models (eds. Stevens, S.E. and Bryant, D.A.) Am.Soc.Plant Physiol. Rockville.* pp. 69–90.

Bryant, D.A. and Stirewalt, V.L. (1990). *FEBS Letters* 259: 273–280.

Bryant, D.A., de Lorimier, R., Lambert, D.H., Dubbs, J.M., Stirewalt, V.L. Stevens, S.E., Porter, R.D., Tam, J. and Jay, E. (1985). *Proc. Natl. Acad. Sci. USA* 82: 3242–3246.

Bryant, D.A., Schluchter, W.M. and Stirevalt, V.L. 1991. *Geme* 98: 169–175.

Burnap, R.L. and Trench, R.K. (1989a). *Proc. R. Soc. Lond.* B 238: 73–87.

Burnap, R.L. and Trench, R.K. (1989b). *Proc. R. Soc. Lond.* B 238: 89–102.

Cantrell, A. and D.A. Bryant (1988). *Photosynth. Res.* 16: 65–81.

Cavalier-Smith, T. (1987). *Evol. Trends Plants* 1: 75–78.

Cavalier-Smith, T. (1990). In: *Endocytobiology IV (eds. Nardon, P. et al.)*, INRA, Paris. pp. 571–574.

Cavalier-Smith, T. (1991). In: *Evolution of Life (eds. Osawa, S. and Honjo, T.) Springer Verlag Tokyo.* pp. 271–304.

Cavalier-Smith, T. and Lee, J.J. (1985). *Protozool.* 32: 376–379.

Chapman, D.J. (1966). *Arch. Mikrobiol.* 55: 17–25.

Chiba, Y. (1951). *Cytologia* 16: 259–264.

Colwell, G.L. and Wickstrom, C.E. (1976). *J. Phycol.* 12 (Suppl.):11.

Dodge, J.D. (1973). *The fine structure of algal cells. Academic Press, London.*

Dodge, J.D. (1980). In: *Endocytobiology I. Endosymbiosis and Cell Biology (eds. Schwemmler, W. and Schenk, H.E.A.) De Gruyter, Berlin.* pp. 33–50.

Drum, R.W. and Pankratz, S. (1965). *Protoplasma* 60: 141–149.

Echlin, P. (1966). In: *Proc. VIth Intern. Conf. Electron Microscopy, Kyoto Vol. 2*, pp. 285.

Evrard, J.-L., Johnson, C., Janssen, I., Löffelhardt, W., Weil, J.-H. and Kuntz, M. (1990a). *Nucleic Acids Res.* 18: 1115–1119.

Evrard, J.-L., Kuntz, M., Straus, N.A. and Weil, J.H. (1988). *Gene* 71: 115–122.

Evrard, J.-L., Kuntz, M. and Weil, J.H. (1990b). *J. Mol. Evol.* 30: 16–25.

Fenwick, M.G. (1966). *Trans. Amer. Microsc. Soc.* 85: 477–480.

Fillat, M.F., Bakker, H.A.C. and Weisbeck, P. (1990). *Nucl. Acids Res.* 18: 7161.

Floener, L. and Bothe, H. (1980). In: *Endocytobiology I (eds. Schwemmler, W. and Schenk, H.E.A.) W. De Gruyter, Berlin, New York.* pp. 541–552.

Floener, L. and Bothe, H. (1982). *Planta* 156: 78–83.

Floener, L. and Bothe, H. (1983). In: *Endocytobiology II (eds. Schenk, H.E.A. and Schwemmler, W.), De Gruyter, Berlin, New York.* pp. 471–474.

Floener, L., G. Danneberg and H. Bothe (1982). *Planta* 156: 70–77.

Gebhart, U.B., Maier, T.L., Stevanovic, S., Bayer, M.G. and Schenk, H.E.A. (1992). *Protein Expression and Purific.* 3: 228–235.

Geitler, L. (1959). In: *Handbuch der Pflanzenphysiologie (ed. Ruhland) Vol. XI.* 530–545. Springer, Berlin.

Gibbs, S.P. and Gillot, M.A. (1980). In: *Endocytobiology I. (eds. Schwemmler, W. and Schenk, H.E.A.) De Gruyter, Berlin.* pp. 737–743.

Giddings, T.H., C. Wasmann and L.A. Staehelin (1983). *Plant Physiol.* 71: 409–419.

Giovannoni, S.J., Turner, S., Olsen, G.J., Barns, S., Lane, D.J. and Pace, N.R. (1988). *J. Bacteriol.* 170: 3584–3592.

Gray, M.W. (1989). *Trends in Genetics* 5: 294–299.

Gray, M.W. and Doolittle, W.F. (1982). *Microbiol. Rev.* 46: 1–42.

Griffiths, B.M. (1915). *Ann. Bot.* 29: 423–432.

Häder, D.P. (1985). *Arch. Microbiol.* 143: 100–104.

Halanych, K.M. (1991). *Mol. Biol. Evol.* 8: 249–253.

Hall, W.T. and Claus, G. (1963). *J. Cell Biol.* 19: 551–563.

Hall, W.T. and Claus, G. (1967). *J. Phycol.* 3: 37–51.

Heimann, K., Reize, I.B. and Melkonian, M. (1989). *Protoplasma* 148: 106–110.

Heinhorst, S. and Shively, J.M. (1983). *Nature* 304: 373–374.

Heinz, G. (1973). *Versuche zur Isolierung der lysozymempfindlichen Stützmembran von Cyanocyta korschikoffiana*, der Endocyanelle aus *Cyanophora paradoxa* KORSCH. Dipl. Arb., University of Tübingen.

Herdman, M. and Stanier, R.Y. (1977). *FEMS* 1: 7–12.

Holm-Hansen, O., Prasad, R. and Lewin, R.A. (1965). *Phycologia* 5: 1–14.

Hori, H. and Osawa, S. (1987). *Mol. Biol. Evol.* 4: 445–472.

Ikan, R. and Seckbach, J. (1972). *Phytochem.* 11: 1077–1082.

Jakowitsch, J., Brandtner, M., Löffelhardt, W., Gebhart, U.B., Bayer, M.G., Maier, T.L., Schenk, H.E.A., Michalowski, C.B. and Bohnert, H.J. (1991). In: *Program and Abstracts. III. Intern. Congr. of Plant Molecular Biology. Tucson, 6–11 October 1991.*

Jakowitsch, J., Bayer, M.G., Maier, T.L., Gebhart, U.B., Lüttke, A., Brandtner, M., Hamilton. B., Neumann-Spallart, C., Michalowski, C.B., Bohnert, H.J., Schenk, H.E.A. and Löffelhardt, W. (1993). *Plant Mol. Biol.* 21: 1023–1033.

Jansen, T., Reiländer, H., Steppuhn, J. and Herrmann, R.G. (1988). *Curr. Genet.* 13: 517–522.

Jánssen, I., Mucke, H., Löffelhardt, W. and Bohnert, H.J. (1987). *Plant Mol. Biol.* 9: 479–484.

Janssen, I., Jakowitsch, J., Michalowski, C., Bohnert, H.J. and Lüffelhardt, W. (1989). *Curr. Genet.* 15: 335–340.

Jehn, U. and Zetsche, K. (1988). *Planta* 173: 58–60.

Jensen, T.E. (1991). *Endocytobiosis & Cell Res.* 8: 1–16.

Jeon, K.W. and Jeon, M.S. (1976). *J. Cell Physiol.* 89: 337–344.

Karplus, P.A., Walsh, K.A. and Herriott, J.R. (1984). *Biochemistry* 23: 6576–6583.

Kenyon, C.N., Rippka, R. and Stanier, R.Y. (1972). *Arch. Mikrobiol.* 83: 216–236.
Keyhani, E. (1983). In: *Endocytobiology II (eds. Schenk, H.E.A. and Schwemmler, W.) De Gruyter, Berlin, New York.* pp. 129–145.
Kies, L. (1976). *Protoplasma* 87: 419–446.
Kies, L. (1979). *Ber. Deutsch. Bot. Ges.* 92: 445–454.
Kies, L. (1980). In: *Endocytobiology I. (eds. Schwemmler, W. and Schenk, H.E.A.) De Gruyter, Berlin.* pp. 33–50.
Kies, L. (1984a). In: *Compartments in Algal Cells and Their Interaction (eds. Wiessner, W. et al.)* Springer, Berlin, Heidelberg. pp. 191–199.
Kies, L. (1984b). *Biol. Rdsch.* 22: 145–157.
Kies, L. (1988). *Endocytobiosis & Cell Research* 5: 361–372.
Kies, L. (1989). *Pl. Syst. Evol.* 164: 65–73.
Kies, L. (1992). In: *Algae and Symbioses (ed. Reisser, W.) Biopress Limited, Bristol.* pp. 353–377.
Kies, L. and Kremer, B.P. (1986a). *BIUZ* 16: 106–112.
Kies, L. and Kremer, B.P. (1986b). *Taxon* 35: 128–133.
Kies, L. and Kremer, B.P. (1990). In: *Handbook of Protoctists (eds. Margulis, L. et al.)*, Jones and Bartlett Publ., Boston. pp. 152–166.
Kirk, J.T.O. (1971). In: *Autonomy and biogenesis of chloroplasts and mitochondria (eds. Boardman, N.K. et al.)* North-Holland Publishing Co., Amsterdam. pp. 267–276.
Klein, S.M., Jaynes, J.M., Kent, S.S. and Vernon, L.P. (1981). *Plant Physiol.* 68: 407–410.
Klein, R.M. and Cronquist, A. (1967). *Q. Rev. Biol.* 42: 105–296.
Kleinig, H., Beyer, P., Schubert, C., Liedvogel, B. and Lütke-Brinkhaus, F. (1986). *Z. Naturforsch.* 41c: 169–171.
Kleinschmidt, M.G. and McMahon, V.A. (1970). *Plant Physiol.* 46: 286–289.
Ko, K., Jaynes, J.M. and Straus, N.A. (1985). *Plant Science* 42: 115–123.
Korschikov, A.A. (1924). *Russian Archive for Protistology* 3: 57–74.
Kowallik, K.V. (1992). In: *Symbiogenesis, Prochlorophytes and the origins of plastids (ed. Lewin, R.A.). Chapman & Hall, New York.* pp. 223–263.
Kraus, M., Götz, M. and Löffelhardt, W. (1990). *Plant Mol. Biol.* 15: 561–573.
Kremer, B.P. (1982). *Br. Phycol. J.* 17: 51–61.
Kremer, B.P., Kies, L. and Rostami-Rabet, A. (1979). *Zeitschr. für Pflanzenphysiol.* 92: 303–317.
Kuntz, M., Crouse, E.J., Mubumbila, M., Burkard, G., Weil, J.H., Bohnert, H.J., Mucke, H. and Löffelhardt, W. (1984). *Mol. Gen. Genet.* 194: 508–512.
Kuntz, M., Evrard, J.L. and Weil, J.H. (1988). *Nucleic Acids Res.* 16: 8733.
Lambert, D.H., Bryant, D.A., Stirewalt, V., Dubbs, J.M., Stevens, S.E. and Porter, R.D. (1985). *J. Bacteriol.* 164: 659–664.
Law, R. and Lewis, D.H. (1983). *Biol. J. Linnean Soc.* 20: 249–276.
Lefort, M. (1965). *C.R. Hebd. Seances Acad. Sci.* D 261: 233–236.
Lefort-Tran, M. (1981). *Ber. Deutsch. Bot. Ges.* 94: 463–476.
Lemaux, P.G. and Grossmann, A. (1984). *Proc. Nat. Ac. Sci. USA* 80: 4100–4104.
Lewis, D.H. (1985). In: *The Biology of Mutualism: Ecology and Evolution (ed. Boucher, D.H.) Croom-Helm, London.* pp. 29–39.
Löffelhardt, W., Mucke, H. and Bohnert, H.J. (1980). In: *Endocytobiology I. (eds. Schwemmler, W. and Schenk, H.E.A.) De Gruyter, Berlin.* pp. 523–530.
Löffelhardt, W., Mucke, H., Crouse, E.J. and Bohnert, H.J. (1983). *Curr. Gen.* 7: 139–144.
Löffelhardt, W., Kraus, M., Pfanzagl, B., Götz, M., Brandtner, M., Markmann-Mulisch, U., Subramanian, A.R., Michalowski, C. and Bohnert, H.J. (1990). In: *Endocytobiology IV (eds. Nardon, P. et al.)* INRA, Paris. pp. 561–564.
Löffelhardt, W., Michalowski, C.B., Kraus, M., Neumann-Spallart, C., Jakowitsch, J., Flachmann, R. and Bohnert, H.J. (1993). In: *Endocytobiology V (eds. Sato, S., Ishida, M.R. and Ishikawa, H.) Tübingen University Press. Tübingen.* pp. 189–194.
Lowe, R.L., Rosen, B.H. and Fairchild, G.W. (1984). *J. Phycol.* 20 (Suppl.): 24.
Lüttke, A. (1991). *Endocytobiosis & Cell Res.* 8: 75–82.
Lüttke, A. and Fuchs, R. (1992). *CABIOS* 8: 591–594.
Maid, U., Valentin, K. and Zetsche, K. (1990). *Curr.Gen.* 17: 255–259.
Mangeney, E. and Gibbs, S. P. (1987). *Europ. J. Cell Biol.* 43: 65–70.
Margulis, L. (1970). *Origin of Eukaryotic Cells, Yale University Press, New Haven and London. 349 pp.*
Margulis, L. (1981). *Symbiosis in Cell Evolution, Freeman and Co.,San Francisco. 419pp.*
Margulis, L. (1988). *Endocytobiosis & Cell Res.* 5: 133–162.
Marten, S., Brandt, P. and Wiessner, W. (1981). *Ber. Deutsch. Bot. Ges.* 94: 727–732.
Marten, S. and Brandt, P. (1984). In: *Compartments in algal cells and their interaction (eds. Wiessner, W. et al.)* Springer, Berlin. pp. 69–75.
Maxwell, E.S., Liu, J. and Shively, J.M. (1986). *J. Mol. Evol.* 23: 300–304.
McCracken, D.A., Nadakavukaren, M.J. and Cain, J.R. (1980). *New Phytologist* 86: 39–44.
Menzel, K. and Wild, A. (1989). *Z. Naturforsch.* 44c: 743–748.
Mereschkowsky, C. (1905). *Biol. Zentralbl.* 25: 593–604.
Mereschkowsky, C. (1910). *Biol. Zentralbl.* 30: 278–303, 321–347, 353–367.
Metzner, H. (1952). In: *Biol. Zbl.* 71: 257–272.
Michalowski, C., Schmitt, J.M. and Bohnert, H.J. (1989). *Plant Physiol.* 89: 817–822.
Michalowski, C., Bohnert, H.J. and Löffelhardt, W. (1990a). *Nucleic Acids Res.*18: 2186.
Michalowski, C., Pfanzagel, B., Löffelhardt, W. and Bohnert, H.J. (1990b). *Mol. Gen. Genet.* 225: 222–231.
Michalowski, C., Bohnert, H.J. and Löffelhardt, W. (1991a). *J. Biol. Chem.* 266: 11866–11870.
Michalowski, C., Flachmann, R., Löffelhardt, W., and Bohnert, H.J. (1991b). *Plant Physiol.* 95: 329–330.
Mucke, H., Löffelhardt, W. and Bohnert, H.J. (1980). *FEBS Lett.* 111: 347–352.
Nakamura, H. (1993). In: *Endocytobiology V (eds. Sato, S., Ishida, M.R. and Ishikawa, H.) Tübingen University Press, Tübingen.* pp. 337–344.

Nass, S. and Nass, M.M.K. (1963). *J. Cell Biol.* 19: 613–629.
Neumann-Spallart, C., Brandtner, M., Kraus, M., Jakowitsch, J., Bayer, M.G., Maier, T.L., Schenk, H.E.A. and W. Löffelhardt (1990). *FEBS Lett.* 268: 55–58.
Neumann-Spallart, C., Jakowitsch, J., Kraus, M. Brandtner, M., Bohnert, H.J. and Löffelhardt, W. (1991). *Curr. Genet.* 19: 313–315.
Neu-Müller, M. and Schenk, H.E.A. (1983). *Z. Naturforsch.* 38c: 990–995.
Newman, B.J. and Gray, J.C. (1988). *Plant Mol. Biol.* 10: 511–520.
Nichols, B.W. and Appleby, R.S. (1969). *Phytochem.* 8: 1907–1915.
Palmer, J.D. (1990). *TIG* 6: 115–120.
Park, M.S. and Jeon, K.W. (1988). *Endocytobiosis & Cell Res.* 5: 215–224.
Parthier, B. (1974). (P. Bohley, ref.) *Biol. Rundsch.* 12:357, and (1975), *Nova Acta Leopoldiana, NF*, 42, 218: 223–239.
Parthier, B., Krauspe, R., Munsche, D. and Wollgiehn, R. (1975). In: *The chemistry and biochemistry of plant proteins. (eds. Harborne, J. and Van Sumere, C.F.) Academic Press, London, New York.* pp. 167–210.
Pascher, A. (1914). *Ber. Deutsch. Bot. Ges.* 32: 339–352
Pascher, A. (1929). *Jahrb. Wiss. Bot.* 71: 386–462.
Prasad, B.M. (1961). *Bull. Bot. Soc., Univ. of Saugar* 13: 44–45.
Pringsheim, E.G. (1958). *Plant. Physiol.* (Praha): 165–184.
Provasoli, L. and Pintner, J. (1952) *News Bull. Phycol. Soc. Amer.* 5: 7.
Raff, R.A. and Mahler, H.R. (1972). *Science* 177: 575 582.
Raven, P.H. (1970). *Science* 169: 641–645.
Ris, H. and Plaut, W. (1962). *J. Cell Biol.* 13: 383–391.
Robinson, D.G. and Preston, R.D. (1971a). *J. Exp. Bot.* 22: 635–643.
Robinson, D.G. and Preston, R.D. (1971b). *Brit. Phycol. J.* 6: 113–128.
Rogerson, A., Polne-Fuller,M., Trench, R.K. and Gibor, A. (1989). *Symbiosis* 7: 229–241.
Sagan, L. (L. Margulis) (1967). *J. Theor. Biol.* 14: 225–275.
Sager, R. and Ishida, M.R. (1963). *Proc. Nat. Acad. Sci. U.S.* 50: 725.
Saitou, N. and Nei, M. (1987). *Mol. Biol. Evol.* 4: 406–425.
Schaede, R. (1962). *Die Pflanzlichen Symbiosen (ed. Meyer, F.H.). Fischer, Stuttgart.*
Schaffner, J.H. (1922). Ohio J. Sci. 22: 129–139.
Schenk, H.E.A. (1970). *Z. Naturforsch.* 25b: 640+656.
Schenk, H.E.A. (1971). *Z. Physiol. Chem.* 352: 321–324.
Schenk, H.E.A. (1973). *Chloroplasten Evolution in biochemischer Sicht, 205 pp. Habilitation. University of Tübingen.*
Schenk, H.E.A. (1977). *Arch. Protistenk.* 119: 274–300.
Schenk, H.E.A. (1990). In: *Endocytobiology IV (eds. Nardon, P. et al.)* INRA, Paris. pp. 199–209.
Schenk, H.E.A. (1992a). In: *The Prokaryotes. A Handbook of ecophysiology, isolation, identification, application. (eds. Ballows, A. et al.)* Springer, New York. Vol. 4, pp. 3819–3854.
Schenk, H.E.A. (1992b). *Endocytobiosis & Cell Res.* 8: 215–222.
Schenk, H.E.A. and Hofer, I. (1972). In: *Proc. of the 2nd Intern. Congress on Photosynthesis Res., Stresa 1971 (eds. Forti, G. et al.)* W. Junk Publ., The Hague. pp. 2095–2100.
Schenk, H.E.A., Hanf, J. and Neu-Müller, M. (1983). *Z. Naturforsch.* 34c: 972–977.
Schenk, H.E.A., Poralla, K., Härtner, T., Deimel, R. and Thiel, D. (1985). *Endocytobiosis & Cell Res.* 2: 233–238.
Schenk, H.E.A., Bayer, M.G. and Maier, T. (1987a). *Endocytobiosis & Cell Res.* 4: 167–176.
Schenk, H.E.A., Bayer, M.G. and Zook, D. (1987b). *Ann. NY. Ac. Sci.* 503: 151–167.
Schenk, H.E.A., Bayer, M.G., Maier, T.L., Lüttke, A., Gebhart, U.B. and Stevanovic, S. (1992). *Z. Naturforsch.* 47c: 387–393.
Schimper, A.W.F. (1883). *Bot. Ztg.* 41: 105–114.
Schlichting, R., Zimmer, W. and Bothe, H. (1990). *Bot. Acta* 103: 392–398.
Schluchter, W.M. and Bryant, D.A. (1992). *Biochemistry* 31: 3092–3102.
Schmidt, B., Kies, L. and Weber, A. (1979). *Arch. Protistenk.* 122: 164–170.
Schnepf, E. (1965). *Planta* (Berlin) 67: 213–224.
Schnepf, E. (1966). In: *Funktionelle und morphologische Organisation der Zelle (ed. Sitte, P.) Springer, Berlin, Heidelberg, New York.* pp. 372–393.
Schnepf, E. and R.M. Brown (1971). In: *Origin and Continuity of Cell Organelles (eds. Reinert, J. and Ursprung, H.) Springer, Berlin, Heidelberg, New York.* pp. 299–322.
Schnepf, E. and Koch, W. (1966). *Z. Pflanzenphysiol.* 55: 97–109.
Schnepf, E., Koch, W. and Deichgräber, G. (1966). *Arch. Mikrobiol.* 55: 149–174.
Schnepf, E. and Elbrächter, M. (1988). *Bot. Acta* 101: 196–203.
Schuster, W. and Brennicke, A. (1988). *Plant Science* 54: 1–10.
Schwemmler, W. (1979). *Mechanismen der Zellevolution. De Gruyter, Berlin/New York.*
Schwemmler, W. (1989). Symbiogenesis a macromechanism of evolution. De Gruyter, Berlin/New York.
Schwemmler, W. (1991). *Symbiogenese als Motor der Evolution. Verlag Paul Parey, Berlin/Hamburg.*
Scott, O.T., Castenholz, R.W. and Bonnett, H.T. (1984). *Arch. Microbiol.* 139: 130–138.
Scott, O.T. (1987). *Ann. NY. Acad. Sci.* 503: 555–558.
Seckbach, J. (1992). In: *Algae and Symbioses (ed. Reisser, W.) Biopress Limited, Bristol.* pp. 399–426.
Seckbach, J., Fredrick, J.F. and Garbary, D.J. (1983). In: *Endocytobiology II (eds. Schenk, H.E.A. and Schwemmler, W.) De Gruyter, Berlin.* pp. 947-962.
Sitte, P. (1987). In: *Beobachtung, Experiment und Theorie in Naturwissenschaft und Medizin (ed. Lüst, R.) Wiss. Verlagsges. Stuttgart.* pp. 431–446.
Sitte, P. (1990). *Biol. Rundsch.* 28: 1–18.
Sitte, P. (1991). *BIUZ* 21: 85–92.
Skuja, H. (1954). In: *Syllabus der Pflanzenfamilien (eds. Melcher, H. and Werdemann, E.), Borntraeger, Berlin.* Vol. 1, pp. 56–57.
Smith, D.C. and Douglas, A.E. (1987). *The Biology of Symbiosis, 302 pp. Arnold, London.*
Sneath, P.H.A. and Sokal, R.R. (1963). *Numerical Taxonomy Freeman, San Francisco.*
Soldo, A.T. (1983). *Int. Rev. Cytol., Suppl.* 14: 79–109.
Starnes, S.M., Lambert, D.H., Maxwell, E.S., Stevens, S.E., Porter, R.D. and Shively, J.M. (1985). *FEMS Microbiol. Lett.* 28: 165–169.
Starr, M.P. (1975). In: *Symbiosis* (eds. Jennings, D.H. and Lee,

D.L.). *Symposia of the Soc. for Experimental Biology* 29:1–20. Cambridge University Press, Cambridge.
Stevanovic, S., Bayer, M.G., Tröger, W. and Schenk, H.E.A. (1989). *Endocytobiosis & Cell Res.* 6: 219–226.
Stirewalt, V.L. and Bryant, D.A. (1989a). *Nucl. Acids Res.* 17: 10095.
Stirewalt, V.L. and Bryant, D.A. (1989b). *Nucl. Acids Res.* 17: 10096.
Taylor, F.J.R. (1974). *Taxon* 23: 229–258.
Tell, G. (1979). *Rev.Algol.,N.S.* XIV: 39–48.
Tomas, R.N. and Cox, E.R. (1973). *J. Phycol.* 9: 304–323.
Trench, R.K. (1981). *Ann. N.Y. Acad. Sci.* 361: 341–355.
Trench, R.K. (1982). In: *Progress in Phycol. Res. Vol.1 (eds. Round, F.E. and Chapman, D.J.) Elsevier Biomedical Press. Amsterdam.* pp. 257–288.
Trench, R.K., Pool, R.R., Logan, M. and Engelland, A. (1978). *Proc. Royal Soc. London* B202: 423–443.
Trench, R.K. and Ronzio, G.S. (1978). *Proc. Royal Soc. London* B202: 445–462.
Trench, R.K. and Siebens, H.C. (1978). *Proc. Royal Soc. London* B202: 473–482.
Ueda, K. (1961). *Cytologia.* 26: 344–358.
Valentin, K. and Zetsche, K. (1989). *Curr. Genet.* 16: 203–209.
Valentin, K. and Zetsche, K. (1990a). *Plant Mol. Biol.* 15: 575–584.
Valentin, K. and Zetsche, K. (1990b). *Mol. Gen. Genet.* 222: 425–430.
Van den Eynde, H., De Baere, R., De Roeck, E., Van de Peer, Y., Vandenberghe, A., Willekens, P. and de Wachter, R. (1988). *J. Mol. Evol.* 27: 126–132.
Van Wisseling, C. (1920). *Z. indukt. Abstamm.- u. Vererbungsl.* 22: 65.
Wasman, C.C., Löffelhardt. W. and Bohnert, H.J. (1987). In: *The Cyanobacteria (eds. Fay, P. and van Baalen, C.) Elsevier Sc. Publ.* pp. 303–324.
Whatley, F.R. and Whatley, J.M. (1980). In: *Endocytobiology I. (eds. Schwemmler, W. and Schenk, H.E.A.) De Gruyter, Berlin, New York.* pp. 849–865.
Willison, J.H.M. and Brown, R.M. Jr. (1978). *Planta* 141: 51–58.
Wolters, J., Erdmann, V.A. and Stackebrandt, E. (1990). In: *Endocytobiology IV (eds. Nardon, P. et al.)* INRA, Paris. pp. 545–552.
Wood, B.J.B., 1974. In: *Algal Physiology and Biochemistry (ed. Stewart, W.D.P.) Blackwell, Oxford.* pp. 236–265.
Yao, Y., Tamura, T., Wada, K., Matsubara, H. and Kodo, K. (1984). *J. Biochem.* 95: 1513–1516.
Zillig, W., Klenk, H.-P., Palm, P., Leffers, H., Pühler, G., Gropp, F. and Garrett, R.A. (1989). *Endocytobiosis & Cell Res.* 6: 1–25.
Zook, D., Schenk, H.E.A., Frank, H.G., Thiel, D., Poralla, K. and Härtner, T. (1986). *Endocytobiosis & Cell Res.* 3: 99–103.
Zook, D. and Schenk, H.E.A. (1986). *Endocytobiosis & Cell Res.* 3: 203–211.

Chapter 3

Alternative pathway (cyanobacteria to eukaryota)

THOMAS E. JENSEN

Department of Biological Sciences, Lehman College of City University of New York, Bronx, NY 10468, USA

Outlines:

1. Introduction
2. State of methods for ultrastructural studies
3. Results of ultrastructural investigations of cyanobacteria
   3.1 Inclusions of regular occurrence
      3.1.1 Thylakoids
      3.1.2 Polyhedral bodies
      3.1.3 Glycogen granules
      3.1.4 Lipid inclusion
      3.1.5 Cyanophycin granules
      3.1.6 Polyphosphate bodies
   3.2 Inclusion of irregular occurrence
      3.2.1 Membrane monolayer limited inclusions
      3.2.2 Membrane limited or composed
      3.2.3 Tubules, filaments and sheets
      3.2.4 Crystalloid inclusions
      3.2.5 Storage polymer-like inclusions
      3.2.6 Intrathylakoidal inclusions
4. Perspectives and conclusions
   4.1 Origin of nucleus, mitochondria and chloroplasts
   4.2 Origin of the eukaryotic cell
References

1. Introduction

It is quite evident from the fossil record that prokaryotic cells were the first cell type evolved. After a long interval of nearly two billion years eukaryotic cells appeared. There is little doubt that this cell type arose from the bacterial cell. How this occurred has been the subject of much speculation. In this chapter we will present evidence that supports the idea that the eukaryotic cell arose by slow gradual change in a prokaryotic cell. This process of arriving at the eukaryotic cell has been termed the 'Autogenous Theory' (Klein and Cronquist 1967; Klein 1970; Cavalier-Smith, 1975; Taylor 1976).

The theory generally suggests an ancient ancestor termed the 'Uralga'. In this cell all of the inclusions evolved by slow gradual selection. Some more recent theories adopt some of these tenants but then include aspects of the endosymbiont theory. This theory is presented in various formats with the one by Margulis (1970) being the most complete. A variety of objections have been raised against the endosymbiotic theory. These are very well summarized by Dillon (1981).

More recently a number of authors have suggested that the cell interior is really similar to an ecological habitat (Richmond and Smith 1979). In spite of its small area the same forces operate here as in the larger environment. In the natural habitats most ecologists agree that a niche can only be occupied by a single species. Some define a species on this basis. In the environment of the cell the same kind of competition occurs and only one inclusion will survive for one specific function (nitch). This hypothesis has been presented in various forms such as the 'Locked Cell Hypothesis' of Keyhani (1983). Taylor and Harrison (1983) point out that homeostasis is a fundamental property of living systems and it is to be expected that the cytocosmic (cytoplasmic) environment must also maintain homeostasis. In any theory of eukaryotic cell origin there is little debate about what appears to be the fact that it arose from a prokaryotic cell. How this came about, i.e. autogenously, endosymbiotically, or a combination of the two is the real question. Over the last nearly 30 years we have examined the ultrastructure of a group of bacteria formally called blue-green algae but now generally called cyanobacteria. We and others have investigated in detail perhaps 200 isolates of the approximately 2,000 to 4,000 species in this group. In no other group of bacteria has this many isolates been examined at the electron microscope level.

We will present evidence for how the nucleus, chloroplasts and mitochondria may have originated. We will also suggest based on the

J. Seckbach (ed.), Evolutionary Pathways and Enigmatic Algae: Cyanidium caldarium *(Rhodophyta) and Related Cells,* 53–66, 1994.

cyanobacterial group how the other organelles and inclusions could have arisen in a prokaryotic cell and then in combination with the nucleus mitochondria and chloroplast given rise to the eukaryotic cell.

2. State of methods for ultrastructural studies

Fixation of the bacterial cell for electron microscopy was and still is a problem. The best fixation seems to be with the procedure developed by Kellenberger and his collaborators (Kellenberger *et al.* 1958). This procedure should be supplemented with a glutaraldehyde-osmium fixation as it uses osmium as the staining and fixing agent. More recently freeze-substitution of cells has also been used and provides good cellular preservation (Graham and Beveridge 1990). Using these procedures the cellular organization of prokaryotes can be elucidated in most cases. Most groups of bacteria have been looked at but not one group has been investigated in great depth. Many studies at the ultrastructural level are not really studies at this level. In many cases appropriate fixation is not achieved or cells were examined at very low magnification so all inclusion bodies may not have been observed. Thus a large number of studies of the bacteria at the electron microscope level fail to really be fine structural investigations. If proper fixation is not achieved no real studies at this level can be carried out. We should also keep in mind that in studies using thin sections, even though most blue-greens are quite small, it will still take a number of sections to get through a cell. For example, if a cell is 5 μm in diameter and the sections are 60 nm it will take 83.3 thin sections to section through a cell. If we use an example we can see what this means for cellular inclusions which may not be seen in each cell in thin section. If a cell this diameter has a single body 0.5 μm in diameter it will take 8.3 sections to cut through the body. It took 83.3 sections to cut through the cell, thus 10% of the sections should show the inclusion body. One can then extrapolate to other situations such as if only one 0.2 μm body is found in every 10 cells how often will it be observed in thin sections. Some investigators will dismiss as unimportant a body which is seen so rarely and it will not be considered as a regular part of the cells mix of inclusion bodies. If one were to think in terms of a pair of centrioles in an animal cell and how often they will be seen in thin sections and then their relative importance in the cell, we can get some of the blue-greens inclusions of irregular occurrence into proper perspective.

3. Results of ultrastructural investigations of cyanobacteria

In order to establish that the cyanobacteria have a large and diverse number of inclusion bodies in their cells we will first examine the inclusions in this bacterial group. Studies have shown that most cyanobacteria have a common set of inclusions that are generally found in all cells of the various isolates. This group of inclusions is called 'Inclusions of Regular Occurrence.' In addition a second group of inclusions is found either sporadically in different blue-green bacterial isolates, or under certain growth conditions or in a certain group. We have termed this group of inclusions 'Inclusions of Irregular Occurrence (Jensen 1984; Jensen 1985; Jensen 1990). Some of these inclusions are illustrated in Figures 1 and 2.

3.1 Inclusions of regular occurrence

3.1.1 Thylakoids

These sac like membrane components have associated with them all of the pigments necessary for photosynthesis.

3.1.2 Polyhedral bodies

These polygonal bodies are found in all cyanobacteria. They are also found in colorless sulfur bacteria, nitrate and ammonia-oxidizing bacteria and the purple and green bacteria (Codd and Marsden 1984). These membrane monolayer limited bodies are from 0.2 μm to 0.7 μm in diameter with from 6 to 40 found per cell. They contain mainly ribulose 1,5-biphosphate carboxylase/oxygenase (Lanaras and Codd 1981).

3.1.3 Glycogen granules

They are located between the thylakoids and are the immediate storage product of photosynthesis.

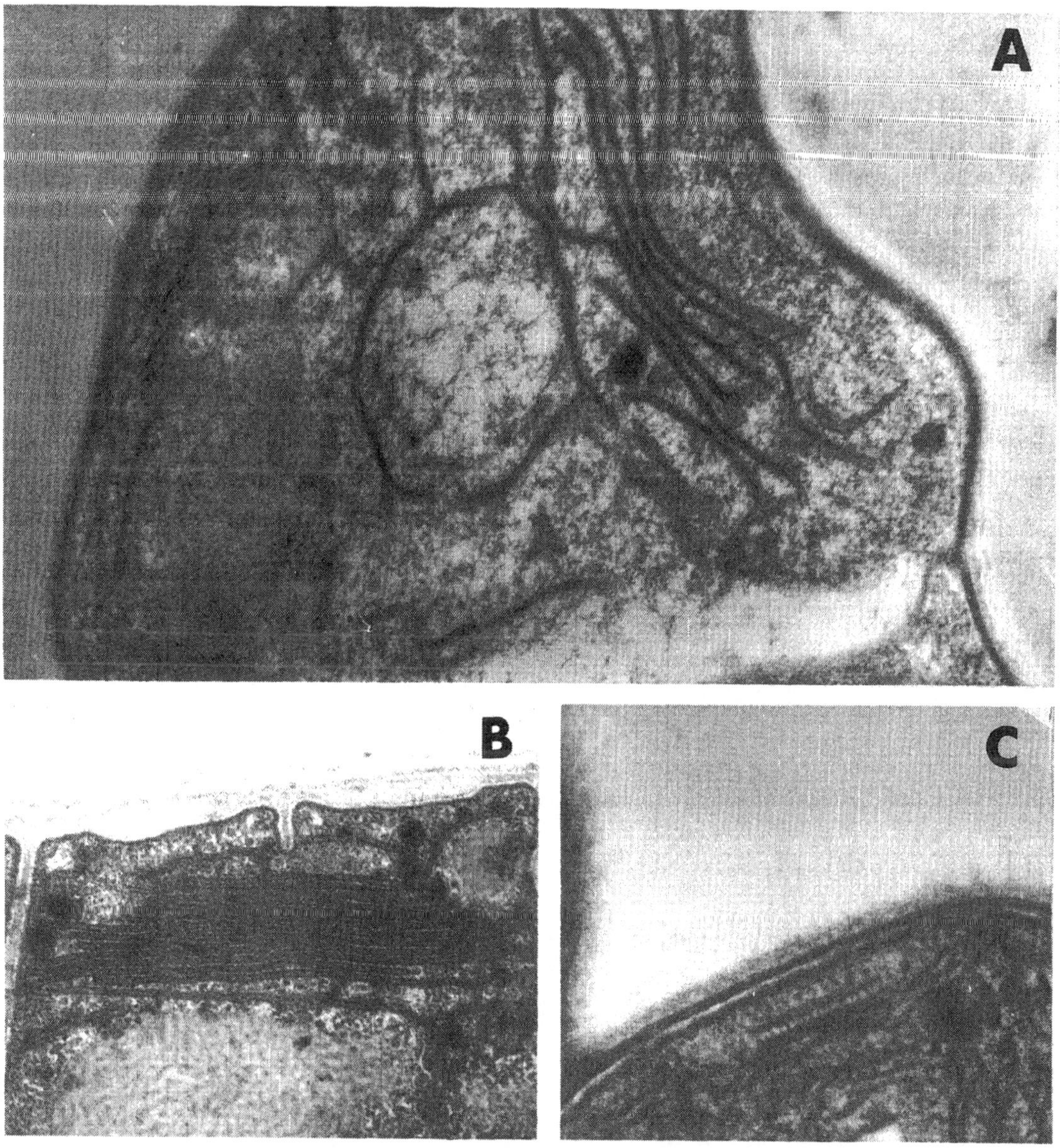

Fig. 1. (A) Thin section of a cell of *Oscillatoria* sp. showing thylakoids contorted around an area of DNA. 49,000X. (B) Part of a cell of *Microcoleus vaginatus* showing a membrane stack. 85,000X. (C) Part of a cell of *Anabaena catenula* showing a tri-lamellar body. 85,000X.

3.1.4 Lipid inclusions
They are spherical, electron dense and about 30 nm in diameter. Each cell has many of these bodies.

3.1.5 Cyanophycin granules
These roughly spherical bodies have been shown to be co-polymers of arginine and aspartic acid (Simon 1971).

3.1.6 Polyphosphate bodies
Cyanobacteria almost always have these spherical electron dense inclusion in their cytoplasm. The bodies are composed of polyphosphate, an organic component and Mg, Ca and K (Jensen 1985).

3.2 Inclusions of irregular occurrence

3.2.1 Membrane monolayer limited inclusions
3.2.1.1 Gas vacuoles. The gas vacuole is a complex body consisting of a number of gas vesicles. The gas

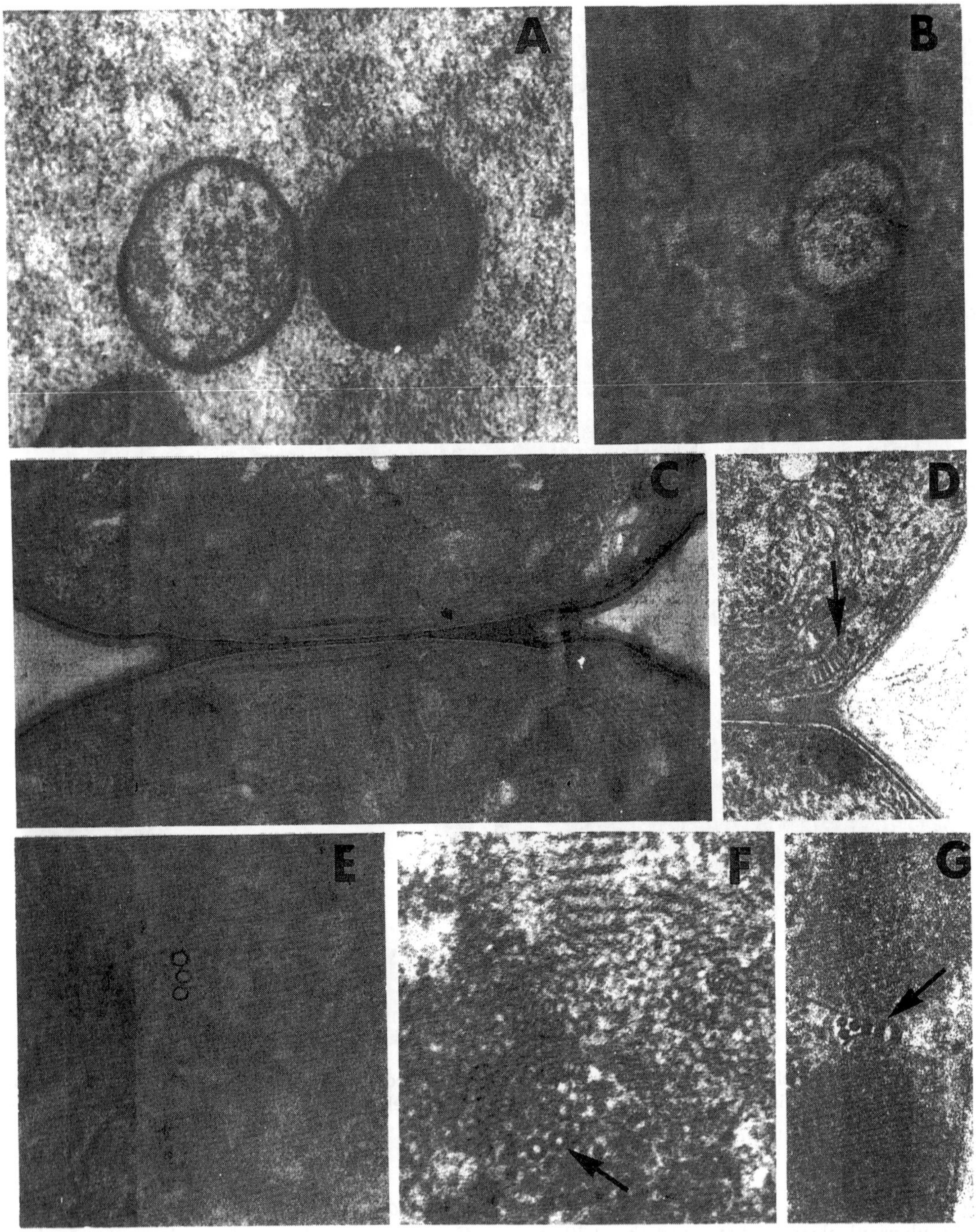

Fig. 2. (A) Intrathylakoidal granules from *Nostoc commune*. Note the limiting unit membrane. 130,000X. (B) Part of a cell of *Anabaena minutissima* showing a membrane limited crystalline inclusion. 130,000X. (C) Part of a cell of *Nostoc pruniforme* showing a plate filament array. 70,000X. (D) Part of a cell of *Anabaena* sp. showing a microplate-microtubule array (arrow). 45,000X. (E) Part of a cell of *Nostoc* sp. showing variable diameter microtubule-like inclusion. 95,000X. (F) Portion of a cell of *Anabaena minutissima* showing striated microtubules. Note substructure in the wall can be seen in the tubule at the arrow. 110,000 X. (G) Polyhedral body associated tubules in *Anabaena* sp. (arrow). 180,000 X.

vesicles are limited by a membrane monolayer. They function in cell density regulation.

3.2.1.2 Poly-β-hydroxybutyrate granules. These spherical membrane monolayer limited inclusions are generally 0.5 μm in diameter. They seem to function as a carbon reserve.

3.2.1.3 Polymorphic bodies. These bodies are found only in certain *Nostoc* and *Microcystis* isolates. They are spherical and limited by a membrane monolayer.

3.2.1.4 Vesicles and vesicular crystalloids. The vesicles are spherical, limited by a membrane monolayer and are about 33 nm in diameter. In some isolates they align to form a crystalloid.

3.2.2 Membrane limited or composed

3.2.2.1 Mesosomes. These bodies are complicated invagination of the plasma membrane.

3.2.2.2 Prolamellar body-like inclusions. These inclusions consist of a complicated twisting of a section of the thylakoids.

3.2.2.3 Membrane limited crystalline inclusions. These bodies contain needle like crystalline material or granules. They are probably composed of calcite, apatite or hydroxyapatite.

3.2.2.4 Intrathylakoidal granules. These spherical bodies are formed by the end of the two oppressed membranes of a thylakoid dilating out at the end.

3.2.2.5 Scroll-like membrane system. They consist of a 'membrane' sheet which is rolled up like a scroll.

3.2.2.6 Membrane spheres. These bodies consist of a 'membrane' rolled into a sphere about 0.6 μm in diameter.

3.2.2.7 Membrane stacks. These inclusions consist of membrane sheets stacked on top of each other. In some cases the ends are dilated.

3.2.2.8 Peripheral lamellar sheets. They consist of membrane segments which provide a layer completely around the cell just inside the plasma membrane.

3.2.2.9 Tri-lamellar bodies. These bodies consist of 3 membranous discs stacked on top of each other. They are generally located at the junction of the cross and longitudinal walls.

3.2.2.10 Cylindrical bodies. They seem to originate from tubular thylakoids. In cross section a series of osmiophilic lines are observed.

3.2.3 Tubules, filaments and sheets

3.2.3.1 Striated microtubules. These 18 nm diameter tubules have a wall composed of 11 to 13 subunits. They show osmiophilic striations at right angles to the longitudinal axis.

3.2.3.1 Variable diameter microtubule-like inclusions. These occur in groups of up to 5 and are 9.5 to 22 nm in diameter. Lengths up to 1 μm have been observed.

3.2.3.3 Plasma membrane associated microtubule-like inclusions. These 15 nm diameter tubules are orientated with their long axis perpendicular to the plasma membrane.

3.2.3.4 Polyhedral body associated tubules. These occur in groups and are always associated with one face of the polyhedral bodies.

3.2.3.5 Sleeve bodies. They consist of a short cylinder 32 nm in diameter with a rod-like body inside the sleeve.

3.2.3.6 Microtubule-like inclusions. These are located in the cytoplasm and are in the size range of eukaryotic microtubules.

3.2.3.7 Plate-filament array. The plates are always located close to the crosswall and parallel to it. Projecting at right angles to the plate are filaments 14 nm in diameter.

3.2.3.8 Microplate-microtubule array. They consist of an arched plate about 0.2 μm long with tubules 12 nm in diameter at right angles.

3.2.3.9 Microfilaments. These inclusions occur in bundles and are about 8 nm in diameter.

3.2.3.10 Wall bodies. These inclusions traverse the length of the cell. They seem to be an internal scaffolding to maintain the shape of the large cells in which they occur.

3.2.4 Crystalloid inclusions

3.2.4.1 Paracrystalline inclusions. They typically have osmiophilic striations with a 80 A periodicity.

3.2.4.2 Segmented rods. These bodies are about 100 nm in diameter and have osmiophilic striations when viewed in longitudinally section.

3.2.5 Storage polymer-like inclusions

3.2.5.1 Granular areas. These bodies consist of granules aligned in rough rows. The bodies are located in the thylakoid area.

3.2.5.2 Spherical bodies. These bodies are about 0.4 to 0.8 μm in diameter. They resemble other storage polymers.

3.2.5.3 Convoluted inclusions. In these bodies individual elements are arranged in three-dimensional whorled and convoluted patterns.

3.2.5.4 Spherical spoke inclusions. They are about 88 nm in diameter and resemble lipid inclusions.

3.2.6 Intrathylakoid inclusions

3.2.6.1 Rosettes. The bodies consist of a central core 30 nm in diameter surrounded by a number of similar elements.

3.2.6.2 Spheroids. They occur in clusters, are roughly spherical and about 30 nm in diameter.

4. Perspective and conclusions

Above we have briefly surveyed the inclusion bodies in the cyanobacteria. Details relating to these inclusions can be found in the various reviews (Lang 1968; Wolk 1973; Stanier and Cohen-Bazire 1977; Allen 1984; Jensen 1985; Shively *et al.* 1988; Jensen 1990).

4.1 Origin of nucleus, mitochondria and chloroplasts

A number of ideas have been presented on how the nucleus could have arisen. Mereschkowsky (1910) suggested that an endosymbiont gave rise to the nucleus. Recently, this idea has been again put forward but most have discounted it. Cavalier-Smith (1975) and Margulis (1970) have proposed that the nuclear envelope arose by the invagination of the plasma membrane. Taylor (1976) has suggested that either the endoplasmic reticulum or the thylakoids associated to form the membranes of the eukaryotic nucleus. More recently, Lake (1989) has used evolutionary parsimony to suggest which group of prokaryotes gave rise to the nucleus. We have recently described how thylakoids arrange themselves around the nucleoid in aged cells of two *Oscillatoria* isolates (Jensen 1989). This kind of arrangement is shown in Figure 1a. In this paper we suggest that it is perhaps a basic property of some membranes when appropriate proteins are present to contort around the nuclear material in a manner which can be seen in the eukaryotic nucleus. If such an arrangement provided some selective advantage to the cell, especially if it would have maintained this configuration upon returning to a more normal environment, then it might tend to be perpetuated. That is, a cell or group of cells with this kind of membrane system around the nuclear area would be maintained. Of course, no nuclear pores were observed in this study. These structures would not appear until proteins were obtained to provide a primitive structure and then await any selective advantage a cell may obtain by having these structures.

Many different prokaryotic cells contain intracellular membranes, the blue-green bacteria of course have thylakoids, all of the nitrogen fixers have lamellae, some in great numbers, the methane-oxidizing bacteria (Weaver and Dugan 1975), and many other bacteria have membranes but in lesser amounts such as the methylotrophic bacteria (Corpe *et al.* 1986). In fact, some sort of internal membrane system is present in a great many bacteria. Sometimes these take on complicated arrangements such as is shown in *Methylococcus capsulatus* (Hyder *et al.* 1979)., in others such as a *Thiothrix* sp. and a type 021N bacterium they arrange themselves to form double membrane limited parts of the cytoplasm (Williams *et al.* 1987). *Nitrosolobus multiformis* has invagination of the plasma membrane which segregate the DNA area by a double membrane system (Watson *et al.* 1971). These authors suggest that this may be the means of transition from the prokaryotic to the eukaryotic cell. In *Chromatium buderi* they from double membrane-bound organelles (Remsen and Truper 1973). In many cyanobacteria the thylakoids closest to the nuclear area surround it and in essence isolate it from the rest of the cytoplasm. Good examples of this are seen in *Pseudanabaena* (Guglielmi and Cohen-Bazire 1984) and *Anacystis nidulans* (Lawry and Jensen 1979).

Lake (1989) based on evolutionary parsimony suggests that a group he calls eocytes gave rise to the eukaryotic nucleus. In this group are *Thermoproteus tenax*, *Sulfolobus salfataricus* and *Desulfurococcus mobilis*. It would be interesting to see if the internal membranes in these organisms can be induced to form nuclear envelope-like configurations as described in the two *Oscillatoria* isolates.

I have observed in long term studies of cells of tomato root tips exposed to colchicine, that membranes which appear to be nuclear envelope segments re-associate but do not necessarily enclose the chromosomes (Jensen 1988). This suggests that perhaps a protein on the surface of the nuclear envelope binds to one or more of the DNA associated proteins. Colchicine allows this protein to be associated with other cellular structures and then the nuclear envelope segments associate around any structure with these proteins. Witkus *et al.* (1982) described a similar phenomenon in colchicine-treated *Allium cepa* root tip cells. In this case, two double membranes associated around each other creating an internal space surrounded by four membranes. Whaley *et al.* (1964) reported convoluted membranes in root cap cells of *Zea mays* after colchicine treatment. This suggests that membrane segments can under appropriate conditions associate in a similar way whether in a prokaryote or in a eukaryote.

In some reports, it is suggested that the nucleoplasmic area in prokaryotic cells is not a discrete structure but is diffuse. This is not the case. The DNA is bound together as a distinct unit. This is shown by the fact that in certain blue-green bacteria and other bacteria the nucleoplasm can be displaced through the cytoplasm without becoming mixed with it. This suggests that it is a separate unit and that it possesses considerable stability (Beams and Kessel 1987).

Various studies have shown that bacterial DNA does have proteins associated or bound to it. This subject has recently been reviewed by Drlica and Rouviere-Yaniv (1987). They report that bacteria have at least five proteins associated with the DNA. Histone-like HU is a small basic, very abundant DNA binding protein which is capable of wrapping DNA. IHG is a histone-like protein capable of wrapping DNA which has considerable amino acid homology with HU protein. Three other proteins are considered to be histone-like, *viz* H,HI and FirA. In the archaebacterium *Thermoplasma acidophilum* HU protein is reported to be similar to the eukaryotic histone proteins H2A and H3. In this bacterium protein Hta binds DNA and prevents thermal denaturation of the DNA.

Thus bacteria have typically five proteins associated with their DNA. These undoubtedly form a discrete structural entity and this is the reason why the DNA moves as a unit when cells are centrifuged. Under certain conditions some of these proteins may bind to components in membranes as has been described.

Based on these observations a new scenario for the origin of the nucleus, and a scheme whereby the different codes become located in the nucleus, chloroplast and mitochondria was proposed (Jensen 1989). We proposed a variation on the 'Cluster-clone' hypothesis of Bogorad (1975). He suggested that a part of the genome from the nucleus moved into the cytoplasm and then became enclosed by membranes, thus forming the mitochondrion and chloroplast. We have seen that in aged cyanobacteria the thylakoids associate and segment-off DNA areas of the cell. It is suggested that at least three different areas were segmented off and that each contained the entire genome. Reproduction of the primitive nucleus would be by invagination (pinching in two) of the nucleus as is found in many lower eukaryotic organisms. This would allow the cell to maintain all three genomes in these primitive nuclei. Selection pressure would now come into play.

Early on a large portion of the genomes in the future mitochondria and chloroplast were probably lost. However, certain genes which gave the cell a reproductive advantage over other cells would be maintained in the primitive chloroplast and mitochondria. Thus, over the span of time this selection process would operate.

We can see evidence for this in where the code for RuBPCase resides. All reports on the code of RuPBCase in higher plants and green algae describe the code for RuBPCase as being separated with a small subunit coded for in the nucleus. See also the chapter of Zetsche and references there, in this volume . The recent observation that *Porphyridium aerugineum*, a red alga (Rhodophycophyta) (Steinmüller *et al.* 1983; Van den Eynde *et al.* 1988), and *Olisthodiscus luteus*, a yellow-green alga (Chrysophycophyta) (Reith and Cattolico 1986), have the code for both subunits in the chloroplast is very interesting. The observation can be explained easily with the proposed nucleus, chloroplast, mitochondria origin. Originally the code was located in these three places in the cell. Selection operated to keep the entire code in one place, the chloroplast, while it was selected against in the other two locations. In other cell lines and the one giving rise to the chlorophyte algae and higher plants, the code for RuBPCase was separated. In the selection process only the code most suited to a particular location would tend to be retained. It would be a waste of energy to maintain in the future mitochondria a complete genome and it probably would not function as efficiently. Only after loss of all genetic material, except that which is necessary for its function, would the cell be using its energy supply most efficiently. It is essential to keep in mind the immense time period involved. The selection pressure inside the cell would operate in the same way we observe it operating at the organismic level. This pressure could have led to the eukaryotic cell by the method discussed.

A member of molecular studies link chloroplasts directly to blue-green bacteria. However molecular data can be interpreted in a number of different ways. Recently, Bremer and Bremer (1989) have challenged the data on which a number of workers using molecular data have suggested that cyanobacteria gave rise to chloroplasts. Using cladistic analysis of blue-green prokaryote interrelationships and chloroplasts origin based on 16S rRNA oligonucleotide catalogues, they conclude that little or nothing can be said about interrelationship on the basis of these particular catalogues, except the *Prochloron* may be more closely related to the blue-green bacteria than to chloroplasts.

The use of ribosomal RNA data to show phylogeny in the animal kingdom has also recently been challenged by Neilsen (1989) who stated, 'That it is important to point out that molecular data do not provide unequivocal phylogenetic trees and must be treated with just as much criticism, care, and tact as the traditional morphological characters.' Recently the use of DNA hybridization technique to show the relationships among birds and the human/ape group have come under severe criticism (Lewin 1988a,b), and again the validity of these data are questioned (Lewin 1988b) in regard to the genetic distance between gorilla and human/chimpanzee: 'Are these techniques missing something? Or is there something consistently amiss with DNA hybridization?' We can see from the above that complete reliance on molecular data can lead to erroneous results. These data must be considered a part of all data from all sources in sorting out relationships.

To summarize, it is suggested that the nuclear envelope arose from internal cellular membranes in a prokaryotic cell based on the observation of two *Oscillatoria* isolates. In colchicine-treated tomato root cells, the nuclear envelope reforms around various cell organelles. This suggests that membranes under appropriate conditions whether prokaryotic or eukaryotic associate together to form nuclear membrane-like configurations. If this arrangement in a prokaryote were to provide a reproductive advantage, it would be retained, and with modifications over time, become the nuclear envelope with its associated structures. It is proposed that this event occurred in one cell to produce three double membrane limited structures each with a complete genome. Selection pressure then operated to produce the nucleus, with most of the genetic material, a chloroplast with a small set of genetic information and the mitochondria with the least genetic information.

Data presented in this paper may be used to support the ideas put forward by Taylor (1976) on the origin of the nuclear membranes. The internal membranes may be thylakoids, lamellae or any of the other various membranes reported in the prokaryotes. It is suggested that most membranes will behave in a similar manner under similar conditions. This does not mean that the nucleus arose in the cyanobacterial group or in any other particular group although the cyanobacteria seem to be best suited as progenitor. Indeed, membrane behavior may be the same whether in cyanobacteria, archaebacteria, eocytes (Lake 1989) or some other group.

4.2 Origin of the eukaryotic cell

Above we have discussed how the nucleus, chloroplast and mitochondria may have arisen. We will now discuss how the eukaryotic cell with its set of organelles and inclusion could have arisen. Many writers in discussing evolution of cells suggest that prokaryotic cells are very simple in structure. They suggest they only have ribosomes, DNA and perhaps some storage material and thus there are no structures on which evolutionary selection could have come to bear to produce the various organelles which we see today. This statement is far from a description of the true nature of the cellular organization of living prokaryotes. Again, let us take the cyanobacteria as an example.

We can see in this one group of prokaryotes that, in addition to the usual set of inclusions, thirty two additional inclusions occur in various isolates.

Other inclusions will surely be found if they are searched for in properly fixed cells. Based on the excellent review of inclusion bodies in other bacteria by Shively (1974), the same situation holds true here. This subject should be reviewed again as numerous papers have been published in this area, especially in the Archaebacteria. There is no reason to doubt that the different groups of bacteria had similar sets of inclusions before the eukaryotic cell arose.

It is not suggested that only the cyanobacteria should be investigated for the origin of eukaryotic cellular inclusions, since it is likely that this same diversity is found in many of the bacterial groups. For example, microtubule-like structures have been reported in nine different blue-green bacteria (Jensen 1984, 1985) and in many other bacteria (Shively 1974). If one of these were to give the cell possessing them some selective reproductive advantage, they would tend to be retained. The same applies to other inclusion bodies in prokaryotes. The function of these inclusions is slowly being elucidated.

In the cyanobacteria gas vacuoles and gas vesicles have received much attention (Walsby 1978) as have polyhedral bodies (Codd and Marsden 1984). In all cases where an inclusion body has been investigated in depth it has been found to have a specific function. The same is true in the eubacteria. For example, flagellated bacteria generally have a polar organelle which is closely associated with the flagellar apparatus and is probably a part of it (Tauschel 1985). Tauschel and Drews (1969) suggest that basal bodies of the flagellum originate from it by a morphogenetic process. It is generally found associated with the flagellar insertion site. Recently, Tauschel (1985) reported ATPase activity associated with the inner boundary and central layer of the organelle. He also reported cytochrome oxidase activity in the fine fibrils which traverse the structure. These bodies resemble the tri-lamellar body reported by Jensen (1984) in various *Anabaena* and *Nostoc* isolates.

The function of other complicated bacterial inclusions, such as the R bodies found in various *Pseudomonas* and *Caedibacter* isolates (Lalucat et al. 1979, 1986; Bird *et al.* 1986; Gibson 1984; Wells and Horne 1983), is also being investigated. These bodies, which were at first a curiosity associated with killer bacteria strains of *Paramecium aurelia*, have now been found in free-living bacteria and recently correlated with leaf blight on oats, corn, maize and barley, due to *Pseudomonas avenae* (Wells and Horne 1983).

As more investigators work on the bacterial inclusion structure-function relationships, a situation may be found such as that in eukaryotic cells where nearly every structure described at the electron microscope level has a function. We are still at the stage in the bacterial world where new cellular inclusions are generally dismissed as unimportant because they are found in only one or a few isolates and are suggested as being probably due to a mutant virus. Even if some of the inclusions were the result of a mutant virus, providing that they conferred some reproductive advantage to the cell, they would be retained.

Thus as work proceeds, we will probably come to understand the function of the numerous microtubule-like structures in prokaryotes. Consider the effort still going on to understand the structure-function relationships in eukaryotic filamentous systems. In any event, the prokaryotes have produced fibrillar and tubular structures of many morphologies. Surely this happened also in the billion plus year interval between prokaryotes and eukaryotes and again they would be selected for, if reproductive advantage were thereby obtained.

Periodically because of a new arrangement of genetic material, a protein or proteins will be formed. If this new protein or structure which it forms allows the cell to reproduce at a more rapid rate than other cells of this kind, this cell type will soon dominate. If on the other hand, the new structure slows the rate of reproduction by utilizing too much energy to have it produced or in some way slows the energy and reproductive processes, this cell line will lose out to its more efficient neighbors. This view of the cell as a very small ecological niche has been discussed by others (Richmond and Smith 1979; Keyhani 1983; Taylor and Harrison 1983). We should accept that competition is a part of natural selection at cell levels. It can be within cells or among cells as well as at the familiar organism level. Michaelson has recently suggested the cellular competition begins in the embryo and continues through the life of the organism (Angier 1991). The origin of the code for the various inclusions in cells may be due to what has been called 'neutral mutation random drift hypothesis' (Jukes and Kimura 1984). This theory holds that mutations occur at a set rate and that most are neutral. They do not confer selective advantage to a cell. However if we consider this theory in relation to the cyanobacteria we can see how neutral mutations may become beneficial. For example, a series or a neutral mutation could result in a protein forming a tube instead of a pleated sheet. If this tube then allowed the cell to reproduce at a higher rate it would be selected for. Perhaps some of the inclusion in prokaryotes are of this nature. Selection is at all levels in a cell. Lake (1985) shows that the shape of ribosomes varies between archaebacteria, eubacteria, eocytes and eukaryotes.

We can assume that various prokaryotic groups now and in the past have a similar set of inclusion bodies to those found in the cyanobacteria. Based on this assumption and using the blue-green bacteria as an example, we can speculate on how the other organelles arose in the precursor prokaryotic cell. In the cyanobacteria nine or ten different kinds of tubules and filaments have been described. Some of these are illustrated in Figure 2. Several of these could have evolved into the microtubules, microfilaments, intermediate filaments and fine filaments now found in eukaryotic cell. The tri-lamellar bodies or the membrane stacks (Fig. 1) could have evolved into the Golgi apparatus. Lysosomes are derived from the Golgi apparatus and this could have evolved from the system. Peroxisomes seem to originate from endoplasmic reticulum. The intrathylakoidal granules (Fig. 2A) originate from a thylakoid in a similar manner. The membrane limited crystalline inclusions are also limited by a unit membrane (Fig. 2B). Either of these could be precursors to peroxisomes (microbodies). The microplate-microtubule array may be the basal body precursor. At 0.2 μm it is in the right size range and the projecting tubules suggest the basal body microtubules. We can see that structurally the cyanobacteria have inclusions that could be modified to produce the familiar organelles in eukaryotic cells.

Let us next look at eukaryotic cells to see if any evidence of this kind of competition can be seen. This kind of competition has been observed in growing root tips in studies where chromosome numbers have been doubled by agents such as colchicine. The cells with the increased chromosome number divide slower than their normal diploid counterparts. Because only cells in the meristematic area divide, the slowly dividing cells soon pass into the zone of differentiation and are then no longer part of the meristematic zone. The fast dividing diploid cells have been retained. This kind of selection has been going on for billions of years. It probably explains why in the much studied eukaryotic cells, very few if any structures with no function have been found.

If we look at microtubule organizing centers (MTOC) which are involved in cell division we can see another example of intracellular competition. In the dinoflagellates (Dinophyceae) the chromosomes are attached to the inner membrane of the nuclear envelope. Segregation of the chromosomes is accomplished by either movement of the attachment sites or preferential membrane growth between the attachment sites to separate the chromosomes (Oakley and Dodge 1974). After separation the nuclear envelope constricts to form two nuclei. In yeast (Ascomycota) cells a different system is found. On the inner membrane of the nuclear envelope at opposite poles a dense layer of material is observed. This structure has been termed a nuclear plaque (Moens and Rapport 1971). When cell division occurs the spindle microtubules align to these areas. The chromosomes move to the poles and the nuclear envelope constricts forming two nuclei. In the Basidomycota a different system has been described. In this case a dumbbell shaped dense body sits in the cytoplasm in an invagination of the nuclear envelope (McLaughlin 1971). The spindle microtubules converge on these bodies, one at each pole, and the chromosomes move to the poles. The nuclear envelope which remains intact then constricts forming two nuclei. In one of the diatoms (Chrysophyceae) a large 0.5 μm diameter dense body is normally present in the cell (Tippit and Pickett-Heaps 1977). This MTOC forms two dense plate like bodies. These move to the poles and become the focal point of the spindle apparatus. After cell division these bodies termed polar organelles disappear. The large dense MTOC remains. These MTOC systems stand in contrast to the familiar centriole of many protists and animals. Other variations have been described. We can see from this brief discussion that a number of MTOC's have evolved to participate in chromosomal segregation. This suggests that many such bodies have been produced by eukaryotic cells and depending on selection pressure one has been selected for and the rest eliminated. All of these different morphological forms of MTOC's perform the same function. Others will probably be described as cells in division are studied in different groups.

If we examine microtubule arrangement and numbers in flagella we get a similar scenario. Flagellated protistan cells and most animal cells have the familiar 9 + 2 organization of microtubules. Papers have been written describing the conservatism of natural forces as all flagella seemed to have this kind of organization. However when sperm of organisms in the Phylum Arthroptera were examined many variations have been discovered. In the mealybug *Pseudococcus obscurus* the sperm have microtubules arranged in two and a half concentric rings and total 56 in the active region and in the less active region two or three concentric rings with 28 or 56 microtubules (Ross and Robison 1969). Phillips (1969) describes a 9 + 0 pattern in three species of mayflies, a 9 + 1 pattern in sperm of two mosquito species, a 9 + 7 pattern in two species of caddis flies and a 9 + 9 + 2 pattern in sperm of tree hoppers. Many others patterns have been described such as 12 + 0 in sperm of *Nymphon* (Van Deurs 1973); 9 + 9 + 3 in sperm of *Simuliid deptera* (Baccetti *et al.* 1974b); 170 doublets in sperm of *Monarthropalpus buxi* (Baccetti *et al.* 1974a);and 6 + 0 in the Gregarine *Lecudina tuzetae* (Schrevel and Besse 1975). Recently Dallai and Afzelius (1990) examined sperm from 49 insect species in 20 orders. They found in addition to number variation different numbers of protofilaments (subunits) in the wall of the microtubules. They observed 13, 14, 15, 16, 17 and 19 protofilaments in different microtubules with the number being generally the same in an order of insects.

We can see from these three examples that competition has occurred in the eukaryotic cell. In the case of MTOCs a variety of structures have been selected for. In the case of number of microtubules in sperm a wide range of numbers have been selected for in the insects. We can also see that even in the number of subunits in the wall of microtubules competition has occurred with a variety of numbers being selected for. Dillon (1981) gives other examples.

Another aspect of eukaryotic cellular origin discussed by many authors is that prokaryotic cells are too small to have given rise to the eukaryotic cell. However, in several groups, a number of extant species have very large cells. The following are the cell dimensions of some of these large prokaryotes: *Chroococcus gigantus* 54 to 55 μm; *Chroococcus maximus* up to 45 μm; *Synechococcus aeruginosa* up to 16 μm broad by 30 μm long; *Dermocarpa gigantea* up to 43 μm broad by 76 μm long; *Oscillatoria principes* up to 60 μm broad and 7 μm long (Desikachary 1959; Prescott 1962); *Beggiatoa gigantea* up to 55 μm wide by 13 μm long; *Beggiatoa mirabilis* up to 21 μm wide by 13 μm long; *Achromatium* sp. up to 30 μm wide by 125 μm long (Buchanan and Gibbons 1974); and specifically *Achromatium oxaliferum* shown by DeBoer *et al.* (1971) to be 27.5 x 35 μm. The average eukaryotic cell is about 30 μm in diameter.

The largest prokaryotic cell described so far is a *Beggiatoa*-like organism from a microbial mat in Guaymas Basin deep-sea hydrothermal vent (Jannasch 1984). This organism is 137 μm wide x 100 μm long x 100 μm in depth. We can see from this listing that a number of present day prokaryotes are composed of cells as large or larger than our average eukaryotic cell. It is suggested that many prokaryotic groups probably evolved large cells such as described above. However, in only a few groups does some unknown advantage allow them to thrive at this time.

It should also be kept in mind that some eukaryotes have quite small cells. In *Saccharomyces cerevisiae* the cells are about 6 μm in length and have within them all of the usual eukaryotic organelles. The smallest eukaryotic cells observed until recently were in the fungi. Yeast cells with a diameter of about 6 μm were thought to be the smallest eukaryotes. Recently, however, studies of picoplankton have confirmed early reports of very small eukaryotic cells. Johnson and Sieburth (1982) reported a non-flagellated chlorophyte probably *Nannochloris atomus* with a cell size of 1 x 2 μm. It has a single chloroplast, mitochondria and a nucleus. They also reported a *Chlorella*-like cell with a cell size of 2 to 3 μm and a third chlorophyte with a cell diameter of 2.0 μm. More recently, Fahnstiel *et al.* (1986) described a small flagellate with nucleus, chloroplast and scales from Lake Superior picoplankton. The cell had a diameter of 1.5 μm. Moestrup (1984) reported that *Nephroselmis* (Prasinophyceae) is 0.5 μm by 1.0 μm. Thus, the view that prokaryotic cells are too small to have given rise to any eukaryotic cell organisms, is just not correct. Prokaryotic cells may be very large, and eukaryotic cells which have the usual cellular inclusions may be very small. Compare with small cells of other eukaryotes: *Nanochlorum eucaryotum* and *Cyanidioschyzon merolae* (chapters by Zahn; Seckbach, resp., in this volume).

The last aspect of eukaryotic cellular origin that I wish to discuss is the time factor. The cyanobacteria go back in the fossil record some 3.3 to 3.5 billion year (Schopf and Packer 1987). Thus this group was well established at this time and other prokaryotes probably pre-date the cyanobacteria. The eukaryotes appear in the fossil record about 1.2 billion years ago (Schopf and Oehler 1976).

Many different cellular inclusions undoubtedly came into these prokaryotic cells, based on the fact that so many currently exist, and if they had no selective reproductive value were lost. If today we have so far categorized nearly thirty-five inclusions in the blue-green bacteria alone, we must surmise that the production of these bodies has been going on since the bacteria were established as cellular entities. We can really think of the cytoplasm on ecological terms. It is almost like a world unto itself. If we consider going back from the present, how many biomes existed with their complicated interaction between all the species making up the biome and then of the cell cytoplasm where a similar kind of reproductive struggle is and was going on. For example the code for a microtubule-like structure would, perhaps by chance, come into being, just as we see at least nine different kinds in blue-greens today. If this structure interfered with reproduction, the line would be lost. However, if it had no impact on reproduction, it may be maintained at least for a time. Then if the code for some other component came into being which together with the previous component allowed for reproductive advantage, it would be retained.

Loss of a cell wall in a certain environment could provide reproductive advantages. Enclosing the DNA by a set of membranes in a certain environment may have resulted in a reproductive advantage. As new structures were produced by the cell, for example various parts of the pore complex, then again this would be retained (selected for) if the cell were better able to reproduce itself. Thus slowly the organelles we see in eukaryotes would be selected for. We must keep in mind that this sort of give and take went on for well over a billion years. Perhaps in this time frame, we can think in a new way about the origin of the eukaryotic cell and its individual complexity.

The complexity of present day prokaryotic cells with their many inclusions must be emphasized. Selection pressure for those which would have given the prokaryote a reproductive advantage may have led to formation of the various eukaryotic cell organelles which are evident. I interpret the present data on this topic to support the view, as have many others (see Lee and Frederick 1987, and papers cited therein), that the eukaryotic cell arose by direct filiation or autogenously. Furthermore, I am suggesting how the chloroplast and mitochondria and their genomes could have evolved. After the eukaryotic cell was established endosymbiotic events could occur and must be interpreted as providing some selective advantage giving a cell the advantage to reproduce at a faster rate than competing cells as discussed by Taylor and Harrison (1983). The ideas presented here are not in conflict with evolutionary schemes based on molecular or any other data. It is only a matter of using the proper time scale and what the event was which the scheme leads to.

Acknowledgements

Thanks are expressed to the Editors of Cytobios for permission to reproduce Figure 1A and parts of the discussion from Jensen, T.E. 1989. Cytobios 60: 47–61.

References

Allen, M.M. (1984). *Ann. Rev. Microbiol.* 38: 1–25.
Angier, N. (1991). *Medical Science. New York Times Feb. 12. C-2 to C-3.*
Baccetti, B., Dallai, R., Giusti, F. and Bernini, F. (1974a). *Tissue and Cell* 6: 269–278.
Baccetti, B., Dallai, F., Giusti, F. and Bernini, F. (1974b). *J. Ultrast. Res.* 46: 427–440.
Beams, H. W. and Kessel, R. G. (1987). *Int. Rev. Cytol.* 100: 15–48.
Bird, B. C., Bedingfield, G., Gibson, I. and Horne, R. W. (1986). *Micron Microsc. Acta* 17: 377–384.
Bogorad, L. (1975). *Science* 188: 891–898.
Bremer, B. and Bremer, K. (1989) *J. evol. Biol.* 2: 13–30.
Buchanan, R. E. and Gibbons, N. W. (1974). *Bergey's Manual of Determinative Bacteriology. Eighth ed. The Williams and Wilkins Co., Baltimore, Maryland.*
Cavalier-Smith, T. (1975). *Nature* 256: 463–468.
Codd, G. A. and Marsden, W. J. N. (1984). *Biol. Rev.* 59: 389–422.
Corpe, W. A., Jensen, T. E. and Baxter, M. (1986). *Arch. Microbiol.* 145: 107–112.
Dallai, R. and Afzelius, B. A. (1990). *J. Struct. Biol.* 103: 164–179.
DeBoer, W. E., Riviere La, J. W. M. and Schmidt, K. (1971). *Antonie van Leewenhoek* 37: 553–563.
Desikachary, T. V. (1959). *Cyanophyta. Indian Council of Agricultural Research, New Delhi, India.*
Dillon, L. S. (1981). *Ultrastructure, Macromolecules,* and *Evolution.* Plenum Press, New York and London.
Drlica, K. and Rouviere-Yaniv, J. (1987). *Microbiol. Rev.* 54: 301–319.
Fahnstiel, G. L., Sicko-Goad, L., Scavia, D. and Stoermer, E. F. (1986). *Can. J. Fish. Aquat. Sci.* 43: 235–240.
Gibson, I. (1984). *Micron Microsc. Acta* 15: 253–259.
Graham, L. L. and Beveridge, T. J. (1990). *J. Bacteriol.* 172: 2150–2159.
Guglielmi, G. and Cohen-Bazire, G. (1984). *Protistologica* 20: 377–391.
Hyder, S.L., Meyers, A. and Cayer, M. L. (1979). *Tissue Cell* 11: 597–610.
Jannasch, H. W. (1984). In: *Microbial Mats: Stromatolites. Alan R. Liss, Inc., New York, U.S.A.* pp. 121–131.
Jensen, T. E. (1984). *Cytobios* 39: 35–62.
Jensen, T.E. (1985). *Arch. Hydrobiol.* (Suppl.) 71: 33–73.
Jensen, T. E. (1988). *Proc. 46th Ann. E.M.S.A. San Francisco Press, California, U.S.A.* pp. 306–307.
Jensen, T. E. (1989). *Cytobios* 60: 47–61.
Jensen, T. E. (1990). In: *Phycotalk Vol. 2. (ed. Kumar, H. D.) Rastogi & Co., Subhash Bazar, Meerut-250002, India.*
Johnson, P.W. and Sieburth, J. McN. (1982). *J.Phycol.* 18: 318–327.
Jukes, T. H. and Kimura, M. (1984). *J. Mol. Evol.* 21: 90–92.
Kellenberger, E., Ryter, A. and Sechaud, J.(1958). *J. Biophys. Biochem. Cytol.* 4: 671–678.
Keyhani, E. (1983). In: *Endocytobiology II. (eds. Schenk, H.E.A. and Schwemmler, W.) Walter de Gruyter, Berlin.*
Klein, R. and Cronquist, A. (1967). *Quart. Rev. Biol.* 42: 105–296.
Klein, R. M. (1970). *Ann. N.Y. Acad. Sci.* 175: 623–633.
Lake, J. A. (1985). *Ann. Rev. Biochem.* 54: 507–530.
Lake, J. A. (1989). *Can. J. Microbiol.* 35: 109–118.
Lalucat, J., Meyer, O., Mayer, F., Pares, R. and Schlegel, H.G. (1979). *Arch. Microbiol.* 121: 9–15.
Lalucat, J., Wells, B. and Gibson, I. (1986). *Micron Microsc. Acta* 17: 243–245.
Lanaras, T. and Codd, G.A. (1981). *Planta* 153: 279–285.
Lang, N. J. (1968). *Ann. Rev. Microbiol.* 22: 15–46.
Lawry, N. H. and Jensen, T .E. (1979). *Arch. Microbiol.* 20: 1–7.
Lee, J.L. and Frederick, J.F. (1987). *Endocytobiology* III. *Ann. N.Y. Acad. Sci.* USA. 503: 1–589.
Lewin,R. (1988a). *Science* 241: 1598–1600.
Lewin, R. (1988b). *Science* 241: 1756–1759.
McLaughlin, D. J. (1971). *J. Cell Biol.* 50: 737–745.
Margulis, L. (1970). *Origin of Eukaryotic Cells. Yale Univ. Press, New Haven, Conn. USA.*
Mereschkowsky, C.(1910). *Biol. Centralbl.* 30: 278–303, 321–347, 353–367.
Moens, P. B. and Rapport, E. (1971). *J. Cell Biol.* 50: 344–351.
Moestrup, O. (1984). *Nord. J. Bot.* 4: 409–421.
Neilsen, C. (1989). *Science* 243: 548–551.
Oakley, B. E. and Dodge, J. D. (1974). *J. Cell Biol.* 63: 322–325.
Phillips, D. M. (1969). *J. Cell Biol.* 40: 28–43.
Prescott, G. W. (1962). *Algae of the western Great Lakes area. W.C. Brown Co., Dubuque, Iowa, U.S.A.*
Reith, M. and Cattolico, R. A. (1986). *Proc. Natl. Acad. Sci.* USA. 83: 8599–8603.
Remsen, C. C. and Truper, H. G. (1973). *Arch. Mikrobiol.* 90: 269–280.
Richmond, M. H. and Smith, D. C. (1979). *The Cell as a Habitat. London: Royal Society.*
Ross, J. and Robison, W.G. Jr. (1969). *J. Cell Biol.* 40: 426–445.
Schopf, J. W. and Oehler, D. Z. (1976). *Science,* 193: 47–49.
Schopf, J. W. and Packer, B. M. (1987). *Science* 237: 70–73.
Schrevel, J. and Besse, C. (1975). *J. Cell Biol.* 66: 492–507.
Shively, J. M. (1974). *Rev. Microbiol.* 28: 167–187.
Shively, J. M., Bryant, D. A., Fuller, R. C., Konopka, A. E., Stevens, S. E. Jr. and Strohl, W. R. (1988). *Int. Rev. Cytol.* 113: 35–100.

Simon, R. D. (1971). *Proc. Natl. Acad. Sci.* U.S.A. 68: 265–267.
Stanier, R.Y. and Cohen-Bazire, G. (1977). *Rev. Microbiol.* 31: 225–274.
Steinmüller, K., Kaling,M. and Zetsche, K. (1983). *Planta* 159: 308–313.
Tauschel, H. D. and Drews, G. (1969). *Arch. Mikrobiol.* 66: 166–179.
Tauschel, H. D. (1985). *Arch. Microbiol.* 141: 303–308.
Taylor, F. J. R. (1976). *Taxon* 25: 377–390.
Taylor, F. J. R. and Harrison, P.J. (1983). In: *Endocytobiology II. (eds. Schenk, H.E.A. and Schwemmler, W.) Walter de Gruyter, Berlin.*
Tippit, D. H. and Pickett-Heaps, J. D. (1977). *J. Cell Biol.* 73: 705–727.
Van Den Eynde, A., DeBaere, R., DeRoeck, E., Van dePeer, Y., Vandenberghe, A., Willekens, P. and DeWachter, R. (1988). *J. Mol. Evol.* 27: 126–132.
Van Deurs, B. (1973). *J. Ultrastruct. Res.* 42: 594–598.
Walsby, A. E. (1978). *Symp. Soc.Gen. Microbiol.* 28: 327–358.
Watson, S. W., Graham, L. B., Remsen, C. C. and Valois, F. W. (1971). *Arch. Mikrobiol.* 76: 183–203.
Weaver, T. L. and Dugan, P. R. (1975). *J. Bact.* 127: 704–710.
Wells, B. and Horne, R.W. (1983). *Micron Microsc. Acta* 14: 329–344.
Whaley, W. G., Kephart, J. E. and Mollenhauer, H. H. (1964). In: *Cellular Mechanisms in Development (ed. Locke, T.) Academic Press.* pp. 135–173.
Williams, T. M., Unz, R. F. and Doman, J. T. (1987). *Appl. Environ. Microbiol.* 53: 1560–1570.
Witkus, R., Herold, L. and Vernon, G. M. (1982). *Expl. Cell Biol.* 50: 56–58.
Wolk, C. P. (1973). *Bact. Rev.* 37: 32–101.

Note added in proof

A prokaryotic symbiont of very large size has recently been described from the gut of surgeonfish. This bacterium is up to 576 μm in length (Clements, K.D. and Bullivant, S. (1991) J. Bacteriol. 173: 5359–5362).

Nakamura (Nakamura, H. and Hase, A. (1990–1991) Origins of Life and Evolution of the Biosphere 20: 499–514) has recently published data summarizing the facts which support the Membrane Evolution Theory. He provides excellent coverage of metabolic and molecular aspects of prokaryotic and eukaryotic cells which do not support the endosymbiont theory.

PART II

Green enigmatic algae

Chapter 4

Nanochlorum eucaryotum – a green enigmatic alga: morphology, biochemistry and molecular biology

RUDOLF K. ZAHN
Academy of Science and Literature, Building for Science and Development of the University of Mainz, Obere Zahlbacher Str. 63, D–6500, Mainz, Germany

Abbreviations: DMSO, Dimethylsulfoxide; EDTA, Ethylene-diamine-tetraacetate, disodium salt; kbp, Kilo base pairs, short for: number of base pairs of DNA nucleotides; kdal, Kilo dalton, short for: molecular mass in multiples of the hydrogen atom mass as unity; MOPS, 3(N-morpholino)-2-hydroxy-propanesulfonic acid; N.e, *Nanochlorum eucaryotum*; Tris, Tris (hydroxymethyl)-aminomethane.

Outlines:

1. Introduction
2. Occurrence of the organism
3. Description of the organism
   3.1 Size and shape
   3.2 Subcellular components
      3.2.1 DNA preparation
      3.2.2 Nucleus
      3.2.3 Mitochondrion
      3.2.4 Chloroplast
      3.2.5 Golgi-apparatus
      3.2.6 Cytoplasmic apparatus
      3.2.7 Cytoplasmic granules
      3.2.8 Outer membrane and cell wall
      3.2.9 Immunological identification
4. Isolation and cultivation
   4.1 Isolation
   4.2 Aerobic cultivation
   4.3 Anaerobic cultivation of N.e.
   4.4 Cultivation under nitrogen deficiency
   4.5 Cultivation with methane as sole carbon source
5. Growth and cell division
   5.1 Inhibitor assays
   5.2 Mitotic apparatus
   5.3 Histones and histone-like proteins
   5.4 Autospore formation
6. Physiological properties
   6.1 Salinity
   6.2 pH dependence
   6.3 Light tolerance
References

1. Introduction

In search of a natural systematic locality for algae — and also for other organisms — quite regularly the central question of evolution comes up. This proves as especially pertinent, when the taxonomic position of the most original phytoflagellates has to be considered. In the classical view of the direct filiation theory the ancestral phytoflagellata is positioned at the verge of the prokaryotic to the eukaryotic stage.

Evolution of eukaryotes in its first stages still is not well understood (Margulis 1981), and therefore is heavily discussed in the literature, especially since so little is known so far. This comes among other reasons mostly from the absence of well enough preserved cellular fossils, of which, besides sporopollenin- like remains, only inorganic materials have been left over.

Therefore investigation into cellular evolution to the largest extent has to rely upon recent model cells, which have to fulfill certain requirements (Zahn 1984; Reitz 1986):

1. They have to be unequivocally eukaryotic
2. At the same time they have to show archaic or primitive attributes
3. It must be clear however that they are originally primitive and do not represent reduced forms of higher developed species.

Nanochlorum eucaryotum (N.e.) in the first stages of investigation possibly seemed to be a primitive eukaryotic organism, due to its reduced size and numbers of subcellular elements. This impression was enhanced by the apparent fission type of nuclear division. Consequently, N. e. seemed to represent a possible candidate as a model cell for the study of eukaryotic evolution:

J. Seckbach (ed.), Evolutionary Pathways and Enigmatic Algae: Cyanidium caldarium *(Rhodophyta) and Related Cells,* 69–86, 1994.

a) Its DNA-content is among the lowest known for eukaryotic cells.
b) Histones and nucleosome-like structures could not be detected. They are absent from all of the subcellular particles.
c) Cell division strongly deviates from the types common for eukaryotes. Condensed chromosome figures, spindle apparatus and phragmoplast are missing. N. e. apparently does not dispose of actin-like proteins that are susceptible to common mitotic poisons.
d) The nuclear membrane does not disintegrate during division.
e) The remaining characteristic eukaryotic features are present only in reduced form or number.

However the study has been hampered by the fact, that this small eukaryote has an extremely tough cell wall, composed of one of the toughest material that nature has used either sporopollenin or a substance that closely resembles it.

The absence of any signs of classical mitosis made it especially hard to classify this organism. So far no histones could be found and mitotic poisons did not bring about any significant inhibition unless at very high concentrations.

Preparing large enough DNA molecules for genetic screening etc. at the time of the first detection was difficult, since the *N. eucaryotum* cells were so small and their envelopes, fortified by sporopollenin, proved to be so extremely tough. So special tactic had to be applied. Two categories of DNA have been isolated and their properties were used in the considerations on the placement of the organism in the evolutionary tree.

The sum of all these signs brought up the question whether *N. eucaryotum* represents a primitive organism or an organism in a reduced state derived from a higher developed taxon. The investigation of N. e. could disclose in a model like fashion what the state of a 'marginal eukaryote' may be and how far deviations from a truly eukaryotic state can go without disabling biological fitness.

2. Occurrence of the organism

Nanochlorum eucaryotum (Nano) was detected by shear serendipity, when a bloom of bright green, bacterial sized microorganisms occurred in a sea water aquarium in the institute for physiological chemistry of the university of Mainz, Germany. In this aquarium the organisms were kept in artificial sea water. Its composition is shown in Table 1.

Table 1. Composition of the artificial sea water used in Mainz at the time of the first bloom

No	Chemical	Concentration in g/l
1	NaCl	27
2	$MgCl_2.6H_2O$	12
3	$Na_2SO_4.10H_2O$	10
4	$CaCl_2.2H_2O$	1.3
5	KCl	0.76
6	KNO_3	0.51
7	KBr	0.11
8	KH_2PO_4	0.068
9	$NaHCO_3$	0.022
10	$FeSO_4.7H_2O$	0.004 6
11	$Sr(NO_3)_2$	0.003 6
12	H_3BO_3	0.003 0
13	$MnCl_2.4H_2O$	0.000 39
14	NaF	0.000 36
15	$B(OH)_3$	0.000 31
16	$ZnSO_4$	0.000 055
17	$CuSO_4.5H_2O$	0.000 055
18	$NiSO_4.7H_2O$	0.000 055
19	$Al_2(SO_4)_3$	0.000 055
20	$Co(NO_3)_2.6H_2O$	0.000 055
21	$(NH_4)_6Mo_7O_{24}.4H_2O$	0.000 055
22	KJ	0.000 028
23	KBr	0.000 028
24	LiCl	0.000 028

The color of the bloom seemed to point to chlorophyll a and b and therefore to an eukaryotic organism, except the unlikely possibility of *Prochloron*, which however never had been found to cause a bloom. Since the size was pointing to a prokaryotic species, strong efforts were undertaken to solve the apparent riddle. However the organisms at this time could not be identified as one of the known algal or cyanobacterial species. Other algae species were scarce, as shown in Table 2. During the following few weeks, the bloom decreased and the new organisms gradually disappeared from the aquarium. In the meantime however, the culture of the organism became possible, thus enabling further research (Wilhelm *et al.* 1982).

Since the aquarium had been stocked with organisms of the Northern Adriatic from the vicinity of Rovinj (Istria), Yugoslavia, we decided

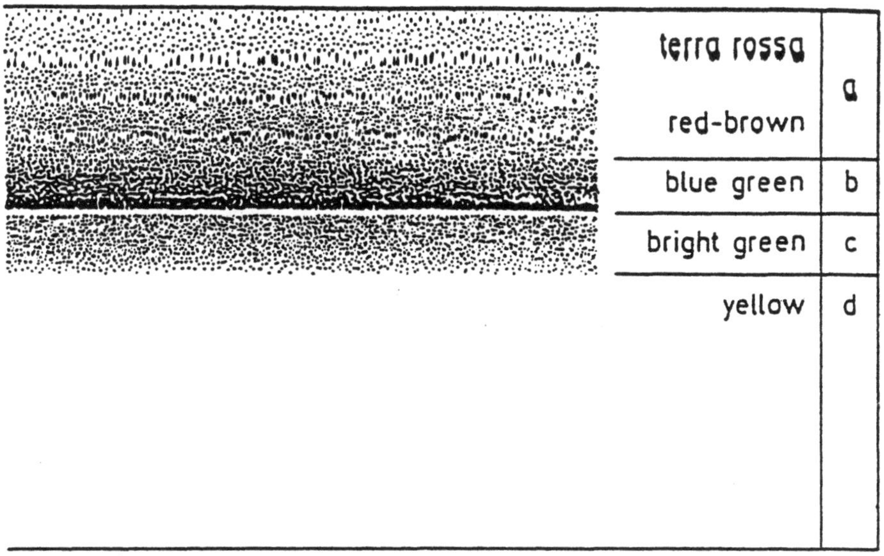

Fig. 1. Location of *Nanochlorum eucaryotum* in a crack of a rock. The sketch gives a view onto the wall of a crack of calcium-carbonate rock from the Location 'Colona', near Bale, Istria, Northern Adriatic. The upper heavy line corresponds to the rock's surface, from there on spreads a 1 cm wide red brown zone (a) coloured by fine soil particles (terra rossa) followed by a dark blue-green band (b) inhabitated by cyanobacteria. Below this follows a bright green, 4mm wide band, from which *Nanochlorum eucaryotum* could be isolated. Below this follows a faint yellow band, about 1.3 cm wide and then fresh fracture faces.

Table 2. Algal composition of the first bloom in the aquarium. (from: Wilhelm 1983).

Species	Number of cells per ml
Nanochlorum e.	10^7
Green Flagellates	10^3
Coccal Green Algae	5×10^4
Diatomeae	10^4
Cyanobacteria	5×10^5
(Ciliata	3×10^2)
(Crustaceae	10^2)

to start searching for the origin of the organism. Thirty eight samples were taken from the Northern Adriatic, from the shores up to 5 nautical miles from the Center for Marine Research at Rovinj (Istria, Yugoslavia). This is the marine stretch which we, during the previous 2 years (15 cruises) have taken samples. We presumed that ultimately the N. e. detected in our sea water aquarium in Mainz must have arrived from these samples.

Thirteen samples were taken from the surface and 13 from the bottom at 24–32 meters. They were filtered by gravity through a 3 μm filter. The filtrate was then put through a 25 mm 0.42 μm filter (Millipore, Neu-Isenburg, FRG). These filters were cut into quarters and put into sterile flasks. To two of them, 10 ml of sterile Adria water was added, to the two others 10 ml of sterile solution according to Table 1. They were plugged and put into a roller apparatus (30 rpm) under a 12 h 2000 lux daylight, 12 h dark cycle, at 22 °C.

Nine samples were rinsed off from macro-algae, shells and sponges, each with 250 ml sterile Adria water. Three samples came from stones. One of them, from 'Colona', the source of a successful preparation, was in a fine crack going 4 cm deep into an otherwise undisturbed, about 12 kg heavy rock of white calcium carbonate. When a piece had been knocked off with a hammer a 3 cm broad colored band became visible (Fig. 1).

Under the microscope, the blue-green zone showed single celled organisms considerably larger than *Nanochlorum* without nuclear structures, whereas in the bright green area among others, small green organisms of the expected diameter (around 1 μm) could be discerned. The organisms of the different layers could be brushed off with sterilized instruments. They were suspended in 250 ml sterile artificial sea water. — Later, when our knowledge was more advanced, a different solution was used (see paragraph 4). — The 250 ml suspension, in 500 ml Erlenmeyer flasks were shaken with 2 strokes per second, at an illumination of 800 lux at 22 °C for 6 weeks. Later, by diluting out and stocking large numbers of roller tubes it became

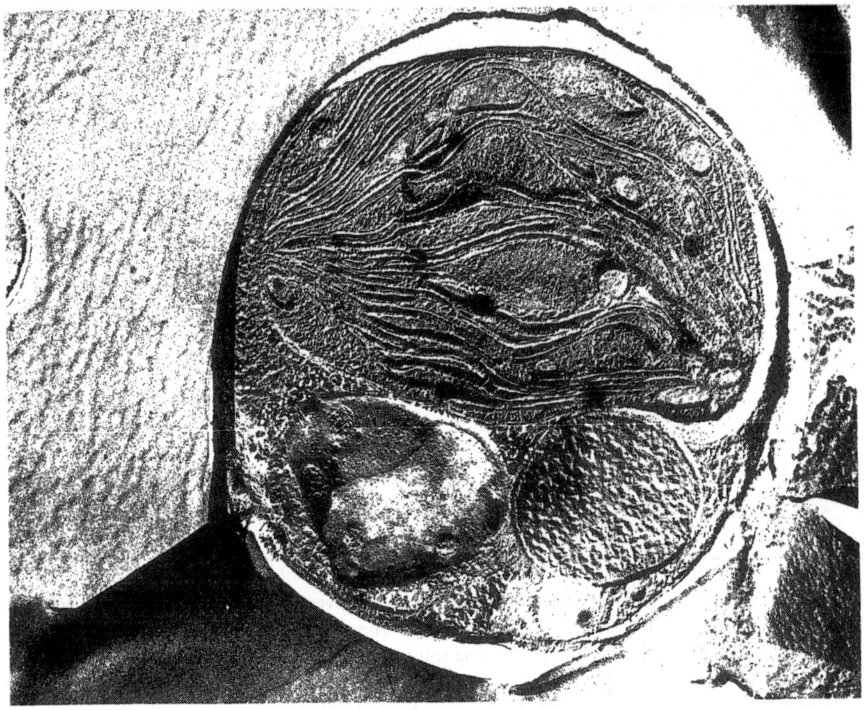

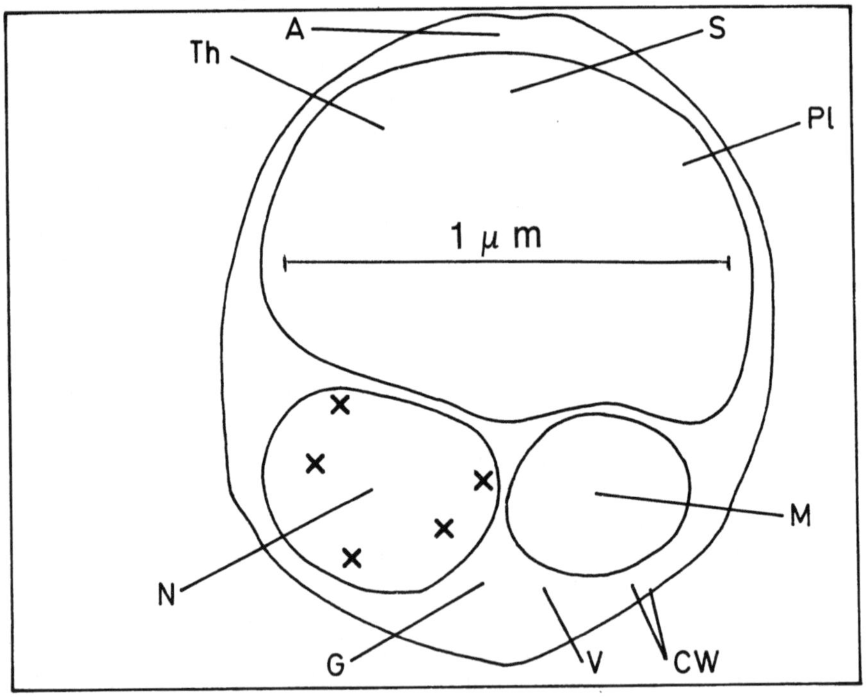

Fig. 2. Freeze fracture of *Nanochlorum eucaryotum* giving essential details of a cell of 1.4μm × 1.6μm. There is an artifactual gap (A) around the cell splitting the thin cell wall (CW) (30 nm). The largest subcellular body, about 60% of total cell volume (TCV), is the plastid with thylakoid bands (Th). Between the thylakoids starch grana (S) and plastoglobuli (Pl) are found in cells approaching the end of the logarithmic phase. The second largest body is the mostly nearly round nucleus (N), occupying 21% TCV, with some nuclear pores (x), which partically has been stripped down to and across its nuclear membrane. The mitochondrion (M), with 12% TCV, which suffered fracture across its inner structure, is also nearly round. The Golgi region (G), mostly with vacuoles (V), is only sometimes visible. (Platinum vapor shadowing was from lower left corner).

possible to make large numbers of the organisms in nearly pure culture to be used in further investigations.

3. Description of the organism

Figure 2 gives a freeze fracture cross section aspect of a *Nanochlorum eucaryotum* cell approaching the end of its logarithmic growth phase.

3.1 Size and shape

The size of N. e. is variable and it depends on the conditions of preparation, on the suspension milieu and on the state of cell cycle. The dimensions of the smallest individuals were little more than 1 μm, in most cases 1.3 to 1.8 μm have been measured. In the largest measurements the cells reached 2.0 μm in their diameter. The size has been found to depend on the iron supply, so that smaller cells may result when iron in its divalent state is ample in supply (Wilhelm 1983). The cells usually are round to slightly elliptic. Shortly before division their size nearly doubles.

3.2 Subcellular components

All cellular components were examined without delay prior to division, they only occur as single structures. Their separation and biochemical investigation as pure entities has not been accomplished since the properties of the cell wall do not allow their isolation in a state of sufficient intactness. There are only two ways in which information about substructures of N. e. can be conveyed: the electron microscopic evaluation and information on different DNA populations.

3.2.1 DNA preparation

Satisfactory DNA preparation however turned out to be rather difficult to achieve since this small organism, even in absence of atmospheric oxygen, continues to produce enough sporopollenin (Geisert *et al.* 1985) for making extremely tough cell wall and thus very hard to break. Therefore, all measures taken for isolation of macromolecular components involved rough handling which caused physical damage of different kinds and to different degrees. Among the different procedures tried, the best results were attained by a glassy form placed into ice and forcing it under controlled conditions through a small orifice (Zahn and Beyer unpublished). The criterium for quality was the molecular length of the DNA as viewed in the electron microscope (Kleinschmidt *et al.* 1960; Zahn *et al.* 1987).

The preparation of N. e. DNA in brief: The harvested cells were centrifuged at about 5 000 g, 20 min, 4 °C. The supernatant was sucked off, including the whitish top layer, which contains cell fragments. The dark green pellet was suspended, 1 g/100 ml in solution A (Na_2SO_4 10 mmol/l; $NaHCO_3$ 2 mmol/l; NaCl 500 mmol/l; KCl 10 mmol/l; EDTA 20 mmol/l; MOPS 50 mmol/l; pH 7.5; 0–4 °C; Proteinase K 20 mg/l), centrifuged as before and an consequent top layer was sucked off. The pellet contains about 4×10^{11} N. e. cells per g. It was suspended using a loose fitting pestle homogenizer by slowly adding 100–300 ml solution B (sodium acetate 1 mol/l; Na_3citrate 50 mmol/l; EDTA 20 mmol/l; Mercaptoethanol 20 mmol/l; DMSO 100 mmol/l; Proteinase K 10 mg/l; pH 8.5, 25 °C) per g of pellet to yield a homogenous suspension.

The suspension has been poured into the 1,2 cm diameter bore of a thick walled steel cylinder with a 1 mm hole in its front, cooled with dry ice. A prechilled tightly fitting pestle is set and via a strong threaded spindle turned by a large wheel two people can exert a pressure of more than 50 tons per cm^2. This forces the ice through the 1 mm opening into a thick walled roller flask containing a phenol mixture for DNA purification. further steps follow one of the usual DNA purifications (e.g., Graham 1978). In other cases different procedures have been used (Schreiner 1990). Such pre-purified DNA has been fractionated by repeated density gradient centrifugation for further characterization of molecular populations containing the programs of the different subcellular components (Weinblum *et al.* 1988; 1991, in preparation). The total DNA had a base ratio $(A+T)/(G+C) = 1.10 \pm 0.15$ and a content of 5-methylcytosine of 1 mol/100 mol bases.

3.2.2 Nucleus

The nucleus occupies about 20% of the cross section. By electron microscopical measurement it has a volume of 0.15 ± 0.06 μ^3. In most cases it is round, with few, large pore complexes. In some cases, the cross section shows a darker portion lying usually in the center. The texture in the sections shows rather invariable fine granulation. No tubuli or other structures ever have been observed. No distinct nucleolus could be detected, except that some parts in the electron microscope seem more dense than others, being very variable in size and in numbers. Freeze edging and sectioning exhibits a surrounding double membrane. Both surfaces of the inner membrane are seeded with numerous particles which are more frequent than on the outer membrane. The nuclear surface is interrupted by 2–8 large pore complexes being at least 20 nm apart.

During division the nuclear double membrane stays intact, while wedge-like protuberances form from it, cutting the nucleus from its equator and finally pinching it in two. Never a spindle apparatus, a centrosome or condensed chromosomes have been observed. In no cases extra nuclear micronuclei surrounding a prophase nucleus and being complexed by endoplasmatic reticulum was observed. Polar fenestrations are also absent. No alterations of the appearance of the inner nuclear substance have been observed except the appearance of same darker region in some cases shortly before cytokinesis.

During the CsCl density centrifugation the main band DNA represents the nuclear DNA with 44% deoxyribonucleotides G + C. Ninety five percent of this DNA is single copy DNA. It has a genetic complexity of $3x10^7$ base pairs. This corresponds a figure five times lower than from *Chlorella fusca* as determined by reassociation kinetics with single-copy DNA from *C. fusca*. Taking into consideration a DNA content of $6x10^{-14}$ g per N. e. cell, this could signify N. e. to be haploid.

3.2.3 Mitochondrion

A single mitochondrion occupies about 20% of the cross section occurs as a smooth round structure but in somewhat variable forms, from circular to kidney shape. It is of the tubular type and is surrounded by a double membrane which is strongly coherent, so that separation by fracturing is difficult. It is often difficult to visualize the inner membrane due to weak contrast.

3.2.4 Chloroplast

This is the largest subcellular organelle in the N. e. cell. It amounts to about 50% or more of the area in the cross sections. The chloroplast is circular in some cases, but often kidney shaped, even sometimes slimly crescent shaped. Its inner space is filled with three to five lammellar thylakoid bands. The surfaces exposed by freeze fracture from the stacked regions of the thylakoids carry considerably more particles than those from unstacked parts. The particle structures can be assigned to the two photosystems I and II as well as to their appropriate light harvesting chlorophyll a/b proteins. While the photosystem II is restricted nearly exclusively to the stacked membrane regions, the system I is distributed rather evenly between stacked and unstacked regions (Bauer 1986). Altogether these points perform an effective system for light exploitation under conditions of weak illumination. In cases of optimal nutrition, starch granule and some plastoglobuli (the end products of the chloroplasts own metabolism) are visible as 30 nm electron dense particles. Pyrenoids have never been observed. In the stroma, one can distinguish threads, most probably DNA molecules. The chloroplast is surrounded by a double membrane. Its outer surface is relatively smooth, while the surfaces exposed by freeze fracture on the inner membrane are occupied by particles considerably more dense than on the outer membrane. Cells illuminated with either 2000 lux or with 500 lux surprisingly show the same ultrastructure.

The chloroplast DNA (C_tDNA) has been isolated, following observation that its fraction counts up to 3% of the total N. e. DNA and has a G+C content of 34%. Therefore this DNA could be separated from the rest by density gradient centrifugation. Restriction fragments of such DNA could be hybridized to different plant chloroplast genes. The size of the chloroplast genome has been determined to be $9x10^4$ base pairs by restriction analysis and reassociation kinetics. So far two segments, one of 8.6 kbp and another one with 15.5 kbp have been mapped. The 8.6 kbp segment comprises the genes for the 32 kdal herbicide binding protein, for the C-terminal end of the ATP- synthase sub-unit, and for the large sub-units of the ribulose bisphosphate carboxylase. The 15.5 kbp segment contains only one single rRNA operon, much in contrast to most other chloroplast

TTTCACGGAG	AGTTTGATCC	TGGCTCAGGA	TGAACGCTGG	CGGCATGCTT
AACACATGCA	AGTCGAACGA	ACATAAAATT	GTGCTTGCAT	AATTTTATAC
TGAGTGGCGG	ACGGGTGAGT	AACACGTAAG	AACCTACCTT	TTGGCGAGGG
ACAACCATTG	GAAACGAATG	GCTAATACCT	CATATTACTG	AGAAGTGAAA
GATGAATAAT	CACCAAGAGA	TGGGCTTGTC	TGATCAGCTT	GTTGGTAAGG
TAATGGCTTA	CCAAGGCAAC	GATCAGTAGC	TGGTCTGAGA	GGATGATCAG
CCACACTGGG	ACTGAGACAC	GGCCCAGACT	CATACGGGAA	GGCAGGCAGT
GAGGAATTTT	CCGCAATGGG	GCGGAAAGCC	GTGACGGAGC	AATGCCGCGT
GAAGGATGAA	TGCCTATGGG	TTGGTAAACT	TCTTTTCTCA	GAGAAGAATT
TTGACGGTAT	CTGAGGAATA	AGCATCGGCT	AACTCTGTGC	CAGCAGCCGC
GGTAAGACAG	AGGATCGAAG	CGTTATCCGG	ATTGATTGGG	CGTAAAGCGT
CTGTAGGTGG	TTTATCCAGT	CTTCTGTCAA	AGTCAGGGCT	TAACCTTGGA
TAGGCAGGAG	AAACTAATAG	ACTAGAGTTC	GGTAGGGGCA	GAGGGAATTC
CCGGTGGAGC	GGTGAAATGC	GTAGAGATCG	GAGAACACCA	AAGGCGAAAG
CACTTCTGCT	GGGCCGCGAC	TGACACTGAG	AGACGAAAGC	GAGGGGAGCA
AATGGGATTA	GATACCCCAG	TAGTCTCGCC	GTAAACGATG	GATACTAGGT
GTTGGGTGAT	ATTACATTCA	GTACCGTAGC	TAACGCGTGA	AGTATCCCGC
CTGGGGAGTA	TGCTCGCAAG	CGTGAAACTC	AAAGGAATTG	ACGGGGGCCC
GCACAAGCGT	GGAGCATGTG	GTTTAATTCG	ATGCAACGCG	AAGAACCTTA
CCAGGGCTTG	ACATGCCACT	TTTTTCTTGA	AAAAGAAAGT	TCCGAGTGGA
CACAGGTGGT	GCATGGCTGT	CGTCAGCCTC	GTGTCTTGAG	ATGTTGGGTT
AAGTCCGCAA	CGAGCGCAAC	CCTTATTTGT	GTTGCTATTT	TAGGAACTCA
AAGACTCCGT	GATAGCGAGA	AGGTGAGGAT	GACGTCAAGT	CAGCATGCCT
GACGCTGGGC	ACACACGTGA	TACAATGGCT	GGGACAAAGA	GATGAACCCC
ACAAGGCAGt	CCAACCTCAA	AAACCCAGTC	TTAGTTCGGA	TTGTAGGCTG
AAACTCGCCT	ACATGAAGCT	GGAATGCTAG	TAATGCGCAG	GTCTATGCCA
TACTGCGGTG	AATACGTTCC	CGGGCCTTGT	ACACACCGCC	GTCACACCAT
GGAAGCTGGC	TATGCCAAAG	TCGTACTCAA	CTTAGGAGGA	GGAGCTAAG1
CAGACTAGTG	ACTAGGGTGA	AGTCGTAACA	AGGTAGCGTA	CTGGAAGGTG
CGCTGGATCA	CCTTAAAG			

Fig. 3. Ncleotide sequence of 16S rDNA of N.e. chloroplasts. (from: Schreiner 190).

genomes. A 5.3 kbp region has been sequenced. It carries the genes for the 16S rRNA (Fig. 3), the 23S rRNA (Fig. 4), the tRNA(Ile) and tRNA(Ala) in structures similar to many chloroplasts and to bacteria. However, a gene for 5S RNA could not be detected.

The 16S rRNA nucleotide sequence has been compared with other available sequences of *Euglena gracilis*, *Pylaiella littoralis*, *Chlamydomonas rheinhardii*, *Marchantia polymorpha*, *Tobacco* and Soy Bean. The phylogenetic evaluation shows, N. e. to have the least consensus with the first two (Schreiner 1990).

3.2.5 Golgi-apparatus

This organelle occupies usually about 5% of all the space and shows about 5–7 tubular crescent-like structures with wider, club-shaped, outer ends. It often can be found in the angle between the mitochondrion near the cell wall.

3.2.6 Cytoplasmic apparatus (the ribosome and rRNA)

There is only a relatively small, however variable space for it. Only very small, few, mostly less than a dozen of structures resembling small ribosomes, can be detected with a size of 16–17 nm. The DNA, coding for the (ribosomal) 18S rRNA was of special interest with respect to the question whether N. e. as a micro alga might represent a primitive or rather a reduced form in evolutionary terms. It has been isolated by the procedure outlined under paragraph 3.2.1. This DNA has been cloned and sequenced (Sargent *et al.* 1988). The sequence of 1796 nucleotides is given in Figure 5. The corresponding rRNA structure is similar to those of other green plants. Sequence similarities are highest, over 90%, with *Chlamydomonas*, slightly lower with some seed plants and drop to 60% with some protists. The results are in favor of evolutionary reduction as the source of minimal features in this organism.

3.2.7 Cytoplasmic granules

In cells grown under optimal conditions, one can observe in the chloroplasts all sizes of round grana,

TCAAATGATA	GAAAGCGTAG	GGTGGATACC	TAGGCACCTA	GAGACGATGA
AGGGCGTCCA	ACGACGATAC	GGTTCGGGGA	GTTGGAAACG	AACTTTGATC
CGAAGATTCC	CGAATGGGGC	AACCCTATAA	ACTTCTTACT	GAATTCATAG
GTAAGAAAGA	GAAAACTTAG	TGAACTGAAA	CATCTTAGTA	GCTAAAGGAA
GAGAAAGCAA	ACGCGATTCC	CTGAGTAGTG	GCGAGCGAAA	TGGGAACAAG
CCTAAACCAA	TATATTTATA	TTGGGGTTGT	GGGAAAACAT	ACTATATTTA
TATATGTATT	AGAAAGATTT	AGACGAAATA	GCTGAATCCT	ATACCAAAGA
AGGTAAAAGT	CTGTAGTTAA	AAAATACTAA	AAAATACGAA	CATATGTTCA
ATATGTTTAT	CCGAGATGCA	TGGGACACGT	GAAATCCGTG	TGATGCGAGG
ACCACCTCGT	AAGGCTAAAT	ACTCTAGGTG	ACCGATAGTG	AAGTAGTACC
GAAAGAAAGG	TGAAAAGAAC	CCCTGATGGG	AGTGAAATAG	AACATGAAAC
CTACGCTGAC	AAACAGTGGG	AGGTACTTTA	GTACTGACCG	CGTACCTGTT
GAAGAATGGG	CCGGCGACTT	AGAAAAAGTG	GCAAGGTTTA	AAGACTCAAA
TCTGGAGCCT	AAGCGAAAGC	AAGTCTGAAT	AGGGCATTTT	AAAGTCATTT
TTTCTAGACC	CGAACCCGGG	TGATCTAACC	ATGACCAGGA	TGAAGCTTGG
GTAATACCAC	GTGAAGGTCC	GAACCGACCG	ATGTTGAAAA	ATCGGCGGAT
GAGTTGTGGT	TAGCGGTGAA	ATACCAGTCG	AACTCGGAGC	TAGCTGGTTC
TCCCCGAAAT	GCGTTGAGGC	GCACGGTCAG	CAAGACTATC	TAGGGGTAAA
AGCACTGTTT	CGGTGCGGGC	CGCGAAAGCT	GTACCAAATC	GTGGCAAACT
CTGAATACTA	GATATGTTAT	TTTCTGGCCA	GTGAGACAGT	GGGGGATAAG
CTTCATTGTA	AGAGGGAAAC	ACCCAGATAC	GCTAAGGCCC	CTAAATGATT
GTTAAGTGGC	AAAGGAGGTG	AGAATGCTCA	GACAACCAGA	AGGTTTGCTT
AGAAGCAGCA	ATCCTTTAAA	GAGTGCGTAA	CAGCTCACTG	GTTAAGCGTT
CTTTCGCCGA	AAATGTCGGG	ACTAAATGAT	CTCCCGAAGC	TGTGGGATAT
TCTAATGAAT	ATCGGTAGGG	GAGCGTTGCT	GCTCTAGGGT	GAAGCAAATA
TGTAAGTATT	TGTGGACGAA	GCAGAAGTGA	GAATGTCGGC	TTGAGTAACG
TAAACATTGG	TGAGAATCCA	ATGCCCCGAA	ACCTAAGGAT	CCTCACCAGG
TCGTCATGGA	GGGTGAGTCA	GGACCTAAGG	CGAGGCTGAA	AGCGTAGACG
TTGGCAAACA	GGTTAATATT	CCTGTACTAT	TTTATATTTG	ATACCAAGGA
CGAAGCAGGT	AGAGATTTGC	TGTTATTGGA	TTTCAGTGTA	AGTTCTCAAG
GCGTTGAGAG	GTAGAAGAAA	AACTATGGTT	GAGCTAAGCA	AACGAACGTG
CAGTACTTCA	TTTTTAAAAT	AACGTATTTC	TTAATCTATC	ATACTTGCAA
CGAAACATGC	TTACGAACTA	TTCACTTAAA	GTAATCAAAA	TGACCTGTAC
CGTAAATCCG	ACACAGGCTA	CAGCGTTAGT	AGAGTATACT	AACGGCGGCG
CGAGATAACT	CTCTCTAAGG	AATCTCGGCA	AAATGACCCG	TACTTCGGAG
AAGGGTGCCC	TCTCTAAAGA	GGCCGCGAGT	ACCAGGCCCA	GGCGACTGTT
TATCAAAAAC	ATAGGTCTCC	GCAAAGTCGT	AACCCATACT	GGGCTGACGC
CTGCCCAGTG	CCGGAAGGTT	AAGGAAGTAG	GTTACCATTT	ATCCAAAGCT
TACGACTGAA	GCCCCGGCGA	ACGGCGGCCG	TAACTATAAC	GGTCCTAAGG
TAGCGAAATT	CCTTGTCGGG	TAAGTTCCGG	CCCGCACGCA	AGGCTTACGT
ACTGGCACTG	TCTCGGAGAG	AGCTCGGTGA	AATAGACTGT	CTGTGAAGAT
CGACTACCTA	CACTTGGACA	GTAAAGGCCC	TATGAAGCTT	GACTGTAGCT
TGTAATTGTG	TCTCGCGCTT	TTTTTGCGCA	GACTAGGTGG	AGGCTAATGA
AGAAGTATTT	TTGGATATTT	TCCAGCCATC	AGTGAGAGTC	CACTCTGAAG
AACGCTTAGA	ATTCTAATGG	CGTTCCTTGA	AGCAGGACGC	TTGGAGTTTC
CAGGTGGGCA	GTTTGACTGG	GGCGGTCACC	TCCTAAAAGG	TAACGGAGGT
GTGCAAAGGT	TCCCTCAGGC	TGGACGGAAA	TCAGCCGTAG	AGTGTAAAGG
CAGAAGGGAG	CTTGATCGCA	AGACCTACAA	GTCGAGCAGG	GGCGAAACGG
GCCTTAGTGA	TCCGACGGTT	CCGTGTGGAG	AACCGTCGTA	AACGGATAAA
AGTTACTCTA	GGGATAACAG	GCTGATCTTC	CCAAGAGTTC	ACATCGACGG
GAAGGTTTGG	CACCTCGATG	TCGGCTCTAC	GCAACCTGGG	CGTAGTACGT
CCAAGGGTTG	GGCTGTCGCC	ATGAAGCGTA	CGTAAGCTCG	GGTCAGAACG
TGAGAGCAGT	TCGGTCCATA	TCCGGTGTAG	GCGCTGGAGC	ATTGAGAGGA
GTCTTCCATA	GTACGAGAGG	ACCTGGAAGA	ACTACACCTC	TGGTATACCA
GTTCTCACGC	CAGTGGGAAA	CGCTGGGTAG	CCAAGTGTGG	AGTGGATAAC
CGGCTAAAGC	ATCTAAGTCG	GAAGCCCACC	TCAAGATGAG	TGCTCCTATG
AGGTCACGGC	AAGACTAGCC	GGTTGATAGG	CAGTAGGTCT	AAGTCTAGCA
ATAGATGTAG	CTGAGCTGTA	CTAATAGACC	CATTGATTTT	GAC

Fig. 4. Ncleotide sequence of 23S rDNA of N.e. chloroplasts. (from: Schreiner 1990).

TACCTGGTTG	ATCCTGCCAG	TAGTCATATG	CTTGTCTCAA	AGATTAAGCC
ATGCATGTCT	AAGTATAAAC	TGCTTTATAC	TGTGAAACTG	CGAATGGCTC
ATTAAATCAG	TTATAGTTTA	TTTGATGGTA	CCTACTTACT	CGGATACCCG
TAGTAATTCT	AGAGCTAATA	CGTGCGCACA	TCCCGACTTC	TGGAAGGGAC
GTATTTATTA	GATAAAAGGC	CGACCGGATT	GTTCCGACTC	GCGGTGACTC
ATGATAACTT	CACGAATCGC	ATGGCCTCGT	GCCGGCGATG	TTTCATTCAA
ATTTCTGCCC	TATCAACTTT	TGATGGTAGG	ATAGAGGCCT	ACCATGGTGG
TAACGGGTGA	CGGAGAATTA	GGGTTCGATT	CCGGAGAGGG	AGCCTGAGAA
ACGGCTACCA	CATCCAAGGA	AGGCAGCAGG	CGCGCAAATT	ACCCAATCCT
GACACAGGGA	GGTAGTGACA	ATAAATAACA	ATACCGGGCC	TTTGGTCTGG
TAATTGGAAT	GAGTACAACC	TAAACACCTT	AACGAGGATC	AATTGGAGGG
CAAGTCTGGT	GCCAGCAGCC	GCGGTAATTC	CAGCTCCAAT	AGCGTATATT
TAAGTTGCTG	CAGTTAAAAA	GCTCGTAGTT	GGATTTCGGG	TGGGGCCTGC
CGGTCCGCCG	TTTCGGTGTG	CACTGGCCGG	GCCCACCTTG	TTGCCGGGGA
CGGGCTCCTG	GGCTTCGCTG	TCCGGGACCC	GGAGTCGGCG	AGGTTACTTT
GAGTAAATTA	GAGTGTTCAA	AGCAGGCCTA	CGCTCTGAAT	ACATTAGCAT
GGAATAACAC	GATAGGACTC	TGGCCTATCC	TGTTGGTCTG	TAGGACCGGA
GTAATGATTA	AGAGGGACAG	TCGGGGGCAT	TCGTATTTCA	TTGTCAGAGG
TGAAATTCTT	GGATTTATGA	AAGACGAACT	ACTGCGAAAG	CATTTGCCAA
GGATGTTTTC	ATTAATCAAG	AACGAAAGTT	GGGGGCTCGA	AGACGATTAG
ATACCGTCCT	AGTCTCAACC	ATAAACGATG	CCGACTAGGG	ATCGGCGGGT
GTTTTTTTGA	TGACCCCGCC	GGCACCTTAT	GAGAAATCAA	AGTTTTTGGG
TTCCGGGGG	AGTATGGTCG	CAAGGCTGAA	ACTTAAAGGA	ATTGACGGAA
GGGCACCACC	AGGCGTGGAG	CCTGCGGCTT	AATTTGACTC	AACACGGGAA
AACTTACCAG	GTCCAGACAT	AGTGAGGATT	GACAGATTGA	GAGCTCTTTC
TTGATTCTAT	GGGTGGTGGT	GCATGGCCGT	TCTTAGTTGG	TGGGTTGCCT
TGTCAGGTTG	ATTCCGGTAA	CGAACGAGAC	CTCAGCCTGC	TAACTAGTCA
CGCGTGCTCC	GGCACGCGGC	GGACTTCTTA	GAGGGACTAT	TGGCGACTAG
CCAATGGAAG	CATGAGGCAA	TAACAGGTCT	GTGATGCCCT	TAGATGTTCT
GGGCCGCACG	CGCGCTACAC	TGATGCATTC	AACGAGCCTA	TCCTTGGCCG
AGAGGTCCGG	GTAATCTTTG	AATCTGCATC	GTGACGGGGA	TAGATTATTG
CAATTATTAA	TCTTCAACGA	GGAATGCCTA	GTAAGCGCAA	GTCATCAGCT
TGCGTTGATT	ACGTCCCTGC	CCTTTGTACA	CACCGCCCGT	CGCTCCTACC
GATTGGGTGT	GCTGGTGAAA	TGCTCGGATT	GGCGGCTTTG	GGCGGTTTCC
GCCCTTGGCT	GTCGAGAAGT	TCATTAAACC	CTCCCACCTA	GAGGAAGGAG
AAGTCGTAAC	AAGGTTTCCG	TAGGTGAACC	TGCGGAAGGA	TCATTG

Fig. 5. Nucleotide sequence coding for 18S rDNA of N.e. (from: Sargent *et al.* 1988).

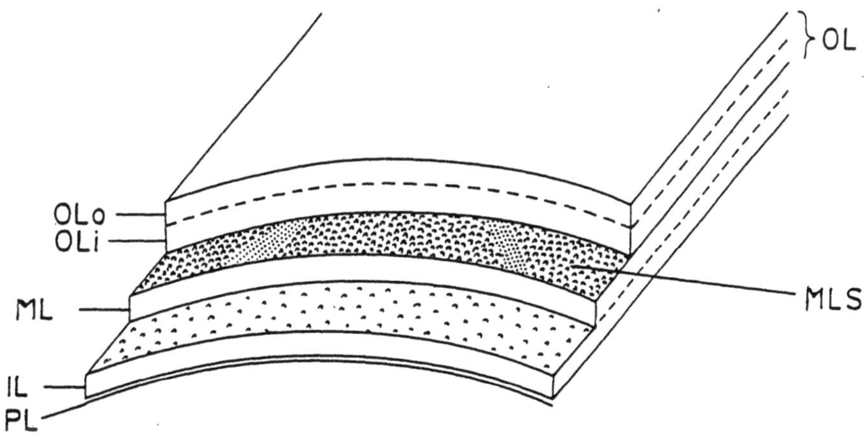

Fig. 6. Cell wall structure of N.e. The cell wall is considered to be trilaminar. The outer layer OL consists of two leafs, an outer leaf, OLo and an inner one, OLi. The inner layer ML shows a rough outer surface MLS with a banded structure, presumably carrying sporopollenin. IL is the inner layer and PL the plasmalemma. (from: Geisert *et al.* 1987).

with diameters between 0.005 μm up to about 0.1 μm. Some of them show inner onion-like lamellation, others are very homogeneously filled with structureless material.

3.2.8 Outer membrane and cell wall

There is a double membrane surrounding the cell. The surfaces exposed by freeze fracture are occupied by particles. They also show elongated impressions. The outer, cytoplasm orientated surface, only rarely can be exposed. The cell wall with its average thickness of 30 nm is though although relatively thin. Slanting freeze fracture exposes several layers with different granulations.

The cell wall is considered as trilaminar, with one outer layer consisting of an outer and an inner leaf, one middle layer, and another one directly adjacent to the plasma membrane. The outer face of the middle layer carries bands of different sizes with regularly arranged small particles which are embedded in a roughly structured surface. Degradation of the cell wall with the enzymes cellulase, proteinase K and zymolyase does not appreciably alter these structures. The appearance of the middle layer face even persists after acetolysis, a treatment that involves concentrated acetic and sulfuric acids. There is only one typical organic substance known with such a stability: sporopollenin. This substance could be isolated, purified and compared with authentic substance from *Lycopodium clavatum* and synthetic material, that had been polymerized from beta-carotene (Geisert *et al.* 1987).

The comparison of the materials shows enough coincidence to justify the assumption that N. e. is protected by a very tough coat containing sporopollenin or compounds of very similar

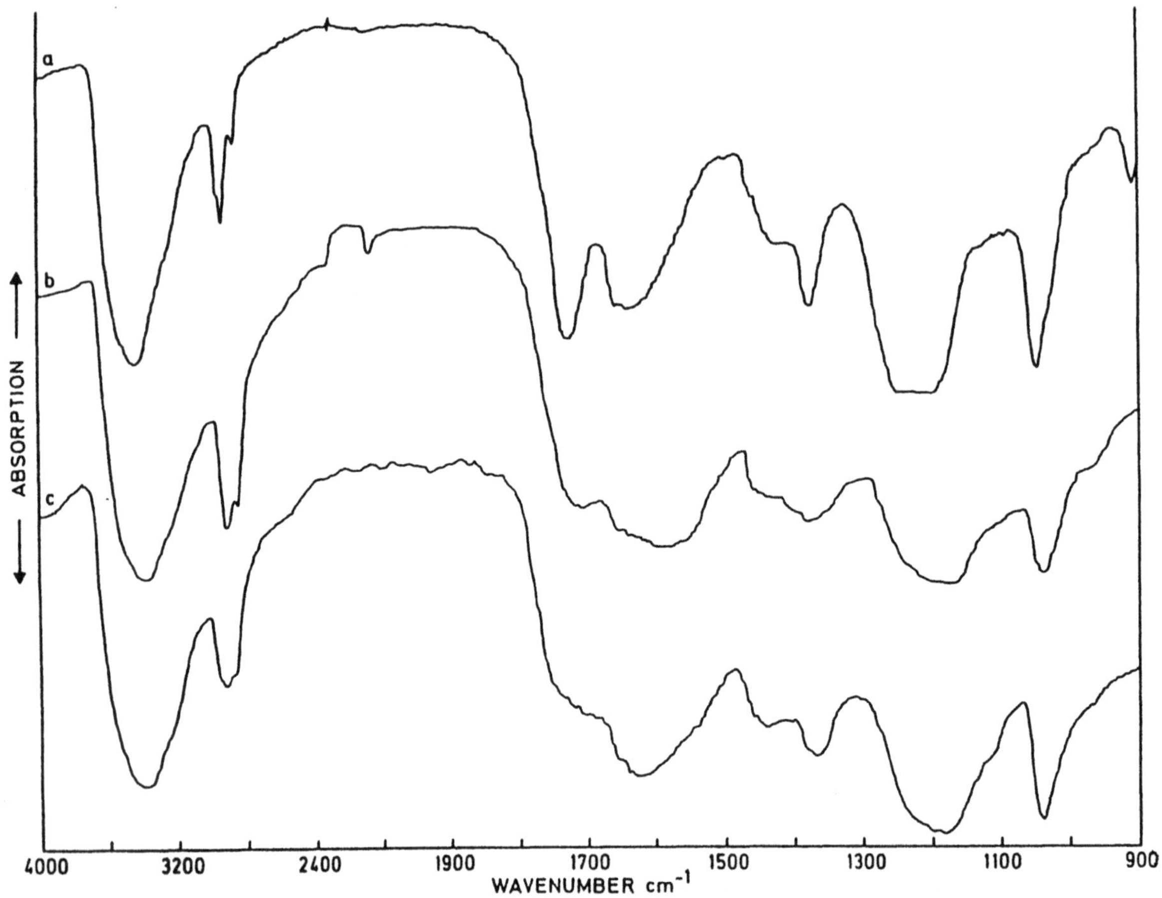

Fig. 7. Comparison of infrared spectra of sporopollenins. Infrared spectra from of acetolysis-resistant material from: a) sporopollenin isolated from *Lycopodium clavatum;* b) sporopollenin isolated from N.e., strain Mainz; c) sporopollenin synthesized from beta-carotene. (from Geisert *et al.* 1987).

Table 3. Elemental analysis of sporopollenins. (from: Geisert *et al.* 1987)

Origin of the Sporopollenin	%C	%H	%O	C/H
Lycopodium clavatum theoretical values[a]	65.2	8.7	26.1	7.5
β-Carotene polymer, theoretical values[b]	63.9	7.7	28.4	8.3
N. eucaryotum strain Mains, values found	63.6	7.7	26.4	8.3
β-Carotene polymer, values found	62.1	6.2	28.1	10.0

[a] Estimated from a theoretical *Lycopodium clavatum* sporopollenin $(C_{30}H_{144}O_{27})_n$
[b] Estimated from a theoretical β-carotene polymer $(C_{90}H_{130}O_{3}o)_n$ as reported by Brooks and Shaw (1973) and Gooday (1981).

resistance. Since the behavior of the cell wall in the course of cell division may offer important taxonomic hints, special consideration has been given to it (see paragraph 5.4).

3.2.9 Immunological identification
For comparison and for securing the identity of N. e. species, white, male NMRI mice were injected with 50 mg of washed, lyophilized algal culture corresponding to about 10^{10} cells, which had been suspended in 0.2 ml of saline. Ten days later the mice were killed, their spleens removed, and torn with tweezers followed by gentle homogenization to liberate the immunocytes . Authentical *Nanochlorum eucaryotum* (Mainz II) cells, were added for the assay. After 2 h incubation at 4°C the rosettes were counted at a multiplicity of 1 spleen cell per more than 10 *Nanochlorum* cells (Nowotny 1979). Using this method *Nanochlorum* can be distinguished from other algae e.g., different *Chlorella* species (Deutsche Algensammlung, Göttingen, FR Germany) even in small volumes in coincidence with the morphological and physiological data.

4. Isolation and cultivation

N. eucaryotum cells can grow in three different modes, 1: autotrophic, aerobically and anaerobically; 2: heterotrophically and 3: under special conditions.

4.1 Isolation

In order to achieve growth of the organisms as pure monoculture the requirements for an optimal medium condition are by far more stringent than for keeping an already established culture going with optimal or near optimal yields. The conditions can even be further relaxed considerably, especially with respect to the trace elements, when other additional prokaryotic microorganisms that can be found in the natural habitat, are permitted. The sufficient and necessary components in this respect have not been as yet fully explored. Near optimal growth has been achieved by the medium given in Table 4.

Table 4. Chemical composition of a optimal growth solution for *Nanochlorum eucaryotum* (from: Zahn 1983)

Number	Component	Mol/l	Concentr. compared to sea w
1	NaCl	1.1	2.3
2	$MgCl_2$	1.5×10^{-1}	2.8
3	Na_2SO_4	8.8×10^{-2}	3.1
4	KNO_3	3.3×10^{-2}	6.9
5	KCl	2.2×10^{-2}	2.2
6	$CaCl_2$	2.2×10^{-2}	2.2
7	KH_2PO_4	1.1×10^{-3}	3.8×10^{2}
8	$NaHCO_3$	1.1×10^{-3}	2.4
9	H_3BO_3	1.0×10^{-3}	2.4
10	KBr	8.0×10^{-4}	0.95
11	LiCl	5.0×10^{-4}	2.0×10
12	$Sr(NO_3)_2$	5.0×10^{-4}	5.6
13	Humic.Subst.	1.0×10^{-4}	1.5×10^{2}
14	NaF	1.0×10^{-4}	1.5
15	Rb_2SO_4	5.0×10^{-5}	3.6×10
16	EDTA	5.0×10^{-5}	--
17	$FeSO_4$	1.5×10^{-5}	2.8×10^{2}
18	$MnCl_2.4H_2O$	5.0×10^{-6}	1.4×10^{2}
19	Ga_2O_3	5.0×10^{-6}	1.2
20	$ZnSO_4$	1.0×10^{-6}	6.7
21	MoO_3	1.0×10^{-6}	1.0×10
22	$Al_2(SO_4)_3$	5.0×10^{-7}	1.3×10
23	$Co(NO_3)_2.6H_2O$	5.0×10^{-7}	7.4×10
24	$CuSO_4.5H_2O$	5.0×10^{-7}	1.1×10
25	$NiSO_4.6H_2O$	5.0×10^{-7}	4.2
26	$BaCl_2$	5.0×10^{-7}	3.3
27	KI	5.0×10^{-7}	9.1×10
28	SeO_2	1.0×10^{-7}	9.1×10
29	$TiOSO_4.H_2O$	1.0×10^{-7}	5.0
30	$ZrOCl_2.8H_2O$	1.0×10^{-7}	3.0
31	As_2O_3	1.0×10^{-7}	2.9
32	V_2O_5	1.0×10^{-7}	2.6
33	$Ce(SO_4)_2.4H_2O$	5.0×10^{-8}	7.1×10^{3}

Table 4. Continued

Number	Component	Mol/l	Concentr. compared to sea w
34	$Y(NO_3)_3$	5.0×10^{-8}	4.6×10^3
35	$La(NO_3)_3$	5.0×10^{-8}	2.3×10^3
36	$Be(NO_3)_2.4H_2O$	5.0×10^{-8}	7.1×10^3
37	$Sc(NO_3)_3$	5.0×10^{-8}	5.6×10^2
38	$BiNO_3$	5.0×10^{-8}	5.0×10^2
39	$TiNO_3$	5.0×10^{-8}	1.0×10^2
40	$WO_3.H_2O$	5.0×10^{-8}	9.3×10
41	$HgCl_2$	5.0×10^{-8}	5.0×10
42	PbOAc	5.0×10^{-8}	3.4×10
43	CsCl	5.0×10^{-8}	2.5×10
44	$CdSO_4$	5.0×10^{-8}	2.2×10
45	$SnCl_3$	5.0×10^{-8}	2.0×10
46	$AgNO_3$	5.0×10^{-8}	1.8×10
47	UO_2Ac	5.0×10^{-8}	3.9
48	$KTeO(OH)_5H_2O$	5.0×10^{-8}	--

The mixture in its final form was put together from several stock solutions. Stock a: components 1–5, Stock b: component 6, Stock c: 7–12, and Stock d: component 13, for which 1 kg of dry peat-moss (Floratorf) + 30 g Kieselgel 'Silica' (Woelm, Eschwege, Germany) were suspended in 3 l of distilled water and kept at 37 °C for 24 h. It was then centrifuged and to the clear supernatant ethylene-diamine-tetraacetate-disodium (EDTA), component 16, has been added. It was divided into two portions. In one of them the components 26–46 were suspended, resp. dissolved, the components, 13, 14 and 16–24 in the other, each one at 1000 fold concentration of the final culture medium. Component 25 was added to the final solution separately. The culture medium in its final composition is heated in an autoclave to 110°C for 30 min. On cooling turbidity arises, which however soon disappears when inoculated with more than 10^4 cells per ml.

Using this solution isolation of N. e. in the course of time became possible from different locations. Cultivation in different modes was easy. Generally the generation times were about 27 h.

4.2 Aerobic cultivation

Five liter flasks with conical bottoms containing 4 l of solution (Table 4) were inoculated with a few milliliters of a pre grown culture, containing a cell clone and gassed with compressed air at 400 l per h at 22 °C. The vessel was illuminated at 1000 lux and sterile solution (Table 4) was pumped continuously into the 5 l flask at 2 l per day. In such a way the volume surpassing 4 l is driven out into a receiving bottle for harvesting.

In the course of 5 years of cultivation of different populations of N. e. different pigment variants have been observed, identical and typical, as outlined in paragraph 3. From all of these pigments, the carotenoids 1–6 could be isolated. In a Colona strain derivative carotenoids 7–10 have been found when cultivated under air gassing.

Table 5. Carotenoids in N.e.

No	Pigment
1	alpha-carotene
2	beta-carotene
3	lutein
4	violaxanthin
5	neoxanthin
6	cryptoxanthin
7	canthaxanthin
8	astaxanthin
9	echinenon (traces)
10	unidentified keto-carotenoid (trace amounts)

4.3 Anaerobic cultivation of N.e.

Since N. e. has grown so well freely in suspension it seemed unlikely that it should exist in its habitat attached to solid surfaces exclusively. Therefore the search in the Adriatic was continued by filtrating large amounts of water first through 4 μm pore filters to keep back larger plankton and then through 0.42 μm filters. These were then incubated in the medium described before and kept in slowly rolling motion for longer times. Most of the water samples from the different places proved to be negative except those from regions where natural gas was bubbling up in the sea, indicating the possibility of partial anaerobic existence. Several samples from such a place yielded algae resembling *Nanochlorum eucaryotum*. This has been verified in two subsequent years. So the suspicion came up that these algae preferred such a habitat. Therefore the anaerobic cultivation of the organisms was initiated. For reasons of economy and for attempting to imitate the natural gas seepage, city gas from the local company, known to be supplied from natural sources mostly has been used instead of air. Air has totally been excluded from the growth vessels. With this source of commercial gas, optimal growth of N. e. could be achieved.

Table 6. Composition of the city gas used for anaerobic cultivation of N.e.

Component	Volume percent
CH_4	82.87
N_2	5.60
C_2H_6	5.50
CO_2	1.86
H_2	1.83
C_3H_8	1.66
C_4H_{10}	0.53
C_5H_{12}	0.10
C_6H_{14}	0.03
C_7H_{16}	0.01
C_8H_{18}	0.01
C_6H_6	0.01

The city gas consisted essentially of natural gas. I am grateful to the Stadtwerke Mainz AG. for their analysis.

Whenever large amounts of cells are needed, air gassing is avoided and the cultures are gassed with natural gas, from the city line (Table 6). This city gas completely substitutes the air. Under these conditions the cells grow at their high doubling rates to higher cell densities. When grown with city gas, besides the carotenoids 1–6, the secondary carotenoids astaxanthin and canthaxanthin (No 7 and 8 in Table 5) could be also isolated.

Table 7 shows that there are considerable differences in pigment composition when the organism is grown with city gas. Especially the content in carotenoids is going up and secondary carotenoids are formed. This causes a rise of the total pigment content per cell, while the ratio of chlorophyll a to chlorophyll b is lowered. The carotenoids, which also apparently occur in association with the cell wall, deserve special consideration since sporopollenin is known to arise by oxidative polymerization from them (Brooks and Shaw 1973).

Table 7. Quantitative pigment composition under gassing with city gas or air.

Pigment	Culture with city gas [a]	Air culture
Total pigment content (chl a + b + carotenoids) (g-cell)	2.72×10^{-14}	1.49×10^{-14}
Carotenoids (g/cell)	6.36×10^{-15}	3.50×10^{-15}
(chl a/chl b)	2.78	3.72
Carotenoids/total chlorophyll (chl a + b)	0.31	0.31

[a] The composition of the city gas is given in Table 6 (from: Geisert *et al.* 1983).

4.4 Cultivation under nitrogen deficiency

N.e. was grown aerobically in the absence of nitrogen containing compounds. Initially only the carotenoids 1–6 (Table 5) could be isolated from the cultures. In the 11th week of nitrogen-free culture however the secondary carotenoids 7–10 appeared. This was coupled with a decrease in chlorophylls. This behavior is common in many green algae (Czygan 1982).

4.5 Cultivation with methane as sole carbon source

N.e. was gassed with ultra pure methane supplemented with 0.1 vol% of oxygen. In this CO_2-free atmosphere the organism did not grow at all. It is to be concluded that N. e. cannot utilize methane as a carbon source contrary to methanotrophic organisms. Its growth is dependent on carbon dioxide.

5. Growth and cell division

5.1 Inhibitor assays

In these experiments cells were grown in 10 ml cultures in sets of ten parallel samples in roller tubes, at 30 rpm, 1000 lux, 22 °C in 12 h light, 12 h dark cycles and daily changes of culture fluid. Cell numbers were evaluated as turbidity difference of the growing culture and its supernatant after centrifugation at 1000 xg, 22 °C, for 10 min. The instrument was calibrated weekly for cell number correlation. Statistical significance evaluation was with the non-parametric X-test (Van der Waerden and Nievergelt 1976). Agents of poor water solubility were dissolved in 10 μl of dimethylsulfoxide (DMSO), which then was added to the 10 ml culture, the assay receiving the vehicle without the inhibitor. The solutions were freshly prepared each time. The experiments lasted up to 96 h with positive values of generation times, otherwise they were stopped at 48 h. The cells were inspected by light microscopy after 48 h for crude morphological changes, for contamination with alien organisms, and for precipitation. Whenever feasible, the highest possible concentration has been tested, as for solubility and the allowed price. The active concentrations were diluted and tested until the effect would disappear.

For the determination of intracellular inhibitor concentration, inhibition of L5178Y cell multiplication has been used. At the end of the experiment the N. e. cells were sedimented at 10 000 xg, 20 min, 3 °C and washed with 1 vol of culture fluid. The dry sediment was suspended in 1 Vol of 50% aqueous DMSO and subjected to a series of 5 cycles of freeze-thawing on dry ice and warming to 40 °C. It was then centrifuged at 15 000 xg, 40 min, 25 °C. The clear supernatant was added as a concentration series to 5 roller tubes each with 3000 L5178Y mouse lymphoma cells per ml growing in suspension culture in Fischer medium with 10% horse serum. They were operated in parallel to a concentration series of the respiratory inhibitor added to 50% DMSO extracts of control N. e. cells. After an incubation time of 72 h the cell numbers and their volume distribution was evaluated in a Coulter counter with size distribution plotter (Coulter Electronics, Hialeah, Fla., USA). This procedure allowed for determination of intracellular inhibitor concentrations within at least one order of magnitude.

In the case of colchicine inhibition, ^{3}H-Colchicine, spec. act. 2.5 Ci/mmol (Amersham-Buchler, Braunschweig, Germany) was used. At low colchicine concentrations the intracellular level was estimated by radioactivity measurements of chloroform extracts of N. e. cells, where quenching has been more than 90%. At $8x10^{-6}$ mol colchicine/l in the growth medium the drug is accumulated by a factor of at least of 10 in 48 h by *Nanochlorum*. With the higher concentration at $8x10^{-3}$ mol/l after 48 h of incubation the N. e. cells were ground and extracted into chloroform. The extract was evaporated to dryness and chromatographed on HPTLC Kieselgel 60, F 254 nm plates (E. Merck, Darmstadt, Germany), using the solvent system ethylacetate-isopropanol aqueous ammonia (25%) (50 : 30 : 2 by vol). Colchicine under these conditions appears as a spot, strongly quenching the fluorescence caused by 254 nm excitation. The colchicine quenching spot stays clear of the N. e. pigments. Its quenching intensity was evaluated using a TLC scanner.

Table 8. Action of inhibitors on cell division. (from: Zahn 1983).

Azide	$1.5.10^{-5}$ – $4.5.10^{-4}$ mol/l inhibition
Diethylpyrocarbonate	2.10^{-6} – 6.10^{-8} mol/l inhibition
Chloramphenicol	3.10^{-5} – 3.10^{-3} mol/l inhibition
Cycloheximide	10^{-7} – 10^{-4} mol/l inhibition
Colchicine [a], [b]	8.10^{-6} – 8.10^{-3} mol/l no inhibition
Colcemid	8.10^{-8} – 10^{-5} mol/l no inhibition b)
Trifluralin	10^{-5} – 3.10^{-4} mol/l no inhibition b)
Cytochalasin B	2.10^{-7} – 2.10^{-5} mol/l no inhibition b)
Lysozyme	0.5 mg/ml no inhibition
Penicillin	1.5 mg/ml 15% inhibition

[a] intracellular concentration determined by chromatography
[b] intracellular concentration determined by testing cell culture inhibition

Azide has been tested, since it is used as a sterilizing agent. It is a strong inhibitor for ATPase and a series of other enzymes at a range comparable to the one for N. e. (Jain 1982). Diethylpyrocarbonate, another sterilizing agent, with the advantage of being easily hydrolyzed and thus reversible in time, known to form amide bonds in proteins between free amino and carboxyl groups, is inhibitory in a range to be expected from literature (Jain 1982). For the two drugs chloramphenicol and cycloheximide reciprocal inhibition has been shown. Thus inhibition of protein synthesis by chloramphenicol in bacteria, 'blue-green algae' (cyanobacteria), and in prokaryotic subcellular organelles of eukaryotes such as: mitochondria and chloroplasts, have been well documented. No inhibition of chloramphenicol was observed in the cytoplasm of eukaryotes as fungi, green algae, higher plants and mammals. The reverse has been reported to be true for cycloheximide. This is the consequence of its action on ribosomes, the chloramphenicol-sensitive 70S prokaryotic ribosomes being insensitive to cycloheximide, the 80S eukaryotic ribosomes being cycloheximide-sensitive and chloramphenicol-insensitive (Smillie and Scott 1969; Jain 1982).

In N. e. the amount of cytoplasm is relatively small, so that not much reserve substances and precursors seem to be present, making the macromolecular syntheses dependable from the metabolic status of the subcellular elements, especially of the large chloroplast. The consequence of this is, besides high sensitivity towards cycloheximide acting on the eukaryotic ribosomes in the cytoplasm, an appreciable inhibition by chloramphenicol acting presumably on the 70S ribosomes of the chloroplast (Table 8). Thus while chloramphenicol is a 20–100 times better inhibitor of typical prokaryotic division at concentrations of $3x10^{-6}$ — $3x10^{-5}$ mol/l than on eukaryotes (Otten *et al.* 1975) the higher value is overlapping with the minimal range in N. e. (Jain 1982). In this context it will be interesting to find out, what the sensitivity will be under heterotrophic conditions.

Cytochalasin B, a drug that is known to inhibit intracellular motility phenomena (Dustin 1978) at concentrations around or below $3x10^{-5}$ mol/l (Jain 1982) does not show measurable effects. Since however especially in this case determination of intracellular concentration may be inexact, not too much weight can be given to this observation. For future experiments however this introduces the interesting question how an apparently marginal eukaryote solves its problems of shifting its subcellular structures. Lysozym and Penicillin which are potential inhibitors for prokaryotic cell wall differentiation do not exert any appreciable effect.

5.2 Mitotic apparatus

This is considered as a leading marker in taxonomic evolution and has been incorporated as such into working hypotheses (e. g. Heath 1980; Margulis 1981). Colchicine is a frequently used poison for the synthesis of the mitotic apparatus and its function. By reaction with one binding site (Bryan 1972) of the tubulin subunits, colchicine prevents their assembly into microtubules. This is the basis of its poisonous action, e.g., leading to mitotic arrest (Dustin 1978). In N. e., under conditions of apparently normal growth and propagation, colchicine is without significant effect, even at high concentrations. Therefore the question may be asked whether colchicine is able to enter the cells under the given conditions. At low concentrations of culture fluid ($8x10^{-6}$ mol/l) the drug is accumulated by a factor of at least 10 in the course of 48 h, as measured with radiolabelled alkaloid. At the higher concentrations ($8x10^{-mol}$/l) it could be demonstrated with the chromatographic method that the outside concentration had been reached. Upon longer incubation periods, the cells will perish, while thin flat, feathery- like crystals appear occasionally at the outer cellular circumference, below the cell wall (Bauer and Zahn, unpublished). This affords an explanation basing on insensitivity of the 'mitotic' apparatus towards colchicine, colcemid and trifluralin at the concentrations tested and at which they had been found active in other taxa (Dustin 1978; DeRosa *et al.* 1978; Jain 1982). This can signify that the colchicine binding site is not accessible in an otherwise functional protein or that such a protein may not be present. At the moment we cannot decide which of the two alternatives is present in N. e.. Investigations to clear up the matter are under way.

As a working hypothesis we assume that a colchicine binding protein is present in N. e. giving rise to crystal formation. Whether such crystals are caused by a microtubular derivative which is not in the doublet form, corresponding to those of the marginal bundle of thrombocytes and of nucleated erythrocytes (Behnke and Forer 1967; Behnke 1970) is at the moment an open question.

It is apparently unlikely that a mitotic apparatus with typically active proteins, like the tubulins of higher eukaryotes exists, since in such a case they ought to be inhibited by the colchicine treatment.

We see deviations from the fully eukaryotic state in the mitotic apparatus and its function in N. e. in the sense that this organism belongs to a taxon with marginal eukaryotic features, characteristics that deserve further experimental efforts in order to be understood.

In this context the apparent absence of typical histones along with the missing nucleosome architecture may lead to the anticipation of alterations in genomic distribution and reshuffling in mitosis. Here the term: 'mitosis' is to be considered in its wider sense (Dustin 1978). This among other considerations raises the question about the fidelity of marker transmission in such organisms, e.g., with the replicated genomes 'attached to the nuclear envelope at a differentiated site' (Pickett-Heaps 1974).

5.3 Histones and histone-like proteins

Electronmicroscopic inspection of 'nucleoprotein' fractions that were prepared by 20 different methods never showed beaded structures. Nuclease digestion of putative chromatin containing preparations never showed periodic structures or fractions, e.g., on electrophoresis. Exposing protein fractions of N. e. to N. e.-DNA at low ionic strength did not yield any characteristic formations either (Zahn and Pondeljak, personal observation). Electrophoretic separation of basic proteins either from acid, or from phenol extracts, or by salt-gradient extraction (Zahn and Stahl 1953, 1955) did not yield any clues to histone-like components. Protein fractions from N. e. where mixed with total N. e.-DNA under conditions of nuclease inhibition and the DNA separated subsequently on Sephadex. From this DNA, proteins could be released by different means. They however did not show histone-like components (personal observation and communication from Schmidt and Beyer). Further attempts where directed to detect the presence of histone gene sequences (Nickel 1986) in the total N. e. DNA, using yeast genes, kindly supplied by M.L. Birnstiel. No satisfactorily stringent hybridizing fractions could be found however (Breter, H. J. personal communication). So the question whether the protein repertoire of N. e. includes any histones has to be answered with 'no', with the stigma of a conclusion derived from negative evidence.

5.4 Autospore formation

Ultrastructure and pigmentation of N. e. is suggestive of the Chlorococcales and by the taxonomic system of Ettl (1980) the tendency was to place it into the subfamily of Chlorelloidea. However the representatives of this taxon reproduce by means of autospores, the relics of which, empty mother cell walls never had been found, until the work of Menzel and Wild (1989). These authors showed the occurrence of free mother cell walls in N. e. cultures as well as autospore formation with trilaminar walls surrounded by the parental trilaminar wall. This rules out simple cell division (Naumann, 1921). This places the species (N.e.) under the genus *Nannochloris*. However as for the genus *Nannochloris*, no further reliable information concerning the physiology of division is available. Thus, at the moment the questions stemming from this problematic issue remain without perspectives.

6. Physiological properties

N. e. is an organism of very remarkable adaptability as far as some aspects of ionic composition of its medium go.

6.1 Salinity

N. eucaryotum is able to grow under a wide range of salinity corresponding to NaCl concentrations between 0.1% to 12%. Only above that concentration cells stop to grow and finally die. Growth optimum occurs around 3.5% to 4% corresponding to the osmotic pressure of sea water (Wilhelm and Wild 1982).

6.2 pH dependence

N. e. can grow between pH 4 to pH 9 with an optimum pH around 6 (Wilhelm and Wild 1982). This differs markedly from the hydrogen ion activities measured at the locations of its isolation. In the crack of a stone, very similar to the one of its first reisolation (Zahn 1984), this pH was 8.9, while at the location of gas bubbling it was 8.1. In other possible habitats the hydrogen ion activity was between pH 8.1 to 8.3. One reason for that may be that the conditions of measured growth differed is so markedly from the ones of its growth in the natural habitat.

6.3 Light tolerance

Contrary to the behavior of plants that can adapt to high light conditions and that under high light reach higher photosynthetic rates, N. e. under such conditions, e.g. 10 000 lux, looses most of its photosynthetic activity. However N. e. when grown under 100 lux shows the same activity as under 2 000 lux. This drastic inhibition of photosynthesis by high light intensities and the absence of a clear adaptation to different low light regimes in the range between 100 to 2 000 lux is considered as a sign of a lacking adaptability to the factor light. This could signify that N. e. usually does not occur in a free living state as a plankton member (Wilhelm and Wild 1982). At least this points out to a rather narrow ecological light niche.

The analysis of the photosynthetic apparatus (see also paragraph 3.2.4) shows a high content of chlorophylls per cell, at a low chlorophyll a to chlorophyll b ratio. This in association with a low photosynthetic capacity along with the ability to grow under extremely low light intensities signifies a very effective light harvesting system, that contains 70–80% of all the chlorophylls (Wilhelm and Wild 1982).

Acknowledgements

The support of the Deutsche Forschungsgemeinschaft, of the Academy of Science and Literature, of the University of Mainz, of the Bundesministerium for Forschung und Technology through the Internationales Burreau Geesthacht, and of the Croatien Authority for Scientific Cooperation is gratefully acknowledged.

I wish to thank for more than five years of common research which were enabled by the dedicated work especially of Winfried Bauer, Rudolf Beyer, Manfred Geisert, Vesna Pondeljak, Monika Schreiner, Dieter Weinblum and Gertrud Zahn. The cooperation with Christian Wilhelm and Aloysius Wild always has been highly appreciated.

The help by the staff members of the Center for Marine Research in Rovinj, Istria has essentially promoted the work, as have discussions with Werner. E.G. Müller, Manfred Reitz, Branko Kurelec and Renato Batel.

References

Behnke, O. and Forer, A. (1967). *J. Cell Sci. 2*: 169–192.

Behnke, O.(1970). *Intern. Rev. Exp. Path. 9*: 1–92.

Bauer, W. (1986). *Untersuchungen zur Ultrastructur von Nanochlorum eucaryotum mit besonderer Berücksichtigung der Zellwand. Ph.D. Disser. FBR Biologie, Universität Mainz.*

Brooks, J. and Shaw, G. (1973). In: *Origin and Development of Living Systems. Academic Press, London.*

Bryan, J. (1972). *Biochem. 11*: 2611–2615.

Czygan, F.-C. (1982). *Arch. Hydrobiol. Suppl. 60*: 470–488.

DeRosa, F., Haber, D., Williams, C. and Margulis, L. (1978). *Cytobios 21*: 37–43.

Dustin, P. (1978). *Microtubules, Springer Verlag, Heidelberg.*

Ettl, H. (1980). *Grundriss der allgemeinen Algologie. Fischer, Stuttgart.*

Geisert, M., Rose, T. and Zahn, R. K. (1985). *Biolog. Chemistry 366*: 761–870.

Geisert, M., Rose,T., Bauer, W. and Zahn, R.K. (1987). *BioSystems 20*: 133–142.

Gooday, G. W. (1981). In: *The Fungal Spore (eds. Turian, G., Hohl, H.R.), Academic Press, London.* pp. 487–505.

Graham, D. E. (1978). *Anal. Biochem. 83*: 609–613.

Heath, J. B. (1980). In: *International Review of Cytology (eds. Bourne,G.H., Danielli, J.F. and Jean, K.W.) Vol.64, Academic Press, New York.*

Jain M.K. (1982). *Handbook of Enzyme Inhibitors, Wiley-Interscience New York.*

Kleinschmidt, A., Lang, D. and Zahn, R. K. (1960). *Naturwiss. 1*: 16–17.

Margulis, L. (1981). *Symbiosis in Cell Evolution. Freeman, San Francisco.*

Menzel, K. and Wild, A. (1989). *Botanica Acta 102*: 117–180.

Naumann, E. (1921). *Archiv für Botanik 16*: 1–19.

Nickel, G. (1986). *Nachweis von Histongenen in der DNA von Nanochlorum eucaryotum durch Hybridisierung mit klonierten Histonfragmenten. Diplomarbeit, Universität Mainz.*

Nowotny, A. (1979). *Basic Exercises in Immunochemistry. Springer-Verlag Berlin, Heidelberg, New York.*

Pickett-Heaps, J. D. (1974). *Biosystems 6*: 37–48.

Reitz, M. (1986). In: *Die Alge im System der Pflanzen. Nanochlorum eucaryotum — eine Alge mit minimalen eukaryotischen Kriterien. Gustav Fischer Verl., Stuttgart, New York.* pp. 190–234.

Rose, T., Bauer, W. and Zahn, R.K.(1986). *Biol. Chem. Hoppe Seyler 367*: 198.

Sargent, M., Zahn, R., Walters, B., Gupta, R. and Kaine, B. (1988). *Nucleic Acids Research 16*: 4156.

Schreiner, M. (1990). *Isolierung und Analyse von Chloroplasten-DNA aus Nanochlorum eucaryotum. Ph.D. Disser., Fachbereich Chemie, Johannes Gutenberg Universität Mainz.*

van der Waerden, B.L. and Nievergelt, E. (1956). *Tables for Comparing Two Samples by X-Test and Sign Test. Springer-Verlag, Berlin.*

Weinblum, D., Geisert, M., Schreiner, M., Güngerich,U. and Zahn, R. K. (1988). *Biol. Chem. Hoppe-Seyler 369*: 941.

Weinblum, D., Geisert, M., Schreiner, M., Güngerich, U. and Zahn, R. K. (1989). *Biol. Chem. Hoppe-Seyler 370*: 956.

Wilhelm, C., Eisenbeis, G., Wild, A. and Zahn, R.K. (1982). *Z. Naturforsch. 37c*: 107–114.

Wilhelm, C. (1983) *II. Nanochlorum eucaryotum, eine neue, marine, sehr kleine, Chlorophyll A und B enthaltende, einzellige Alge. Ph.D. Disser., Fachbereich Biologie, Johannes Gutenberg-Universität Mainz.*

Wilhelm, C. and Wild, A. (1982). *Z. Naturforsch. 37c*: 115–119.

Zahn, R. K. and Stahl, I. (1953). *Hoppe-Seyler's Z. Physiol. Chem. 302*: 1–10.

Zahn, R. K. and Stahl, I. (1955). *Hoppe-Seyler's Z. Physiol. Chem. 302*: 204–216.

Zahn, R. K. (1984). *Origins of Life 13*: 289–303.

Zahn, R. K., Reinmüller, J., Beyer, R. and Pondeljak, V. (1987). *Mech. Age. Devel. 41*: 73–114.

Chapter 5

Enigmatic chlorophycean algae forming symbiotic associations with ciliates

WERNER REISSER
Pflanzenphysiologisches Institut der Universität, Untere Karspüle 2, D – 37073 Göttingen, Germany

Outlines:

1. Introduction
2. General features of symbiotic associations with chlorophycean algae living in aquatic habitats
3. The green *Paramecium*-system
   3.1 Special ultrastructural features of the symbiotic algae
   3.2 Special physiological features of the symbiotic algae
References

1. Introduction

Algae are omnipresent members of the biosphere and settle most different habitats among which symbiotic associations belong to the most interesting ones. There algae get into intimate contact with other organisms and often have to stand peculiar physiological conditions. Those algae are also of interest regarding their taxonomic affiliation. Mainly algae which live in a stable hereditary relationship with their hosts represent taxonomically rather interesting forms which often are not present as free-living specimen (Reisser 1986, 1992a). Sometimes they also show conspicuous features of adaptation to the symbiotic milieu. This is most interesting in regard to mechanisms which may have been responsible for the evolution of autotrophic partners within heterotrophic hosts, eventually resulting in the formation of chloroplasts (Reisser 1992b).

In general, algae of symbiotic associations living in freshwater and terrestrial habitats belong to chlorophycean genera whereas algae of marine associations show a much greater taxonomic diversity including dinoflagellates, diatoms, red algae, and chrysophycean species but comparably few green ones (for an overview see Reisser and Widowski 1992).

Traditionally symbiotic associations are described as either 'endo-' or 'ecto-' symbiotic with reference to the localization of the smaller partner organisms within or outside the host. However, those terms can be misleading and should be used only carefully. In fact, there exist very few really 'ecto'-symbiotic associations. In most cases those relationships should be better called 'epibiotic'. Even in heterologous lichens algae which are situated at the surface of the thallus, are surrounded by hyphae (Honegger 1992). Therefore, in the following, the term 'symbiotic' will be preferentially used.

'Symbiosis' shall also be defined in a general sense as 'living in physical contact and by mutual exchange of signals with a genetically different partner' without any implications as to mutual harm or benefit. Traditionally, the symbiotic partner or the symbiont which is bigger in volume is called 'host'. The term 'endocytobiont' is commonly assigned to a symbiont living within a host cell. By convention hosts which have lost their symbionts are called 'aposymbiotic' and symbionts which have been isolated from their hosts and cultured independently are referred to as 'exsymbiotic'. Organisms which naturally live without a symbiotic partner are called 'free-living' or 'non-symbiotic' (Reisser 1986).

In the following general features of symbiotic associations formed by chlorophycean algae of aquatic habitats will be discussed and examined for potential adaptive properties of the algae which set them apart from free-living species.

2. General features of symbiotic associations with chlorophycean algae living in aquatic habitats

The majority of symbiotic associations of chlorophyceae is formed with freshwater hosts. Endocytobiotic associations are dominating and

J. Seckbach (ed.), Evolutionary Pathways and Enigmatic Algae: Cyanidium caldarium *(Rhodophyta) and Related Cells,* 87–95, 1994.

hosts are either protozoa or digestive cells of *Hydra viridis* or *Spongilla* sp. (Reisser 1992a, Table 1). Regardless to the nature of respective host cells stable endocytobiotic associations of freshwater habitats show an interesting pattern of common features:- The overwhelming amount of chlorophycean partners belongs to the genus *Chlorella* (Table 1).

Table 1. Symbiotic associations of heterotrophic hosts and chlorophycean freshwater algae[a]

Host[b]	Alga	Location of algae	Carbohydrates released by isolated algae	References
Protozoa				
Amoeba				
Acanthocystis turfacea	C	In	M	1, 2, 3, 4, 5
Amphitrema flavum	C	In	/	6, 7
Chaos zoochlorellae	C	In	/	8
Difflugia oblonga	C	In	/	9
Heleopara sphagni	C	In	/	6
Hyalosphenia papilio	C	In	/	6
Mayorella viridis	C	In	/	10
Ciliata				
Acaryophrya sp.	C	In	/	11
Climacostomum virens	C*	In	G,F,X,	3, 4
Coleps hirtus	C*	In	X	3, 4, 12, 13
Coleps spetai	C	In	/	14
Disematostoma bütschlii	C	In	/	11
Euplotes daidaleos	C*	In	F,X	3,4
Frontonia leucas	C	In	/	1, 2, 15, 16
Frontonia vernalis	C	In	/	11, 17
Halteria bifurcata	C	In	/	18
Holosticha viridis	C	In	/	17
Malacophrys sphagni	C	In	/	19
Ophrydium versatile	C	In	/	1, 2, 16, 20
Paramecium bursaria	C*	In	G,M,T	3, 4, 21
Prorodon viridis	C	In	/	17
Prorodon ovum	C	In	/	17
Psilotricha viridis	?c	In	/	16
Spirostomum viridis	C	In	/	22
Stentor polymorphus	C'	In	M	3, 4
Stentor niger	C	In	/	23
Stentor roeseli	C	In	/	16
Teutophrys trisulca	C	In	/	24
Vorticella sp.	C	In	/	25
Vorticella sp.	?d	In	/	25

Table 1. Continued

Host[b]	Alga	Location of algae	Carbohydrates released by isolated algae	References
Porifera				
Ephydatia fluviatilis	C	Ine	G	26
Heteromeyenia sp.	C	Ine	/	27
Spongilla lacustris	C	Ine	/	27, 28, 29
Hydra viridis	C*	Inf	M	30
Hydra magnipapillata	S	Inf	?	31
Mollusca				
Anodonta sp.	C	In,It	/	32, 33
Linnaea sp.	C	It	/	33
Unio pictorum	C	It	/	33
Rotatoria				
Ascomorpha helvetica	C	/	/	33
Cephalodella sp.	C	It	/	34
Turbellaria				
Castrada viridis	C	/	/	35
Dalyellia viridis	C*	In,It	/	36, 37, 38
Phaenocora typhlops	C	/	/	39, 40
Typhloplana viridata	C*	It	/	36

a: after Reisser (1992a); b: listed in alphabetical order, selection of species, for further species, mostly described only as 'green' see Buchner (1930, 1953), Dangeard (1900), Goetsch and Scheuring (1926), Kahl (1930); c: alga reported with stigma (*Chlamydomonas* sp.?); d: *Oocystis* sp.?; e: algae in choanocytes and amoebocytes; f: algae in digestive cells; C: coccoid chlorophycean alga; C*: belongs to *Chlorella vulgaris*-group; S: *Symbiococcum hydrae;* In: Intracellular; It: Intercellular; F: fructose; G: glucose; M: maltose; T: trehalose; X: xylose; /:no data available, 1: Bucher (1930, 1953); 2: Dangeard (1900); 3: Reisser (1984b); 4: Reisser (1986); 5: Matzke *et al.* (1990); 6: Schönborn (1965); 7: Bonnet *et al.* (1981; 8: Willumsen (1982); 9: Chardez (1968); 10: Cann (1981); 11: Berninger *et al.* (1986); 12: Christopher and Patterson (1983); 13: Klaveness (1984); 14: Foissner (1984); 15: Hood (1927); 16: Kahl (1930); 17: Sud (1968); 18: Tamar (1968); 19: Kawakami and Kawakami (1985); 20: Winkler and Corliss (1965); 21: Pardy *et al.* (1989); 22: Hauck (1986); 23: Kawakami (1984); 24: Clémont-Iftode and Versail (1967); 25: Graham and Graham (1978, 1980); 26: Wilkinson (1980); 27: Gilbert and Allen (1973); 28: Williamson (1979); 29: Frost and Williamson (1980); 30: Muscatine *et al.* (1967); 31: Rahat and Reich (1989); 32: Pardy (1980); 33: Goetsch and Scheuring (1926); 34: Droop (1963); 35: Gelei (1927); 36: Douglas (1987); 37: Haffner (1925); 38: Heitkamp (1979); 39: Eaton and Young (1975); 40: Young and Eaton (1975).

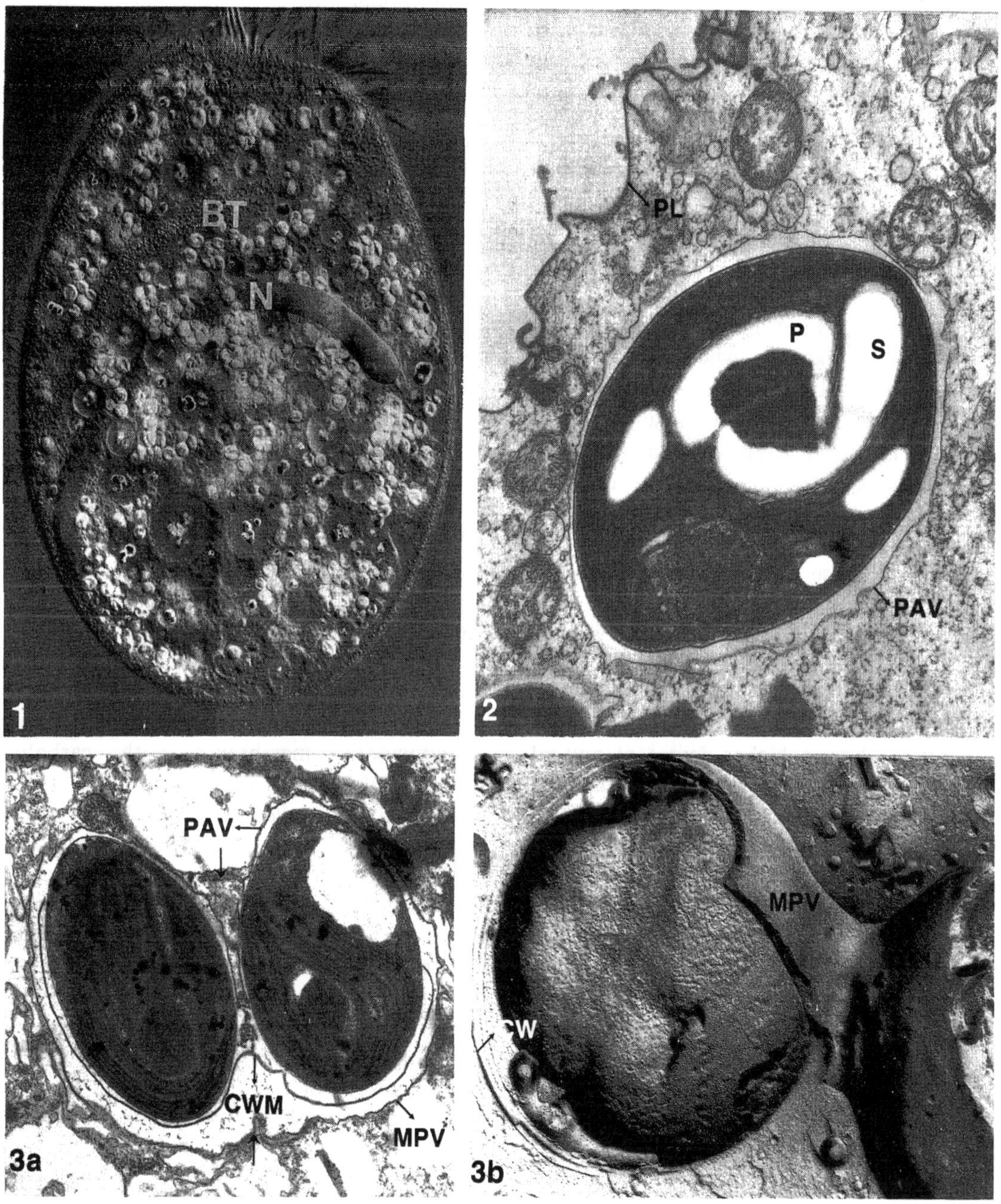

Fig. 1–3. (1) *Climacostomum virens* with symbiotic chlorellae. BT: Buccal tube, N: Nucleus. bar: 50 μm (from Reisser *et al.* 1984); (2) Symbiotic *Chlorella* sp. in a perialgal vacuole of *Paramecium bursaria. P: Pyrenoid, PAV: Perialgal vacuole, PL: Pellicula of the host, S: Starch. Bar: 1 μm (from Reisser and Wiessner 1984); (3) Formation of autospores by a symbiotic Chlorella* in a perialgal vacuole. Around each autospore an individual new perialgal vacuole membrane is formed (arrows). a: Division of a symbiotic *Chlorella* in a perialgal vacuole of *Stentor polymorphus*. CWM: Cell wall of mother cell, MPV: Membrane of perialgal vacuole, PAV: perialgal vacuole, bar: 1 μm (from Reisser 1981). b: Division of a symbiotic *Chlorella* in a perialgal vacuole of *Paramecium bursaria*. CW: Cell wall of autospore, MPV: Membrane of perialgal vacuole, bar: 1 μm (from Reisser and Wiessner 1984).

— Algae are only taken up by host cells which are capable of phagocytosis and thus can form lysosomal compartments (Figs. 1–3).

— Algae are taken up via phagocytosis and are enclosed in special compartments of host cells (so-called perialgal vacuoles) where they are protected against host lytic enzymes. Thus two types of vacuoles exist in host cells, perialgal and digestive.

- Algae are individually enclosed in perialgal vacuoles. Thus a maximum host-symbiont-interface area and a high capacity for exchange of signals are guaranteed.
- The phagocytotic uptake of algae is specific in that only suitable algae are enclosed in perialgal vacuoles, non-suitable ones are sequestered into digestive vacuoles.
- This selection does not only take place during phagocytosis of algae, i.e. when the host feeds on particulate matter, but also when algae in perialgal vacuoles divide and a new protective membrane is formed around each daughter cell.

Endocytobiotic associations with protozoan hosts differ in stability ranging from temporary associations such as the green *Coleps* which can take up and loose algal partners repeatedly during the host's lifetime, to hereditary ones such as the green *Paramecium* in which growth rates of host and algae are coordinated to some extent so that the algae are distributed to daughter cells when hosts divide. Hereditary associations show a higher degree of mutual adaptation of partners, probably because of a longer time of common evolution (Reisser 1986, 1992b). Since this report deals with special adaptive features of symbiotic algae the following will concentrate on algae of those stable hereditary systems. Most attention will be paid to the 'green *Paramecium*' which is a hereditary association of *Paramecium bursaria* (Ciliata) and *Chlorella* sp. (Chlorophyceae) and the best studied endocytobiotic association of the freshwater habitat (Reisser 1986). Additional information will be added from studies on other freshwater associations.

3. The green *Paramecium*-system

Green ciliates are common in freshwater habitats (Table 1, Figs. 1–3). Especially the green *Paramecium* is well known as to ultrastructure, physiology, and behavior (for an overview see Reisser 1992a, 1992c). A pertinent feature of this association is its specificity, i.e. the fact that a stable hereditary ciliate-alga association is formed only with special strains of *Chlorella* sp. Other algal genera and even taxonomically closely related *Chlorella* spp. are not recognized by the ciliate as suitable for symbiosis formation. They do not trigger the formation of a perialgal vacuole but are digested in food vacuoles. Obviously suitable algae show special properties which allow them to enter and to settle the symbiotic habitat. Those properties are not species but probably strain specific. Their very nature is still under study and thus the following can only be a sort of status report. However, data clearly show that special algal features exist on both the ultrastructural and the physiological level.

3.1 Special ultrastructural features of the symbiotic algae

When algae are taken up by the *Paramecium* via phagocytosis a series of cell-to-cell recognition events decides on whether cells are enclosed individually in protective perialgal vacuoles or are sequestered into food vacuoles. It could be shown that the recognition mechanism bases on an interaction of ciliate proteins which are bound to phagocytotic membranes with carbohydrate groups of algal cell wall surfaces. Only the symbiotic *Chlorella* strains and few free-living specimen of the genus *Chlorella* are endowed with the 'proper' carbohydrate structures which trigger the enclosure into protecting perialgal vacuoles. The very nature of those surface groups is unknown. Probably they react with ciliate membrane proteins by a lectin-type mechanism which leads to a close attachment of the ciliate membrane to the algal cell wall surface and thus to an enclosure of an individual alga in a special (perialgal) vacuole (Reisser 1990).

Algal cell wall carbohydrates are different during different developmental stages of the algae since only autospores but not mother cells of the symbiotic *Chlorella* trigger the formation of a perialgal vacuole. This is in accordance with the fact that the cell-to-cell recognition mechanism also works when symbiotic algae within perialgal vacuoles divide by forming autospores which are enclosed immediately by new specific membrane material and thus are enclosed again in a perialgal vacuole (for details see Reisser 1990).

3.2 Special physiological features of the symbiotic algae

Once suitable algae have been enclosed in perialgal vacuoles they can populate the host according to their ability to meet the requirements of their symbiotic milieu. Actually more algae are enclosed in perialgal vacuoles and get into the host than ultimately are successful in staying permanently within it. Usually only the original symbiotic algae are able to reestablish a stable hereditary relationship. All other algae although being enclosed in perialgal vacuoles — are lost again. Their growth rate is too low to keep up with host cell division. At latest under stress conditions such as permanent darkness or high host growth rates most algae are lost again. Obviously the proper algae show some special adaptive physiological features which allow them to stay within the ciliate even under stress conditions (Reisser 1986).

The nature of those features is still a matter of discussion. However, available data strongly indicate that there don't exist any principal qualitative differences between symbiotic chlorophycean algae and their free-living counterparts as far as such features as the potential for photoheterotrophic growth, a requirement of vitamins or organic nitrogen sources, resistance to UV-rays, pH tolerance, and number of autospores produced, are concerned (Reisser and Widowski 1992). However, the metabolism of symbiotic algae differs from the one of free-living algae by some modifications:

In the green *Paramecium* system algae and host are integrated according to the classical autotroph-heterotroph-scheme, i.e. algae use respiratory CO_2 of the host for photosynthesis and release oxygen which in turn supports host's respiration. In addition, algae excrete maltose and glucose and receive ammonia and glutamine as nitrogen compounds (Reisser 1986, Fig. 4).

The phenomenon of sugar excretion by symbiotic *Chlorella* spp. has been studied thoroughly (Schilling *et al.* 1991; Ziesenisz *et al.*. 1984) and illustrates well the nature of modifications separating symbiotic from free-living specimen.

The release of organic material is not uncommon among algae and has been described repeatedly for planktonic species which give off different substances such as glycolic acid, various kinds of amino acids, sugars, etc. However, excretion of organic compounds by symbiotic Chlorellae shows some special modifications (Reisser 1986, 1992a):

(i) Only few organic substances are released but in relatively high amounts: The symbiotic chlorellae of *Paramecium bursaria* excrete maltose, alanine, and traces of glucose which together can amount up to 45 % of the total photosynthetically fixed carbon (Table 2).

Table 2. Execretion of carbon compounds and photosynthetic oxygen production by symbiotic and free-living *Chlorella* spp[a]

	Carbon excreted [% of total photo-synthetically fixed carbon][b]	Type of excreted sugars	Oxygen produced at 6 Wm^{-2} [μMol × 10^{-2} mg Chl-1 × min^{-1}][c]
Symbiotic *Chlorella*			
Acanthocystis turfacea isolate (3.83) (A)	16.68	M	74.21
Climacostomum virens isolate (Cvi) (Ci)	8.10	G,F,X	41.92
Coleps hirtus isolate (Chi) (Ci)	4.21	X	311.32
Euplotes daidaleos isolate (Edi) (Ci)	6.32	F,X	27.09
Paramecium bursaria isolate (Pbi) (Ci)	45.21	G,M	74,52
Stentor polymorphus isolate (Spi) (Ci)	.76	M	70.31
Spongilla fluviatilis isolate (211-40c) (P)	15.10	G	27.23
Hydra viridis isolate (242.80) (Coe)	37.00	M	156.70
Free-living *Chlorella*			
C. vulgaris Beij. (211-11b)	1.92	–	24.59
C. mirabilis Andr. (211-30)	2.12	–	20.21
C. fusca var. vacuolata Sh. et Kr. (211-8b)	0.74	–	82.40

a: modified after Reisser (1984); b: measured at 20.9 Wm^{-2} for 3 h at 22°C, experiments were run at pH values which give maximum excretion, i.e. at pH 5.7 for Spi, pH 6.6 for Cvi, and pH 5.4 for other strains; c: calculated for true photosynthesis; C: *Chlorella;* G: glucose; F: fructose; M: maltose; X: xylose; A: Amoeba; Ci: Ciliata; P: Porifera; Coe: Coelenterata; for numerical strain designations see Schlösser (1982).

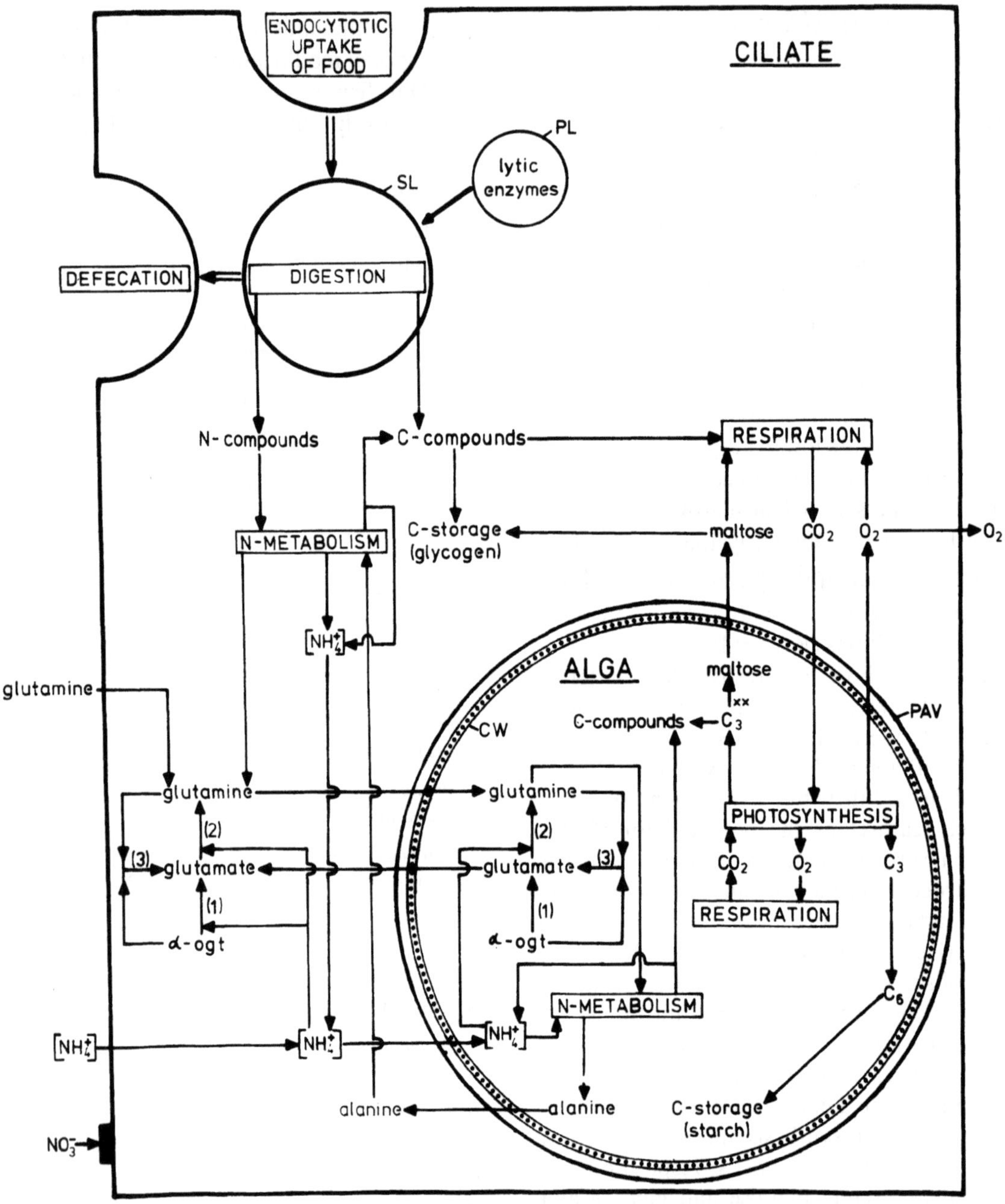

Fig. 4. Schematic drawing of C- and N-metabolism of the endocytobiotic association of *Paramecium bursaria* and *Chlorella* sp. fed with bacteria and grown in the light above the photosynthetic compensation point. **: in darkness maltose is synthesized from starch, CW: Cell wall of symbiotic *Chlorella*, PAV: Perialgal vacuole, PL: Primary lysosome, SL: Secondary lysosome, <=: Cyclosis, (1): NADH-dependent glutamate-dehydrogenase, (2): glutamine synthetase, (3): glutamate synthase, α-ogt: α-oxoglutarate. (from Reisser 1986).

(ii) Excretion is not triggered by any inducing factor and requires energy, i.e. is not due to leakage.

(iii) Amount and type of substances excreted depend on pH, i. e. the more acid the pH of the growth medium the more sugars are excreted (Table 3). At an alkaline pH this excretion nearly stops and mainly amino acids such as alanine are released.

(iiii) Excretion dominates over growth: When the

Table 3. Carbohydrates produced by a symbiotic *Chlorella* isolated from *Paramecium bursaria* and by a free-living *Chlorella* vulgaris at different pH[a]

pH	Carbohydrate produced (CP) [$10^{-6}\mu g \times alga^{-1} \times d^{-1}$]	Carbohydrate in supernatant[b] [% of CP]	[$10^{-6}\mu g \times alga^{-1} \times d^{-1}$]
symbiotic *Chlorella*			
5.2	0.97	91.86	0.89
7.4	.44	35.05	1.56
free-living *Chlorella*			
5.2	5.32	1.30	0.07
7.4	3.90	2.10	0.08

Carbohydrates are measured as maltose-equivalents (+/− 5%; d: day; a: after Reisser and Herbig 1987; b: tested 7 days after start of mass culture, in cultures of the symbiotic *Chlorella* predominantly maltose, in cultures of the aposymbiotic *Chlorella* (strain 211-11b, Schlösser 1982) not identified (cell debris)

Table 5. Carbohydrates produced by a symbiotic *Chlorella* isolated from *Paramecium bursaria* and by a free-living *Chlorella vulgaris* at different light conditions[a]

Illuminance [lux]	Carbohydrate produced (CP) [$10^{-6}\mu g \times alga^{-1} \times d^{-1}$]	Carbohydrate in supernatant[b] [% of CP]	[$10^{-6}\mu g \times alga^{-1} \times d^{-1}$]
symbiotic *Chlorella*			
3500	1.66	96.50	1.60
10000	8.59	68.67	5.90
free-living *Chlorella*			
3500	6.78	2.40	0.16
10000	5.56	1.80	0.10

Carbohydrates are measured as maltose-equivalents (+/− 5%); d: day; a: after Reisser and Herbig 1987; b: tested 7 days after start of mass culture, in cultures of the symbiotic *Chlorella* predominantly maltose, in cultures of the free-living *Chlorella* (strain 211-11b, Schlösser 1982) not identified (cell debris).

photosynthesis of symbiotic chlorellae is limited by a low partial pressure of CO_2 (Table 4) or low illuminance (Table 5) the overall amount of carbohydrates produced decreases — as is also the case in free-living algae — but the percentage of the excreted part increases, i. e. excretion is maintained even under unfavorable conditions. It even can happen that storage carbohydrates are mobilized in order to feed excretion while growth completely stops (Table 4). This 'altruistic' behavior has been observed in symbiotic chlorellae of *Paramecium bursaria*, *Acanthocystis turfacea*, and *Spongilla* sp. (Reisser and Widowski 1992). It is unique and might be the only really 'symbiotic' modifi-

Table 4. Carbohydrate budgets of a symbiotic *Chlorella* isolated from *Paramecium bursaria* and of a free-living *Chorella vulgaris* in an inorganic medium in the light and in the dark under different pCO_2[a]

pCO_2 [Mm]	Carbohydrate produced (+) or consumed (−) [$10^{-8}\mu g \times alga^{-1} \times h^{-1}$]			Carbohydrate budget[b] [$10^{-8}\mu g \times alga^{-1} \times 24h^{-1}$]		
	symb. C. in cell		freel.C.[c] released[d]	symb. C. in cell		freel. C.[c] released[d]
Light						
14.20	+ 16.70	+ 3.10	+ 30.82	+ 146.40	+ 124.80	+ 1030.68
11.50×10^{-3}	− 5.50	+ 7.70	+ 2.96	/	/	+ 85.56
#	− 5.80	+ 6.80	+ 2.74	/	/	+ 78.12
Dark						
#	− 4.50	+ 7.30	− 0.44	/	/	/

Carbohydrates are measured as maltose-equivalents (+/− 5%); a: after Reisser and Herbig 1987; b: calculated for 12 h light and 12 h dark; c: no release of sugars; d: maltose in supernatant; C.: *Chlorella*; /: calculation not possible since cellular car-bohydrates would be completely exhausted; # measured when KOH in the center well of Warburg vessel.

cation of the carbohydrate metabolism of symbiotic chlorophycean algae. It is genetically based and cannot be induced by some putative host factor as is e.g. the case in symbiotic algae living as partners in lichens (Smith *et al.* 1969).

This phenomenon is the more intriguing since the role which excreted sugars play for the survival of the whole symbiotic unit is rather obscure. Measurements on the carbon budget of the green *Paramecium* clearly show that under natural conditions of suboptimal light intensity the host can rely only to a limited extent on the endogenous supply of sugars by algae (Reisser and Herbig 1987). The ciliate always has to take up exogenous nutrients such as bacteria since algal photosynthesis is limited by available CO_2. Actually green paramecia always feed on bacteria or microalgae which also supply them with the necessary nitrogen, phosphorus, sulfur, etc compounds (Fig. 4).

Thus excretion of sugars by the algal partners does not seem to be the evolutionary driving force in establishment of the symbiotic system. Instead data suggest another explanation: Since sugar excretion is not triggered by an external stimulus it continues after isolation of the algae from the host. Thus in a free-living status algae cannot compete with non-symbiotic relatives because of a permanent energy loss. Only within the symbiotic unit they have found a sort of evolutionary shelter place. Actually the overwhelming part of sugar excreting *Chlorella* strains — although belonging to the *Chlorella vulgaris/sorokiniana* group — cannot be unequivocally assigned to any existent free-living *Chlorella* genera but presents special, probably either rather old or transitional taxa (Reisser and Widowski 1992).

The study of endocytobiotic associations of freshwater ciliates and chlorophycean partners gives interesting insights into potential mechanisms and special features of adaptation during the evolution of associations of heterotrophic and autotrophic cells and thus potentially into the evolution of plastids as e.g. of the chlorophyll a, b — type. Algae in special vacuoles are enclosed by two membranes, i.e. algal plasmalemma and perialgal vacuole membrane. These membranes differ by ultrastructure and function and thus show parallels to the outer and inner chloroplast membrane.

In hereditary stable systems, algae within perialgal vacuoles show special features of adaptation, such as low division rate, low compensation point and high rates of excretion of different substances. This might be characteristic for a fundamental principle in the evolution of plastids from algae: Those features of the alga which add a new quality to the heterotrophic partner such photosynthesis are maintained whereas properties which are not essential for the association are lost (for details and a thorough discussion see Reisser 1992b).

References

Berninger, U. G., Finlay, B. J. and Canter, H. M. (1986). *J. Protozool.* 33: 557–563.

Bonnet, L., Brabet, J., Comoy, N. and Guitard, J. (1981). *Protistologica* 17: 225–233.

Buchner, P. (1930). *Tier und Pflanze in Symbiose. Bornträger, Berlin*

Buchner, P. (1953). *Endosymbiose der Tiere mit pflanzlichen Mikroorganismen. Birkhäuser, Basel, Stuttgart.*

Cann, J. P. (1981). *Arch. Protistenk.* 124: 353–360.

Chardez, D. (1968). *Hydrobiol.* 31: 1–6.

Christopher, M. H. and Patterson, D. J. (1983). *Annales de Station Biologique de Besse-en-Chandesse* 17: 278–293.

Clémont-Iftode, F. and Versavel, G. (1967). *Protistologica* 3: 457–464.

Dangeard, P. A. (1900). *Botaniste* 7: 161–191.

Douglas, A. E. (1987). *Br. phycol. J.* 22: 157–161.

Droop, M. R. (1963). In: *The 13. Symposium of the society for general microbiology. Cambridge Uni. Press, Cambridge*. pp 171–199.

Eaton, J. W. and Young, J. O. (1975). *Arch. Hydrobiol.* 75: 50–75.

Foissner, W. (1984). *Stapfia* 12; 1–165.

Frost, T. M. and Williamson, C. E. (1980). *Ecology* 61: 1361–1370.

Gelei, J. V. (1927). *Biol. Zentralbl.* 47: 449–461.

Gilbert, J. J and Allen, H. L. (1973). *Int. Rev. ges. Hydrobiol.* 58: 633–658

Goetsch, W. and Scheuring, L. (1926). *Z. Morph. Ökol. Tiere* 7: 220–253.

Graham, L. E. and Graham, J. U. (1978). *J. Protozool.* 25: 207–210.

Graham, L. E., Graham, J. U. (1980). *Trans. Amer. Micros. Soc.* 99: 160–166.

Haffner, K. V. (1925). *Z. wiss. Zool.* 126: 1–69.

Hauck, A. (1986). *Mikrokosmos* 75: 198–199.

Heitkamp, U. (1979). *Arch. Hydrobiol.* 86: 499–514.

Honegger, R. (1992). In: *Algae and Symbioses, (ed. Reisser, W.), Biopress, Bristol.* pp. 255–275.

Hood, C. L. (1927). *Biol. Bull.* 52: 79–88.

Kahl, A. (1930). In: *Die Tierwelt Deutschlands und der*

angrenzenden Meeresteile und nach ihrer Lebensweise, Vol. 18. (ed. Dahl, F.), Fischer, Jena.
Kawakami, H. (1984). *J. Protozool.* 31: 247- 253.
Kawakami, H. and Kawakami, N. (1985). *7 Int. Congr. Protozool., Nairobi, Abstr.* No 300.
Klaveness, D. (1984). *Protistologica* 20: 335–349.
Matzke, B., Schwarzmeier, R. E. and Loos, E. (1990). *Planta* 181: 593–598.
Muscatine, L., Karakashian, S. J. and Karakashian, M. W. (1967). *Comp. Biochem. Physiol.*, 20: 1–12.
Pardy, R. L. (1980). *Biol. Bull.* 158: 349–355.
Rahat, M. and Reich, W. (1989). *Phycologia* 28: 181–187.
Reisser, W. (1981). *Protoplasma* 105: 273–284.
Reisser, W. (1984). *Br. phycol. J.* 19: 309–318.
Reisser, W. (1986). *Progr. in Protistol.* 1: 195–214.
Reisser, W. (1990). In: *Experimental Phycology, Vol. 1 (eds. Wiessner, W., Robinson, D. G., Starr, R. C.), Springer, Berlin, Heidelberg, New York.* pp. 55–68.
Reisser, W. (1992a). In: *Algae and Symbioses, (ed. Reisser. W.).* pp. 1–19.
Reisser, W. (1992b). In: *Symbiogenesis: Prochlorophytes and the origins of plastids, (ed. Lewin, R. A.) Chapman and Hill, New York.* pp. 27–43.
Reisser, W. (1992c). In: *Algae and Symbioses (ed. Reisser, W.), Biopress, Bristol.*
pp. 657- 674.
Reisser, W. and Herbig, E. (1987). *Endocyt. Cell Res.* 4: 305–316.
Reisser, W. and Widowski, M. (1992). In: *Algae and Symbioses. (ed. Reisser, W.), Biopress, Bristol.* pp. 21–40
Reisser, W. and Wiessner, W. (1984). In: *Encyclopedia of Plant Physiology, New series 17, Intercellular Interactions, (eds. Linskens, H. F., Heslop-Harrison, J.) Springer, Berlin, Heidelberg, New York.* pp. 91–112.
Reisser, W., Fischer-Defoy, D., Staudinger, J., Schilling, N. and Hausmann, K. (1984). *Protoplasma* 119: 93–99.
Schilling, N., Reisser, W. and Dittrich, P. (1991). *Verh. Int. Ver. Limnol.* 24: 2665–2667.
Schlösser, U. G. (1982). *Ber. dt. bot. Ges.* 95: 181–276.
Schönborn, W. (1965). *Limnologica* 3: 173–176.
Smith, D., Muscatine, L. and Lewis, D. (1969). *Biol. Rev.* 44: 17–90.
Sud, G. C. (1968). *J. Protozool.* 15: 605–607.
Tamar, H. (1968). *Acta Protozool.* 6: 175–188.
Wilkinson, C. R. (1980). freshwater sponge Ephydatia fluviatilis. *Hydrobiologia* 75: 241–250.
Williamson, C. E. (1979). *Trans. Amer. Micros. Soc.* 98: 59–77.
Willumsen, N. B. S. (1982). J. *Protozool.* 29: 500.
Winkler, R. H. and Corliss, J. O. (1965). *Trans. Amer. Micros. Soc.* 84: 127- 137.
Young, J. O. and Eaton, J. W. (1975). *Arch. Hydrobiol.* 75: 225–239.
Ziesenisz, E., Reisser, W. and Wiessner, W. (1981). *Planta* 153: 481–485.

PART III

The paradoxical Cyanidiophyceae

Chapter 6

The natural history of *Cyanidium* (Geitler 1933): past and present perspectives

JOSEPH SECKBACH

The Hebrew University of Jerusalem, School for Overseas Students, Mount Scopus, Jerusalem 91905, Israel

Outlines:

1. Succinct features of Cyanidiophyceae
2. Personal background and involvement
   2.1 Following the 'favorite organisms'
3. Milestone of the investigation with *Cyanidium*
   3.1 Past history of this organism
      3.1.1 The progenitor and the founder of current name
      3.1.2 The alkaline *Cyanidium* species
      3.1.3 M.B. Allen — the Western algal pioneer
      3.1.4 Research at Yellowstone Park during the 60's
      3.1.5 The Neapolitan presentations
      3.1.6 Contributions of the Japanese workers
      3.1.7 Recent Cyanidian investigations in Germany
      3.1.8 Russian investigation with Cyanidiophytes
      3.1.9 Current Ott's bibliographic collection
   3.2 Nomenclature of *Cyanidium*
   3.3 The systematic position
   3.4 The transitional features: the algal bridge
4. Proposals towards endosymbiotic association
5. The current status of Cyanidiophyceae
References

1. Succinct features of Cyanidiophyceae

Cyanidium caldarium is an acid hot spring alga which resembles *Chlorella* in its external morphological appearance. During reproduction, this alga divides into four endospores (while other species in this class divide into different number of cyanidiospores). This unicellular organism tolerates very acidic growing media of pH 2–4 at maximum temperature of 57 °C (Doemel 1970) and shows better rates of growth and photosynthesis when cultured with pure CO_2 (Seckbach *et al.* 1970). It has primitive eukaryotic cellular structure and biochemical traits which made it a suitable candidate for a transitional cell between cyanobacteria and primitive eukaryotic algae.

Rosen and Sigesmund (1961) were the first team to investigated the ultrastructural features of *C. caldarium*, and since there have been numerous publications on the electron microscopical studies of this organism (Staehelin 1967, 1968; Seliakh *et al.* 1981; Seckbach 1971, 1972, 1987; Seckbach and Fredrick 1980; Seckbach *et al.* 1981; chapters by Kuroiwa *et al.*, Ueda, and by Seckbach, in this volume). Its chloroplast possesses cyanobacterial pigments i.e., chlorophyll *a*, C-phycocyanin, allophycocyanin and carotenoid (Allen, 1959; Seckbach 1971; see chapter of Troxler in this volume). The thylakoids are arranged in concentric bands or in parallel membranes, they are non-stocked lamellae (i.e., without grana). The phycobilisomes granules are located outside the thylakoids as in blue-green and red algae (Seckbach 1972; Seckbach and Ikan 1972). In spite of the fact that *C. caldarium* is an obligated thermo-acidophilic (pH 2–4, 37 to 57 °C), its internal cytoplasmic environment is close to neutrality (Beardal and Enwisle 1984; Zenwirt *et al.* 1985; Enami *et al.* 1986). *C. caldarium* itself is an obligate autotroph although the larger cyanidiophyte ('Allen strain', or forma *B*, M-8, see Seckbach 1991, 1992) may also grow heterotrophically (in dark) with addition of sugar to the liquid medium (Allen 1959). For further characteristics of the cyanidiophytes see Seckbach 1992, 1987, 1972; Seckbach and Fredrick 1980, 1981; Seckbach *et al.* 1993, 1992, 1990, 1983 and their relevant references).

With time, a great deal of confusion and pandemonium appeared in the literature concerning *C. caldarium* studies. Lewin (1974, p. 26) notice the taxonomic problems with this alga and declared that: '....each new analysis of its biochemistry serving only to complicate the picture further ... certain despairing phylogenists are inclined to assign it to limbo, or in a subclass by itself.' In an attempt to solve this confusion within these acidothermophytes, the Italian School indicated that a natu-

J. Seckbach (ed.), Evolutionary Pathways and Enigmatic Algae: Cyanidium caldarium *(Rhodophyta) and Related Cells,* 99–112, 1994.

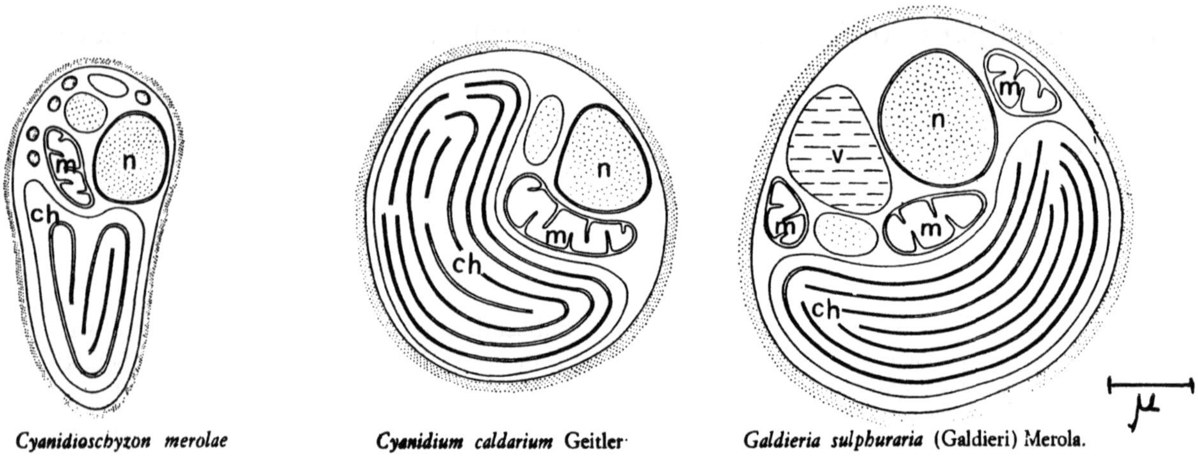

Fig. 1. A schematic sketch showing the ultrastructural features of three thermophilic acidophilic algae (Cyanidiophyceae). Note: n = nucleus; ch = chloroplast; m = mitochondrion; V = vacuole. Modified after Merola *et al.* (Gior. Bot. Ital. 115(4–5): 192). Kind permission by Professors P. DeLuca and R. Taddei (University of Napoli). Bar = 1 μm.

ral population of *Cyanidium* actually contain two, three or more members of the Cyanidiophyceae (Merola *et al.* 1981; see Seckbach 1992; Seckbach *et al.* 1993). Three genera (Fig. 1) have been original inserted into the Cyanidiophyceae: *Galdieria sulphuraria, Cyanidium caldarium* and *Cyanidioschyzon merolae* [for details and the specific features of each one of these organisms, see Merola *et al.* (1981); Review by Seckbach 1992 and 1991; Seckbach *et al.* (1993, 1992, 1990)]. Figure 2 indicates that all algal members of this group possess the same pigment content (Seckbach 1991; Nagashima *et al.* 1993).

Based upon the characteristics of these cyanidiopytes, we have organized these cells in a gradual developmental line and proposed that this algal group may display an evolutionary bridge over the gap between the prokaryotic cyanobacteria and the primitive red algae (Seckbach, 1992; Seckbach *et al.* 1983). We advocated that the first link in this transitional chain towards the Eukaryota is the very primitive eukaryote — *Cyanidioschyzon.* Then comes *Cyanidium* — the intermediate member of this class which is followed by the most advanced cyanidiophyte — *Galdieria* (this cell exhibits already several advanced, Rhodophycean characteristics). Additional information of the Cyanidian bridge is presented below (Seckbach and his coauthors 1993, 1992, 1991, 1990, 1987, 1983).

Recently, Sentsova (see this volume) observed three additional species of *Galdieria*, namely, *G.partita; G.maxima* and *G.daedala* (Sentsova 1991). Currently Fukuda *et al.* (1991) and Nagashima *el al.* (1993) report on additional algae of *Cyanidioschyzon species* which are different from *C.merolae.*

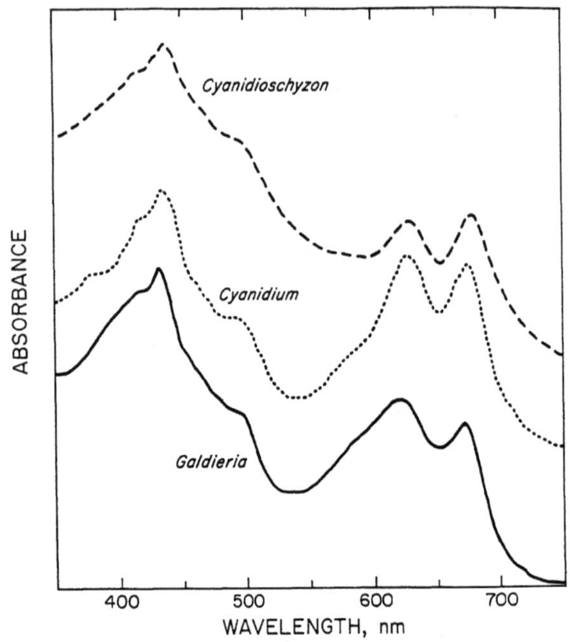

Fig. 2. The *in vivo* absorption spectra of three acidophilic thermophilic algae. The peaks at the blue (436 nm) and red (674 nm) regions are of chlorophyll a. The 'shoulder' at 488 nm is contributed by carotenoid and the peak at 626 nm is of phycocyanin. The base line for the different algae was shifted on the ordinate scale. After Seckbach 1991. Compare to Nagashima *et al.* 1993 (*Endocytobiology V.* p. 282).

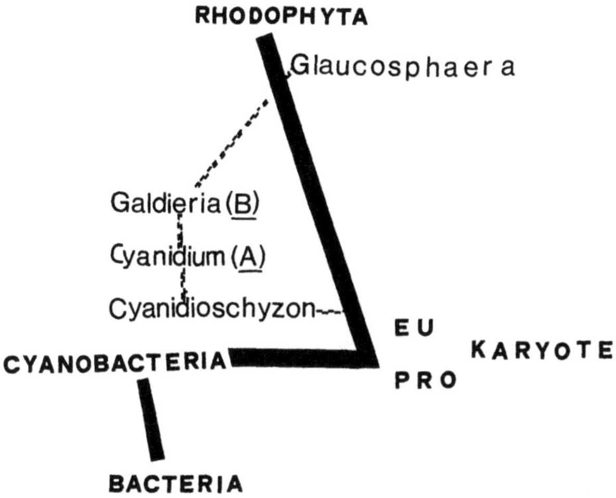

Fig. 3. A possible scheme of evolutionary path from cyanobacteria (Prokaryota) to Rhodophyta (Eukaryota). The transition between the two superkingdoms can be followed via the present-day relics of the Cyanidiophyceae. After Seckbach *et al.* *1983 (Endocytobiology II* p. 958).

The ultrastructure and other cytological and biochemical features of *Cyanidioschyzon* (Fig. 4) may suggest that this alga is the most primitive eukaryote (Fig. 5). Contemporary, fine structural observations on the newly discovered *Cyanidioschyzon*, could not always detect any mitochondrion and other cellular constituents like vacuole, golgi or peroxisome was not observed (Nagashima *et al.* 1993). If this is the real picture than we might have a strong case for the first nucleated cell which shares both prokaryotic and initial eukaryotic traits. The next cell in our transitional algal line is *Cyanidium* which represents an intermediate position in this chain, it shows still simple cellular substructure and possesses a single mitochondrion. Finally, *Galdieria*, the next step in this line, contains already several profiles of mitochondria and chloroplasts, this alga shows a more advanced level of cell differentiation (see Seckbach 1992, 1987).

Lately, Hoffmann (see this volume) reinvestigated two species of cave alkaline *Cyanidium* (Schwabe 1936, 1942), these cells may be of great interest since they do not occupy the 'normal' acidothermophilic habitat. For additional information on these caveous aerophilic cyanidiophytes, see chapter on the Taxonomy and Systematic in this volume.

Cyanidium caldarium has gained lately a great deal of interest, it turned from almost an inconspicuous and obscure alga in the past (with only a few citations in the literature) to a prominent alga with about 950 articles dealing with this fascinating organism (see Ott's Bibliographic list in this book). Today *C. caldarium* is involved in biochemical studies (see chapters by Gross; Troxler; Rigano *et al.*, or Nagashima in this volume) in ultrastructural investigations (see Ueda, Kuroiwa *et al.*, this volume), and in molecular biology (see chapter by Zetsche).

For additional review on these algae see Brock (1978), Seckbach (1971 to 1993), and chapters on Taxonomy and systematic by Ott and Seckbach in this volume. Table 1 and Figure 6 summarize the general and ultrastructural features of the three cyanidian species.

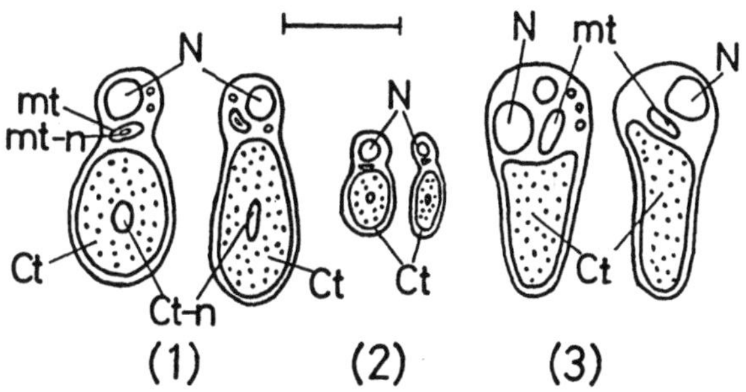

Fig. 4. Schematic diagram showing cell morphology of three strains of *Cyanidioschyzon*: 1. *C. merolae YR-89*; 2. *C. merolae YR-87*; 3. *C. meroale.* Note: N = cell nucleus; Ct = chloroplast; Ct-n = chloroplast nucleoid; mt = mitochondrion; mt-n = mitochondrial nucleoid (after Nagashima *et al.* 1993 in: *Endocytobiology V* p. 284). Bar = 2 μm.

2. Personal background and involvement

2.1 Following the 'favorite organism'

My interest and involvement in this thermophilic alga started about 25 years ago. The late Prof. W.F. Libby (Univ. of California at Los Angeles, UCLA) invited me to explore the possibilities of Life on the planet Venus. Our aim was to demonstrate that terrestrial photosynthetic organisms are able to tolerate and survive extreme conditions, such as were believed then, to exist on Venus polar caps. These organisms had to tolerate extreme conditions, to grow and photosynthesize in an atmos-

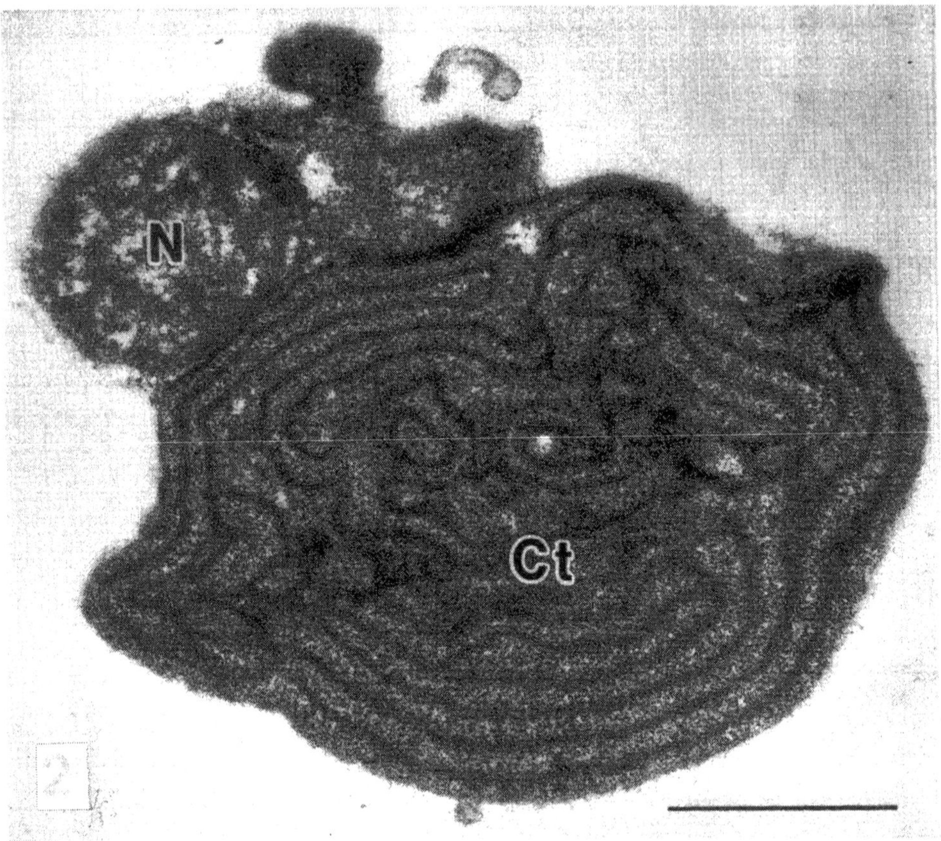

Fig. 5. Higher magnification of an electron micrograph of *Cyanidioschyzon merolae YR-89* fixed with glutaraldehyde-$KMnO_4$. The rabbit shape of this alga has a relatively large chloroplast (Ct) contiguous to the nucleus (N). The thylakoids are single (unstacked) and parallel arranged (similar to cyanobacteria and rhodophytes). After Nagashima *et al.* 1993 (*Endocytobiology V* p. 283). Bar = 0.5 μm.

Table 1. Comparative characteristics of the Cyanidiophyceae (Rhodophyta)

Feature	*Cyanidioschyzon*	*Cyanidium*	*Galdieria*
Cell Morphology			
Shape	oval, club-like	spherical	spherical
Size (μm)	1.5 × 3,5	2–5	3–16
Division	binary fission	endospores	endospores
Daughter cells	2	4	4–16–32
Nuclear DNA (N) ($X10^4$ photon)	104[1]	193[1,4]	378[2]
relative level of N-DNA among algae[5]	53–65%	100%	137–195%
Ultrastructure			
Chloroplast (Ct)	one	one	one, polyprofiled
–Ct shape	polymorph	spherical/cup	multilobed
–Ct DNA form	bead-like	rod shape	ring shape
–Ct DNA region	central	central	periphery
–Ct DNA value ($x10^4$ photon)	8.3[1]	72.8[1,3,4]	231[3,4]
–DNA ratio Ct/N	0.08[1]	0.38[1,2,3]	0.6[2]
Mitochondrion	one	one	several profiles
Vacuoles	absent	absent	several

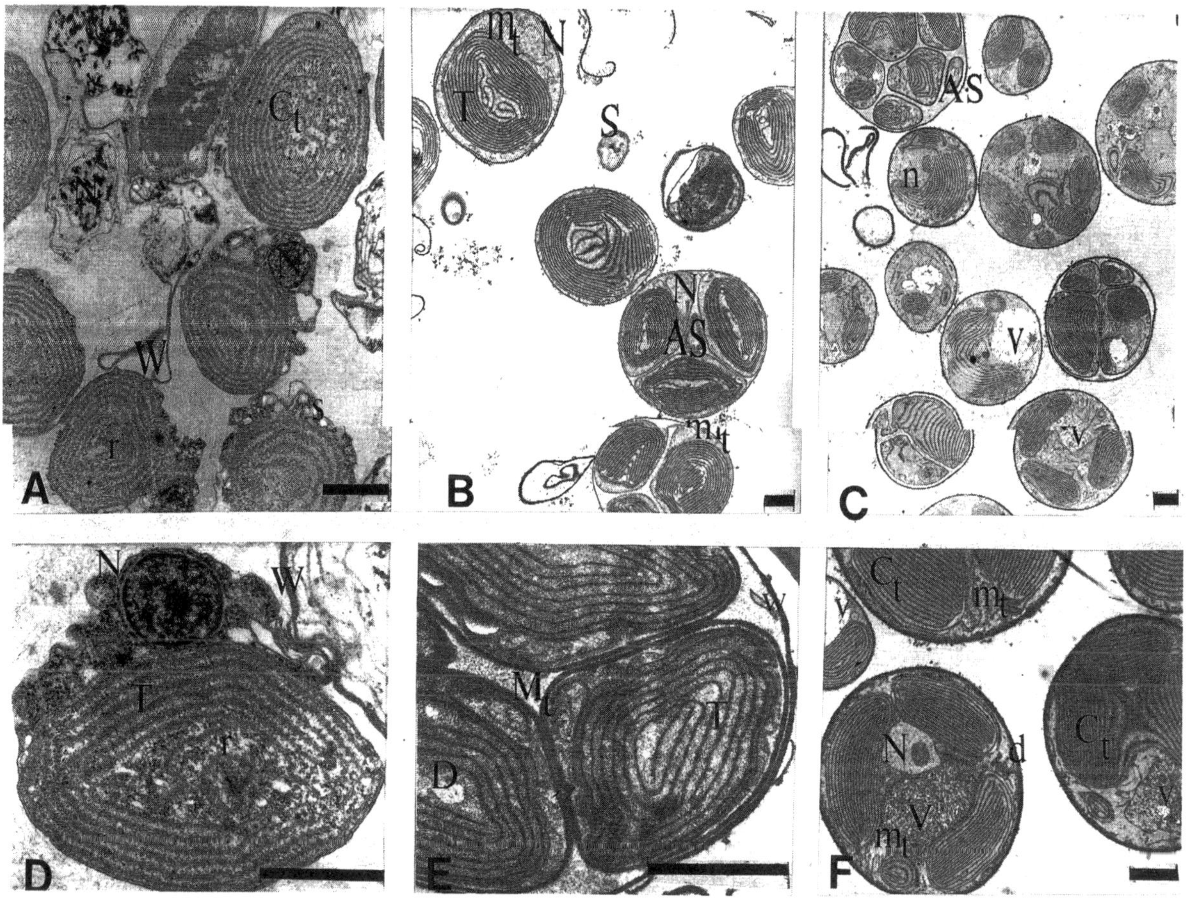

Fig. 6. The tree algae of the Cyanidiophyceae at lower magnification (sections A, B and C) and at higher views (D, E and F). Electron micrographs A and D represent *Cyanidioschyzon merolae* cells, while B and E are of *Cyanidium caldarium*. This alga divides into 4 cyanidiospores (see Parts B and E), in section E are three large autospores within the mother cell, the three mitochondria (M_t) are oriented at a similar position within the spores. The sections C and F are of *Galdieria sulphuraria*. Section C represents single cells, one containing six profiles of autospores (AS) within the mother cell. Many of these cells possess vacuoles (which are not present in the other two members). Most of the *Galdieria's* thylakoids are oriented in a parallel lines which are surrounded by an outermost thylakoid. At part F, the left cell exhibits three profiles of chloroplasts enclosing a nucleus and a dense appearing vacuole, at the right side is present a dictyosome (d). The cell at the right of section F contains a chloroplast which has been sectioned at a position suggesting that these organelles shown in a single cell (of *Galdieria*) are actually one multilobed chloroplast (as was recently demonstrated by the Japanese recent studies with chloroplast-nucleoid).
All bars at the right part represent 1 μm. Note: Ct = chloroplast with Thylakoids (T); N or n = nucleus; W = wall; M_t = mitochondrion; AS = autospores within the mother cell; V = vacuole; d = dictyosome; D = chloroplast nucleoid (DNA); r = plastidial ribosomes. For more details of algal cultures and fixation methods, see Seckbach 1992 (in: Algal Symbioses) and Seckbach *et al.* 1993 (in: Endocytobiology V). Permission from Biopress publisher, Bristol, England.

phere of pure CO_2 and at elevated temperature and pressure, as well as in a very acidic media.

We have treated several microalgae and higher plants but did not reach an encouraging success. That time I welcomed a letter from Dr. Lawrence Bogorad (Harvard U.) who suggested we try his 'favorite organism *Cyanidium caldarium*' as a suitable candidate for our Venusian project. We had several phone conversations with various colleagues on our engaging project, among them with W.N. Doemel (then a graduate student of Dr. T.D.Brock at Indiana University). He was studying the ecological physiology of algae in acid hot springs of Yellowstone National Park. He kindly sent us axenic cultures of *Cyanidium* isolated from Norris Geyser Basin of Yellowstone . The primary results of our algal growth and physiology in high concentration of CO_2, acidic medium and elevated temperatures were reported to the *International Botanical Congress* (Seattle, Wash. 1969) and at the

Table 1. Continued

Feature	*Cyanidioschyzon*	*Cyanidium*	*Galdieria*
Microbodies	–	+/–	+
catalase activity	+	+	+
Physiology			
Storage glucans			
cytosolic 'starch'	phytoglycogen	phytoglycogen	floridean
% α-1,6 glucose-glucose linkage[6]	7.2	7.1	5.8
Iodine reaction	red	red	red-violet
Branching enzyme	3 b.e	3 b.e.	2 b.e., 1 q
Ramification degree	++++	++++	++
Activity-Products			
Linolenic acid	absent	absent	present
NH_4 uptake	yes	yes	yes
NO_3 uptake	yes	yes	no
% salt tolerance	3	3–4	10
Growth conditions	autotrophic	autotrophic	auto/heterotroph

Based on: [1] Nagashima *et al.* 1993; [2] Fukuda *et al.* 1991; [3] Seckbach *et al.* 1992; [4] Fukuda and Nagshima 1987; [5] Suzuki *et al.* 1992; [6] Fredrick 1993; see also Seckbach 1992; Seckbach *et al.* 1990; Sentsova 1991; Ohta *et al.* 1992.

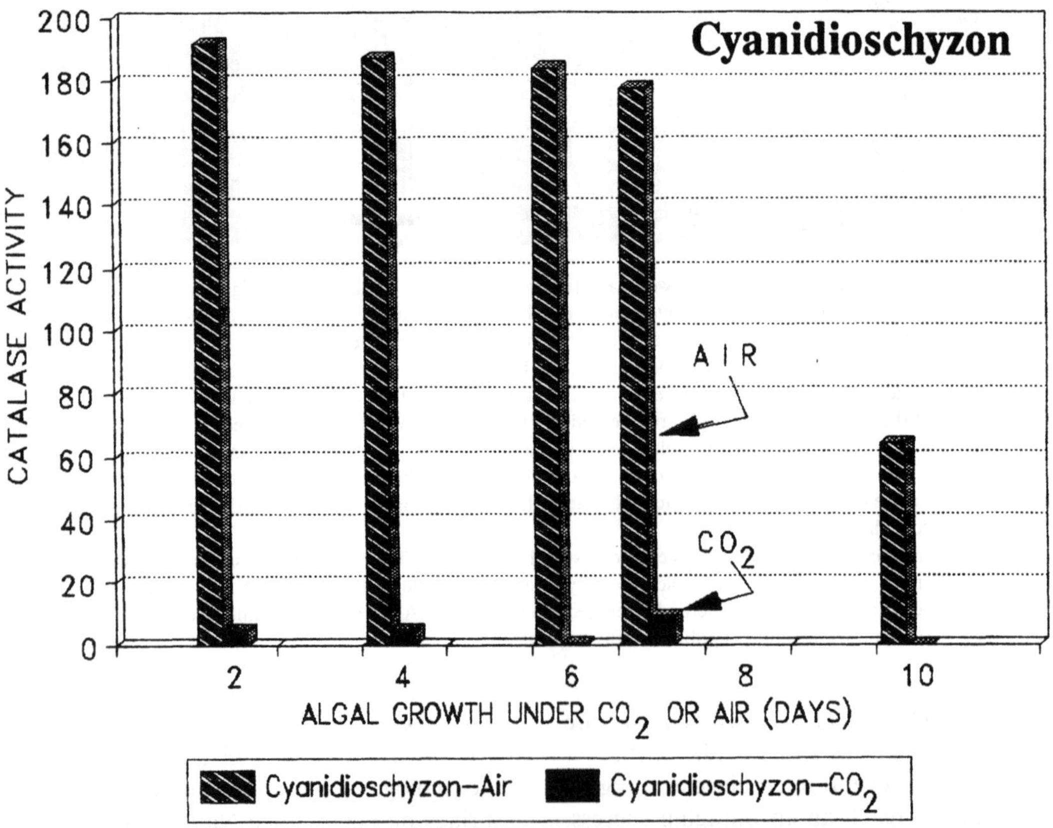

Fig. 7. Peroxisomal activity of *Cyanidioschyzon* cultured under aerated conditions (higher levels) and of cells grown under pure CO_2 (very lower enzymatic level). For more discussion on catalase activity of the other members of this algal family see Seckbach *et al.* 1992. Catalase activity is expressed as μM/min/mg protein.

Symposia of International Astronomic Union #40 at Marfa, Texas (Seckbach and Libby 1970, 1971) and elsewhere (Seckbach and coauthors 1969–1972). At this Botanical Congress I met Dr. Richard M. Klein (U. of Vermont) who was quite excited over our electron micrographs of *Cyanidium*. Thus, Dick Klein inspired me to consider a publication on the ultrastructural observations with a taxonomic approach (Seckbach 1971, 1972). We studied also the C^{13} / C^{12} natural isotopic fractionation in *C. caldarium* and hot spring algal mats (Seckbach and Kaplan 1973). Later, in Jerusalem we examined the lipids (Fatty acids and sterols) of *Cyanidium caldarium* cells obtained from Doemel (Adam *et al.* 1971; Ikan and Seckbach 1972; Seckbach and Ikan 1972). When in the late 70's I was studying cellular fine structure [at Bar Ilan University (Ramat Gan)], Dr. Jerry F. Fredrick (Dodge Chem. Comp. NY) sent me then (at my request) cultures of the larger cells of *Cyanidium caldarium* (Allen strain). This minor cooperation, followed in our future mutual cooperation in searching the biochemical of storage glucans, ultrastructure and on the bridge concept for the *Cyanidium* members (see list of publication). Our new ultrastructural observations were presented to the New York Academy of Sciences at the Symposium on *Origins and evolution of Eukaryotic Intracellular Organelles* (Seckbach *et al.* 1981). One of our ultrastructural pictures decorates the illustrated cover of this volume (Seckbach *et al.* 1981). Another scanning EM micrograph from our *C. caldarium* photos (during its dividing stage), has been chosen to decorate the cover page of *Mada*, the Israel Science Magazine (Seckbach and Fredrick 1981a). At that time I associated also with Prof. I. Fukuda from Tokyo and assisted his bibliographic list of *Cyanidium caldarium*, an early rapport which led afterwards to research cooperation with the Science University of Tokyo group. In 1987, when I visited the University of Naples, Dr. Aldo Moretti drove me, on a warm and muggy Mediterranean afternoon to Solfatara at Pozzuoli Park (in Naples region). There, Dr. Moretti showed me layers of *Cyanidium* under the upper surface of the white warm walls contiguous to hot fumaroles. He scratched and scraped the soft layer of the walls to expose out the blue-green covering

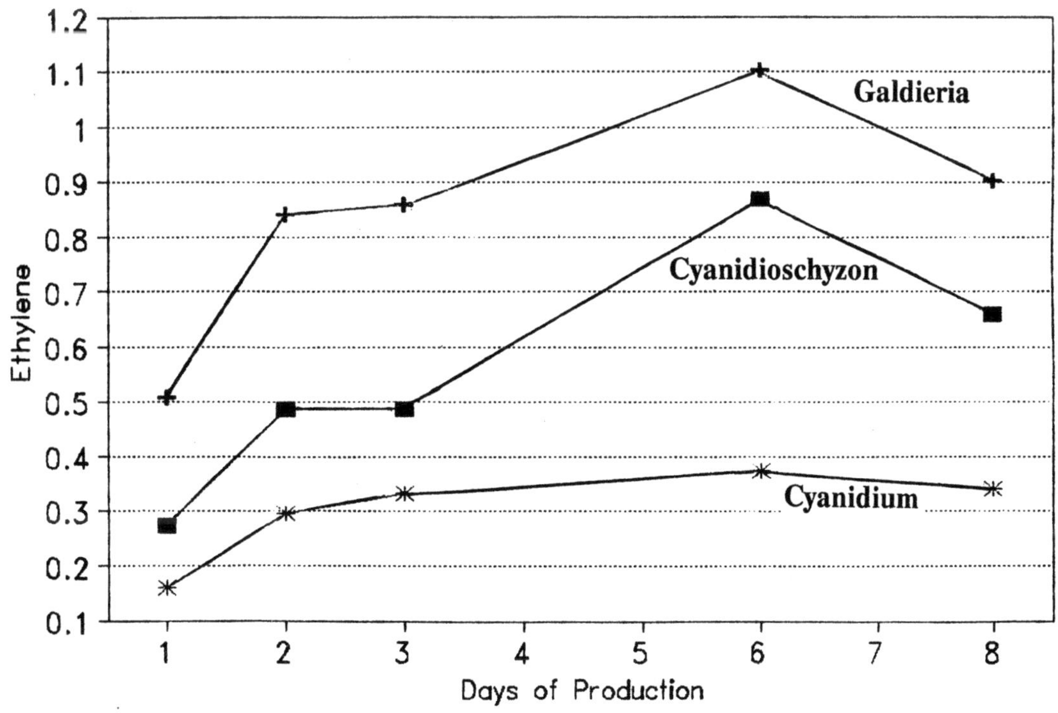

Fig. 8. Ethylene excretion into nutrient solution of the three cyanidiacean. Units of ethylene are expressed as μM/mg algal d. wt.

of *Cyanidium*. During 1989/1990 I spent a sabbatical year at UCLA at D.J. Chapman's laboratory and investigated *Cyanidium* biology from various aspects (Seckbach 1990, 1991, 1992; Seckbach *et al.* 1992; 1993). We observed the cyanidiophycean ultrastructure under various growth conditions, and studied with Dr. Elma González, their peroxisomal (catalase) activity (Seckbach *et al.* 1992) see Fig. 7. Another investigation with David Starrett (a graduate student of Dr. G. Laties), led us to discover in the first time the excretion of ethylene in the Cyanidiophyceae (Fig. 8). The lipids have been reexamined (with the cooperation of Dr. Kaplan's laboratory at UCLA and Drs. David White and Ringelberg from U. of Tennessee/Oak Ridge National Laboratory at Knoxville, TN) and a new sterol spectrum has been being published elsewhere (Seckbach *et al.* 1993, 1993a).

Our main contribution in the biology of *Cyanidium* include, algal life in pure CO_2 (Seckbach *et al.* 1970), ultrastructural aspects of the chloroplast, e.g. phycobilisomes, fatty acids and sterol content (Adam *et al.* 1971; Ikan and Seckbach 1972; Seckbach and Ikan 1972; Seckbach *et al.* 1993, 1993a) and extending the algal Bridge concept with the Cyanidiophyceae and introducing the new taxon of Prerhodophyta (Seckbach 1987; Seckbach *et al.* 1983). Our current studies include new aspects of fine structure, biochemical (ethylene activities) and evolutionary phylogenetic lines of these microalgae. Our recent studies have been published in *Endocytobiology* books (*II–V*) and elsewhere.

3. Milestones of the investigations with *Cyanidium*

The chronological review of *Cyanidium* indicates that this alga changed various names of genera (and species) as well as its systematic position. Ott and Seckbach (in this volume) reviewed the taxonomy, systematic position and proposed a new classification for the species of the genus *Cyanidium* and its ecological cohorts.

3.1 Past history of this organism

Cyanidium caldarium has been labelled at one time or another different names and classified into various systematic positions (see extended chapter by Ott and Seckbach in this volume and in Brock 1978; Doemel 1970). The notable progressive stations in *Cyanidium* material have been recorded by the following investigators:

3.1.1 The progenitor and the founder of current name

The American pioneer in this field, Josephine Tilden (1898) dealt with *Chroococcus varium* and with *Protococcus botryoides* f. *caldarium* (1898). The binomial name of this cell is associated with her name to indicated her contribution made over 95 years ago.

The Austrian, Luthar Geitler, first coined the name *Cyanidium caldarium*, (Gietler 1933, 1959; Gietler and Ruttner 1936). In the same time Copeland (in USA) gave this themoacidophilic alga the euphonious name *Pluto caldarius* (1936) which was rejected because of the name *Cyanidium* had a priority.

3.1.2 The alkaline Cyanidium *species*

Other taxa of *Cyanidium caldarium* have been observed in ecological niches of non acidic environments by Schwabe (1936, 1942) in Chile who described *C. caldarium* var. *chilenses* (1936) and *C. caldarium* var. *rumpens* (1942) growing inside caves. Others workers have later also observed these type of *Cyanidium* in Europe and in Israel (see Hoffmann; Ott and Seckbach's chapters in this volume; Hoffmann and Mayele, 1993).

3.1.3 M.B. Allen — the Western algal pioneer

Mary B. Allen studied the hot springs algae from California. She initiated in the Western world, the modern natural history of *Cyanidium caldarium* by culturing microalgae in axenic nutrient solutions in the laboratory. Her comprehensive article (Allen 1959) is quoted very frequently and the 'Allen' growing medium has received widespread used. In spite of her comprehensive contribution (Allen 1959) she continued to classify *Cyanidium* as an acid *Chlorella* or as 'anomalously pigmented chlorophyte'.

3.1.4 Research at Yellowstone Park during the 60's

The completion of the Yellowstone Park Investigations by Brock's group is presented in his book entitled *Thermophilic microorganisms and Life at high*

Temperatures (Brock 1978). An inclusive chapter in that book is devoted to the genus *Cyanidium* (Brock 1978). Doemel (1970) studied under Dr. Brock the physiological ecology of *C. caldarium* and published his observations during the early seventies (see references in Brock 1978). They confirmed the pH ranges, upper temperature limit of *Cyanidium* and made experiments with light intensities and photosynthesis on populations of this alga. They have investigated only the *C. caldarium* (Tilden) Geitler and have not studied the other species of the cyanidiophyceae [although Brock (1978) mentioned briefly the existence of additional strains of this alga].

3.1.5 The Neapolitan presentations

The Italian school has a long history with *Cyanidium* biology (Meneghini 1839; Galdieri 1899, De Luca and workers). Merola and coauthors have differentiated between the three algae in the *Cyanidium* populations and described *C. caldarium (=forma A), Galdieria sulphuraria* (=*C. caldarium* form *B*) and a smaller, more primitive algae *Cyanidioschyzon merolae* (Merola *et al.* 1981). They have studies the ultrastructure, physiological aspects, lipid content and other aspects of these organisms. Rigano and coworkers have been investigating for several years the nitrogen metabolism of these algae and found differences in the nitrate/ammonium uptake among the cyanidiophyceae (See chapters by Pinto *et al.* and by Rigano *et al.* in this volume).

3.1.6 Contributions of the Japanese workers

A major portion upon the Japanese background in Cyanidian studies is presented in Ott and Seckbach as well as in Fukuda's chapter in this volume. Indeed, Fukuda initiated and explored since 1958 the physiology of *Cyanidium*. His group studied the physiology of acidothermophilic conditions, photosynthesis, e.g., Hill reaction of the *Cyanidium* cells. Last decade, Fukuda compiled and published 3 pamphlets with the main bibliography of *C. caldarium* which covered the literature citations until 1981. Currently Fukuda and Nagashima as well as Kuroiwa and associates (Ohta *et al.* 1992; Suzuki *et al.* 1992) are examining the DNA content and shape of nucleoids of the cyanidiophyceae (within the organelles and nucleus) and discovered recently additional species of *Cyanidioschyzon* (Nagashima *et al.* 1993). Kuroiwa *et al.* (in this volume) presents the evolutional tree of algae on the basis of chloroplast nucleoid's profile and orientation. In their demonstration, the evolutionary sequence of the cyanidiophyte members correspond to our scheme (e.g., Seckbach *et al.* 1983; Seckbach 1992), namely, *Cyanidioschyzon* → *Cyanidium* → *Galdieria*. Suzuki *et al.* (1992) determined the cellular DNA distributed in the nucleus, mitochondrion and chloroplast of these thermophilic algae. They demonstrated (Suzuki *et al.* 1992) the increasing amounts withing the nuclear DNA for *Cyanidioschyzon* ($8x10^3$ kbp), *Cyanidium* ($13x10^3$ kbp) and for *Galdieria* ($18x10^3$). The electron microscopical illustrations of Ueda and of Kuroiwa *et al.* (in this volume) presents the recent ultrastructure of *C. caldarium* (Ueda and Yokochi 1981; Ueda and Chida 1987: Kuroiwa 1989; Kuroiwa *et al.* 1989). Their EM techniques reveal new aspects of cellular detail, e.g., the chloroplast double membrane (Ueda and Yokoci 1981, Ueda and Chida 1987) or the newly discovered stages which take place in *Cyanidium* during cell division (Kuroiwa *et al.* 1989). Additional groups in Japan performed ultrastructural studies (see chapter by Fukuda in this volume). The chapter on natural products by Nagashima (in this volume) describes various cytomolecules and the contribution of the Japanese workers [e.g., ferredoxin and phylogeny (Hase *et al.* 1978), lipids, polyamines (Hamana *et al.* 1990)]. Currently, the group of Kuroiwa (see separate chapter) is studying the gene structure and expression which adds more support to the presumption that *C. caldarium* is a primitive eukaryote. They notice in their molecular studies that *Cyanidium* (Ohta *et al.* 1992) and *Cyanidioschyzon* (Suzuki *et al.* 1992) possesse the smallest amount of nuclear DNA and its chloroplast *trnK* gene lacks introns (Ohta *et al.* 1992).

3.1.7 Recent Cyanidian investigations in Germany

A decade ago Kremer and coauthors (1978, 1979, 1982, 1983) became involved in biochemical and a symbiotic approach for *C. caldarium*. Zetsche's group is involved currently in the molecular aspects of various cellular components in *Cyanidium caldarium* (actually *Galdieria*, see Seckbach, 1991), their origin of genome coding and sequences (see his chapter). Maid and Zetsche (1990) studied the nucleotide sequence of the plastid 16S gene in

Cyanidium. They found (Maid *et al.* 1990) that Rubisco (key enzyme in Photosynthesis) is completely encoded in the rhodoplast genome of *Cyanidium* (similar to the origin of this enzyme in cyanelles of *Cyanophora paradoxa*, *Porphyridium* and other rhodophytes as well as in two chromophytes examined). On the other hand, in the higher plants and in chlorophytes the small subunits of Rubisco are encoded in the nucleus and are imported into the chloroplast. These authors suggested some endosymbiotically, phylogenetic implications out of their observations.

Additional molecular nucleic acids studied of *Cyanidium* were performed by Maid *et al.* 1990, 1992; Abu Elela and Nazar 1992; Lopes *et al.* 1985 and by Stempka and Richter 1978.

3.1.8 Russian investigations with Cyanidiophytes

During the last 20 years there have been published several research papers from the USSR (see Bibliographic list) describing the ultrastructure, physiology and biochemistry of *Cyanidium* members. Unfortunately, for the so-called Western world, almost all these publications have been published in Russian language and are not adequately circulated among the western phycologists. An important contribution has been presented by Dr. Olga Sentsova (see her chapter in this volume) who discovered very recently three new species of the genus *Galdieria* (Sentsova 1991).

3.1.9 Current Ott's bibliographic collection

Dr. Franklyn Ott has been devoted a great deal of time to the Literature of the non-marine rhodophytes, including *Cyanidium*. He has compiled for almost 30 years the world's Literature on non-marine rhodophytes. His inclusive list of the *Cyanidium* Bibliographic chapter has a very useful key for sorting the references into subfields. The total list in this volume contains about 950 references. In his other articles we present the chronological history of the nomenclature, taxonomy and systematic classification of *Cyanidium caldarium*. The paper (Ott and Seckbach, in this book) on the new taxonomic proposal is practical and may help to alleviate confusion among the '*Cyanidium*' workers (Seckbach 1991) in the future. In Ott's new taxonomic approach, we propose to include all the species of *Cyanidium* and *Galdieria* algae in one genus and family (Cyanidiaceae), all species have been placed in the single genus (*Cyanidium*). The most primitive alga within Cyanidiaceae, i.e., *Cyanidioschyzon merolae* (and its newly discovered additional species — see Nagashima *et al.* 1993) should be removed from this family because they do not share with the Cyanidia important features, and they have been placed with the Porphyridiaceae (Ott and Seckbach, in this volume).

3.2 Nomenclature of Cyanidium

The initial study on thermophilic acidophilic algae (which we recognize today as *Cyanidium* material) has been performed @ 155 years ago by the Italian investigator Meneghini who investigated *Coccochloris osiniana* (1839). With time several other algal names have been bestowed upon this acido thermophilic alga such as *Chroococcus; Chroococcopsis; Pluto; Dermacarpa; Entophysalis; Protococcus; Rhodococcus; Cyanidium caldarium (= C.c. formae A and B; RK-1 and M-8)*, *Galdieria and Cyanidioschyzon* (cf. Allen 1959; Doemel 1970; Brock 1978).

The *Cyanidium* data has shown various contradictions and inconsistencies among the results from its investigators. These confusions have guided the Neapolitan workers to distinguish among the '*Cyanidium*' members at least three separate organisms. They have termed the smaller cell (dividing to four endospores) 'C.c. forma *A*', while the larger cell was tagged as 'C.c. forma *B*'. On the other hand the Japanese workers called these cyanidian members as: RK-1 (for Cc forma A) and M-8 (for Cc forma *B*). We (Seckbach *et al.* 1981) also noticed the different between the cells of this organism and designated the larger one strain A (cells originated from Allen culture) while to the smaller one we designated strain B (culture obtained originally from Brock's laboratory). Later, the Italian School renamed the three Cyanidiophyceae as following: the smaller, *Cyanidium caldarium* (Geitler) [= C.c. forma *A* or identified to the Japanese terminology RK-1], while the larger organism was named *Galdieria sulphuraria* [= C.c. forma *B*, or to the Japanese 'M-8']. The smallest alga in this group was named *Cyanidioschyzon merolae* (Merola *et al.* 1981; Nagashima *et al.* 1993).

We are proposing a reasonable and straight forward key to classify the Cyanidiaceae, in our new scheme (Ott and Seckbach, see 'New Classification', in this volume) all the cyanidian members (except *Cyanidioschyzon)* have been placed in the genus *Cyanidium* (including also the *Galdieria* various species). It is our hope that this new, sensible, taxonomic revision will soon be recognized and accepted. For further information see features of these algae in Seckbach 1991, 1992, 1987 and Seckbach and Fredrick 1980, 1981; Seckbach *et al.* 1993, 1993a, 1992, 1990, 1983; Chapters by Ott in this volume; Fredrick 1976, 1987; Fredrick and Seckbach 1983, 1986.

3.3 The systematic position

Since the early pioneering observations on *C. caldarium*, there have been several attempts to classify it into various systematic and phylogenetic categories. The initial systematic placements of *C. caldarium* (because of its color and limitation of optical observations) were into the blue–green algae, i.e., Cyanophycophyta (cyanobacteria). Into this prokaryotic division was placed the first recognized alga *C. caldarium*, which has been placed in various genera, such as: *Coccochloris*, *Chroococcus*, *Chroococcopsis*, *Pluto*, *Dermocarpa*, *Entophysalis*), for further data see Ott and Seckbach in this volume. Unfortunately, a few contemporary authors still inaccurately place it to the Cyanobacteria (see Seckbach 1992), while other confuse name of these species (e.g., Suzuki *et al.* 1992). Following the cytological recognition of its eukaryotic features, *Cyanidium* was shifted into the green algae (Chlorophycophyta). For example genera like *Palmella, Protococcus, Prototheca* or *(acid) Chlorella* (all older names of the genus *Cyanidium*) have been classified as Chlorophyceae.

Others suggested placing *C. caldarium* in the Cryptophyceae (Fogg 1956; Lewin 1961) or as a 'genus of an uncertain taxonomic position' (Silva, 1962; Lewin and Gibbs 1982), Chrysophyta (see Doemel 1970) or Glaucophyta (see Allen 1959; Brock 1978; Seckbach 1992; Seckbach *et al.* 1993).

Hirose (1958) reported the presence of the nucleus and several Rhodophycean features in *C. caldarium* and suggested it be placed in the Porphyridiaceae (Rhodophycophyta) and changed its name into *Rhodococcus caldarius*. Later Chapman (1974) placed *Cyanidium* in the family Cyanidiaceae, in the subclass Porphyridiales among the more primitive Rhodophycophyta belonging to the Bangiophyceae. Ultrastructural and various biochemical studies (Seckbach and Fredrick 1980) as well as molecular analyses of ferredoxin (Hase *et al.* 1978) have confirmed this systematic pigeonhole for *Cyanidium* within the Rhodophycophyta. Further analyses of the biochemical and the primitive ultrastructural features of all cyanidiophytes suggested to Seckbach (1987) that these Cyanidiaceae should be placed in a new taxon i.e., Prerhodophyta.

3.4 The transitional features: algal bridge

The unusual ultrastructural, biochemical features of these acido-thermophilic algae have generate during the sixties the concept of transitional alga or bridge cell between various algal groups. Klein and Cronquist (1967) intensively advocated for the

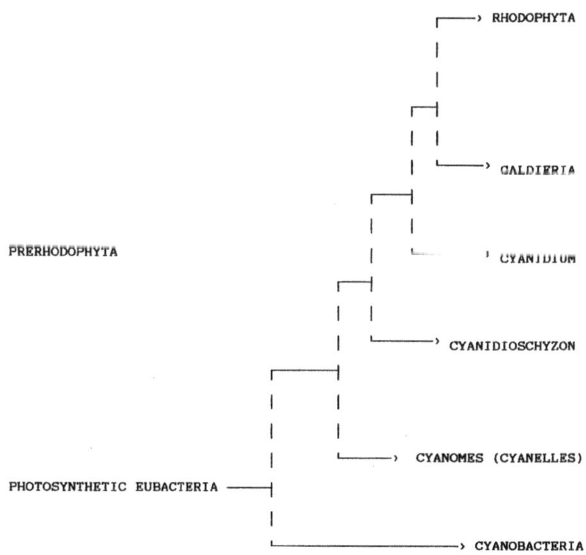

Fig. 9. Cyanidiophyceae – the classical (non symbiotic) model for eukaryogenesis (based on morphological, cytological and biochemical investigations).

Bridge position of this 'Uralga' and placed *Cyanidium* between the Cyanophyceae and the eukaryotic (green or red) algae. Klein (1970) and Fredrick (1976) supported the bridge and placed it between the blue-green and the red algae (cf. Seckbach and Fredrick 1981). Seckbach (1987) and coauthors (1983) extended the algal Bridge concept for the three Cyanidiophycean members and established these cells in a gradual evolutionary order between the cyanobacteria and the more primitive rhodophytes (Fig. 3). Kuroiwa *et al.* (in this volume) studied the organization of chloroplast-DNA among various algae and observed that the cyanidiophytes have a sub-evolutionary line in the algal evolutionary tree in the same order as our Prerhodophyta proposal (Seckbach 1992, 1991, 1987; Seckbach *et al.* 1983). Although there have been additional studies which support the intermediate position of the Cyanidiophyceae we feel that additional molecular sequences are needed to sustain or to infirm the bridge view for the Cyanidiophyceae. Figure 9 exhibits the possible evolutionary relationships from bacteria towards the red algae via the Cyanidian members.

4. Proposals towards endosymbiotic association

Castaldo (1968) suggested that the fine cellular structure of *Cyanidium* chloroplast may present symbiotic cyanelles harbored within this [host] alga. Margulis pointed out that *Cyanidium* presents an anomalous symbiosis (1968). Raven (1970) supported the above hypothesis that the multilobed chloroplast of *C. caldarium* is actually a blue green alga that entered into symbiotic association an ancestral organism and became stabilized as a photosynthetic organelle. Edwards and Mainwaring (1973) indicated in their abstract that the chloroplast of this alga presents a symbiotic prokaryotic cyanophyte within the apochloric (eukaryotic) host. Kremer with his colleagues (1978, 1979, 1982, 1983) rejected the taxonomic position of *Cyanidium* among the rhodophytes. They proposed on their biochemical observations (absence of floridiosides and the presence of monosaccharides), that *Cyanidium* is an endosymbiont similar to *Cyanophora paradoxa* (and to other members of Glaucophycophyta). The above biochemical criteria led Kremer and coauthors to the belief that the plastids of *C. caldarium* were cyanelles (and the cell is a cyanome, i.e., a protistan carrying endosymbiotic cyanobacteria). This concept has been challenged by others (De Luca and Moretti 1983; Nagashima and Fukuda 1983; Reed 1983). These workers have observed contradictory data in *C. caldarium* which argue against Kremer's symbiotic proposal.

Furthermore, Maid and Zetsche (1989), Maid *et al.* (1990) reported that they have obtained sequence data for the rubisco operon of *Cyanidium* which did not support that its chloroplast is a cyanelle like *C. paradoxa* (cf. Seckbach 1991). It seems that these conflicting data may weaken the Kremer's symbiotic notion for *Cyanidium.*

Another attempt to associate *C. caldarium* as a dual cell association was made by Castenholz (1979) who composed a symbiotic ecological origin for his model. He proposed that a thermophilic cyanobacterium entered a mesophilic-thermoacidophilic *Chlorella* or *Prototheca* (a colorless form). Such an association would have been an advantage to the prokaryotic endosymbiont in an intolerable acidic environment. It would now be protected by a cytoplasm regulated near neutrality. The thermophilic cyanobacterium would 'rise' the temperature limits of its host from 42 °C (the upper limit of the mesophilic *Chlorella*) to 57 °C — the current upper temperature range of *Cyanidium* (Castenholz 1979). Fredrick (1979, 1981) investigated the storage glucans of various algae and noticed that in *Cyanidium* this deposition is completely different from that of *Prototheca* (or *Chlorella*). These algae vary in their type of storage carbohydrate (and the enzymatic system involved in their synthesis) and also accumulate their carbohydrates in different location within the cellular space (Seckbach and Fredrick 1981; Seckbach *et al.* 1983).

These different opinions will also emphasize the idea that *Cyanidium* is a descendent of a colorless (apochloric) green alga which had originally been associated with a cyanobacterium as proposed by Castenholz (1979) and Kremer (see above) or by others.

There has been opposition to intracellular symbiotic concept in general (e.g., Nakamura 1993; Nakamura and Hase 1988; Jensen 1989, 1991; see also their chapters in this volume) and specifically for *Cyanidium* by the autogenous proponents (Allen 1959; Mercer *et al.* 1962; Klein and Cronquist 1967; Alsopp 1969; Chapman 1974; Seckbach *et al.* 1981, 1983, 1990, 1992; Seckbach and Fredrick 1980, 1981). See opposite concepts in Margulis (1981).

5. The current status of Cyanidiaceae

We propose herein (Seckbach *et al.* 1993) that the most primitive genus *Cyanidioschyzon*, a cohort of the other cyanidiophytes, (which we propose to move into the Porphyridiaceae — see new Classification by Ott and Seckbach in this volume) has perhaps evolve from a symbiotic association (Fig. 10). The progenitors of *Cyanidioschyzon* may present an archaebacterium *Thermoplasma*-like cell (displaying the host-nucleo-cytoplasm) which merged with a cyanobacterium (acting as chloroplast). Discussion on this new hypothesis has been recently presented (Seckbach *et al.* 1993).

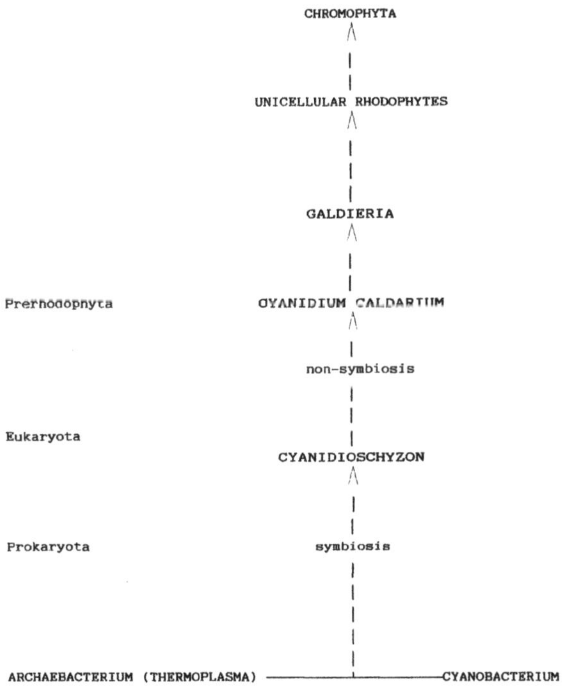

Fig. 10. Cyanidiophyceae-symbio-autogenous model: the exogenous origin (endosymbiotic view) for *Cyanidioschyzon* and autogenous development of the higher cyanidia (Prerhodophyta). See Seckbach *et al.* (1993) for further discussion.

Acknowledgements

The author who has followed the route of *Cyanidium caldarium* for a number of years wishes to thank all who took an interest and cooperated with him in his studies of these enigmatic organism. Special appreciations are due to Dr. and Ms. F.D. Ott for proof reading of this manuscript and to his colleagues who gave permissions to use their published illustrations.

References

Abu Elela, S. and Mazar, R.N. (1992). *Biochim. Biophys. Act.* 130: 339–342.

Adam, B.L., McMahon, V. and Seckbach, J. (1971). *Bichem. Biophys. Res. Commun.* 42: 359–365.

Allen, M. B. (1959). *Arch. für Mikrobiol.* 32: 270–277.

Allsopp, A. (1969). *New Phytol.* 68: 591–612.

Beardal, J. and Enwisle, L. (1984).*Phycologia* 23: 397–399.

Brock, T.D. (1978). *Thermophilic microorganisms and life at high temperatures.* New York, Springer-Verlag. Chapter 9: The genus *Cyanidium.* 255–302.

Castaldo, R. (1968). *Delpinoa* (Napoli) 8/9 (1966–1967): 135–145.

Castenholz, R.W. (1979). In: *Strategies of Microbial Life in Extreme Environments.* (ed. Shilo, M.) Verlag Chemie Weinheim N.Y. pp. 373–392.

Chapman, D.J. (1974). *Nova Hedwigia* 25: 673–682.

Copeland, J. J. (1936). *Ann. N.Y. Acad. Sci.* 36: 1–229.

De-Luca, P. and Moretti, A. (1983). *J. Phycol.* 19: 368–369.

Doemel, W.N. (1970). Ph.D. dissertation, Indiana University, Bloomington. USA.

Edwards, M.R. and Mainwaring, J.D. (1973). 31st Ann. Proc. Electron Microscopy Soc. Am., New Orleans, Louisiana, USA, ed. C.J. Arceneaux. p.462 + 1 pl.

Enami, I., Akutsu, H. and Kyogoku, Y. (1986). *Plant Cell Physiol.* 27: 148–165.

Fogg, G. E. (1956). *Bacter. Rev.* 20: 148–165.

Fredrick, J. F. (1976). *Plant & Cell Physiol.* 17: 317–322.

Fredrick, J. F. (1979). *Phytochem.* 18: 1823–1825.

Fredrick, J. F. (1981). *Ann. N.Y. Acad. Sci.* 361: 426–434.

Fredrick, J. F. (1987). *Ann. N.Y. Acad. Sci.* 503: 438–448.

Fredrick, J.F. (1993). In: *Endocytobiology* V (eds. Sato *et al.*) Tübingen University Press, Tübingen, FRG. pp. 475–479.

Fredrick, J. F. and Seckbach, J. (1983). *Phytochem.* 22: 1155–1157.

Fredrick, J. F. and Seckbach, J. (1986). *Phytochem.* 25: 363–365.

Fukuda, I and nagashima, H. (1987). *Ann. N.Y. Acad. Sci.* 503: 575–577.

Fukuda, I., Nagashima, H. and Seckbach, J. (1991). *Program & Abstracts of Internatl. Symbiosis Congress,* Jerusalem, Israel. November 1991. p. 70.

Galdieri, A. (1899). *Rend. Accad. Sci. Mat. Napoli Ser.* 3[a] (Napoli) 5: 160–164.

Geitler, L. (1933). *Arch. Hydrobiol. Suppl.* 12: 622–634.

Geitler, L. (1959). *Östrr. Bot. Zeitschr.* (Wien) 106: 172–173.

Geitler, L. and Ruttner, F. (1936). *Arch. Hydrobiol. Suppl.* 14: 553–715.

Hamana, K., Matsuzuki, S., Niitsu, M., Samejima, K. and Nagashima, H. (1990). *Phytochem.* 29: 377–380.

Hase, T., Wakabayashi, S., Wada, K., Matsubara, H., Jütner, P., Rao, K., Fry, I. and Hall, D.O. (1978) *FEBS Lett.* (Fed. Eur. Biochem. Soc. Amsterdam) 96: 41–44.

Hirose, H. (1958). *Bot. Mag., Tokyo* 71(844): 347–352.

Hoffmann, L. and Mayele, B. (1993). *Phycologia* 32: 307–309.
Ikan, R. and Seckbach, J. (1972). *Phytochem.* 11: 1077–1082.
Jensen, T.E. (1989). *Cytobios* 60: 47–61.
Jensen, T.E. (1991). *Endocytobiosis & Cell Res.* 8: 1–16.
Klein, R.M. (1970). *Ann. N.Y. Ac. Sci.* 175: 623–633.
Klein, R.M. and Cronquist, A. (1967). *Quart. Rev. Biol.* 42: 105–296.
Kremer, B.P. (1982). *Br. Phycol. J.* 17: 51–61.
Kremer, B. P. (1983). *Endocytobiology II.* (eds. Schenk H. and Schwemmler W.), pp 963–970.
Kremer, B.P. and Feige, G.B. (1979). *Z.Naturforsch Sect C Biosci.* 34: 1209–1214.
Kremer, B.P., Feige, G.B. and Schneider, H.A.W. (1978). *Naturwissenschaften* 65: 157.
Kuroiwa, T. (1989). *Bot. Magaz. Tokyo* 102: 291–329.
Kuroiwa, T., Nagashima, H. and Fukuda, I. (1989). *Protoplastma* 149: 120–129.
Lewin, R. A. (1961). *Phycol. News Bull.* 14: 6–7.
Lewin, R. A. (1974). In: *Algal Physiology and Biochemistry,* (ed. Stewart, W.I.P.) Blackwell Sci. Publ. pp 1–39.
Lewin, R.A. and Gibbs, S.P. (1982). In: *Selected papers in Phycology II.* (eds. Rosonski J. R. and Parker, B.C.) publish. by Americ. Soc. Phycol. pp 659–662.
Loppes, R., Denis, C. and Bairiot-Delcoux, M. (1985). *J. Gen. Microbiol.* 131: 1743–1751.
Maid, U., Valentin, K. and Zetsche, K.(1990) *Curr. Genet.* 17: 25–229.
Maid, U., Steinmüller, R. and Zetsche, K. (1992). *Curr. Genet.* 21: 521–525.
Maid, U. and Zetsche, K. (1989). *Curr. Genet.* 16: 203–209.
Maid, U. and Zetsche, K. (1990). Nucleic Acids Res. 18: 3996.
Margulis, L. (1968). *Science* 161: 1020–1022.
Margulis, L. (1981). *Symbiosis in Cell Evolution.* W.H. Freeman and Company, San Francisco. XXII+ (2) + 419 pp.
Meneghini, G. (1839). *Nuovo Giorn. Lett. (Pisa)* 39: 67–68.
Mercer, F.V., Bogorad, L. and Mullens, R. (1962). *J. Cell Biol.* 13: 393–403.
Merola, A., Castaldo, R., De-Luca, P., Gambardella, R., Musacchio, A. and Taddei, R. (1981). *Giorn. Bot. Ital.* 115: 189–195.
Nagashima, H. and Fukuda, I. (1983). *Phytochem.* 22: 1949–1951.
Nagashima, H., Ueki, K., Fukuda, I and Seckbach, J. (1993). In: *Endocytobiology V* (eds. Sato, S. et al.) Tübingen University Press. Tübingen, Germany. pp. 279–285.
Nakamura, H. (1993). In: *Endocytobiology V* (eds. Sato, S. et al.) Tübingen University Press. Tübingen, Germany. pp 337–344.
Nakamura, H. and Hase, A. (1988). *Mem. Konan University Sci. Sec.* 35: 1–15.
Ohta, N., Nagashima, H., Kawano, S. and Kuroiwa, T. (1992). *Plant and Cell Physiology* 33: 657–661.
Raven, P.H. (1970). *Science* 169: 641–646.
Reed, R.H. (1983). *Phycologia* 22: 351–354.
Rosen, W.G. and Siegesmund, K.A. (1961). *J. Biophys. Biochem. Cytol.* 9: 910–914.
Schwabe, G.H. (1936). *Verh. Deutsch. Wissenschaftl. Verein Santiago Chile (Valparaiso), N.F.* 3: 113–146.
Schwabe, G.H. (1942). *Mitt. Deutsch. Ges. Nat.- Voelk. Ostasiens (Tokyo, Leipzig)* 33: C1-C67.
Seckbach, J. (1969). *Plant Physiol.* Suppl. 44: 15.
Seckbach, J. (1971). *Israel J. Bot.* 20: 302–310.
Seckbach, J. (1971a). *Limno. Oceanogr.* 16: 567–572.
Seckbach, J. (1972). *Microbios* 5: 133–142.
Seckbach, J. (1987). *Ann. N.Y. Acad. Sci. (Endocytobiology III).* 503: 424–437.
Seckbach, J. (1990). *J. Phycol.* (*suppl.*) 26 (2): 18.
Seckbach, J. (1991). *J.Phycol.* 27: 794–796.
Seckbach, J. (1992).In: *Algae and Symbioses: Plants, Animals, Fungi, Viruses, Interactions Explored* (ed. Reisser, W.) Biopress Ltd. pp. 399–426.
Seckbach, J., Baker, F.A. and Shugarman, P.M. (1970). *Nature* 227: 744–745.
Seckbach, J. and Fredrick, J.F. (1980). *Microbios* 29: 135–147.
Seckbach, J. and Fredrick, J.F. (1981). In: *Origin of Life* (ed. Wollman, Y. ed.) Reidel Pub. Comp. pp. 567–574.
Seckbach, J. and Fredrick, J.F. (1981A). *Mada* (Sci. Magazine in Hebrew) 25: 122–125.
Seckbach, J., Fredrick. J.F. and Garbary, D. J. (1983). In: *Endocytobiology II.* (eds. Schenk H. and Schwemmler W.) Walter de Gruyter and Comp, Berlin. pp 947–962.
Seckbach, J., González, E., Wainwright, I. M. and Gross, W. (1992). *Nova Hedwigia* 55: 99–109.
Seckbach, J., Gross, H. and Nathan, M.B. (1971). *Israel J. Bot.* 20: 84–90.
Seckbach, J., Hammerman, I.S. and Hanania, J. (1981). *Ann. N.Y. Acad. Sci.* 361: 409–425.
Seckbach, J. and Ikan, R. (1972). *Plant Physiol.* 49: 457–459.
Seckbach, J., Ikan, R., Nagashima, H. and Fukuda, I. (1993). In: *Endocytobiology V* (eds. Sato, S. et al.) Tübingen University Press, Tübingen. pp. 241–254.
Seckbach, J., Ikan, R., Ringelberg, D. and White, D. (1993A). *Phytochem.* 34 (5): 1345–1349.
Seckbach, J. and Kaplan, I.R. (1973). *Chem. Geol.* 12: 161–169.
Seckbach, J. and Libby, W. F. (1971). In: *Planetary Atmospheres* (eds. Sagan, C. et al.) Reidel Publ. Comp. pp 62–83. also in *Space Life Sciences* 2: 121–143 (1970).
Seckbach, J., Nagashima, H. and Fukuda, I. (1990). In: *Endocytobiology IV* (eds. Nardon, P., et al.) INRA, Paris. pp 575–578.
Seliakh, I.O., Romanova, N.I., Baulina, O.I., Mineeva, L.A. and Gusev, M.V. (1981). *Sov. Plant Physiol.* (New York) 28: 880–886. see also *Fiziol. Rast. (Moskva)* 28: 1210–1215.
Sentsova, O. (1991). *Bot. Zhurn. (Moskva, Leningrad)* 76: 69–79.
Silva, P.C. (1962). In: *Physiology and Biochemistry of Algae.* (ed. Lewin, R.A.) Acad. Press. New York, pp 827–837.
Staehelin, L. A. (1967). *Nature*(London) 214: 1158.
Staehelin, L.A. (1968). *J. Ultrast. Res. (N.Y.)* 22: 326–247.
Stempka, R. and Richter, G. (1978). *Arch. Microbiol.* 119: 187–196.
Suzuki, K., Ohta, N. and Kuroiwa, T. (1992). *Protoplasma* 171: 80–84.
Tilden, J. (1898). *Bot. Gaz.* 25: 89–105.
Ueda, K. and Chida, Y. (1987). *Br. Phycol. J.* 22: 61–66.
Ueda, K. and Yokochi, J. (1981). *Bot. Mag. Tokyo* 94: 159–164.
Zenwirt, D., Voloka, M. and Kaplan, A. (1985). *Plant Physiol.* 77: 237–239.

Chapter 7

A review on the taxonomic position of the algal genus *Cyanidium* Geitler 1933[1] and its ecological cohorts *Galdieria* Merola in Merola *et al.* 1981[1] and *Cyanidioschyzon* De Luca, Taddei and Varano 1978[2]

FRANKLYN D. OTT[1] and JOSEPH SECKBACH[2]
[1] *317 Osborne Avenue, Catonsville, MD 21228, USA*
[2] *The Hebrew University of Jerusalem, School for Overseas Students, Mount Scopus, Jerusalem 91905, Israel*

Outlines:
1. Introduction
2. Taxonomy of *Cyanidium*
   2.1 The pioneering studies
   2.2 Tilden's epithet *'Caldaria'*
   2.3 Galdieri and his *Pleurococcus*
   2.4 Naming game proceeds during the turn of century
   2.5 Geitler's preliminary treatment of Tilden's *'Caldaria'*
   2.6 Molisch - A western explorer of Japanese hot springs
3. Geitler confirms Cyanidium in Japan
   3.1 Negoro — The prominent algal investigator in Japan
   3.2 Geitler's contribution
   3.3 Copeland and his *Pluto caldarius*
4. Schwabe and the non-acidic *Cyanidium*
5. Is *Cyanidium* a *Chlorella*?
   5.1 Schwabe and the more alkaline Cyanidiaceae
   5.2 Negoro — First isolation and culturing of axenic *Cyanidium*
   5.3 Schwabe's taxonomic reduction
6. Hirose — Towards the contemporary view of Eukaryotic nature
   6.1 Returning to Solfatara de Pozzuoli
7. Mary Belle Allen's 'acid *Chlorella*'
8. Hirose placing *Cyanidium* into Rhodophycophyta
9. Allen's article of 1959 and the algal temperature range
10. The alkaline Cyanidia — Schwabe's report
   10.1 Alkaline *Cyanidium* from Jerusalem and Italian caves
11. Lewin's attempt to classify *Cyanidium*
12. Ultrastructural investigations
13. De Luca's two strains of *Cyanidium*
14. Ecological physiology of *Cyanidium*
15. Comparative studies of the two strains
16. *Cyanidioschyzon* — A third alga in the Cyanidian habitat
17. Fatty acids analyses of the three members
18. From formae to new genus of *Galdieria*
   18.1 The *Galdieria* genus proliferates — Sentsova's observations
19. New classification of all Cyanidian members
References

1. Introduction

The alga presently known as *Cyanidium caldarium* (Tilden 1898a) Geitler 1933 has received a great deal of attention in the last 25 years both from the more classical phycologists who are concerned with phylogenetic considerations (e.g., Klein and Cronquist 1967) and from the algal physiologists who are attempting to define more precisely the physiological parameters permitting organisms to thrive in moderate to extreme environmental conditions (see Doemel 1970; Brock 1978). These two areas of endeavor are not exclusive of each other, but are mutually complimentary and indivisible.

The phylogenetic considerations essentially turn on the evolutionary niche of *C. caldarium* as a possible link between the prokaryotic Cyanophyceae (=cyanobacteria) and the eukaryotic Rhodophycophyta. The interest here has long revolved about a number of characteristics common to both of these algal divisions and to *C. caldarium*, such as: 1) the complete lack of flagellated cells, 2) the presence of phycobilins as the major accessory photosynthetic pigments and 3) at least for some taxa of both algal divisions and *C. caldarium*, an asexual reproduction involving only the formation of endospores. We may now also emphasize similarities derived from the more modern types of investigation: 1) the non-stacked, photosynthetic thylakoids, without grana, arranged in concentric rings or parallel running lamellae, which are found in the cyanophycean cell, in the rhodophycean chloroplast (rhodoplast) and in the plastids of *C. caldarium*, 2) the phycobilisomes are attached to the

[1] Rhodophycophyta, Porphyridiales, Cyanidiaceae Geitler 1933 emed. D.J. Chapman 1974
[2] Rhodophycophyta, Porphyridiales, Porphyridiaceae Kylin [1937] ex Skuja 1939.

J. Seckbach (ed.), Evolutionary Pathways and Enigmatic Algae: Cyanidium caldarium *(Rhodophyta) and Related Cells,* 113–132, 1994.

outer surface of the thylakoids, 3) the branching molecular structure of carbohydrate storage products, ∝ 1–6 glucosyl linkages, 4) the storage of the glucans outside of the photosynthetic organelles, i.e., in the cytoplasm, 5) at least three types of glucotransferases, namely phosphorylases, synthetases and branching isoenzymes types, and 6) current molecular sequencing investigation of a few genes and their products.

The environmental conditions which have attracted much attention to *C. caldarium* are two, the relatively high temperature and low pH under which it thrives. Its tolerance to extremely low pH values, pH 1 in the natural habitat (see Brock 1978) and pH 0.24 under strictly controlled laboratory conditions, has permitted it to colonize what supposedly otherwise would be a rather depauperate ecological niche. Because *C. caldarium* shows little or no growth at >pH 5 values (Doemel 1970) it must be recognized as an obligate acidophile and, as such, one should not expect it to be found in neutral and alkaline waters. Hand in hand with this obligate acidophile goes also its thermophily. Its acclaimed thermophily, initially reported to be 75–80 °C (Copeland 1936; Allen 1959) has now been shown (Doemel 1970; Brock 1978) to be only in the range of 56–57 °C with an optimal temperature of only 45 °C. These ranges do not permit *C. caldarium* to be ranked among the extreme thermal autotrophs such as may be represented by *Synechococcus lividis* Copeland (Cyanophycophyta) which can tolerate (but only in neutral or alkaline waters) temperature up to 74 °C (see Brock 1978). With an optimal growth temperature of 45 °C, *C. caldarium* has to be considered only moderately thermophilic.

Seckbach and his colleagues (Seckbach, 1969, 1971; Seckbach and Libby 1969, 1970; Seckbach *et al.* 1970, 1971) have demonstrated that *C. caldarium* may be grown in pure carbon dioxide (CO_2) and under elevated atmospheric pressures. Because of these attributes and those enumerated above, the Seckbach school has suggested that *C. caldarium* might well represent a very primitive eukaryote having possibly evolved when the young earth was very acidic, the atmosphere abundantly rich with CO_2 (ca. 90%) and possessing an atmospheric pressure in excess of 14 atm (Seckbach and Libby 1971). Also based upon these studies, it has been suggested (Seckbach 1971; Seckbach and Libby 1969, 1970, 1971) that similar life forms might possibly exist on other planets, such as Venus. This hot Cytherian planet is known to possess both high atmospheric pressures and high CO_2 levels. Seckbach and Libby (1971) also emphasized the danger of contaminating or colonizing other planets with terrestrial microbes.

Against this rich and intellectually stimulating background, the following review on the taxonomic (and systematic) history of the algal genus *Cyanidium* Geitler 1933 is presented to help place in proper perspective the various attributes of these disciplines as they may apply to *Cyanidium*. In so doing, the painfully slow evolution of our understanding of *Cyanidium* will become obvious.

2. Taxonomy of *Cyanidium*

2.1 The pioneering studies

The earliest binomial which was assigned to material of *Cyanidium caldarium* (which can be determined with reasonable certainty) is *Coccochloris orsiniana* Meneghini 1839. The algal material was collected by Antonio Orsini de Ascoli from the sulfuric acid rich thermal waters of the spa of Acquasanta (Ascoli Piceno, Italia). According to De Luca and Taddei (1976, p. 199) Meneghini described, in his 1839 paper, the cells of *C. orsiniana* as being spherical and falling within the size range of 0.2–2.0 μm. Shortly after this, Meneghini [1842 (actually published in 1841)] emended and expanded this 1839 description. In this emended description Meneghini (1842, p. 63) gave new cell dimensions ranging from 0.1 to 1.0 μm with some cells reaching 10.0 μm in diameter (De Luca and Taddei 1976, p. 199). A few years later, Kützing [1845–1859 (the pertinent section actually published in 1846) p. 12 Tab. 16, fig. III] places Meneghini's *C. orsiniana* in the genus *Palmella* establishing the new combination *P. orsiniana* (Meneghini 1839 emend. 1842) Kützing 1845. The genus *Palmella*, is today known to be a member of the Chlorophycophyta.

A *nomen nudum, Oncobyrsa orsiniana*, attributed to Bergen (1897), can be found in the paper by Profili and Maini (1950 p. 320). However, as far as we can determine at this writing, the only paper published by Bergen in 1897 reported on an alga which he had collected in the Solfatara at Pozzuoli, Italy, and which had been identified by G.W. Farlow as Meneghini's *Coccochloris orsiniana*; no mention of *O. orsiniana* can be found in this 1897 paper by Bergen.

Chronologically, the next binomial that was applied to material of *C. caldarium* was *Chroococcus varius* A. Braun in Rabenhorst 1876, a species established by A. Braun (Rabenhorst 1876) for a cyanophycean taxon which was initially found 'An der Wand eines Warmhouses zu Poppelsdorf bei Bonn, gesammelt in Sept. 1875 von C. Bouchè'. It was to this taxon that Josephine E. Tilden (Tilden 1896, Cent. 2, No. 198) erroneously identified material of *C. caldarium* that she had collected on June 27, 1896, 'On rocks. Near vent of geyser. Sometimes heated. Norris Geyser Basin, Yellowstone National Park'. *Chroococcus* is a cyanophycean genus presently characterized by possessing free-living spherical or hemispherical cells with distinct, colorless, gelatinous sheaths forming amorphous colonies usually composed of less than 50 cells; the formation of endospores was not reported but the formation of nanocysts (endospores) has been reported. *Protococcus vulcanicus* Cesati in Balsamo 1892 is a *nomen nudum* which V. Cesati (Balsamo 1892) had applied to material which undoubtedly was composed of *C. caldarium*. Cesati had collected the material in 'Marzo 1869' in the 'Solfatara di Pozzouli'. Balsamo gives only '*Pr. vulcanicus* Ces. *inéd.* (sine descriptione necicone)'. The genus *Protococcus* is eukaryotic, belonging to the algal division Chlorophycophyta, and its cells are characterized by possessing a single parietal, laminate chloroplast that is more or less lobed at its margins and, depending on the species, this chloroplast may or may not possess pyrenoids. Reproduction is exclusively by means of cell division forming packets of two, three, four, or more cells and never reproducing via zoospores.

As herein alluded to previously, Bergen (1897) reported the presence of *Coccochloris orsiniana* Meneghini as the dominant alga among those which he had collected in the Solfatara di Pozzuoli; W.G. Farlow, Harvard University, Cambridge, Massachusetts, identified this dominant alga as *C. orsiniana*. The algae had been collected in the main stream outlet of the Solfatara, locally referred to as the 'Bacca'. According to Bergen, the aqueous environment supporting the algal populations ranges from 20 °C – 55 °C. Bergen further reported that the stream issuing from the rock crevices, at *c.* 100 °C, was highly charged with sulfur dioxide and that delicate needles of sublimed sulfur were to be found about all the crevices or *fumaroles*; the mixed algal populations grew up to the very orifices of these fissures. In this early paper, it is of interest to note that the reported upper temperature range of 55 °C for the aqueous habitat supporting the growth of *C. orsiniana* is very close to that recently reported for *C. caldarium*. Allen (1959), Ascione *et al.* (1966) and Doemel and Brock (1970) have demonstrated, using more modern techniques, that the upper temperature tolerance for *C. caldarium* as determined both on natural populations and on cultured materials is in the range of 56–57 °C. While Bergen did not report the acidity of the Pozzuoli habitat, it must be acknowledged that the stream, heavily charged with sulfur dioxide, certainly could have produced an exceptionally low pH on condensing in or on the algal habitat; for example: 0.1 N sulfurous acid solution has a pH value of *c.* 1.5 [cf. Handb. Chem. Phys. 40th ed. (1958) p. 1721],

2.2 Tilden's epithet 'Caldaria'

It was in 1898 that the specific epithet of *Cyanidium caldarium* was estublished. In that year Tilden (1898b, p. 94) published *Protococcus botryoides* Kützing 1845 emend. Kirchner 1878 forma *caldarium* Tilden 1898 to accommodate a new algal form which she had collected from a thermal spring. Tilden felt that this new alga was identical to *P. botryoides* except for the habitat. She gave the following brief description of this new form: 'Aquatic, vegetative or yellowish green: cells globose, 4–6 μ in diameter, for the most part solitary; membrane thin, homogeneous, hyaline; cell contents homogeneous, vegetative green'. In the same paper, Tilden (1898b, p. 104) reported the occurrence also in Yellowstone National Park of an alga which she had mistakenly identified as *Chroococcus varius* A. Braun in Rabenhorst 1876. Tilden gave four collecting sites for this alga of which site (b) was almost identical to the site she gave for *P. botryoides* f. *caldarius*; Tilden described site (b) as follows: 'With *Microspora amoena* var. *thermalis*, lying in overflow from spring. Temperature 41 °C Frying Pan Basin. Yellowstone National Park. J.E.T. 7 Jl 1896'. Tilden gave the following very brief description of the material which she had identified as *C. varius*: 'Cells spherical, single or in small groups, 1.6–5.6 μ in diameter; cytoplasm pale aeruginous.'

Tilden's material of *C. varius* was subsequently placed into synonymy along with her *P. botryoides* f. *caldarius* by Setchell when he made the new combination of *Pleurocapsa caldaria* (Tilden 1898a) Setchell in Collins, Holden and Setchell 1901. In the same year, she (Tilden 1898a, Cent. III, No. 283) issued exsiccate specimens of *P. botryoides* f. *caldarius*. Tilden (1902, Cent. VI, No. 600) issued an additional exsiccate specimen of *C. varius* which had been collected by 'Mr. David Griffiths 17 S 1898' at Lo Lo Hot Springs, Lo Lo, Montana. For this exsiccate specimen she gives the following brief description: 'Cells somewhat spherical, isolated or in twos or fours, 2.5–5 μm in diameter; protoplasm pale aeruginous. In hot water'. Tilden here made no mention of Setchell's *Pleurocapsa caldaria*. However, in a later work (Tilden 1910, p. 47) accepted Setchell's new combination of *P. caldaria* as well as the reduction therewith to synonymy of both her *P. botryoides* f. *caldarius* and *C. varius*. In passing, it is interesting to note, Tilden's use of 'cytoplasm pale aeruginous' and 'protoplasm pale aeruginous' respectively in 1898 and 1902 for her *C. varius* represents, as far as we can determine, the first references to the blue-green color of *C. caldarium*. Because *Protococcus* and *Chroococcus varius* are respectively members of the Chlorophycophyta and the Cyanophycophyta and because both these taxa have been previously treated above, we will dwell no further on these taxa at the present time.

2.3 Galdieri and his Pleurococcus

The year after Tilden's publication on *P. botyoides* f.*caldarius* and *C. varius*, Galdieri published his description of *Pleurococcus sulphurarius* Galdieri 1899. The material serving as the basis of this description was again collected in the Solfatara de Pozzuoli (in the vicinity of Naples). From his description it is obvious that Galdieri had before him an eukaryotic alga possessing a parietal, laminated, cup-shaped, blue-green colored chloroplast. The cells lacked a gelatinous sheath and reproduced only by the formation of 2,4,8 or more endospores which were subsequently released by the splitting of the sporangial wall. The cells of *P. sulphurarius* had a size range of 3–12 μm. In the future, De Luca and Taddei (1976 p. 205) will consider that Galdieri's *P. sulphurarius* as identical to their *Cyanidium caldarium* forma *B*; this particular aspect will be discussed later in more detail in this paper. Galdieri's assignment of his material to the genus *Pleurococcus* (Galdieri 1899, p. 162) as circumscribed in Engler and Prantl by Wille (1897 p. 56) is somewhat puzzling since as described by Wille this genus did not reproduce via the formation of endospores. Wille (1897 p. 56) describes *Pleurococcus* as possessing cells which are spherical, or polyhedral by mutual compression, lacking a gelatinous sheath, and possessing several chlorophyll containing structures which appear to fuse into a single, cup-shaped, parietal chloroplast and, depending on the species, may or may not possess a pyrenoid; cell division is in three planes with up to 32 or more cells being produced which may remain united together after cell division for a period of time.

2.4 Naming game proceeds during the turn of the century

The next binomial to be applied to material of *C. caldarium* is *Pleurocapsa caldaria* Setchell in Collins *et al.* 1901. The following is given concerning the habitat of the collection (Collins *et al.* 1901, Fasc. XVIII, No. 851): "Forming an emerald green, rather thick coating on steaming rocks above the 'Devil's Kitchen', Geysers, Sonoma County, California, June 4, 1900". On establishing this new combination, Setchell stated the following basis for so doing: 'Clearly a member of the Cyanophyceae and forming a limited number (4) of schizospores which seem to relate it most closely to the genus *Pleurocapsa*'. As has previously been indicated above in the present paper, Setchell cited as synonymous with his *P. caldaria* the materials Tilden (1898b, p. 94, p. 104) identified and described as *Protococcus botryoides* forma *caldarius* and *Chroococcus varius*. Later Tilden (1910 p. 47) will accept Setchell's *P. caldaria* and the reduction therewith of these two taxa. *Pleurocapsa* as a member of the prokaryotic Cyanophycophyta.

Palmellococcus thermalis G. S.West 1904 was mentioned by Hirose (1950 p.110) as a possible synonym of *C. caldarium*. The material on which *P.*

thermalis is based had been collected in Dominica by Mr. A. Howard from a hot spring having a temperature of *c.* 32 °C. Because West (1904 pp 287–288) did not indicate an acid character for the spring, De Luca and Taddei (1976 p. 205) suggested that *P. thermalis* was incorrectly placed into synonymy with *C. caldarium* but, on stronger grounds, they also argued that the possession of more than one chloroplast was sufficient to exclude West's *P. thermalis* as being a bonafide synonym of *C. caldarium*. In any event, *Palmellococcus* is a member of the Chlorophycophyta and was characterized as an alga whose cells possessed a single laminated chloroplast without pyrenoids, which divided by the formation of 8, 16, 32, or 64 endospores that are subsequently released from the sporangium by the splitting of the sporangial wall; the cells formed a thin mucous coat on the substrate. Because *P. thermalis* possessed 2–3 chloroplasts per cell, West proposed that the generic description should be altered to accommodate this species having more than one chloroplast per cell. In a later work, West (1916, p. 194) will make the following statement: "Many of the recently described 'genera' of unicells found in cultures are referable to *Chlorella*"; Among these unicells found in culture, West listed *Palmellococcus*.

2.5 Geitler's preliminary treatment of Tilden's 'Caldaria'

In a short synoptic review of the systematic and morphological characteristics of the Cyanophyceae, Geitler (1925 p. 241) made the new combination, *Chroococcopsis caldaria* (Tilden 1898a) Geitler 1925. Geitler gave no reason for including *Pleurocapsa caldaria* (Tilden) Setchell in the genus *Chroococcopsis*, giving only, at this time, the generic description. *Chroococcopsis* is a member of the Cyanophycophyta. Geitler did acknowledge that *Chroococcopsis* may be related to *Pleurocapsa*: 'Ob es sich nich doch nur um ein extremes Stadium von *Pleurocapsa*-Arten handelt, müssen weitere Untersuchungen zeigen'. In passing, it is of interest to note that on establishing the genus *Cyanidium* in 1933, Geitler made no mention of his *Chroococcopsis caldaria* of 1925.

2.6 Molisch – a western explorer of Japanese hot springs

In 1924, Molisch visited numerous Japanese thermal springs in addition to other areas of botanical interest. Subsequently, Molisch (1926) published the results of his Japanese journey including physical data and the composition of the algal flora of these thermal springs. Among the numerous thermal springs he discussed was the 'Thermen von Noboribetsu-Onsen'. Concerning this spring, Molisch (1926, p. 90) made his observation on the strong acidic reaction and the temperature (40 °C) where one *Chroococcus* sp. (Solfataric alga) dwells. Geitler and Ruttner (1936 pp 705–706) made the following observation and prediction pertaining to Molisch's identification of *Chroococcus* sp. from Japanese thermal, acid springs: 'Bei Temperaturen zwischen 37° und 40.5° fand er in mehreren Fällen eine als *Chroococcus* sp. bezeichnete Cyanophyceen. Es scheint uns durchaus möglich zu sein, dass auch dort unser *Cyanidium* vorlag'.

3. Geitler confirms *Cyanidium* in Japan

3.1 Negoro – the prominent algal investigator in Japan

In 1944, Negoro confirmed the presence of *C. caldarium* in the Noboribetsu-Onsen and the 1936 prediction of Geitler and Ruttner (1936 pp 705–706), based purely on ecological considerations, that Molisch's *Chroococcus* sp. was probably material of *C. caldarium*, was substantiated as valid. Negoro's additional observations (Negoro, 1944 pp 302–303; see Fukuda's chapter in this volume) concerning the presence of *C. caldarium* in the Noboribetsu springs are resolved from his German above paper: 1. Molisch investigated the Japanese algae and published his observation in his book *Pflanzenbiologie in Japan* and indicated in this volume that a species of *Chroococcus* has been found in the sulfuric acid rich water at a temperature range of 37–40.5 °C. 2. Geitler and Ruttner proposed that this alga indicates the cyanophyte *C. caldarium*. 3. It was Yonosuke Okada who sent such Japanese algal material to Geitler in Austria and has demonstrated for the first time the appearance of *C. caldarium* in Japan.

3.2 Geitler's contribution

Up to this point in time, materials of *C. caldarium* had been placed in a number of already described genera falling either to the Cyanophycophyta or Chlorophycophyta. *Cyanidium*, the preset, universally accepted genus designation, having the type species, *C. caldarium*, was established in 1933 by L. Geitler to accommodate an unicellular alga which was present among the liquid-preserved (formalin) cyanophycean materials collected during the 1928–1929 Deutsche Sunda-Expedition to Java, Bali, and Sumatra. The species epithet, *caldarium*, was based upon *Protococcus botryoides* (Kützing 1846) Kirchner 1878 forma *caldarium* Tilden 1898. Geitler's original description of *Cyanidium caldarium* (Tilden) Geitler was published in the Latin language (Geitler 1933 p. 624).

While initially appointing *C. caldarium* to the prokaryotic Cyanophycophyta (cyanobacteria), Geitler did so somewhat begrudgingly, clearly acknowledging his suspicions that indeed *C. caldarium* might well possess a cup-shaped chloroplast and a nucleus. In that event, the possession of these two organelles, along with the phycocyanin, would rank *C. caldarium* among the 'Bangiaceen' the Bangiophycidae of present day phycologists, and which, in turn, is ranked as a subclass or primitive or reduced forms of the Rhodophycophyta. It was not possible for Geitler in 1933 to resolve these suspicions with the preserved material available to him. Geitler's doubts as to the cyanophycean affinities of *C. caldarium* have been published in Geitler and Ruttner, 1935 (p 389).

Supposedly, *C. caldarium* differed from all other theretofore described cyanophytes in that it multiplied only by the formation of four, and only four, endospores which were subsequently released from the unicellular sporangium (mother cell) by either gelatinization or splitting of the sporangial wall. To systematically accommodate this new genus, as well as the simultaneously described *Chroococidiopsis* Geitler 1933. Geitler circumscribed the family Cyanidiaceae, essentially for unicellular, globose, cyanophytes showing no *cellular polarity* and possessing no vegetative cell division, i.e., reproducing only by endospores. In turn, the family, Cyanidiaceae, was placed in the Chamaesiphonales, a well established order for those cyanophycean taxa which, up to this point in time, had been described as unicellular, attached, possessing *some degree of cellular polarity*, and having no vegetative cell division, *i.e.*, reproducing only via endospores or exospores.

3.3 Copeland and his Pluto caldarius

Copeland (1936 p. 72) apparently unaware of Geitler's works (Geitler 1933; Geitler and Ruttner 1935) on *C. caldarium* established the new genus, *Pluto*, with *P. caldarius* (Tilden 1898a) Copeland 1936 as the type species. Copeland retained *P. caldarium* in the Cyanophycophyta as a member of the Dermocarpaceae sensu Geitler 1925. In his description, of *Pluto, gen. nov.* Copeland mentioned the following points: 1. the plants are unicellular, yellowish to blue-green or arsenic green, spherical with thin gelatinous tegument, vegetation division is absent, multiplication through 4 endospores per each cell, cellular size is 4–6 μ in diameter. 2. The genus *Pluto* is based on the single species *Pleurocapsa caldaria* (Tilden) Setchell, but they are totally different in morphological and by mode of cell division and number of endospores. 3. Because this species is found in acid , hot waters in Yellowstone and Roaring Mountain or in mud volcanos, it was dedicated to the God of the Underworld, Pluto. 4. The pH of the best growth is below 5.0 ranging from 2.6 to 5.4 and the temperature range 34 to 80.2 °C. (Copeland stated that '*Pluto caldarius* can endure water with a temperature as high as 90 °C without injury'.

Future workers (Allen 1959; Doemel 1970) will confirm Copeland's observations that *P. caldarius* (i.e., *C. caldarium*) thrives in waters whose acidity is below pH 5. In fact *C. caldarium* will be found to be an obligate acidophile (Doemel 1970). However, Copeland's report that *C. caldarium* grew at temperatures of 82.2 °C appears to be erroneous as stated above in the light of modern investigations utilizing both natural populations as they occurred in the thermal springs and laboratory cultures under strictly controlled conditions (Doemel 1970).

4. Schwabe and the non-acidic *Cyanidium*

In the same year in which Copeland described *Pluto*, Schwabe described a second species of *Cyanidium*, namely *Cyanidium chilense* Schwabe

1936. This species of *Cyanidium* is, or perhaps more appropriate epipelagic (epilithic), having been collected from damp, rock walls of caves located along the Pacific coast of central Chile. Schwabe's highlights (1936 pp 119–121) are as follows; 1) *Cyanidium chilense* n. sp. occurs in caves of the coasts of Chile. 2) The cells are spherical and blue-green with a diameter of 3–6 μ. 3) The 4 endospores are born in cells of 4.5–6 μ. 4) Their membrane is different than that of *C. caldarium* Geitler. Schwabe discussed the ecological conditions of this species (light conditions inside the caves, humidity and other atmospheric conditions). For further information on cave *C. caldarium* consult Hoffmann's chapter in this volume.

5. Is *Cyanidium* a *Chlorella*?

Inman (1940 p. 664) who was investigating chlorophylls of thermal algae reported the following for a unicellular alga which was unquestionably material of *C. caldarium*; the identification to *Chlorella* was made by F. Drouet (1943) who reported the absence of blue-green algae from these collection sites. This lack of cyanobacteria was attributed both to the warm temperatures, up to 55–57 °C and the acidness, pH 5 and below. Both of these factors work in concert to exclude the typical Cyanophycean genera of more alkaline thermal waters (Doemel and Brock 1970, 1971). The identification of the preserved, collected material as *Chlorella* sp. is obviously an error in identification, an error to be repeated quite often in subsequent years.

Notwithstanding this identification of material of *C. caldarium* by Drouet for Inman (1940 p 665, footnote no. 2) to *Chlorella* sp., Drouet (1940 p. 665) proceeded to make a new combination, *Dermocarpa caldaria* (Tilden 1898a) Drouet 1943. While Drouet (1943 p. 672) cited the place and material used in Inman's 1940 paper, i.e. ['the Geysers, Sonoma County, O.L. Inman, July 1939 (as *Chlorella* sp. reported by Inman. *Journ. Gen. Physiol. 23: 664 1940)']* Drouet gave no comments as to why this material which he had earlier identified as the eukaryotic alga, *Chlorella* sp. was now placed into synonymy with his prokaryotic *D. caldaria*. In this same article, Drouet (1943 p. 672) also acknowledges he had seen Copeland's 1936 paper, but again he gave no reasons for rejecting *Pluto* whose cells showed no polarity over *Dermocarpa* whose cells showed distinct polarity. Drouet, in this 1943 paper, made no mention of Geitler's works on *C. caldarium*. We will further note later on, in the present paper, that Drouet continues to identify, on various occasions, material of *C. caldarium* as a green alga, as a species of *Chlorella*, or as a growth form of *Entophysalis rivularis* (Kützing) Drouet 1943. The genus *Dermocarpa* is a member of the Cyanophycophyta and is characterized as an alga forming crustose thalli in the form of parenchymatous discs without distinct filaments and with the cells showing distinct polarity. In the time honored, classical sense (cf.: Geitler, 1932), *Dermocarpa* is a member of the Dermocarpaceae, a family belonging to the order Chamaesiphonales.

Because Copeland failed to provide a Latin language diagnosis for *Pluto*, Drouet (1943 p. 672, footnote no. 2) did so on placing *Pluto* into synonymy with his *Dermocarpa caldaria* (Tilden 1898a) Drouet 1943; the exact reasons why Drouet provided a Latin language diagnosis for *Pluto* at this time (1943 p. 672, footnote no.2), when he was placing it into synonymy with *D. caldaria*, remain elusive to these writers. In later works Drouet (Drouet and Daily, 1956, 1957 pp. 161 and 222 resp.) will cite '*Pluto caldarius* Copeland ex Drouet' as a synonym for *Chlorella caldaria* (Tild.) M.B. Allen!

5.1 Schwabe and the more alkaline Cyanidiaceae

In 1942, Schwabe reduced his *Cyanidium chilense* to a variety of *C. caldarium* (Schwabe 1936; Schwabe 1942, pp C41-C42). In this paper Schwabe also described a new form of *C. caldarium* var. *chilensis*, namely *Cyanidium caldarium* (Tilden 1898a) Geitler 1933 var. *chilensis* Schwabe 1936) Schwabe 1942 forma *rumpens* Schwabe 1942. He presented a complete description of this additional organism (Schwabe 1942, pp C42-C43) and described the critical points used to separate out these taxa from the typical acidothermophilic *C. caldarium*. In so doing, Schwabe also gave the information concerning the habitat of the aerophilic and epipelagic *C. caldarium* var. *chilensis* and *C. caldarium* var. *chilensis* f. *rumpens*.

5.2 Negoro — first isolation and culturing of axenic Cyanidium

At this point in time, the reader's attention should be drawn to Negoro's paper (Negoro 1943 pp 308–310) in which he described *Chroococcidiopsis thermalis* Geitler 1933 var. *nipponica* Negoro 1943. As the reader may recall in the above paragraph of the present paper, the genus *Chroococcidiopsis* was simultaneously published in the same paper (Geitler 1933 p. 625) as was *Cyanidium*. The monotypic genera, *Chroococcidiopsis* and *Cyanidium*, were both placed in Geitler's newly created Cyanidiaceae (Geitler, 1933 p. 624). The generic distinction between *Chroococcidiopsis* and *Cyanidium* resided in the number of endospores per sporangium (mother cell) respectively produced by each genus, eight to thirty-two endospores in *Chroococcidiopsis* and only four endospores in *Cyanidium*. The type species of *Chroococcidiopsis, C. thermalis* (var. *thermalis*) was collected 'aus den Frauenteich (Temp. 37 °C, pH 6.8) der warmen Quellen von Kadjaj, Mittel-Sumatra' (Negoro, 1944, p. 307). Negoro's *C. thermalis* var. *nipponica* was separated from Geitler's 1933 *C. thermalis* because it produced smaller sporangia (mother cells). The type material of *C. thermalis* var. *nipponica* was collected in the 'Thermen von Kusatu, Gumma Prafektur, Japan' intermixed with *C. caldarium* at temperatures ranging from 35–50 °C and at a pH of 1.6. Negoro (1943 p. 311) also reported the occurrence of *C. thermalis* var.*nipponica* in the 'Kawayu-Thermen, Hokkido' in waters with a temperature of 32.2 °C and a pH of 1.7. As will be noted later on in the present paper, Nagashima and Fukuda (1981a, b) will report, based on electron microscopic studies, that *C. thermalis* var. *nipponica* is an eukaryotic alga which they feel is identical to *Cyanidium caldarium* forma *B* of De Luca and Taddei (1970).

Negoro in this same paper (Negoro 1943 pp 307–310), also reported on his detailed cytological studies on living cells of *C. caldarium* utilizing both naturally collected material and living cultures. Hirose presents later (1950 pp 107–208) an English language summary of Negoro's observations.

The contemporary interest in the so-called western world in *C. caldarium* stems from the reports of M.B. Allen; 1) on an *Acid Alga* initially given in her paper, *The cultivation of the Myxophyceae* (Allen 1952); 2) in her subsequent presentation before the Huitième Congrès International de Botanique, Paris (Allen 1954 a, b) 3) and in her more detailed treatment (Allen 1959) on what she referred as 'an anomalously pigmented chlorophyte'. Allen, generally, has been given credit for first isolation into axenic culture of *C. caldarium* and with the subsequent utilization of these cultures in various physiological investigations. However, such credit rightfully should be given to the Japanese worker, K. Negoro, who as early as 1943 reported in the Japanese language (cf. Hirose, 1950 p.107) that he had utilized cultures of *C. caldarium* (Negoro 1943). However, the next year Negoro (1944, pp 353–354), writing in the German language, gave the following detailed discussion of his techniques yielding axenic (Bakterien Frei) cultures of *C. caldarium*.

The sterile mineral liquid medium used by Negoro (1944) had the following content:

$Ca(NO_3)_2$	0.50 g
KH_2PO_4	0.20 g
$MgSO_4$	0.20 g
$Al_2(SO_4)_3$	0.50 g
$FeCl_2$	0.01 g
H_2O	1000.00 ml

He also recommended a solid medium made of 2% agar dissolved in the original nutrient solution. Maintaining the experimental cultures at 35 °C Negoro carried out a series of experiments to determine the pH which would allow optimal growth of *C. caldarium* and determined that there is no growth over pH 5.4 while the optimum is in the range between 1.5 and 3.1.

These reports by Negoro and Hirose also represented the first modern indication that *C. caldarium* was not a prokaryotic cyanophyte, but indeed possessed a typical organelle, i.e., a chloroplast. However, for an indication of such in an earlier period, the reader should recall that Galdieri (1899 pp 161–162) published descriptions and drawings which clearly demonstrate the presence of a chloroplast in *Pleurococcus sulphurarium*, a binomial which De Luca and Taddei (1976 p.205) also considered to be synonymous with their *C. caldarium* forma *B*. Negoro's drawings (Negoro 1943, p. 308, fig. 8) showing these colored and colorless portions in *C. caldarium* are very reminiscent of the drawings by Geitler (Geitler and Ruttner 1935, p. 391, fig. 16). Shortly after Negoro's publication, Hirose (1947) published the first formal paper in which Negoro's 'colored' portions were definitely called a plastid (i.e., a chloroplast). Hirose's drawings (Hirose 1947, p. 45) clearly show the perforated, laminated, parietal chloroplast; these drawings by Hirose, made from visual observation using the light microscope, are excellent.

5.3 Schwabe's taxonomic reduction

1944 witnessed another publication by Schwabe in which he reduced his *C. caldarium* var. *chilensis* to *C. caldarium* forma *chilensis*. We were not able to secure a complete copy of this work which had been issued in Shanghai in 1944 during the height of World War II. This journal is extremely rare, but we were able to secure a few pages pertaining to *Cyanidium* through the kindness of Dr. W. Doemel. In the pages available to us, Schwabe gives no reason for reducing his var. *chilensis* to forma *chilensis*. A complete copy of this rare paper was finally secured in September 1992 through the kindness of Prof. Wolfgang Becker, Tübingen.

6. Hirose — towards the contemporary view of eukaryotic nature

In 1950, Hirose published a more extensive treatment of *C. caldarium*. In this paper he essentially repeated, in English, his 1947 Japanese language publication on the chloroplast of *C. caldarium*. In addition, he also gave the results of his investigations on the fat soluble pigments of this alga. Hirose (1950 p.110) reported that *C. caldarium* contained '... chlorophyll *a*, carotin and xanthophyll, but no chlorophyll *b*...,' however, he made no mention of the water-soluble phycocyanins. Based upon these observations Hirose (1950, p. 109) gave the following diagnosis of *Cyanidium* which was the first to emphasize the eukaryotic nature of this alga: 'Genus *Cyanidium* Geitler, Family Oocystaceae, Order Chlorococcales, Class Chlorophyceae. Plants unicellular, spherical or roundish triangular, always containing single chromatophore within; chromatophore parietal along the inner wall of cell, containing chlorophyll *a*, carotin and xanthophyll, but no chlorophyll *b*; endospores 4 in a tripartite group'.

He pointed out that the plant mass of *Cyanidium caldarium* (Tilden) Geitler was pulverulent on sand, stone rock or in and out or submerged wooden fibers, hygrophytic, thermobiotic; cells are up to 5 – 6μ in diam., the single chromatophore is laminated or discoid which is often perforated. In this same paper, Hirose (1950 p. 110) proceeded to suggest that *C. caldarium* might well be a species of *Chlorella* and he even went so far as to make the combination: *Chlorella caldaria* (comb. nov.), i.e., *Chlorella caldaria* (Tilden 1898a) Hirose 1950, but in so doing, Hirose acknowledged his reluctance to establish this new combination because he could find no chlorophyll *b* in *C. caldarium*. In that paper Hirose conveys his feelings on the possible taxonomic position of *C. caldarium* (Hirose 1950, p. 110).

6.1 Returning to Solfatara de Pozzuoli

In the same year which saw the publication of Hirose's significant contribution to our understanding of the eukaryotic nature of *C. caldarium*, the scene again returns to the Solfatara de Pozzuoli (Naples). In a somewhat confusing contribution Profili and Maini (1950) reported on their investigations on the algae growing in 'la Solfatara de Pozzuoli'. These authors concluded that the principle alga growing there was *Entophysalis rivularis* (Kützing) Drouet forma *puteolana* Profili and Maini 1950, and gave a large number of synonyms which they broke down into two groups 'Eusinonimi' and 'Parasinonimi'.

For those who may wish to persue further these opinions, the reader is referred to the 1950 paper by Profili and Maini.

7. Mary Belle Allen's 'acid *Chlorella*'

In 1952, M. B. Allen published a significant paper dealing with her investigations on the cultivation of Myxophyceae. Among those with which she worked was an unidentified unicellular blue-green alga, an alga which in the future would be identified as *C. caldarium*. This isolate was sufficiently unique, from her other cyanophycean isolates, to warrant the following comments (Allen, 1952, pp 41–42). "...One culture in the list [Unidentified unicellular alga. 'Lemonade Spring', The Geysers Sonoma County, Calif. grows in acid solution] requires special mention. Its taxonomic position is uncertain, since it possesses the mode of reproduction (nannocyte formation).....It will tolerate and grow well in extremely acid solutions (up to 1N sulfuric acid); it grows on organic media in the dark at a rate comparable to its rate of growth in the light; and when grown in the dark it loses its chlorophyll and phycocyanin and becomes a light straw yellow in color. It will be referred to as the 'acid alga' in this paper". In the summary to this paper, Allen adds that the 'acid alga' is very similar in its morphology and developmental history to *Chlorella variegata* Beijerinck. However she noticed that the 'acid alga' has pigments characteristic of a blue-green alga (phycocyanin and chlorophyll *a*).

During the Huitième Congrès International de Botanique, Paris, Allen (1954b, pp 41–42) presented additional data on her acid alga such as 1) the absence of α, ϵ-diaminopimelic acid, 2) the presence of only chlorophyll *a* which substantiated Hirose's 1950 report, and 3) the ability to grow oxidatively in the dark on media containing sugars. These characteristics along with those previously published such as 1) tolerance to low pH; this acid alga would grow in nutrient media which contained 1N H_2SO_4 and also would show some growth at pH 7, and 2) reproduction only by endospore formation led Allen to make the combination *Chlorella caldaria* (Tilden 1898a) Allen 1954. Evidently, at this time, Allen was unaware of Hirose's paper in which he had already made this new combination (Hirose 1950 p. 110). However, in response to Allen's report that the acid alga was to be considered a species of *Chlorella*, a chlorophycean genus, Bourrelly (Allen 1954a, p. 39) suggested that this alga may be a member of the Xanthophyceae.

It was also in 1954 that Bourrelly communicated to Allen (via personal ommunication, cf. Allen 1959, p. 275) his opinion that *C. caldarium* might be a coccoid cryptomonad. Fogg (1956, p. 157) held a similar opinion in suggesting that this alga might be classed as a coccoid representative of the Cryptophyceae.

Drouet and Daily in 1956 published their *Revision of the Coccoid Myxophyceae* and in their section therein entitled *Nomina Excludenda*; these authors excluded *C. caldarium* from the Cyanophycophyta. But in so doing, they concluded (Druoet and Daily, 1956 p. 161) that *C. caldarium* was a species of *Chlorella*. The next year, Drouet and Daily (1957) published a short paper in which they gave 'additions and Corrections' to the above cited 1956 paper. They (Drouet and Daily 1957, p. 222) made the following change in the above quoted passage: "P. 161, L. 1–4 from bottom: delete all after 'MIN)' and insert, = *Chlorella caldaria (Tild.) M. B. Allen.*"

8. Hirose placing *Cyanidium* into Rhodophycophyta

In 1958, Hirose published further studies with *C. caldarium* (Hirose 1958), among his new data we should mention that: 1) he demonstrated by Feulgen techniques that cells of *Cyanidium* possessed a nucleus and 2) he determined that the storage assimilatory polysaccharide was close to floridean starch and glycogen. Hirose (1958 pp 349–350) summarizes in detail the traits of *C. caldarium* and concluded that it must be placed in the Rhodophyceae as a new genus of the Porphyridiaceae. In this synopsis he made the following points: 1) the plant body is unicellular and spherical; 2) this cell reproduced merely by means of endospore-formation; 3) it has a definitely shaped chloroplast, which is parietal laminate, without pyrenoid; 4) it contains only chlorophyll *a* and never has chlorophyll *b*; 5) phycoerythrin is not included, but merely phycocyanin exists as a sort of phycobilin; 6) it never produces starch as an assimilatory product, but a kind of soluble polysaccharide which comes nearer to floridean starch or dextrine; 7) chromatins aggregate to form a mass which may be called nucleus. Based his view on characters No. 5 and 6 (existence of phycocyanin and absence of starch resp.) he concluded the alga must be transferred from Chlorophyceae to either Cyanophyceae or Rhodophyceae. And based on the characteristics No. 3 and 7 (existence of chloroplast and nucleus resp.), the systematic position of the alga must fall to Rhodophyceae.

Because Hirose belabored under the impression that the changing of *C. caldarium* (from the Cyanophyceae) to the Rhodophyceae required a new genus to be established to accommodate *C.*

caldarium as a rhodophyte, he (Hirose 1958 p. 351) made the new combination: *Rhodococcus caldarius* (Tilden 1898a) Hirose 1958. Geitler (1959 p. 351) promptly agrees with Hirose that *C. caldarium* should be considered a Bangialean alga but pointed out that the name *R. caldarius* was inadmissable according to the rules of nomenclature. In the same year, Hirose (1959 p. 348) also realized that *R. caldarius* was an illegitimate name under the *International Code of Botanical Nomenclature* and acknowledged that the correct name should be *C. caldarium*.

9. Allen's article of 1959 and the algal temperature range

Also in 1959, Allen published a more detailed treatment of her studies on *C. caldarium*. As a result of these studies, Allen realized that her combination of *Chlorella caldaria* (Tilden 1898a) Allen 1954 was untenable and concluded (Allen 1959 p. 271): 'Further study has indicated that the characteristics of this alga are sufficiently unusual to deserve generic rank, and this suggested name is withdrawn in favor of Geitler's name, *C. caldarium*,...'. In this paper Allen showed that the phycocyanin of *C. caldarium* was *C*-phycocyanin and indistinguishable from the *C*-phycocyanin of the Cyanophycophyta. Notwithstanding this determination, Allen retained *C. caldarium* as a member of the Chlorophycophyta and referred to it as an anomalously pigmented one. While Allen cited Hirose's 1950 paper on *C. caldarium*, apparently she was not aware of his 1958 paper in which Hirose proposed rhodophycean affinities for *C. caldarium*. Allen further reported that *C. caldarium* showed optimal growth in the temperature range of 45–50 °C. In the future, this optimal temperature will be confirmed by Ascione *et al.* (1960), Doemel (1970) and Doemel and Brock (1970) to be 46 °C. However, because one of Allen's isolations had been secured from a thermal spring which reportedly had a temperature of 75–80 °C and because Copeland (1936) had reported the occurrence of *C. caldarium* in thermal springs supposedly having temperatures of 80.2 °C, Allen argued that her isolates of *C. caldarium* had lost some of their thermal tolerance because they had been maintained at lower temperatures for a period of time. Doemel (1970) and Doemel and Brock (1970) will show that (in natural habitats and in laboratory cultures) no thriving natural populations of *C. caldarium* occurred in temperatures above 57 °C. These authors concluded that the reported higher temperatures of *ca.* 80 °C were in error due to instrumentation or could be attributed to an unusually high discharge from the thermal springs which may have temporarily raised the temperature at the collecting sites to *c.* 80 °C.

10. The alkaline Cyanidia — Schwabe's report

In an extremely interesting paper entitled: *Zur autotrophen Vegetation in ariden Boden*, Schwabe (1960 pp 304–305) gave detailed additional information on the habitat of the aerophilic and epipelagic *C. caldarium* forma *chilensis* and *C. caldarium* forma *rumpens*. Before proceeding on with this taxonomic review, We wish, at this point in time, to call to the readers' attention several reports of the occurrence of *C. caldarium* in supposedly neutral and alkaline habitats and to place these populations of *C. caldarium* in proper taxonomic perspective. It has been amply demonstrated by various workers that *C. caldarium* is an acidothermophilic alga requiring for its optimal growth a temperature of 45 °C and a pH of *c.* 3.0–3.5. Both observations on naturally occurring populations and on cultures maintained under strictly controlled laboratory conditions have confirmed that *C. caldarium* shows little or no growth at pH values above five. However, the publications of G. H. Schwabe (1936, 1942, 1944, 1960) which have been reviewed up to this point in the present paper, have given considerable data on habitats, for varieties and formae of *C. caldarium*, which cannot be interpreted as strictly thermoacidic. It is obvious that Schwabe's recognition of the peculiar ecological niche which his forma *chilensis* and forma *rumpens* occupy was indeed correct. We only question whether or not the designation of *forma* is appropriate, since obviously the ability to produce dense populations outside of the thermoacidic environment is a genetic character and not an ecological expression for which the term *forma* is usually reserved. It may well be that varietal or even specific rank should be given to Schwabe's formae *chilensis* and *rumpens*. While Schwabe retained these formae in the Cyanophycophyta, he was clearly aware of their uncertain systematic position and acknowledged their eukaryotic nature with the following statement (Schwabe 1942, p. C42): 'Da die zellen einen Kern zu besitizen scheinin, ist, wie schon Geitler betont, die Stellung der Art in System unsicher.'

10.1 Alkaline Cyanidium *from Jerusalem and Italian caves*

Additional non-acidothermic sites for *C. caldarium* have also been reported. Leiv-Hetzroni (1965) has described the occurrence of *C. caldarium* in alkaline environments of caves in and about Jerusalem. Friedmann (1964) also alludes to the presence of *Cyanidium* in these caves but gives no sites of the collections. [Also personal communication from I. Friedmann to W. Doemel (Doemel 1970 p. 69); personal communication from I. Friedmann per letters from J. Seckbach to F.D. Ott on 23/III/82 and on 21/VI/82].

Friedmann [personal communication to W.Doemel (cf. Doemel 1970, p. 69)] had also collected an alga similar to *C. caldarium* in the mythical Sybil Cave at Cuma, Italy. Doemel (1970 p. 69) reported that T.D. Brock had also collected material of *C. caldarium* from this same cave. While no viable cultures could be established from these cave collections, Doemel (1970, pp 69–70) made some observations based upon Brock's collections in 1969 (from dimly lighted recesses of the Cave of Sybil at Cuma) like: 1) the pH of scrapings from the cave at Cuma was 7.2; 2) no growth was obtained at culturing these cells at acidic or neutral media; 3) the *in vivo* spectrum, in glycerol, demonstrated peaks of 640 and 680 nm, corresponding to phycocyanin and chlorophyll *a* respectively; 4) the similar morphology and pigmentation of this alga to *C. caldarium* posed an interesting problem. 'If however, *C. caldarium* is defined as existing strictly in acid habitats as studies of the thermal habitats indicate, then the alga found in this cave may be another species of *Cyanidium*, but cannot be *C. caldarium*. The solution to this problem however, awaits the isolation of this alga in pure culture.'

Skuja (1970) reported the occurrence of *C. caldarium* among the *Alghe cavernocole nelle zone illuminate delle Grotte de Catellana (Murge di Bari)* in Italy. Skuja analyzed 22 samples which had been collected in various lighted parts of the cave. Two of the samples contained *C. caldarium* and Skuja reported the alga to be abundant in both samples (Skuja 1970, p. 198). Skuja (1970) gives no information in the text of his paper on the *C. caldarium* material, but in the 'Spiegazione della Tavola X,' under figure 14 he gave cytological and ecological observations (Skuja 1970, p. 202). For more information the reader is guided to Hoffmann's chapter in this volume.

11. Lewin's attempt to classify *Cyanidium*

Returning the reader's attention back to the taxonomic survey Lewin (1961) reviewed the various systematic positions for which *C. caldarium* had been assigned and rejected. He offered additional arguments to strengthen the suggestion of Fogg (1956, p. 157) to rank *C. caldarium* among the Cryptophyceae, and in particular to the Cryptococcales. Lewin further cited the blue-green colored, acidophile, *Tetragonidium verrucatum* Pascher as a possible close associate of *C. caldarium* . The absence of chlorophyll *c* and α-carotene seem not to support Lewin's arguments for placing *C. caldarium* in the Cryptococcales.

12. Ultrastructural investigations

The 1960's will see a series of papers on the ultrastructure of *C. caldarium* which will have in the future years a significant bearing on the taxonomy of *C. caldarium* but which was not evident during the decade of their publication. In 1961, Rosen and Siegesmund published the first electron micrographs of *C. caldarium*. These micrographs show the presence of a nucleus, vacuoles and a chloroplast whose limiting membrane is enclosed in a parallel array of lamellae. Rosen and Siegesmund (1961, p. 914) conclude: "The *Cyanidium* chloroplast appears to be transitional between the 'primitive' photosynthetic apparatus of the photosynthetic bacteria and blue-green algae, and the more elaborate chloroplasts of higher algae and vascular plants." The culture of *C. caldarium* utilized by Rosen and Siegesmund was from Dr. L. Bogorad who had in turn secured it from M.B. Allen. Rosen and Siegesmund (1961, p. 914) reported that this isolate reproduced by the formation of up to eight endospores per sporangium (mother cell). They further reported: 'Each endospore contains a single small chloroplast which appears to have been produced by division of the chloroplast of the mother cell.' Mercer, Bogorad and Mullens (1962) using this same isolate published a somewhat more extensive treatment of the fine structure of *C. caldarium*.

These authors reported the occurrence of vacuoles, a nucleus and a chloroplast similar to those reported by Rosen and Siegesmund (1961). However, they were at variance with Rosen and Siegesmund on the behavior of the chloroplast on endospore formation. In addition to these organelles, Mercer *et al.* (1962) also reported the presence of an endoplasmic reticulum and mitochondria. Clusters of membranes were occasionally observed but lacked sufficient detail to be identified with certainty as Golgi bodies. These authors arrived at the following conclusions (Mercer *et al.* 1962, p. 403): ... *Cyanidium* is not a blue-green alga, not a symbiotic association, not a red alga similar to *Porphyridium*, and not a green alga similar to *Chlorella* or *Chlamydomonas*. It is less differentiated that *Chlorella* and may be more primitive. On the basis of the structure of its chloroplast, *Cyanidium* resembles both *Nostoc* and *Chlorella*. The fine structure of the lamellae is similar to that of the photosynthetic membranes in *Nostoc*...., while in the arrangement of lamellae into a definite organelle *Cyanidium* more closely resembles *Chlorella*...Whether it should be included in the Chlorophyceae remains a matter of conjecture. For further initial ultrastructural aspects and classification status, see Seckbach and Fredrick (1980) and Seckbach review article (1992).

Castaldo (1968) working with material collected in its natural habitat from 'terreni fumarolici acidi dei Campi Flegrei', published her results on the ultrastructure of *C. caldarium*. She essentially confirmed the organelles and membrane systems as reported in the two previous ultrastructural studies, but she also clarified that the few mitochondria possessed an almost elementary system of cristae and, for the first time, she definitely demonstrated the presence of Golgi bodies. She (Castaldo 1968, p. 139) concluded that her material collected at Campi Flegrei was essentially identical to the American isolate used by the previous workers (Rosen and Siegesmund 1961, Mercer *et al.* 1962) in their ultrastructural studies with the only apparent differences being in the average cell size and the number of endospores produced per sporangium. The following sums up these two differences (Castaldo, 1968, p. 145): For such morphological characteristics the *Cyanidium* growing at Campi Flegrei is closely similar to the Yellowstone Park one. The only differences between the two strains are: 1) the slight[l]y larger size of the cells (6–8,5 μ versus 3–5 μ of the American strain); 2) the number of endospores for each mother cell — up to 32 in the Neapolitan strain; no more than 8 in the Yellowstone Park strain. Castaldo was working with materials collected from the natural habitat and, hence, there was no assurance that the collections were unialgal. Indeed, her electron micrographs clearly show cells possessing chloroplasts with two different arrangements of the photosynthetic thylakoids. One type of thylakoid arrangement was identical to that reported previously by Rosen and Siegesmund (1961) and by Mercer *et al.* (1962). However, there was also present in the collected material smaller cells *c.* 3 μm in diameter. The chloroplasts of these smaller cells possessed a different thylakoid arrangement which essentially consisted of four or more thylakoids concentrically arranged within the limiting chloroplast membrane. In this 1968 paper Castaldo held that this thylakoid arrangement was reminiscent of symbiotic 'cianellae'. Recently, Seckbach (1992) reviewed the additional EM studies performed with *Cyanidium*.

13. De Luca's two strains of *Cyanidium*

De Luca and Taddei (1970) recognized that morphological, ecological and physiological investigations utilizing various isolates of *C. caldarium*, which had been secured from diverse localities throughout the world, failed to yield identical results. Among these differences one may mention the conflicting reports on cell dimensions and the number of endospores produced per sporangium. Because of such differences, De Luca and Taddei initiated a comparative study on a large number of isolates which they had made from the *Cyanidium* populations at Campi Flegrei. From their studies of these isolates, De Luca and Taddei (1970, p. 88) clearly established that two different eukaryotic algae were present in the naturally occurring population of *Cyanidium caldarium* forma *A* and *C. caldarium* forma *B*.

C. caldarium f. *A* was characterized by the following: 1) possessing chloropphyll *a* and *C*-phycocyanin; 2) forming 4 endospores per mother cell (sporangium); 3) having the optimal growth temperature of 38 °C; 4) having the optimal growth pH of 2.2. On the other hand, *C. caldarium* f. *B* was characterized by the following: 1) possessing chlorophyll *a* and *C*-phycocyanin; 2) forming up to 32 endospores per sporangium; 3) having the optimal growth temperature of 45 °C; 4) having the optimal growth pH of 1.5.

As a result of these studies, De Luca and Taddei attributed the conflicting results of various investigators to the probability that their isolates were not identical, but fell to one or the other of these two forms. In the future, additional contrasting characters will be assigned to forma *A* and *B* of *C. caldarium*. Castaldo (1970) published the results of her comparative ultrastructural studies of these two forms of *C. caldarium* based upon cultures supplied by De Luca and Taddei. The electron micrographs in this work are excellent and clearly demonstrate the reported differences.

14. Ecological physiology of *Cyanidium*

The year 1970 also saw the publication of a significant work on the physiological ecology of *C. caldarium* by Doemel (1970). The following review constitutes Doemel's summary (Doemel 1970, p. 183–185): 1) '... *C. caldarium* does not exist at a pH value greater than 5.0.... 2) It is an obligate acidophile.... 3) ..It does not exist at temperature greater than 55–57°C.. 4) It is a moderate thermophile. 5) It is an extreme acidophile. 6) At temperatures greater that 35 °C, *C. caldarium* is the sole photosynthetic organism [in its habitat]. 7) Unshaded populations are adapted to high light intensity and photosynthesis is not inhibited by full sunlight (>10^6 foot-candles). 8) Shaded populations have increased amount of chlorophyll *a* and are inhibited by full sunlight. 9) Because mats of *C. caldarium* are quite dense (10^{12} cells/m^2) and *c.* 0.5 cm thick,.... in areas of reduced light or where carbon dioxide is limiting, *C. caldarium* may be growing either mixotrophically or heterotrophically on photosynthetic organic carbon. Therefore even though at temperatures greater that 35 °C the population is essentially a pure culture of *C. caldarium*, the population is energetically more diverse and thus inherently more stable than a purely autotrophic population.'

15. Comparative studies of the two strains

In 1972 , De Luca and Taddei (1972 p. 8) reported that *C. caldarium* f. *A* utilized nitrates while *C. caldarium* f. *B* could not. De Luca *et al.* (1972a, p. 27) reported that *C. caldarium* f. *A* does not utilize glucose in the dark but may remain alive for over six weeks under such conditions. *C. caldarium* f. *B* utilized glucose both in the light and in the dark producing three times as much growth as realized under purely autotrophic conditions. These authors concluded: '....*C. caldarium* forma *A* is, like most Cyanophyta, an obligatory autotrophic alga, while *C. caldarium* forma *B*, like most eukaryotic algae, is a facultative heterotrophic one'.

Continuing with these comparative nutritional studies, De Luca *et al.* (1972b, p. 122) further reported that the ratio of chlorophyll to phycocyanin is definitely higher in *C. caldarium* forma *A* than is the ratio in forma *B*. These authors further report that the carotenoid composition of the two formae are remarkably different; forma *A* contained 30% β-carotene, 48% zeaxanthin and 22% lutein while forma *B* contained 55% β-carotene, 32% zeaxanthin and 13% lutein. They also reported that when forma *B* is grown on glucose in the dark it synthesized neither phycocyanin nor chlorophyll. They further noted that the β-carotene concentration, under these conditions, may drop to only 17% of the total carotenoid complement. In 1976, De Luca and Taddei reviewed the taxonomic binomials which had been applied to material of *C. caldarium* throughout the years. Their first observation was the fact that *C. caldarium* forma *A* and *C. caldarium* forma *B* were two distinct organisms. The second observation was about the heterogeneity of the *Cyanidium* populations at Aquasanta, Italy. The type locality of Meneghini's *Coccochloris orsiniana* and collection at the fumaroles of Yellowstone Nat. Park and of Tilden's *Protococcus botryoides* forma *caldarius*, could not been identified with certainty as *C. caldarium* Geitler. The third, and last significant observation to be emphasized here is the following (De Luca and Taddei 1976, p. 215): Galdieri pointed out near the Solfatara of Pozzuoli just one alga, and his descriptions and drawing unequivocally indentify his *Pleurococcus sulphurarius* as being identical to *C. caldarium* forma *B* which had been isolated by De Luca and Teddei from the identical same site.

Accepting this statement as an appropriate and correct identification, it follows, therefore, in our opinion, that the remaining forma, i.e., *C. caldarium* forma *A*, is indeed representative of the other alga present also in the type localities of Meneghini's *Coccochloris orsiniana*, Tildens's *Protococcus botryoides* forma *caldarius*, Setchell's *Pleurocapsa caldaria* and Copeland *Pluto caldarius*. Certainly, the identification of *C. caldarium* forma *B* to *Pleurococcus sulphurarius* Geitler leaves no other described candidate in the supposedly mixed populations of Tilden's collections for the *C. caldarium* forma *A*.

However, in all fairness to Tilden who collected her material in 1898, collections made 80 years after hers *only testify* to the fact that in 1976, the samples examined by De Luca collected from supposedly the same general area, were indeed composed of two distinct species; they have failed to demonstrate that Tilden's collections were heterogeneous in nature (see later for more detail).

Indeed, concerning these collections housed at the University of Minnesota, Department of Botany, Minneapolis, USA, De Luca and Taddei (1976) stated 'The sample of *C. varius* (inv. no. 79) is practically useless. In the sample of *P. botryoides* f. *caldarius* (inv. no. 2385) there can be recognized cells with a diameter varying between 2 and about 8 μm. Only the walls of these cells are left and inside them no structure is identifiable'. The reader's attention should also be called to the fact that as early as 1901, Setchell (Collins, Holden and Setchell, 1901, Fasc XVIII, no 851) stated, on placing Tilden's *Protococcus botryoides* forma *caldarius* into the genus *Pleurocapsa*: 'clearly a member of the Cyanophyceae and forming a limited number (4) of schizospores which seem to relate it most closely to the genus *Pleurocapsa*'. Tilden (1910, pp 47–48) accepted Setchell's transfer of her taxon to *P. caldaria*. Furthermore, she even quotes the above from Setchell, thus clearly accepting not only the transfer to *P. caldaria* but also his additional information, without comment, on the production of four endospores! Surely, there is nothing, to the present authors' knowledge, which requires the rejection of the specific epithet each time a new bit of knowledge accrues concerning the plant.

We find untenable the premises of De Luca and Taddei (1976) that Tilden's varietal epithet is to be rejected because 1) samples collected in supposedly the same general area 80 years later were found to be composed of two distinct species; they have failed to demonstrate that Tilden's material, collected in 1896, was also heterogeneous as to specific composition. Because there has been presented no information which indicates that Tilden's 1896 collection is composed of mixed populations corresponding to their *C. caldarium* forma *A* and forma *B* the application of Article 70 of the International Code of Botanical Nomenclature (Stafleu, 1972) appears to be unjustified in this case and 2) because Tilden failed to cite the number of endospores produced during each vegetative cell division even though Galdieri, the next year (1899) may have done so for his *Pleurococcus sulphurarius*, especially untenable since Galdieri made no mention of Tilden's taxon as a possible point of contention; Setchell (1901) indicated that four endospores were produced by Tilden's taxon, a fact which she acknowledged on quoting Setchell's observation (Tilden 1910, p. 48).

Additional comparative studies between the two formae of *C. caldarium* continue to appear. Boenze *et al.* (1977, pp 131–133), on investigating the fatty acids in forma *A* and forma *B*, concluded that the linolenic acid is present in *C. caldarium* form *B* and not in *C. caldarium* forma *A*. In fact, all eukaryotic vegetables and part of the Cyanophyceae possess trienoic acids; therefore *C. caldarium* forma *A* is the only eukaryotic plant lacking these fatty acids. Boenze *et al.* (1977) also concluded that *C. caldarium* forma *A* is a very primitive alga, related to Cyanophyceae more that *C. caldarium* forma *B*. The following year Bresciano *et al.* (1978 p. 23) published the results of their immunological comparative studies of these two formae and concluded that amount of γ-globulins response was markedly different in the two groups of experimental rabbits. This study was followed by the report of Musacchio *et al.* (1978 p. 43) of the respective salt tolerance of the two forms of *C. caldarium*; they established that forma *B* is more salt-tolerant than forma *A*.

From the above review of comparative studies (1970–1978), it became progressively obvious that the *Cyanidium* population at Campi Flegrei was composed of two distinct eukaryotic algae. This realization mandated that several taxonomic decisions be made among which the most important two were: 1) which of the two formae of *C. caldarium*, *A* or *B*, should be designated as *C. caldarium sensu stricto* and 2) what taxonomic binomial should be applied to the remaining forma? To these ends, De Luca, Moretti, Musacchio and Taddei (1974) examined the *Cyanidium* populations growing in the acid thermal springs of Acquasanta located near Ascoli Piceno (Italy); the reader may recall that these thermal springs were the source of the material on which Meneghini (1839) based his *Coccochloris orsiniana.*

In the same year, De Luca, Moretti and Taddei (1974) examined additional localities in Sicily and southern Italy containing populations of *Cyanidium*. Their analyses revealed that seven of these populations contained both forms of *C. caldarium*. Similarly, Pinto and Taddei (1975 p. 135) reported the distribution of these two formae in several additional sulfur thermal collecting sites. In about a dozen sites, they observed also a third cell, *Cyanidioschyzon merolae*. This survey of acid thermal springs was extended to north America by De Luca, Gambardella and Merola (1979) including springs in Mexico, El Salvador and the United States. Those springs in the United States which were investigated included the collecting sites of Tilden's 1898 *Protococcus botryoides* forma *caldarius*, Setchell's 1901 *Pleurocapsa caldaria* and Copeland's 1936 *Pluto caldarius*. De Luca, Gambardella and Merola (1979, pp 421–425) concluded that two species of algae (and not just one, as reported until then), are living in nearly all hot, acid springs of North and Central America. The two species live in mixed populations where the pH is very low. The ratio between the species depends on environmental conditions. There are some hot, acid springs where only one species is present. The high salinity of the water or the high temperature of the water, may make the selection for the one organism.

Gambardella, Moretti and Musacchio (1980, p. 44) carried out a series of autoecological and ultrastructural studies and concluded the uniformity among all the strains of each species coming from the American and European Continents. De Luca, Musacchio and Taddei (1981 pp 7–8) noticed that their results of autoecological and physiological tests have shown a remarkable identity of behavior among the strains of *Pleurococcus sulphurarium* coming from Java, from Pozzuoli Solfatara and from Yellowstone National Park. They reported 'The presence of *C. caldarium* Geitler and of *P. sulphurarius* Galdieri near the European acid fumaroles (Pozzuoli Solfatara, Italy), the American (Yellowstone Park, U.S.A.) and in Asia (Mt. Lawu, Indonesia) suggests that these algae are cosmopolitan and occur in particular environments often in extreme conditions of temperature and acidity'. For additional data on these algae, see Seckbach (1991, 1992).

16. *Cyanidioschyzon* — a third alga in the Cyanidian habitat

As De Luca, Taddei and Varano (1978) were investigating the acidothermophilic flora at Campi Flegrei, they also discovered a third alga among the populations of *C. caldarium* forma *A* and forma *B*. This third alga possessed the same photosynthetic pigments as did *Cyanidium*, namely chlorophyll *a* and *C*-phycocyanin (see absorption curves of the Cyanidian algae in Seckbach 1991; Nagashima *et al.* 1993). It differed, however, from both forms of *C. caldarium* in possessing only vegetative cell division and never dividing by the formation of endospores. This new alga was named *Cyanidioschyzon merolae*, it seemed to occupy the same general ecological niche as did the two forms of *C. caldarium*. The abstract of De Luca *et al.* (1978 p. 37) describes this newly observed alga.

The same year, De Luca, Moretti and Taddei (1978) also reported the presence of *C. merolae* among the *Cyanidium* populations in the thermal springs at Yellowstone National Park, Wyoming and in those located on Mount Lawu (Java, Indonesia). Thus, the supposedly unialgal populations in acido-thermal springs having a pH below 5 and temperatures above 35 °C in reality now may be composed of a mixed algal population containing up to three taxa, namely, *C. caldarium* f. *A*, *C.*

caldarium f. *B* and *Cyanidioschyzon merolae*. Recently Nagashima *et al.* (1993) observed a couple of additional species of *Cyanidioschyzon* which will be treated elsewhere.

17. Fatty acids analyses of the three members

Because there appeared to be contrasting data on the fatty acid composition of *C. caldarium* as had been reported by Kleinschmidt and McMahon (1970 a,b), Allen, Good and Holton (1970), Adams *et al.* (1971), Ikan and Seckbach (1972), Boenzi *et al.* (1977) and Bedord *et al.* (1978). Moretti and Nazzaro (1980) felt that these reported differences might be due to the fact that the above investigators were utilizing different strains (i.e., clones) of *C. caldarium* representing *C. caldarium* forma *A* or *C. caldarium* forma *B*. Moretti and Nazzaro, therefore, undertook comparative investigations on the fatty acid composition utilizing axenic strains of these two taxa as well as the eukaryotic, thermoacidophilic *Cyanidioschyzon merolae*. Their results (Moretti and Nazzaro 1980, p. 8) indicated that there are qualitative differences among the lipids of the 3 species tested. At different temperature ranges (20 versus 55 ° C) of the cultures there were differences within a single species. Linolenic acid is present only in *C. caldarium* forma *B*, while it is lacking in *C. caldarium* forma *A* and in *C. merolae* (both of the latter cells are the only eukaryotes not having trienoic acids. For futher information on the lipid and sterol components, please see, in this compendium, Chapter 6 by Seckbach and Chapter 16 by Nagashima.

18. From formae to new genus of *Galdieria*

Nagashima and Fukuda (1981a, b) described their comparative studies on the two strains of *C. caldarium* from Japan. They proposed that their cell designated as RK-1 is *C. caldarium* (Tilden) Geitler forma *A*, while their M-8 strain may belong to the genus *C. caldarium* forma *B. [= Chroococcidiopsis* Geitler]. In the same year, Merola, Castaldo, De Luca, Gambardella, Musacchio and Taddei (1981), published their taxonomic investigations of *Cyanidium*. This was a follow-up to the preliminary conclusions of their work published by De Luca and Taddei (1976). In a recent paper Suzuki *et al.* (1992) analyzed the DNA content of the cyanidian members, unfortunately, some taxonomic inaccuracies have permeated this interesting article. In the interim, since the publication of 1976 paper, De Luca, Musacchio and Taddei (1981) had the opportunity to investigate material collected from the *locus classicus* at Mount Lawu in Java, the same site on which was based Geitler's 1933, *sic* 1935! description of *Cyanidium*, the conclusions of this investigation have already been summarized above in this review [De Luca, Musacchio and Taddei (1981, pp 7–8)]. In the present 1981 paper, Merola *et al.* concluded: 1) that *C. caldarium* (Tilden 1898a) Geitler 1933, *sic* 1935! would best be cited as *C. caldarium* Geitler 1933. They had earlier concluded that 1) Tilden's basionym '*caldarius*' was based upon a heterogeneous population and, thus, was to be rejected. *C. caldaria* Geitler 1933 was to accommodate their *C. caldarium* forma *A* (dividing into 4 endospores); 2) that for their *C. caldarium* forma *B* a new genus was to be described and a new combination was to be made, namely *Galdieria sulphuraria* (Galdieri 1899) Merola ex Castaldo, De Luca, Gambardella, Musacchio and Taddei 1981 (forma *B* reproduced by the formation of 4, 8, or 32 endospores); 3) that the ecological cohort of the above two taxa, *Cyanidioschyzon merolae* De Luca, Taddei et Varano 1978, which divides by binary fission, should be included in the same family as *C. caldarium* which reproduces by the formation of four endospores, namely Cyanidiaceae Geitler 1933 emend.

While Merola *et al.* (1981) emend the family description to accommodate these two widely separated genera, it appears to us that the new circumscription of the family is quite untenable in view of the classical taxonomy usually employed to classify unicellular algae falling to the Rhodophycophyta. De Luca and Moretti (1983) reported on the floridosides of *Cyanidium caldarium*, *Galdieria sulphuraria* and *Cyanidioschyzon merolae*. The inclusion of these algae in the Rhodophycophyta by these authors had been done with some hesitation; but based upon these studies, they determined that the presence of floridosides and isofloridosides, as storage products in these taxa, were compatible with their assignment of these algae to the Rhodophycophyta.

18.1 The Galdieria *genus proliferates — Sentsova's observations*

In terminating this extensive review of the Cyanidiaceae, the authors wish to call to the reader's attention the recent paper by Sentsova (1991) in which three new species of *Galdieria* were described. They all were isolated from acidothermal springs of the USSR. Sentsova reported that one of the cultures isolated by the late M.B. Allen, from material collected in Yellowstone National Park, USA, fell to one of these newly described species, namely, *G. partita.* This alga is characterized by having cell dimensions of 3–11 μm, producing four to thirty-two endospores on division, and by having a single, laminated, parietal, cup-shaped chloroplast. *G. daedala* Sentsova has cell dimensions of 3–11 μm, producing two, four, or eight endospores on division, and by having a single, laminated, parietal, irregularly lobed chloroplast and with the edges of the chloroplast wavy. *G. maxima* Sentsova has cell dimensions of 6.5 to 16.5 μm in diameter, producing four, eight, or sixteen cells per division, and by typically having a single, laminated, irregularly to oval shaped chloroplast, on occasion, up to four chloroplasts per mother cell. See also Sentsova's chapter in this volume.

19. New classification of all Cyanidian members

We have proposed (Ott and Seckbach, in this volume) to reclassify all the *Cyanidium* and *Galdieria* species into one genus, namely, *Cyanidium*. We feel that *Cyanidioschyzon* should be removed from of the Cyanidiaceae and placed to the Porphyridiaceae (Bangiophytes). We hope this new, logical, taxonomic arrangement will alleviate further complications.

References

Allen, M. B. (1952). *Arch. Mikrobiol.* (Berlin, Heidelberg) 17: 34–53.

Allen, M. B. (1954). *Arch. Mikrobiol.* (Berlin, Heidelberg) 32: 270–277.

Allen, C. F., Good, P. and Holton, R. W. (1970). *Plant Physiol.* 46: 748–751.

Ascione, R., Southwick, W. and Fresco, J. R. (1966). *Science* 153: 752–755.

Balsamo, F. (1892). *Boll. Soc. Nat. Napoli.* Ser. I (Napoli) 6: 77–79.

Bedord, C.J., McMahon, V. and Adams, B. (1978). *Arch. Biochem. Biophys.* 181: 15–20.

Bergen, J. Y. (1897). *Bot. Gaz.* 23: 198–199.

Boenze, D., De Luca, P. and Taddei, R. (1977). *Giorn. Bot. Ital.* 111: 129–134.

Bresciano, E., Pinto, G. and Vuotto, M.L. (1978). *Delpinoa* 18/19(1976–7): 17–24.

Brock, T.D. (1978). *Thermophilic microorganisms and life at high temperatures.* Springer verlag. Chapter 9: *The genus Cyanidium.* pp, 255–302.

Castaldo, R. (1968). *Delpinoa* (Napoli) 8/9(1966–1967): 135–145.

Castaldo, R. (1970). *Delpinoa* (Napoli) 10/11(1968–1969): 91–109.

Chapman, D. J. (1974). *Nova Hedwigia* 25: 673–682.

Collins, F. S., Holden, I. and Setchell, W. A. (1901). *Phycotheca Borealis American. A collection of Dried Specimens of the Algae of North America. Fascicle XVIII.* Malden, Massachusetts. November, 1901. (2) + Specimens Nos. 851–900.

Copeland, J. J. (1936). *Ann. N.Y. Acad. Sci.* (New York) 36: 1–229.

De Luca, P., Gambardella, R. and Moretti, A. (1979). *Bot. Gaz.* 140: 418–427.

De Luca, P. and Moretti, A. (1983). *J. Phycol.* 19: 368–369.

De Luca, P., Moretti, A., Musacchio, A. and Taddei, R. (1974). *Delpinoa* 14/15(1972–1973): 3–11.

De Luca, P., Moretti, A. and Taddei, R. (1974). *Delpinoa* 14/15(1972–1973): 49–60.

De Luca, P., Moretti, A. and Taddei, R. (1978). *Delpinoa* 18/19(1976–1977): 69–76.

De Luca, P., Musacchio, A. and Taddei, R.(1972a). *Delpinoa* 12/13(1970–71): 19–27.

De Luca, P., Musacchio, A. and Taddei, R.(1972b). *Delpinoa* 12/13(1970–71): 115–123.

De Luca, P., Musacchio, A. and Taddei, R. (1981). *Giorn. Bot. Ital.* 115: 1–9.

De Luca, P. and Taddei, R. (1970). *Delpinoa* 10/11(1968–1969): 79–89.

De Luca, P. and Taddei, R. (1972). *Delpinoa* 12/13(1970–1971): 3–8.

De Luca, P. and Taddei, R. (1976). *Webbia* 30: 197–218.

De Luca, P., Taddei, R. and Varano, L. (1978). *Webbia* 33: 37–44.

Doemel, W. N. (1970). *Ph.D. The Physiological Ecology of Cyanidium caldarium, Dissertation, Indiana University, Bloomington.* x + 222 pp + (1).

Doemel, W. N. and Brock, T. D. (1970). *Arch. Mikrobiol.* 72: 326–332.

Doemel, W. N. and Brock, T. D. (1971). *J. Gen. Microbiol.* 67: 17–32.

Drouet (1943) *Am. Midl. Nat.* (Notre Dame) 30: 671–674.

Drouet, F. and Daily, W.A.(1956). *Butler Univ. Bot. Stud.* (Indianapolis)12: 1–218.

Drouet, F. and Daily, W. A. (1957). *Trans. Am. Microsc. Soc.* 76: 219–222.

Fogg, G. E. (1956). *Bacter. Rev.* 20: 148–165.

Friedmann, I. (1964). *Int. J. Speleol.* (Weinheim) 1: 29–33.

Galdieri, A. (1899). *Rend. Accad. Sci. Fis. Mat. Napoli* Ser. 3ª 5: 160–164.

Gambardella, R., Moretti, A. and Musacchio, A.(1980). *Delpinoa* 21: 35–45.

Geitler, L. (1925). *Beih. Bot. Zentralbl.* (Kassel) 41, Abt. II: 163–294.

Geitler, L. (1932). *Dr. L. Rabenhorst's Kryptogamen-Flora von Deutschland, Österreich und der Schweiz. Zweite, völlstading neu bearbeitete Auflage. Vierzehnter Band. Die Algen (Herausgegeben von Prof. Dr. R. Kolkwitz). Cyanophyceae von Europa unter berücksichtigung der andern Kontinento von Dr. Lothar Geitler. Akademische Verlagsgesellschaft m.b. H., Leipzig. VI+1196 SS.*

Geitler, L. (1933). *Arch. Hydrobiol. Suppl.* (Stuttgrart) 12: 622–534.

Geitler, L. (1959): *Österr. Bot. Zeitschr.* 106: 172–173.

Geitler, L. and Ruttner, F. (1935). *Arch. Hydrobiol. Suppl.* 14: 371–482.

Geitler, L. and Ruttner, F. (1936). *Arch. Hydrobiol. Suppl.* 14: 553–715.

Hirose, H. (1947). *Kagaku To Kagaku Kyoiku (Sci. Scient. Educ.)* 1: 43–45.

Hirose, H. (1950). *Bot. Mag.* (Tokyo) 63: 107–111.

Hirose, H. (1958). *Bot. Mag.* (Tokyo) 71: 347–352.

Hirose, H. (1959). *Bot. Mag.* (Tokyo) 72: 348.

Ikan, R. and Seckbach, J. (1972). *Phytochem.* 11: 1077–1082.

Inman, O. L. (1940). *J. Gen. Physiol.* (New York) 23: 661–666.

Kirchner, O. (1878). *Kryptogamen-Flora von Schlesien. Im Namen der Schlesischen Gesellschaft für vaterländische Cultur herausgegeben Prof. Dr. Ferdinand Cohn. Zweiter Band. Erste Hälfte. Algen, bearbeited von Dr. Oskar Kirchner.* J.U. Kern's Verlag (Max Müller), Breslau. IV + Sachliche Abkürzungen (1) + Abkurzungen der Autoren-Namen (1) + 284 SS.

Klein R. M. and Cronquist, A. (1967). *Q. Rev. Biol.* 42: 105–296.

Kleinschmidt, M. G. and McMahon, V. (1970a). *Plant Physiol.* 46: 286–289.

Kleinschmidt, M. G. and McMahon, V. (1970b). *Plant Physiol.* 46: 290–293.

Kützing, F. T. (1845). *Phycologia germanica, d.i. Deutschlands Algen in Bündigen Beschreibungen. Nebst einer Anleitung zum Untersuchen und Bestimmen dieser Gewächese für Anfänger. Bearbeitet von Friedrich Kützing. Zu finden bei Wilh. Köhne, Nordhaussen. VIII+340 SS+(2)+SS IX–X (Verzeichniss).*

Kützing (1845–1859). *Tabulae Phycologicae oder Abbildungen der Tange.* Herausgegeben von Friedrich Traugott Kützing. I band. Mit 100 Tafeln. Gedruckt auf Kosten des Verfassers (In Commission Bie Wilhelm Köhne), Nordhausen. VI + 54 SS + Tab. 1–100.

Kylin, H. (1937). *Fysiogr. Sällsk Förhandl.* 7: 119–123.

Leiv-Hetzroni, O. (1965). *Algae in Israel Caves.* Master of Science Thesis, Hebrew University of Jerusalem.

Lewin, R.A. (1961). *News Bull. Phycol. Soc. Am.* 14: 6–7.

Meneghini, G. (1839). *Nuovo Giorn. Lett.* 39: 67–68.

Meneghini, G. (1842). *Monographia Nostochinearum Italicarum Addito Specimine de Revulariis. Auctore Professore Iosepho Meneghinio. (Iconibus ab Auctore Dipictis).* Ex Officina Regia, Augustae Taurinorum. (2) + 143 pp.

Mercer, F.V., Bogorad, L. and Mullens, R. (1962). *J. Cell Biol.* 13: 393–403.

Merola, A., Castaldo, R., De Luca, P., Gambardella, R., Musacchio, A. and Taddei, R. (1981). *Giorn. Bot. Ital.* 113: 189–195.

Molisch, H. (1926). *Pflanzenbiologie in Japan auf Grund eigener Beobachtungen.* Verlag von Gustav Fischer, Jena. X +270 SS nut 84 Abbildungen in Text.

Moretti, A. and Nazzaro, R. (1980). *Delpinoa* N. Ser. (Napoli) 21: 1–10.

Musacchio, A., Pinto, G., Sabato, S. and Taddei, R. (1978). *Delpinoa* N. Ser. (Napoli) 18/19(1976–1977)L 37–44.

Nagashima, H. and Fukuda, I. (1981a). *in Abstracts of the XIII International Botanical Congress, Sydney, Australia, 21–28 August 1981,* p. 288.

Nagashima, H. and Fukuda, I. (1981b). *Jpn. J. Phycol.* 29: 237–242.

Nagashima, H., Ueki, K., Fukuda, I. and Seckbach, J. (1993). In: *Endocytobiology V (eds. Sato S. et al.) Tübingen University Press, Tübingen, Ver.* pp 279–285.

Negoro, K. (1943). *Bot. Mag.* (Tokyo). 57: 302–312.

Negoro, K. (1944). *Sci. Rep. Tokyo Bunrika Daigaku Sect. B.,* 6: 231–374.

Pinto, G. and Taddei, R. (1975). *Delpinoa* N. Ser., 18/19(1976–1977): 125–139.

Profili, P. L. and Maini, P. S. (1950). *Ann. Ist. Super. Sci. Lett. 'S. Chiara'* (Napoli) XVII [1949–1950 (2)]: 317–327.

Rabenhorst, L. (1876). *Die Algen Europa's mit Berücksichtigung des ganzen Erdballs. Gesammelt und bearbeitet von Professor Dr. A. Braun. Herausgegeben von Dr. L. Rabenhorst. Dec. CCXLVI–CCXLVIII.* Druck von C. Hienrich, Dresden. (2) + Specimens Nos. 2151–2480.

Rosen, W. G. and Siegesmund, K. A. (1961). *J. Biophys. Biochem. Cytol.* (New York) 9: 910–914.

Schwabe, G. H. (1936). *Verh. Dtsch. Wissenschaftl. Verein. Santiago Chile N.F.* (Valparaiso) 3: 113–174.

Schwabe, G. H. (1942). *Mitt. Dtsch. Ges. Nat.- Völk. Ostasiens* (Tokyo, Leipzig) 33(Teil C): C1-C67.

Schwabe, G. H. (1944). *Mitt. Dtsch. Ges. Nat.- Völk. Ostasiens* (Schanghai) supplementband 21: 1–239.

Schwabe, G. H. (1960). *Schwiez. Zeitschr. Hydrol.* (Basel) 22: 759–792.

Seckbach, J. (1969). *Plant Physiol.* 44(Suppl.) 15.

Seckbach, J. (1971). *Isr. J. Bot.* (Jerusalem) 20: 302–310.

Seckbach, J. (1991). *J. Phycol.* 27: 794–796.

Seckbach, J. (1992). In: *Algae and Symbioses: Plants, Animals, Fungi, Viruses, Interactions Explored* (ed. Reisser, W.) Biopress Ltd. pp 399–426.

Seckbach, J. and Fredrick, J.F. (1980). *Microbios* 29: 135–147.

Seckbach, J. and Libby, W.F. (1969). *Plant Physiol.* 44(Suppl.): 15–16.

Seckbach, J. and Libby, W.F. (1970). *Space Life Sci.* 2: 121–143.

Seckbach, J. and Libby, W.F. (1971): In: *Planetary Atmospheres. International Astronomical Union, Symposium, Marfa, Texas, USA, Oct. 1969.* pp 62–83.

Seckbach, J., Baker, F. A. and Shugarman, P.M. (1970). *Nature* 227: 744–745.

Seckbach, J., Gross, H. and Nathan, M.B. (1971). *Isr. J. Bot.* 20: 84–90.

Sentsova, O. (1991). *Bot. Zhurn. (Moskva, Leningrad)* 76: 69–79.

Skuja, H. (1939). *Latvijas Univ. Bot. Darzu Raksti (Acta Horti Bot. Univ. Latviensis, Riga)* 11/12: 23–38.

Skuja, H. (1970). *La Grotte d'Italia* (Trieste)2[(1968–1969) (4)]: 193–202.

Stafleu, F. A. (1972), *International Code of Botanical Nomenclature. Adapted by the Eleventh International Botanical Congress, Seattle, August 1969.* Published with Financial Assistance of IUBS-ICSU-UNESCO by A. Oosthoek's Uitgeversmaatschppij N.V. for the International Association for Plant Taxonomy, Utrecht, the Netherlands. (Regnum Vegetabile, Volume 82) 426 pp.

Suzuki, K., Ohta, N. and Kuroiwa, T. (1992). *Protoplasma* 171: 80–84.

Tilden, J. (1896). *American Algae: Century II.* Minneapolis, Minnesota. (2) + Nos. 101–200 + Specimens Nos. 12b, 12c, 32b, 35b, 47b, 92b on five page leaves.

Tilden, J. (1898a). *American Algae: Century III.* Minneapolis, Minnesota. (2) + Specimens Nos. 201–300 + Specimen No. 65b.

Tilden, J. (1898b). *Bot. Gaz.* 25: 89–105 + pls VIII–X.

Tilden, J. (1902). *American Algae: Century VI.* Minneapolis, Minnesota. (2) + Specimens Nos. 501–600 + Specimen No. 154b and 498b.

Tilden, J. (1910). *Minnesota Algae. Volume I. The Myxophyxeae of North America and Adjacent Regions Including Central America, Greenland, Bermuda, The West Indies and Hawaii.* Report of the Survey, Botanical Series VIII, Minneapolis, Minnesota. IV + 328 pp + pls 1–20.

West, G. S. (1904). *J. Bot. Br. Foreign* (London) 42: 281–294 + pl. 464.

West, G. S. (1916). *Algae. Volume I. Myxophyceae, Peridinieae, Bacillarieae, Chlorophyceae, Together with a Brief Summary of the Occurrence and Distribution of Freshwater Algae.* At the University Press, Cambridge. VIII + (2) + 475 pp.

Wille. N. (1897). Pleurococcaceae. SS 54–60. In: Engler, A. and Prantl, K. *Die natürlichen Pflanzenfamilien nebst ihren Gattungen und wichtigeren Arten insbesondere den Nutzpflanzen unter Mitwirkung zahlreicher hervorragender Fachgelehrten begründet von A. Engler und K. Prantl fortgesetzt von A. Engler.* Verlag von Wilhelm Engelmann, Leipzig. XII + 580 SS.

Chapter 8

Systematic position and phylogenetic status of *Cyanidium* Geitler 1933

JOSEPH SECKBACH[1] and FRANKLYN D. OTT[2]
[1] *The Hebrew University of Jerusalem, School for Overseas Students, Mount Scopus, Jerusalem 91905, Israel*
[2] *317 Osborne Avenue, Catonsville, MD 21228, USA*

Outlines:

1. Introduction to the genus *Cyanidium* and its allies
   1.1 The Cyanidiophyceae: a brief inspection
   1.2 Nomenclature, systematic and phylogenetic problems
   1.3 Additional species of cyanidiophytes
2. Assignment to various algal divisions
   2.1 Placement within the Cyanophycophyta and Chlorophycophyta
   2.2 Affiliation with the Rhodophycophyta
   2.3 Assignment to some chromophytes
3. *Cyanidium* and red algae as primitive characteristic organisms
4. The transitional (algal Bridge) view
5. Discussion on symbiotic association
6. Summary of the systematic and phylogenetic notion
References

1. Introduction to the genus *Cyanidium* and its allies

1.1 The Cyanidiophyceae: a brief inspection

The alga known in the literature as *Cyanidium caldarium* is an acido-thermophilic organism distributed ubiquitously throughout the world. This alga resembles *Chlorella*, is unicellular, eukaryotic and exhibits a rather simple ultrastructure (contains a nucleus, a single mitochondrion and chloroplast which are embedded the ground cytoplasm) surrounded by a thick proteinaceous cell wall. The fine structure, physiology and biochemical properties such as its ability to thrive under pure carbon-dioxide, are being reviewed elsewhere in this volume.

1.2 Nomenclature, systematic and phylogenetic problems

Cyanidium caldarium has a long history of occupying many 'Musical chairs' of names (genera and species — see, Ott and Seckbach in this volume). In addition, this alga has suffered from its floating taxonomic position. At one time or another *Cyanidium* has been assigned into the following algal divisions: first to the prokaryotic Cyanophycophyta (Cyanobacteria) then to the Chlorophycophyta, Rhodophycophyta, Chromophyta, Cryptophyta, Glaucophycophyta. Other biologists proposed that indeed this enigmatic alga does not possess a stable systematic position. Some phylogenetic proponents pointed out that it represents a transitional cell between the Cyanophyceae and the eukaryotic algae. Such an algal Bridge was supported by ultrastructural and biochemical studies.

During the last 25 years there were also several attempts to consider this blue-green eukaryotic alga as a symbiotic association between a apochloric host cell and a blue-green alga (which took over the function of the photosynthetic organelle). There have been controversial discussions pro and con these systematic and phylogenetic propositions. Currently, the general accepted view is that *C. caldarium* (Cyanidiaceae) is a primitive rhodophytan (Bangiophyceae, Rhodophycophyta). It may serve as a transitional organism between the Prokaryotic and Eukaryotic worlds (Seckbach *et al.* 1983 and later on in this chapter). In addition to the above enumerated taxa, *C. caldarium* has also been assigned to several well established genera and species either by mistaken identification or by failure to recognize the total pertinent characters possessed by *C. caldarium*. For the very early period, Doemel (1970, p. 159) called to our attention that material in the Azores which undoubtedly was *C. caldarium* was identified by Moseley (1875) as *Chroococcus* sp., a member of the Cyanophycophyta. In the same year, Archer (1875) referred to the same material as *Botryo-*

J. Seckbach (ed.), Evolutionary Pathways and Enigmatic Algae: Cyanidium caldarium *(Rhodophyta) and Related Cells,* 133–143, 1994.

coccus braunii Kützing, then considered by most phycologists as a member of the Chlorophycophyta, but in contemporary classification schemes, *B. braunii* is considered to be a yellow-green alga, the Class Xanthophyceae, and thus, belonging to the Chrysophycophyta.

1.3 Additional species of cyanidiophytes

Contemporarily, there have been observed and isolated a few additional algae which are different but related one to the other, all affiliated with the Bangiophyceae — the lower ('primitive' unicellular) rhodophytes. These algae ('Cyanidia') share common basic features, like, unusual habitat and identical pigment content (see absorption spectra in Nagashima *et al.* 1993; Seckbach 1991). The algae in question are at least three genera of the Cyanidiophyceae. Despite the fact that all the cyanidiophytes survive under extreme conditions, these members exhibit interesting and possibly significant differences in their subcellular organization, DNA (nuclear and organelles) content and biochemical characteristics. These differences, described elsewhere, suggest a gradual development of features from the most primitive to more advanced condition among these members of the Cyanidiaceae (Seckbach 1987, 1991, 1992; Seckbach *et al.* 1983, 1990, 1992; Suzuki *et al.* 1992 and Kuroiwa *et al.* in this volume).

Seckbach *et al.* (1992) placed all three genera within the primitive Rhodophycophyta in the following hierarchical systematic arrangement:

Division: Rhodophycophyta
- Class: Bangiophyceae
  - Subclass: Cyanidiophyceae
    - Family: Cyanidiaceae
      - Genera: *Cyanidioschyzon, Cyanidium, Galdieria.*

Recently Nagashima *et al.* (1993) have isolated new strains of *Cyanidioschyzon* which differ morphologically from *C. merolae* (the first published species). Furthermore, Sentsova (1991, see her chapter enclosed this volume) has described recently three additional species of the genus *Galdieria* from eastern Soviet Union. Hoffmann (this volume) characterizes two *Cyanidium* species living in alkaline epipelagic caves.

The cellular features of the Cyanidian members are sometimes difficult to interpret and their systematic classification has involved taxonomic and phylogenetic problems (Seckbach *et al.* 1983, 1990, 1992, Seckbach 1991, 1992). Because of the primitive (rhodophytic) algal features of the Cyanidiaceae, Seckbach (1987) suggested that they be classified as a new taxon (Prerhodophyta). This author and his coworkers (Seckbach, 1987, 1992; Seckbach *et al.* 1983, 1992) recommend to place the cyanidiophytans in the following hierarchical systematic arrangement: *Cyanidioschyzon* (the most primitive) → *Cyanidium caldarium*, the middle stage in this progression → *Galdieria sulphuraria* (which is very closely related to the lower rhodophytes).

Seckbach *et al.* (1983, 1990, 1992) proposed that the Cyanidian members in the above sequence may serve as an evolutionary ascending ladder between the anucleated cyanobacteria and the eukaryotic primitive Rhodophyta. Further aspects of these algae is presented in this volume by Ott and Seckbach (see review on the Taxonomy and New Classification). Recent investigations on the molecular data (nucleoid content of DNA and its shape of the Cyanidian organelles) have led Suzuki *et al.* (1992) and Kuroiwa *et al.* (in this book) to place these algae in the above evolutionary line.

2. Assignment to various algal divisions

2.1 Placement within the Cyanophycophyta and Chlorophycophyta

From the chapters dealing with the binomials which have been applied to material of *Cyanidium caldarium* (Ott and Seckbach in this volume), it is obvious that the majority of these fall to the Cyanophycophyta (cyanobacteria) which possesses chlorophyll *a* as the *primary* photosynthetic pigment. However, in addition to chlorophyll *a*, *C. caldarium* also contains as its *major* photosynthetic pigment *C*-phycocyanin which is similar, if not identical, to the typical phycobilin of the Cyanophycophyta (see chapter of Troxler in this volume). It is probably the blue-green color of the *C*-phycocyanin which, at least in the earlier years, caused many investigators (who studied liquid or dried material prepared in the field) to place specimens of *C. caldarium* into cyanophycean genera. Such material, from field harvesting, obviously, may not always reflect, on examination in the laboratory, their typical eukaryotic cytology. Or, if examined in the living state, poor quality of the employed optical instruments may have prevented the true eukaryotic nature from being recognized. However, in regards to this last possibility, these writers, on one hand, cannot ignore at an early date the excellent observations and drawings by Galdieri (1899) on *Pleurococcus sulphurarius* which De Luca and Taddei (1970) considered to be their *C. caldarium* forma *B*, while on the other hand, we acknowledge the excellent drawings of the chloroplast of *C. caldarium* by Hirose (1947) which clearly testify to the utilization of excellent optical instruments and also to the keen eye of an accomplished phycologist. Returning our attention to the taxonomic summary (see Ott and Seckbach, in this volume), material of *C. caldarium* has been assigned to various Cyanophycean genera such as *Coccochloris* by Meneghini (1839), *Chroococcus* by Tilden (1898), *Chroococopsis* by Geitler (1925), *Pluto* by Copeland (1936), *Dermocarpa* by Drouet (1943) and *Entophysalis* by Profili and Maini (1950). It may be seen from the above summary that various authors have also assigned material of *C. caldarium* to Chlorophycean genera such as *Palmella* by Kützing (1845–1849), *Protococcus* by Balsamo (1892) and by Tilden (1898a), *Pleurococcus* by Galdieri (1899), *Pleurocapsa* by Setchell in Collins, Holden and Setchell (1901), *Palmellococcus* by West (1904) and *Chlorella* by Hirose (1950) and to the 'acid *Chlorella*' by Allen (1954b).

Based upon the misidentification of Drouet, Inman (1940) reported material of *C. caldarium*, collected in the Yellowstone National Park, as *Chlorella* sp., a member of the Chlorophycophyta. Shortly after this identification, Drouet (1943) placed the material of this same collection into the Cyanophycean genus, *Dermocarpa*. Slightly later, Drouet (cf. Profili and Maini 1950) considered material of *C. caldarium* collected at the Solfatara di Pozzuoli, to be only a stage of *Entophysalis rivularis*, a member of the Cyanophycophyta. Then Drouet reverted back to identifying material of *C. caldarium* as *Chlorella* (Allen 1952, 1954b; Drouet and Daily 1956). It appears that even Allen (1959) could not break with the established and traditional concepts in spite of the fact that she acknowledged that the characters of *C. caldarium* were sufficiently unique as to rank generic recognition; she continued to refer to *C. caldarium* as an anomalously pigmented chlorophyte! Gusev (1961a, b) included *C. caldarium* in a treatment of the blue-green algae but clearly acknowledged, on several occasions, the eukaryotic nature of its cytology. Goodwin (1966) considered *C. caldarium* to be a green alga containing zeaxanthin and not lutein as the main xanthophyll.

2.2 Affiliation with the Rhodophycophyta

Hirose (1958) was the original investigator to recognize that the characters displayed by *C. caldarium* showed strong Rhodophycean affinities and he established, based upon this resemblance, the Rhodophycean genus *Rhodococcus* to accommodate *C. caldarium* in the Rhodophycophyta. Since it is not required by the botanical code that a species be assigned to a new genus just because it is transferred to a new systematic position, the genus *Rhodococcus* (Hirose) is untenable as has noted elsewhere (see Ott and Seckbach on Taxonomic Review, this volume). Geitler (1959) agreed with Hirose (1958) that *C. caldarium* should be considered a member of the Bangiales but that Hirose's *Rhodococcus* was taxonomically an unjustifiable genus. In the same year, Hirose (1959) reached the same conclusion on the taxonomic validity of *Rhodococcus*.

In the next year, Geitler (1960) stated: '*Cyanidium* ...ist eine Bangiaceen-artige Alge mit Zellkern und Chromatophor'. Nicols and Bogorad (1960) reported the presence of allophycocyanin in *C. caldarium* (see chapter by Troxler in this volume).

As a result of his extensive studies on the carotenoids of the flagellated Cryptophyceae Chapman (1965, p. XI) concluded that assignment of *C. caldarium* to the Rhodophycophyta appears justified and consequently a new order Cyanidiales within this division has been proposed. Further on, Chapman (1965, p. 220) noted that there are three possible solutions to assigning *Cyanidium* to an appropriate taxon. These are inclusion in the Chlorophycophyta, Rhodophycophyta or the erection of a new taxon. He suggested the following classification: 'Sub-class: Bangiophycidae. Order: Cyanidiales. Family: Cyanidiaceae'. De Luca, Musacchio and Taddei (1972) investigated the photosynthetic pigment complex in *C. caldarium*. Based upon these investigation, these authors concluded (1972b, p. 121) that *C. caldarium* is a primitive red alga similar to *Porphyridium aerugineum*.

The year 1972, also witnessed the publication of Seckbach's comprehensive review on the taxonomic position of *C. caldarium*. Based upon results of his ultrastructural studies, Seckbach reported the following results which apparently eluded previous workers: Seckbach has cultivated *C. caldarium* in a purely mineral medium lacking the glucose previous workers had utilized. Seckbach's (1972, p. 139) results follow: 'This study allowed us to determine the normal location of phycobiliprotein and to conclude that the phycobiliproteins are closely associated with the photosynthetic lamellae. Figures 4 and 5 show electron dense granules in chloroplasts of *Cyanidium* embedded in the stroma side of the thylakoids and are identical to the phycobilisomes.....(p. 140–141). *Cyanidium* is not closely related to the chlorophytes because of its more primitive type of plastid, its pigment composition and the absence of other components found in the green alae...Thus it seems that *Cyanidium* is not a candidate for the Cryptophyta as suggested elsewhere.....*C. caldarium* is a red alga....Chapman *et al.* (1968)..stated that the Rhodophyceae are primitive eukaryotes and that the taxonomic position of *C. caldarium* is among the more primitive members (Bangiophyceae)'. The investigations on the sterols of *C. caldarium* by Ikan and Seckbach (1972), Seckbach and Ikan (1972, pp 457–458) have suggested that this paradoxical alga is a transitional form between the procaryotic blue-green and more advanced algae (it resembles closer that of the Rhodophyta). However, recent sterol analyzes from *Cyanidium* and *Galdieria* have shown that the previous conclusion is more complicated as suggested earlier (Seckbach *et al.* 1993). Also the ultrastructural investigations of *C. caldarium* which are described (e.g., the rhodoplast, thylakoidal orientation, the outermost lamella, phycobilisomes, storage grains in cytoplasm) are similar to the rhodophytes. The conclusion of these authors is: *C. caldarium* is a blue-green colored rhodophyte closely related to the unicellular red algae (Seckbach and Ikan 1972).

Chapman (1974) concluded that Geitler's 1933 family Cyanidiaceae, is sufficiently appropriate to accommodate *C. caldarium* and transferred Geitler's Cyanidiaceae from the Cyanophycophyta to the Rhodophycophyta. He (Chapman 1974) advised that *Cyanidium* must be assigned to the Porphyridiales or to a new order within the Bangiophycidae. He compares *Cyanidium* with *Rhodospora*, both of these Porphyridiales lack pyrenoids, both reproduce by endospores. The thick cell wall of *C. caldarium* is probably an adaption to the acid hot springs environment (and should not be considered a major taxonomic feature). It is interesting that Margulis (1981) agrees with Chapman (1974) and states that *C. caldarium* is a eukaryote highly specialized to hot, acid water. In his paper Chapman (1974) canceled the order Cyanidiales (which he himself has proposed earlier) and placed the family Cyanidiaceae under the order Porphyridiales.

Hara and Chihara (1974) published the results of their extensive comparative study on chloroplast ultrastructure in the Rhodophycophyta which was based upon ultrastructural examination of 116 taxa. They have divided their taxa into seven types on the basis of fine structural features and considered *C. caldarium* belonging to the type of *Polysiphonia* chloroplast. These authors have based their divisional proposal upon some special algal cellular features [lack of pyrenoid, variation in chloroplast shape and architecture e.g., outermost thylakoid (girdle band)].

The year 1980 also witnessed the publication of an excellent ultrastructural study on *C. caldarium* by Tôyama. The following constitutes the main points of Tôyama's (1980, pp 788–789) paper: the taxonomic position has been undetermined, contains a defined nucleus, a large chloroplast composed of a single envelope, single thylakoids closely resembling those of red algae, phycobilisome, etc. The algal cell is enveloped by a thick cell wall made from at least three dense sub-layers. Only one dumbbell shaped mitochondrion was located between the nucleus and chloroplast. Other organelles seen in eukaryotic cells (golgi bodies, microbodies, etc.) were not observed. From the standpoint of cell ultrastructure, *C. caldarium* may be a primitive rhodophyte, although from the chloroplast characteristic (of fine structure), this cell shows a transitional form from blue-green to red algae.

Seckbach, Hammerman and Hanania (1981) presented a detailed comparison of various features of *C. caldarium* and those of other algal groups. They concluded that the Rhodophyta seems to be the group most similar to *Cyanidium* in all the aspects compared... The red algae are followed by the blue-green algae, while the Cryptophyceae are the most remote group, at least for these features (Seckbach *et al.* 1981 p. 410). They stressed that 'With all the above phylogenetic and ultrastructure data in mind, we propose the *C. caldarium* is a blue-green rhodophyta. This is supported also by others with different taxonomic approaches.....We do hope that the problem with this fascinating alga will soon be settled, so that this black sheep among the algae will turn out to have been only an ugly duckling' (Seckbach *et al.* 1981, pp 419–420).

2.3 Assignment to some chromophytes

Because *C. caldarium* lacked chlorophyll *b*, Bourelly suggested in 1954 (cf. Allen 1954a) that this alga might possibly be a member of the Xanthophyceae in the event that starch also was found to be lacking. In this connection, he also called to our attention that the Chromophyceae often contained blue colored pigments such as those found in certain dinoflagellates and in some members of the Cryptophyceae (Allen 1954a). However, at the same period, Schmidt (1954) in the 12th revision of A. Engler's *Syllabus der Pflanzenfamilien* retained *C. caldarium* in the Cyanophycophyta. Fogg (1956) suggested, since the Cryptophyceae contained taxa possessing blue-green and red colored pigments, that the taxonomical difficulties which *C. caldarium* was posing might be resolved if it were to be classified as a coccoid Cryptophyceae. Lewin (1961) tentatively suggested that *C. caldarium* might possibly be a member of the Cryptococcales, but his recognition that *C. caldarium* lacked chlorophyll *c* and α-carotene seem to weaken this contention. On placing *C. caldarium* in the Cryptophyta, Silva (1962) argued that because of its pigmentation and the presence of starch, this cell should be assigned to the Class Cryptomonads rather than in the Class Cryptophyceae.

Christensen (1962) established the order, Cyanidiales, which he ranked, with some question, under the Cryptophyceae, a class of the Chromophyta. Later (Christensen 1966) in his second edition of '*Alger*' still retained his Cyanidiales in the Chromophyta but now indicated that it might be better placed in the Porphyridiales (Rhodophycophyta).

3. *Cyanidium* and red algae as primitive characteristic organisms

Allsopp (1969, p. 593) considered *C. caldarium* to be a rhodophyte and referred to it as '....the simplest eukaryotic organism hitherto described.....' Lee (1974) studied the chloroplast structure and the starch grains as phylogenetic indicators in lower (unicellular) rhodophytes. He proposed that *C. caldarium* (lacking a pyrenoid) possibly originated from an ungerminated spore of an advance rhodophytan (such as *Porphyra* or *Bangia* 'that maintained their unicellular condition'). This means that *Cyanidium* should be regarded as advanced and reduced alga rather than presenting a primitively simple stage. It seems that this notion has not gained popularity.

Demoulin (1980) called to our attention that the red algae could be among the most primitive eukaryotes and pointed out additional criteria which might be considered in evolutionary treatments of the origin of the primitive eukaryotic cell out from the blue-green algae. He discusses the known similarities between the blue-green and red algae and pointed out (1980, p. 179): '...some Bangiophycidae are hardly distinct from blue-green algae without excellent microscopy.... The preliminary [molecular] data available already show a distinctly primitive character of red algae: smallest sized eukaryotic ribosomal RNAs and N-terminus of mitochondrial cytochrome *c* closer to that of prokaryotic than those of the protozoa *Euglena* and *Tetrahymena* are'.

In his current review, Seckbach (1992 pp 415–419) discussed that the rRNA sequencing studies may be ambiguous, the origin of Rhodophycophyta (and therefore of *C. caldarium*) is either very ancient, on the basis of 5S rRNA, or these algae have evolved more recently as based on the 16s or 28S rRNA. Further molecular studies have to be accomplished in order to determine the origin and 'age' of *C. caldarium*.

Ohta *et al.* (1992) reported that *C. caldarium* strain RK-1 [= forma A] has the smallest amount of (eukaryotic) nuclear DNA. The chloroplast DNA includes a *trnK* sequence that shows a high degree of homology to the chloroplast of land plants but the *Cyanidium* gene lacks a typical intron, i.e., the non coding area in the gene (see also Fukuda in this volume). In the above research group Suzuki *et al.* (1992) have published molecular data on *Cyanidioschyzon* and the other Cyandian cells.

4. The transitional (algal bridge) view

Allen, French and Brown (1960) felt it was best to consider *C. caldarium* as a transitional form since it did not fit well in the Chlorophycophyta, the Rhodophycophyta or the cryptomonads. Goodwin (1960), based upon carotene analyses which revealed 55% β-carotene, 32% xanthin and 13% of an unidentified xanthophyll, suggested that *C. caldarium* might be a transitional form between the cryptomonads and the green algae.

Rosen and Siegesmund (1961), based upon electron microscopy studies, stated "the *Cyanidium* chloroplast appears to be transitional between the 'primitive' photosynthetic apparatus of the photosynthetic bacteria and blue-green algae and vascular plants". Mercer, Bogorad and Mullens (1962), based upon their ultrastructural studies, concluded that *C. caldarium* was not a blue-green alga, not a symbiotic association, not a red alga and not a green alga.

After reviewing the current literature, Klein and Cronquist (1967, p. 220) concluded that they were inclined to place *Cyanidium* between the blue-green and the green or red algae, as a more or less separate line from the postulated 'Uralga'. Their justification is related to the presence of the pigmentation and cellular constituents which were then known in the literature. They insisted in this, rather 'classical' review, that the transitional status is more suggestive rather than is the possibility that the organism is reduced from a more advanced alga (cf. Lee 1974).

In 1970, Klein discussed the supposed relationship between the blue-green and eukaryotic algae and indicated the possibility that *C. caldarium* might be an extant representative of the postulated Ur-eukaryote serving as the non-flagellated bridge between the blue-green and red algae. He concluded: "*Cyanidium* fulfills all these [of the bridge alga] criteria...the carbohydrate metabolism of *Cyanidium* is 'correct' for the possible bridge" (Klein 1970, pp 629–630).

Fredrick (1971) reported on his studies on the electrophoretic patterns of polyglucoside synthesizing isozymes and concluded that *C. caldarium* may be a 'bridge' form between the cyanobacteria and the Rhodophyceae. In the above paper he states: 'It would seem probable that the blue-green algae gave rise to the red algae with *C. caldarium* as a probable extant transition form'. Fredrick (1975, pp 312–313) discusses the phylogenetic position of *C. caldarium* following his immunological studies with algal enzymes. His investigation supports the proposed pathway from a Cyanophycean ancestor for the red algae, while '*C. caldarium* fits even more firmly into the niche of a transition form between the prokaryotic cyanophytes and the primitive eukaryotic rhodophytes'.

Castaldo and Ligrone (1978) reported the presence of triplet thylakoids in *C. caldarium* forma *A* and pointed out that similar ultrastructural arrangements were also known of certain taxa of the Cyanophycophyta. These authors (Castaldo and Ligrone 1978, pp 32–33) stated that the rhodophytes do not possess such 'triple' structure for the lamella, however some affinities exist between the rhodoplasts of red algae, and those of *C. caldarium* and of the Cyanophyta. They pointed out also that: 'to include *C. caldarium* in a possible group of very primitive eukaryotic algae, constituted by intermediate forms between Cyanophyta and Rhodophyta'.

Hase *et al.* (1978) investigated the ferredoxin present in *C. caldarium* and compared this compound with those from other plant groups. They concluded that ferredoxin of *C. caldarium* might possibly be a red algal type (cf. to the chapter of Nagashima in this volume). Their comparative results suggest that (Hase *et al.* 1978, pp 43–44):'...*C. caldarium* may be a member of red algae. However, more complete and extensive sequence study of red algal ferredoxins is required to confirm that *Cyanidium* forms an evolutionary link between the blue-green and red algae'.

Seckbach and Fredrick (1980, p. 135) suggested the following six characteristics for an organism which might occupy a link between the Prokaryota and the Eukaryota: 1) simple eukaryotic compartmentalization (cellular constituents with primordial stages of differentiation; 2) asexual reproduction by forming endospores; 3) lack of prokaryotic specific cell wall constituents ; 4) the chloroplast should possess pigmentation and synthesize assimilated products, and be constructed of both blue-green and (primitive) eukaryotic red algae; 5) an intermediate biochemical status in the storage polyglucans and of the enzymatic system involved in their synthesis should be apparent; and finally, 6) the ecological physiology of such a transitional cell should survive in and tolerate environments similar to the habitats of the blue-green algae, e.g. in elevated temperatures and under high concentrations of carbon dioxide. These authors showed that *C. caldarium* possesses many of these desirable traits for being a 'bridge' between the prokaryotic and the eukaryotic algae. " Some features of *Cyanidium* chloroplast.... point to the primeval status of this early-evolved eukaryote which is considered a 'living fossil'. It seems to be the true transitional organism between the anucleated and nucleated algae and is unlikely to be an endosymbiotic association of two or more cells, as speculated elsewhere" (Seckbach and Fredrick 1980 pp 135–147).

5. Discussion on symbiotic association

Endocytobiotic origin (i.e., growth of alien organelles within the cytoplasm of a host organism) has been reported for *Cyanidium caldarium* (see Seckbach 1972, 1987, 1991, 1992; Seckbach and Fredrick 1980, 1981; Seckbach *et al.* 1983, 1990, 1992). This current review will feature the development of this notion for *Cyanidium*. As was stated under the taxonomic section, Castaldo (1968) indicated that a small unicellular organism was also present in a natural collection of *C. caldarium*. This unicellular organism possessed an ultrastructure which was reminiscent of symbiotic '*Cianelle*'. Margulis (1968, pp 1020–1021) expressed her opinion on the phylogenetic status of *C. caldarium*. She agrees that this alga is eukaryotic and expects that perhaps (it lost its original symbiotic plastids and) re-established new symbioses with somewhat different forms of blue-green algae. She states: 'Such anomalous symbioses could include not only *Cyanidium* but also *Cyanophora paradoxa* and *Glaucocystis nostochinearum*' (Margulis 1968).

In 1970, Raven published a comprehensive paper on a multiple origin for plastids and mitochondria and declares 'that chloroplasts of red algae, flagellated cryptophyceae and *Cyanidium* are in fact blue-green algae that entered into symbiotic relationships with the ancestors of these organisms'. He also supported in his text books (Biology of Plants, see Seckbach *et al.* 1983) the symbiotic association for *Cyanidium* which he considered a green alga. Raven continues (1970): 'There appears to be no valid reason not to regard the large lobed chloroplast of *C. caldarium* as a symbiotic blue-green algal cell that has become stabilized in this role. These chloroplasts contain chlorophyll a, carotenoids, and phycobilin, thus closely approximating the biochemistry of living blue-green algae, to which they are likewise similar in structure. If this is the case, then biochemical evidence from the chloroplasts should not be taken into account in deciding where *Cyanidium* should be placed among the groups of organisms'.

Margulis (1981, p. 43) in her treatment of *Symbiosis in Cell Evolution* made the following comment on *Cyanidium*: 'Some authors have claimed that *Cyanidium* or *Cyanophora*, eukaryotes with pigment systems like cyanobacteria, make good candidates for the uralga. However, *Cyanophora*, at least, is probably an anomalous symbiotic partnership between cryptomonads and cyanobacteria, whereas *Cyanidium* may be a red alga highly specialized and adapted to hot acid waters. In fact, the search for the uralga has led to increased recognition of the fundamental discontinuity between the cyanobacteria and all the nucleated algae'.

Kremer and his associates (1978, p. 157, 1982, 1983, 1979; cf. Seckbach, 1992) proposed a 'new' systematic position for *C. caldarium*. They stressed that *C. caldarium* is not a rhodophyte (as suggested from the structural properties). When the soluble low molecular weight compound of *Cyanidium* and *Cyanophora* were extracted and analyzed, these workers could not detect heterosides typical of Rhodophyceae (such as glycerol galactoside and mannosido glyceride). In contrast to the unicellular red algae, free glucose and fructose (and short chain oligosaccharides, polysaccharides) are found in *C. caldarium*. They conclude that in this regard too, *Cyanidium* resembles the Cyanophyceae more that the red algae. Their final solution is that *C. caldarium* is a endocyanome and represents a colorless, apoplastidal eukaryotic alga. This association contains an endosymbiotic blue-green alga (cyanelle) providing the function of a normal chloroplast. Its phylogenic aspect is similar to symbioses of *Cyanophora, Glaucocystis* and *Gloeochaete* which are well documented. Indeed, the concept of the 'cyanelles' (or 'cyanoplast', see Schenk's paper) which organisms like *Cyanophora paradoxa* possess have been considered to be the beautiful intermediate between a free living cyanobacterium and an organelle (chloroplast). Very recently these postulates regarding cyanelles have been demonstrated as untenable following molecular sequencing studies (see Seckbach 1992; Schenk's chapter in this volume).

Castenholz (1979) composed a similar endosymbiotic ecological combination for *C. caldarium* (see Seckbach 1987, 1992; Seckbach *et al.* 1983). He proposed that a mesophilic apochlorotic (acidophilic) *Chlorella* engulfed a thermophilic cyanophyte and as a result of such union the combined cellular association lives in acidic external thermophilic medium. However, this endosymbiont cyanobacterium (originated from alkaline waters) which serves as a chloroplast, is now protected in an internal neutral cytoplasmic environment (Castenholz 1979).

There have been arguments against the symbiotic evolution of the eukaryotic cell (the reader is referred to the chapters of Jensen and of Nakamura in this volume). Disputes against *Cyanidium* being a symbiotic association have been raised by several investigators (Mercer *et al.* 1962; Allen 1959; Seckbach 1972, 1987; Seckbach and Fredrick 1980; Seckbach *et al.* 1983, 1990).

Fredrick (1979, p. 1823) reported on his research on the storage glycogen biosynthesis in *Cyanidium*, *Chlorella* and *Prototheca* and concluded there was no evidence that '...*C. caldarium* was actually a colorless *Chlorella*, probably a *Prototheca*, that has acquired blue-green algae, and that these symbiotic cyanobacteria served as the functional chloroplast for this thermophilic alga'.

Fredrick summarizes: '.....*C. caldarium* differs markedly from even the high temperature strain of *Chlorella*, as to the storage glucan it forms the types of branching isozymes present and the site of formation of the storage glucan. Because of these differences, it seems highly unlikely that *Cyanidium* represents a *Prototheca*-like *Chlorella* that has acquired cyanobacteria and established an endosymbiotic relationship with these blue-green alge so that they now function as chloroplasts of their host'. Thus, also Fredrick's above conclusion may weaken the proposals of Castenholz (1979) and Kremer *et al.* (1978). Additional rebuttal concerning the symbiotic notion for *Cyanidium* has been reported elsewhere (Seckbach 1987, 1992; Seckbach and Fredrick 1981; Seckbach *et al.* 1983). Furthermore, repeated investigations on the rhodophytic photosynthetic products (floridosides, isofloridosides, free monosaccharides) in *C. caldarium* have demonstrated contradictory results from those obtained by Kermer *et al.* (1978). These results may weaken such a symbiotic view for this organism (Nagashima and Fukuda 1983; De Luca and Moretti 1983; Reed 1983; see also Seckbach 1992 p. 418). Maid, Valentin and Zetsche recently (1990) have isolated plastid-DNA from *C. caldarium* and sequenced the cloned psbA-gene. In their summary they stated: '...the overall sequence homology does not support a direct relationship between the plastids of *Cyanidium*, cyanelles [of *Cyanophora*] and cyanobacteria'.These authors continue (Maid *et al.* 1990, p. 255) :' *Cyanidium* has been interpreted by some authors as an association between an uncolored eucyte and an endosymbiotic cyanobacterium......We have recently obtained sequence data for the RubisCO-operon of *Cyanidium* which did not support this hypothesis. In fact we found a directly higher homology between the RubisCO-operon of *Cyanidium* and the red alga *Porphyridium aerugineum* than between *Cyanidium* and cyanobacteria..'

During the preparation of Seckbach's 1991 paper to *Journal of Phycology*, Dr. K. Zetsche replied to Dr. Lembi (the editor of J. Phycol.) with the following: '... there is a totally wrong interpretation of our data, ... [that our studies, Maid *et al.* 1990] do not support the endosymbiotic hypothesis for *Cyanidium*... In contrary, our dates [=data] support the endosymbiotic origin of the plastids of *Cyanidium* and the red algae from a cyanobacteria like organism. Our statement is, that the plastids of *Cyanidium* and the red algae originate from another group of cyanobacteria[-]like prokaryotes as the chloroplasts of green algae and higher plants. Furthermore our data exclude the hypothesis that *Cyanidium* may be an association between a cyanelle as described for the alga *Cyanophora paradoxa*'.

Additional discussion on the speculation of the

symbiotic origin of *Cyanidium* and its classification with the glaucophytes has been published recently (Seckbach 1992, pp 409–419; Seckbach *et al.* 1983, pp 953–953).

6. Summary of the systematic and phylogenetic notion

It is obvious from the above review that *C. caldarium* has been placed, by various workers, in a wide range of families, orders, classes and divisions of algae. Often these taxa are quite disjunct and, thus, have little in common with one another. Perhaps the most striking and deplorable, characteristic of this systematic, 'musical chairs' is the utilization of a single bit of new information as the foundation for a new systematic scheme, hardly a firm foundation and certainly not a very scientific approach.

Chapman (1974) seems to be the most rational in his reasons for placing *C. caldarium* in the Porphyridiales *sensu* Kylin 1956 within the family, Cyanidiaceae. It appears to us that the overwhelming weight of new information which has accrued since 1974 testifies to the correctness of Chapman's scheme. The acceptance of recognition of this systematic position for *C. caldarium* in the present scheme of things certainly should not distract from those premises which hold that perhaps *C. caldarium* may represent the most primitive, photosynthetic, eukaryotic organism presently known to science and, as such, may occupy a missing link, perhaps one among many, between the prokaryotic and eukaryotic biospheres. The systematic position (pigeonhole) which *C. caldarium* may occupy is part of a man-made scheme and, as such, has not been handed down to us on stone tablets.

The systematic position occupied by *Cyanidium*, and indeed all organisms, should make a statement of their evolutionary role or niche and on their relationship among other forms, hopefully closely related. Since the supposedly evolutionary (systematic) scheme is a product of man's intellect, since the evolutionary development or product was not witnessed, i.e., recorded by man, and since the evolutionary event cannot be reproduced at will for others to observe, it appears, at this stage of our intellectual and scientific embryology, that we should not lose sight of the fact the late Professor Dr. R. H. Thompson, repeatedly emphasized to his students (and one of us – F.D.O.), that systematic and evolutionary positions and discussions simply reflect our intellectual curiosity and the primitive state of our knowledge and understanding of the history and composition of our biosphere.

References

Allen, M. B. (1952). *Archiv. Mikrobiol.* 17: 34–53.

Allen, M. B. (1954a). In: *Huitième Congrès International del Botanique, Paris, 1954. Comptes-Rendus des Sèance et Rapports et Communications déposés lors du Congrès dans la Section 17 (Phycologie).* Pierre André, Imprimeur, Paris. p. 39.

Allen (1954b). In: *Huitième Congrès International de Botanique, Paris, 1954. Comptes-Rendus des Séance et Rapports et Communications déposés lors du Congrès dans la Section 17 (Phycologie).* Pierre André, Imprimeur, Paris. pp 41–42.

Allen, M. B. (1959). *Archiv. Mikrobiol.* 32: 270–277.

Allen, M. B., French, C. S. and Brown, J. S. (1960). In: *Comparative Biochemistry* (ed. Allen, M. B.) pp 33–52.

Allsopp, A. (1969). *New Phytol. Br. Bot. J.*, (London) 68: 591–612.

Archer, W. (1875). *J. Linn. Soc. London Bot.* (London) 14: 328–340.

Balsamo, F. (1892). *Boll. Soc. Nat. Napoli* ser. I (Napoli) 6: 77–97.

Bourelly, P. (1954). *Proc. Int. Bot. Congr.* 17:117–118. (cited from Klein and Conquist, 1967).

Castaldo, R. (1968). *Delpinoa* N. er. (Napoli) 8/9 (1966–1967): 135–147.

Castaldo, R. and Ligrone, R. (1978). *Delpinoa* N. Ser (Napoli) 18/19(1976–1977): 25–36.

Castenholz, R. W. (1979). In: *Strategies of Microbial Life in Extreme Environments* (ed. Shilo, M.) Dahlem Konferenzen, Verlag Chemie Berlin. pp 373–393.

Chapman, D. J. (1965). Ph.D. Dissertation, Department of Marine Biology, University of California, San Diego. XI + 169 pp.

Chapman, D. J. (1974). *Nova Hedwigia* 25: 673–382.

Christensen, T. (1962). In: Böcher, T. W., Lange, M. and Sørensen, T. (Redig.). *Botanik. Bind II. Systematisk Botanik. Nr. 2.* I Kommission hos Munksgaard, København. 178 pp.

Christensen. T. (1966). In: Böcher, T. W., Lange, M. and Sørensen, T. (Redig.). *Botanik. Bind II. Systematisk Botanik. Nr. 2. 2. Udgave.* I Kommission hos Munksgaard, København. 180 pp.

Collins, F.S., Holden, I. and Setchell, W. A. (1901). *Phytotheca Borealis Americana Fasc. 18, No. 851–900* Cited by Tilden, 1910.

Copeland, J. J. (1936). *Ann. N.y. Acad. Sci.* (New York) 36: 1–229.

De Luca, P. and Moretti, A. (1983). *J. Phycol.* 19: 368–369.

De Luca, P., Musacchio, A. and Taddei, R. (1972). *Delpinoa* N. Ser. (Napoli) 12/13(1970–1971): 115–123.

De Luca, P. and Taddei, R. (1970). *Delpinoa* N. Ser. (Napoli) 10/11(1968–1969): 79–89.
Demoulin, V. (1980). In: *Abstracts, Second International Congress of Systematic and Evolutionary Biology. The Uni. of British Columbia, Vanxouver, Canada.* p. 179.
Doemel, W.N. (1970). Ph.D. Dissertation, Department of Microbiology, Indiana University, Bloomington. Z + 222 pp + Vita (1).
Drouet, F. (1943). *Am. Midl. Nat.* (Notre Dame) 30: 671–674.
Drouet, F. and Daily, W. A. (1956). *Butler Univ. Bot. Stud.* (indianapolis) 12: 1–128.
Fogg, G. E. (1956). *Bacter. Rev.* 20: 148–165.
Fredrick, J. F. (1971). *Phytochem.* (Oxford) 10: 395–398.
Fredrick, J. F. (1975). In: *Isozymes IV. Genetics and Evolution.* (ed. Markert, C.) *The Third International Conference on Isozymes Held at Yale University, New Haven, Connecticut, April 18–20, 1974.* Acad, Press. N.Y. pp 307-321.
Fredrick, J. F. (1979). *Phytochem.* 18(11): 1823–1825.
Galdieri, A. (1899). *Rend. Accad. Sci. Fis. Mat. Napoli* Ser. 3[a] (Napoli) 5: 160–164.
Geitler, L. (1925). *Beih. Bot. Zentralbl.* (Kassel) 41 Abt II: 163–294.
Geitler, L. (1933). *Arch. Hydrobiol. Suppl.* (Stuttgart) 12: 622–234.
Geitler, L. (1959). *Österr. Bot. Zietschr.* (Wein) 106: 172–173.
Geitler, L. (1960). In: *Handbuch der Pflanzenanatomie. Band VI, Teil I. Abteilung: Spezieller Teil.* (Zimmermann, W. and Ozenda, O. (Hrsg.)] Gebrüder Borntraeger, Berlin-Nikolasse. VII + (1) + 131 SS.
Goodwin, T. W. (1960). In: *Comparative Biochemistry of Photoreactive Sysytems* (ed. Allen, M. B.) Academic Press, New York. pp 1–10.
Goodwin, T. W. (1966). In: *Comparative Phytochemistry* (ed. Swain, T.) Academic Press, New York. pp 121–137.
Gusev, M. V. (1961a). *Mikrobiologiia* (Moskva) 30: 1108–1128.
Gusev, M. V. (1961b). *Microbiology* (U.S.S.R.; New York) 30: 897–911.
Hara, Y. and Chihara, M. (1974). *Sci. Rep. Tokyo Bunrika Daigaku Sect. B* (Tokyo) 13(232): 209–235 + pls I–XV.
Hase, T., Wakabayashi, S., Wada, K., Matsubara, H., Jüttner, F., Rao, K. K. Fry, I. and Hall, D. O. (1978). *FEBS Lett. (Fed. Eur. Bioch;em. Soc.* Amsterdam) 96: 41–44.
Hirose, H. (1947). *Kagaku To Kagaku Kyoiku (Sci. Scient. Educ.)* 1: 43–45.
Hirose, H. (1950). *Bot. Mag.* Tokyo 63(745/746): 107–111.
Hirose, H. (1958). *Bot. Mag.*, Tokyo 71: 347–352.
Hirose, H. (1959). *Bot. Mag.*, Tokyo 72: 348–352.
Ikan, R. and Seckbach, J. (1972). *Phytochem.* 11: 1077–1082.
Inman, O. L. (1940). *J. Gen. Physiol.* (New York) 23: 661–666.
Klein, R. M. (1970). *Ann. N.Y. Acad. Sci.* (New York) 175: 623–633.
Klein, R. M. and Cronquist, A. (1967). *Q. Rev. Biol.* 42: 105–296.
Kremer, B.P. (1982). *Br. Phycol. J.* (London) 17: 51–61.
Kremer, B. P. (1983). In: *Endocytobiology II* (eds. Schenk, H.E.A. and Schwemmler, W.) W. de Gruyter, Berlin, New York. pp 963–970.
Kremer, B. P. and Feige, G. B. (1979). *Zeitschr. Naturforsch. Teil C, Biosci.* 34: 1209–1214.
Kremer, B.P., Feige, G.B. and Schneider, A.W. (1978). *Naturwissenschften* 65: 157.
Kützing, F. T. (1845–1849). *Tabulae Phycologicae oder Abbildungen der Tange. Herausgegeben von Friedrich Traugott Kützomg. I. Band. Mit 100 Tafeln.* Nordhausen. VI + 54 SS + Tab. 1–100.
Kylin, H. (1956). GWK Gleerups Förlag, Lund. Frontispiece + XV + 673 SS.
Lee, R. E. (1974). *Br. Phycol. J.* 9: 291–295.
Lewin, R. A. (1961). *News Bull. Phycol. Soc. Am.* (Urbana) 14: 6–7.
Maid, U., Valentin, K. and Zetsche, K. (1990). *Curr. Genet. 17*: 255–259.
Margulis, L. (1968). *Science* 161: 1020–1022.
Margulis, L. (1981). *Symbiosis in Cell Evolution. Life and Its Environment on the Early Earth.* W. H. Freeman and Company. San Francisco. XXII + (2) + 419 pp.
Meneghini, G. (1839). *Nuovo Giorn. Lett.* (Pisa) 39: 67–68.
Mercer, F. V., Bogorad, L. and Mullens, R. (1962). *J. Cell Biol.* (New York) 13: 393–403.
Moseley, H. N. (1875). *J. Linn. Soc. London Bot.* (London) 14: 321–325.
Nagashima, H. and Fukuda, I. (1983). *Phytochem.* 22: 1949–1951.
Nagashima, H., Ueki, K., Fukuda, I. and Seckbach, J. (1993). In: *Endocytobiology V (eds. Sato S. et al.). Tübingen University Press, Tübingen, Germany.* pp 279–285.
Nicols, K. E. and Bogorad, L. (1960). *Nature* 188: 870–872.
Ohta, N., Nagashima, H. and Kawano, S. and Kuroiwa, T. (1992). *Plant Cell Physiol.* 33: 657–661.
Profili, P. L. and Maini, P. S. (1950). *Ann. Ist. Super. Sci. Lett. 'S. Chiara'* (Napoli) XVII[1949–1950 (2)]: 317–327.
Raven, P. H. (1970). *Science* 169: 641–646.
Reed, R. H. (1983). *Phycologia* 22: 351–354.
Rosen, W. G. and Siegesmund, K. A. (1961). *J. Biophys. Biochem. Cytol.* (New York) 9: 910–914.
Schmidt, E. W. (1954). In: *A. Engler's Syllabus der Pflanzenfamilien mit besonderer Berücksichtigung der Nutzpflanzen nebst einer Übersicht über die Florenreiche und Florengebiete der Erde. I. Band. Allgemeiner Teil. Bakterien bis Gymnospernem. (Melchior, H. and Werdermann, E.) Zwölfte, völlig neugestaltete Auflage. Gebrüder Borntraeger, Berlin-Nikolassee. pp 50–55.*
Seckbach, J. (1972). *Microbios* (London) 5: 133–142.
Seckbach, J. (1987). *Ann. N.Y. Acad. Sci.* (New York) 503: 424–437.
Seckbach, J. (1991). *J. Phycol.* 27: 794–796.
Seckbach, J. (1992). In: *Algae and Symbioses (ed. Reisser, W.) Biopress Ltd. pp 399–426.*
Seckbach, J. and Fredrick, J. F. (1980). *Microbios* 29: 135–147.
Seckbach, J. and Fredrick, J. F. (1981). In: *Origin* of Life (ed. Wolman, Y.) Reidel Publ. Comp., Dordrecht, Holland. pp 567–574.
Seckbach, J., Fredrick, J. F. and Garbary, D. J. (1983). In: *Endocytobiology II (eds. Schenk H. E. A. and Schwemmler, W.)* Walter de Gruyter, Berlin, New York, pp 947–962.
Seckbach, J., González, E., Wainwright, I. M. and Gross, W. (1992). *Nova Hedwigia* 55: 99–109.
Seckbach, J., Hammerman, I. and Hanania, J. (1981). *Ann. N.Y. Acad. Sci.* (New York) 381: 409–425.
Seckbach, J. and Ikan, R. (1972). *Plant Physiol.* 49: 457–459.

Seckbach, J., Ikan, R., Ringelberg, D. and White, D. (1993). *Phytochem.* 34(6): 1345–1349.
Seckbach, J., Nagashima, H. and Fukuda, I. (1990). In: *Endocytobiology IV (eds. Nardon, P. et al.)* INRA, Paris. pp 575–578.
Sentsova, O. (1991). *Bot. Zhurn. (Moskva, Leningrad)* 76: 69–79.
Silva, P. C. (1962). In: *Physiology and Biochemistry of Algae* (ed. Lewin, R. A.) Acad. Press, New York and London, pp 827–837.
Suzuki, K., Ohta, N. and Kuroiwa, T. (1992). *Protoplasma* 171: 80–84.
Tilden, J. (1898). *American Algae: Century III. Minneapolis, Minnesota. (2) + Specimens Nos. 201–300 + Specimen No. 65b.*
Tileden, J. (1898a) *Bot. Gaz.* 25: 89–105 + pls VIII–X.
Tôyoma, S. (1980). *Cytologia (Tokyo)* 45: 779–790.
West, G. S. (1904). *J. Bot. Br. Foreign* (London) 42: 2810294 +pl. 464.

Chapter 9

New classification for the genus *Cyanidium* Geitler 1933

FRANKLYN D. OTT[1] and JOSEPH SECKBACH[2]
[1] *317 Osborne Avenue, Catonsville, MD 21228, USA*
[2] *The Hebrew University of Jerusalem, School for Overseas Students, Mount Scopus, Jerusalem 91905, Israel*

Outlines:
1. Summary of the taxonomic review
2. Highlights of the chronological line of nomenclature and taxonomy
3. Seven selected members of the Cyanidiaceae
   3.1 *Cyanidium caldarium*
   3.2 *Galdieria sulphuraria*
   3.3 *Cyanidium caldarium* var. *chilense*
   3.4 *Cyanidium caldarium* var. *chilensis* f. *rumpens*
   3.5 *Galdieria partita*
   3.6 *Galdieria maxima*
   3.7 *Galdieria daedala*
4. Basic justification for reclassification
5. The enigma of '*Cyanidium caldarium*'
6. New classification
7. Key to species
8. Synonym list
9. *Cyanidioschyzon* to be placed in the Porphyridiaceae
10. Conclusions
References

1. Summary of the taxonomic review

The taxonomic and systematic chapters (Ott and Seckbach in this volume) gave the following binomials (and where applicable their respective formae) that have been applied at various times throughout the years to material presently recognized as the Cyanidaceae Geitler 1933. After each listed binomial there will be found, in parentheses, the modern name of the respective algal division to which the binomial would have been assigned at the time of its initial circumscription. The names of these respective algal divisions here employed were those recommended by Papenfuss (1955) with the single exception that the name for the blue-green algal division has been taken from Bold (1967).

2. Highlights of the chronological line of nomenclature and taxonomy

1) *Coccochloris orsiniana* Meneghini: *Nuovo Giorn. Lett.* (Pisa) XXXIX: 67. 1839 (Cyanophycophyta).
2) *Coccochloris orsiniana* Meneghini 1839 emend. Meneghini: *Monogr. Nostoch. Ital.* (Augustae Taurinorum) 63, Tab. VIII, figs 1, 1a, 1b. 1842. (Cyanophycophyta).
3) *Palmella orsiniana* (Meneghini 1839 emend. Meneghini 1842) Kützing: *Tab. phycol.* (Nordhausen) I, 16, Tab. 16, fig. III. 1846. (Chlorophycophyta).
4) *Chroococcus varius* A. Braun in Rabenhorst: *Alg. Europ.* (Dresden) Dec. CCXLVI–CCXLVIII, Nos 2451 et 2452. 1876. (Cyanophycophyta).
5) *Protococcus volcanicus* Cesati in Balsamo: *Boll. Soc. Nat. Napoli* (Napoli), ser. I, VI: 94. 1892. (Chlorophycophyta).
6) *Protococcus botryoides* (Kützing 1846) Kirchner 1878 forma *caldarius* Tilden: *Bot. Gaz.* (Crawfordsville) XV: 94, Pl. VIII, fig. 18. 1989. (Chlorophycophyta).
7) *Pleurococcus sulphurarius* Galdieri: *Rendiconti Reale Accad. Sci. Fis. Mat.* (Napoli), ser. 3[a], V: 162, figs A-I on p. 161. 1899. (Chlorophycophyta).
8) *Pleurocapsa caldaria* (Tilden 1898) Setchell in Collins, Holden et Setchell: *Phycoth. Bor. Amer.* (Malden) Fasc. XIII, No. 851. 1901. (Chlorophycophyta).
9) *Palmellococcus thermalis* G.S. West: *Jour. Bot.* (London) LXII (502): 287, Pl. 464, fig. 21. 1904. (Chlorophycophyta).
10) *Chroococcopsis caldaria* (Tilden 1898) Geitler: *Beih. Bot. Centralbl.* (Kassel) XLI, Abt. II: 241, 1925. (Cyanophycophyta).

J. Seckbach (ed.), Evolutionary Pathways and Enigmatic Algae: Cyanidium caldarium *(Rhodophyta) and Related Cells,* 145–152, 1994.

11) *Cyanidium caldarium* (Tilden 1898) Geitler: *Arch. Hydrobiol.* (Stuttgart) Suppl. XII: 624. 1933. (Cyanophycophyta).
12) *Pluto caldarius* (Tilden 1898) Copeland: *Ann. New York Acad. Sci.* (New York) XXXVI: 72, fig. 33 in p. 73. 1936. (Cyanophycophyta).
13) *Cyanidium chilense* Schwabe: *Verhandl. Deutsch. Wissenschaftl. Ver. Santiago Chile* (Valparaiso) N.F. III: 119–120, abb. 5. 1936. (Cyanophycophyta).
14) *Cyanidium caldarium* (Tilden 1898) Geitler 1933 var. *chilensis* (Schwabe 1936) Schwabe: *Mitt. Deutsch. Ges. Natur-Völkerk. Ostasiens* (Tokyo) XXXIII (Teil C): C41-C42. 1942. (Cyanophycophyta).
15) *Cyanidium caldarium* (Tilden 1898) Geitler 1933 var. *chilensis* (Schwabe 1936) Schwabe 1942 forma *rumpens* Schwabe: *Mitt. Deutsch. Ges. Natur- Völkerk. Ostasiens* (Tokyo) XXXIII (Teil C): C42-C43. 1942 (Cyanophycophyta).
16) *Chroococcidiopsis thermalis* Geitler 1933 var. *nipponica* Negoro: *Shkubutsugaku Zasshi* [*Bot. Mag.* (Tokyo)] LVII: 310, fig. 9. 1943. (Cyanophycophyta).
17) *Dermocarpa caldaria* (Tilden 1898) Drouet: *Amer. Midl. Nat.* (Notre Dame) XXX: 672. 1943. (Cyanophycophyta).
18) *Entophysalis rivularis* (Kützing 1833) Drouet: *Amer. Midl. Nat.* (Notre Dame) XXX: 671. 1943 (Cyanophycophyta).
19) *Pluto caldarius* (Tilden 1898) Copeland ex Drouet: *Amer.Midl. Nat.* (Notre Dame) XXX: 672. 1943. (Cyanophycophyta).
20) *Cyanidium caldarium* (Tilden 1898) Geitler 1933 forma *rumpens* (Shwabe 1942) Schwabe: *Mitt. Deutsch. Ges. Natur- Völkerk. Ostasiens* (Shanghai) *Supplement band* XXI: 135. 1944 (Cyanophycophyta).
21) *Cyanidium caldarium* (Tilden 1898) Geitler 1933 forma *chilensis* (Schwabe 1936) Schwabe: *Mitt. Deutsch. Ges. Natur- Völkerk, Ostasiens* (Shanghai) Supplement band XXI:135, Abb. 42a. 1944. (Cyanophycophyta).
22) *Chlorella caldaria* (Tilden 1898) Hirose: *Shokubutsugaku Zasshi [Bot. Mag.* (Tokyo)] LXIII: (745–746): 110. 1950. (Chlorophycophyta).
23) *Entophysalis rivularis* (Kützing 1833) Drouet 1943 forma *puteolana* Profili et Maini: *Ann. Ist. Super. Sci. Lett. 'S. Chiarara'* (Napoli) XVII: 327. 1950 (Cyanophycophyta).
24) *Oncobyrsa orsiniana* (Meneghini 1839 emend. Meneghini 1842) Bergen in Profili et Maini: *Ann. Ist. Super. Sci. Lett. 'S. Chiara'* (Napoli) XVII: 230. 1950. (Cyanophycophyta).
25) *Chlorella caldaria* (Tilden 1898) M.B. Allen: *Rapp. comm. Parvenue avant le Congrés, Huit. Congr. Internat. Bot.* (Paris) XVII: 43. 1954. (Chlorophycophyta).
26) *Rhodococcus caldaria* (Tilden 1898) Hirose: *Shokubutsugaku Zasshi [Bot. Mag.* (Tokyo)] LXXI (844): 351. 1958. (Rhodophycophyta).
27) *Cyanidium caldarium* (Tilden 1898) Geitler 1933 forma *piccola* (A) De Luca et Taddei: *Delpinoa* (Napoli) n. ser., X-XI: 84. 1970. (Rhodophycophyta). (this entry is not a formal synonym, but is here given because it has been widely cited as such by a number of workers since 1970).
28) *Cyanidium caldarium* (Tilden 1898) Geitler 1933 forma *grande* (B) De Luca et Taddei: *Delpinoa* (Napoli) n. ser. X-XI: 84. 1970. (Rhodophycophyta). (this entry is not a formal synonym, but is here given because it has been widely cited as such by a number of workers since 1970).
29) *Galdieria sulphuraria* (Galdieri 1899) Merola in Merola, Castaldo, De Luca, Gambardella, Musacchio et Taddei: *Giorn. Bot. Ital.* (Firenze) 115:(4–5). 191, 192, fig. C. 1981. (Rhodophycophyta).
30) *Cyanidium caldarium* Geitler 1935, sic. 1933, sensu Merola, Castaldo, De Luca, Gambardella, Musacchio et Taddei: *Giorn. Bot. Ital.* (Firenze). 115 (4–5). 191. 1981. (Rhodophycophyta) non *Cyanidium caldarium* (Tilden 1898) Geitler: *Arch. Hydrobiol.* (Stuttgart) Suppl. XII: 624. 1933. (Cyanophycophyta).
31) *Galdieria partita* Sentsova 1991: *Bot. Zhur.* (Moskva, Leningrad) 76(1): 73, Ris. 1, figs. A1-A14. 1991. (Rhodophycophyta).
32) *Galdieria maxima* Sentsova 1991: *Bot. Zhur.* (Moskva, Leningrad) 76(1): 75, Ris 1, figs. G1-G9. 1991. (Rhodophycophyta).
33) *Galdieria daedala* Sentsova 1991: *Bot. Zhur.* (Moskva, Leningrad) 76(1): 75, Ris 1, figs. B1-B11. 1991. (Rhodophycophyta).

Table 1. New classification of the species within the (family) Cyanidiaceae

Species of *Cyanidium*	Number of endospores	Diameter cell (μm)	Special features (C+ = chloroplast)	Synonym. list
C. caldarium	4	4 – 6	acid pH 5↓ max. 57C	many names
C. sulphurarius	4, 8, 16 (32)	3 – 11	multi-lobed Ct acid pH 5↓, 54C	Cc f. B, *Galdieria*
C. chilense	4	3 – 6	neutral medium pH 5→	cave-Cc var. *Chilense*
C. rumpens	8 or more	4.5 – 6	neutral medium pH 5→	cave-Cc var. *Chilense f. rumpens*
C. partitium	4 -16- 32	3 – 11	cup shaped Ct pH 5↓ 54C	*G. partita*
C. maximum	4-8-16	6.5-16.5	up to 4 oval Ct (larger cells) pH 5↓ 54C	*G. maxima*
C. daedalium	2, 4, 8, 16	3 – 11	Ct lobed & wavy edges pH 5↓	*G. daedala*

– (order) Cyanidiales – (subclass) Bangiophycidae/Cyanidiophyceae – (class) Rhodophyceae – (division) Rhodophycophyta. *Cyanidium* natural history started in 1838, there were 33 taxa registered. Ct = chloroplast; Cc = *Cyanidium caldarium.*

3. Seven selected members of the Cyanidiaceae

Based upon the above 33 listed taxa, we suggest that at least the following seven taxa (Table 1) ought to be considered valid by students of the genus *Cyanidium* Geitler 1933, namely:

1) *Cyanidium caldarium* (Tilden 1898) Geitler 1933.
2) *Galdieria sulphuraria* (Galdieri 1899) Merola in Merola, Castaldo, De Luca, Gambardella, Musacchio et Taddei 1981.
3) *Cyanidium chilense* Schwabe 1936.
4) *Cyanidium caldarium* (Tilden 1898) Geitler 1933 var. *chilensis* (Schwabe 1936) Schwabe 1942 forma *rumpens* Schwabe 1942.
5) *Galdieria partita* Sentsova 1991.
6) *Galdieria maxima* Sentsova 1991.
7) *Galdieria daedala* Sentsova 1991.

3.1 Cyanidium caldarium

The type species of the genus, *Cyanidium caldarium*, reproduces by the formation of four endospores which are released by the splitting of the mother cell (sporangial) wall. It typically lives in habitats below pH 5, often approaching pH 1, and at elevated temperatures up to 57 °C. Thus *C. caldarium* is a moderate thermophile and an extreme acidophile. It appears to be widely distributed in such habitats throughout the world. The presence of only one mitochondrion, a chloroplast (having a club-like nucleoid), the absence of vacuoles, and trienoic fatty acids (Merola *et al.* 1981, p. 191) or the utilization of nitrates, are best to be considered only of secondary, taxonomic importance and, in any event, are scarcely significant to be the basis of generic and familial distinctions.

3.2 Galdieria sulphuraria

The second taxon, *Galdieria sulphuraria* (Galdieri) Merola *et al.* 1981 reproduces by the formation of four, eight, sixteen and thirty-two endospores. The cells possess a single, parietal, cup-shaped or multilobed chloroplast (having a ring-link nucleoid), the endospores possess a similar plastid type. This alga typically lives in habitats below pH 5 and at temperatures up to 54 °C. This taxon is distributed throughout the world in such habitats. As was the case of *C. caldarium*, the presence here of several mitochondria, one vacuole, possession of trienoic fatty acid (Merola *et al.* 1981, p. 191), or its incompatibility to use nitrate, are best to be considered only of secondary, taxonomic importance and, in any event, are scarcely significant to be the basis of generic and familial distinctions.

3.3 C. caldarium *var.* chilense

The third taxon, *C. caldarium* (Tilden) Geitler var. *chilense* Schwabe was first described from material 'in Brandungshöhlen (caves caused by surf) von Ramuntcho bei Talcahuan Mittelchilie and in Küstenhölen (coastal caves) bei Zapallar (Chilie)'.

This aerophilic, or perhaps better, lithophilic, alga reproduces by the formation of four endospores which are released by the splitting of the mother cell (sporangial) wall. Its typical (epipelagic) habitat is on dimly lighted walls and floors near the openings of caves. *C.caldarium* var. *chilense* receives most of its moisture from water condensation from the damp atmosphere (Schwabe 1936, pp 120–121). The pH of the collecting site was not determined but, Schwabe later (1942, p. 213) indicated that it was above pH 5. This rarely encountered, cavernous species, as has been previously noted, is known (reported as *C. caldarium*) from Israel (Leiv-Hetzroni, 1965), Italy (Skuja, Missouri, U.S.A. (personal communication from Dr. Russel Rhodes to Dr. F.D. Ott), and France (Leclerc, Couté and Dupuy, 1983). The alga reported by Cribb (1976) as *C. caldarium*, collected from a shady ledge below a waterfall in Australia, probably falls to this taxon. For more details see Ott and Seckbach (on Taxonomy and Systematics) and Hoffmann's chapter, in this volume.

3.4 C. caldarium *var.* chilensis *forma* rumpens

The fourth taxon, *C. caldarium* (Tilden) Geitler var. *chilensis* Schwabe forma *rumpens* Schwabe was described from material collected along the precipitous, Chilian, Pacific coast in the vicinity of Vina del Mar and Valparasio. This alga is also aerophilic or lithophilic. Forma *rumpens* is known only from its type locality. The material was found in narrow and extremely fine fissures in granite rock. Indeed, later, Schwabe (1960, p. 304) employed the term 'Endolithische' to describe this habitat and attributed the growth of forma *rumpens* as the cause of scaling and breakdown of the rock serving as the substrate (Schwabe, 1960, p. 303), hence the varietal epithet '*rumpens*'.

This taxon is characterized by the formation of eight or more endospores which have been produced from several cells whose cell walls remain intact. This is accomplished by the failure of the mother cell to release its endospores before they proceed with endospore formation for the next generation. Thus, several successive generations may be retained within the initial mother cell wall prior to the final release of the several endospore generations by the splitting of the mother cell walls. The populations received ocean spray during extreme high surf and as a result they were often encrusted with salt crystals. At other times, freshwater from rain moistened the populations. The pH of the collecting sites was not determined, but considering the strong buffering properties of seawater, certainly the pH must not have been strongly acidic and perhaps even about neutral. For more detail see Hoffmann's chapter; Ott and Seckbach reviews in this volume.

3.5 Galdieria partita

The fifth taxon, *G. partita* Sentsova, has recently been described from material collected in the acido-thermal springs at Uzon Caldara, Kamtchatka, USSR and in the Yellowstone National Park, USA. The habitat of the collections is the typical for these acido-thermal strains with pH below 5 and temperatures up to 54 °C. The cells possess a single parietal, laminated, lobed, or partitioned chloroplast which occupies most of the periphery of the cell wall. In the endospores, the chloroplasts appear to be parietal and are ribbon, band, or belt shaped. The alga typically reproduces by the formation of four to sixteen endospores (Sentsova, 1991, p. 73). See chapter by Sentsova in this volume.

3.6 Galdieria maxima

The sixth taxon, *G. maxima* Sentsova, is characterized by the cells possessing more that one (perhaps up to four) parietal, laminated, lobed to oval, chloroplasts; the endospores reportedly contain four parietal plastids. This alga typically reproduces by the formation of either two, four, or eight endospores (Sentsova, 1991, p. 75). *G. maxima* is known only from the Kunashir Islands, USSR, from acido-thermal springs. See chapter by Sentsova in this volume.

3.7 *Galdieria daedala*

The seventh, and the last remaining taxon to be recognized by the present authors, is *G. daedala* Sentsova. This taxon is known also only from acido-thermal springs on the Kunashir Islands, USSR. It is characterized by the cells possessing typically one, or occasionally two, parietal, laminated, irregularly lobed chloroplasts; the plastid margins are wavy. The endospores have a single, parietal chloroplast. It reproduces by the formation to two, four, eight, or sixteen endospores. See chapter by Sentsova in this volume.

4. Basic justification for reclassification

Because these seven algae 1) reproduce vegetatively by the formation of endospores and by no other means, 2) are photosynthetic with *C*-phycocyanin as their dominant photosynthetic accessory pigment and 3) possess parietal, laminated chloroplasts not possessing a pyrenoid, we think they should best be handled by retaining all seven of these taxa in the genus *Cyanidium* Geitler 1933. The primary basis for separation into species resides in the number of endospores produced and in the morphology of the chloroplast.

As in this case, when species can be distinguished by purely morphological and cytological characters easily discernable by the use of the light microscope, the use of biochemical and ultrastructural characteristics appears to be superfluous and certainly should not be the basis for generic and familial distinctions. While such information is surely important, it is best to be considered only of secondary taxonomic importance at this point in time.

5. The enigma of '*Cyanidium caldarium*'

The conflicting data observed in *Cyanidium caldarium* is not new, but have been known for several years (see Lewin 1974; Seckbach 1992, 1987; Brock 1978). These conflicts revolve about 1) the application of the scientific name for a given organism and 2) the use of a single binominal for more than one taxon as applied to *C. caldarium* forma *A*, *C. caldarium* forma *B* and *Galdieria sulfuraria* by various investigators. Some investigators have employed *C. caldarium* for the smaller size algae as represented by *C. caldarium* forma *A* and Strain RK-1 (of Dr. Fukuta's Laboratory, University of Tokyo, Tokyo). Other investigators have applied *C. caldarium* for the larger size alga represented by *C. caldarium* forma *B*, Strain M-8 of the Institute of Applied Microbiology, Tokyo) and the Allen Strain (Strain No. C-14.1.1 of the Laboratory of Comparative Biology, Kaiser Foundation Research Institute, Richmond, California) of which the latter has also been designated *Galdieria sulfuraria* by the Italian School. Still other investigators may have reported results of their investigations under the binominal, *C. caldarium*, while in actuality they may have been investigating mixed populations of the larger and smaller size algae; this latter case may be especially pertinent to studies which have been carried out on naturally occurring populations. Seckbach (1991) has recently discussed this enigma and offered some solutions concerning the disorder between these algae carrying the same genus and species names.

6. New classification

Therefore, we advocate the following classification:

Division RHODOPHYCOPHYTA Papenfuss 1946
- *Class* Rhodophyceae *Ruprecht in Middendorff 1891*
  - *Subclass* Bangiophycidae *De Toni 1897 orth. mut. Newton 1953*
    - *Order* Cyanidiales *Christensen, T. in: Böcher, Lange et Sørensen 1962*
      - *Family* Cyanidiaceae *Geitler 1933 emended D.J. Chapman 1974*

Type species:

1) *Cyanidium caldarium* (Tilden 1898) Geitler 1933.

Other species:

2) *Cyanidium sulphurarium* (Galdieri 1899) F.D. Ott, comb. nov. Specific basionym: *Pleurococcus sulphurarius* Galdieri: *Rendiconti Reale Accad. Sci. Fis. Mat.* (Napoli) ser 3ª V. 162, figs A-I on p. 161. 1899.
3) *Cyanidium chilense* Schwabe 1936.
4) *Cyanidium rumpens* (Schwabe 1936) F.D. Ott, comb. nov.

Specific basionym: *Cyanidium caldarium* (Tilden 1898) Geitler 1933 var. *chilensis* (Schwabe 1936)
Schwabe 1942 forma *rumpens* Schwabe: *Mitt. Deutsch. Ges. Natur-Völkerk. Ostasiens* (Tokyo) XXXIII(Teil C). C42-C43. 1942.

5) *Cyanidium partitum* (Sentsova 1991) F.D. Ott, comb. nov. Specific basionym: *Galdieria partita* Sentsova: *Bot. Zhur.* (Moskva, Leningrad) 76(1): 73. Ris 1, figs A1-A14. 1991.
6) *Cyanidium maximum* (Sentsova 1991) F.D. Ott, comb. nov. Specific basionym: *Galdieria maxima* Sentsova: *Bot. Zhur.* (Moskva, Leningrad) 76(1): 75. Ris 1, figs. G1-G19. 1991.
7) *Cyanidium daedalum* (Sentsova 1991) F.D. Ott, comb. nov. Specific basionym: *Galdieria daedala* Sentsova: *Bot. Zhur.* (Moskva, Leningrad) 76(1): 75. Ris 1, figs. B1-B11. 1991.

7. Key to the species

1) populations typically found in acidothermal habitats 2
1) populations typically in non-acidothermal habitats 6
  2) cells typically reproducing by the formation of *only* four endospores, with mother cells possessing only a single laminated, parietal cup-shaped chloroplast, cells 4–6 μm in diameter .... *C. caldarium*
  2) cells typically reproducing by the formation of fewer than, or more than four, endospores, as well as four endospores 3
3) cells typically reproducing by the formation of two, four, or eight endospores, with the mother cell possessing a single, laminated, parietal, irregularly lobed chloroplast and with the edges the chloroplast wavy, the mother cell wall mildly protuberant, conforming to the shape of the enclosed endospores, cells 3–11 μm in diameter .... *C. daedalium*
3) cells typically reproducing by the formation of four to sixteen or more endospores 4
  4) cells typically reproducing by the formation of four to thirty-two endospores, with the mother cell possessing a single, laminated, parietal, cup-shaped chloroplast, cells 3–11 μm in diameter .... *C. partitium*
  4) cells typically reproducing by no more than sixteen endospores 5
5) cells typically reproducing by the formation of four, eight, sixteen, or thirty-two endospores, with mother cell possessing a single, laminated, parietal, cup-shaped chloroplast, cells 3–11 μm in diameter .... *C. sulphurarius*
5) cells typically reproducing by the formation of four, eight, or sixteen endospores, with the mother cell typically possessing a single, laminated, parietal, irregular to oval shaped chloroplast, on occasion, up to four chloroplast per mother cell, cells 6.5–16.5 μm in diameter.......................... *C. maximum*
  6) cells typically reproducing by the formation of four endospores, with the mother cells possessing single, laminated, parietal cup-shaped chloroplast, cells 3–6μm in diameter .... *C. chilense*
  6) cells typically reproducing by the formation of eight or more endospores which are retained in successive mother cell walls for several generations of endospore formation prior to their release, with mother cells possessing a single, parietal, laminated, cup-shaped chloroplast, cells 4.5–6 μm in diameter .... *C. rumpens*

8. Synonym list

From the above listed 33 binomials of *Cyanidium caldarium* (*sensu lato*) in paragraph 2 of this article, it is now possible to distribute these binomials among the seven taxa recognized in this contribution; the parenthesized number after each entry refers to the number in that list:

1. *Cyanidium caldarium* Geitler:
Coccochloris orsiniana Meneghini (1), *C. orsiniana* Meneghini emen. Meneghini (2), *Palmella orsiniana* (Meneghini) Kützing (3); *Chroococcus varius* A. Braun in Rabenhorst (4), *Protococcus vulcanicus* Casati in Balsamo (5), *P. botryoides* (Kützing) Kirchner forma *caldarius* Tilden (6), *Pleurocapsa caldaria* (Tilden) Setchell in Collins, Holden et Setchell (8), *Palmellococcus thermalis* G.S. West (9), *Chrooccopsis caldaria* (Tilden) Geitler (10), *Pluto caldarius* (Tilden) Copeland (12), *Dermocarpa caldaria* (Tilden) Drouet (17), *Entophysalis rivularis* (Kützing) Drouet (18), *Pluto caldarius* (Tilden) Copeland ex Drouet

(19), *Chlorella caldaria* (Tilden) Hirose (22), *Entophysalis rivularis* (Kützing) Drouet forma *puteolana* Profili et Maini (23), *Oncobyrsa orsiniana* (Meneghini emend. Meneghini) Bergen ex Profili et Maini (24), *Chlorella caldaria* (Tilden) M.B. Allen (25), *Rhodococcus caldaria* (Tilden) Hirose (26), *Cyanidium caldarium* (Tilden) Geitler forma *piccola* (A) De Luca and Taddei (27), *C.caldarium* Geitler non *C. caldarium* (Tilden) Geitler sensu Merola *et al.* (30).

2. *Cyanidium sulphurarium* (Galdieri) F.D. Ott. *Pleurococcus sulphurarius* Galdieri (7), *Chroococcidopsis thermalis* Geitler var *nipponica* Negoro (16), *Cyanidium caldarium* (Tilden) Geitler forma *grandi (B)*(28), *Galdieria sulphuraria* (Galdieri) Merola in Merola *et al.* (29).
3. *Cyanidium chilense* Schwabe: *C. caldarium* (Tilden) Geitler var. *chilense* (Schwabe) Schwabe (14). *C. caldarium* (Tilden) Geitler forma *chilense* (Schwabe) Schwabe (21).
4. *Cyanidium rumpens* (Schwabe) F.D. Ott *C. caldarium* (Tilden) Geitler var. *chilensis* (Schwabe) Schwabe forma *rumpens* Schwabe (15), *C. caldarium* (Tilden) Geitler forma *rumpens* (Schwabe) Schwabe (20).
5. *Cyanidium partitum* (Sentsova) F.D. Ott *Galdieria partita* Sentsova (31).
6. *Cyanidium maximum* (Sentsova) F.D. Ott *Galdieria maxima* (Sentsova) (32).
7. *Cyanidium daedalum* (Sentsova) F.D. Ott *Galdieria daedala* Sentsova (33)

9. *Cyanidioschyzon* to be placed in the Porphyridiaceae

The alga genus *Cyanidioschyzon* Merola in Merola *et al.* 1981, with its type species, *C. merolae*, constitutes only an ecological cohort of the thermo-acidophilic *Cyanidium* species. These authors placed *C. merolae* in their newly circumscribed family Cyanidiaceae, among the endospore reproducing *Cyanidium* species. However, since it does not reproduce by the formation of endospores, but only by binary fission, it has no taxonomic place here among these endospore, reproducing Cyanidiacean (sensu D.J. Chapman 1974) algae. Suzuki *et al.* (1992) have currently investigated the molecular data of *C. merolae* and indicated the amount of cell-nuclear DNA (54,9%), mitochondrion DNA (10.6%) and chloroplast DNA (34.5%). They concluded that *C. merolae* has the smallest amount of cell-nuclear DNA among all eukaryotes (see Suzuki *et al.* 1992, Tables 1 and 2). *C. merolae* would be best transferred to the Porphyridiaceae among other unicellular members of the Rhodophycophyta reproducing vegetatively only by binary fission. Recently Nagashima *et al.* (1993) have observed additional species of this genus which differ in their ultrastructural and in other characteristics. *Cyanidioschyzon*, with its two new, additional species, would best be placed among the Porphyridiaceae. Further, Seckbach *et al.* (1993) have postulated a new (endosymbiotic) evolutionary pathway for *Cyanidioschyzon* which may develop into the other *Cyanidium* species *via* autogenous (non-symbiotic) evolutionary lines.

10. Conclusions

We believe that the above, straight forward, new classification of the various *Cyanidium* species will avoid further mystification and disorder among the biologists (and phycologists) who use these fascinating organisms for their explorations. Furthermore, our new proposal of the genus *Cyanidium* may bring consolation and relief for those who used *Galdieria* and labelled it also *Cyanidium* (see Seckbach 1991).

References

Bold, T. D. (1967). *Morphology of Plants. Second Edition.* Harper & Row, Publishers, New York, Evanston and London, XXIX + (3) + 541 pp.

Brock, T. D. (1978). *Thermophilic microorganisms and life at high temperatures. Springer verlag. Chapter 9: the genus Cyanidium.* pp. 255–302.

Chapman, D. J. (1974), *Nova Hedwigia* 25: 673–682.

Christensen, T. (1962). In: *Böcher, T.W., Lange, M. and Sørensen (Reding.) Botanik. Bind II. Systematisk Botanik. Nr. 2. København* (11) + pp. 12–178.

Cribb, A. B. (1976). *Queensl. Nat. (Brisbane)* 21: 132–133.

De Toni, J. B. (1897). In: *Sylloge Algarum Omnium Husque Cognigarum.* Vol. IV. *Florideae.* Section I. Familiae I-IX. pp. 2–388.

Leclerc, J. C., Couté, A. and Dupuy, P. (1983). *Cryptogam., Algol.* (Paris) 4(1–2): 1–19.

Leiv-Hetzroni, O. (1965). Algae from Israel Caves. Final Thesis, Master of Natural Sciences. Hebrew University of Jerusalem, Jerusalem. 36 pp. + References (3).

Lewin, R. A. (1974). In: *Algal Physiology and Biochemistry (ed. Steward, W.D.P.) Blackwell Sci. Pub.* pp. 1–39.

Nagashima, H., Ueki, K., Fukuda, I. and Seckbach, J. (1993). In: *Endocytobiology V (eds. Sato S. et al.) Tübingen University Press, Tübingen, Germany.* pp. 279–285.

Newton, L. M. (1953). *Marine Algae* In: *The John Murray Expedition 1933–1934. Scientific Reports. Volume IX Zoology and Botany. No. 5. pp. 394–420. Printed by Order of the Trustees of the British museum, London.*

Papenfuss, G. F. (1955). In: *A Century of Progress in the Natural Sciences 1853–1953. Published in Celebration of the Centennial of the California Academy of Sciences.* California Acad. of Sciences, San Francisco. XII + 807 pp.

Ruprecht, F. J. (1851). *Tange des Ochotskischen Meeres.* SS 191–435. In: *A.T. von Middendorff. Reise in der äussersten Norden und Osten Sibiriens während der Jahre 1842 und 1844 mit Allerhöchster Genehmigung auf Veranstaltung der Kaiserlichen Akademie der Wissenschaften zur St. Petersburg ausgeführt und in Verbindung mit vielen Gelehrten ausgegeben von Dr. A. Th. v. Middendorff. Erster Band. Einleitung. Klimatologie. Geognosie. Botanik. Theil 2. Botanik. Bearbeitet von E.R.v. Trautvetter, F.T. Ruprecht, C.A. Meyer, E. und G. Borszczow. Buchdrucherei der Kaiserlichen Akademie der Wissenschften, St. Petersburg.* IV-(IX) + 435 Ss + Taf. 1-19.

Schwabe, G. H. (1960). *Österr. Bot. Zeitschr.* 107: 281–309.

Seckbach, J. (1987). *Ann. N.Y. Acad. Sci. (New York)*: 503: 424–437.

Seckbach, J. (1991). *J. Phycol.* 27: 794–796.

Seckbach, J. (1992). In: *Alage and Symbioses: Plants, Animals, Fungi, Viruses, Interactions Explored (ed. Reisser, W.) Biopress Ltd.* pp. 399–426.

Seckbach, J., Ikan, R., Nagashima, H. and Fukuda, I. (1993). In: *Endocytobiology V (eds. Sato et al.) Tübingen University Press, Tübingen, Germany.* pp. 241–254.

Suzuki, K., Ohta, N. and Kuroiwa, T. (1992). *Protoplasma* 171: 80–84.

Chapter 10

Cyanidium investigations in Japan

IKUJIRO FUKUDA
Department of Biology, Faculty of Science, Science University of Tokyo, Kagurazaka, Shinjuku-ku, Tokyo, 162 Japan

Outlines:

1. The study of Cyanidiophyceae in Japan
 1.1 Background studies of Negoro with *Cyanidium* material
 1.2 Hirose's contributions
 1.3 Modern pioneering and current investigations
  1.3.1 Physiological studies
  1.3.2 Structural investigations
  1.3.3 The Japanese phycological volume
  1.3.4 Chemical composition and chloroplast DNA analysis
  1.3.5 Phylogenetic explorations
References

1. The study of Cyanidiophyceae in Japan

1.1 Background studies of Negoro with Cyanidium *material*

The history on study of this enigmatic alga, *Cyanidium caldarium* in Japan started during the 1940's. The first full scale publication assigned for this subject can be considered his article 'Untersuchungen über die Vegetation der mineralogen-azidotrophen Gewässer Japans' by Negoro (1944). This article was written in German and published in 'Tokyo Bunridai Kiyô' (Academic Reports of Tokyo College of Science and Literature). The bulk of this report consists from five parts about 140 pp. and includes: 1. Introduction, 2. Living habitat, 3. Culture, 4. Summarized conclusion, and 5. Literature survey.

Innately Japan is one of the famous volcanic regions, and the occurrence of *Cyanidium* in hot springs has been known earlier to that publication (Negoro 1935, 1943). In his papers, Negoro described not only *Cyanidium caldarium* but also *Synechococcus minutissimus*, *Chroococcidiopsis thermalis* and related algae. During his period *C. caldarium* was classified as a member of the Cyanophyceae (i.e., cyanobacteria) because of the algal blue-green color and due to some primitive observations. See chapter on Taxonomy and Systematic survey by Ott and Seckbach in this volume. After persevering endeavor, Negoro succeeded in culturing *C. caldarium* in liquid medium and his cultures were growing in steady progressing conditions. In these experiments, neither CO_2 bubbling procedure nor bacterial contamination was regarded. It should be mentioned here, that the initial credit for *Cyanidium* contemporaneous studies is given to M. B. Allen, who developed laboratory methods of for cultivation blue-green algae (Allen 1952) and for *Cyanidium* (Allen 1959). Brock declares in his book (1958 pg. 255) that: 'The modern history of *Cyanidium* begins with the late Mary Belle Allen, who developed methods...and isolated an unidentified unicellular alga from Lemonade Spring...California'. However, chronological, Negoro is deserving to be recognized as the pioneer in this line.

1.2 Hirose's contributions

After these investigations, until the sequelae of the 2nd World was over, the existence of this interesting algal problem was neglected and forgotten. Later, in the 1950's H. Hirose and his group had some queries on the taxonomic position of this alga (Hirose 1950, 1958). First, Hirose classified *Cyanidium* also into prokaryotic cyanophyceae (cyanobacteria) and then reclassified it into Chlorophyceae (green algae). He described microscopic structure of *Cyanidium* cells and spectroscopic observations of the algal pigments and concluded that this alga belong to *Chlorella* group. But, later on, his new nomenclature has not been accepted according to the International Botanical Code.

J. Seckbach (ed.), Evolutionary Pathways and Enigmatic Algae: Cyanidium caldarium *(Rhodophyta) and Related Cells,* 153–155, 1994.

1.3 Modern pioneering and current investigations

1.3.1 Physiological studies

Physiological study of *Cyanidium* has been started by Fukuda and his group upon the photosynthetic activity, Hill reaction, acidophilic and thermophilic conditions (Fukuda 1958, 1962; Enami and Fukuda 1975). Their early determinations on physiological activity started with classical manometric method which gradually changed to oxygen electrode measurements. The technical apparatus used by Fukuda extended to ESR techniques (Fukuda and Ishisone 1965), polarography (Enami *et al.* 1978) and into other biophysical fields.

1.3.2 Structural investigations

In parallel with these physiological studies, also morphological ultrastructural investigations by electron microscopic technique have been performed by Enami *et al.*(1975), Toyama (1974, 1980; Nagashima and Fukuda (1981), Ueda and coauthors (1981, 1987), Mita *et al.* (1986) and their teams. See also chapter of Ueda in this volume.

1.3.3 The Japanese phycological volume

During these investigations the proponents of *C. caldarium* in Japan increased gradually, and their topics also changed a constant stream day after day. One of the introductory books concerning studies of Cyanophyceae has been published in Japan ('Advance of Phycology in Japan') edited by Tokida and Hirose (1975). In this book they introduced several original papers on Cyanidiaceae presented by Japanese researchers.

The bulk volume of 'illustrations of the Japanese Fresh Water Algae' edited by Hirose and Yamagishi was published in 1977. In this Monograph *C. caldarium* has been introduced exactly by Kumano and Hirose, within the Family Cyanidiaceae, order Porphyridiales, Subclass Bangiophycidae, Class Rhodophyceae. They have listed in the Cyanidiaceae three genera: *Cyanidium, Chroococcidium* and *Chroococcidiopsis*. In addition, they illustrated these cells in various magnifications.

1.3.4 Chemical composition and chloroplast DNA analysis

Chemical components of *Cyanidium* have also been investigated, for example, carbohydrates (Nagashima and Fukuda 1981, 1983), hydrocarbons and fatty acids (Nagashima *et al.* 1986), polyamines (Hamana *et al.* 1990) and other compounds which represent these studies.

Chloroplast DNA studies of *Cyanidium* was carried out by Fukuda and his group since 1984 (Nagashima *et al.* 1984; Fukuda and Nagashima 1987, 1990; Nagashima and Fukuda 1987; Kuroiwa *et al.* 1989; Ohta *et al.* 1992; Suzuki *et al.* 1992; see chapters by Kuroiwa *et al.*, and by Nagashima, in this book). Recently, Ohta *et al.* (1992) isolated chloroplast-DNA from *C. caldarium* (RK-1) and sequenced the trnK gene (its product is Lysine-tRNA). It turned out that among the eukaryotic unicellular organisms, *C. caldarium* (Ohta *et al.* 1992) and *Cyanidioschyzon* (Suzuki *et al.* 1992) possess the smallest amount of nuclear DNA. In addition, they also measured the relative amount of DNA from *Cyanidium*'s nucleus, mitochondrion and chloroplast. Their data indicated that the distribution of cellular DNA is 85%, 3% and 11% respectively (Ohta *et al.* 1992). Suzuki *et al.* (1992) analyzed the cellular DNA distributed within the smallest eukaryotes *Cyanidioschyzon merolae* (Cyanidiophyceae) and detected similar patterns of DNA: 54.9% in the nucleus, 10.6% in the mitochondrion nucleoid and 34.5% in chloroplast 'nucleus'. It is interesting that chloroplast DNA from *C. caldarium* contains a trnK gene that shows a high degree of homology to those chloroplasts of land plants but it lacks a typical intron (Ohta *et al.* 1992). See also chapter by H. Schenk in this volume.

1.3.5 Phylogenetic explorations

An additional area of interest has been a study in phylogenesis and following our studies the paper 'Amino acid sequence of ferredoxin isolated from *Cyanidium caldarium* strain RK-1' has been published by Nagashima *et al.* (1989). Phylogenical implications have been discussed several times by Fukuda and his group (Fukuda 1984; Fukuda *et al.* 1986). The discussion of this area is now most interesting topic among the Japanese phylogenical researchers. Most of us believed now that Cyanidiophyceae belong to the most primitive 'Prerhodophyta' and are serving as a phylogenetic bridge linking prokaryota to eukaryota as proposed original by Seckbach (1987). The proposed evolutionary path moves from the cyanobacteria to

Cyanidioschyzon → *Cyanidium* → *Galdieria* → Rhodophyta (Seckbach 1987, 1992).

This scheme was confirmed by morphology of chloroplast DNA in these algae (see Kuroiwa in this volume) and by the increasing amounts of nuclear DNA in the cyanidiaceae (Suzuki *et al.* 1992).

References

Allen, M.B. (1952). *Arch. Mikrobiol.* 17: 34–53.

Allen, M.B. (1959). *Arch. Mikrobiol.* 32: 270–277.

Enami, I. and Fukuda, I (1975). *Plant & Cell Physiol.* 16: 211–220

Enami, I. Matsubayashi, T. and Fukuda, I. (1978). *Photosynthetica* 12: 83–86.

Enami, I. Nagashima, H. and Fukuda, I. (1975). *Plant & Cell Physiol.* 16: 221–231.

Fukuda, I. (1958). *Bot. Mag. Tokyo* 71: 79–86.

Fukuda, I. (1962). *Bot. Mag. Tokyo* 75: 349–355.

Fukuda, I. (1984). *Viva Origino* 12: 95–96.

Fukuda, I. and Ishisone, S. (1965). *Japan Electron Optic News* 44: 4–5.

Fukuda, I. and Nagashima, H. (1987). *Ann. N.Y. Acad. Sci.* 503: 575–577.

Fukuda, I., Nagashima, H. and Kuroiwa, T. (1986). *Viva Origino* 14: 130–131.

Fukuda, I., Nagashima, H. and Seckbach, J. (1990). *Endocytobiology IV:* 585–588.

Hamana, K., Matsuzaki, S., Niitsu, M., Samejima, K. and Nagashima, H. (1990). *Phytochem.* 29: 377–380.

Hirose, H. (1950). *Bot. Mag. Tokyo* 63. 107–111.

Hirose, H. (1958). *Bot. Mag. Tokyo* 71: 347–352.

Kumano, S. and Hirose, H. (1977). In: *Illustrations of the Japanese Fresh Water Algae.* (eds. Hirose, H. and Yamagishi, T.) pp. 160–161. Pub. Uchida-Rohkakuho Comp. Tokyo, Japan.

Kuroiwa, T., Nagashima, H. and Fukuda, I. (1989). *Protoplasma* 149: 120–129.

Mita, T., Kanbe, T. Tanaka, K. and Kuroiwa, T. (1986). *Protoplasma* 130: 211–213.

Nagashima, H. and Fukuda, I. (1981). *Phytochem.* 20: 439–442.

Nagashima, H. and Fukuda, I. (1983). *Phytochem.* 22: 1949–1951.

Nagashima, H. and Fukuda, I. (1987). *Plant & Cell Physiol.* 28: 315–321.

Nagashima, H., Kuroiwa, T. and Fukuda, I.(1984). *Experientia* 40: 563–564.

Nagashiman, H., Matsumoto, G. and Fukuda, I. (1986). *Phytochem.* 25: 2335–2341.

Nagashima, H., Sakai, K., Kamo, M. and Tsugita, A. (1989).*Protein Sequence & Data Analysis* 2: 457–460.

Negoro, K. (1935). *Bot. Mag. Tokyo* 49: 457–460.

Negoro, K. (1943). *Bot. Mag. Tokyo* 57: 302–312.

Negoro, K. (1944). *Acad. Reports of Tokyo College of Sci. and Literature* 101: 232–274.

Ohta, N., Nagashima, H., Kawano, S. and Kuroiwa, T. (1992). *Plant Cell Physiology* 33: 657–661.

Seckbach, J. (1987). *Ann. N.Y. Acad. Sci. (Endocytobiology III)* 503: 424–437.

Seckbach, J. (1992). In: *Algae and Symbioses* (ed. Reisser, W.) Biopress Ltd. pp. 399–426.

Suzuki, K., Ohta, N. and Kuroiwa, T. (1992). *Protoplasma* 141: 80–84.

Tokida J. and Hirose H. (ed). (1975). *Advance of phycology in Japan.* Ver. Gustav Fischer, Jena. pp 355.

Tôyama, S. (1974). *Jpn. J. Bot.* 20: 413–438.

Tôyama, S. (1980). *Cytologia (Tokyo)* 45: 779–790.

Ueda, K. and Yokochi, J. (1981). *Bot. Mag. Tokyo* 94: 159–164.

Ueda, K. and Chida, Y. (1987). *Br. phycol. J.* 22: 61–65.

Chapter 11

Italy's contribution to the systematics of *Cyanidium caldarium 'sensu lato'*

GABRIELE PINTO[1], P. ALBERTANO[2] and A. POLLIO[1]
[1] *Dipartimento di Biologia Vegetale, Università 'Federico II', Naples, Italy*
[2] *Dipartimento di Biologia, Università 'Tor Vergata', Rome, Italy*

Outlines:
1. Introduction
2. Historical background
3. The revival of Italian research on *Cyanidium*
   3.1 Research at the Solfatara of Pozzuoli near Naples
   3.2 Acquasanta and new stations in Italy
   3.3 The *locus classicus* of *Cyanidium caldarium* Geitler and extra-European habitats
4. The new class Cyanidiophyceae
5. Current research and perspectives
References

1. Introduction

Microorganisms from extreme acidic environments are interesting because of the physiological or biochemical peculiarities retained or developed in the colonization of such habitats. Among the algae inhabiting acidic and thermal environments, the *C. caldarium* – group has attracted the attention of biologists because of its unique characteristics, whereas the systematics of the group have been often ignored. Recently, Seckbach (1991, see also this volume) discussed the implications that this gap in our knowledge had on molecular biology studies and on phylogenetic concepts.

Most of the data on the systematics of *C. caldarium* were obtained in Japan and in Yellowstone National Park in the U.S.A. where there are wide extensions of impressive volcanic phenomena. In South Italy, besides the classical Solfataras arising from secondary volcanisms, there is an abundance of acidic waters and soils (pH < 3.0) originating from the oxidation of large sulphur deposits dating back to the gypsum-sulphur formation of the Upper Miocene.

In the light of earlier works carried out in the last century, from 1965 Italian researchers began to focus on thermo-acidophilic algae and on the systematics of '*C. caldarium*'; their findings suggest that the taxonomy of these organisms be revised according to autoccological, morphological, physiological and ultrastructural data.

2. Historical background

Giuseppe Meneghini (1839) was the first to describe the presence of thermo-acidophilic algae in the sulphurous hot spring of Acquasanta (Ascoli Piceno, Italy). Following a suggestion by an acqauintance, Antonio Orsini, he observed mucilaginous bright green colonies among the sulphate crystals spread on the ceiling of the cave housing the waters, and described the extreme acidity of the site 'contact with which burns hands and clothes'. From observations of small and very small green 'globules' (0.2–2 μm) in the colonies he proposed, without any diagnosis the new species *Coccochloris orsiniana* (Cyanophyta). Later, he gave a detailed but confused description of *C. orsiniana* (1841) (Fig. 1) and, in contrast with his previous work, reported cell dimensions from 0.1–1 up to 10 μm.

Before the end of 19th century this species was cited by Kützing (1849) who transferred it to the genus *Palmella* (Chlorophyta) on the sole basis of Meneghini's description, and by Bergen (1897) who recognized the species near the fumaroles of the Solfatara near Pozzuoli (Naples, Italy) on the basis of Farlow's determination of his samples.

Two years later, Agostino Galdieri, from the Naples Real Orto Botanico, found an alga on the rocks near the fumaroles of the Solfatara. The morphological observations of the green-bluish round-shaped cells, which were 3–12 μm in diameter, resulted in the publication of the new palmellacean species *Pleurococcus sulphurarius* (Chlorophyta)

J. Seckbach (ed.), Evolutionary Pathways and Enigmatic Algae: Cyanidium caldarium *(Rhodophyta) and Related Cells,* 157–166, 1994.

AUCTORE PROF. IOSEPHO MENEGHINIO 63

4. Coccochloris Orsiniana, Mgh. (Nuova sp. di Alga ecc.).

C. *fronde gelatinosa, diffusa, viridi, globulis minimis, subrotundo-angulosis viridibus, vesiculis sphaericis intensius coloratis, granula pauca magnitudine varia, angulosa includentibus.*

Hab. Ad Thermas hydrosulphuricas sic dictas *Acquasanta* legit Cl. Orsini.

Descript. Substantia gelatinosa late effusa, demersa laete virescit, conglomerata selenitica emersa obducens fuscescit. Oscillariis et Confervis in iisdem fontibus degentibus immixta varios sumit adspectus. Demersa quoque innumeros gypsi crystallos sinu fovet et tenuissimo facillimeque secedenti muco globulos includit minimos, magnitudine tamen summopere varios a decimillesima ad millesimam millimetri partem, subrotundos vel angulosos, vesiculasque globosas diametro usque centesimam millimetri partem adipiscentes, granula parce numerosa egregie angulosa intense viridia gerentes. Maiores globuli formam evidentius angulosam praeseferentes granulari quidem substantia vitris maxime augentibus foeti conspiciuntur, granulis vero ipsis minutissimis et vix distinguendis. Vesiculae quamquam granulis inclusis non omnino farciantur omnino tamen virides conspiciuntur, nec ullo margine diaphano praecinctae, ex quo suspicari potest ipsas praeter granula illa maiora, minora quidem et propter parvitatem inconspicua includere.

Obs. Praeter eximios characteres quibus species haec a caeteris omnibus differt, physiologorum attentionem etiam promeretur propter conditiones in quibus degit: evolvitur enim et vegetat in ipsis fontibus ubi quaecumque alia organica corpora enecantur et dissolvuntur.

Tab. VIII. fig. 1. Conglomeratum seleniticum Coccochloride Orsiniana obductum.
1. *a.* Eiusdem fragmentum, centies auctum.
1. *b.* Portiuncula quadringenties aucta.

Fig. 1. Diagnosis and iconotypes of *Coccochloris orsiniana* (from Meneghini 1841).

(1899). This alga was extremely resistant to high temperatures and sulphuric acid concentrations, and reproduced by 2–8 or more daughter cells (Fig. 2). Galdieri evidently did not know Meneghini's work, and he referred to Tilden (1898) in discussing the thermoresistance of *P. sulphurarius*. However, he failed to recognise the similarity with the alga collected by Tilden in Yellowstone National Park and classified as *Chroococcus varius*.

At the beginning of this century the last Italian citation on acidophilic algae was reported in 'Silloge algarum omnium hucusque cognitarum' by Achille Forti (1907) who included *Pleurocapsa caldaria* (Tilden) Setch. in the species *insufficienter descripta*, while he accepted the new combination *Palmella orsiniana* (Menegh.) Kützing for the alga found at the thermal spring of Acquasanta. Thereafter, italian phycologists lost interest in these algae and the microflora of sulphuric and thermal environments was largely ignored for more than 50 years.

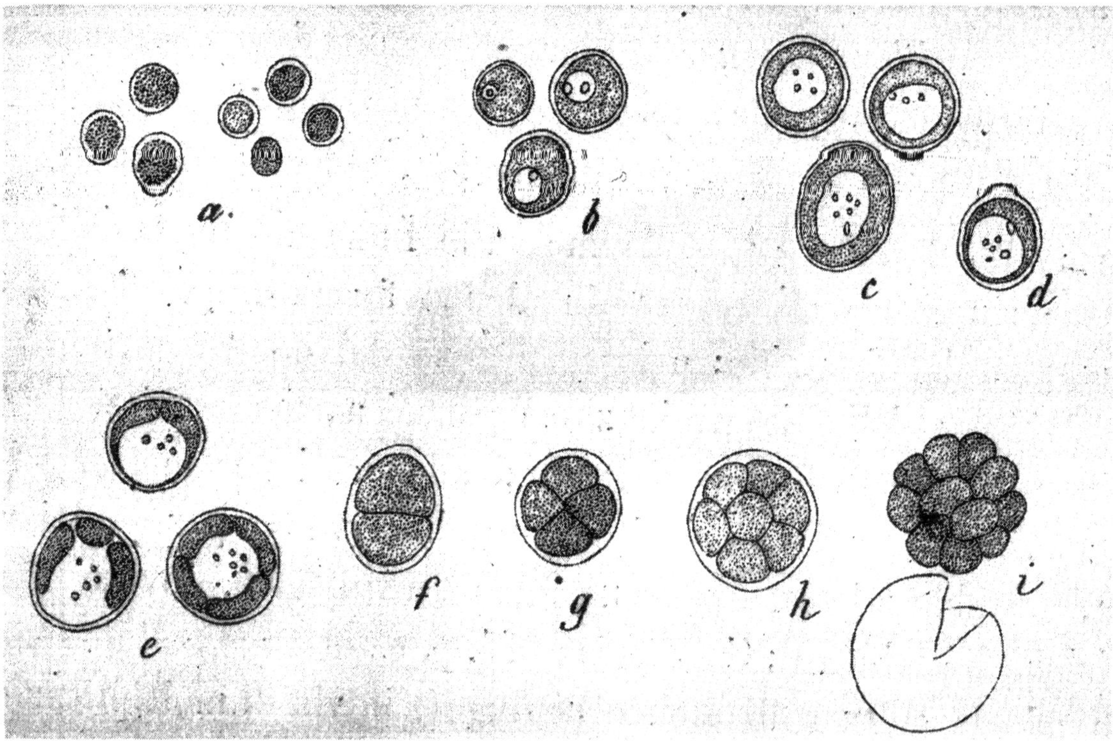

Fig. 2. Iconotype of the algae isolated by Galdieri (from Galdieri 1899).

3. The revival of italian research on *Cyanidium*

3.1 Research at the Solfatara of Pozzuoli near Naples

After the International Phycological Symposium held at the Stazione Zoologica Anton Dohrn di Napoli in 1963, Aldo Merola, Professor of Botany at the University of Naples, and his colleagues began to study the organisms inhabiting the volcanic areas in the vicinity of the town. Rigano was the first to report the presence of *C. caldarium* in the European continent (1965), and two years later he recorded the ecological data of this organism and the presence of an unidentified alga associated to *C. caldarium* (1967). The auto-ecological studies focused first on the growth limits with respect to pH (Rigano and Taddei 1968). In the same year the ultrastructure of *C. caldarium* from the Solfatara of Pozzuoli was described by Castaldo (1968) who established for the first time the occurrence of dictyosomes, reproduction by up to 32 endospores and confirmed the presence of a single plastidial membrane already reported for the Yellowstone Park strain (Rosen and Siegesmund 1961). In addition, Castaldo described the unknown associated alga reported by Rigano (1965). This alga had a cell diameter of 3 μm and a single plastidial membrane; it differed from earlier species in that it possessed only one mitochondriom and no vacuole (Fig. 3).

De Luca and Taddei (1970) compared morphometrical analyses of field samples and laboratory cultures and confirmed the occurrence in the same habitat of two different algae; these they provisionally named *C. caldarium forma A* and *C. caldarium forma B* (Fig. 4). In the same year Castaldo (1970) conducted a detailed ultrastructural description of the two forms of *C. caldarium* and identified the *forma B* with that previously observed by Rosen and Siegesmund (1961) and Mercer *et al.* (1962), and *C. caldarium forma A* with the smallest one. She confirmed earlier observations on the *forma B*, and found dictyosomes and a second membrane around the plastid of *forma A*, which suggested the alga had an endosymbiotic origine. Subsequently, the tripartite thylakoid structure was interpreted as a functional analogy with the partitions extant into the plastids of most photosynthetic organisms (Castaldo and Ligrone 1977).

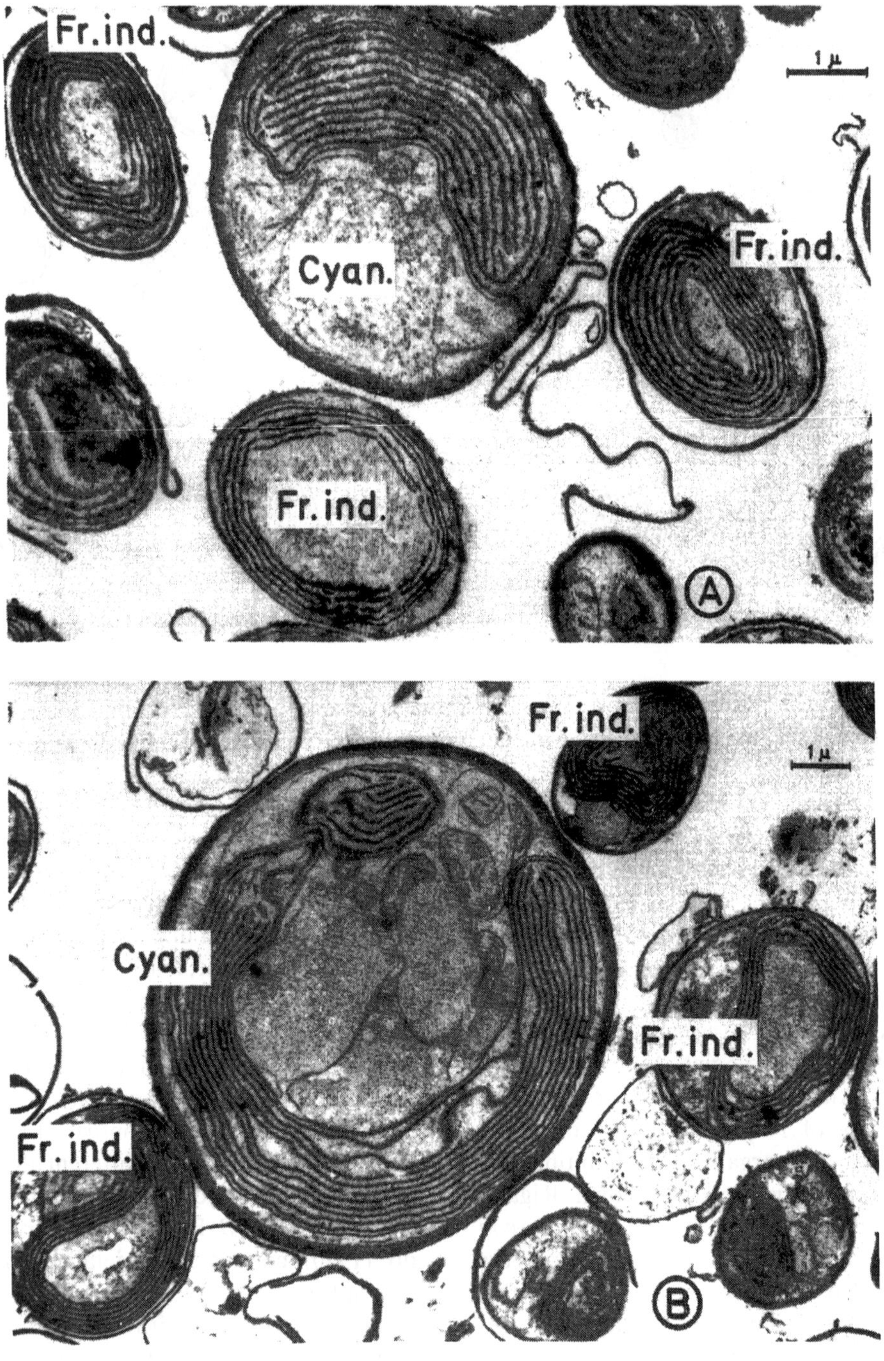

Fig. 3. The first micrograph of *Cyanidium caldarium* reported as Indetermined form (= Fr. ind.) (from Castaldo 1968). The larger alga identified as *Cyanidium caldarium* (= Cyan.) was subsequently classified as *Galdieria sulphuraria*.

Following the isolation of the two algae (De Luca and Taddei 1970), physiological studies showed the ability of *C. caldarium forma A* to utilize nitrate or ammonium as a nitrogen source, whereas *C. caldarium forma B* metabolized only NH_4 and survived prolonged periods of nitrogen starvation. Moreover, only the latter alga grew heterotrophically in the dark on glucose and glutamate and did not synthetize chlorophyll *a* or phycocyanin under these conditions (De Luca and Taddei 1972; De Luca *et al.* 1972a, 1972b). Furthermore, the wide variability in the cell dimension of *C. caldarium forma B* in relation to pH convinced De Luca and co-workers (1975) of the significant differences between this alga and the types of Tilden (1898) and Geitler (1935). Therefore, it became necessary to compare the two forms with the type description of *C. caldarium* (Tilden) Geitler. However Geitler established the binomium *C. caldarium* on Indonesian material, and recognised the synonymy with *Protococcus botryoides* f. *caldarium* Tilden and *Pleurocapsa caldaria* (Tilden) Setchell isolated from Yellowstone Park, but ignored the report on the acidophilic *C. orsiniana* by Meneghini (1839, 1841) and the citation by Berger (1897).

The difficulties in comparing the original descriptions and the iconotypes given by subsequent authors prompted investigators to collect living 'original' acidophilic algae from the *loci classici* in Italy and in other areas.

3.2 Acquasanta and new stations in Italy

De Luca *et al.* (1974a) examined Meneghini's station of Acquasanta (Ascoli Piceno, Italy) and found algal colonies on the walls of the cave housing the thermal spring. The algae were found in the same place and had characters similar to those originally described by Meneghini (1839). Moreover, the site yelded two algae identifiable with the *C. caldarium forma A* and *C. caldarium forma B* previously found at the Pozzuoli Solfatara. This observation led a series of samplings in sulphuric acidic environments throughout Italy, and *Cyanidium caldarium 'sensu lato'* was identified in more than one hundred stations in central and southern regions of the country (De Luca *et al.* 1974b; Pinto and Taddei 1975, 1977). These field studies revealed numerous other acidophilic or acid-tolerant algae in highly acidic soils and waters (Pinto and Taddei 1977; Albertano 1993; Pinto 1993). In addition, the occurrence of *Cyanidium caldarium 'sensu lato'* in acidic and non-thermal habitats (i.e., cold springs, sulphur mines, etc.) was reported for the first time (Taddei 1977; Pinto and Taddei 1977).

During that period a third primitive rhodophycean alga, closely associated with *Cyanidium caldarium forma A* and *C. caldarium forma B*, was isolated and described as the new genus and species *Cyanidioschyzon merolae* De Luca, Taddei et Varano. This species was characterized by small dimension (1.5 x 3–4 μm), reproduction by binary fission, one mitochondrion and a single plastid membrane (De Luca *et al.* 1978) (Fig. 5). In addition to the reference strain, which was isolated from the Solfatara of Pozzuoli, about ten other isolates were obtained from different Italian stations (Pinto and Taddei 1977).

3.3 The locus classicus of Cyanidium caldarium *(Tilden) Geitler and extra-european habitats*

Following the Italian explorations, reports from an expedition to Mount Lawu in Java (Indonesia) established the concomitant presence of the three rhodophytes in Geitler's *locus classicus* and the identity of the alga observed by Geitler with *Cyanidium caldarium forma A* found on the crater of Tjitrodromoko (De Luca *et al.* 1981). Therefore, the deterioration of the iconotypes and the absence of geological and anthropological changes in the environment 45 years after the first collection justified the comparison made by the authors. Subsequent expeditions to Yellowstone National Park, Playon de Ahuachapan (El Salvador), Los Azufres and Cerro Prieto (Mexico) confirmed the occurrence of the three algae in all these sites (De Luca *et al.* 1977, 1979; Gambardella *et al.* 1980). Moreover, the isolation of strains from extra-European environments allowed a comparison of the ecophysiological features and showed stronger similarities between *Cyanidium caldarium forma A* and *Cyanidioschyzon merolae* than between these and *Cyanidium caldarium forma B* (De Luca *et al.* 1977).

4. The new class Cyanidiophyceae

The collection of strains from Europe, Asia and America suggested that unicellular rhodophytes were present in all thermo-acidophilic environments. The different immunoresponses to cell surface antigens of *Cyanidium caldarium forma A* and *C. caldarium forma B* (Bresciano *et al.* 1977), the halo-tolerance approaching 7% and 10% respectively (Musacchio *et al.* 1977), and the different fatty acids composition of *C. caldarium forma B*

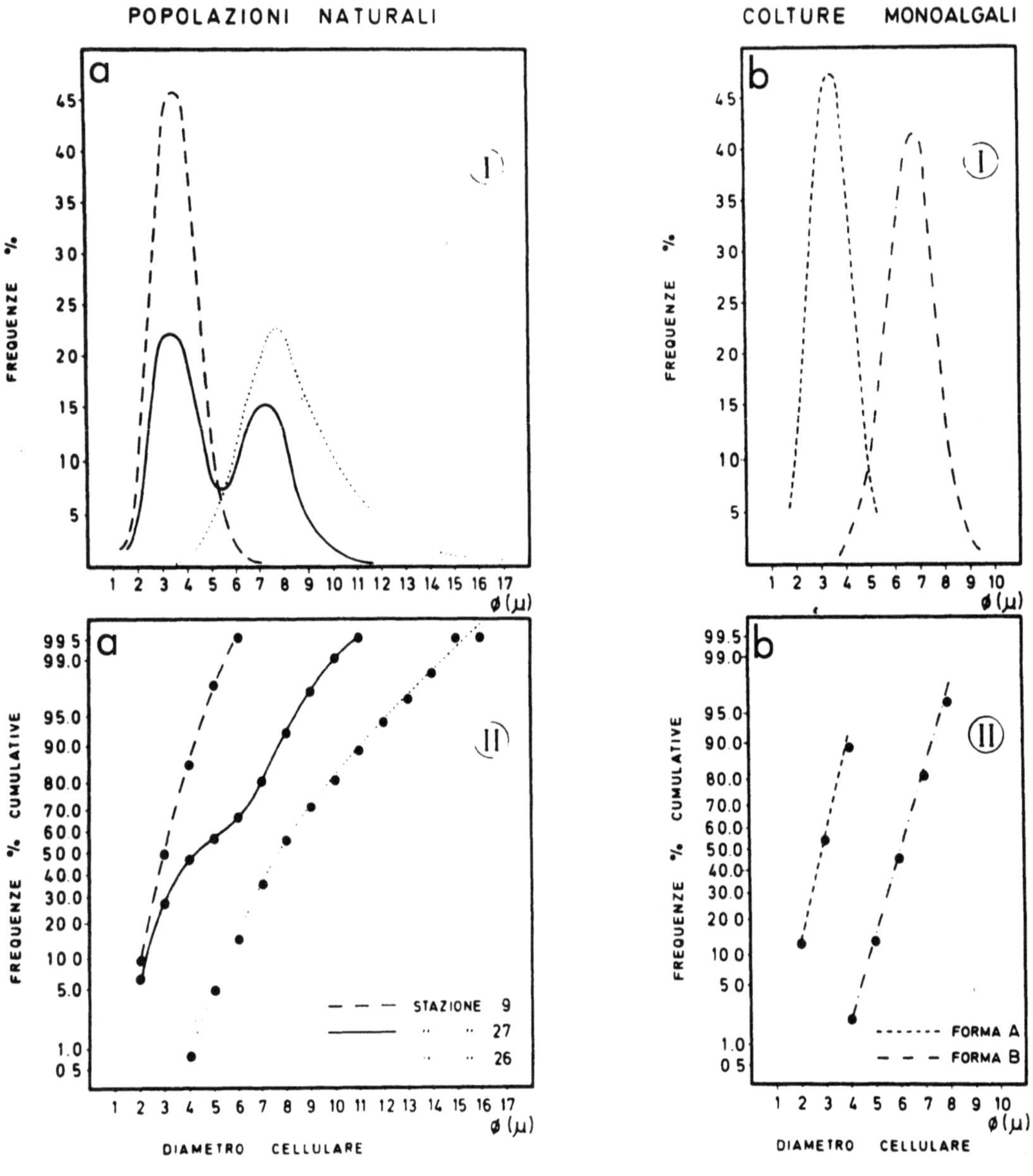

Fig. 4. Frequency (%)(I) and cumulative frequency (II) of the cell diameter in natural populations of (a) '*C. caldarium*' and two isolated clones (b), showing the occurrence of two different algae in the same population. (c) pH and temperature optima of the two isolated algae (*C. caldarium forma A* and *forma B*) (from De Luca and Taddei 1970).

compared to *C. caldarium forma A* and *Cyanidioschyzon merolae* (Boenzi *et al.* 1977; Moretti and Nazzaro 1981) confirmed the need for a systematic revision of the thermal acidophilic alga *C. caldarium* (Tilden) Geitler.

De Luca and Taddei (1976) listed the discrepancies among the morphological, ultrastructural and physiological data available and analyzed all the extant diagnoses of the thermo-acidophilic algae. According to the International Code of Nomenclature (1972), they proposed the specific epithet *sulphurarius* Galdieri to replace *caldarium* in the case of *C. caldarium forma B*, but left unsolved the nomenclatural question of *C. caldarium forma A*.

The definitive revision of the genus *Cyanidium* by Merola *et al.* (1981) invalidated the early description of *C. caldarium* either sub *Coccochloris orsi-*

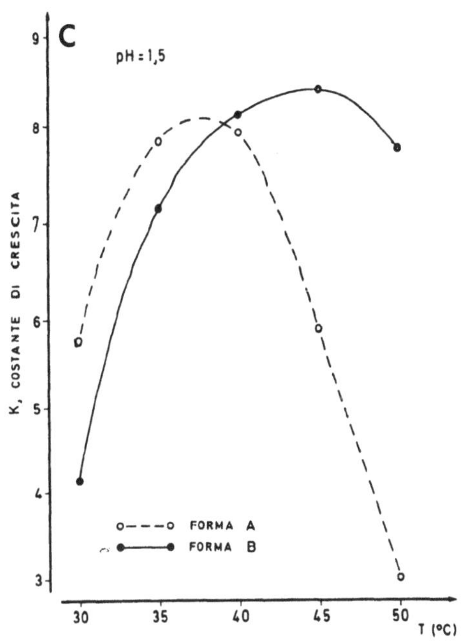

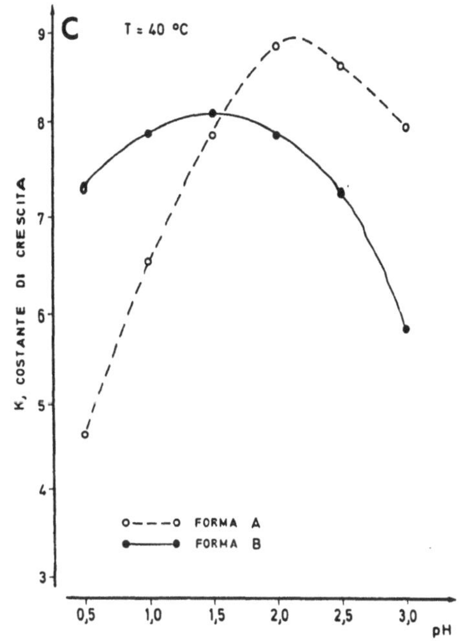

Fig. 4. Continued.

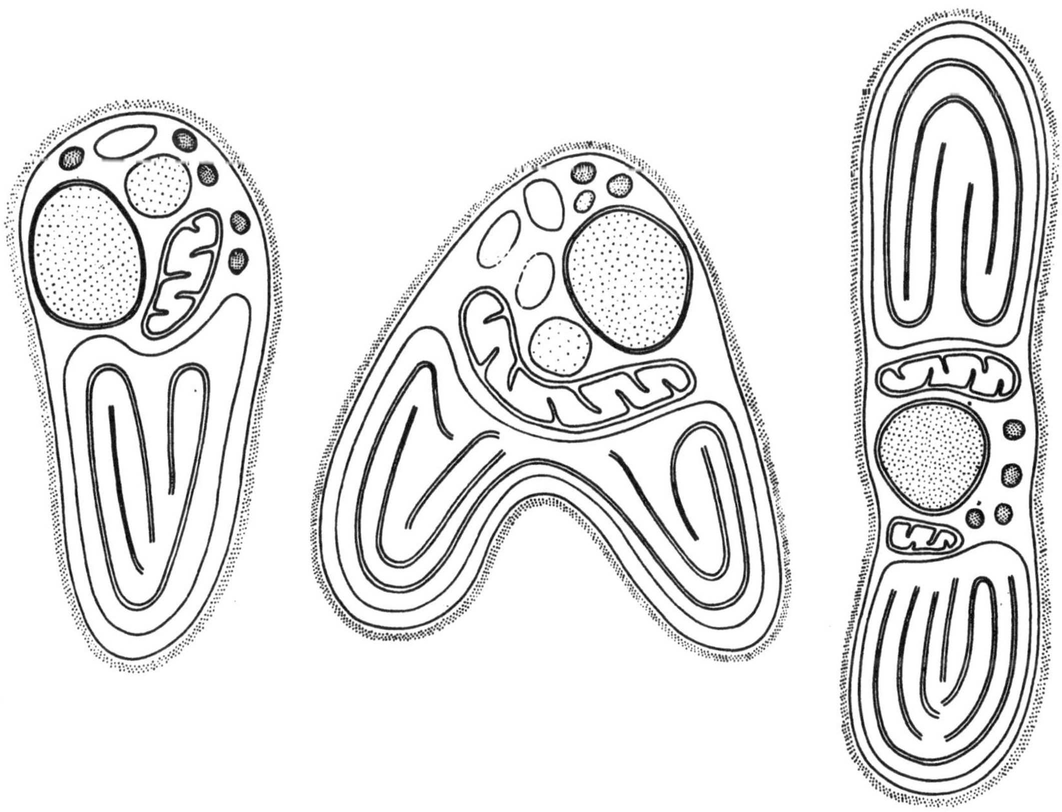

Fig. 5. Sketch of cell division in *Cyanidioschyzon merolae* De Luca, Taddei & Varano. x ca. 16,000 (from De Luca *et al.* 1978).

niana (Meneghini 1839, 1841), or sub *Protococcus botryoides* f. *caldarius* (Tilden 1898) or sub *Pleurocapsa caldaria* (Tilden 1910) because the diagnoses were referred to mixed populations. According to Merola and co-workers, the first valid description unequivocally given for one of the three species was that of *Pleurococcus sulphurarius* (Galdieri 1899), and they identified it with *C. caldarium forma B* according to Galdieri's diagnosis. The attribution of this alga to the genus *Pleurococcus* Meneghini (Chlorophyta) was rejected on the basis of pigment content, and the new genus *Galdieria* with the unique species *G. sulphuraria* (Galdieri) comb. nova was instituted. Geitler's diagnosis of *C. caldarium* (Geitler and Rutner 1935) was held to be the second valid description in which *C. caldarium forma A* was included. Moreover, the identity between *C. caldarium* and *Pleurocapsa caldaria* (Tilden) Setchell was rejected and the binomium *C. caldarium* Geitler non (Tilden) Geitler proposed. Consequently, the new class Cyanidiophyceae Merola was instituted; this incorporated the *familia* Cyanidiaceae Geitler emend. including the genus *Cyanidium* Geitler emend. and *Cyanidioschyzon* De Luca, Taddei and Varano, and the *familia* Galdieriaceae Merola including *Galdieria* Merola (Fig. 6).

Further chemiotaxonomic studies on the three species confirmed their attribution to the rhodophytes and revealed various primitive characters with respect to *Porphyridium* species (Moretti and Musacchio 1982; De Luca and Moretti 1983).

5. Current research and perspectives

In spite of the systematic revision by Merola *et al.* (1981), the taxonomic as well the phylogenic position of the three species is still the object of intense research and discussion. Several aspects have not yet been sufficiently clarified, mainly because of the heterogeneity of the strains experimentally employed. In particular, the question of the single or double plastidial membrane will help to confirm or refute the symbiotic hypothesis (for a review, see Seckbach 1992).

The contradictory ultrastructural observations already reported by several authors resulting from different fixation techniques can be resolved only by a comparative study of the type strains of *Cyanidium caldarium* IBN182 *Cyanidioschyzon merolae* IBN199 and *Galdieria sulphuraria* IBN002 housed in the Culture Collection of the Dipartimento di Biologia Vegetale (University of Naples) (Pinto *et al.* 1992). The use of new fixation and stain procedures in electron microscopy investigations of these strains is providing evidence for a double plastidial membrane in *Cyanidioschyzon merolae* and *Cyanidium caldarium* (unpublished data from our laboratory) in contrast with the first observations by De Luca *et al.* (1978), and could confirm

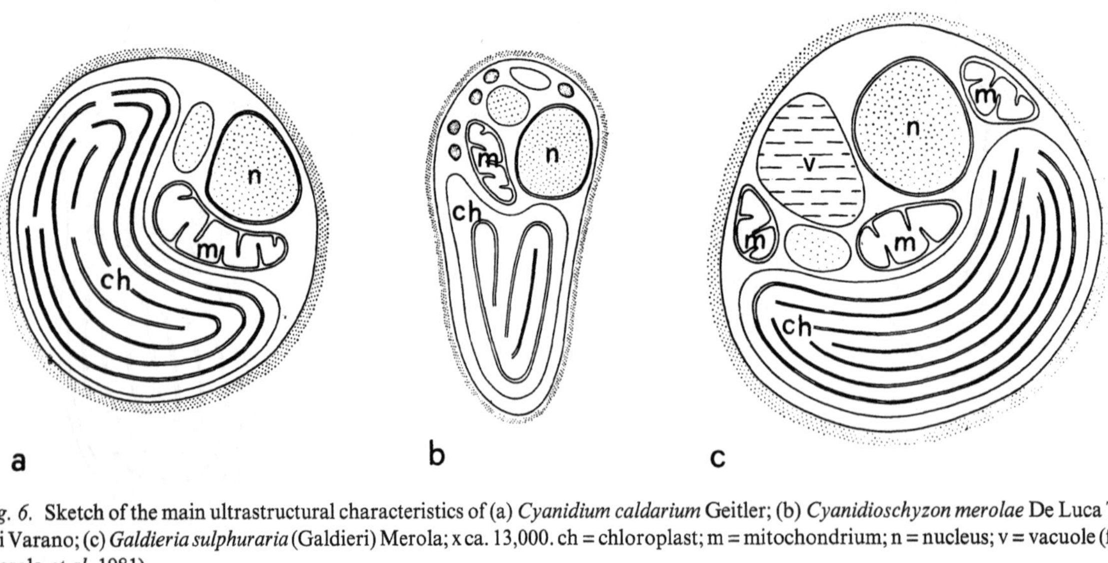

Fig. 6. Sketch of the main ultrastructural characteristics of (a) *Cyanidium caldarium* Geitler; (b) *Cyanidioschyzon merolae* De Luca Taddei Varano; (c) *Galdieria sulphuraria* (Galdieri) Merola; x ca. 13,000. ch = chloroplast; m = mitochondrium; n = nucleus; v = vacuole (from Merola *et al.* 1981).

the presence of dictyosomes in *C. caldarium* (Castaldo 1968).

The ecology of these algae in Italian sulphureous acidic environments has yet to be investigated in detail. Most of the climatological and ecophysiological data collected over the last twenty years concern pH, temperature and halo-tolerance limits for growth, while little is known about microclimatic conditions, substrate composition, light requirements, primary production and relationships with other prokaryotic and eukaryotic microorganisms inhabiting the same sites.

The occurrence of several acidophilic strains of *C. caldarium* in caves and non-thermal habitats of Italy (Pinto and Taddei 1977) and reports of non-acidic cave-*Cyanidium* (see Hoffmann this volume) suggest either a specific diversity of these taxa or a wide ecological range of the same species. In addition, because of the high concentration of heavy metals due to the volcanic emissions (i.e. Hg, As) in soils of the Pozzuoli Solfatara, and the availability of such metals at extremely low pHs, we were able to study the effects of toxic elements on the primary metabolism of *C. caldarium* (Aliotta and Pollio 1982) and on the type of resistence to different heavy metals (Albertano *et al.* 1980; Albertano and Pinto 1986). There is evidence that in *G. sulphuraria* a gradual exposure to low concentration of mercury leds to adaptation to lethal amounts of this element (Pinto *et al.* 1988).

Finally, encouraging results were recently obtained on biotransformation of steroidic compounds by *C. caldarium*, *C. merolae* and *G. sulphuraria* (unpublished data from our laboratory). In fact, the ability of these microorganisms to thrive in extreme conditions (low pHs, high temperatures and CO_2 saturated atmosphere) can be exploited both for biotechnological applications, because of the low probability of contamination of other organisms and for exobiological purposes.

References

Albertano, P. (1993). In: *Algae, man and environments* (Wiessner, W. ed.). Chapman and Hall publ. (in press).

Albertano, P., Pinto, G. and Taddei, R. (1980). *Delpinoa* 21: 23–34.

Albertano, P. and Pinto, G. (1986). *Boll. Soc. Naturalisti in Napoli* 95: 319–328.

Aliotta, G. and Pollio, A. (1982). *Giorn. Bot. Ital.* 116: 123–129.

Bergen, J.Y. (1897). Algae in the Solfatara at Pozzuoli, Italy. *Bot. Gaz.* 23: 198–199.

Boenzi, D., De Luca, P. and Taddei, R. (1977). *Giorn. Bot. Ital.* 111: 129–134.

Bresciano, E., Pinto, G. and Vuotto, M.L. (1977). *Delpinoa* 18–19: 17–24.

Castaldo, R. (1968). *Delpinoa* 8–9: 135–147.

Castaldo, R. (1970). *Delpinoa* 10–11: 91–109.

Castaldo, R. and Ligrone, R. (1977). *Delpinoa* 18–19: 25–36.

De Luca, P., Gambardella, R. and Merola, A. (1979). *Bot. Gaz.* 140: 418–427.

De Luca, P. and Moretti, A. (1983). *J. Phycol.* 19: 368–369.

De Luca, P., Moretti, A., Musacchio, A. and Taddei, R. (1974a). *Delpinoa* 14–15: 3–11.

De Luca, P., Moretti, A. and Taddei, R. (1974b). *Delpinoa* 14–15: 49–60.

De Luca, P., Moretti, A. and Taddei, R. (1975). *Delpinoa* 14–15: 141–152.

De Luca, P., Moretti, A. and Taddei, R. (1977). *Delpinoa* 18–19: 69–76.

De Luca, P., Musacchio, A. and Taddei, R. (1972a). *Delpinoa* 12–13: 19–27.

De Luca, P., Musacchio, A. and Taddei, R. (1972b). *Delpinoa* 12–13: 115–123.

De Luca, P., Musacchio, A. and Taddei, R. (1981). *Giorn. Bot. Ital.* 115: 1–9.

De Luca, P. and Taddei, R. (1970). *Delpinoa* 10–11: 79–89.

De Luca, P. and Taddei, R. (1972). *Delpinoa* 12–13: 3–7.

De Luca, P. and Taddei, R. (1976). *Webbia* 30: 197–218.

De Luca, P., Taddei, R. and Varano, L. (1978). *Webbia* 33: 37–44.

Forti, A. (1907). In: *Sylloge Algarum omnium hucusque cognitarum* (De Toni, G.B. ed.) 5: 123–124.

Galdieri, A. (1899). *Rend. della Real Acc. Scienze Fis. Mat. Napoli* 6: 160–164.

Gambardella, R., Moretti, A. and Musacchio, A. (1980). *Delpinoa* 21: 35–45.

Geitler, L. and Ruttner, F. (1935). *Arch. Hydrobiol., suppl.* 14: 371–481.

Kützing, F.T. (1849). *Tabulae phycologicae. 1, Nordhausen.*

Meneghini, G. (1839). *Nuovo Giornale de' Letterati* 39: 67–68.

Meneghini, G. (1841). *Monographia Nostochinearum italicarum. Acc. R. delle Scienze.* Torino, Mat. e Fis., ser.2, 5: 1–143.

Mercer, F.V., Bogorad, L. and Mullens, K. (1962). *J. Cell. Biol.* 13: 393–403.

Merola, A., Castaldo, R., De Luca, P., Gambardella, R., Musacchio, A. and Taddei, R. (1981). *Giorn. Bot. Ital.* 115: 189–195.

Moretti, A. and Nazzaro, R. (1981). *Delpinoa* 21–22: 3–11.

Moretti, A. and Musacchio, A. (1982). *Giorn. Bot. Ital.* 116: 269–274.

Musacchio, A., Pinto, G., Sabato, S. and Taddei, R. (1977). *Delpinoa* 18–19: 37–44.

Pinto, G., Pollio, A. and Taddei, R. (1988). *Delpinoa* 29–30: 57–62.

Pinto, G., Pollio, A. and Taddei, R. (1992). *Boll. Soc. Adriatica di Scienze*, LXXII, suppl. 1: 5–24.

Pinto, G. and Taddei, R. (1975). *Delpinoa* 14–15: 125–139.

Pinto, G. and Taddei, R. (1977). *Delpinoa* 18–19: 77–106.

Pinto, G. (1993). *Giorn. Bot. Ital.* 127: 400–406.

Rigano, C. (1965). *Delpinoa* 6–7: 277–284.

Rigano, C. and Conforti, T. (1967). *Delpinoa* 8–9: 1–11.
Rigano, C. and Taddei, R. (1968). *Delpinoa* 8–9: 57–63.
Rosen, W.G. and Siegesmund, K.A. (1961). *Journ. Biophys. Biochem. Cytol.* 9: 910–914.
Seckbach, J. (1991). *J. Phycol.* 27: 794–796.
Seckbach, J. (1992). In *Algae and Symbioses*: (Reisser, W. ed.), Biopress Limited, Bristol, England. pp 399–426.
Taddei, R. (1977). *Annuario Speleologico* 4: 27–32.
Tilden, J. (1898). *Bot. Gaz.* 26: 89–105.
Tilden, J. (1910). *Minnesota algae*. Minneapolis.
International Code of Botanical Nomenclature (1972). 1–426.

Chapter 12

The study of Cyanidiophyceae in Russia

Algae of genus *Galdieria*: diversity, characterization and occurrence in mixed populations with *Cyanidium caldarium*

OLGA YU. SENTSOVA
Moscow State University, Biology Faculty, Moscow, Russia
Present address: Mir Publishers, 2 Pervy Rizhsky Pereulok, Moscow, 1-110 GSP, Russia

Outlines:

1. Introduction
 1.1 The taxonomy and classification of cyanidiophytes
2. Cyanidiophycean investigations in Russia
 2.1 Algal materials and the directions of studies
 2.2 Revision of laboratory cultures based on new taxonomy
 2.3 Characteristics of isolated clones and factors of species dominance in mixed cultures
 2.4 Study of natural populations
3. The description of new *Galdieria* members
 3.1 The new species *G. partita*
 3.2 The new species *G. maxima*
 3.3 The new species *G. daedala*
4. Comparative studies of the new *Galdieria* species
References

1. Introduction

1.1 The taxonomy and classification of cyanidiophytes

The eukaryotic unicellular acidothermophilic algae Cyanidiophyceae occupy a special place in the algal systems since they combine distinct characters of cyanophytes, unicellular chlorophytes and rhodophytes on the one hand, and some unique features on the other hand (see Seckbach 1992; Seckbach *et al.* 1992). For a long time these organisms were erroneously referred to the same taxon *Cyanidium caldarium*, although in reality they represent a significant range of morphological and metabolic diversity. This was established by Neapolitan (Italian) researchers who revised '*Cyanidium caldarium*' species and classify three different acidothermophilic algae (Merola *et al.* 1981). These organisms were identified as three species of three different genera and two families formed a new class Cyanidiophyceae. Accordingly, it became necessary to re-identify available strains of *C. caldarium* based on the new taxonomy. This would be of special importance for certain fields of investigations including comparative molecular biology studies of cyanidiophytes (Seckbach 1991). However, the identification up to a species level would entail certain difficulties for the strains isolated from different world regions since there is only one species in each genus in this taxonomy. Meanwhile, the occurrence of cyanidiophytes in the regions so remote from each other (Allen 1959; Brock 1978; De Luca *et al.* 1972/1973; Nagashima and Fukuda 1981) would suggest any diversity within their genera. Correct distinction of new forms or species requires the morphological criteria which in turn may be determined after a detailed examination of different strains and samples.

This article describes the studies of laboratory cultures, natural samples and isolates of cyanidiophytes from different geographical regions which permitted to introduce three new species of the genus *Galdieria* and to reveal certain regularities of their growth and coexistence with *C. caldarium* in mixed populations.

2. Cyanidiophycean investigations in Russia

2.1 Algal materials and the directions of studies

Cyanidiophyceae members have been observed in the territory of Far Eastern regions of Russia. A few individual strains and mixed cultures have been isolated from samples of natural populations inhabiting acid hot springs (Gromov *et al.* 1979;

J. Seckbach (ed.), Evolutionary Pathways and Enigmatic Algae: Cyanidium caldarium *(Rhodophyta) and Related Cells,* 167–174, 1994.

Gerasimenko 1983). The studies of these strains and cultures as well as the 'Allen strain' of *Cyanidium caldarium* from North America (originally isolated by M.B. Allen) were concerned with the following: changes of cell morphology and ultrastructure during the life cycle, mode of nutrition (Gerasimenko *et al.* 1972; Selyakh *et al.* 1984); cell morphology, DNA and pigment content (Gromov *et al.* 1979); influence of pH on the growth rate, pigment content and photosynthesis rates (Sentsova *et al.* 1982; Ignatievskaya and Mineeva 1982) and the gas exchange in the algal cultures (Gerasimenko and Miller 1985).

These recent studies enabled the revealing of different members of Cyanidiophyceae within the laboratory cultures as well as in natural populations. Numerical ratios of *C. caldarium* and *Galdieria* sp. were determined in these two kinds of populations. Three new species from Kamchatka Peninsula and Kunashir Island are being described.

2.2 Revision of laboratory cultures based on new taxonomy

Comparative microscopical examination of *C. caldarium* culture from Kamchatka (isolated and kindly presented to the author by L.M. Gerasimenko), as well as the *C. caldarium* strain Allen indicated significant differences in cell morphology between them. The Kamchatka culture consisted of small cells containing no vacuoles, while *C. caldarium* strain Allen had cells of larger dimensions (maximal diameter) and displayed vacuoles. According to the modern classification of acidothermophilic algae developed by Merola *et al.* (1981) it was reasonable to identify the cells from Kamchatka culture as *C. caldarium* Geitler non (Tilden) Geitler emend. Whereas the *C. caldarium* strain Allen is actually *Galdieria* sp. (cf. Seckbach 1991). However, after the transfer of Kamchatka culture of *C. caldarium* from autotrophic (AT) to mixo- (MT) or heterotrophic (HT) growth conditions, these cells acquired the morphological characteristics (cell distribution according to the size) of *Galdieria* sp. strain Allen (Fig. 1). In order to clarify the situation, spray plating of cultures grown under various conditions were made on agar medium containing glucose and acidified with H_2SO_4. The results of plating showed that Kamchatka algal culture was heterogeneous. It consisted of two cell types forming correspondingly two colony types. There were small dot-like blue-green colonies of *C. caldarium* cells and large soft blue-green colonies of *Galdieria* sp. cells (Fig. 2). On the Petri dishes with Allen strain of *Galdieria* sp. all the colonies were large and had green color.

When the Kamchatka culture subcultured for a long time on mineral medium had been plated, colonies of *C. caldarium* predominated in number (ratio about 1000:1), but when MT- or HT-grown culture (first passage) had been plated the relative amount of *Galdieria* colonies was a hundred fold higher (number ratio *C. caldarium*: *Galdieria* sp. approximately 10:1). When mineral solid medium was used for plating, results were the same, but differences in colony sizes were less significant because *Galdieria* sp. cells formed smaller colonies in comparison with MT-grown ones.

Axenic monocultures were isolated from the colonies of both types. A repeated plating of the *C. caldarium* culture has been required for purifying *Galdieria* sp. clones. Clones originated from large colonies had no differences between themselves. Since there was known only one species in the taxonomy of this genus, these isolated clones were identified as one strain of *G. sulphuraria* (Sentsova *et al.* 1988). Such identification was later proved to be incorrect in respect to a species (as will be discussed further). On the other hand, all clones of *C. caldarium* were referred to one strain.

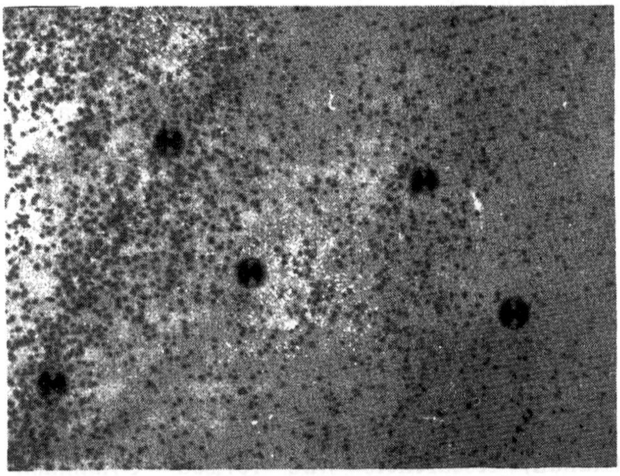

Fig. 2. Colonies of two types on the solid medium containing glucose as a result of plating of *C. caldarium* accumulative culture grown under AT- conditions. Incubation at 30 °C in the light for the period of 30 days.

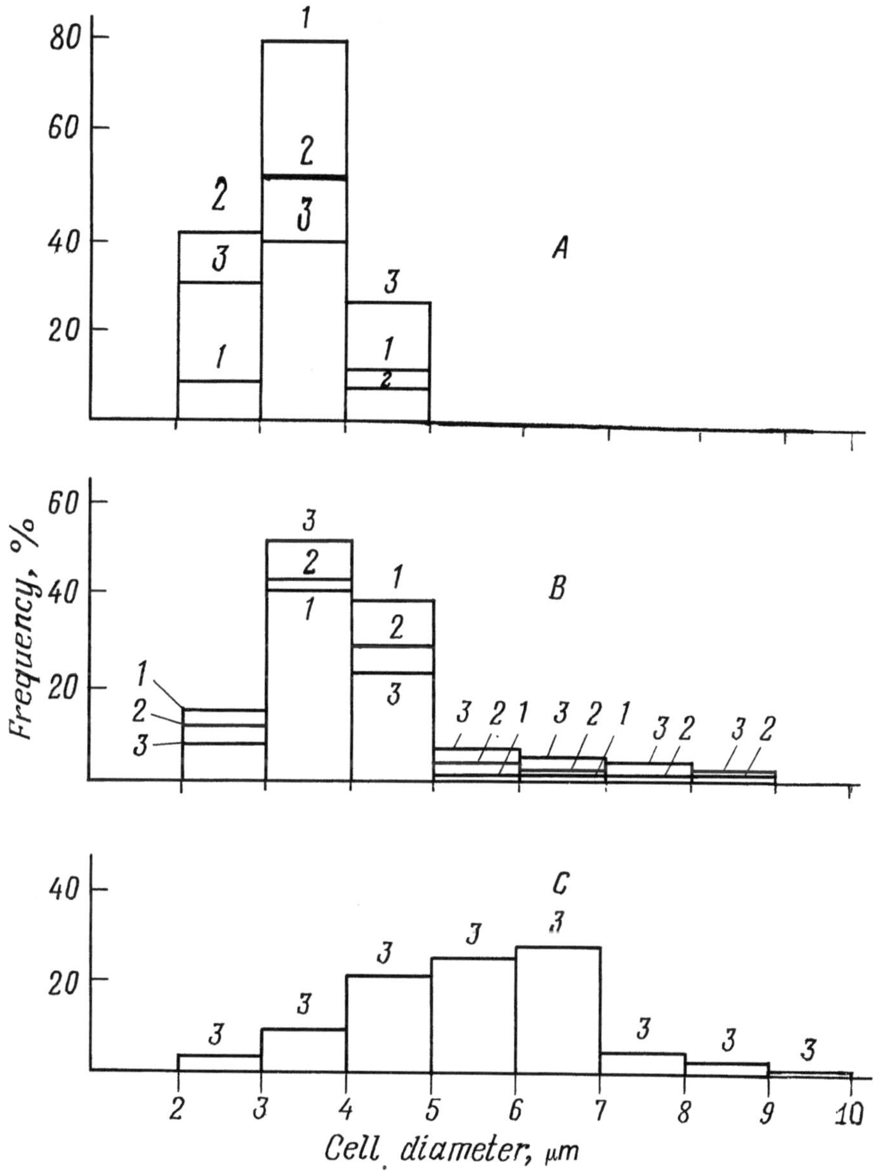

Fig. 1. Cell distribution according to sizes in accumulative culture of *C. caldarium* under AT- (A), MT- (B) and HT- (C) µm conditions. Numbers refer to the age of culture: 3–5 days (1), 7–9 days(2) and 14–16 days (3). Incubation at 35 °C, in the shaker, in the light (for A+B).

In order to account for the characteristics of the mixed culture, the features of its components such as growth rate, cell sizes and pigment content were estimated using various cultivation conditions.

2.3 Characteristics of isolated clones and factors of species dominance in mixed cultures

Under AT-conditions cells of *C. caldarium* were 2 to 5 µm and that of *Galdieria* 2.5 to 8 µm in diameter. The growth rate and pigment level were essentially higher in *C. caldarium*. The supplement of glucose to the medium had no effect on any characteristic examined in the light and didn't support growth in the dark in *C. caldarium*. However, in *Galdieria* strain, cultured both in the light and the in dark, glucose caused an essential increase of growth rate. The cells of *Galdieria* cultured on the medium with glucose, showed an increase of maximal cellular size (up to 10 – 11 µm in diameter) and decline of pigment content which was more significant in the dark (Fig. 3, see also Sentsova *et al.*

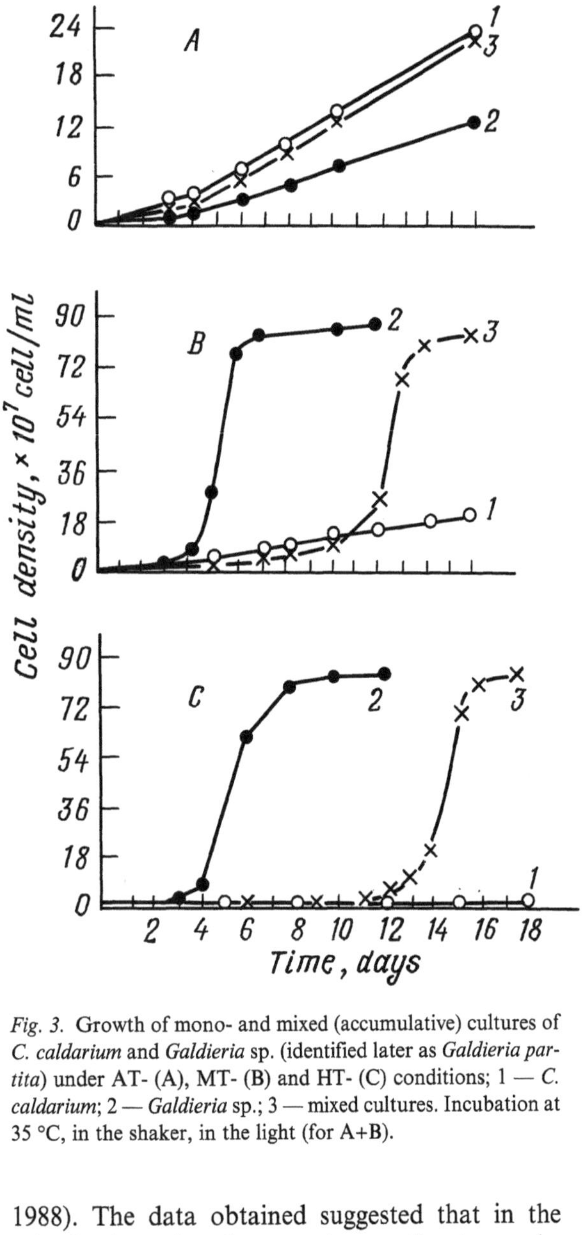

Fig. 3. Growth of mono- and mixed (accumulative) cultures of *C. caldarium* and *Galdieria* sp. (identified later as *Galdieria partita*) under AT- (A), MT- (B) and HT- (C) conditions; 1 — *C. caldarium*; 2 — *Galdieria* sp.; 3 — mixed cultures. Incubation at 35 °C, in the shaker, in the light (for A+B).

1988). The data obtained suggested that in the mixed culture the advantage in growth rate may be considered as the main cause of *C. caldarium* number predominance under AT-conditions and of *Galdieria* cells predominance under MT- or HT-conditions. Thus, the dominance of one or other alga in mixed culture apparently was determined by either higher photoautotrophic potential of *C. caldarium* or by the capacity of *Galdieria* cells for fast mixotrophic or chemoheterotrophic growth according to whether CO_2 + light were the sole carbon and energy sources or organic substrate was available, correspondingly.

In order to elucidate whether the observed numerical ratio of the two algae in accumulative culture is typical for natural mixed population, and to look for already known, or any new members of the Cyanidiophyceae [in addition to those described by Merola *et al.* (1981)], we studied the cyanidiophycean natural populations.

2.4 Study of natural populations

Several samples from five hot acidic springs of Kunashir Island (Kuriles) were kindly brought by S.N. Gorjachev at the request of the author. Most of the samples were relatively dense cell suspensions, blue-green in color. Microscopic examinations indicated the abundance of small cells like *C. caldarium* and certain amount of larger cells with blue-green pigmentation. In the samples there were also few other unicellular algae (green and diatoms) as well as some zooplankton organisms and numerous bacteria.

Spray plating from the dilutions of the samples or from autotrophic accumulative cultures (derived from some samples) onto a solid medium with glucose confirmed the abundance of *C. caldarium* in all samples and cultures. This alga formed numerous dot-like colonies or almost continuous growth on the plates. In addition, large colonies of several other, thus far unknown members of the Cyanidiophyceae, were observed on the plates. These colonies were few in comparison with *C. caldarium* colonies (Table 1). Growth of bacteria and fungi was also observed on the plates. On the basis of cell sizes, presence of vacuoles and ability for heterotrophy, clones of chlorophyll *a* and phycocyanin containing algae isolated from the large colonies, could be referred only to the genus *Galdieria* Merola among all the described genera of the Cyanidiophyceae .

In order to identify these clones as well as earlier isolated *Galdieria* strain from Kamchatka and the Allen strain, detailed analysis of cell and chloroplast morphology was carried out. All the clones were grouped mainly according to the following criteria: cell sizes, shape and number of chloroplasts (depending on the life-cycle), autospore number in maternal cell, and empty envelopes of these cells after autospores release. At least three groups of algae could be separated according to this set of characters (Table 2). They rep-

Table 1. Growth of Cyanidiophycean (expressed as small/large colonies) from Kunashir samples on the solid media [a]

#	Hot Spring	Sample	Dilution	Medium	Growth of algae [b] (colonies per dish)	
					Small	Large
1	pH 4 41.5°C	natural population	–	mineral	continuous growth	5 to 10
			10^{-3}	mineral	0	10 to 30
			10^{-3}	+glucose	2.5×10^3	5 to 7
2	pH 4 38°C	natural population	–	mineral	continuous growth	0
			10^{-3}	mineral	50	0
			10^{-3}	+glucose	10^3	10 to 30
3	pH 4 33°C	accumulative culture (1st passage) [c]	–	mineral	10^3	0
4	pH 3 40°C	natural population	–	mineral	continuous growth	0 to 3
5	pH 4 37°C	accumulative culture (1st passage)[c]	–	mineral	10^3	0 to 7

[a] For conditions see Fig. 2.
[b] Some of the large colonies grew on the surface of large cream-colored bacterial colonies while growth of small colonies was often observed on the surface of black-colored fungus growth.
[c] For growth conditions see Fig. 1.

resented well defined taxons in terms of morphological criteria. None of them corresponded to the description of *G. sulphuraria* (Merola *et al.* 1981; Seckbach 1992).

The characters mentioned were taken as decisive systematic features and used as the criteria for the determination of three new species of *Galdieria*. Original descriptions accomplished by Latin descriptions of each species of the genus *Galdieria* were formally published (Sentsova 1991).

Taking into consideration the new species, the genus *Galdieria* Merola emend. Sentsova includes unicellular acidothermophilic Cyanidiophycean algae of 2.5 to 8; 3.0 to 11; 6.0 to 16.5 μm in diameter depending on species, some of which are characterized by the increase of maximal cell size on transfer from AT- to MT- or HT- growth conditions. Chloroplasts are of variable shape depending on the species and on the stage of life cycle. The form and number of chloroplasts are of great importance for the identification of species. Autospores are of the same sizes and may be present within maternal cells in numbers 4–8, 2–4–8, 4–8–16–32 or 4–8–16 in different species. Within these limits their number depends on growth conditions. After the release of autospores, empty envelopes, lentil-shaped or bursiform in different species, remain in the culture. Table 2 compares cellular characteristics of the *Galdieria* members.

3. The description of new *Galdieria* members

3.1 The new species Galdieria partita

The *Galdieria* strain from Kamchatka (type strain) and that from North America (strain Allen) formed a new species *G. partita* Sentsova. Only one difference in cell sizes was observed between two strains (maximal diameter of strain Allen under all growth conditions was 10–11 μm), all the other characteristics were similar. The parietal chloroplast is single cup-shaped in autospores (like that in mature *G. sulphuraria* which is demon-

Table 2. A partial comparison of cellular sizes, number of autospores and chloroplast shape of newly discovered *Galdieria* species and *G. sulphuraria* (Cyanidiophyceae). All algae have been grown autotrophically

<table>
<tr><th rowspan="2">Galdieria species</th><th rowspan="2">Cell sizes (in μm)</th><th rowspan="2">Autospore numbers</th><th colspan="4">Chloroplasts within[a]</th></tr>
<tr><th>1:</th><th>2:</th><th>3:</th><th>4:</th></tr>
<tr><td>G. partita</td><td>2.5–8.0</td><td>4, 8</td><td>single parietal cup-like</td><td>single parietal belt-shape</td><td>single parietal lobed or several</td><td>numerous oval</td></tr>
<tr><td>G. maxima</td><td>6.0–16.5</td><td>2, 4, 8</td><td colspan="4">two or more, parietal</td></tr>
<tr><td>G. daedala</td><td>2.5–8.0</td><td>4, 8, 16</td><td colspan="2">single parietal variable shape</td><td colspan="2">single parietal multilobed or several</td></tr>
<tr><td>G. sulphuraria[b]</td><td>3.0–11.0</td><td>4, 8, 16 or 32</td><td colspan="4">single cup-like</td></tr>
</table>

[a] Chloroplasts were observed in the: 1 = autospores; 2 = young cells; 3 = mature cells; 4 = old cells.
[b] After Merola *et al.* (1981).

strated by Merola *et al.* 1981), belt-shaped in young cells, and lobed or divided into parts, occupying almost all the cell periphery, in mature cells. Senescent cells contain numerous oval chloroplasts. Autospores develop in amount of 2–4–8. The empty envelopes of maternal cells are lentiform or bursiform (Fig. 4A).

3.2 The new species Galdieria maxima

The second new species *G. maxima* Sentsova was introduced to identify several strains from Kunashir Island. Among the Cyanidiophycean members, thus far known, this strain is characterized by the largest cells, up to 16.5 μm in diameter, which does not depend on growth conditions. The cells contain more than one chloroplast already on the stage of autospores (Fig. 4B).

3.3 The new species Galdieria daedala

Two other Kunashir strains were described as the third new species *G. daedala* Sentsova. This species resembles *G. partita* in its cellular dimensions (including increase of cell diameter under MT- or HT-growth conditions), but differs from this and all the other species of *Galdieria* in the morphology of chloroplasts and shape of the cells with autospores. Chloroplasts are single, parietal, daedal or of variable shape in autospores. In mature cells, the chloroplast is either single, laciniate (multilobed or carved in ornamental manner) or one can observe that more than one chloroplast are present. In *G. daedala* maximal number of autospores (up to 16) may be higher than in *G. partita* and *G. maxima*. In addition, specific feature of *G. daedala* is the shape of maternal cells that have bulges conforming the outlines of autospores (Fig. 4C).

4. Comparative studies of the new *Galdieria* species

The study of type strains of these three species showed that the growth rate as well as the amounts of chlorophyll *a* and phycocyanin under the used culturing conditions are similar in *G. partita* and *G. daedala*, but lower in *G. maxima* (Sentsova 1991).

In type strains of all these species pigment level under AT-growth conditions was lower than in *C. caldarium* from Kamchatka and from Kunashir. On the transfer to MT- or HT-growth conditions, pigment content decreased in all *Galdieria* species.

The metabolic peculiarity of *G. partita* strain Allen was the ability to use citric acid as the source of carbon and energy. This strain was proved to be heterogeneous by this feature as it contained cit+ and cit− cells. Significant population shifts were observed in this strain cultures in relation to the predominance of cit^+ or cit^- cells depending on the presence or the absence of citric acid in the medium (Sentsova 1991). It looks like that cit^+ cells con-

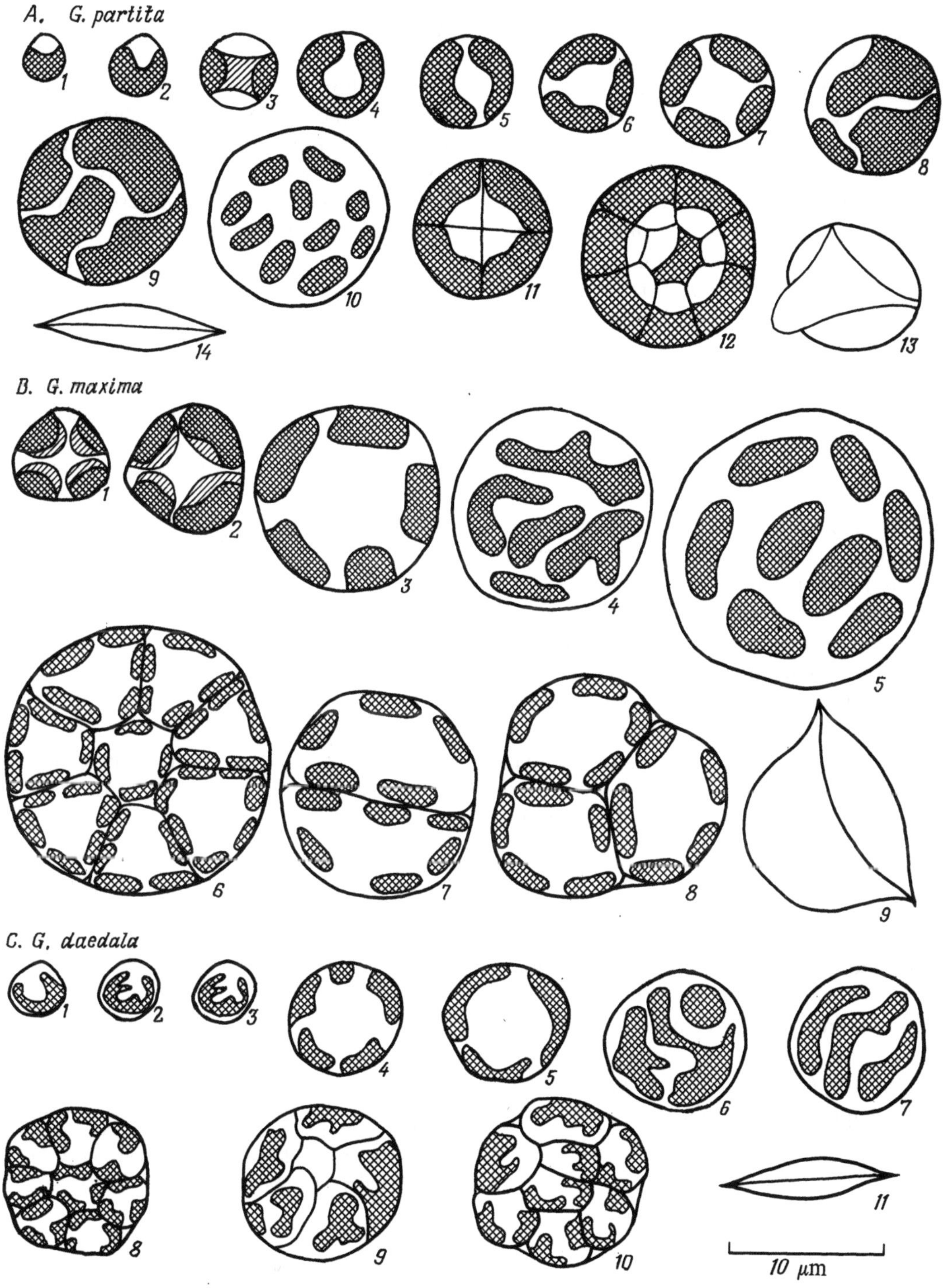

Fig. 4. Morphology of *Galdieria* species (shaded regions indicate chloroplasts). A. *G. partita*: 1,2 — autospores after releasing; 3,4 — young cells: 5–7 — optical cross-sections of different cells: 8–10 — mature cells (surface view); 11–12 — cells with 4 or 8 autospores; 13–14 — empty envelopes of maternal cells. B. *G. maxima*; 1,2 — autospores after releasing; 3 — optical cross-section of a cell: 4,5 mature cells (surface view): 6–8 — cells with 2, 4, and 8 autospores; 9 — empty envelopes of maternal cell. C. *G. daedala*; 1–3 autospores after releasing; 4,5 — optical cross-sections of cells; 6,7 — mature cells (surface view); 8,9 — cells with 16 autospores; 11 — empty envelope of maternal cells.

tained the degradation plasmid, transmitted to cit$^-$ cells after the addition of citric acid. Moreover, *G. partita* strain Allen differed from type strain of this species in that it was unable to grow on nitrate.

As a result of this investigation, different algae of the genus *Galdieria* were discovered. The genus *Galdieria* includes now four species (*G. sulphuraria, G. partita, G. daedala* and *G. maxima*) and represents a single genus of Cyanidiophyceae that is characterized by the diversity of members. Unlike *C. caldarium* who is widely spread throughout the world, separate species of *Galdieria* were found in different world regional areas. Some of the *Galdieria* species may be endemic to certain regions. The species *G. sulphuraria* was shown to inhabit the hot spring of Italian Peninsula (Merola *et al.* 1981). The species *G. partita* occur in two mainlands — Asia (Kamchatka) and North America. The species of *G. daedala* and *G. maxima* were found in the same populations from Kunashir Island (Sentsova 1991).

In the natural populations of Cyanidiophyceae from Kunashir as well as in the mixed culture of these algae from Kamchatka, *C. caldarium* has been predominated in number so that species of *Galdieria* represented minor components of communities, apparently because they have lower growth rate than *C. caldarium*.

Further studies of a diversity of Cyanidiophyceae in their places of inhabitance and geography of each species might be of importance for a clearer understanding of the history and biology of these algae, as well as for their adequate taxonomy and classification. In addition, molecular biological studies (e.g., sequence of nucleic acids and proteins) may also shine light on intra-relations of these species. For further taxonomic proposal of the Cyanidiophyceae, see Ott and Seckbach (chapt. 9) in this volume.

Acknowledgements

The author is very grateful to Drs. L.M. Gerasimenko and S.N. Gorjachev for providing the culture and natural samples; V.M. Andreeva for valuable remarks on taxonomy; and to Dr. J. Seckbach for editing the manuscript and shaping it into the required form.

References

Allen, M.B. (1959). *Arch. Mikrobiol.* 32: 270–277.

Brock, T.D. (1978). In: *Thermophilic microorganisms and life at high temperatures.* Springer Verlag, N.Y., Heidelberg, Berlin. 465 pp.

De Luca P., Moretti A. and Taddei R. (1972/1973). *Delpinoa* 14–15: 49.

Gerasimenko L.M. (1983). In: *Thermophilic microorganisms in nature and practice. Short communications.* Moscow p. 44.

Gerasimenko, L.M. and Miller, Yu. M. (1985). *Mikrobiologiya* 54: 693–698.

Gerasimenko L.M., Pusheva M.A. and Goryunova S.V. (1972). *Mikrobiologia* 41: 324–326.

Gromov B.V., Avilov I.A., Andreev L.A. and Lyutikov, V.N. and Nikitina, V.N. (1979). *Biol.Nauki* (Moscow). 10: 78–82.

Ignatievskaya M.A. and Mineeva, L.A. (1982). *Physiologya Rastenii* (Moscow) 29: 586–590.

Merola, A., Castaldo, R., De Luca, P., Gambardella, R., Musacchio, A. and Taddei, R. (1981). *Giorn. Bot. Ital.* 115(4–5): 189–196.

Nagashima H. and Fukuda I. (1981). *Japan J. Phycol.* 29: 237–242.

Seckbach, J. (1991). *J. Phycol.* 27: 794–796.

Seckbach, J. (1992). In: *Algae and symbioses.* (ed. Reisser, W.) *Biopress, Bristol. England,* pp 399–426.

Seckbach, J., Gonzáles, E., Wainwright, I. M. and Gross, W. (1992). *Nova Hedwigia.* 55(1–2): 99–109.

Selyakh, I.O., Mineeva L.A. and Gusev M.V. (1984). *Izvestia Akad. Nauk SSSR.* Ser.Biol. 1: 74–81.

Sentsova O. Ju. (1991). *Botanichesky J.* St. Petersburg 76: 69–79.

Sentsova O. Ju., Romanova N.I., Mineeva L.A. and Gusev M.V. (1982). *Doklady Akad. Nauk.* SSSR. 266: 762–765.

Sentsova O. Ju., Kravtsova T.R. and Gusev M.V. (1988). *Physiologia Rastenii. (Moscow)* 35: 516– 525.

Chapter 13

Cyanidium-like algae from caves

LUCIEN HOFFMANN
University of Liège, Department of Botany, B.22, Sart Tilman, B-4000 Liège, Belgium

Outlines:
1. Introduction
2. Reports of the genus *Cyanidium* from caves
3. Life cycle and morphology
4. Pigments
5. Ecology
6. Taxonomic position
References

1. Introduction

Cyanidium Geitler and the related genera *Cyanidioschyzon* De Luca, Taddei & Varano and *Galdieria* Merola are typical representatives of hot, acidic ecosystems; a large amount of data concerning their biology has accumulated over the last decades. The genus *Cyanidium* was, however, also reported by several authors from non thermal, non acidic habitats such as limestone caves. The taxonomic position of this enigmatic aerophytic alga, called 'cave *Cyanidium*' in the following, is not resolved; in this chapter the literature data will be reviewed and some recently obtained (preliminary) results on its ultrastructure and ecophysiology will be presented.

2. Reports of the genus *Cyanidium* from caves

Caves have a special microclimate characterized by generally even temperatures, a constant level of humidity and a pronounced light gradient. Cave algae are reported from all over the world (Hoffmann 1989). Some algae like the blue-green algae *Geitleria calcarea* Friedmann, *G. floridana* Friedmann, *Scytonema julianum* (Meneghini ex Frank) Richter and the red alga *Phragmonema sordidum* Zopf are typical of these low light environments. In this ecological niche, a small unicellular alga also sometimes forms large, often almost unialgal, blue-green coatings. This alga was first discovered in five caves in Chile by Schwabe (1936) who described it as a new blue-green algal species, *Cyanidium chilense*. Because of its great morphological similarity to the hot spring alga *C. caldarium* (Tilden) Geitler from which it mainly differs by the ecology, Schwabe (1942) transferred *C. chilense* to *C. caldarium* as *C. caldarium* var. *chilense*. In the same paper he also described a new, morphologically slightly different, form, *C. caldarium* var. *chilense* f. *rumpens* from fine rock cracks of the Pacific coast of Chile. In a later paper (Schwabe 1944), these two described aerophytic taxa were considered by the author as forms of *C. caldarium*. After these studies by Schwabe, aerophytic *Cyanidium* cells were only reported from three locations (Fig. 1), where they were always found in the same type of environment, that is in caves. In 1964 Friedmann mentioned the discovery of a *Cyanidium* species during the biological exploration of caves in Israel. This taxon was isolated in culture by Leiv-Hetzroni (1965) who considered it as related to *C. caldarium*. Skuja (1970) reported *Cyanidium caldarium* from caves in Southern Italy. Leclerc *et al.* (1983) described the microclimatic conditions of a limestone cave in central France in which live populations of an alga which they placed tentatively in the genus *Cyanidium*.

3. Life cycle and morphology

The cave *Cyanidium* is a unicellular alga which carpets the walls of caves with a thin, bright blue-green layer. In the case of a large development up to 1 mm thick, greenish black coatings are built up (Schwabe 1942). The plant mass is generally more or less powdery and granular, never gelatinous or

J. Seckbach (ed.), Evolutionary Pathways and Enigmatic Algae: Cyanidium caldarium *(Rhodophyta) and Related Cells,* 175–182, 1994.

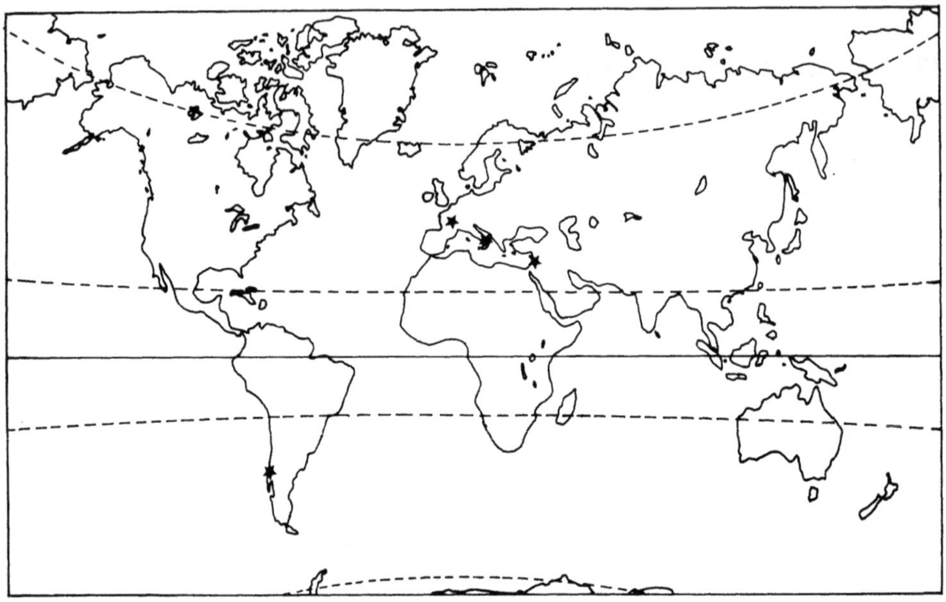

Fig. 1. Geographical locations of cave *Cyanidium* reports.

slimy. The cells are spherical (Fig. 2) and their sizes vary between 3.0–6.0 μm (Schwabe 1936), 3.5–7.5 μm (Leiv-Hetzroni 1965), and 3.6–5.0 μm (Fig. 3). In culture the cells may be slightly larger than in the corresponding field population (Fig. 3). The cell is surrounded by a rigid and homogeneous cell wall. Mature cells show a double layered cell wall with an outer electron denser and granular layer. The alga multiplies through the formation of generally 4 autospores which form a tetrad inside the mother cell wall; occasionally 2 (Skuja 1970) or 8 (Schwabe 1942 for f. *rumpens*) autospores develop. Auto-

Fig. 2. Photomicrographs of the cave *Cyanidium* (field material from a cave at Angles, France). The bar represents 10 μm.

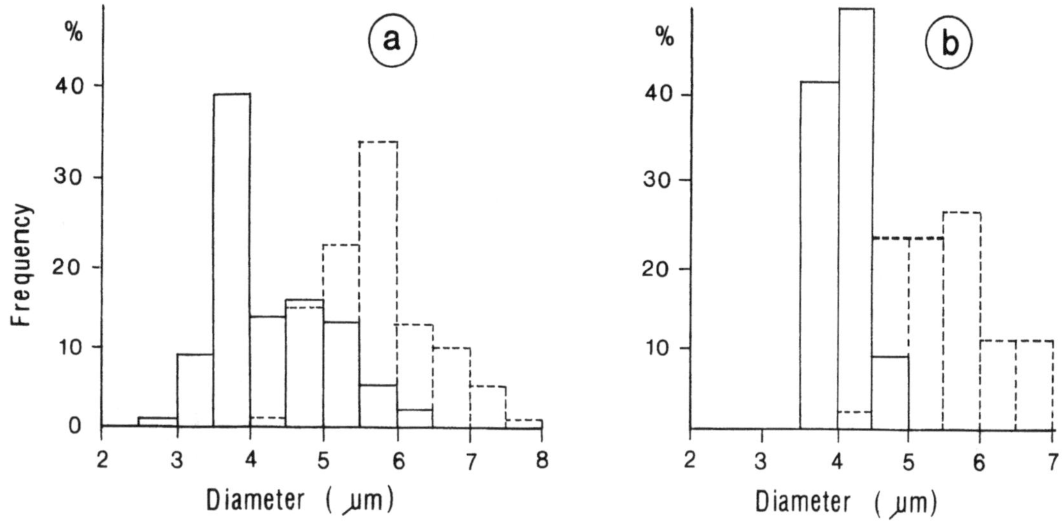

Fig. 3. Frequency distribution of the cell (—) and tetrad (--) diameters; a) culture material (liquid BG 11, 20°C; 4 µE x/m.^{2}s); b) sample from a cave at Angles, France.

spore formation takes place in 4.5–6.0 µm (Schwabe 1936), 4.5–8.0 µm (Leiv-Hetzroni 1965), 4.4–7.0 µm (Fig. 3) wide mother cells. The autospores form their cell wall inside the mother cell; they are released by the breakage of the mother cell wall and have a diameter of 1.8–3.0 µm (Schwabe 1942), 1.5–3.2 µm (personal observation). A single, spherical or cup-shaped, blue-green chloroplast generally occupies the greater part of the cell (Fig. 4); the chloroplast lacks a pyrenoid. The rest of the cell is hyaline except that in some cases one or several dark particles are to be seen by light microscopy in the cytoplasm; these probably correspond to lipid bodies (Leiv-Hetzroni 1965). The chloroplast is surrounded by a closely pressed pair of membranes and contains in the granular matrix up to 15 single, non-stacked thylakoids. A girdle band runs around and parallel

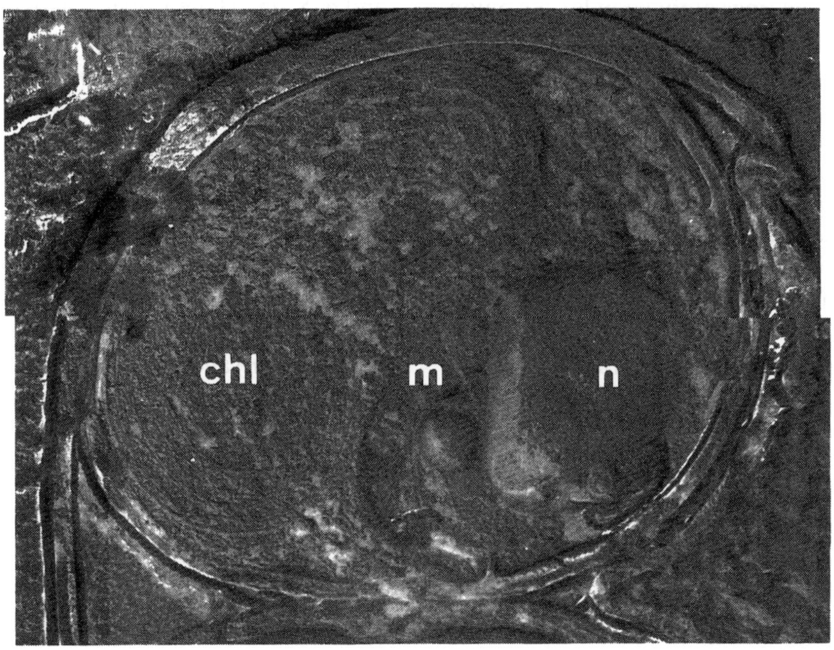

Fig. 4. Electron micrograph of a cross-fractured cell of the cave *Cyanidium* (chl: chloroplast, n: nucleus, m: mitochondrion ?) (15000 x).

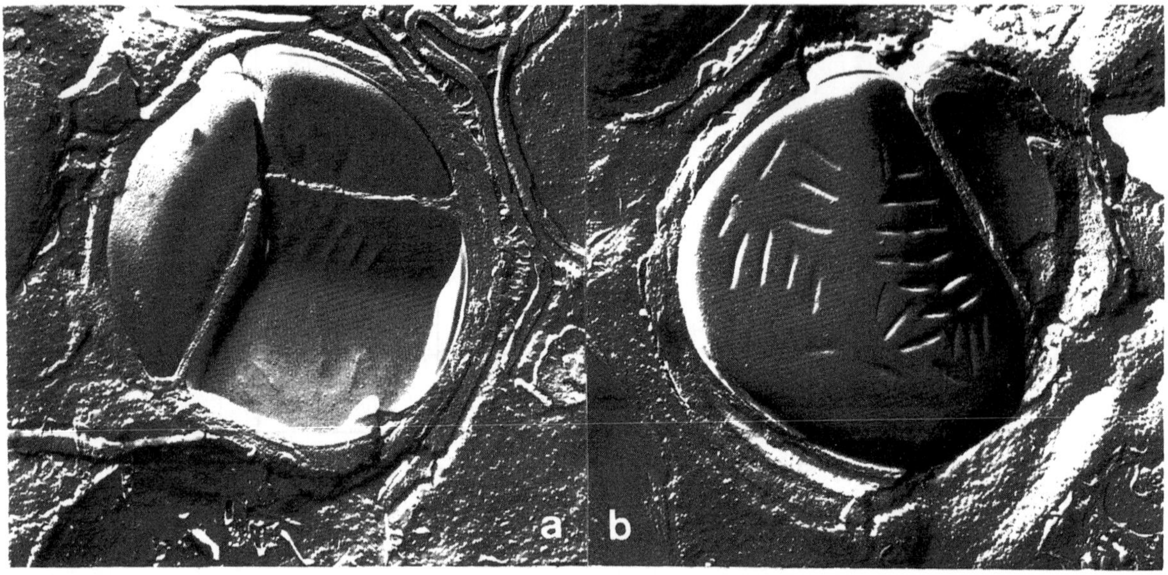

Fig. 5. Different aspects of cross-fractured plasmalemmas (field material from a cave at Angles, France) (10000 x).

to the periphery of the plastid membrane. Six to twelve separated thylakoids run parallel with the chloroplast envelope, each composed of a pair of adjacent membranes joined at their ends; they are open circular or U-shaped. The center of the chloroplast is often occupied by parallel rows of straight cisternae parallel with each other. Another pattern observed consists of nearly parallel, concentric, non stacked rings of thylakoids (like onion rings). A characteristic feature of the chloroplast is the association of electron dense particles, the phycobilisomes, with the thylakoids; as usual these structures are only located on the outside surface of each thylakoid as described for blue-green and red algae. Storage grains are absent from the plastid. The cell possesses a definite nucleus in which electron dense material may occasionally be seen. Other cellular inclusions typical for eukaryotic cells like mitochondria, vacuoles, dictyosomes could so far not been demonstrated with certainty.

During the life cycle of the cell, the plasmalemma develops a number of differentiated structures which can be demonstrated with the freeze-fracturing technique (Hoffmann and Mayele 1993). The most frequently observed structures are narrow ridges of different lengths covered with particles (Fig. 5). Similar differentiations of the plasmalemma were also observed for the thermoacidophilic *Cyanidium caldarium* (Staehelin 1968, Figs. 13, 17).

4. Pigments

The absorption spectrum of intact cells of the cave *Cyanidium* shows major absorption maxima at 625 and 675 nm. These values compare favorably with those reported for the phycobiliprotein C-phycocyanin and chlorophyll-a. Fourth derivative spectra of cave *Cyanidium* cells (Leclerc *et al.* 1981) also reveal the presence of a second phycobiliprotein, allophycocyanin (with a peak at 652 nm). The presence of these pigments is confirmed by the fluorescence emission spectra (Fig. 6) with peaks at 688 and 697 nm (chlorophyll a-protein complex of photosystem II), at 726 nm (chlorophyll a-protein complex of photosystem I), at 646 nm (C-phycocyanin) and a shoulder at 657 nm (allophycocyanin). The cave *Cyanidium* has thus the same major pigments as the thermoacidophilic *Cyanidium* (Nichols and Bogorad 1962).

5. Ecology

Few data are available concerning the ecology and

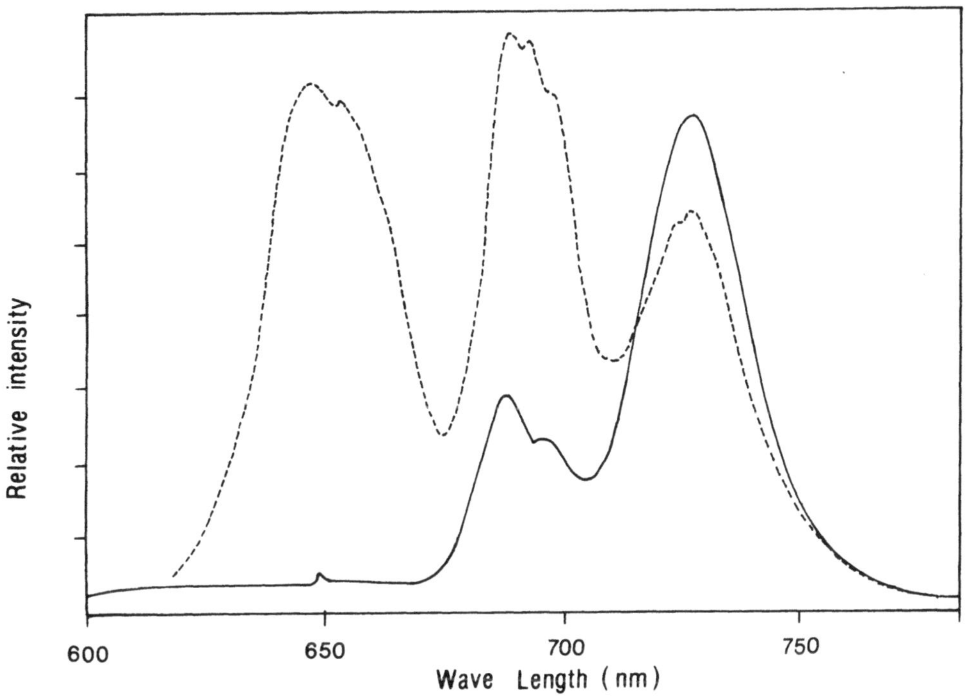

Fig. 6. Fluorescence emission spectra of intact cells taken at −186 °C (—: emission spectrum by exciting at 440 nm; - -: emission spectrum by exciting at 578 nm).

ecophysiology of the cave *Cyanidium*, but from what is known, this aerophytic alga seems to develop best in ecological niches characterized by:

a) low light intensities: The best development of the alga is found in the interior of caves, up to 40 m away from the entry (Leclerc *et al.* 1983), where light intensity is very low. Photoinhibition may already be observed at 10 $\mu E/m^2.s$ (Leclerc 1985). A certain amount of light is however necessary as the alga was never observed on cave walls receiving no light at all. Leclerc *et al.* (1983) measured the light intensity received by populations of the cave *Cyanidium* at Angles, France (Table 1). In this cave the light intensity received by the algae in winter and spring is higher than in summer because the leaves of the trees in front of the cave entry retain most of the light during the latter period. This cave is particular in that its *Cyanidium* populations receive direct sunlight during a few minutes for a couple of days within a year. Generally, however, the algae receive less than 20 lux which is comparable to the light climate of another cave alga, *Geitleria calcarea*, but which is incompatible with the survival of most photosynthetic organisms. In culture the cave *Cyanidium* grows at a light intensity of 4 $\mu E/m^2.s$ (personal observation).

b) stable 'low' temperatures: Caves have generally stable temperatures all over the year. In the cave at Angles, the temperature varied between 4 and 10 °C; daily variations did not exceed 0.5 °C (Leclerc *et al.* 1983). The forma *rumpens* described by Schwabe (1942) seems however able to survive larger temperature variations; it is found in cracks of rocks exposed to full sunlight and so to important daily temperature oscillations. In culture the alga grows at 20 °C (personal observation).

c) constant and high air humidity: The alga is only found on dry, never on permanently wet cave walls. It seems, however, to prefer an atmosphere with a constant and high air humidity (Schwabe 1942). In the cave at Angles the air humidity varied between 89 and 95 %. In culture the alga is growing as well in liquid as on solidified media (personal observation).

d) slightly acidic to alkaline pH : So far the alga

Table 1. Duration of the different light intensities (in lux) received by the *Cyanidium* colonies at Angles, France (after Leclerc *et al.* 1983)

Light	0.5–1	1–2	2–5	5–10	10–20	20–50	50–200	200–2000	2000–10000	>10000
28.1.82	–	–	3h30′	1h10′	15′	–	–	–	–	–
13.3.81	–	–	–	–	4h10′	2h10′	50′	13′	12′	6′
19.6.81	7h	3h15′	10′	–	–	–	–	–	–	–

was only found in habitats with a pH of at least 5.0 (Schwabe 1944); in limestone cave habitats, the pH is higher than 7.0. In culture the alga is growing in neutral or slightly alkaline (pH up to 8) media (Leiv-Hetzroni 1965, personal observation).

These habitat characteristics, especially the low light tolerance, eliminate most of the photosynthetic organisms so that the cave *Cyanidium* generally forms unialgal populations (Schwabe 1942; Leiv-Hetzroni 1965; Leclerc *et al.* 1983) towards the depth of caves. In less extreme light conditions, *e.g.* at the entrance of caves or in the vicinity of artificial light sources it is mixed with other aerophytic algae, like *Pleurococcus* sp. (Schwabe 1944), *Pseudocapsa dubia* Erceg., *Pleurochloris cavernosa* Skuja, *Gloeocapsa punctata* Naeg., or *Denticula tenuis* (Kütz.) Hust. (Skuja 1970).

6. Taxonomic position

On the basis of the properties of the cave alga described above, it is clear that it is a eukaryote and that it bears a striking resemblance to red algae. The features that assure this alga a place in the red algae are: the absence of flagella, the presence of single, non stacked thylakoids, the presence of phycobilisomes and of the pigments phycocyanin and allophycocyanin. Within the red algae, this alga can be assigned to the Phragmonemataceae Skuja or the Cyanidiaceae Geitler, if a narrower family concept is used. It shares with the members of the latter family the production of autospores, the presence of a single chloroplast per cell, the presence of a cell wall and the absence of pyrenoids. Within the Cyanidiaceae, it is morphologically most closely related to the genus *Cyanidium* (Table 2) which has the same chloroplast morphology and also reproduces by the formation of 4 autospores.

The major morphological, ultrastructural, biochemical and ecological characteristics of the cave *Cyanidium* and the thermoacidophilic *C. caldarium* are summarized in Table 3. The two algae have the same morphology and life cycle, as well as the same major pigments. A number of data are still missing for the ultrastructure and biochemistry of the cave *Cyanidium*; this is due to the fact that so far no fully satisfying fixation and inclusion procedure could be worked out and that it is not yet in axenic culture. From what is known so far from the observation of freeze-fractured and thin-sectioned cells, the two algae seem to have the same ultrastructural organization: thus, especially two chloroplast envelopes exist as it seems to be the case in the hot spring alga (e.g. Staehelin 1967; Ueda and Chida 1987) and similar differentiations of the cytoplasmic membrane are observed. Although the two algae are thus very similar in their attributes, they live in widely different environments. The most striking environmental variables which differ in these habitats are :

a) Availability of water: the cave *Cyanidium* is an epilithic alga adopted to live on generally dry cave walls. The thermoacidophilic *Cyanidium caldarium*, by contrast, often lives in running water and shows a marked sensitivity to drying

Table 2. Comparison of the major diacritical characteristics of the cave *Cyanidium* and the other Cyanidiaceae

Character	Cave *Cyanidium*	*Cyanidioschyzon*	*Cyanidium*	*Galdieria*
Size (μm)	3–7.5	1.5 x 3.5	2–6	3–11
Shape	spherical	claviform to oval	spherical	spherical
Division	(2–)4(–8) autospores	binary fission, 2 autospores	4 autospores	4–32 autospores
Chloroplast	spheric or cup shape	polymorph	spheric or cup shape	multilobed

Table 3. Comparison of the characteristics of cave *Cyanidium* and the thermoacidophilic *Cyanidium caldarium*

	Cave *Cyanidium*	Thermoacidophilic *C. caldarium*
Cell size (μm)	3–7.5	2–6
Cell shape	spherical	spherical
Reproduction	(2–)4(–8) autospores	4 autospores
Colour	blue-green	blue-green
Photosynthetic pigments	chlorophyll-a C-phycocyanin allophycocyanin	chlorophyll-a C-phycocyanin allophycocyanin
Chloroplast number	1	1
Chloroplast morphology	(hemi-)spheric or cup-shape	spheric or cup-shape
Envelopes of chloroplast	2	(1?),2
Pyrenoid	absent	absent
Vacuole	?	absent
Mitochondrion	1?	1
Dictyosome	?	absent
Differentiation of the cytoplasmic membrane	present	present
Cell wall	?	high protein content
pH range	?-5–8-?	2–4(–6)
Temperature range (°C)	?–4–20–?	(12–)35–57
Light intensity	generally low	variable
Storage glucan	?	phytoglycogen
Linolenic acid	?	absent

(Smith and Brock 1973). In soil the lowest water potential at which it was found was −4 bars. The fact that the most desiccation resistant populations were found in soils with the lowest water potential suggest, however, that the alga can adapt to the water conditions of the environment.

b) pH : The cave *Cyanidium* was only found in habitats with a pH above 5.0 and in culture it grows at a pH of 7.0 to 8.0. In both aquatic and soil environments, thermophilic *C. caldarium*, by contrast, is generally restricted to habitats with a pH less than 5.0 (Doemel and Brock 1971) and is only occasionally found in waters with a pH up to 6.0 (Seckbach 1987). Doemel and Brock (1971) showed that in culture no growth occurs above pH 5.0, although the alga is still able to photosynthesize. The lack of growth at pH above 5.0 is apparently associated with the inability of the alga to regulate its internal pH effectively once the pH approaches neutrality (Beardall & Entwisle 1984).

c) Temperature : Available data from one cave indicate that the cave *Cyanidium* is living at temperatures between 4 and 10 °C; the temperature range is however probably considerably broader as it grows at 20 °C in culture. The thermoacidophilic *Cyanidium caldarium*, although having a temperature optimum for growth and photosynthesis between 45 and 50 °C, exists at considerably lower temperatures (Doemel and Brock 1971) and is a eurythermal alga.

d) Light: The cave *Cyanidium* has its best development in very low light environments which prevents the growth of most photosynthetic organisms. It seems, however, to survive exposure to full sunlight, at least for a short duration. The thermoacidophilic *Cyanidium* grows under a broad range of light intensities, from very low intensities in deep layers of mats to full sunlight. The cells living in shaded environments seem to adapt by modifying their cellular chlorophyll content (Doemel & Brock 1971).

Finally, although the two algae have their optimum development in widely different extreme environments, their ecological spectrum seems to overlap for most of the environmental parameters (water availability, light, temperature), except for pH. *C. caldarium* is indeed an obligate acidophile, whereas the cave *Cyanidium* is at most acidotolerant, growing in media with a pH above 7.0. In view of this overall similarity between the two algae, the question arises whether to consider the cave *Cyanidium* as a separate species, as a subspecific taxon of *C. caldarium* or merely as a physiological race of the latter. In many genera of the green algal order of the Chlorococcales, for example, acid tolerance is a species specific character (Kessler 1982) and phenotypically similar taxa proved to be genetically very diverse (Huss *et al.* 1989). In order to decide whether the two *Cyanidium* members are conspecific the application of molecular techniques allowing the detection of natural relationships between nucleic acids and proteins will be required. The chapter on 'new classification' for *Cyanidium* by Ott and Seckbach (in this volume) proposes to include these cave algae as: *C. chilense* Schwabe and as *C. rumpens* (Schwabe) F.D. Ott.

Acknowledgements

I thank Dr. V. Demoulin for reading the manuscript, Dr. F. Franck for taking the fluorescence spectra, B. Mayele for freeze-fracturing the cells, R. Gago for preparing the plates, M. Lambion for typing the manuscript, and M. Wolff for translating the thesis of Leiv-Hetzroni. The author is research associate of the Belgian National Fund for Scientific Research.

References

Beardall, J. and Entwisle, L. (1984). *Phycologia* 23: 397–399.
Doemel, W.N. and Brock, T.D. (1971). *J. Gen. Microbiol.* 67: 17–32.
Friedmann, I. (1964). *Int. J. Speleol.* 1: 29–33.
Hoffmann, L. (1989). *Bot. Rev.* 55: 77–105.
Hoffmann, L. and Mayele, B. (1993). Phycologia 32: 307-309.
Huss, V.A.R., Huss, G. and Kessler, E. (1989). *Pl. Syst. Evol.* 168: 71–82.
Kessler, E. (1982). *Progress in Phycological Research* 1: 111–135.
Leclerc, J. C. (1985). *Comad. J. Bot.* 63: 1893–1899.
Leclerc, J.C., Couté, A. and Dupuy, P. (1983). *Cryptogamie, Algologie* 4: 1–19.
Leclerc, J.C., Couté, A. and Hoaran, J. (1981). In: *Photosynthesis VI. Photosynthesis and Productivity, Photosynthesis and Environment (ed. Akoyunoglou, G.)* Balaban International Science Services, Philadelphia, 443–453.
Leiv-Hetzroni , O. (1965). *Thesis in Natural Sciences, Hebrew University at Jerusalem, Israel,* 23 pp. (in Hebrew).
Nichols, K.E. and Bogorad, L. (1962). *Bot. Gaz.* 124: 85–93.
Schwabe, G.H. (1936). *Verh. Deutsch. Wiss. Ver. Santiago de Chile N.F.* (Valparaiso) 3: 113–146.
Schwabe, G.H. (1942). *Mitt. Deutsch. Ges. Nat.- Völk. Ostasiens* (Tokyo, Leipzig) 33(Teil C): C1-C67.
Schwabe, G.H. (1944). *Mitt. Deutsch. Ges. Nat.-Völk Ostasiens* (Schanghai), Supplementband 21: 1–239.
Seckbach, J. (1987). *Ann. N.Y. Acad. Sci.* 503: 424–437.
Skuja, H. (1970). *Le Grotte d'Italia* (Trieste) 10(1968–1969): 193–202.
Smith, D.W. and Brock, T.D. (1973). *J. Gen. Microbiol.* 79: 219–231.
Staehelin, L.A. (1967). *Nature* 214: 1158.
Staehelin, L.A. (1968). *Proc. Roy. Soc. B.* 171: 249–259.
Ueda, K. and Chida, Y. (1987). *Br. phycol. J.* 22: 61–65.

Chapter 14

The function of peroxisomes in the Cyanidiaceae

WOLFGANG GROSS
Institut für Pflanzenphysiologie und Mikrobiologie, Freie Universität Berlin, Königin-Luise-Str. 12–16a, 14195 Berlin, Germany

Outlines:
1. Introduction
2. Evidence for peroxisomes from ultrastructural studies
3. Enzymatic capabilities of the peroxisomes
   3.1 Isolation of peroxisomes
   3.2 Peroxisomal enzymes in the Cyanidiaceae
   3.3 Cytosolic localization of the hydroxypyruvate reductase
   3.4 Peroxisomal enzymes in *Cyanidioschyzon merolae*
4. Peroxisomes in the Cyanidiaceae and other algae
5. *Cyanidium* as an endocyanome
6. Summary
References

1. Introduction

The term microbody was coined by Rhodin in 1954 to describe structures in mammalian cells. Although in the following years similar structures had been observed in plant tissue, the term microbody was adopted not before 1966 when Mollenhauer *et al.* showed the similarities of these structures in animals and plants. Microbodies are ultrastructural vesicles measuring usually from 0.2 to 0.6 μm, in some instances, however, up to 5 μm in diameter. On electron micrographs these organelles are visible as ovoid or elongated shaped vesicles, surrounded by a single membrane which usually encloses electron dense material. Any ultrastructural objects that shows these structural features can be called a microbody without implying any particular function. If it has been established that such a microbody contains the enzyme catalase and at least one H_2O_2-producing oxidase it is referred to as peroxisome. In higher plants all microbodies examined contain catalase and an oxidase, thus the terms microbody and peroxisome are generally used as synonyms.

Careful biochemical examination of the peroxisomes in various plant tissues mainly in the laboratories of H. Beevers and of N. E. Tolbert revealed that plant peroxisomes can be divided into three types according to their function in cell metabolism: In fat-storing tissue of germinating seeds peroxisomes participate in the conversion of fat to carbohydrates. The characteristic pathway of these organelles is the glyoxylate cycle, hence they are termed glyoxysomes (Breidenbach and Beevers 1967). In photosynthetically active cells, peroxisomes are involved in photorespiration, converting glycolate to glycine and serine to glycerate (Fig. 1). This type is termed leaf-peroxisome (Tolbert *et al.* 1968). In other plant tissues peroxisomes of the so-called 'unspecialized' type are present, which contain the enzymes catalase and uricase as well as enzymes of fatty acid β-oxidation.

Although in mammals and higher plants the function of the microbodies is fairly well understood, the enzymatic capabilities of these organelles have been studied only in a few algal species. From the data obtained so far, it has already become apparent that in many algae the function of these organelles is different from those known from higher plants and animals. Despite this diversity, the peroxisomal enzyme complement appears very conservative within certain algal groups.

To obtain information about the function one has to get data on the enzymatic capabilities of the peroxisomes. Cytochemical studies offer only a very limited view on the enzymes present in peroxisomes. Therefore, it is usually necessary to isolate the organelles and test for the presence of various enzymes.

In recent years the alga *Cyanidium caldarium* has been the subject of considerable ultrastructural and biochemical studies mainly because this alga is regarded as one of the most primitive eukaryotic organisms. *Cyanidium* appears to be related to unicellular rhodophytes, although it also contains

J. Seckbach (ed.), Evolutionary Pathways and Enigmatic Algae: Cyanidium caldarium *(Rhodophyta) and Related Cells,* 183–191, 1994.

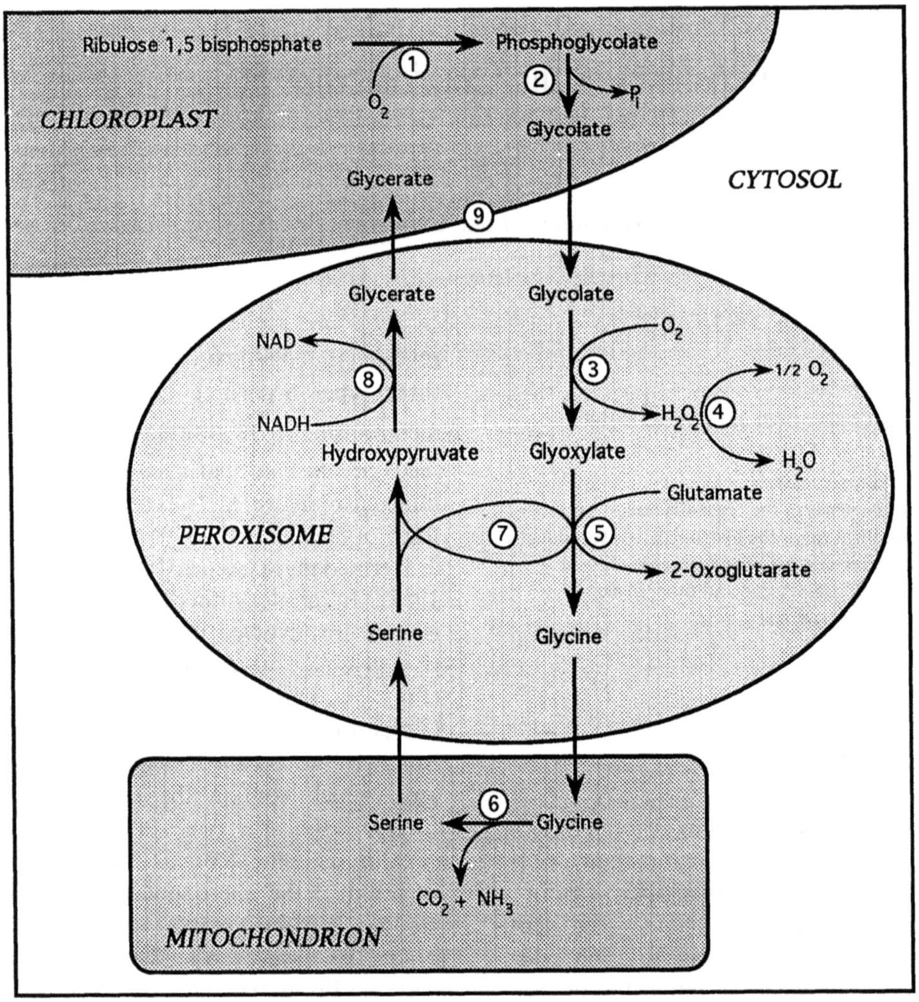

Fig. 1. Cooperative interaction of organelles in the photorespiratory pathway in higher plants and the Charophyceae. 1: Ribulose 1,5-diphosphate carboxylase/oxygenase; 2: phosphoglycolate phosphatase; 3: glycolate oxidase; 4: catalase; 5: glutamate-glyoxylate aminotransferase; 6; serine hydroxymethyltransferase; 7: serine-glyoxylate aminotransferase; 8: hydroxypyruvate reductase; 9: glycolate-glycerate transporter.

characteristics of the cyanobacteria (see Seckbach 1987). The enzyme profile of peroxisomes has been suggested as a phylogenetic marker for algal groups (Gross 1987), therefore, information on the function of peroxisomes in *Cyanidium* could be valuable in determining the taxonomic affiliation of this alga.

2. Evidence for peroxisomes from ultrastructural studies

Several reports have been published that mention the occurrence of microbody-like structures on electron micrographs from *'Cyanidium'*-cells (DeLuca *et al.* 1979; Ueda and Yokochi 1981; Ford 1984). However, Merola *et al.* (1981) could show that the *Cyanidium* cultures did not consists of one but of two or three related but distinct species termed *Cyanidioschyzon merolae*, *Cyanidium caldarium*, and *Galdieria sulphuraria*. These three species have since been characterized both ultrastructurally and biochemically (for review see Seckbach 1987), but unfortunately has not received general attention. Therefore, it was, and often is, not taken into account which species was used when *'Cyanidium'* cells are described. Re-examining the ultrastructural studies, one can often find

Table 1. Subcellular distribution of peroxisomal enzymes in Cyanidiaceae and other photosynthetic organisms

	Higher plants[a]	Charophyceae[b]	Chlorophyceae[b]	Xanthophyta[c]	*Cyanidium*[d] *Galdieria*
Catalase	Perox	Perox	Perox	Perox	Perox
Glycolate oxidoreductase	Perox	Perox	Mito[e]	Perox	Perox
Glu.-glyoxylate-aminotransferase	Perox	Perox	Mito	Perox	Perox
Ser.-glyoxylate-aminotransferase	Perox	Perox	Mito	absent	Soluble
Hydroxypyruvate reductase	Perox	Perox	Mito	absent	Soluble
β-oxidation	Perox[f]	Perox	Mi/Per	Mito	Mito
Carnitine-acetyltransferase	absent	absent[g]	--	Mi/Per	absent
Uricase	Perox	--	Perox	Perox	absent

Perox: peroxisomes; Mito: mitochondria; Mi/Per: dual localization;
Soluble: soluble fraction; absent: not detected; —: not assayed.
a: see Gerhardt (1978) and (1983);
b: after Stabenau (1984);
c: after Gross (1987) and Gross *et al.* (1985);
d: after Gross and Beevers (1989);
e: glycolate dehydrogenase;
f: but see Masterson *et al.* (1990);
g: Gross, unpublished.

that not *Cyanidium*, but *Galdieria sulphuraria* (Ueda and Yokochi 1981; Ford 1984), a more advanced member of the Cyanidiaceae, was examined. In addition, much of the controversy whether or not microbodies are present in *Cyanidium* is caused by the problems concerning the preparation of the cells for electron microscopy. No microbodies have been reported from the most primitive member of the Cyanidiaceae, *Cyanidioschyzon merolae*. All electron micrographs of *Cyanidioschyzon* published so far, are of very poor resolution making it usually difficult identifying any cell organelle except the chloroplast. The fixation and preservation of *Cyanidium caldarium* (e.g. from unialgal cultures) for electron microscopy is usually much better. However, reports on the occurrence of microbodies in this alga are controversial. DeLuca *et al.* (1979) described microbodies in *Cyanidium* as well as in *Galdieria*. In contrast, Toyoma (1980) did not observe these structures in *Cyanidium*. Taken in toto, it seems that only few microbodies are present in cells of *Cyanidium* and *Cyanidioschyzon*. From biochemical studies it can be confirmed that microbodies are present at least in *Cyanidium* and *Galdieria* .

3. Enzymatic capabilities of the peroxisomes

3.1 Isolation of peroxisomes

Although the technique to isolate peroxisomes from *Cyanidium* has been described elsewhere (Gross and Beevers 1989), there are some general problems when one is isolating organelles from these microalgae. The cell size of the three members of the Cyanidiaceae ranges from 2 to 10 μm (Seckbach 1987) and *Cyanidium* and *Galdieria* possess a strong cell wall. Because of the small size, the cells cannot be broken by conventional tissue homogenizers. Even grinding with mortar and pestle is time consuming and gives only low yield of intact organelles. The best results are obtained by grinding the cells in the presence of small glass beads (0.2 mm) with a Teflon impeller (VirTis Co.) After homogenization for 30–60 sec at medium to high speed in a VirTis homogenizer about 50 % of the cells are broken and about 30–50 % of the released mitochondria and peroxisomes are intact. Intact chloroplasts or nuclei, however, can not be obtained by this technique. Grinding at lower speed for longer duration (3–6 min) always gives low recovery of intact organelles. Short time (15 sec) high speed grinding breaks only 10–20 % of the cells; the recovery of intact organelles, however, increases to about 70 %. After cell disruption, the homogenate must be filtered through Nylon mesh (50 μm) to remove the glass beads and the filtrate is centrifuged for 10 min at 1,500 x g to remove unbroken cells, cell walls, and chloroplast fragments. The resulting supernatant is carefully collected with a pasteur pipette and centrifuged again for 10 min at 2,000 x g. The second centrifugation step is necessary to remove the bulk of the chloroplast fragments. The supernatant is collected again with a pipette and can be loaded directly onto a density gradient of 30–60 % sucrose. However, the separation of the organelles is usually very poor due to high trapping effects by the chloroplast fragments. During centrifugation these fragments move rapidly to a density of about 1.20 g/cm^3, forming a dense layer. Peroxisomes, heavier but migrating much slower, eventually enter this layer of chloroplast fragments and the majority of peroxisomes is trapped in this band. This problem can be overcome if the particles are isolated by flotation from high sucrose densities. For this so-called flotation gradient centrifugation, the 2000 x g supernatant is stirred gently on ice and brought to 60 % (w/w) sucrose by adding slowly powdered sucrose. This solution is then poured into a centrifuge tube and a sucrose gradient from 60 to 30 % (w/w) is carefully layered on top. The gradient is centrifuged for about 12 h at 65,000 x g and then fractionated. Although this method gives a fairly good separation, it is still very important to remove most of the chloroplast fragments without removing too many mitochondria and peroxisomes before the sample is loaded onto the flotation gradient. For example, the chloroplast fragments of *Cyanidioschyzon merolae* sediment only slightly faster than the mitochondria. This makes it impossible to remove a sufficient amount of chloroplast fragments and results in high trapping effects even when a flotation gradient is used.

3.2 Peroxisomal enzymes in the Cyanidiaceae

Mitochondria and peroxisomes from *Cyanidium* and *Galdieria* were separated using the above mentioned flotation gradient system. Figure 2 shows the distribution of marker enzymes and peroxisomal enzymes from *Galdieria* within a flotation gradient. Similar to *Cyanidium* (Gross and Beevers 1989), the peroxisomes in *Galdieria* contain one key enzymes for glycolate metabolism, glycolate oxidase, in the peroxisomes (Fig. 2b) and the other key enzyme, hydroxypyruvate reductase (HPR), is found in the soluble fraction (Fig. 2d). In addition to glycolate oxidase, the peroxisomes contain the enzymes catalase and glutamate-glyoxylate aminotransferase. The term soluble fraction refers to the cytosolic component plus soluble enzymes released from broken organelles. Therefore, it can only be concluded that the HPR is not localized in the peroxisomes and mitochondria. However, a cytosolic localization of this enzymes seems most likely. The mitochondria reached their equilibrium density at 1.19 g/cm^3 as indicated by the distribution of citrate synthase activity (Fig. 2c). The presence of mitochondrial membranes is indicate by the peak of cytochrome c oxidase and malate dehydrogenase activity. Although malate dehydrogenase is usually regarded as a soluble matrix enzyme of mitochondria, in *Cyanidium* it has been shown (Gross 1990a) that the mitochondrial malate dehydrogenase is completely membrane-bound. Coinciding with the activity of the mitochondrial citrate synthase is the activity of an enzyme of fatty acid β-oxidation, enoyl-CoA hydratase (Fig. 2d). Unlike peroxisomes from higher plants, enzymes for β-oxidation of fatty acids were not present in the peroxisomes of *Cyanidium* (Gross 1989) and *Galdieria.* This pathway is, therefore, confined to the mitochondria in these two algal species.

3.3 Cytosolic localization of the hydroxypyruvate reductase

In higher plants a cytosolic HPR, in addition to the peroxisomal HPR, has been reported (Kleczkowski and Randall 1988). This cytosolic enzyme, however, is specific for NADPH and not for NADH as the peroxisomal or the *Cyanidium* enzyme. The NADPH-preferring HPR has been proposed to

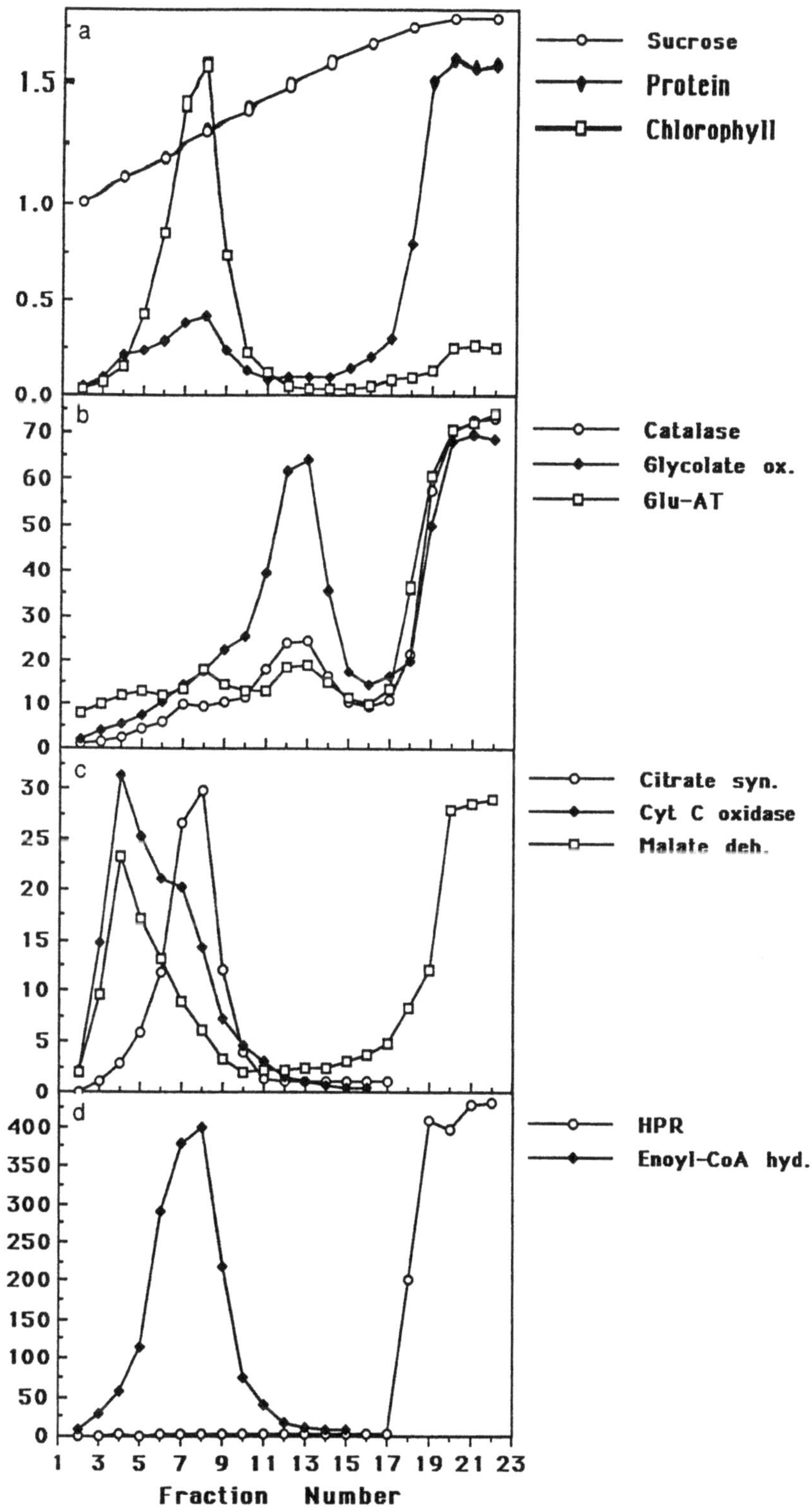

Fig. 2. Separation of organelles from a crude homogenate of *Galdieria sulphuraria* on a linear sucrose flotation gradient. Ordinate value x factor = nmoles · min^{-1} · ml $fraction^{-1}$; % sucrose; mg protein · ml $fraction^{-1}$; mg chlorophyll · ml $fraction^{-1}$. Sucrose, x 30; protein, x 1; chlorophyll, x 0.7; catalase, x 1.10^3; glycolate oxidase, x 0.2; glutamate-glyoxylate aminotransferase, x 1; citrate synthase, x 1; cytochrome C oxidase, x 13; malate dehydrogenase, x 35; hydroxypyruvate reductase, x 1; enoyl-CoA hydratase, x 1.

deal with hydroxypyruvate leaking out from the peroxisomes, thereby indirectly involved in photorespiration. Although HPR is present in *Galdieria* and *Cyanidium* it is not unquestionable that this enzyme is connected to photorespiration. It has been found in other organisms that glyoxylate can be converted to C^3-compounds via the glyoxylate cycle, tartronic acid semialdehyde pathway (Badour and Waygood 1971) or the β-hydroxyaspartate pathway (Kornberg and Morris 1963). However, no activity of the key enzymes of these pathways could be detected in *Cyanidium* (data not shown). Furthermore, under growth conditions that suppress photorespiration (high CO_2, low O_2), the activity of all enzymes of the glycolate metabolism including HPR are repressed (Gross and Beevers 1989). This indicates that all five enzymes tested participate in the same pathway. As mentioned earlier, the total activity of HPR is about 20-times higher than the total glycolate oxidase activity. This is not surprising if one considers that the activity of HPR per volume of cytosol should be about equal to the activity of the glycolate oxidase per volume of peroxisomal matrix. Although this does not take into account the difference in kinetic characteristics of these two enzymes, it offers one way to explain the different total activity of these enzymes.

In higher plants and animals a three amino acid motif (ser-lys-leu) is suspected to function as a topogenic signal for the import of proteins into peroxisomes (Gould *et al.* 1989). Although, there are still open questions about this concept, a cytosolic location of a typical peroxisomal enzyme could be explained easily by assuming that the target signal is no longer functional in this protein. In other words, the HPR in *Cyanidium* and *Galdieria* lost this signal or has never acquired an import signal. Sequencing the HPR from *Cyanidium* could help to clarify this problem.

It is interesting to note that in the peroxisomes of *Cyanidium* and *Galdieria* no enzymes seem to be present that use NAD(H) as a cosubstrate. In contrast, peroxisomes from higher plants contain the β-oxidation enzyme, hydroxyacyl-CoA dehydrogenase, as well as hydroxypyruvate reductase and malate dehydrogenase. Therefore, a system is necessary to regenerate NAD from NADH via a malate-aspartate shuttle (Mettler and Beevers 1980) or via a membrane-bound redox system (Hicks and Donaldson 1982). Both systems are apparently missing in the peroxisomes of the Cyanidiaceae.

3.4 Peroxisomal enzymes in Cyanidioschyzon merolae

Peroxisomes could not be isolated from the most primitive member of the Cyanidiaceae, *Cyanidioschyzon*, due to poor separation in sucrose gradients. However, the enzymes serine-glyoxylate-aminotransferase and hydroxypyruvate reductase as well as all enzymes found in the peroxisomes of the other two species, could be demonstrated in crude extracts of *Cyanidioschyzon*. In addition, two enzymes of the β-oxidation pathway, enoyl-CoA hydratase and hydroxyacyl-CoA dehydrogenase, are not part of a bifunctional enzyme as in peroxisomes but are separate proteins as it is typical for the mitochondrial pathway. This indicates that *Cyanidioschyzon* possesses only mitochondrial β-oxidation (Gross 1989). Further discussion on peroxisomal function in the Cyanidiophyceae and their phylogenetic relation was recently published by Seckbach *et al.* (1992).

4. Peroxisomes in the Cyanidiaceae and other algae

As mentioned earlier, the Cyanidiaceae seem to be closely related to unicellular rhodophytes. In this respect it is interesting that a glycolate oxidase is present in the unicellular red alga *Porphyridium purpureum* (Suzuki *et al.* 1991). In addition, the enzymes phosphoglycolate phosphatase, catalase, glutamate-glyoxylate aminotransferase, and HPR are present in several red algae (Gross 1990c; Gross, unpublished). Thus, red algae and the Cyanidiaceae show a similar profile of enzymes of glycolate metabolism. However, peroxisomes, although present in rhodophytes, have not been isolated from red algae making predictions about the function of the peroxisomes in this group highly speculative.

Glycolate oxidase is not only present in the Rhodophyta and higher plants but also in the Chrysophyta (Suzuki *et al.* 1991), Xanthophyta, Phaeophyta (Gross 1990b; Suzuki *et al.* 1991), and the green algal group of the Charophyceae

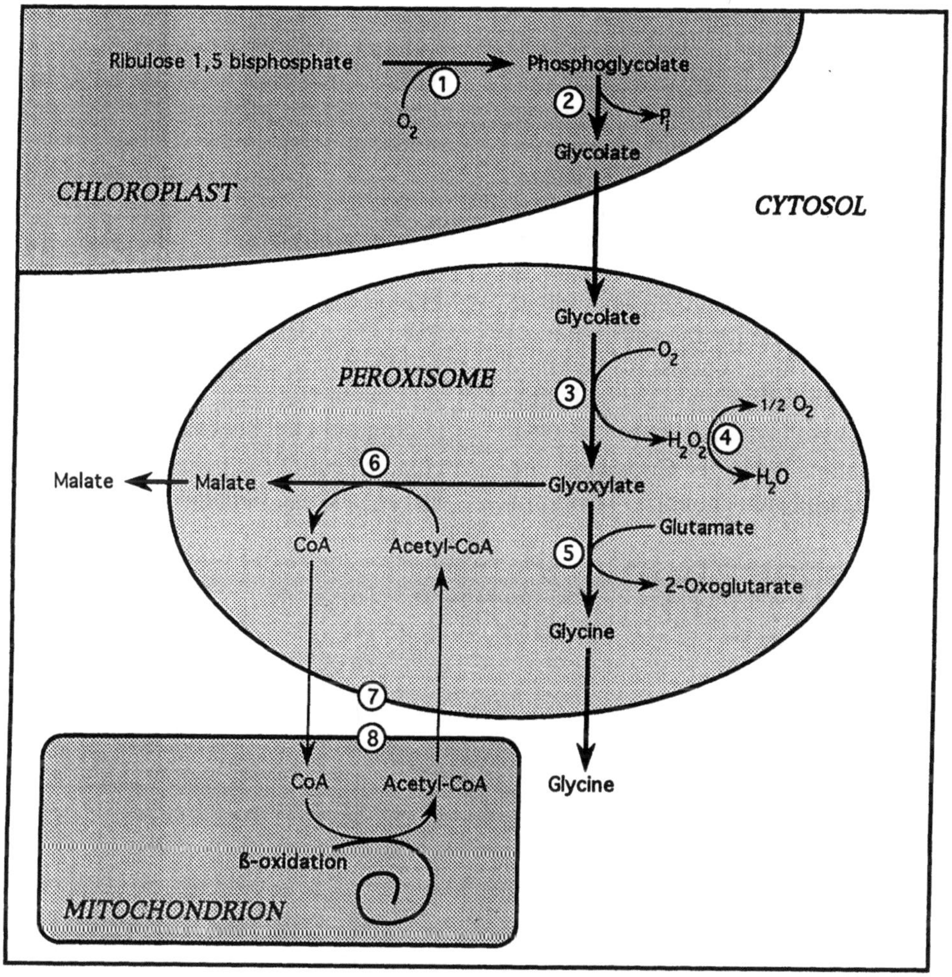

Fig. 3. Conversion of glycolate to malate in the Xanthophyta. 1: Ribulose 1,5-diphosphate carboxylase/oxygenase; 2: phosphoglycolate phosphatase; 3: glycolate oxidase; 4: catalase: 5: glutamate-glyoxylate aminotransferase; 6: malate synthase; 7: peroxisomal carnitine acetyltransferase; 8: mitochondrial carnitine acetyltransferase.

(Stabenau 1984) Table 1. However, in the Xanthophyta (Gross *et al.* 1985; Gross 1987) and the Chrysophyta (Gross, unpublished), the second key enzyme of glycolate metabolism, hydroxypyruvate reductase, is missing. For the Xanthophyta it has been shown that glyoxylate, produced by the glycolate oxidase, can be converted to malate (Fig. 3).

For the Rhodo- and Phaeophyta, the subcellular location of glycolate oxidase and HPR is not yet known, making a comparison with the peroxisomes in the Cyanidiaceae difficult. The Charophyta, like the higher plants, possess peroxisomal glycolate oxidase and HPR (Stabenau 1984) and thereby differ from the Cyanidiaceae. The distribution of glycolate oxidase within the algae not necessarily indicates that this enzyme is of polyphyletic origin because the occurrence of the glycolate oxidase should be independent of the type of plastid. All algae could have originated from a common host cell which contained a peroxisomal glycolate oxidase. The separation into different algal lines occurred only by the type of endosymbiont the cells harbor as plastids. This would imply that algae containing no glycolate oxidase but glycolate dehydrogenase lost the oxidase secondarily. Sequencing the glycolate oxidase from mammalian and various algal sources would help to clarify the validity of this scenario.

5. *Cyanidium* as an endocyanome

In the past it has been suggested by several authors that *Cyanidium* represents a cyanelle-possessing alga. According to Raven (1970) and Castenholz (1979) the most likely candidate for the *Cyanidium*-host cell is a colorless green alga whereas the plastid evolved from a cyanobacterium. If this theory is correct then where did the glycolate oxidase in *Cyanidium* come from? Only a colorless charophyte could have served as a host for the endocyanome because all other green algae possess no glycolate oxidase but a mitochondrial glycolate dehydrogenase. However, charophytes have a peroxisomal fatty acid β-oxidation while in *Cyanidium* this pathway is located in the mitochondria. Therefore, assuming that *Cyanidium* represents an endocyanome requires a number of assumptions concerning the change of metabolic pathways. This in turn makes any other alga as likely or unlikely as a candidate for such a symbiosis.

6. Summary

The biochemical characterization indicate that all three Cyanidiaceae, *Cyanidioschyzon*, *Cyanidium*, and *Galdieria*, contain peroxisomes although not yet demonstrated for *Cyanidioschyzon* on electron micrographs (cf. Seckbach *et al.* 1992). The enzyme complement of the peroxisomes seems to be identical at least in *Cyanidium* and *Galdieria* and has not

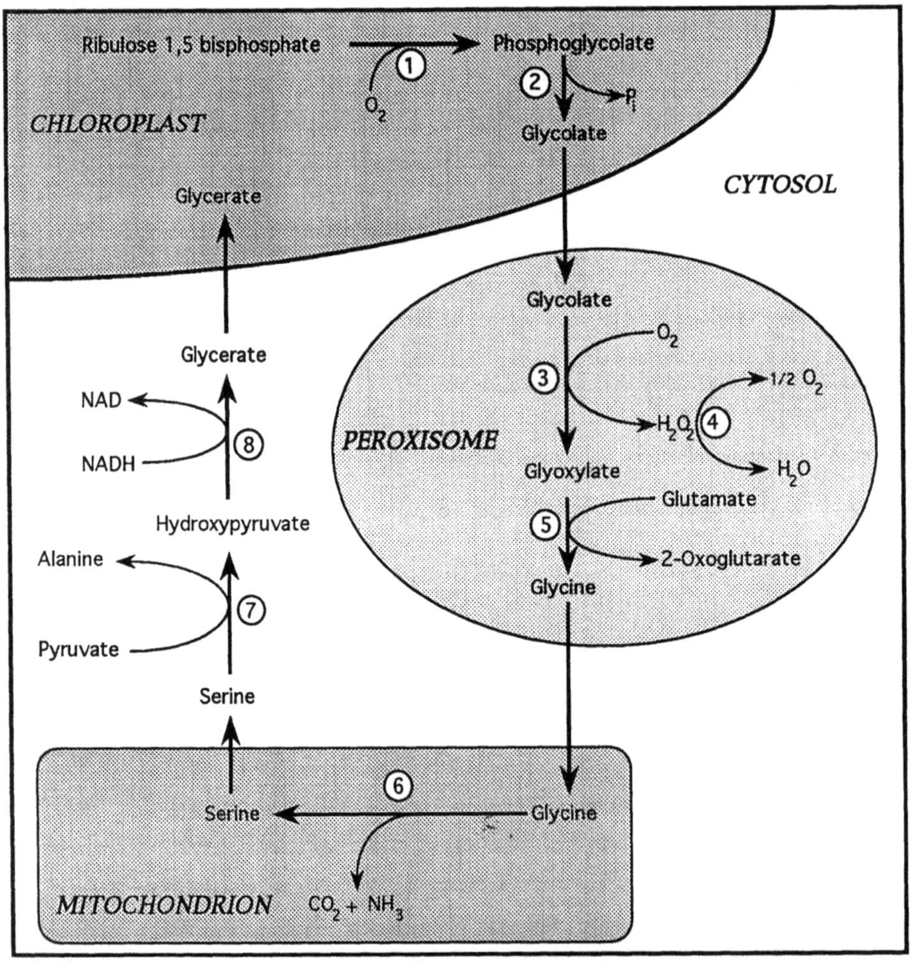

Fig. 4. Subcellular distribution of enzymes of the glycolate pathway in *Cyanidium* and *Galdieria*. 1: Ribulose 1,5-diphosphate carboxylase/oxygenase; 2: phosphoglycolate phosphatase; 3: glycolate oxidase; 4: catalase: 5: glutamate-glyoxylate aminotransferase; 6: serine hydroxymethyltransferase; 7: serine-pyruvate aminotransferase; 8: hydroxypyruvate reductase.

been reported for peroxisomes from other organisms. The major difference to higher plants and the Charophyceae is the separation of the glycolate pathway and the glycerate pathway in two different compartments as illustrated in Figure 4.

Since all enzymes participating in the glycolate/glycerate pathway are present in extracts of various red algae, there is no evidence for differences in the function of peroxisomes in the Rhodophyta and Cyanidiaceae. This assumption needs to be confirmed by studies on isolated peroxisomes from red algae. If the Cyanidiaceae are indeed primitive organisms and not the result of a secondary reduction due to their environment, then the occurrence of glycolate oxidase can no longer be considered as a characteristic of advanced organisms (Tolbert 1976). Furthermore, the distribution of certain enzymes alone can often not be used as a phylogenetic marker unless their subcellular localization is taken into account.

Acknowledgements

I thank Dr. H. Beevers for his help and encouragement during these studies and the Alexander-von-Humboldt foundation (Feodor-Lynen-program) for financial support.

References

Badour, S. S. and Waygood, E. R. (1971). *Biochim. Biophys Acta.* 242: 493–499.

Breidenbach, R. W. and Beevers, H. (1967). *Biochem. Biophys. Res. Commun.* 27: 462–469.

Castenholz, R.W. (1979). In: *Strategies of microbiol life in extreme environments. (ed. Shilo, M.).* Verlag Chemie, Weinheim, FRG. pp 373–392.

De Luca, P., Gambardella, R. and Merola, A. (1979). *Bot. Gaz.* 140: 418–427.

Ford, T. W. (1984). *Ann. Bot.* 53: 285–294.

Gerhardt, B. (1978). *Microbodies/Peroxisomen pflanzlicher Zellen. Cell Biology Monographs,* Vol. 5, Springer Verlag, Vienna.

Gerhardt, B. (1983). *Planta* 159: 238–246.

Gould, S. J., Keller, G. A., Hosken, N., Wilkinson, J. and Subramani, S. (1989). *J. Cell Biol.* 108: 1657–1664.

Gross, W. (1987). *Untersuchungen zur Funktion der Peroxisomen in verschiedenen Xanthophyceen. Doctoral thesis, Oldenburg, F.R.G.*

Gross, W. (1989). *Plant Physiol.* 91: 1476–1480.

Gross, W. (1990a). *Phytochemistry* 29: 3081–3085.

Gross, W. (1990b). *J. Phycol.* 26: 381–383.

Gross, W. (1990c). *J. Phycol. (Suppl.)* 26: S-16.

Gross, W., Winkler, U. and Stabenau, H. (1985). *Plant Physiol.* 77: 296–299.

Gross, W. and Beevers, H. (1989). *Plant Physiol.* 90: 799–805.

Hicks, D. B. and Donaldson, R. P. (1982). *Arch. Biochem. Biophys.* 215: 280–288.

Kleczkowski, L.A. and Randall, D.D. (1988). *Biochem. J.* 250: 145–152.

Kornberg, H. L. and Morris, J. G. (1963). *Nature* 197: 456–457.

Masterson, C., Wood, C. and Thomas, D. R. (1990). *Planta* 182: 129–135.

Merola, A., R. Castaldo, P. DeLuca, R. Gambardella, A. Musacchio and Taddei, R. (1981). *Giorn. Bot. Ital.* 115: 189–195.

Mettler, I. J. and Beevers, H. (1980). *Plant Physiol.* 66: 555–560.

Mollenhauer, H. H., Morré, D.J. and Kelly, A. G. (1966). *Protoplasma* 62: 44–52.

Raven, P. H. (1970). *Science* 169: 641–646.

Rhodin, J. (1954). *Correlation of ultrastructural organization and function in normal and experimentally changed proximal convoluted tubule cells of mouse kidney. Doctoral thesis, Stockholm, Sweden.*

Seckbach, J. (1987). *Ann. N. Y. Acad. Sci.* 503: 424–437.

Seckbach, J., Gonzáles, E., Wainwright, I. M. and Gross, W. (1992). *Nova Hedwigia* 55(1–2): 99–109.

Stabenau, H. (1984). In: *Compartments in algal cells and their interaction. (eds. Wiessner, W., D. Robinson and Starr, R. C.). Springer Verlag, Berlin.* pp. 183–190.

Suzuki, K., Iwamoto, K., Yokoyama, S. and Ikawa, T. (1991). *J. Phycol.* 27: 492–498.

Tolbert, N.E. (1976). *Aust. J. Plant Physiol.* 3: 129–132.

Tolbert, N. E., Oeser, A., Kisaki, T., Hageman, R. H. and Yamazaki, R. K. (1968). *J. Biol. Chem.* 243: 5179–5184.

Toyoma, S. (1980). *Cytologia* 45: 779–790.

Ueda, K. and Yokochi, J. (1981). *Bot. Mag.Tokyo* 94: 159–164.

Chapter 15

Nitrogen assimilation in a thermophilic acidophilic alga

CARMELO RIGANO, AMODIO FUGGI*, VITTORIA DI MARTINO RIGANO and VINCENZA VONA
Dipartimento di Biologia Vegetale, Università di Napoli Federico II, Napoli, Italy
** Istituto di Chimica Agraria e Forestale, Università di Reggio Calabria, Gallina di Reggio Calabria, Italy*

Outlines:
1. Introduction
2. Enzymes of nitrate and ammonium metabolism
   2.1 Nitrate reductase and glutamine synthetase
   2.2 Experiments with methionine sulphoximine
   2.3 Synthesis of glutamine synthetase and nitrate reductase
   2.4 Biochemical properties of inactive nitrate reductase
3. Enzymes from *Cyanidium* and other sources
4. Uptake of inorganic nitrogen compounds
   4.1 Nitrate uptake
   4.2 Nitrite uptake
   4.3 Ammonium uptake
5. Nitrogen assimilation and carbon – nitrogen interaction
   5.1 Ammonium assimilation by N-limited cells
   5.2 Effect of removal of CO_2 on NH_4+ assimilation in N-limited cells
References

1. Introduction

Thermophilic eukaryotic algae are rare in acid hot environments in nature. The sole photosynthetic organisms living in thermal alkaline environments are almost exclusively cyanobacteria. Among algae, only the members of the class Cyanidiophyceae [including the families Cyanidiaceae and Galdieriaceae (Merola *et al.* 1981)] were found to live in thermal acidic habitats up to 57 °C (Doemel and Brock 1970). The family Cyanidiaceae includes the alga *Cyanidium caldarium*. For a larger spectrum on these algae, see chapters by Seckbach, Ott and by Sentsova in this volume.

The occurrence of *C. caldarium* is strictly dependent on acidic pH, and usually it is grown in our laboratory at pH 1.9. Among unicellular green algae, *Chlorella sorokiniana* is able to grow at higher pH levels in waters up to 40 °C. Thus it is not surprising that studies on nitrogen metabolism in thermophilic algae have been only carried out with these two organisms. Noteworthy is the fact that both *C. caldarium* and *C. sorokiniana* share characteristics with respect to control of enzymes of inorganic nitrogen metabolism which distinguish them from the mesophilic algae. *Galdieria* (Galdieriaceae) was unable to use nitrate as nitrogen source (Allen 1959; cf. Sentsova's chapter in this volume). Ott and Seckbach (in this book) proposed a new classification of the *Cyanidium* and *Galdieria* members. *C. caldarium* is a strictly (obligatory) autotrophic alga, being unable to use organic substrate as carbon and energy source. It can use glutamate, nitrate and ammonium as nitrogen source. Glutamate, indeed, is considered to bc a poor nitrogen source (Rigano *et al.* 1974a).

2. Enzymes of nitrate and ammonium metabolism

We have being studying nitrogen metabolism in *C. caldarium* at the biochemical level in cell-free extract by testing activity levels and characteristics of enzymes involved in ammonium and nitrate metabolism. Our investigations include, in particular nitrate reductase and glutamine synthetase. At the physiological level, we assayed the patterns of ammonium and nitrate uptake in intact cells.

The alga possesses an assimilatory nitrate reductase (Rigano 1971) and assimilates ammonium through a glutamine synthetase/glutamate synthase pathway. These observations have been supported by experiments with methionine-D,L-sulphoximine, and by analysis of the intracellular amino acid pools variations (Rigano *et al.* 1977; Di Martino Rigano *et al.* 1982; Di Martino Rigano *et al.* 1989). *C. caldarium* possesses also a NADPH glutamate dehydrogenase (Rigano *et al.* 1979a).

J. Seckbach (ed.), Evolutionary Pathways and Enigmatic Algae: Cyanidium caldarium *(Rhodophyta) and Related Cells,* 193–200, 1994.

2.1 Nitrate reductase and glutamine synthetase

The *Cyanidium* nitrate reductase can use *in vitro* both NADH and NADPH as physiological reductants, as well as the non physiological dithionite-reduced viologens and flavins (Rigano 1971; Rigano and Aliotta 1975). The NAD(P)H nitrate reductase activity of *Cyanidium*, like nitrate reductase from mesophilic algae, is more sensitive to mild treatment with protein denaturing agents (heating, urea, mercurials) than is the activity dependent on viologens or flavins. FAD (0.01 mM) protects the enzyme against the effect of heat (Rigano and Violante 1972a, 1972b; Rigano *et al.* 1980). *Cyanidium* nitrate reductase is, however, more stable to heat treatment than nitrate reductases from other sources (Rigano 1967). Moreover, typical of *Cyanidium* nitrate reductase is the finding that the benzyl viologen activity can be *in vitro,* up to 4-fold activated by heat treatment, as observed in crude extract prepared with cells grown in a batch in sufficient nitrate (Rigano 1971). It was demonstrated later that heat treatment activated an alternative form of nitrate reductase which occurred inactive in the cell, and that the occurrence of this inactive enzyme may be related to some regulatory phenomena which control the rate of nitrate reduction through reversible inactivation of nitrate reductase (Rigano and Violante 1973). Thus it was supposed that the inactivation of nitrate reductase was stimulated (through an unknown mechanism) by some nitrogenous compound formed by nitrate reduction. This proposal is based on the fact that nitrate reductase was immediately transformed into the inactive species upon addition of ammonium to the cells which shared an active enzyme. Then this enzyme was once more transformed into the active form upon ammonium removal, so that the cells remain in condition of nitrogen starvation (Rigano *et al.* 1974b).

In addition to nitrate reductase, also glutamine synthetase of *Cyanidium* has been proven to be a regulatory enzyme.

2.2 Experiments with methionine sulfoximine

The regulatory role of reversible inactivation of nitrate reductase, observed *in vivo* in response to the nitrogenous state of the cell, was established in several ways (Rigano *et al.* 1978, 1979b, 1981). The reversible results obtained in this respect with methionine-DL-sulfoximine (MSX), a structural analogue of glutamate, have been incredible (Di Martino Rigano *et al.* 1982).

MSX is a strong inhibitor of glutamine synthetase and in *C. caldarium* it inhibits assimilation of ammonium and, consequently, the growth (Rigano *et al.* 1977). However the growth inhibition was observed only in cultures inoculated with N-limited cells, or with cells grown on glutamate as the sole nitrogen source, that were permeable to MSX. The MSX added to cells of *C. caldarium* causes a loss of glutamine synthetase activity and prevents inactivation of nitrate reductase. This occurred in fully active form also in the presence of large amount of ammonium (Rigano *et al.* 1979b). If a cell suspension of *C. caldarium*, with added MSX, was supplied with nitrate, there was a significant excretion of ammonium into the external medium. These results support the view that the presence of an active glutamine synthetase is a necessary prerequisite for the ammonium-stimulated inactivation of nitrate reductase. The excretion of ammonium from nitrate-supplied cells, on the other hand, suggests that it is the inability of nitrate reductase to attain the inactive state that, in the presence of MSX, makes nitrate reduction an uncontrolled process leading to a continuous production of ammonium which is not assimilated (Rigano *et al.* 1979b; Di Martino Rigano *et al.* 1982). This contention was supported by experiments *in vivo* at pH 7,5 and 9,4, a range which is not in physiological condition for *Cyanidium* cell which are able to grow only at pH lower than 5 (Rigano *et al.* 1978).

2.3 Synthesis of glutamine synthetase and nitrate reductase

The synthesis of nitrate reductase and glutamine synthetase was studied by culturing *C. caldarium* under different conditions of growth (Rigano and Violante 1973; Rigano *et al.* 1974b, 1981). Glutamine synthetase activity is low in cells grown under conditions of excess ammonium, and it is 2-, 5- and up to 10-fold greater in cells grown under conditions of excess nitrate, limited ammonium or limited nitrate, respectively. The data also indicate that excess activity of ammonium nitrate reductase is almost undetectable, whereas it is present under conditions of excess nitrate, limited nitrate, or limited ammonium, as well as in cells grown on glutamate. It is significant that under all conditions allowing the synthesis of nitrate reductase, comparable values of activity occur. It is thus evident that in *C. caldarium* glutamine synthetase is consistently repressed in excess ammonium, significantly repressed in excess nitrate and derepressed under the limited nitrogen conditions where the growth is seriously restricted. Nitrate reductase, in contrast, is fully repressed in excess ammonium, or it is fully derepressed in other growth conditions. However, nitrate reductase occurs mainly in a catalytically active form under growth rate resulting from nitrogen limiting conditions. In excess nitrate this enzyme is in catalytically inactive form. However under conditions of phosphate limitation, even in the presence of excess ammonium nitrate reductase is fully derepressed. Since it occurs in the catalytical inactive form, we suggest that the signals for repression or inactivation of the enzyme are distinct (Di Martino Rigano *et al.* 1984).

The occurrence of nitrate reductase in the inactive form and the significant repression of glutamine synthetase in excess nitrate, as well as the occurrence of nitrate reductase in the active form and derepression of glutamine synthetase under conditions of nitrogen limitation, suggest that the key effector(s) of nitrate reductase inactivation and of glutamine synthetase repression are to be found in the intracellular pools of metabolites that we assume are still present in excess nitrate, but severely depleted in limited nitrogen. In this respect it is noteworthy that inactive nitrate reductase can *in vivo* be transformed into the active one by transferring the cells to an N-free medium. Also, that the active enzyme can be rapidly transformed into the inactive one by adding ammonium to the cells (Rigano *et al.* 1974b).

2.4 Biochemical properties of inactive nitrate reductase

Due to the regulatory significance of the activation and the inactivation of nitrate reductase, the biochemical characteristics of the active and inactive enzyme were studied to establish the molecular basis of the control mechanism. Thus, we tested the effect of mild treatment (with the protein denaturing agents, heat, urea, mercurials, and ethylene glycol) in crude extracts (Rigano 1971; Rigano *et al.* 1972a, 1972b, 1980). All procedures to purify the inactive enzyme were unsuccessful.

It was observed that heating the inactive nitrate reductase in the range of 45 to 57 °C increases significantly (benzyl)viologen dependent activity. In the presence of FAD (0.01 mM), however, the increase in activity was shifted to the range of 57 to 62 °C (Rigano *et al.* 1980). Concomitantly with this activation of benzyl viologen activity the NAD(P)H dependent activity was destroyed.

The treatment of inactive nitrate reductase with urea brought about a fourfold increase of both benzyl viologen and NAD(P)H activities. At urea concentrations above 3 M, loss of NAD(P)H activity occurred, and above 5 M also loss of benzyl viologen activity. Mersalyl (10–40 μM) increased four-fold both NAD(P)H and benzyl viologen activity of the inactive enzyme. Higher concentrations of mercurial inhibited the NAD(P)H activity by 100% and benzyl viologen activity only by 50%. Treatments of the inactive nitrate reductase with ethylene glycol activated both NAD(P)H and benzyl viologen activities. The maximum extent of activation was found at 25 vol% glycol concentration for benzyl viologen activity, and at 35 vol% for NAD(P)H activity. Higher concentrations produced similar denaturing effects in both active and inactive enzyme.

These results suggest that the *Cyanidium* nitrate reductase molecule contains a control moiety involved in the maintenance of the inactive state of the enzyme. This part, in response to some presently unknown biological signal(s) can contribute to promotion and maintenance of the inactive state of the enzyme. Thus, urea and mersalyl, under sufficiently mild conditions, act on this control region producing activation of the nitrate reductase. Only at higher concentrations they act progressively on the part of the enzyme molecule bearing the site for NAD(P)H and ultimately on the part bearing the sites for benzyl viologen and nitrate. Ethylene glycol, different from the other denaturants, acts simultaneously on the controlling part and on the catalytic part. With respect to the latter action, it seems that ethylene glycol acts on the catalytic part involved in the electron flow from either NAD(P)H and reduced benzyl viologen.

Heat differs from the other denaturants in that it

produces an activation of only benzyl viologen activity and concomitantly a loss of NAD(P)H activity. This means that the integrity of the entire nitrate reductase molecule is required for the maintenance, and perhaps promotion, of the inactive state of nitrate reductase, and not only of its controlling part. The occurrence of a regulatory region in the enzyme is suggested also by *in vivo* experiments: in N-limited cells the inactivation of the enzyme was delayed and, after ammonium addition, protein synthesis was required to allow ammonium to inhibit nitrate assimilation (Fuggi *et al.* 1981b).

3. Enzymes from *Cyanidium* and other sources

It would be useful to compare the characteristics of nitrate reductase and other enzymes of nitrogen metabolism of *Cyanidium* with those of other nitrate reductases from other algae. The nitrate reductase of *Cyanidium* differs somewhat from the enzyme of another red alga *Porphyridium aerugineum* (Rigano *et al.* 1979a). In fact: (1) *Porphyridium* enzyme is functional with NADH only and not with NADPH; (2) it exhibits similar activity with NADH and benzyl viologen, whereas in *Cyanidium* the activity with NAD(P)H is only 25% of that with benzyl viologen ; (3) the enzyme of *Porphyridium* is more sensitive to heat treatment than that of *Cyanidium*; (4) the activity of *Porphyridium* enzyme with benzyl viologen *in vitro* is linear with time, whereas the activity of the *Cyanidium* enzyme declines from linearity after 12 min of incubation. However, it is noteworthy that common characteristics of *Cyanidium* and *Porphyridium* nitrate reductases are that the benzyl viologen activity of both algae can be activated by heat. Perhaps also in *Porphyridium* heat activates some inactive enzyme, and that the same regulatory mechanism may operate in both *Cyanidium* and *Porphyridium.* A striking difference, however, between the two enzymes is that the enzyme of *Cyanidium* is stable in nitrogen-starved cells, whereas that of *Porphyridium* requires nitrate to be stabilized *in vivo*; in fact, cells kept under nitrogen- free conditions lose some of their nitrate reductase activity in less than 1h (unpublished data). In this respect the enzyme of this red alga *Porphyridium* resembles that of *Chlorella sorokiniana*, a green alga. Also in the latter alga nitrate reductase is a regulatory enzyme, but its regulatory patterns are different from those of *Cyanidium* and of other green algae. In *C. sorokiniana* nitrate was necessary to produce and maintain an active enzyme. In an other Chlorellae, e.g., *Chlorella* 211/12, the synthesis of nitrate reductase requires nitrate for induction (Di Martino Rigano *et al.* 1983).

Glutamate dehydrogenase (another enzyme concerning the nitrogen metabolism) was 5–6 fold lower in *Cyanidium* than in *Porphyridium aerugineum* (Rigano *et al.* 1979a). On the other hand, glutamine synthetase occurred at very low level in *Porphyridium* (unpublished).

Studies on enzymes of inorganic nitrogen metabolism of thermal organisms can provide clues on evolutionary aspects of these algal enzymatic systems. Also, they can elucidate, by supplying alternative models, related control mechanisms operating in photosynthetic organisms.

4. Uptake of inorganic nitrogen compounds

In *Cyanidium caldarium*, as in other organism, efficient uptake systems are available at the plasmalemma to concentrate inorganic nitrogen compounds within the cell and sustain growth. Being strictly autotrophic and acidophilic, it exchanges basically only mineral ions with the medium. The acidic conditions, in particular, allowed also to evaluate the proton exchange dependent on uptake and metabolism of nitrogen compounds. These compounds are not masked by the proton exchange dependent on photosynthesis and respiration (Fuggi *et al.* 1981a).

4.1 Nitrate uptake

In *Cyanidium caldarium* nitrate has been taken up by two uptake systems: a high affinity and a low affinity transport system (Fuggi *et al.* 1984). The high affinity system is not constitutive in this alga: it was derepressed in cells transferred to low nitrate media or to nitrogen limitation. Nitrate was not necessary for induction, a feature found in several marine algae and cyanobacteria, but not in green algae, land plants and fungi (Syrett 1981; Ullrich 1983; Schloemer and Garrett 1974). Protein synthesis was a requirement for its expression.

The high affinity nitrate uptake system is able to scavenge trace amounts of nitrate in the medium: it was operative in acidic but not in neutral conditions. Stoichiometric data of proton taken up with pulse amounts of nitrate in transitory conditions and the kinetics of nitrate uptake at various medium pHs fitted in with the mechanism of nitrate-proton symport (Fuggi 1985). This carrier operates with a stoichiometry of 2 protons per nitrate and involves a tight coupling of protons and nitrate independent of the medium pH. The hypothesis that protons could be involved in taking up nitrate has been proposed in the last years, essentially from electrophysiological experiments (Ullrich 1987).

The mechanism is electrogenic as evidenced by the determination of simultaneous proton uptake and potassium ion release. The H^+ and K^+ exchange upon addition of pulses of HNO_3 to a cell suspension approached to a stoichiometry of 2 protons taken up and 1 potassium ion released per nitrate, but only in low chloride medium. Potassium ion release was suppressed in a medium supplemented with 40 mM Cl^-, whereas nitrate uptake suffered an inhibition below 10%. These findings display that K^+ release was not tightly coupled to the carrier, but was related to the charge balance and could be replaced by the Cl^- inflow or/and H^+ outflow. The H^+ extrusion at nitrate exhaustion evidenced that the proton pump, driven by changes of the cytoplasmic pH and/or membrane potential (nitrate uptake through the carrier should depolarize the membrane and acidify the cytoplasm), restored the membrane potential and the charge balance for the overall assimilation of the added nitrate.

The low affinity nitrate transport system has not been completely characterized. It was operative both in cells grown on nitrate and in nitrogen limiting conditions. It was saturable and not affected by changes of the medium pH (Fuggi *et al.* 1984). A NO_3^-/anion antiport mechanism remained questionable. However, this system has been suggested to work as a leak system in regulating the net nitrate uptake in *Cyanidium* (Fuggi 1989a).

4.2 Nitrite uptake

In *C. caldarium* two transport systems for nitrite uptake have been evidenced: one system has been determined by the passive diffusion of HNO_2 across the cell membranes. In very acidic media the uptake rate fitted in with the assumption that HNO_2 was the only permeating species (Fuggi, 1993). The pH gradient across the plasmalemma (the cell sap is neutral) allowed to trap NO_2^- within the cell to a high level, even in ammonium grown cells where assimilation of nitrite does not occur. In conditions in which the uptake through the nitrous acid diffusion was negligible (pH > 5 at low nitrite concentration) the assimilation rate of nitrite was comparable to that of nitrate while nitrate was a strong competitive inhibitor (Fuggi 1989b, 1990). The data suggested that the high affinity nitrate uptake system can take up nitrate and nitrite, nitrate having an affinity higher than nitrite.

Nitrite assimilation in *C. caldarium* did not suffer a short term inhibition by ammonium as occurred for nitrate assimilation.

4.3 Ammonium uptake

Fuggi *et al.* (1981a, 1984) have described an ammonium transport system (ATS) with a high affinity for ammonium in *C. caldarium*. This ATS was active in acidic media and in aerobic conditions. Its activity was increased when ammonium grown cells were transferred to a nitrate- or nitrogen-limited medium and protein synthesis was required for its expression (unpublished). An electrophoretic mechanism for this high affinity ATS could be suggested by analyzing the transitory K^+ and H^+ exchange dependent on ammonium uptake. Ammonium transport across the plasmalemma, in fact, transfers a positive charge and acidifies the cytoplasm (Bertl *et al.* 1984). Potassium and H^+ release from the cells took place just upon ammonium addition. The K^+ concentration in the medium rapidly reached a maximum that maintained until ammonium was depleted. The H^+ extrusion, on the other hand, exhibited a biphasic pattern: the former was simultaneous to potassium release, the latter was delayed from ammonium addition and was dependent on further ammonium assimilation (Fuggi 1990).

The ratio between released K^+ and the corresponding H^+ extruded in the first stage for a certain suspension and a given addition of NH_4^+ was dependent on several factors: in cells resuspended in a low potassium medium and at low chloride, K^+ was the main ion released and H^+ had only a minor contribution. Addition of chloride changed this figure, increasing the release of H^+ and decreasing that of K^+. Methylammonium, an ammonium analogue taken up at a rate lower than ammonium, favored H^+ efflux versus K^+ when added at comparable concentrations. In microaerobiosis also H^+ was the main ion released, the contribution of K^+ being strongly reduced. These results support that membrane potential is the driving force for ammonium uptake: K^+ efflux could be partially replaced by Cl^- influx for charge balance. The H^+ extruded could be related to activation of the proton pump at the plasmalemma driven by the decrease of cytoplasmic pH dependent on NH_4^+ — NH_3 and/or H^+/Cl^- symport or other mechanisms.

Ammonium uptake through ammonia diffusion in *Cyanidium* has been evidenced in basic media (Fuggi *et al.* 1984). In acidic media the trapping due to the low pH suggests that ammonia diffusion can determine a partial leak of the ammonium taken up or produced by nitrate reduction.

5. Nitrogen assimilation and carbon — nitrogen interaction

Nitrogen assimilation requires carbon skeleton and energy and, thus, strongly interacts with carbon metabolism. In photosynthetic organisms it was affected by light and dark conditions and by the CO_2 availability. Carbon dioxide being not only necessary for photosynthesis, but also for dark CO_2 fixation related to the anaplerotic replenishment of Krebs cycle intermediates used for ammonium assimilation and amino acid synthesis .

Assimilation of ammonium of *C. caldarium* cells grown in batch under excess ammonium (N-sufficient cells), was almost immediately inhibited by removal of light and CO_2, and was immediately restored upon light and CO_2 resupply. These facts support the view that ammonium assimilation requires active photosynthesis (Di Martino Rigano *et al.* 1986, 1987a, 1987b, 1990). This is in agreement with findings in other algae (Tacker and Syrett 1972; Elrifi and Turpin, 1986) that the Calvin cycle, can provide a source of triose phosphate for replenishment of C_4 and C_5 carbon compounds suitable for ammonium assimilation. The strict dependence on light and CO_2 indicates that in N-sufficient cells the Calvin cycle represents the sole source of carbon for ammonium assimilation.

5.1 Ammonium assimilation by N-limited cells

N-limited chemostat grown cells of *C. caldarium* assimilated ammonium either in the light or in darkness at higher rates than N-sufficient illuminated cells. On the other hand, addition of ammonium to N-limited cells produces a sudden inhibition of photosynthesis and stimulation of respiration (Rigano *et al.* 1990; Di Martino Rigano *et al.* 1991). However, the rate of assimilation in the light was higher than in the dark. It is noteworthy that (i) the difference between the rate of ammonium assimilation measured in the light and in the dark, was similar to the rate at which ammonium was assimilated in the light by N-sufficient cells, and that (ii) light-dependent part of ammonium assimilation in N-limited cells, like the sole light-dependent activity of N-sufficient cells, was suddenly inhibited upon removal of light, and was immediately restored upon light resupply. This suggests that: (i) in N-limited cells carbon skeletons for ammonium assimilation must be either from the Calvin cycle (light assimilation) or from starch breakdown (dark assimilation); (ii) in *C. caldarium* both ways can operate simultaneously; (iii) each of these metabolic pathways alone is inadequate to supply all the carbon needed for a full ammonium assimilation. Analysis of amino acid pools shows that in N-limited cells ammonium is firstly assimilated into glutamine. Then an increase of the other amino acids is observed as well as of other nitrogen compounds including 5-aminolevulinic acid, a precursor to chlorophyll and phycocyanin (Di Martino Rigano *et al.* 1989).

5.2 Effect of removal of CO_2 on NH_4^+ assimilation in N-limited cells

In suspension of N-limited chemostat cells of *C. caldarium*, in contrast to N-sufficient cells, ammonium assimilation was not immediately inhibited by CO_2 deprivation. The inhibition behaved like a time-dependent phenomenon both in light and darkness (Rigano *et al.* 1990). During an initial period of about 30 min, however, CO_2 deprivation was without effect on ammonium assimilation in darkness, but it produced a 75% inhibition in light. This suggests a light-dependent inhibition promoted by the absence of CO_2. Thus, under CO_2-free air, not only light-dependent ammonium assimilation (which, as referred above, is strictly dependent also on CO_2) was prevented, but there occurred also a light-dependent decrease of dark ammonium assimilation. Resupply of CO_2 stimulated ammonium assimilation in the light but was without such effect in darkness. Thus, as a working hypothesis, we are investigating whether the light-dependent decline of ammonium assimilation in the light in the absence of CO_2 could be promoted by photogenerated ATP and reductans which, being not used in the Calvin cycle, attain intracellular concentrations suitable to trigger control mechanisms operating at level of starch breakdown. Another level of control would be at the glycolytic pathway which ultimately constitutes the primary source of substrates for dark respiration and, consequently, for the generation of carbon substrates for ammonium assimilation. This would be consistent with the finding that dark respiration of leaves and algae is partially inhibited in the light as a consequence of a light dependent increase in the cytoplasm ratio of ATP/ADP (Heber 1974). If so, it could provide a control mechanism by which the cell would be able to economize carbohydrate reserves not replenished in the absence of CO_2. However, the stimulation of respiration after the supply of ammonium suggests that some endogenous CO_2 is available. Thus, alternatively, it might be suggested that under conditions of limited CO_2 supply, in the light, the photosynthetic carboxylation reaction will compete with the dark carboxylation reaction for the available CO_2. Hence it is likely that the production of C_4 and C_5 precursors for amino acid biosynthesis will be slower in the light than in the dark and, as a consequence, ammonium assimilation will be slower, too.

In *C. caldarium* N-limitation brings about a 10 fold increase of glutamine synthetase activity (Rigano *et al.* 1981). This accounts for the increased capacity of the N-limited cells to assimilate ammonium. However, the rate of ammonium assimilation which strictly depends on light and CO_2, as calculated by the difference between ammonium assimilation by N-limited cells in light and darkness, did not increase upon N-limitation. Thus the increase of the capacity to assimilate ammonium upon N-limitation, besides derepression of glutamine synthetase activity, is, apparently, based also on derepression of regulatory enzyme(s) of dark respiratory metabolism. This enables the N-limited cells to face the major demand of carbon skeletons for NH_4^+ assimilation from mobilization of polysaccharides reserve.

Thus, ultimately the variation in the nitrogen status and, in turn, the nitrogen demand of the cell, through a modulation of glutamine synthetase activity and, perhaps, of other enzymes of nitrogen metabolism, could provide a control signal for coordinated regulation of carbon and nitrogen metabolism.

Acknowledgements

This work has been supported by M.V.R.S.T. and Consiglio Nazionale delle Ricerche of Italy.

References

Allen, M.B. (1959) *Arch. Mikrobiol.* 32: 270–277.

Bertl, A., Felle, H. and Bentrup, F.W. (1984). *Plant Physiol.* 76: 75–78.

Di Martino Rigano, V., Vona, V., Fuggi, A. and Rigano, C. (1982). *Physiol. Plant.* 54: 47–51.

Di Martino Rigano, V., Vona, V., Fuggi, A., Di Martino, C. and Rigano, C. (1983). *Plant Sci. Letters* 28: 265–272.

Di Martino Rigano, V., Vona, V., Fuggi, A., Di Martino, C., Martello, A. and Rigano, C. (1984). *Biochim. Biophys. Res. Commun.* 119: 259–264.

Di Martino Rigano, V., Vona, V., Di Martino, C. and Rigano, C. (1986). Physiol. Plant. 68: 34–38.

Di Martino Rigano, V., Vona V., Di Martino, C. and Rigano, C. (1987a): *New Phytol.* 105: 247–254.

Di Martino Rigano, V., Vona, V., Manzo, L. and Rigano, C. (1987b). *New Phytol.* 107: 507–512.

Di Martino Rigano, V., Di Martino, C., Vona, V., Esposito, S. and Rigano, C. (1989). *Phytochemistry* 28: 2891–2895.

Di Martino Rigano, V., Vona, V., Esposito, S., Di Martino, C. and Rigano, C. (1990). *J. Phycol.* 26: 69–72.

Di Martino Rigano, V., Vona V., Esposito, S., Di Martino, C. and Rigano, C. (1991). *J. Phycol.* 27: 220–223.

Doemel, W.N. and Brock, T.D. (1971). *J. Gen. Microbiol.* 67: 17–32.

Elrifi, I.R. and Turpin, D.H. (1986). *Plant Physiol.* 81: 273–279.

Heber, U. (1974). *Ann. Rev. Plant Physiol.* 25: 393–421.

Fuggi, A., Di Martino Rigano, V., Vona, V. and Rigano, C. (1981a). *Plant Sci. Lett.* 23: 129–138.

Fuggi, A., Di Martino Rigano, V., Vona, V. and Rigano, C. (1981b). *Arch. Microbiol.* 130: 349–352.

Fuggi, A., Di Martino, C. and Rigano, C. (1984). *Giorn. Bot. Ital.* 118: (Suppl. 1): 60–61.

Fuggi, A., Vona, V., Di Martino Rigano, V., Di Martino, C., Martello, A. and Rigano C. (1984). *Arch. Microbiol.* 137: 281–285.

Fuggi, A. (1985). *Biochim. Biophys. Acta* 815: 392–398.

Fuggi, A. (1989a). *Biochim. Biophys. Acta* 974: 141–148.

Fuggi, A. (1989b). *Plant Physiol. Biochem.* 27: 563–568.

Fuggi, A. (1990). In: *Inorganic Nitrogen in Plants and Microrganisms (eds. Ullrich et al.), Springer Verlag, Berlin-Heidelberg*, pp. 66–72.

Fuggi, A. (1993) Uptake and assimilation of nitrite in the acidophilic red alga *Cyanidium caldarium* Geitler. *New Phytol.* 125: 361–360.

Merola, A., Castaldo, R., De Luca, P., Gambardella, R., Musacchio, A. and Taddei R. (1981): *Giorn. Bot. Ital.* 115: 189–195.

Rigano, C. (1967). *Delpinoa* 8–9: 75–83.

Rigano, C. (1971). *Arch. Mikrobiol.* 76: 265–276.

Rigano, C. and Violante, U. (1972a): *Biochim. Biophys. Acta*, 256: 524–532.

Rigano, C. and Violante, U. (1972b). *Biochem. Biophys. Res. Commun.* 47: 372–379.

Rigano, C. and Violante, U. (1973). Arch. Mikrobiol. 90: 27–33.

Rigano, C., Aliotta, G. and Violante, U. (1974a). *Plant Sci.Letters* 2: 277–281.

Rigano, C., Aliotta, G. and Violante, U. (1974b). Arch. Microbiol. 99: 81–90.

Rigano, C. and Aliotta, G. (1975). *Biochim. Biophys. Acta.* 384: 37–45.

Rigano, C., Di Martino Rigano, V., Vona, V., Aliotta, G. and Fuggi, A. (1977). *Rend. Accad. Nat. Lincei.* 53: 141–148.

Rigano, C., Di Martino Rigano, V., Vona, V., Fuggi, A. and Aliotta, G. (1978). *Plant Sci. Letters* 13: 301–307.

Rigano, C., Vona, V., Di Martino Rigano, V. and Fuggi, A. (1979a). *Plant Sci. Letters* 15: 203–209.

Rigano, C., Di Martino Rigano, V., Vona, V. and Fuggi, A. (1979b). *Arch. Microbiol.* 121: 117–129.

Rigano, C., Vona, V., Di Martino Rigano, V. and Fuggi, A. (1980). *Biochim. Biophys. Acta* 613: 26–33.

Rigano, C., Di Martino Rigano V., Vona V. and Fuggi, A. (1981). *Arch Microbiol* 129: 110–114.

Rigano, C., Di Martino Rigano ,V., Vona, V., Esposito, S. and Di Martino Catello, C. (1990). In: *Inorganic Nitrogen in Plants and Microrganisms. Ullrich et al. Eds., Springer Verlag, Berlin-Heidelberg.* pp. 131–136.

Schloemer, R.H. and Garrett, R.H. (1974). *J. Bacteriol.* 118: 259–269.

Syrett P.J. (1981). *Can. Bull. Fish. Aquat. Sci.* 210: 346–205.

Thacker, A. and Syrett, P.J. (1972). *New Phytol.* 71: 423–433.

Ullrich W.R. (1983). In: *Inorganic Plant Nutrition. Encyclopedia of Plant Physiology, NS vol 15A. (eds. Lauchli, A, and Bieleski, R.L.) Springer Verlag, Berlin-New York* pp. 376–397.

Ullrich W.R. (1987). In: *Inorganic Nitrogen Metabolism. (eds. Ullrich, W.R., Aparicio, P.J., Syrett, P.J., and Castillo, F.) Springer Verlag, Berlin* pp. 32–38.

Chapter 16

Natural products of the Cyanidiophyceae

HIDEYUKI NAGASHIMA

Department of Biology, Faculty of Industrial Science and Technology, Science University of Tokyo, 2641 Yamazaki, Noda-shi, Chiba 278, Japan

Outlines:
1. Introduction
2. Low molecular weight carbohydrates
   2.1 Algal carbohydrates
   2.2 Floridosides
   2.3 Physiological role of floridosides
   2.4 Sugars
3. Lipids
   3.1 Algal fatty acids
   3.2 Fatty acids in the Cyanidiophyceae and their relation to growth temperatures
   3.3 Hydroxy acids
   3.4 Hydrocarbons
   3.5 Sterols
4. Polyamines
   4.1 Algal polyamines
   4.2 Polyamines in the Cyanidiophyceae
5. Summary and conclusion
References

1. Introduction

Algae contain some characteristic substances that are usually not present in higher plants. The study of natural storage products such as carbohydrates, lipids and other organic substances may be useful also for algal phylogeny. Although studies of algal chemotaxonomy have been carried out over the years, one must still keep in mind that natural compounds change quantitatively under various physiological states and also according to the algal strains used.

Some strains of a hot spring alga '*Cyanidium caldarium*' isolated previously have been found to be actually a mixed culture of 'true' *C. caldarium* (RK-1 type, or forma A) and other algae, i.e., *Galdieria sulphuraria* (*C. caldarium* M-8 type, or forma B), see De Luca and Taddei 1976, Nagashima and Fukuda 1981b. *Galdieria* cells were formerly called various names, such as, *Protococcus sulphurarius* (De Luca *et al.* 1979), *Pleurococcus sulphurarius* (Moretti and Nazzaro 1980) or *Chroococcidiopsis* sp. (Nagashima and Fukuda 1981b). During the previous studies, a new eukaryotic, blue-green colored alga, *Cyanidioschyzon merolae* was found in the same habitat of acid hot springs (De Luca *et al.* 1978). In this chapter a number of strains and their origins are being discussed in relation to the algal natural products.

2. Low molecular weight carbohydrates

The low molecular weight carbohydrate here includes monosaccharides, disaccharides, polyols and their derivatives. Floridosides, glycerol galactosides, are mainly discussed in this section. Polysaccharides such as amylopectin-type polyglucan found in Cyanidiophyceae will be separated treated.

2.1 Algal carbohydrates

Most neutral low molecular carbohydrates are photo-assimilatory reserve products of algae (Lindberg 1956; Meeuse 1962; Craigie 1974; Nagashima 1976). Higher plants and green algae possess sucrose, glucose and fructose as the main products, while blue-green (Cyanophyta, cyanobacteria), brown, and red algae contain different sugars (Table 1). The cyanobacteria are divided into those containing glucose, fructose, or sucrose, and those which contain trehalose (Nagashima and Fukuda 1981a). Brown algae (Phaeophyceae) contain mainly mannitol, but no reducing sugars or sucrose. Red algae (Rhodophyta) contain mainly floridoside (2–O-glycerol-α-D-galactopyranoside) as demonstrated in Figure 1. Iso-floridoside (1–O-glycero-α-D-galctopyranoside), the isomer

J. Seckbach (ed.), Evolutionary Pathways and Enigmatic Algae: Cyanidium caldarium *(Rhodophyta) and Related Cells,* 201–214, 1994.

Table 1. Major low molecular weight carbohydrates in algae

Algae	Glu*	Fru	Man	Flo	IsF	Suc	Tre	Mal	MaG
Cyanidium RK-1				+	+				
Galdieria M-8				+					
Porphyridium	+			+				+	
Cyanobacteria	+	+				+	+		
Rhodophyta				+	+		+		+
Chromophyta									
Phaeophyceae			+						
Chrysophyceae**	+				+	+			
Bacillariophyceae	+								
Euglenophyta							+	+	
Chlorophyta									
Chlorophyceae	+	+				+			
Prasinophyceae			+						

Data from Lindberg (1956), Meeuse (1962), Craigie (1974), Nagashima (1976), Nagashima and Fukuda (1981a). *Cyanidium* = *C. caldarium; Galdieria* = *G. sulphuraria; Porphyridium* = *P. cruentum (Rhodophyta).*

* Glu = glucose; Fru = fructose; Man = mannitol; Flo = floridoside; IsF = iso-floridoside; Suc = sucrose; Tre = trehalose; Mal = maltose; MaG = mannoglyceric acid.

** *Ochromonas malhamensis* only, see Kauss (1967).

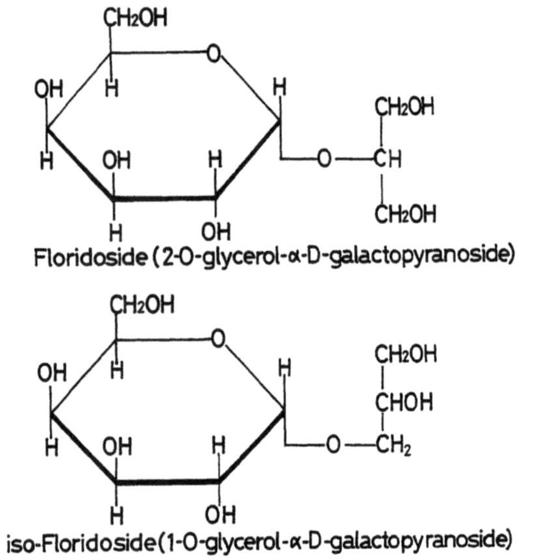

Fig. 1. Structure of floridosides. These compounds are usually found in Rhodophyta.

of floridoside, is abundantly present in the order Bangiales (Rhodophyta). Mannoglyceric acids (2–O-α-D-mannopyranosyl-D-glyceric acid) are found in the order Ceramiales (Rhodophyta) which contain a small amount of floridoside (Impellizzeri *et al.* 1975; Kirst 1980). A few exceptions to these rhodophytans are that a chrysophycean alga *Ochromonas malhamensis* shows iso-floridoside (Kauss 1967).

2.2 Floridosides

Cyanidium caldarium strain RK-1 isolated from Yumoto-spa, Japan contains floridoside and iso-floridoside (Nagashima and Fukuda 1981a), constituting 88.3% of total the low molecular weight carbohydrate fraction (Table 2). These compounds have been reported to be the main constituents of most red algae (Nagashima 1976). Sucrose, glucose, and fructose which were absent in red algae could not be detected in *C. caldarium*. Therefore, Nagashima and Fukuda (1981a) proposed that *C. caldarium* belongs to Rhodophyta.

On the other hand, Kremer and his coworkers pointed out that '*C. caldarium*', similar to *Cyanophora paradoxa*, contains fructose and glucose, but not floridosides (Kremer *et al.* 1978; Kremer and Feige 1979, and Kremer 1982). Therefore, they proposed that *C. caldarium* is an endocyanome (i.e., a symbiotic association between two different cells).

We examined further the carbohydrate of other '*Cyanidium*' strains, like *C. caldarium* strains RK-1, KS-1 (Kusatsu-spa, Japan) and 001 (Campi Flegrei, Italy) and observed that they contain mainly floridoside and iso-floridoside. On the other hand strains M-8 (Noboribetsu-spa, IAM culture collection, Japan) and 002 (Campi Flegrei, Italy) contain only floridoside (Nagashima and Fukuda 1983). The latter two strains (M-8 and 002)

Table 2. Low molecular weight carbohydrates under various culture conditions in the Cyanidiophyceae*

Algae		Condi-tion	Amount (% total carbohydrate)		
			Flo	iso-F	Other
Cyanidium caldarium	RK-1	L**	81.2	7.1	3.0
	RK-1	L	81.3	8.3	10.4
	KS-1	L	82.0	14.8	3.0
	001	L	80.8	8.8	10.9
Galdieria sulphuraria (*Cyanidium caldarium* M-8)		L	95.8	0.0	4.2
		DG	88.9	0.0	11.1
Galdieria sulphuraria (*Cyanidium caldarium* 002)		L	95.0	0.0	5.0
		DG	88.6	0.0	11.4
Cyanidioschyzon merolae		L	0.02	0.01***	–

* Shaking cultures at 38°C. L = inorganic medium, pH 3.0, at 2000 lux; DG = the medium plus 0.5% glucose in the dark; Flo = floridoside; iso-F = iso-Floridoside; Other = other unidentified non-reducing carbohydrates, see Nagashima and Fukuda (1983).

** Aeration culture at 35°C in inorganic medium at 3000 lux, see Nagashima and Fukuda (1981a).

*** Percent of dry weight of the alga, see De Luca and Moretti (1983).

have been found to be a different genus from *Cyanidium*, and are assigned to *Galdieria sulphuraria* (Merola *et al.* 1981). Floridosides were also found in *C. caldarium* and *G. sulphuraria* isolated from Campi Fregrei, Naples by De Luca and Moretti (1983). The presence of floridoside in *C. caldarium* strain 1355/1 (= *G. sulphuraria*) is also confirmed by Reed (1983). Unicellular red algae, *Porphyridium cruentum* (= *P. perpureum*) and *P. aerugineum* are primitive, because of their asexual production, they contain, similar to *Cyanidium*, floridosides and a small amount of iso-floridosides (Nagashima and Fukuda 1981a; Moretti and Musacchio 1982). From these results we conclude that strains (RK-1, KS-1, 001) and strains (M-8, 002) are surely different species, and may belong to the primitive red algae.

Cyanidioschyzon merolae was later found in acid hot-springs (De Luca *et al.* 1978). This alga is similar to the other '*Cyanidium*' members in its habitats, and in photosynthetic pigments (Chlorophyll a, phycocyanin), but it differs in cell shape and cell division. *C. merolae* is club-shaped and multiplies in binary fission. It contains a small amount of floridoside and iso-floridoside (Table 2), similar to *Cyanidium caldarium*, and may belong to primitive red algae (De Luca and Moretti 1983).

2.3 *Physiological role of floridosides*

Floridosides and floridean starch are the main assimilated products of photosynthesis in red algae (Majak *et al.* 1966) and they may play similar physiological roles to sucrose and starch in most higher plants (Nagashima *et al.* 1969). In addition, these compounds relate to osmotic regulation in most red algae (Kauss 1968; Kirst and Bisson 1979). Similar results were obtained with iso-floridoside in *Ochromonas malhamensis*, Chrysophyceae (Kauss 1967).

In *C. caldarium* floridoside may be one of the main photoassimilated reserve substances, also, because only under light conditions could the alga grow and store floridoside (Nagashima and Fukuda 1983). The floridoside content showed tenfold increase when the cells of *C. caldarium* strain 1355/1 (= *G. sulphuraria*) were exposed to a range of NaCl concentration from 0 to 320 mmol at 25 °C (Reed 1983). *C. caldarium* strain M-8 (= *G. sulphuraria*) can accumulate floridoside at about the same ratio in heterotrophic as well as in autotrophic conditions (Nagashima and Fukuda 1983). Therefore, floridoside may play an important role in osmotic balance, at least in *Galdieria*.

2.4 *Sugars*

No detectable amounts of sucrose and reducing sugars such as glucose, galactose and fructose could be detected in all *C. caldarium* and *G. sulphuraria* strains tested (Nagashima and Fukuda 1983). This fact indicates that these algae may be closely related to red algae, because most rhodophytes contain sugar alcohols such as laminitol (methyl-inositol), but not reducing sugars and sucrose (Craigie 1974; Nagashima 1976). In contrast to that observation, Reed (1983) reported small amounts of galactose, glucose, sucrose and trehalose in *C. caldarium* strain 1355/1 (= *Galdieria sulphuraria*).

3. Lipids

Many works on algal lipids have been carried out from the biochemical and chemotaxonomical points of view (Patterson 1971; Wood 1974; Weete 1976). In regards to lipid composition, fatty acid has been widely studied in algae, because it may be a useful marker for algal phylogeny (Holton *et al.* 1968; Kenyon and Stanier 1970) and physiology, especially in relation to growth temperature (Holton *et al.* 1964). In this section we will discuss mainly fatty acid composition and also hydrocarbons, hydroxy acids and sterols.

3.1 Algal fatty acids

Major fatty acid distribution in various algal classes was early summarized by Wood (1974) and is shown in Table 3. Cyanobacteria (Cyanophyta) have been most extensively investigated for fatty acid analysis. Green algae (Chlorophyta) and diatoms (Bacillariophyceae) have been also fairly investigated, but in other algal classes only a small number of species have been analyzed. Fatty acid distribution seems to be characteristic to algal classes, but the pattern is considerably different in each species belonging to the same algal class. C16 and C18 series of fatty acids are ubiquitously present in various algae. C18 unsaturated fatty acids such as oleic (C18:1) and linoleic (C18:2) acids are also widely distributed. However, linolenic acid (C18:3) is rather confined within Chlorophyceae, Euglenophyta, Cryptophyceae and cyanobacteria. α-Linolenic acid (C18:3 ω 9) is common in these algae, and γ-linolenic acid (C18:3 ω 6) is present within some species of cyanobacteria such as *Synechococcus* 6801, *Microcystis* 7005 and *Aphanocapsa* 6702 (Wood 1974). C18:4 is present only in Cryptophyta and Dinophyta. The presence of C20:4 and C20:5 are restricted in some species of Rhodophyta, Cryptophyta, Chromophyta and Haptophyta. These C18:4, C20:4 and C20:5 acids may be a part of the criteria for some algal taxa.

Table 3. Major fatty acids in algae

Algae (number of spp. surveyed)	Fatty acid*												
	14:0	16:0	16:1	17:0	18:0	18:1	18:2	18:3 (α)	18:3 (γ)	18:4	20:4	20:5	>20
Cyanidioschyzon	+	+	+	+	+		+						
Cyanidium		+		+	+	+	+						
Galdieria		+			+	+	+	+					
Cyanobacteria (39)		+	+			+	+	+	+				
Rhodophyta (5)		+				+	+				+	+	
Cryptophyta (10)	+	+						+		+		+	
Chromophyta													
Phaeophyceae (3)		+				+					+		
Chrysophyceae (5)	+	+	+				+					+	
Bacillariophyceae (12)	+	+	+									+	
Xantophyceae (2)		+	+									+	
Dinophyta (6)	+	+								+		+	+**
Euglenophyta (4)		+						+					+**
Haptophyta (6)		+	+			+	+					+	
Chlorophyta													
Chlorophyceae (13)		+				+	+	+					
Prasinophyceae (1)***		+	+									+	

Data from Wood (1974), Moretti and Nazzaro (1980) and others. *Cyanidioschyzon* = *C. merolae*; *Cyanidium* = *C. caldarium; Galdieria* = *G. sulphuraria.*

* Number of carbons: numer of unsaturated bonds; >20 = more than 20 carbons.

** 22:6 in Dinophyta; unsaturated fatty acids with 20 to 24 carbon chains in Euglenophyta.

*** *Heteromastix rotunda*

3.2 Fatty acids in the Cyanidiophyceae and their relation to growth temperatures

The strains of '*Cyanidium caldarium*' used for lipid analysis since 1970, caused some confusion. Boenzi *et al.* (1977) distinguished in such populations two algal species forma A and forma B used for the lipid analysis. Boenzi and co-workers stated that *C. caldarium* forma A contained no linolenic acid (C18:3), but *C. caldarium* forma B (= *Galdieria*) contained 10.6% linolenic acid in a 20 °C cultures (Table 5).

Kleinschmidt and McMahon (1970a, 1970b) found that the lipid composition of '*Cyanidium*' was very variable according to growth temperature (Table 4). In a 20 °C grown culture, 30% of the fatty acids was linolenic acid (C18:3), while in a 55 °C culture, this fatty acid could not be detected in the cells. Ratios of unsaturated to saturated fatty acids in these cells decreased 3-fold as temperature increased from 20 °C to 55 °C. These results show that growth temperature affects fatty acids, especially unsaturated fatty acid composition. However, judging from linolenic acid content which is absent in true *Cyanidium* (forma A) in a 20 °C culture (Boenzi *et al.* 1977); Kleinschmidt and McMahon (1970a, 1970b) used most probably the other alga, i.e., *Galdieria sulphuraria*. The Glycolipids and phospholipids contents of this alga (*Galdieria*) were also analyzed by them at different temperature ranges.

Table 4. Major fatty acids (%) at different temperatures in *G. sulphuraria*

Fatty acid	Amount (%)			
	20°C*	38°C**	42-45°C***	55°C*
16:0	18	33.5	43.1	53
18:0	7	1.7	3.2	5
18:1	34	33.3	29.5	21
18:2	11	23.7	20.7	21
18:3	30	7.3	3.5	0
TUFA****	75	64.3	53.7	42

All the '*C. caldarium*' strains correspond to *G. sulphuraria* (see Text).
* Kleinschmidt and McMahon (1970a).
** Nagashima *et al.* (1986).
*** The sample used was Lot II (Allen *et al.*, 1970). 18:3 = α-linolenic acid.
**** Total unsaturated fatty acids (%).

Allen *et al.* (1970) reported slightly different results from those by Kleinschmidt and McMahon (1970a) as shown in Table 4. According to their report (Allen *et al.* 1970), one of the three '*Cyanidium*' strains, isolated from Lot III of hot springs at Yellowstone National Park (U. S. A.)

Table 5. Fatty acids (%) at different temperatures in the Cyanidiophyceae

Fatty acid	*C. caldarium*			*G. sulphuraria*			*C. merolae*	
	20°C*	20°C	55°C	20°C*	20°C	55°C	20°C	55°C
10:0	5.3	1.9	1.8	6.6	t**	t	t	7.7
12:0	2.7	0.8	0.6	2.0	t	t	0.0	t
12:1	12.9			7.3				
12:2	9.5			9.0				
14:0	t	0.8	2.8	t	0.5	3.0	0.8	8.0
14:1	3.3			2.2				
16:0	10.1	20.6	26.1	15.6	23.7	27.8	24.8	9.0
16:1	t	0.8	3.5	t	0.2	3.4	0.2	6.9
16:2	2.8			t				
17:0	t	4.6	8.3	t	0.2	2.6	1.6	11.0
18:0	2.7	5.7	15.3	t	3.4	10.7	7.6	29.7
18:1	5.7	12.6	13.1	15.9	24.3	22.7	5.7	t
18:2	41.8	51.0	28.0	23.5	13.6	22.7	58.9	21.9
18:3***	–	–	–	10.6	34.0	7.3	–	–
TUFA****	78.8	70.0	49.8	76.2	72.6	57.8	66.6	36.1

see Moretti and Nazzaro (1980). '*Pleurococcus sulphurarius*' in the text corresponds to *Galdieria sulphuraria*. * Boenzi *et al.* (1977). *Cyanidium caldarium* forma A and Forma B in the text correspond to *C. caldarium* and *G. sulphuraria* respectively. ** trace; *** it may be α-18:3 (α-linolenic acid); **** Total unsaturated fatty acids (%).

lacks linolenic acid, while two other strains contain it in a 45 °C culture. This shows that the alga (Lot III) lacking linolenic acid has been most probably true *Cyanidium* [compare to Ikan and Seckbach, 1972] and the others (Lots I and II) have been *G. sulphuraria*, or mixed with *Cyanidium*, because of containing α-linolenic acid. Similar results as in Lot II were obtained by Adams *et al.* (1971). That is, linolenic acid was very low in the '*Cyanidium*' (= *Galdieria*) cells in a 45–50 °C culture, but abundant in the cells in a 20 °C culture. This shows that the algae they used [except for Lot III (Allen *et al.* 1970)] were also not *Cyanidium*, but rather *G. sulphuraria* or this latter alga mixed with *C. caldarium* (Table 4).

The comparison between true *C. caldarium* (forma A) and '*C. caldarium* forma B' (= *G. sulphuraria*) with regards to their fatty acid composition in a 20 °C culture was reported first by Boenzi *et al.* (1977), as stated above (Table 5). The fatty acid composition of two strains of '*Cyanidium caldarium*' (RK-1 and M-8) isolated from hot springs in Japan was also compared when they were cultured at 38 °C (Nagashima *et al.* 1986). RK-1 and M-8 strains correspond to *C. caldarium* and *G. sulphuraria* respectively. *Cyanidium* RK-1 has 17:0 and 18:0 fatty acids considerably, in contrast with other algae (Table 3). There is no alga showing similar fatty acid composition to *C. caldarium* RK-1 among other algal taxa. On the other hand, the fatty acid composition of M-8 strain is very close to those of the primitive red alga, *Porphyridium aeruginosa* (Wood 1974) as well as to the green alga, *Chlorella vulgaris* (Nichols 1965), but it is different from those of red algae, *Porphyra yezoensis* (Araki *et al.* 1986) and *Porphyridium purpureum* (= *P. cruentum*), see Nyberg and Koskimies-Soininen (1984). These data show that two algae (*Cyanidium* RK-1 and M-8 strains) are distinguishable clearly in their linolenic acid content and show that they may be apart from each other phylogenetically.

Morretti and Nazzaro (1980) reported fatty acid compositions among three thermoacidophilic algae in 20 °C and 55 °C cultures (Table 5). They found that *C. caldarium* and *Cyanidioschyzon merolae* contain mainly linoleic acid (C18:2), but no α-linolenic acid (C18:3), and that *Pleurococcus sulphurarius* (= *G. sulphuraria*) contains much linolenic acid in a 20 °C culture, but decreases in quantities in a 55 °C culture. They suggest that there is an affinity between *C. caldarium* and *C. merolae* and these two algae are more primitive than *P. sulphurarius* (= *G. sulphuraria*).

In summary, the growth temperature affects the fatty acid composition of these algae. As the temperature raised, total unsaturated fatty acids decreased in all these algae (Table 5). In *Cyanidium* and *Cyanidioschyzon* linoleic acid (C18:2) were particularly decreased, on the other hand in *Galdieria* linolenic acid (C18:3, probably α-C18:3) was most strikingly decreased with increasing temperature. The fact that unsaturated fatty acids decrease with increasing temperature may be recognized generally in most algae, but the specific unsaturated fatty acid decrease varies among algae. In the cyanobacterium, *Anacystis nidulans*, palmitic acid (C16:0) and a hexadecenoic acid (C16:1) comprised approximately 90% of the total fatty acids at all temperatures, but the ratio of the C16:1 to C16:0 decreased with increasing temperature (Holton *et al.* 1964).

3.3 Hydroxy acids

Hydroxy acids are divided into 2-hydroxy (α-hydroxy), 3-hydroxy (β-hydroxy), and other acids by the position of hydroxy residues in fatty acids. They are considered to be intermediates in the α- and β-oxidation pathways of fatty acid degradation (Eglinton *et al.* 1968). Hydroxy acids, which occur in bacteria, fungi, yeasts and higher plants (Burton and Carter 1964; Stodola *et al.* 1967), have been studied mainly as geochemical indicators (Cardoso and Eglinton 1983). Hydroxy acids were also found in algae (Matsumoto *et al.* 1984; Matsumoto and Nagashima 1984), as well as in cyanobacteria (Weckesser *et al.* 1974).

Cyanidium caldarium strain M-8 (= *G. sulphuraria*) and *C. caldarium* strain RK-1 contain 41.6% of 2-hydroxyhexacosanoic acid (nC26) and 40.3% of 2-hydroxynonadecanoic acid (nC19) respectively in the 2-hydroxy acid fraction (Matsumoto *et al.* 1984), as shown in Table 6. The composition of 3-hydroxy acid in the two algae are fairly similar to each other, see Table 6 (Matsumoto and Nagashima 1984). That is, *C. caldarium* M-8 (= *G. sulphuraria*) contains mainly nC12 (58.8%) and nC14 (37.7%), while *C. caldarium* RK-1 contains mainly nC12 (41.8%), nC13

(20.6%) and nC14 (20.4%). A small amount of branched (iso and anteiso) 3-hydroxy acids (C16 and C18) are present exclusively in the RK-1 strain. These results support the idea that, phylogenetically, these algae may be considerably distantly related (Nagashima *et al.* 1984). More chemotaxonomical data are needed for further phylogenetic postulation. Matsumoto and Nagashima (1984) and Matsumoto *et al.* (1984) also suggested that microalgae are important sources of hydroxy acids in the natural environment, because all 8 samples (*C. caldarium*, *G. sulphuraria*, *Chlamydomonas reinhardii*, *Chlorella pyrenoidosa*, *Anacystis nidulans*, *Phormidium foveolarum*, *Anabaena variabilis* and *Oscillatoria* sp.) tested contain 2- and 3-hydroxy acids.

Table 6. Hydroxy acids (%) in the Cyanidiophyceae

Carbon Chain	2-Hydroxy acid		3-Hydroxy acid	
	Cyanidium	*Galdieria*	*Cyanidium*	*Galdieria*
10	–	–	1.0	0.2
11	–	–	4.2	–
12	–	–	41.8	58.8
13	–	–	20.6	1.4
14	–	–	20.4	37.7
15	–	–	2.9	–
16	4.5	2.5	2.0	1.2
17	1.0	0.1	1.3	–
18	8.1	9.4	2.2	0.3
19	40.3	2.1	0.4	–
20	36.0	33.5	–	–
21	1.4	0.5	–	–
22	0.4	3.0	–	–
23	0.1	0.4	–	–
24	1.4	2.6	–	–
25	3.3	4.3	–	–
26	3.5	41.6	–	–
br14*	–	–	0.2	–
br16	–	–	1.4	–
br17	–	–	0.3	–
br18	–	–	1.0	–

* br = branched; *Cyanidium* = *C. caldarium; Galdieria* = *G. sulphuraria. Data from Matsumoto et al.* (1984) and Matsumoto and Nagashima (1984).

3.4 *Hydrocarbons*

Cyanobacteria are often suggested as source material for organic matter associated with ancient sediments, such as hydrocarbons (Winter *et al.* 1969). A golden-brown alga (Xanthophyceae), *Botryococcus braunii* containing long chain hydrocarbons (C25-C31) has been well studied from a geochemical significance, especially in relation to oil sources (Gelpi *et al.* 1968; Brown and Knight 1969; Murray and Thomson 1977). Weete (1976) has summarized the distribution of hydrocarbons in microalgae (Table 7) and showed that its pattern in algae is related to taxonomy, although the growing conditions and physiological state can affect the hydrocarbon composition (Patterson 1967; Brown and Knights 1969).

Most algae have n-C17 or C21:6 as the major hydrocarbon, but some brown algae (Phaeophyceae) have n-C15 as their principal alkane. Some cyanobacteria have n-C17, C19:1 or branched C18, and certain green algae have n-C17, unsaturated C17, C23 or unsaturated C27, and red algae have n-C17 as their major hydrocarbons. Green algae and cyanobacteria show the greatest diversity in the hydrocarbon distribution (Winter *et al.* 1969; Gelpi *et al.* 1970; Lee and Loeblich 1971; Cranwell *et al.* 1990).

Cyanidium RK-1 has 46.3% n-C17 alkane and 20.4% C19:2 alkene, while *Cyanidium* M-8 (= *G. sulphuraria*) has 43.4% C19:1 and 19.6% C21:1 alkenes at the culture temperature of 38 °C (Nagashima *et al.* 1986). This result shows that *Cyanidium* RK-1 does not fit into any of these algal classes, but *Cyanidium* M-8 resembles some cyanobacteria such as *Coccochloris elabens* and *Agmenellum quadruplicatum* which contain mainly C19:1 (Winter *et al.* 1969).

When culture temperatures were raised from 20 °C to 45 °C, two facts emerge 1) a decrease in the ratio for C19:2 and an increase in the ratio for C17 in *Cyanidium* RK-1, and 2) a decrease in the ratio for C19:1 and an increase in the ratio for C19 in *Cyanidium* M-8 (= *G. sulphuraria*) were observed, as shown in Fig. 2 (Nagashima *et al.* unpublished data). This fact shows that algal unsaturated hydrocarbons are related to the growth temperature as was also noticed in the algal composition of fatty acid which were related to growth temperature.

3.5 *Sterols*

The distribution of sterol in algae are summarized by Patterson (1971, 1991), Goodwin (1974) and

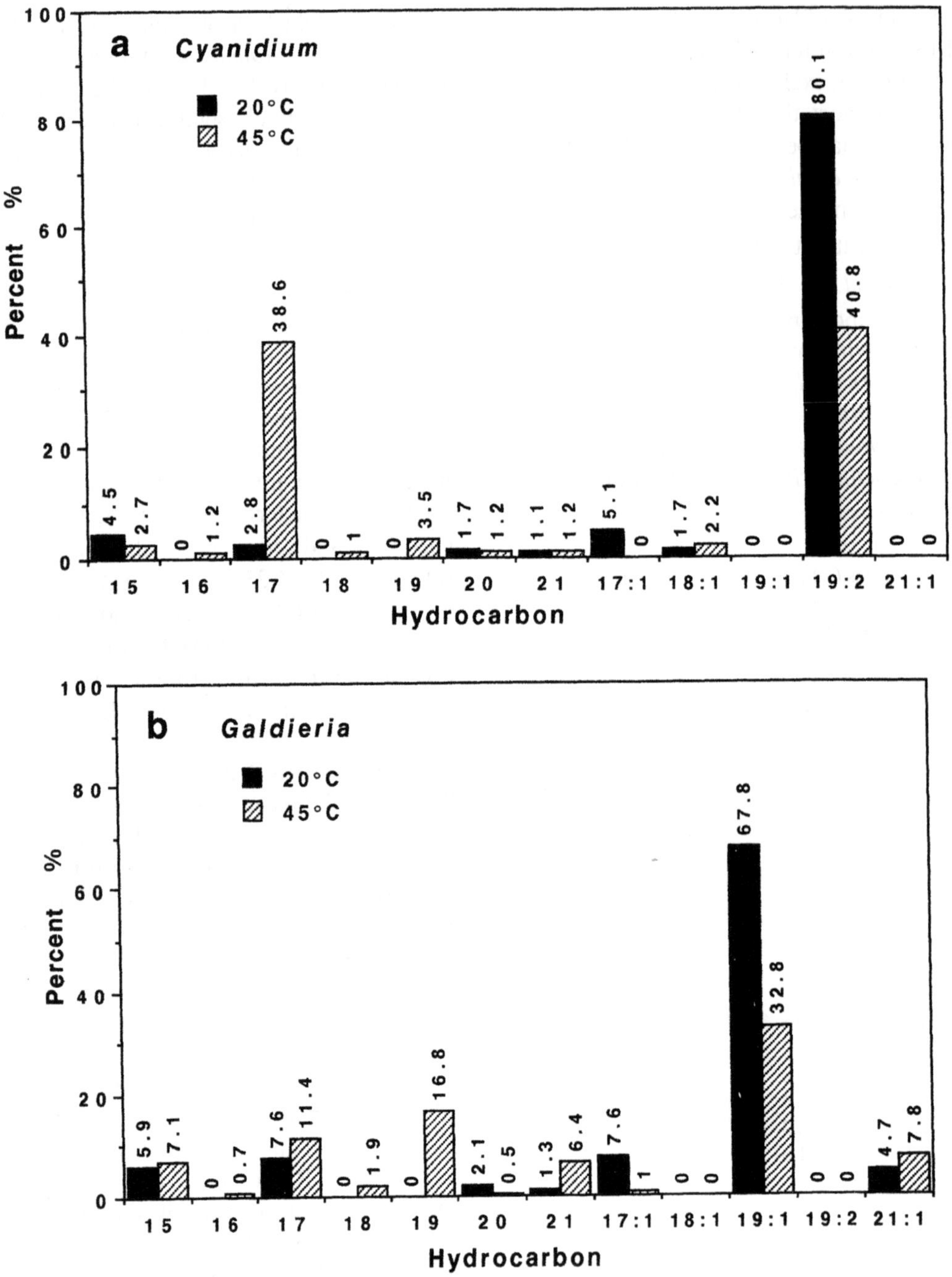

Fig. 2. Hydrocarbons in *Cyanidium*(a) and *Galdieria*(b) at 20 °C and 45 °C (Nagashima *et al.* unpublished data). Major hydrocarbons C19:2 and C19:1 decrease in *Cyanidium* and *Galdieria* respectively at elevated temperature.

Weete (1976). Early studies pointed out that the sterols of algal classes, families and sometimes even species were characteristic of those particular taxa, and that sterol composition could be useful in phylogenetic comparison (Patterson 1991). Cyanobacteria contain small amounts of sterols such as cholesterol and sitosterol (24-ethylcholesterol)(Kohlhase and Pohl 1988). Red algae contain primarily cholesterol (Chardon-Loriaux *et al.* 1976). Fucosterol (24(E)-ethylidenecholesterol) is

Table 7. Major hydrocarbaons in algae

Algae (No. of species surveyed)	Hydrocarbon*										
	15:0	17:0	17:1	19:1	19:2	21:1	21:6	Δ23	Δ25	Δ27	Oth**
Cyanidium RK-1		+			+						+
Galdieria M-8		+	+	+		+		+	+		
Cyanobacteria (18)		+	+	+					+	+	+
Rhodophyta (4)		+									
Cryptophyta (2)							+				
Chromophyta											
Phaeophyceae (8)	+						+				
Chrysophyceae (1)***		+									
Bacillariophyceae (6)							+				+
Xanthophyceae (1)***										+	+
Dinophyta (6)							+				
Euglenophyta (2)							+				
Haptophyta (3)							+				
Chlorophyta											
Chlorophyceae (14)		+	+	+			+	+	+	+	+
Prasinophyceae (1)***							+				

Data from Weete (1974), Lee and Loeblich (1971), Nagashima *et al.* (1986) and Cranwell *et al.* (1990). *Cyanidium* = *C. caldarium; Galdieria* = *G. sulphuraria.*

* Number of carbons : number of unsaturated bonds: Δ = unsaturated; Oth = other hydrocarbons.

** 18 in *Cyanidium* RK-1, branched 18 in cyanobacteria, 29 in Bacillariophyceae, Δ31 in Xanthophyceae, and 23 in Chlorophyceae.

*** *Syracosphaera carterae* (Chrysophyceae), *Botryococcus braunii* (Xanthophyceae), which is classified by some authors in the Chlorophyceae, *Platymonas* sp. (Pracinophyceae), respectively.

the dominant sterol of brown algae (Ikekawa *et al.* 1968). The sterols of green algae are much more varied and complex than those of other groups of algae. Sitosterol, campesterol (24-methylcholesterol), poriferasterol (24-ethylcholesta-5,22-dienol), ergosterol (24-methylcholesta-5,7,22-trienol) and other sterols are found in green algae (Patterson 1971, 1991). The sterol composition of some fresh water green algae was broadly consistent with the respective taxonomic position of each organism (Cranwell *et al.* 1990). Members of the Chromophyta appear to have sterols which are related to their taxonomic position (Patterson and Valkenburg 1990).

Cyanidium caldarium contained ergosterol + 5,6-dihydroergo-sterol (55.0%), β-sitosterol (38.9%) and campesterol (5.6%), and a small amount of cholesterol (0.3%) (Ikan and Seckbach 1972). Seckbach and Ikan (1972) suggested that the presence of cholesterol and campesterol in the alga may be similar to that in red algae. However, its sterol composition is somewhat inexplicable on these grounds (Goodwin, 1974). Recently Seckbach *et al.* (1993) examined active grown *C. caldarium* and *G. sulphuraria* harvested in their log phase. They have observed that ergosterol, Δ^7 ergosterol (24-methyl-cholesta-7-enol), $\Delta^{7,22}$ ergosterol (24-methylcholesta-7,22-dienol), chondrillasterol (24-ethylcholesta-7,22-dienol) and Δ^7 chondrillasterol are present in these two algae. In addition, a new unusual sterol (24-methylcholesta-5-7-22–24 (28)tetraen-3β-ol) was established. However, they could not find any 5,6-dihydroergosterol (24-methylcholesta-5,6-dienol), cholesterol, β-sitosterol and campesterol, those sterols found in the previous analysis of the older cultures of *C. caldarium* (Ikan and Seckbach 1972). Further studies will be needed for phylogenetic implication of these facts.

4. Polyamines

Polyamines are polycationic organic compounds that are ubiquitously present in living organisms and have been implicated in a wide variety of biological reactions, including synthesis of proteins and nucleic acids (Tabor and Tabor 1985).

Table 8. Major polyamines in algae

Algae	Polyamine*					
	C4	C3C4	C3C4C3	C3	C3C3	C3C3C3
Cyanidium RK-1	+++**	+++	++	–	–	–
Galdieria M-8	+++	+++	+++	+	++	+++
Porphyridium	+++	++	+	t	–	–
Glaucocystis	+++	+++	–	–	–	–
Cyanophora	+++	++	–	+	–	–
Cyanobacteria						
N_2-fixing	+	t	t	–	–	–
N_2-nonfixing	+	+++	t	–	–	–
Rhodophyta						
Bangioideae	+	+	++	t	–	–
Florideae	+++	t	+/–	–	+/–	+/–
Pyrrophyta***	+++	++	–	–	+	++
Chrysophyta	+++	++	+/–	–	+/–	–
Phaeophyta	+++	+	t	t/–	+/–	t/–
Euglenophyta	++	+++	+/–	+/–	++	+++
Chlorophyta	++	++	t/–	+/–	++	+/–

Data from Hamana and Matsuzaki (1985) and Hamana *et al.* (1990).

Cyanidium = *C. caldarium; Galdieria* = *G. sulphuraria; Porphyridium* = *P. cruentum; Glaucocystis* = *G. nostochinearum; Cyanophora* = *C. paradoxa.*

* C4 = putrescine; C3C4 = spermidine; C3C4C3 = spermine; C3 = diaminopropane; C3C3 = norspermidine; C3C3C3 = norspermine.

** +++ = abundant; ++ = moderate; + = present; t = trace; +/– = present or absent.

*** *Peridinium* sp. only.

4.1 Algal polyamines

Fig. 3 shows the chemical structure of the major polyamines in regards to this discussion. Distribution of polyamines have been studied in many algae and bacteria in order to elucidate the chemotaxonomy among these organisms (Hamana and Matsuzaki 1982; Hamana and Matsuzaki 1985). As shown in Table 8, putrescine (C4) is the most ubiquitous polyamine, and spermidine (C3C4) is also widely distributed in algae. Other polyamines are distributed in various patterns among algal phyla.

Chemical Structure of Polyamines

Putrescine(C_4)	$NH_2(CH_2)_4NH_2$
Spermidine(C_3C_4)	$NH_2(CH_2)_3NH(CH_2)_4NH_2$
Spermine($C_3C_4C_3$)	$NH_2(CH_2)_3NH(CH_2)_4NH(CH_2)_3NH_2$
Diaminopropane(C_3)	$NH_2(CH_2)_3NH_2$
Norspermidine(C_3C_3)	$NH_2(CH_2)_3NH(CH_2)_3NH_2$
Norspermine($C_3C_3C_3$)	$NH_2(CH_2)_3NH(CH_2)_3NH(CH_2)_3NH_2$

Fig. 3. Structure of polyamines found in Cyanidiophyceae.

4.2 Polyamines in Cyanidiophyceae

Polyamines were first analyzed in *Cyanidium* M-8 *(=Galdieria sulphuraria)* by Hamana and Matsuzaki (1985). Spermidine (C3C4) is the most prominent, constituting a quite high concentration (3.0 μmol/g wet weight); putrescine (C4), spermine (C3C4C3) and Norspermine (C3C3C3) are also abundant. Norspermidine (C3C3) and diaminopropane (C3) are present in small quantities. This pattern of polyamine content could not be found in any other algae tested. Hamana *et al.* (1990) later compared the distribution of polyamines between two species of '*Cyanidium caldarium*' (Fig. 4). *C. caldarium* RK-1 contains 1.02 μmole of putrescine, 1.08 μmole of spermidine and 0.27 μmole of spermine in a 38 °C culture, but not norspermine, norspermidine and diaminopropane. Similar results were obtained from the cells of *C. caldarium* KS-1 isolated from Kusatsu-spa, Gunma, Japan.

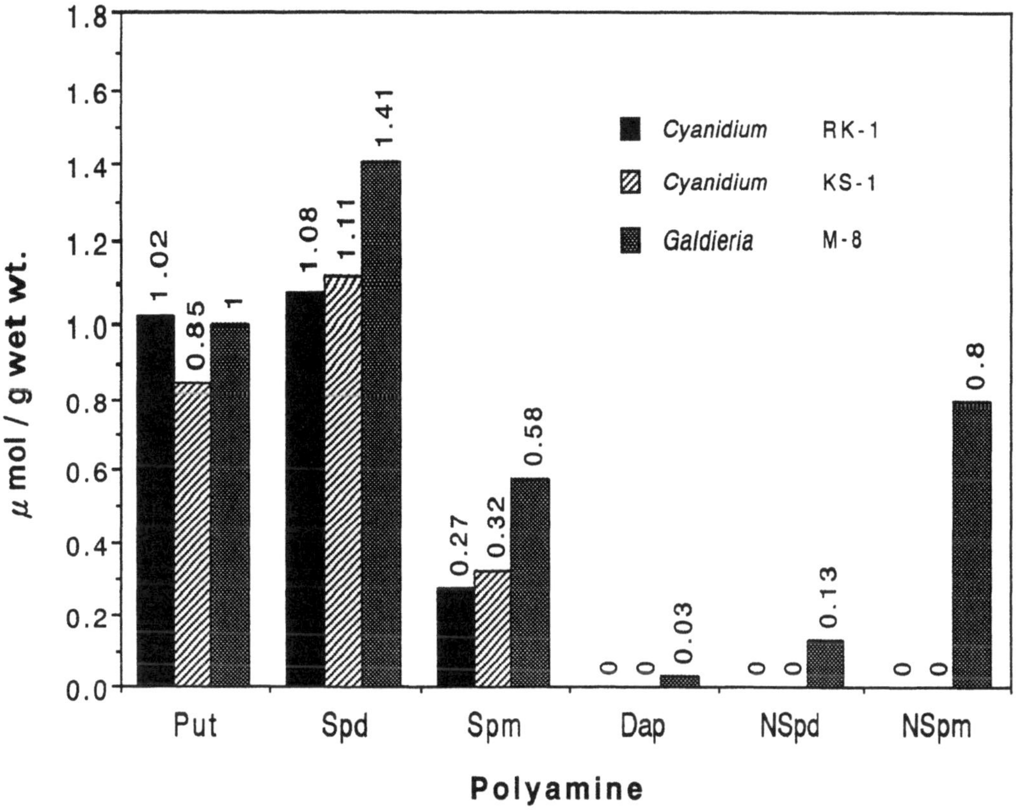

Fig. 4. Polyamine content in *Cyanidium* RK-1 and KS-1 strains, and *Galdieria* M-8 strain (Data from Hamana *et al.* 1990). Put= putrescine; Spd= spermidine; Spm= spermine; Dap= diaminopropane; NSpd= norspermidine; NSpm= norspermine. Dap, NSpd and NSpm are found only in *Galdieria* M-8. Algal cultures were carried out at 38 °C.

Clearly, the RK-1 and KS-1 strains were different from *C. caldarium* M-8 (= *G. sulphuraria*) in their pattern of polyamine content as already stated above.

The unicellular red alga, *Porphyridium cruentum* (= *P. purpureum*) contains putrescine, spermidine and a small amount of spermine (Hamana and Matsuzaki 1985), thus it shows a very similar pattern to those of *C. caldarium* RK-1 and KS-1 in polyamine distribution. As reported elsewhere, *P. purpureum* shares close features with *C. caldarium* in floridoside content and primitive cell structure (Nagashima and Fukuda 1981a; Moretti and Musacchio 1982). These results stated above suggest that *C. caldarium* may be closely related to *P. purpureum*. In addition, *Glaucocystis nostochinearum* and *Cyanophora paradoxa* (Glaucophyta, endosymbionts containing 'cyanelle') contain mainly putrescine and spermidine, as reported for *C. caldarium*. It shows the similarity between *C. caldarium* and *Glaucophyta*. On the other hand, *Galdieria sulphuraria* shows the different pattern of polyamine distribution from any other algal taxa, which suggests that the alga is situated in a specified position in algal phylogeny.

When the temperature was raised from 20 to 50 °C in the *Cyanidium* RK-1 culture, putrescine, spermidine and spermine increased, and the total polyamine content tripled (Fig. 5a). In *Cyanidium* M-8 (= *Galdieria sulphuraria*), the total polyamine content became more than one and a half times as much as that at 20 °C; spermine and norspermine (C3C3C) increased in particular, but putrescine and spermidine decreased (Hamana *et al.* 1990) (Fig. 5b). These results show that polyamines, especially spermine in *Cyanidium* RK-1 or spermine and norspermine in M-8 strain may be related to the thermophilic properties of these algae.

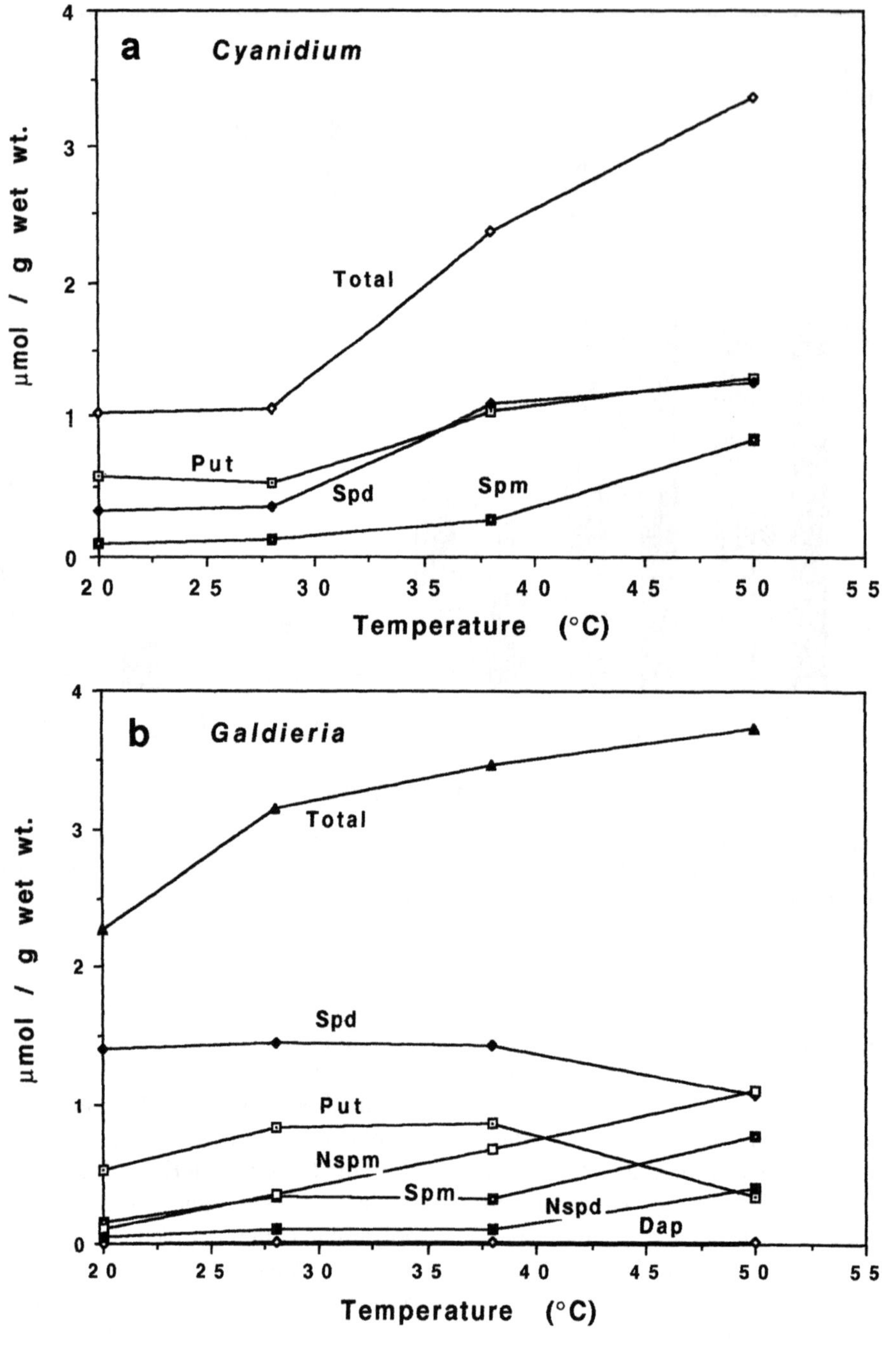

Fig. 5. Changes in polyamine content in *Cyanidium* (a) and *Galdieria* (b) at elevated temperature (Data from Hamana *et al.* 1990). See Fig. 4 captions. When culture temperature raised all polyamines increased in *Cyanidium* RK-1, while in *Galdieria* M-8 only spermine (Spm), norspermine (Nspm) and norspermidine (Nspd) increased.

5. Summary and conclusion

This chapter discusses low molecular weight carbohydrates, fatty acids, hydrocarbons, hydroxy acids, sterols and polyamines in the Cyanidiophyceae. The distribution of these compounds in other algae are briefly summarized, while the phylogeny of *Cyanidium* and its related algae are presented from a chemotaxonomical point of view. Among these substances the low molecular weight carbohydrate is one of the most reliable and useful criterion in the chemotaxonomy of Cyanidiophyceae. Physiological roles of some of these substances are also described here.

References

Adams, B. L., McMahon, V. A. and Seckbach, J. (1971). *Biochem. Biophys. Res. Commun.* 42: 359–365.

Allen, C. F., Good, P. and Holton, R. W. (1970). *Plant Physiol.* 46: 748–751.

Araki, S., Sakurai, T., Omata, T., Kawaguchi, A. and Murata, N. (1988). *Jpn J. Phycol. (Sorui)* 34: 94–100.

Boenzi, D., De Luca, P. and Taddei, R. (1977). *Giorn. Bot. Ital.* 111: 129–134.

Brown, A. C. and Knights, B. A. (1969). *Phytochem.* 8: 543–547.

Burton, A. J. and Carter, H. E. (1964). *Biochem.* 3: 411–418.

Cardoso, J. N. and Eglinton, G. (1983). *Geochim. Cosmochim. Acta* 47: 723–730.

Chardon-Loriaux, I., Morisaki, M. and Ikekawa, N. (1976). *Phytochem.* 15: 723–725.

Cranwell, P. A., Jaworski, G. H. M. and Bickley, H. M. (1990). *Phytochem.* 29: 145–151.

Craigie, J. S. (1974). In: *Algal Physiology and Biochemistry (ed. Stewart, W. D. P.). Blackwell Scientific Publications, London*, pp. 206–235.

De Luca, P. and Taddei, R. (1976). *Webbia* 30; 197–218.

De Luca, P., Taddei, R. and Varano, L. (1978). *Webbia* 33: 37–44.

De Luca, P., Gambardella, R. and Merola, A. (1979). *Bot. Gaz.* 140: 418–427.

De Luca, P. and Moretti, A. (1983). *J. Phycol.* 19: 368–369.

Eglinton, G., Hunneman, D. H. and Douraghi-Zadeh, K. (1968). *Tetrahedron* 24: 5929–5941.

Gelpi, E., Oro, J., Schneider, H. J. and Bennett, E. O. (1968). *Science* 161: 700–701.

Gelpi, E., Schneider, H., Mann, J. and Oro, J. (1970). *Phytochem.* 9: 603–612.

Goodwin, T. W. (1974). In: *Algal Physiology and Biochemistry (ed. Stewart, W. D. P.). Blackwell Scientific Publications, London*, pp. 206–235.

Hamana, K. and S. Matsuzaki (1982). *J. Biochem.* 91: 1321–1328.

Hamana, K. and S. Matsuzaki (1985). *J. Biochem.* 97: 1311–1315.

Hamana, K., Matsuzaki, S. Niitsu, M., Samejima, K. and Nagashima, H. (1990). *Phytochem.* 29: 377–380.

Holton, R. W., Blecker, H. H. and Onore, M. (1964). *Phytochem.* 3: 595–602.

Holton, R. W., Blecker, H. H. and Stevens, T. S. (1968). *Science* 160: 545–547.

Ikan, R. and Seckbach, J. (1972). *Phytochem.* 11: 1077–1082.

Ikekawa, N., Morisaki, N., Tsuda, K. and Yoshida, T. (1968). *Steroids* 12: 41–48.

Impellizzeri, G., Mangiafica, S., Oriente, G., Piattelli, M. and Sciuto, S. (1975). *Phytochem.* 14: 1549–1557.

Kauss, H. (1967). *Z. Pflanzenphysiol.* 56: 453–465.

Kauss, H. (1968). *Z. Pflanzenphysiol.* 58: 428–433.

Kenyon, C. N. and Stanier, R. Y. (1970). *Nature* 227: 1164–1166.

Kirst, G. O. (1980). *Phytochem.* 19: 1107–1110.

Kirst, G. O. and Bisson, M. A. (1979). *Aust. J. Plant Physiol.* 6: 539–556.

Kleinschmidt, M. G. and McMahon, V. A. (1970a). *Plant Physiol.* 46: 286–289.

Kleinschmidt, M. G. and McMahon, V. A. (1970b). *Plant Physiol.* 46: 290–293.

Kohlhase, M. and Pohl, P. (1988). *Phytochem.* 27: 1735–1740.

Kremer, B. P., Feige, G. B. and Schneider, Hj. A. W. (1978). *Naturwissenschaften* 65: S, 157.

Kremer, B. P. and Feige, G. B. (1979). *Z. Naturf.*, 34c: 1209–1214.

Kremer, B. P. (1982). *Br. Phycol. J.* 17: 51–61.

Lee, R. F. and Loeblich III. (1971). *Phytochem.* 10: 593–602.

Lindberg, B. (1956). In: *Second Internatl. Seaweed Symp. (eds. Braarud, T. and Sørensen, N. A.), Pergamon Press, London*, pp. 33–38.

Matsumoto, G. I., Shioya, M. and Nagashima, H. (1984). *Phytochem.* 23: 1421–1423.

Matsumoto, G. I. and Nagashima, H. (1984). *Geochim. Cosmochim. Acta* 48: 1683–1687.

Majak, W., Craigie J. S. and McLachlan, J. (1966). *Can. J. Botany* 44: 541–549.

Meeuse, B. J. D. (1962). In: *Physiology and Biochemistry of Algae (ed. Lewin, R. A.), Acad. Press, New York*, pp. 289–313.

Moretti, A. and Nazzaro, R. (1980) *Delpinoa* 21: 4–11.

Moretti, A. and Musacchio, A. (1982). *Giorn. Bot. Ital.* 116: 269–274.

Merola, A., Castaldo, R., De Luca, P., Gambardella, R., Musacchio, A. and Taddei, R. (1981). *Giorn. Bot. Ital.* 115: 189–195.

Murray, J. and Thomson, A. *Phytochem.* 16: 465–468.

Nagashima, H., Ozaki, H., Nakamura, S. and Nisizawa, K. (1969). *Bot. Mag. Tokyo* 82: 462–473.

Nagashima, H. (1976). *Bull. Jpn. Soc. Phycol.* 24: 103–110.

Nagashima, H. and Fukuda, I. (1981a). *Phytochem.* 20: 439–442.

Nagashima, H. and Fukuda, I. (1981b). *Jpn J. Phycol.* 29: 237–242.

Nagashima, H. and Fukuda, I. (1983). *Phytochem.* 22: 1949–1951.

Nagashima, H., Kuroiwa, T. and Fukuda, I. (1984). *Experientia* 40: 563–564.

Nagashima, H., Matsumoto, G. I. and Fukuda, I. (1986). *Phytochem.* 25: 2339–2341.

Nichols, B. W. (1965). *Biochim. Biophys. Acta* 106: 274–279.

Nyberg, H. and Koskimies-Soininen, K. (1984). *Phytochem.* 23: 2489–2495.

Patterson, G. W. (1967). *J. Phycol.* 3: 22–23.
Patterson, G. W. (1971). *Lipids* 6: 120–127.
Patterson, G. W. and Valkenburg, S. D. V. (1990). *J. Phycol.* 26: 484–489.
Patterson, G. W. (1991). In: *Physiology and Biochemistry of Sterols (eds. Patterson, G. W. and Nes, W. D.) American Oil Chemists' Society, Champaign, Ilinois.* pp. 118–157
Reed, R. H. (1983). *Phycologia* 22: 351–354.
Seckbach, J. and Ikan, R. (1972). Plant Physiol. 49: 457–459.
Seckbach,J., Ikan, R., Ringelberg, D. and White, D. (1993). *Phytochem.* 34: 1345–1349.
Stodola, F. H., Deinema, M. H. and Spencer, J. F. T. (1967). *Bacteriol. Rev.* 31: 194–213.
Tabor, C. W. and Tabor, H. (1985). *Microbiol. Rev.* 49: 81–89.
Weckesser, J., Katz, A., Drews, G., Mayer, H. and Fromme, I. (1974). *J. Bacteriol.* 120: 672–678.
Weete, J. D. (1976). In: *Chemistry and Biochemistry of Natural Waxes (ed. Kolattukudy, P. E.) Elsevier, Amsterdam.* pp. 349–418.
Winter, K., Parker, P. L. and Baalen, C. V. (1969). *Science* 163: 467–468.
Wood, B. J. B. (1974). In: *Algal physiology and biochemistry (ed. Stewart, W. D. P.) Blackwell Scientific Publications, Oxford.* pp. 236–265,

Chapter 17

Ultrastructure of unicellular red algae

SHARON T. BROADWATER and JOSEPH L. SCOTT
Department of Biology, College of William and Mary, Williamsburg, VA 23187, USA

Outlines:
1. Introduction
2. Ultrastructure
   2.1 Cell walls
   2.2 Endoplasmic reticulum
   2.3 Golgi apparatus
   2.4 Mitochondria
   2.5 Chloroplasts
   2.6 Vacuoles, vesicles and microbodies
   2.7 Nucleus
   2.8 Cytoskeleton
3. Cell division
4. Summary
References

1. Introduction

The phylum Rhodophyta is an enigmatic assemblage of organisms whose phylogenetic relationships with other eukaryotes are shrouded in uncertainty. The group was long held to be among the most primitive eukaryotes due to the lack of a centriole complex and the similarities between the red algal chloroplast and cyanobacteria (Garbary and Gabrielson 1990). This view was given further impetus when the analysis of red algal 5s ribosomal RNA indicated an early divergence from other eukaryotes (Lim et al. 1986; Hori and Osawa 1987). However, there is equally compelling evidence that red algae are not more primitive than other algal phyla. Acceptance of the endosymbiotic origin of chloroplasts renders the resemblance of red algae and cyanobacteria to be superficial (Gabrielson *et al.* 1985) while ultrastructural studies show typical eukaryotic cell organization (Pueschel 1990). In addition, recent studies of the 18s ribosomal subunit (Bhattacharya *et al.* 1990) indicate divergence of red algae to be at the same time as that of 'plants, fungi, animals and most protists'.

Regardless of the time of rhodophytan emergence, at least one major question impacting an understanding of red algal unicells remains unanswered. What were the characteristics of the ancestral red alga? Was it a unicell? The red algae are thought to be monophyletic, but no extant group of organisms has been shown to have a close relationship to the Rhodophyta (Gabrielson *et al.* 1985; Gabrielson and Garbary 1986; Garbary and Gabrielson 1990). At different times, phylogenetic affinity has been proposed for the Cyanobacteria, higher fungi (Ascomycetes and Basidiomycetes), Cryptophyta, Chlorophyta and Glaucophyta, but none of these hypotheses have withstood subsequent investigations (Gabrielson *et al.* 1985; Garbary and Gabrielson 1990).

The lack of an outgroup is partially responsible for the taxonomic uncertainties in this group. This uncertainty along with new kinds of data generated within the last two decades is reflected in the flux of taxonomies over the last ten years such that the number of orders has varied from 9 to 18 or more (Gabrielson and Garbary 1986; Woelkerling 1990). One of the few orders which has been accepted by all, albeit reluctantly, is the Porphyridiales even though this order is considered by many to be 'the most problematic order of red algae in terms of classification and potential relationships' (Garbary and Gabrielson 1990).

According to Garbary and Gabrielson (1990), the Porphyridiales is defined solely on the basis of a unicellular or palmelloid habit and is considered by many to be both polyphyletic (having different ancestors) and paraphyletic (having descendants in different orders). Garbary and Gabrielson argue against the primitive nature of these unicells since sexual reproduction has not been found in this order. In order for the Porphyridiales to serve as the outgroup, one must assume that sexuality arose at least twice among eukaryotes. Such an event is

J. Seckbach (ed.), Evolutionary Pathways and Enigmatic Algae: Cyanidium caldarium *(Rhodophyta) and Related Cells,* 215–230, 1994.

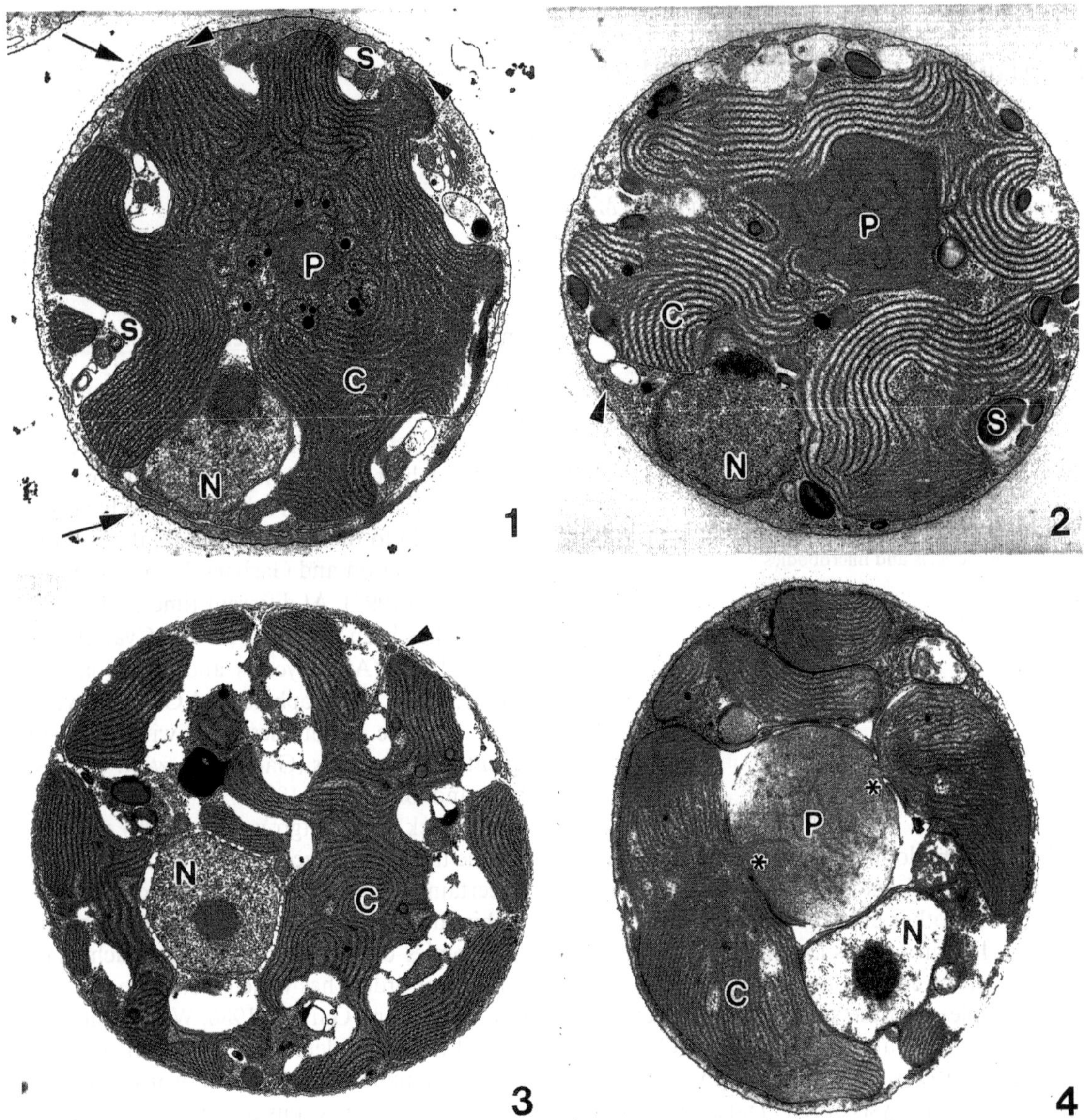

Fig. 1–4. (1) *Porphyridium purpureum* has a peripheral nucleus (N) and a central pyrenoid (P) with thylakoids in a stellate-like chloroplast (C) without a peripheral encircling thylakoid. Starch (S) appears randomly scattered in sparse cytoplasm. A peripheral ER system with perpendicular tubules extending to the plasmalemma (arrowheads) can be detected as can a thin mucilage sheath (arrows). 12,550 X. (Schornstein and Scott 1982, with permission); (2) *Porphyridium aerugineum* contains nearly the same assemblage and disposition of organelles as *P. purpureum*. N, nucleus; P, pyrenoid; C, chloroplast; S, starch; arrowhead, peripheral ER. 14,050 X. (Gantt *et al.* 1968, with permission); (3) *Flintiella sanguinaria* possesses an eccentric nucleus (N) and a variously branched chloroplast (C) without a peripheral encircling thylakoid or an obvious pyrenoid. Arrowhead, peripheral ER system. 7,600 X. (Scott 1986, with permission); (4) *Rhodosorus marinus* has an eccentric nucleus (N) in a cup-shaped chloroplast (C) without a peripheral encircling thylakoid. The central pyrenoid (P) contains random thylakoids and is said to be stalked although the presence of two connections to the chloroplast (asterisks) undermines this assumption. 12,600 X. (Courtesy of B. Saunders).

highly unlikely, especially in light of meiotic studies in the red alga *Dasya baillouviana* which show that the meiotic process in this alga fits easily within the general eukaryotic pattern (Broadwater *et al.* 1986a, 1986b). This consideration gives further credence to the speculation that some unicells may be reduced forms of multicellular red algae (Garbary *et al.* 1980; Gabrielson and Garbary 1986; Scott and Gabrielson 1987; Garbary and Gabrielson 1990).

If the Porphyridiales are not a natural assemblage, then no taxonomy can accurately reflect the true phylogenetic relationships and must instead be a pragmatic solution (Garbary *et al.* 1980; Garbary and Gabrielson 1990). This chapter, therefore, will make no attempt to classify the various red algal unicells; instead we will show the diversity of types with respect to ultrastructure and discuss some enigmatic algae with regard to their affinity to the Rhodophyta.

2. Ultrastructure

Only eight red algal unicells have been studied significantly for details of ultrastructure and will be included in this review: *Porphyridium purpureum* (Fig. 1) (Brody and Vatter 1958; Gantt and Conti 1965; Neushul 1970; Oakley and Dodge 1974, 1966; Lin *et al.* 1975; Bronchart and Demoulin 1977; Schornstein and Scott 1980, 1982; Wanner and Köst 1980), *Porphyridium aerugineum* (Fig. 2) (Gantt *et al.* 1968; Neushul 1970; Ramus 1972; Pekárková *et al.* 1989), *Flintiella sanguinaria* (Fig. 3) (Scott 1986), *Rhodosorus marinus* (Fig. 4) (Giraud 1962, 1963; Ford 1984), *Rhodella cyanea* (Fig. 5) (Billard and Fresnel 1986; Fresnel and Billard 1987), *Dixoniella grisea* (Fig. 6) (Deason *et al.* 1983; Broadwater and Scott 1986; Fresnel and Billard 1987; Pekárková *et al.* 1988; Fresnel *et al.* 1989; Scott *et al.* 1992), *Rhodella maculata* (Fig. 7) (Evans 1970; Evans *et al.* 1974; Callow and Evans 1979; Fresnel and Billard 1987; Patrone *et al.* 1991), and *Rhodella violacea* (Fig. 8) (Wehrmeyer 1971; Fresnel and Billard 1987; Patrone *et al.* 1991). Because of possible phylogenetic affinity to the Rhodophyta, the three thermophilic algae, *Cyanidium caldarium*, *Galdieria sulphuraria* and *Cyanidioschyzon merolae* (Merola *et al.* 1981) and the alga, *Glaucosphaera vacuolata* (Fig. 9) (Richardson and Brown 1970; McCracken *et al.* 1980; Trench 1982) are also included where relevant.

Since other authors in this volume will cover the thermophilic algae in great detail, we will make no attempt to include all the literature. Neither will we discuss the differences in ultrastructural details among these three algae since some of the ultrastructural work was done prior to the realization that three separate species exist. However, most red algal systematists do consider the Cyanidiophytes to belong to the Rhodophyta (Gabrielson *et al.* 1985; Gabrielson and Garbary 1986; Garbary *et al.* 1980; Hara *et al.* 1989); therefore, we include them here. Similarly, the taxonomic position of *Glaucosphaera vacuolata* is uncertain, but it is thought by some to be a red alga (McCracken *et al.* 1980; Melkonian 1982).

Organelles not mentioned in the following discussion can be assumed either to be the same as found in all eukaryotes or unstudied. Table 1 is a compilation of the data herein.

2.1 Cell walls

All non-controversial unicellular members of the Porphyridiales (Chapman 1974) and *G. vacuolata* (Richardson and Brown 1970) lack a cell wall, and instead have a mucilagenous sheath whose thickness varies with culture conditions (Figs. 1, 10, 13). The mucilages are complex, of high molecular weight and variable both in composition and in proportion of similar components (Craigie 1990). The presence of a thick, highly proteinaceous cell wall (Bailey and Staehelin 1968) and absence of a mucilagenous sheath in the thermophilic algae is one factor which has caused taxonomic difficulties with this group. However, this condition may be an adaptation to the acid hot springs habitat of these algae and, consequently, may not represent an impediment to its inclusion in the Rhodophyta (Chapman 1974).

Some unicells exhibit gliding motility which may be due to secretion of mucilage at the trailing end of the cell (Lin *et al.* 1975). Now that actin filaments have been identified in red algae (Broadwater *et al.* 1992; Garbary *et al.* 1992), it will be interesting to see whether unicells may have specialized arrangements of these filaments to enable movement.

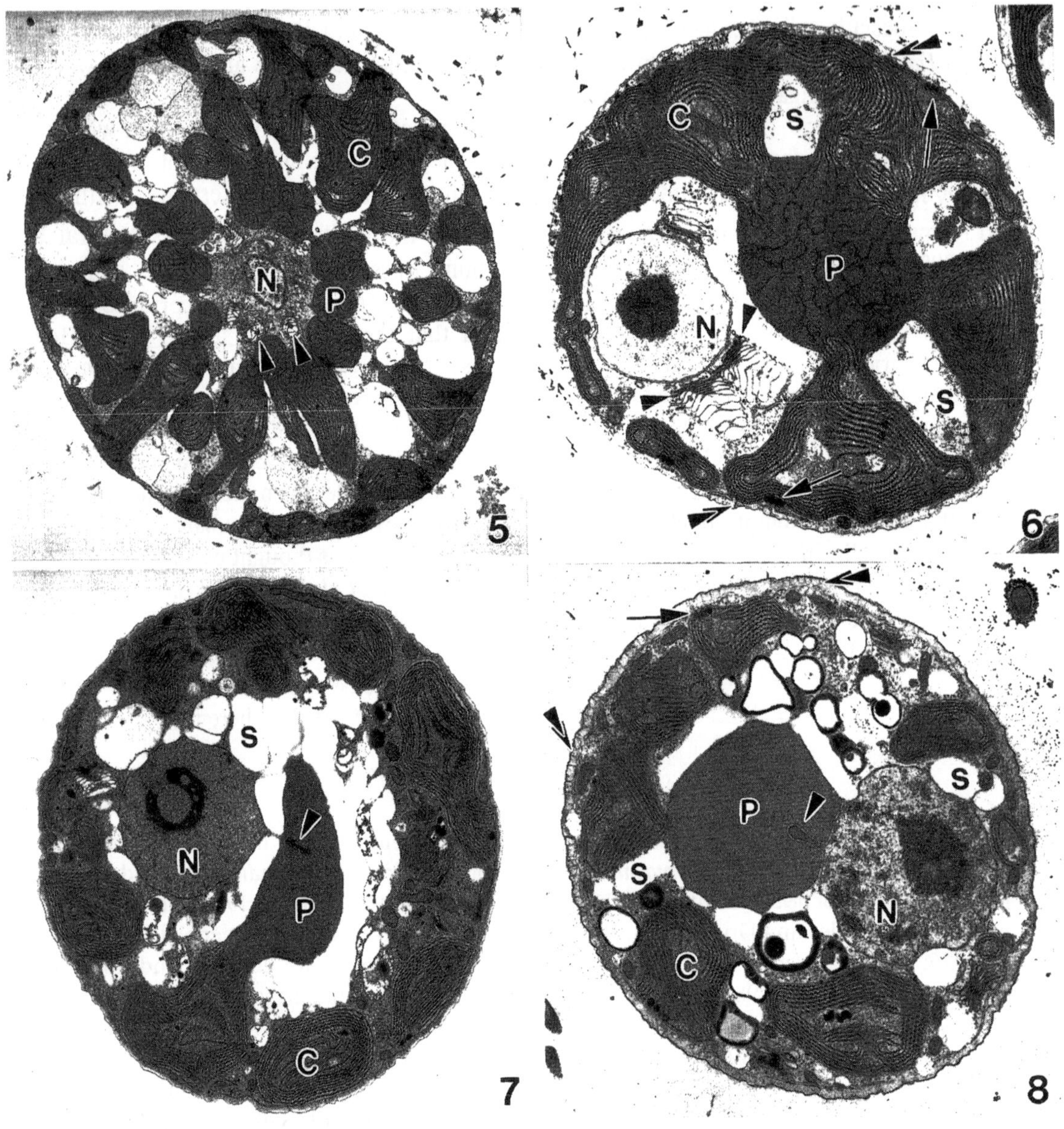

Fig. 5–8. (5) *Rhodella cyanea* possesses a central nucleus (N) to which are juxtaposed the dictyosomes (arrowheads), giving this alga a Golgi-nucleus association. The branched chloroplast (C) fuses to form a perforated shell around the nucleus. The thylakoid arrangement in this region is irregular and has the appearance of a pyrenoid (P). 4050 X. (Billard and Fresnel 1986, with permission); (6) *Dixoniella grisea* contains an eccentric nucleus (N), central pyrenoid (P) with traversing thylakoids, and a branching chloroplast (C) with a peripheral encircling thylakoid and plastoglobuli (arrows). Perinuclear dictyosomes (arrowheads) make this one of only two red algae with a Golgi-nucleus association. A peripheral ER system with perpendicular tubules is present (double arrowheads). S, starch. 10,800 X. (Scott *et al.* 1992, with permission); (7) *Rhodella maculata* has an eccentric nucleus (N), central pyrenoid (P) without thylakoids, and a branching chloroplast (C) with a peripheral encircling thylakoid. This unicell is one of two containing a nuclear projection into the pyrenoid (arrowhead). Starch (S) congregates around the areas of the pyrenoid exposed to the cytoplasm. 6,450 X. (Patrone *et al.* 1991, with permission); (8) *Rhodella violacea* is like *R. maculata* except that the chloroplast (C) is less branched and the pyrenoid (P) is more closely associated with the nucleus (N). Plastoglobuli (arrow) and peripheral ER with tubules (double arrowheads) are visible. S, starch; arrowhead, nuclear projection. 10,000 X. (Patrone *et al.* 1991, with permission).

Table 1. Characteristics of red algal unicells

Red algal unicells	Cell size (μm)	Color (log phase)	Habitat	Cell wall	Peripheral ER and tubules	Golgi associations	Chloroplast number	Chloroplast shape	Peripheral thylakoid	Pyrenoid	Thylakoids in pyrenoid	Pyrenoid location	Nu projection into pyrenoid	Nucleus location	Chromosome number	Kinetochore type	NAO type	NAO size (nm) (width × height)	Metaphase pole	Relative IZM length
Flintiella sanguinaria	6–14	red	fresh water	no	yes	ER/M	one	peripheral	no	no?	no?	peripheral?	no	eccentric	?	small, indistinct	polar ring	30 × 70	gap	long
Porphyridium aerugineum	5–13	blue-green	fresh water	no	yes	ER/M	one	stellate-like	no	yes	yes	central	no	peripheral	2, LM	?	?	?	?	?
Porphyridium purpureum	5-13	red	high salinity	no	yes	ER/M	one	stellate-like	no	yes	yes	central	no	peripheral	2, LM 8, EM	small, indistinct	distal disc prox. granule	d: 70 × 50 g: 140 × 55	gap	long
Rhodella cyanea	22–40	blue-green	marine	no	yes	Nu	one	radial, fusing around nucleus	no	yes?	yes?	perinuclear?	no	central	?	?	?	?	?	?
Dixoniella grisea	8–32	olive	marine	no	yes	Nu	one	stellate-like deeply lobed	yes	yes	yes	eccentric	no	eccentric	?	small, indistinct	polar ring	85 × 40	gap	short
Rhodella maculata	7–24	pink	marine	no	yes	ER	one	stellate-like deeply lobed	no	yes	no	eccentric	yes	eccentric	?	small, indistinct	polar ring	16 × 35	gap	short
Rhodella violacea	8–30	violet	marine	no	yes	ER	one	stellate-like deeply lobed	no	yes	no	eccentric	yes	eccentric	?	small, indistinct	polar ring	16 × 35	gap	short
Rhodosorus marinus	5–15	red	marine	no	yes?	ER/M	one	cup-shaped	no	yes	yes	eccentric stalked?	no	peripheral	?	?	?	?	?	?
Problematic Unicells																				
Cyanidioschyzon merolae	1.5–4	blue-green	thermophilic	yes	?	?	one	polymorphic	yes	no	N/A	N/A	N/A	peripheral	?	?	?	?	?	?
Cyanidium caldarium	2–6	blue-green	thermophilic	yes	yes	ER	one	cup-shaped	yes	no	N/A	N/A	N/A	peripheral	?	?	?	?	?	?
Galdieria sulphuraria	3–11	blue-green	thermophilic	yes	yes	ER	one	multilobed	yes	no	N/A	N/A	N/A	eccentric	?	?	?	?	?	?
Glaucosphaera vacuolata	14–22	blue-green	fresh water	no	yes	Nu	one	peripheral, multilobed	yes	no	N/A	N/A	N/A	central	?	small, trilaminar	N/A	N/A	gap	short

Abbreviations used: d, disc; EM, electron microscopy; ER, endoplasmic reticulum; g, granule; IZM, interzonal midpiece; LM, light microscopy; M, mitochondria; N/A, not applicable; NAO, nucleus associated organelle; Nu, nucleus.

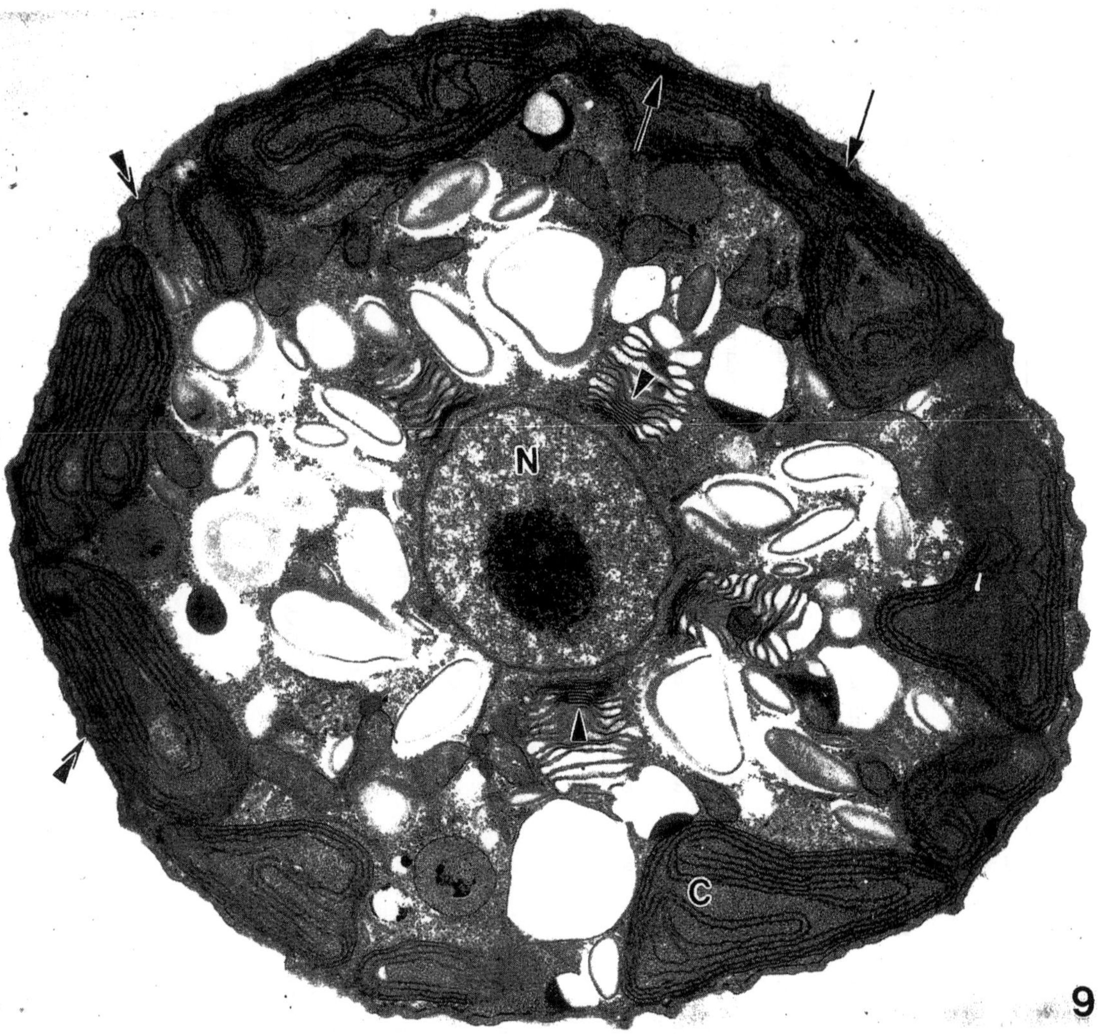

Fig. 9. Glaucosphaera vacuolata possesses large numbers of dictyosomes with appressed cisternae (arrowheads) arranged around the periphery of the central nucleus (N). A variously branched chloroplast (C) with a peripheral encircling thylakoid and plastoglobuli (arrows) is present. Double arrowheads, peripheral ER. 17,300 X. (Courtesy of J. Thomas).

2.2 Endoplasmic reticulum

Both smooth and rough endoplasmic reticulum (ER) appear to be present in all unicells although not usually in large quantities. One characteristic shared by all of the unicells studied (Pueschel 1990; Patrone *et al.* 1991; Scott *et al.* 1992) and *G. vacuolata* (Fig. 9) is a system of smooth ER which underlies the entire plasma membrane (PM) (Figs. 1, 3, 6, 8, 9, 10, 14–16). The 100–150 nm wide region between this peripheral ER system and the plasma membrane is free of major organelles, including ribosómes, but does appear to contain a fibrous substance which is likely actin in nature (Broadwater *et al.* 1992). At irregular intervals, tubules arise at right angles from the ER toward the PM (Figs. 1, 6, 8, 10, 14–16); some appear to fuse with the PM suggesting open communication between the ER and the cell exterior (Pueschel 1990). Evans *et al.* (1974) suggest that this system may play a role in mucilage production although multicellular algae with a peripheral ER system do not appear to have the perpendicular tubules even though some do produce mucilage (Patrone *et al.* 1991). If the system is primarily engaged in mucilage production, it should not be expected in

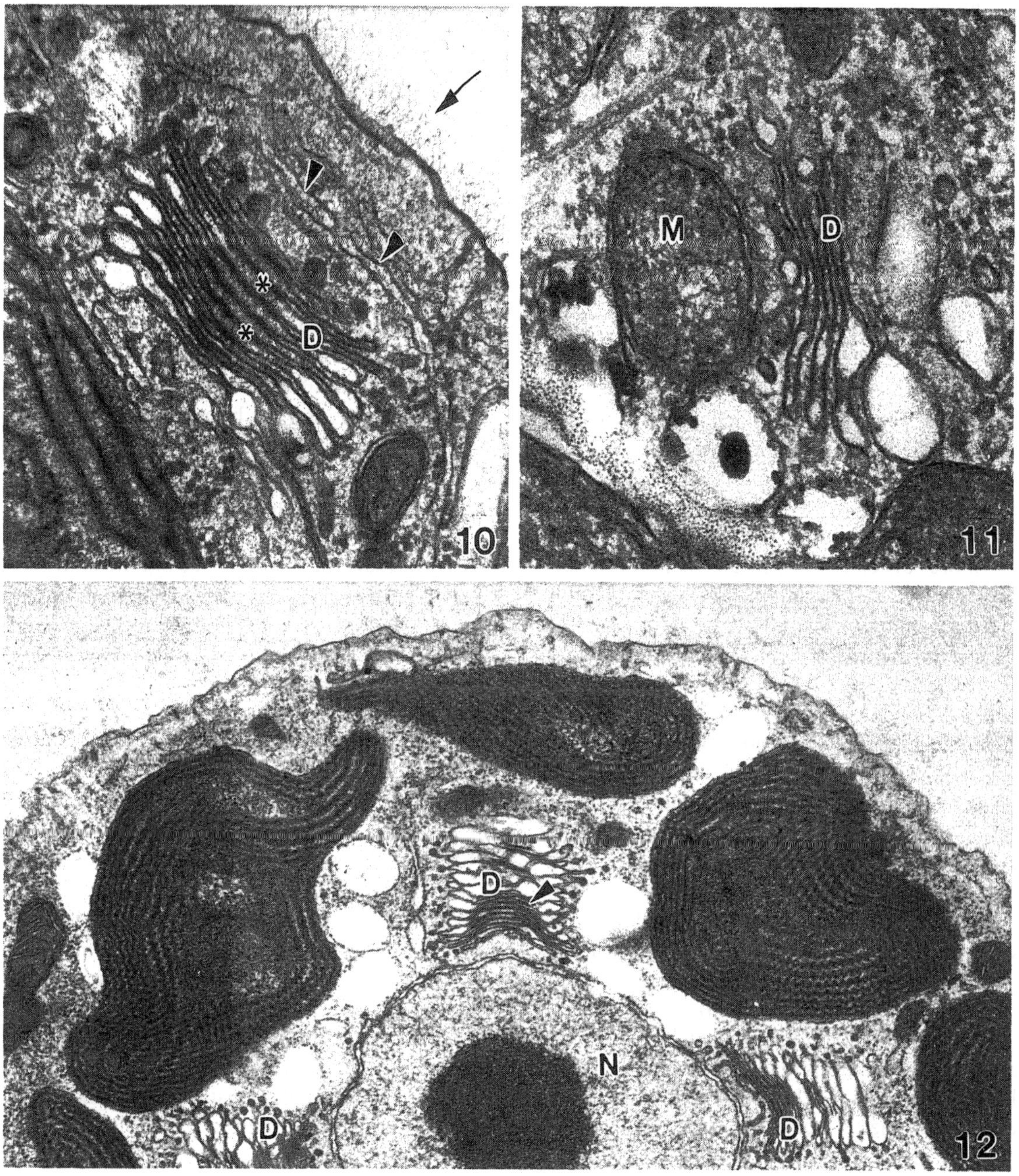

Fig. 10–12. (10) This dictyosome (D) in *R. violacea* is associated only with rough ER (arrowheads) and represents one of the types of Golgi associations present in unicells. Note the appressed cisternae (asterisks). Arrow, mucilage sheath. 73,000 X. (Patrone *et al.* 1991, with permission); (11) This dictyosome (D) and mitochondrion (M) from *P. purpureum* exemplifies the second of three types of Golgi association in the unicells. 76,500 X; (12) The dictyosomes (D) encircling the nucleus (N) of *D. grisea* represent the third type of Golgi association present in unicells. Note appressed cisternae (arrowhead). 26,000 X. (Scott *et al.* 1992, with permission).

these algae or other organisms not producing mucilage.

2.3 Golgi apparatus

All red algae contain a typical eukaryotic Golgi apparatus (GA) (Pueschel 1990) as does *G. vacuolata* (Richardson and Brown 1970; McCracken *et al.* 1980). The presence of GA in all of the thermophilic algae is still questioned by some (Seckbach 1987), but has been demonstrated in *Cyanidium* (De Luca *et al.* 1979; Ueda and Yokochi 1981) and *Galdieria* (De Luca *et al.* 1979; Ueda and Yokochi 1981; Seckbach 1987, 1992).

GA numbers, size and morphology may vary with the cell cycle or culture conditions. For example, log phase cells of *Porphyridium* have higher numbers of dictyosomes (components of GA) and the dictyosomes are greater in size and have more distended cisternae than stationary phase cells (Ramus and Robins 1975, personal observation). During log phase growth, the dictyosomes of some red algal unicells display a unique morphology in which the mid-regions of adjacent cisternae are very closely apposed or fused to each other (Figs. 10, 12) as does *G. vacuolata* (Fig. 9). This condition is found in multicellular red algae only in certain developmental stages of sporangia (Patrone *et al.* 1991; Scott *et al.* 1992) and, to our knowledge, has not been shown to occur in any other eukaryotic cells. The presence of this feature only in unicells and developing sporangia is interesting in light of the possibility that unicells are reduced forms of reproductive cells (in Garbary and Gabrielson 1990).

An important taxonomic indicator for the red algae is the associations made between the GA and other cell organelles. Three different associations are possible. The cis face of the GA may be associated with ER only (Fig. 10), with both ER and a mitochondrial profile (Fig. 11) or with the nuclear envelope only (Figs. 5, 6, 9, 12). GA associations with both ER and mitochondria are the most common condition in red algae (Scott 1984; Pueschel 1990; Patrone *et al.* 1991). Except for the Porphyridiales, all orders of red algae are consistent with regard to type of GA association and exhibit only one pattern throughout the entire order. All three patterns, however, can be found among the unicells (Table 1). The heterogeneity of this character among the Porphyridiales supports the widely held contention that this group is polyphyletic (Garbary *et al.* 1980; Gabrielson and Garbary 1986; Scott and Gabrielson 1987; Garbary and Gabrielson 1990).

2.4 Mitochondria

All red algae have typical mitochondria with flattened cristae (Pueschel 1990) as does *G. vacuolata* (Richardson and Brown 1970) and at least some of the thermophilic algae (Ueda and Yokochi 1981; Seckbach 1987). Determining the exact number and shape of mitochondria per cell requires serial section analysis and has been done completely only for *D. grisea* (as *R. reticulata*, Broadwater and Scott 1986). This alga was found to contain a single anastomosing mitochondrion with occasional small, unattached segments thought to result from dynamic rearrangements of the organelle. It is quite possible that all red algal unicells contain only one to a few highly branched mitochondria as this condition is often typical of unicellular organisms in general (Broadwater and Scott 1986). The thermophilic algae are reported to have from one to several mitochondria (Seckbach 1987), but only serial-section analysis can confirm the actual configuration of the chondriome.

2.5 Chloroplasts

The chloroplast and its complement of pigments is the single most important organelle in differentiating red algae from other organisms. All red algae (Pueschel 1990) and *G. vacuolata* (McCracken *et al.* 1980) have a combination of chloroplast features, including a double-membrane envelope and single, unstacked thylakoids with phycobilisomes containing phycobiliproteins and starch deposition outside the chloroplast, that taken together, are unique to the red algae. Chloroplasts of thermophilic algae apparently share all these features. Early studies of thermophilic algae showing a chloroplast envelope comprised of a single membrane led some investigators to propose these as bridge algae from Cyanophytes to Rhodophytes (Seckbach and Fredrick 1980; Seckbach *et al.* 1983; Seckbach 1987). However, fixation with glutaraldehyde rather than potassium permanganate (Ueda and

Yokochi 1981) and freeze-substitution (Ueda and Chida 1987) show two membranes around the chloroplast.

One unique characteristic of *C. caldarium* is the presence of a 'plastid-dividing ring' which functions to divide the chloroplast (Kuroiwa 1989). The 'plastid-dividing ring', which is apparently composed of actin filaments, contracts, furrowing into and dividing the chloroplast. Nothing analogous has been reported in red algae although similar structures have been seen in other algae, mosses and land plants (Kuroiwa 1989; Kuroiwa *et al.* in this book).

Those characteristics of the chloroplast which vary among the unicells are chloroplast number, shape, thylakoid disposition, presence of pyrenoid, pyrenoid position, presence and arrangement of thylakoids in the pyrenoid, associations with starch deposits, and presence of pyrenoid-nucleus associations (Table 1). Plastoglobuli (Figs. 6, 8, 9), sometimes called 'eyespots', are commonly found in unicellular red algal chloroplasts but are not taxonomically valuable since they appear in plastids of almost all red algae (Pueschel 1990).

The eight unicellular red algal genera considered here appear to have a single plastid (Garbary *et al.* 1980) as does *G. vacuolata* (McCracken *et al.* 1980, personal observation) and the three thermophilic algae (Seckbach 1992; Seckbach *et al.* 1992). However, the unicellular genus *Rhodospora* is thought to contain several chloroplasts (Bold and Wynne 1985) and Hara (personal communication) has discovered a new red algal unicell with many discoid chloroplasts. The chloroplast varies considerably in morphology (Table 1) with various genera having chloroplasts which are cup-shaped (*R. marinus*, Fig. 4; some thermophilic algae), stellate-like (*P. purpureum*, Fig. 1; *P. aerugineum*, Fig. 2), or variously branching (*F. sanguinaria*, Fig. 3; *D. grisea*, Fig. 6; *R. maculata*, Fig. 7; *R. violacea*, Fig. 8; *G. vacuolata*, Fig. 9). Perhaps the most unusual is that of *R. cyanea* (Fig. 5) whose chloroplast has ramifying branches which fuse to form a perforated shell around the nucleus.

The presence of a peripheral encircling thylakoid (Figs. 5–9) is variable among the unicells (Table 1), but the significance of this variability as a taxonomic feature is somewhat questionable since some multicellular red algae have this feature in one stage of the life cycle but lack it in another (Cole and Conway 1975; Pueschel 1990).

The importance of the presence of a pyrenoid can not be fully analyzed at this time since the absence of a pyrenoid in *F. sanguinaria* is questionable (Scott 1986) as is the presence of a pyrenoid in *R. cyanea* (Scott *et al.* 1992).

However, if both species are subsequently shown to have pyrenoids, the absence of pyrenoids in *G. vacuolata* and the thermophilic algae would have greater significance. The presence and arrangement of thylakoids in the pyrenoid and the position of the pyrenoid (whether completely surrounded by thylakoids or partially juxtaposed to the chloroplast envelope) appears to be consistent at the generic level. The presence and position of the pyrenoid has an effect on starch disposition (Lee, 1974; Pueschel 1990). In general, starch will aggregate most heavily near the pyrenoid if regions of the pyrenoid border the cytoplasm (Figs. 6–8); if the pyrenoid is absent or deeply embedded in the chloroplast, starch deposition is more random (Fig. 1).

Two red algal unicells, *R. maculata* and *R. violacea* (Patrone *et al.* 1991), have startling pyrenoid-nucleus associations which have brought into question the present species composition of the genus *Rhodella* (Scott *et al.* 1992). Throughout interphase, the nucleus of *R. maculata* and *R. violacea* sends long finger-like projections into the pyrenoid (Fig. 7–8). These appear to withdraw during mitosis, but an intimate association of pyrenoid and nucleus is maintained by the cupping of the pyrenoid around the nucleus during later stages of mitosis. The function of this unusual association is obscure.

2.6 Vacuoles, vesicles and microbodies

Red algal unicells apparently do not have the large osmotically active vacuoles seen in some multicellular species, probably because those cells without rigid cell walls are less able to use the osmotic-regulation properties of the vacuole (Pueschel 1990). Vacuoles of differing sizes and contents are sometimes seen in unicells (Richardson and Brown 1970; Patrone *et al.* 1991) but their function is unknown. Differently-sized vesicles of varying content are also present, but little cytochemical work has been done to establish

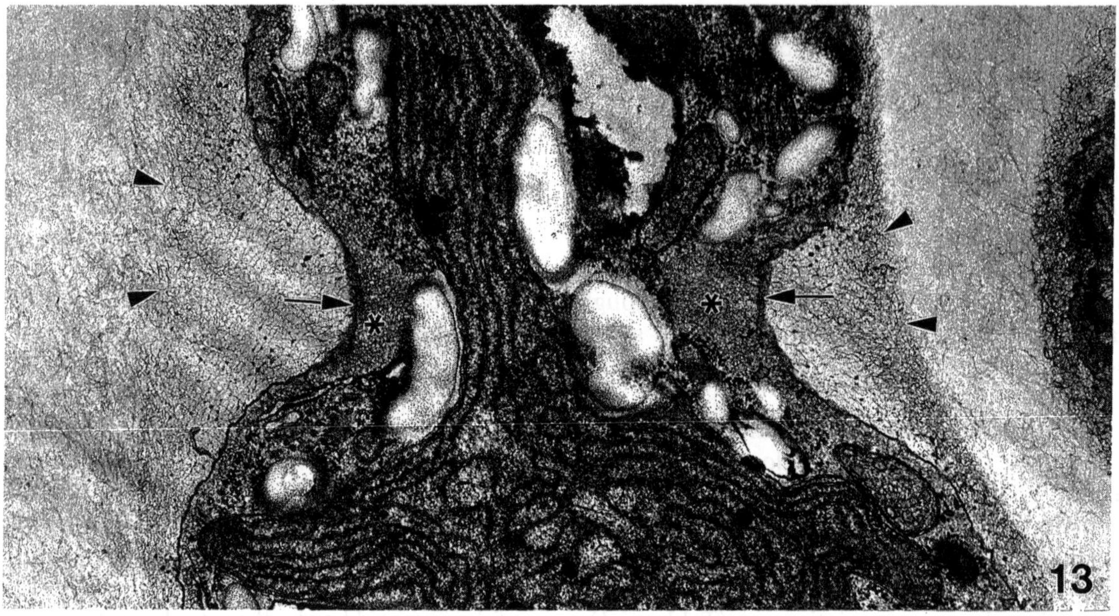

Fig. 13. Cytokinesis in some unicells appears to be mediated by a contractile ring of presumptive actin filaments (asterisks) seen here in *P. purpureum*. Arrows, cleavage furrow; arrowheads, mucilage sheath. 21,600 X.

either the content, longevity, or route of these vesicles. Microbodies with peroxidase activity were shown to be present in *P. purpureum* (Fig. 14) (Oakley and Dodge 1974). Microbodies were also seen at the poles of this alga during nuclear division (Bronchart and Demoulin 1977; Schornstein and Scott 1980, 1982) and were thought by the latter investigators to be utilizing the spindle to ensure equal allocation to daughter cells. In this instance, the microbodies do have taxonomic significance since no other red algae has been shown to have microbodies at the spindle pole (Scott and Broadwater 1990). Although not noted at spindle poles, microbodies have been reported for *G. vacuolata* (Seckbach 1992) and some thermophilic algae (Ueda and Yokochi 1981, Seckbach 1992). Seckbach (1992) also reports catalase activity present in all three thermophilic algae.

2.7 Nucleus

Red algal nuclei are eukaryotic in appearance with typical membranes, nuclear pores, nucleoli and chromatin (Pueschel 1990) (Figs. 1–9, 12, 14–19). Very little information is available on chromosome numbers among red algal unicells with counts available only for *P. aerugineum* (2) (Cole 1990) and *P. purpureum* (2–8) (Sommerfeld and Nichols 1970; Cole 1990). Part of the reason for the difficulty in obtaining counts and in the disparate number in *P. purpureum* may be due to poor condensation or aggregation of chromatin in the unicells leading to low counts (Scott and Broadwater 1990); the higher number in *P. purpureum* was due to counting kinetochores rather than using traditional light microscopic methods (Schornstein and Scott 1982). Despite possible differences in chromatin condensation between the unicells and the more elaborate multicellular taxa of red algae, chromosomes organization appears to be perfectly normal as inferred from work with *P. aerugineum* (Barnes *et al.* 1982). In this alga, chromatin was shown to have nucleosome core particles of the same size as other eukaryotes. In addition, all unicells studied for cell division have been shown to have kinetochores although these are always small and indistinct (Fig. 14) (Scott and Broadwater 1990, Scott *et al.* 1992).

The position of the nucleus is variable among the unicells but appears to be consistent at the level of genus. Most unicells have either eccentric or peripheral nuclei; only two, *R. cyanea* (Billard and

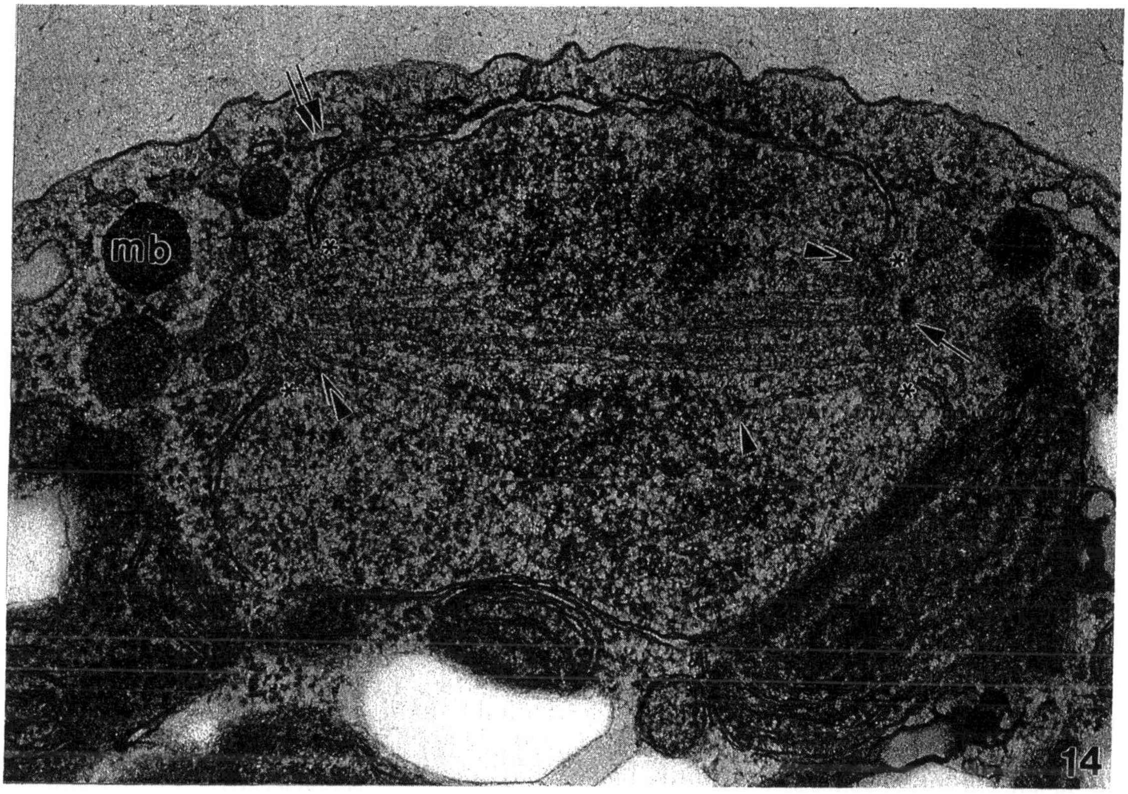

Fig. 14. The metaphase nucleus in the unicells, exemplified by *P. purpureum*, retains the nuclear envelope except for large gaps at the pole (delimited by asterisks). All microtubules abut electron-dense material at the nucleoplasm-cytoplasm interface (double arrowheads); some end at the small indistinct kinetochores (arrowhead). Arrow, distal portion of Nucleus Associated Organelle (NAO); double arrows, peripheral ER system; mb, microbody (found only in *P. purpureum*). 45,000 X (Schornstein and Scott 1982, Broadwater *et al.* 1992, with permission).

Fresnel 1986; Fresnel and Billard 1987) and *G. vacuolata* (Richardson and Brown 1970 and McCracken *et al.* 1980), have central nuclei.

One characteristic found in all undisputed red algae studied for details of cell division is a nucleus associated organelle (NAO) (Scott and Broadwater 1990). This organelle will be discussed in section 3.

2.8 Cytoskeleton

One of the most fundamental features serving to unite the Rhodophyta is the complete lack of centrioles or flagellar apparatus in all members. Although there have been occasional reports of reduced flagella or centrioles, these sightings have invariably been shown to be misinterpretations (in Pueschel 1990).

Microtubules and actin filaments, in general, are not well preserved in ultrastructural investigations of red algal cells. However, studies in other areas are proving more rewarding. Using molecular techniques, the multicellular red alga *Gracilaria tikvahiae* has been shown to contain beta tubulin, actin, and calmodulin (Dassler *et al.* 1990). In addition, fluorescence microscopy has demonstrated extensive arrays of cytoplasmic microtubules and actin filaments in *Griffithsia pacifica*, a filamentous red alga (Broadwater *et al.* 1992; Garbary *et al.* 1992). No unicells have been studied with these techniques. One might expect differences due to their simple morphology and lack of rigid walls, but one should not expect them to exhibit primitive cytoskeletal characters in light of the similarities in spindle that they share with multicellular taxa.

Microtubules are known to be present in the spindle, and a contractile ring of presumptive actin filaments appeared responsible for cytokinesis in the unicells *P. purpureum* (Fig. 13) (Schornstein and Scott 1982), *F. sanguinaria* (Scott 1986) and two species of *Rhodella* (Patrone *et al.* 1991). The presumption of an actin ring has been strengthened by recent immunofluorescence studies with rhodamine-phalloidin which have shown a contractile ring of actin filaments during cytokinesis of *G. pacifica* (Broadwater *et al.* 1992; Garbary *et al.* 1992).

3. Cell division

Cell division, along with chloroplast structure and pigment composition, is a leading area of research for understanding relationships in the Porphyridiales. Five unicellular red algae have been studied for ultrastructural details of cell division: *P. purpureum* (Schornstein and Scott 1982), *F. sanguinaria* (Scott 1986), *R. maculata* and *R. violacea* (Patrone *et al.* 1991) and *D. grisea* (Scott *et al.* 1992). In addition, a study in our laboratory of cell division in *Glaucosphaera vacuolata* is nearly complete and will be included in this discussion where appropriate.

As in all red algae, mitosis is accomplished through interdigitating half spindles which appear to interact with chromosomes at the kinetochore in typical eukaryotic fashion and undergo the normal sequence of stages from prophase to telophase. Spindles are enclosed within the nuclear envelope which remains intact throughout mitosis except at each pole where a single NAO is found (Scott and Broadwater 1990).

As a group, unicells do not have any mitotic characteristics that are not shared by other multicellular taxa. Surprisingly, however, considering the polyphyletic nature of this group, there is a fairly consistent pattern among the unicells whenever a character shows variability within the Rhodophyta as a whole. For example, among red algae, polar openings vary from large gaps, produced when preceding invaginations of the nuclear envelope break down, to openings comprised of many small fenestrations, not preceded by invaginations; all five unicells have large polar gaps (Fig. 14). Kinetochores vary from small and indistinct to discrete trilaminar structures; the five unicells all have the former (Fig. 14). Interestingly, *G. vacuolata* appears to have very definitive trilaminar kinetochores (unpublished information). Perinuclear ER may or may not be present in red algae but has not been observed in unicells. Perinuclear ER is thought to be a mechanism to maintain the environment of the nucleus after partial nuclear envelope breakdown at the poles. In general, it is more highly developed in those genera which are multinucleate or have some multinucleate cells (discussed in Scott and Broadwater 1990).

The only two major differences noted among the unicells were the length of the interzonal spindle and the appearance and subsequent behavior of the NAO (Scott *et al.* 1992). Two of the five unicells (Table 1) have long interzonal spindles (Fig. 15) that result in a much greater separation of daughter nuclei, but the significance of this difference is uncertain. Differences in NAO type and behavior, however, are certain to be phylogenetically important.

Except for *P. purpureum* (Schornstein and Scott 1980, 1982), all NAOs in red algae are ring-shaped organelles of variable diameter (Scott and Broadwater 1990; Patrone *et al.* 1991; Scott *et al.* 1992) (Table 1). In *P. purpureum*, the prophase NAO is bipartite with a small, distal (50 x 70 nm) and large, proximal (55 x 140 nm) portion (Fig. 16); in the four other unicells examined, the prophase NAOs are also bipartite, but consist of two stacked, hollow cylinders with the longitudinal axis usually perpendicular to the nuclear envelope (Figs. 17). The largest of the NAOs observed in these species is approximately the size of the small portion of the *P. purpureum* NAO. Even though cell division in *G. vacuolata* is very similar to that described for red algal unicells, no NAOs have yet been found despite extensive investigation (Figs. 18–19).

NAO behavior is also quite different in *P. purpureum* compared with other species (Schornstein and Scott 1980, 1982; Scott and Broadwater 1990; Patrone *et al.* 1991; Scott *et al.* 1992). In *P. purpureum*, the large proximal portion disappears at prometaphase, concomitant with the breakdown of the nuclear envelope at the poles. The smaller distal portion then moves to the nucleoplasm-cytoplasm interphase (Fig. 14) where it remains until late telophase. In the other four unicells, the NAOs behave similar to each other but unlike the NAOs of *P. purpureum*. In very early prophase the NAOs appear close to or attached to the nuclear envelope. From midprophase through prometaphase, the

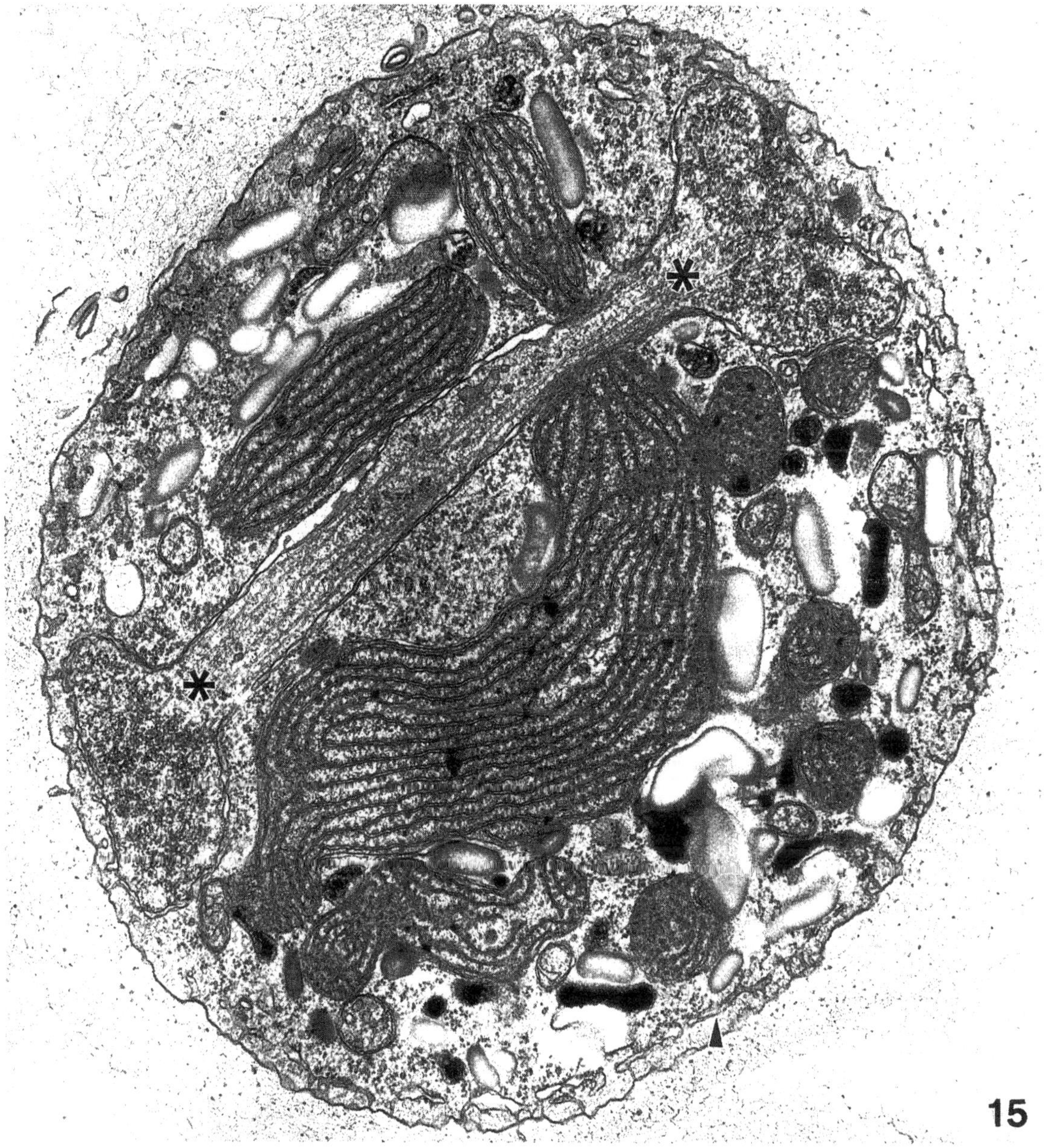

Fig. 15. Anaphase in unicellular red algae, demonstrated here by *P. purpureum*, results in an interzonal midpiece of varying lengths (delimited by asterisks). Arrowhead, peripheral ER system. 28,000 X. (Schornstein and Scott 1982, Scott and Broadwater 1990, with permission).

NAO is located a distance away from the nucleus in a ball of material of variable density. By metaphase the NAO moves to the nucleoplasm-cytoplasm interface, and the electron dense material flattens, partially plugging the gap.

At first analysis, the NAO of *P. purpureum* appears radically different from all other red algal NAOs. However, it is possible that the small distal portion of the *P. purpureum* NAO is analogous to other red algal NAOs, and the large proximal portion is analogous to the electron dense material that is thought to be microtubule organizing material. Although attractive, this hypothesis is not entirely satisfactory since the amount of electron dense material varies significantly and the NAO of *P. purpureum* is associated with some material prior to the disappearance of its proximal portion; also the distal portion is not a hollow cylinder. For now,

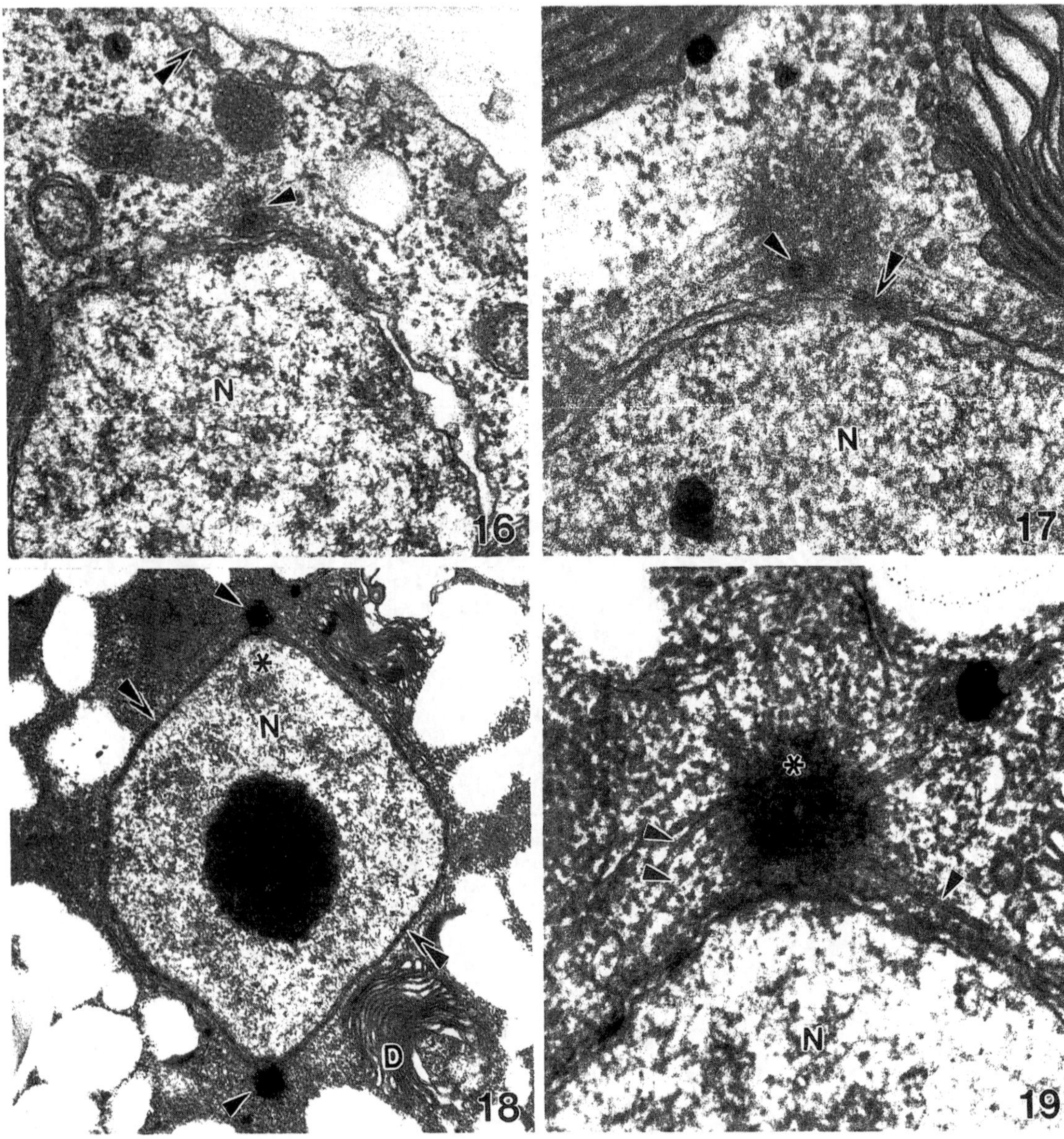

Fig. 16–19. (16) The NAO of P. purpureum (arrowhead) consists of a large amorphous proximal portion and a distal portion which is smaller, more compact, and disc-like. At prometaphase the proximal portion disappears and the distal portion moves to the nucleoplasm-cytoplasm interface (Fig. 14). N, nucleus; Double arrowhead, peripheral ER. 52,500 X. (Schornstein and Scott 1982, Scott and Broadwater 1990, with permission); (17) The NAO of *D. grisea* (arrowhead) has the shape of two rings stacked on top of each other and is similar in morphology to the NAOs of almost all red algae. During mitosis, NAOs are surrounded by electron-dense material which differs in density from one stage to another and from one alga to another. N, nucleus; Double arrowhead, nuclear pore. 79,500 X. (Scott and Broadwater 1990, Scott *et al.* 1992, with permission). (18) Mitosis in *G. vacuolata*, exemplified here by prophase, differs little from that of red algal unicells examined. The electron-dense material at the spindle poles (arrowheads) is very prominent and undergoes morphological changes during mitosis most similar to those observed in *D. grisea*. A major difference is the lack of a NAO in the former. D, dictyosome; N, nucleus; Double arrowheads, nuclear pores; Asterisk, spindle pole enlarged in figure 19. 15,850 X. (Courtesy of S. Goss); (19) In the enlarged spindle pole region of *G. vacuolata* from figure 18, the electron-dense material (asterisk) is thought to be a microtubule organizing center (MTOC). Microtubules (arrowheads) radiate from this region, but no NAO is visible in this or adjacent sections. N, nucleus. 67,200 X. (Courtesy of S. Goss).

one must accept two very different NAO types within the Rhodophyta and even the absence of an NAO if *G. vacuolata* is accepted as a red alga. Unfortunately, neither cell division nor NAOs have been seen in the thermophilic algae. Finding NAOs of either type would be strong evidence for the inclusion of these algae in the Rhodophyta.

4. Summary

The ultrastructural evidence strongly supports the polyphyletic nature of the Porphyridiales as well as the inclusion of *G. vacuolata* and the thermophilic Cyanidiophyceae in the Rhodophyta. Neither *G. vacuolata* or the thermophilic algae, however, appear closely related to the other unicells discussed, which is not surprising in light of the heterogeneity of this group of organisms. The thermophilic algae, in particular, have the potential to show some unique ultrastructural characters as evidenced by the 'plastid-dividing ring'. Further ultrastructural evaluation of these problematic algae as well as studies of red algal unicells not yet examined are necessary in order to get a fuller understanding of the phylogeny and taxonomy of this group of organisms.

Acknowledgements

We wish to thank Bill Saunders and Jewel Thomas for their help in the preparation of this manuscript and also Steve Goss, Jewel Thomas and Bill Saunders for use of their micrographs. Financial support was provided by NSF grant BSR 90-08078.

References

Bailey, R. W. and Staehelin, L. A. (1968). *J. Gen. Microbiol.* 54: 269–276.

Barnes, K. L., Craigie, R. A., Cattini, P. A. and Cavalier-Smith, T. (1982). *J. Cell Sci.* 57: 151–160.

Bhattacharya, D., Elwood, H. J., Goff, L. J. and Sogin, M. L. (1990). *J. Phycol.* 26: 181–186.

Billard, C. and Fresnel, J. (1986). *C. R. Acad. Sc. Paris* 302: 271–276.

Bold, H. C. and Wynne, M. J. (1985). *Introduction to the Algae. Prentice Hall, Inc. Englewood Cliffs,* N.J. 720p.

Broadwater, S. and Scott, J. (1986). *J. Cell Sci.* 84: 213–219.

Broadwater, S. T., Scott, J. L. and Garbary, D. J. (1992). In: *The Cytoskeleton of the Algae (ed. Menzel, D.) CRC Press, Boca Raton, Florida.* pp. 93–112.

Broadwater, S., Scott J. and Pobiner, B. (1986a). *J. Phycol.* 22: 490–500.

Broadwater, S., Scott J. and Pobiner, B. (1986b). *J. Phycol.* 22: 501–512.

Brody, M. and Vatter, A. E. (1958). *J. Biophysic. and Biochem. Cytol.* 5: 289–294.

Bronchart, R. and Demoulin, V. (1977). *Nature* 268: 80–81.

Callow, M. E. and Evans, L. V. (1979). *Br. Phycol. J.* 14: 327–337.

Chapman, D. J. (1974). *Nova Hedw.* 25: 673–682.

Cole, K. M. (1990). In: *Biology of the Red Algae (eds. Cole, K. M. and Sheath, R. G.) Cambridge University Press, Cambridge.* pp. 73–101.

Cole, K. and Conway, E. (1975). *Phycologia* 14: 239–245.

Craigie, J. S. (1990). In: *Biology of the Red Algae (eds. Cole, K. M. and Sheath, R. G.) Cambridge University Press. Cambridge.* pp. 221–257.

Dassler, C. L., Scott, J. and Salisbury, J. L. (1990). *J. Cell Biol.* 111: 182.

Deason T. R., Butler, G. L. and Rhyne, C. (1983). *J. Phycol.* 19: 104–111.

De Luca, P. Gambardella. R. and Merola, A. (1979). *Bot. Gaz.* 140: 418–427.

Evans, L. V. (1970). *Br. Phycol. J.* 5: 1–13.

Evans, L. V., Callow, M. E., Percival, E. and Fareed, V. (1974). *J. Cell Sci.* 16: 1–21.

Ford, T. W. (1984). *Annals. Bot.* 53: 285–294.

Fresnel, J. and Billard, C. (1987). *Cryptogam. Algol.* 8: 49–60.

Fresnel, J., Billard, C., Hindák, F. and Pekárková, B. (1989). *Plant Syst. Evol.* 164: 253–262.

Gabrielson, P. W. and Garbary, D. (1986). *CRC Crit. Rev. Plant Sci.* 3: 325–366.

Gabrielson, P. W., Garbary, D. J. and Scagel, R. F. (1985). *Biosys.* 18: 335–346.

Gantt, E. and Conti, S. F. (1965). *J. Cell Biol.* 26: 365–381.

Gantt, E., Edwards, M. R. and Conti, S. F. (1968). *J. Phycol.* 4: 65–71.

Garbary, D. J. and Gabrielson, P. W. (1990). In: *Biology of the Red Algae (eds. Cole, K. M. and Sheath, R. G.) Cambridge University Press, Cambridge.* pp. 477–498.

Garbary, D. J., Hansen, G. I. and Scagel, R. F. (1980). *Nova Hedw.* 33: 145–166.

Garbary, D. J., McDonald, A. R. and Duckett, J. G. (1992). *New Phytol.* 120: 435–444.

Giraud, G. (1962). *J. Microscopie* 1: 251–274.

Giraud, G. (1963). *Physiol. Vég.* 1: 203–255.

Hara, Y., Nagumo, T. and Katou, S. (1989). In: *Recent Advances in Microbial Ecology (eds. Hattori, T., Ishida, Y., Maruyama, Y., Morita, R. and Uchida, A.) Proc. 5th Intern. Symp. Microbial Ecol.* pp. 292–296.

Hori, H. and Osawa, S. (1987). *Mol. Biol. Evol.* 4: 445–472.

Kuroiwa, T. (1989). *Bot. Mag. Tokyo* 102: 291–329.

Lee, R. E. (1974). *Br. Phycol. J.* 9: 291–295.

Lim, B. L., Kawai, H., Hori, H. and Osawa, S. (1986). Jap. J. Genet. 61: 169–176.

Lin, H., Sommerfeld, M. R. and Swafford, J. R. (1975). *J. Phycol.* 11: 452–457.
McCracken, D. A., Nadakavukaren, M. J. and Cain, J. R. (1980). *New Phytol.* 86: 39–44.
Melkonian, M. (1982). In: *Progress in Botany, Fortschritte der Botanik (eds. Ellenberg, K., Esser, K., Schnepf, E. and Ziegler, H.) Springer-Verlag, Berlin Heidlelberg.* pp. 315–344.
Merola, A., Castaldo, R., De Luca, P., Gambardella, R., Musacchio, A. and Taddei, R. (1981). *Giorn. Bot. Ital.* 115: 189–195.
Neushul, M. (1970). *Amer. J. Bot.* 57: 1231–1239.
Oakley, B. R. and Dodge, J. D. (1974). *Protoplasma* 80: 233–244.
Patrone, L., Broadwater, S. and Scott, J. (1991). *J. Phycol.* 27: 742–753.
Pekárková, B., Hindák, F. and Šmarda, J. (1988). *Arch. Protistenkd.* 135: 69–83.
Pekárková, B., Šmarda, J. and Hindák, F. (1989). *Pl. Syst. Evol.* 164: 263–72.
Pueschel, C. M. (1990). In: *Biology of the Red Algae (eds. Cole, K. M. and Sheath, R. G.) Cambridge University Press, Cambridge.* pp. 7–41.
Ramus, J. (1972). *J. Phycol.* 8: 97–111.
Ramus, J. and Robins, D. M. (1975). *J. Phycol.* 11: 70–74.
Richardson, F. L. and Brown, T. E. (1970). *J. Phycol.* 6: 165–171.
Schornstein, K. L. and Scott. J. (1980). *Nature* 283: 409–410.
Schornstein, K. L. and Scott. J. (1982). *Can. J. Bot.* 60: 85–97.
Scott, J. (1984). *J. Phycol.* 20: (suppl.), 6.
Scott, J. (1986). *Can. J. Bot.* 64: 516–524.
Scott, J. and Broadwater, S. (1990). In: *Biology of the Red Algae (eds. Cole, K. M. and Sheath, R. G.) Cambridge University Press, Cambridge.* pp. 123–145.
Scott, J., Broadwater, S., Saunders, B., Thomas, J. and Gabrielson, P. (1992) *J. Phycol.* 28: 649–660.
Scott, J. and Gabrielson, P. W. (1987). *J. Phycol.* 23: (suppl.), 12.
Seckbach, J. (1987). *Ann. N.Y. Acad. Sci.* 503: 424–448.
Seckbach, J. (1992). In: *Algae and Symbioses (ed. Reisser, W.) Biopress, Bristol, England.* pp. 399–426.
Seckbach, J. and Fredrick, J. F. (1980). *Microbios* 29: 135–147.
Seckbach, J., Fredrick, J. F. and Garbary, D. J. (1983). *Endocytobiology II (eds. Schenk, H. and Schwemmler, W.) Walter der Gruyter & Co. Berlin,* pp. 947–962.
Seckbach, J., Gonzalez, E. Wainwright, I. M. and Gross, W. (1992). *Nova Hedw.* 55: 99–109.
Sommerfeld, M. R. and Nichols, H. W. (1970). *J. Phycol.* 6: 67–78.
Trench, R. K. (1982). In: *Progress in Phycological Research, vol 1 (eds. Round, F. E. and Chapman, D. J.) Elsevier Biomedical Press, New York.* pp. 257–288.
Ueda, K. and Chida, Y. (1987). *Br. Phycol. J.* 22: 61–65.
Ueda, K. and Yokochi, J. (1981). *Bot. Mag. Tokyo* 94: 159–164.
Wanner, G. and Köst, H.-P. (1980). *Protoplasma* 102: 97–109.
Wehrmeyer, W. (1971). *Arch. Mikrobiol.* 75: 121–139.
Woelkerling, W. J. (1990). In: *Biology of the Red Algae (eds. Cole, K. M. and Sheath, R. G.) Cambridge University Press, Cambridge.* pp.1–6.

Chapter 18

Ultrastructure of cytoplasmic organelles in *Cyanidium caldarium*

KATSUMI UEDA
Biological Laboratory, Nara Women's University, 630 Nara, Japan

Outlines:
1. Introduction
2. Organelles in the cell
3. Chloroplasts
   3.1 Envelope membranes of the chloroplast
   3.2 Thylakoids and phycobilisomes
4. Plasma membrane
5. Golgi bodies and the trans Golgi network
6. Lysosomes
7. Possible taxonomic position of *Cyanidium caldarium*, as deduced from its ultrastructure
References

1. Introduction

The organelles of plants in each phylum have their specific ultrastructure as first suggested by Ueda (1961) who examined the structures of chloroplasts in 25 species from 9 phyla. He found that species that belong to a particular phylum have the same type of chloroplast in terms of the arrangement of thylakoids (Ueda 1961). Gibbs (1962) reported similar results from an examination of chloroplasts in 13 species from 9 phyla. Ueda (1961) pointed out a very exiting (and early) observation in the ultrastructural perspective of the chloroplast-like granules from *Glaucocystis nostochinearum* which may represent endosymbionts belonging to the Cyanophyceae. Thus, it became possible to classify an unspecified plant species into the appropriate group from the ultrastructure of its chloroplasts. Since then, many detailed studies on the ultrastructure of chloroplasts have been carried out in connection with taxonomic investigations (Dodge 1973; Mattox and Stewart 1984).

In addition to the arrangement of thylakoids, the envelopes of chloroplasts have also been recognized as important features for the classification of plant species (Gibbs 1970). The number of the envelope membranes ranges from two to four and are constant in species that belong to the same phylum. The structure of pyrenoid is also an important feature for classification. Dodge and Crawford (1971) described 5 types of pyrenoid in the Dinophyceae.

Structures of organelles other than chloroplasts may also serve as markers for the classification of plants. The presence or absence of scales (Manton 1966), the position of eye spots (Hibberd and Leedale 1972), and the arrangement of microtubules during mitosis (Pickett-Heaps 1972) have been discussed in relation to the classification of plants.

In this chapter, details of the ultrastructure of the cytoplasmic organelles in *Cyanidium caldarium* are described. They should serve as effective tools in attempts to resolve the taxonomic position of this enigmatic alga.

2. Organelles in the cell

The most suitable method of preparation for visualization of the ultrastructure of organelles in *Cyanidium* involves the freeze-substitution technique. Cells are rapidly frozen at −190 °C in cooled propane, ice in cells is replaced by acetone at −85 °C, and finally cells are embedded in Spurr resin (Ueda and Noguchi 1986). Distribution patterns of organelles are shown in Figures 1 (prepared by freeze-substitution) and 2 (fixed with potassium permanganate). Both figures show the chloroplasts at the periphery of the cells. The chloroplasts are situated under the cell wall remaining a narrow space between the cell wall and the chloroplasts. Seckbach (1987) and De Luca *et al.* (1979) reported that the chloroplasts of *C. caldarium* are cup-shaped. Three-dimensional reconstruction from serial sections would reveal the entire profile of the chloroplast.

J. Seckbach (ed.), Evolutionary Pathways and Enigmatic Algae: Cyanidium caldarium *(Rhodophyta) and Related Cells,* 231–238, 1994.

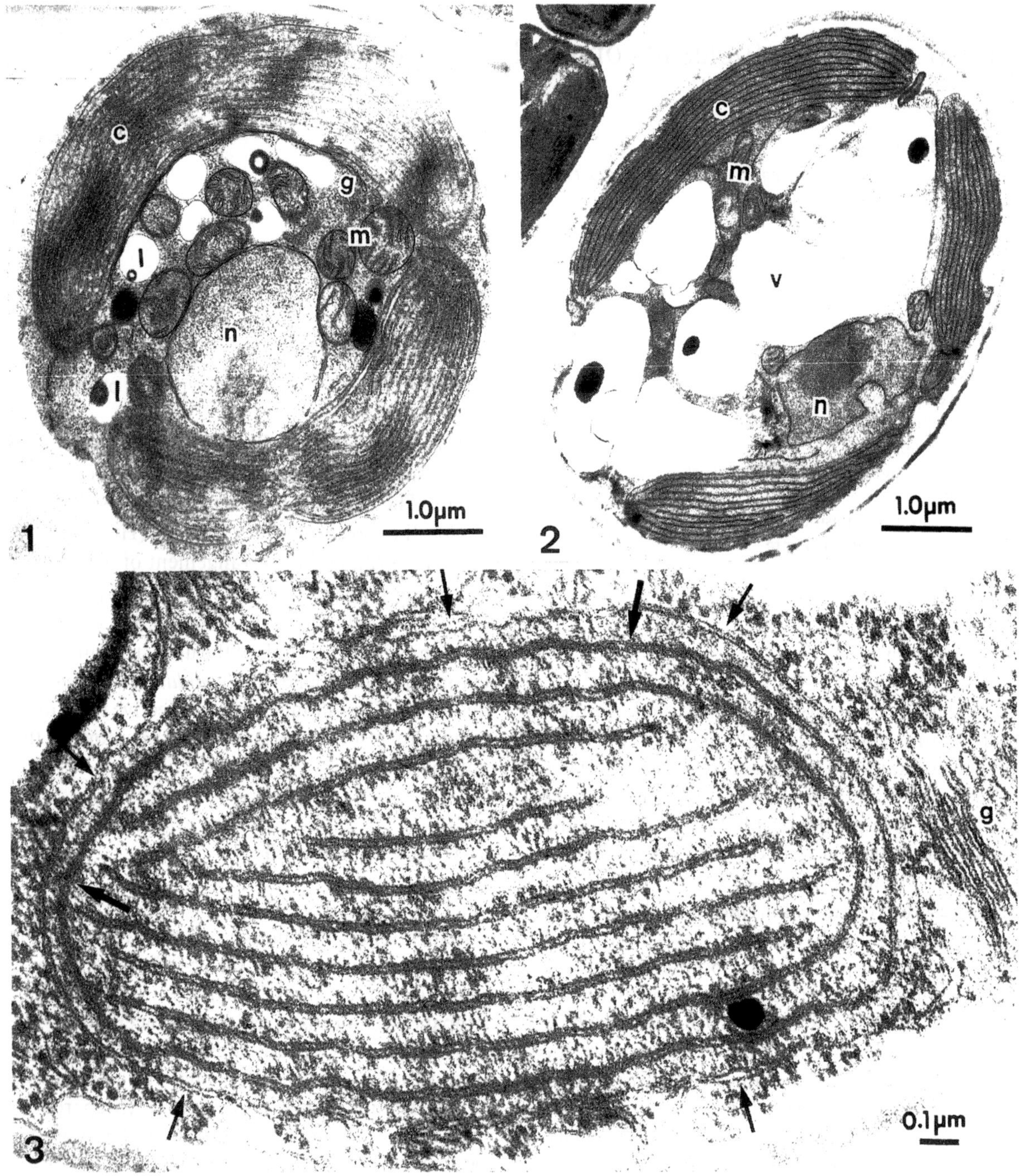

Fig. 1–3. (1) Freeze-substituted of *Cyanidium* cell. c = chloroplast, n = nucleus, m = mitochondria, l = lysosomes, g = Golgi bodies; (2) Potassium permanganate fixed cell. The outline of the nucleus (n) is distorted. c = chloroplast, m = mitochondria, v = vacuole; (3) Chloroplast with unstacked thylakoids which are surrounded by a girdle thylakoid (thick arrows). The double membrane of the envelope (thin arrows) is evident. Fixed with glutaraldehyde and osmium tetroxide. g = Golgi bodies.

A nucleus covered with a double nuclear membrane is located in the cytoplasmic cavity surrounded by the chloroplast ('n' in Figs. 1 and 2). In each nucleus there is a large nucleolus. Freeze-substituted nuclei are spherical or ovoid with smooth outlines, but nuclei in cells fixed with potassium permanganate have wavy outlines as repeatedly reported elsewhere (Seckbach 1987; De Luca *et al.* 1979). The wavy outlines may be an artifact derived from the originally smooth outlines, because cells and organelles inevitably shrink during the dehydration process, even after fixation with glutaraldehyde and osmium tetroxide (Ueda and Noguchi 1986). Accordingly, the shape of the freeze-substituted nucleus (Fig. 1, 'n') is assumed to be the real one.

Many profiles of mitochondria are also visible in the cytoplasmic area that is surrounded by the chloroplast. Several cristae are visible in the mitochondria. It has not yet been determined whether a single or many mitochondria are distributed within a single cell. Sectioning of a single undulating mitochondrion or a complicated network of mitochondria could give rise to many mitochondrial profiles as reported in *Euglena* (Pellegrini 1978) and in *Hydrodictyon* (Hatano and Ueda 1988). Images from serial sections should be reconstructed three-dimensionally to clarify the entire outline of the mitochondrion.

The small cells of *Cyanidium caldarium* RK-1 are different in shape and in the pattern of distribution of organelles (Mita and Kuroiwa 1988) from the cells that we have studies. The RK-1 strain was isolated by Nagashima and Fukuda (1981) and its morphological properties have been investigated in detail by Fukuda's group. Cells of this strain (i.e., RK-1) multiply by the formation of 4 autospores. Each cell contains a spherical to rod-shaped chloroplast without any lobes, a nucleus, and a single rod-shaped mitochondrion (Mita and Kuroiwa 1988). Seckbach (1987, 1991) suggested that *Cyanidium caldarium* with large cell size, which produces 4–32 autospores, should be classified as *Galdieria sulphuraria*. See chapters by Seckbach and Ott in this volume on the classification and taxonomy of the cyanidiophytes.

Examination of the mature cells reveals that they contain many vacuoles. A few vacuoles are large and occupy the central parts of the cells. Some of these vacuole-like inclusions are small and numerous (Fig. 1, 'I'). They seem to develop from the lysosomes. These small vacuoles may also resemble cytoplasmic storage granules in these cells (see Seckbach 1972).

Cisternae of the ER and the Golgi bodies are seen in the cytoplasm, and microbodies have also been reported (De Luca *et al.* 1979; Ueda and Yodochi 1981; see Seckbach *et al.* 1992).

3. Chloroplasts

3.1 Envelope membrane of the chloroplast

Rosen and Siegesmund (1961) were the first to examine the ultrastructure of chloroplast in *C. caldarium* and reported that the envelope of the chloroplast was a single membrane. This observation in *Cyanidium* has been further investigated and discussed because an organism with such a single membrane may provide evidence about the evolutionary origin of chloroplasts. Seckbach (1972, 1987), Seckbach and Fredirck (1980), Tôyama (1980) and De Luca *et al.* (1979) also reported the envelope to be a single membrane (see also Seckbach 1992). The *Cyanidium* cells that the previous investigators examined were fixed with potassium permanganate. In 1981, Ueda and Yokochi clearly demonstrated that a double membrane is enveloping the chloroplasts of *C. caldarium* cells that have been fixed with glutaraldehyde and osmium tetroxide. They also fixed samples of the same algal cells with potassium permanganate, and found that the typical double membranes of various organelles, such as chloroplasts, nuclei and mitochondria, appeared to exhibit just a single membrane. They interpreted their results as evidence that potassium permanganate destroys the outer membrane of the organelle bounded originally with a double membrane. Only the inner membrane is left and demonstrated in the electron micrograph. The presence of the double membrane of the envelope in this alga was subsequently confirmed by Ford (1984) who used glutaraldehyde and osmium tetroxide as fixatives, and by Ueda and Chida (1987) who used the freeze-substitution method.

Figure 3 shows a chloroplast that was fixed by the conventional method with glutaraldehyde and osmium tetroxide. The double membrane of the envelope is clearly visible (indicated by small arrows in Fig. 3). The thickness of the double membrane is about 20 nm and the thickness of each membrane is about 7 nm. Freeze-substituted chloroplast show the double membrane of the envelope more clearly, and the outer one is thicker than the inner one.

3.2 Thylakoids and phycobilisomes

Similar observations have been reported by many workers with respect to the arrangement of thylakoids in the chloroplasts of *Cyanidium caldarium*. Within the envelope, 7 to 12 thylakoids are usually seen. One and sometimes two or three girdle thylakoids enclose other thylakoids (larger arrow in Fig. 3). Thylakoids are never stacked together but are individually separated; they extend from one end to the other end of the chloroplast. In sections cut perpendicularly to the long axis of the chloroplasts, a concentric arrangement of thylakoids is visible. Taking this observation into consideration, we can interpret the three-dimensional arrangement of thylakoids in the chloroplast (Fig. 3) as follows: within the envelope, 5 cylindrical thylakoids are arranged concentrically, and the outermost cylinder of thylakoid is completely closed at both ends of the cylinder, inclosing the other 4 thylakoids. The next outer thylakoid is closed only at one end of the thylakoid, with the other end remaining open. the inner 3 cylindrical thylakoids are never closed at their terminations, they need not always form a cylinder and may be divided lengthwise in certain regions.

Thylakoids that are very well preserved by freeze-substitution reveal that their thickness is similar to that of the envelope, and each of their two constituent membranes appears to be triple-layered (Fig. 4). The unit membranes of various kinds of organelle are seen as triple layers, when they are prepared by freeze-substitution, because their lipid components, in the form of bimolecular leaflets, are dissolved by acetone and the proteins that are inserted into the lipid layers remain, after the disappearance of the lipids, as two electron-dense layers (Howard and Aist 1979; Ueda and Noguchi 1986).

Phycobilisomes are located on the outer surfaces of thylakoids (Fig. 4). They are usually destroyed by potassium permanganate and are easily dissolved out from chloroplasts during the dehydration process that follows glutaraldehyde fixation. Phycobilisomes are cylindrical in shape with a diameter of about 30 nm and a length of 50–500 nm. The cross-sectional images of phycobilisomes are not circular but rather they are star-shaped with several radiating arms. These images are quite in accord with the Gantt's scheme (1981) of phycobilisomes which is derived from biochemical studies. In this scheme there are central allophycocyanin moieties and the radiating phycocyanin and phycoerythrin moieties.

Phycobilosomes are anchored deeply in the membrane of the thylakoids. Electron-dense spots in the inner lipid layers of bimolecular leaflets may be images of such anchors. Thus, the anchors, namely, portions of phycobilisomes, appear to penetrate the outer lipid layer and are actually embedded in the inner lipid layer of the thylakoid membrane. Anchors of phycobilisomes have been reported by Ueda and Chida (1987). Mita *et al.* (1986) assumed such spots to be the connecting elements between the pair of phycobilisomes located on the opposite sides of the thylakoid membrane. The facts that phycobilisomes do not always appear in pairs but they always have the spots at feet suggest they may be the anchors in nature rather than the connecting elements.

4. Plasma membrane

Beneath the cell wall, there is a plasma membrane which surrounds the entire protoplasm. The plasma membrane invaginates into the cytoplasm in many places (Fig. 5). The invaginations are not very deep, usually less than 150 nm, and they never fuse with vesicles or ER. Sometimes, they are seen in contact with the envelopes of chloroplasts (Fig. 3, bottom).

5. Golgi bodies and the trans Golgi network

Golgi bodies were not be detected in early studies and it was assumes that they were absent in *Cyanidium caldarium* (Rosen and Siegesmund 1961; Mercer *et al.* 1962). De Luca *et al.* (1979) described the Golgi bodies in *C. caldarium* to be in lesser maturity of the cisternal stack, while Ueda

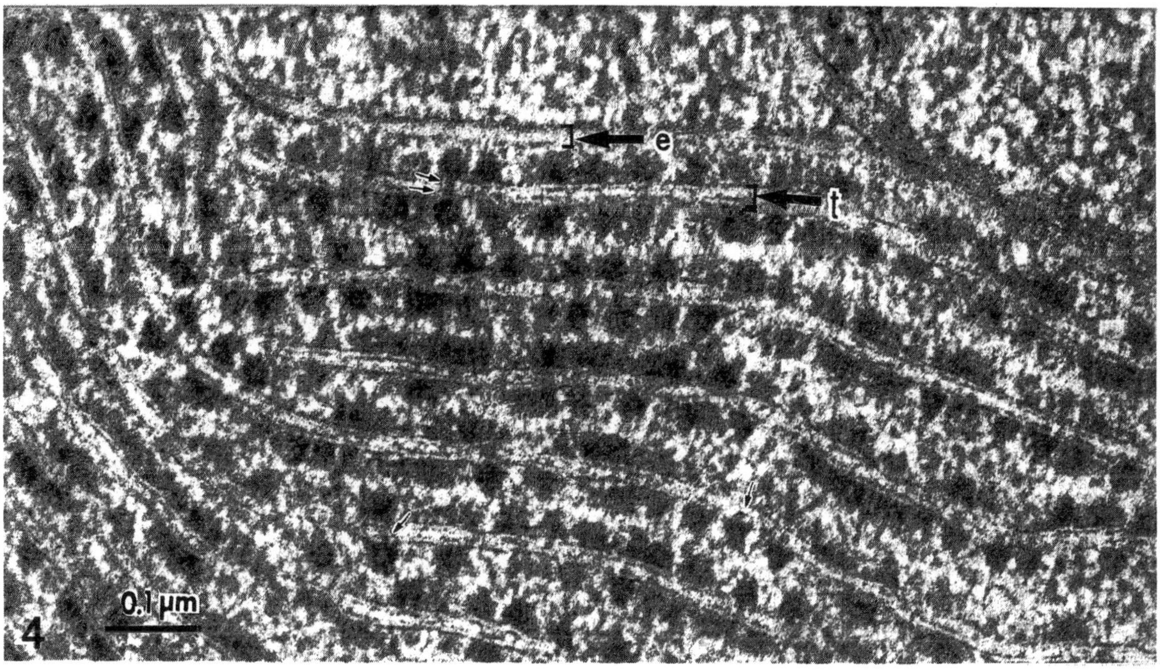

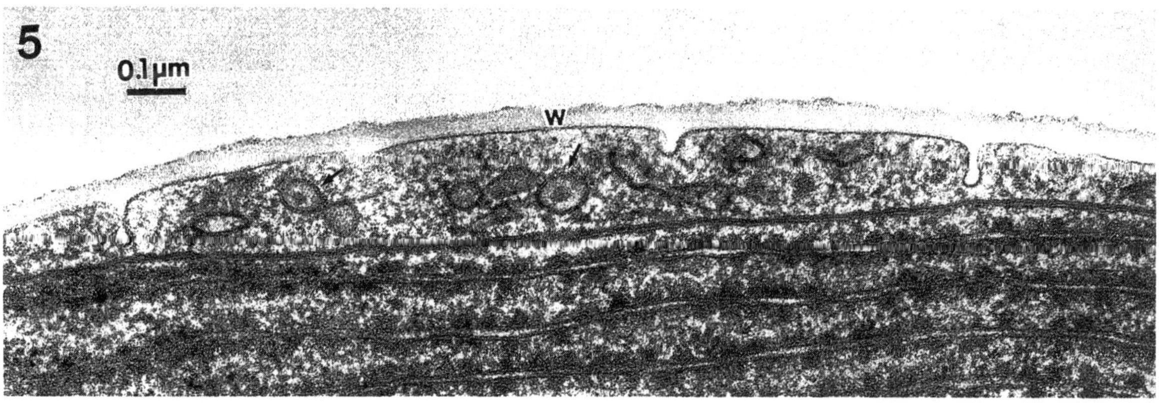

Fig. 4–5. (4) Part of a freeze-substituted chloroplast. The double-membrane envelope (e) is similar in thickness to the thylakoid membrane (t). Phycobilisomes are located on the outer surface of the thylakoids. The small arrows point to the anchors of phycobilisomes; (5) Invagination of the plasma membrane and the trans Golgi network with electron-dense inclusions (arrows). Freeze-substituted. w = cell wall.

and Yokochi (1981) found them to be as well-developed as those in other plant cells. Seckbach (1987) interpreted that *Galdieria sulphuraria* contains a primitive type of Golgi body.

Figures 6 and 7 show the Golgi bodies. A Golgi body is always accompanied by a cisterna of ER at the cis-face. This association was first reported in a green alga, *Chlorogonium elongatum* by Ueda (1966) and has now been widely recognized in many plant cells. Ribosomes are attached on the surface of the ER. The ER produces transitional vesicles of moderate electron-density near the cis-face of the dictyosomes (Figs. 6 and 7, small arrows). Inclusions of these vesicles may fuse to the cisternae of the Golgi body and be processed chemically in the cisternae. The cisterna of the cis-side labeled by the number 1 in Figure 7 is thick and the cisternae labeled 2 and 3 become thinner towards the trans-side. These characteristics are common features of the cisternae in Golgi bodies in green algae (Ueda and Noguchi 1986) and also in higher plants (Staehelin *et al.* 1990). No sign of immaturity in the Golgi bodies of *Cyanidium caldarium* can be recognized.

The cisterna labeled by the number 5 in Figure 7 is thicker that the cisternae labeled 1 to 4. This

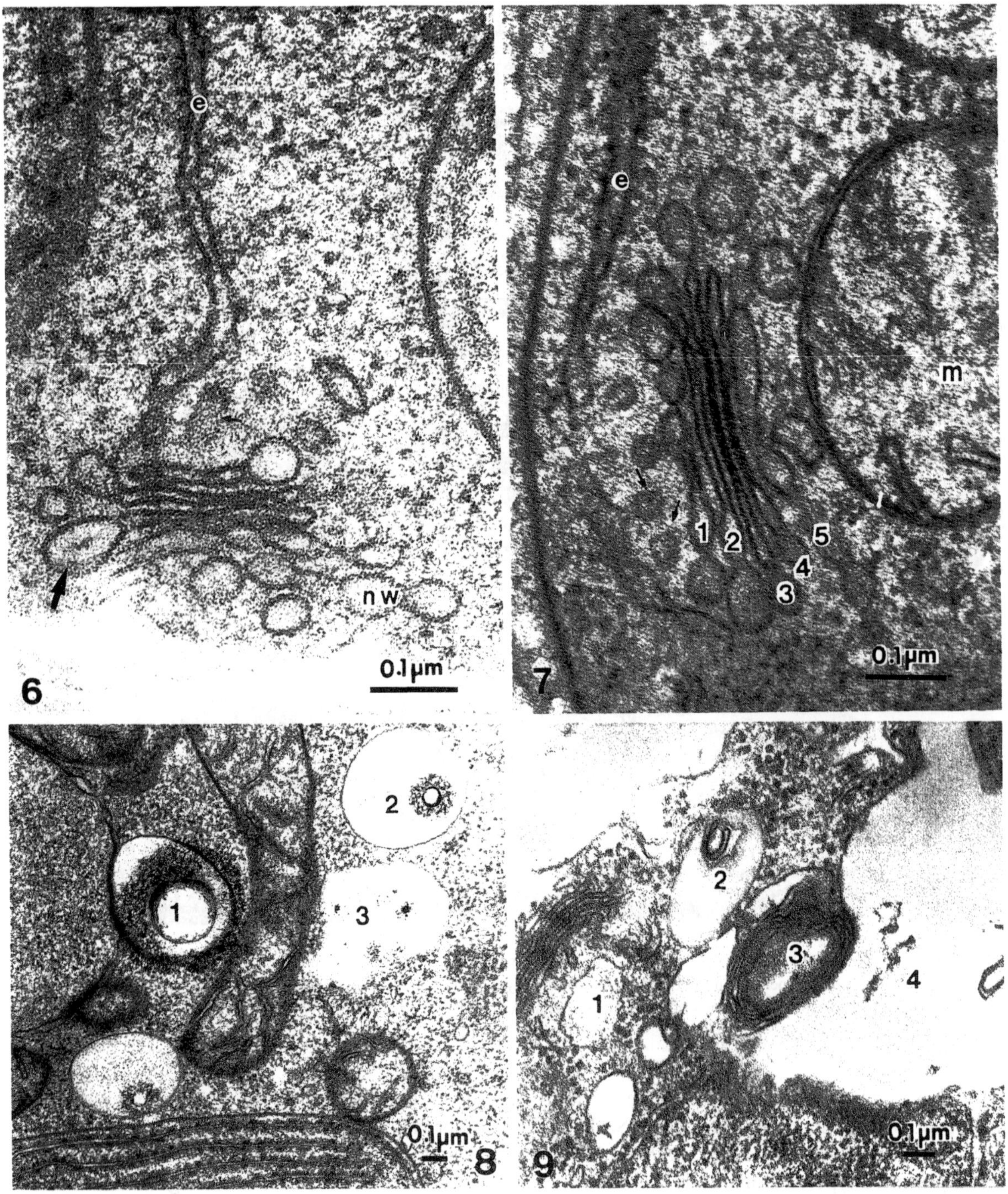

Fig. 6–9. (6–7) Golgi bodies accompanied by cisternae of the ER (e). Transitional vesicles (small arrows) are formed from the ER. The cisterna (1) at the cis-side is thick and those at the trans-side (3 and 4) are thin. Trans Golgi networks (nw and 5) are in close contact with the Golgi bodies. Freeze-substituted. m = mitochondrion; (8) Lysosomes (1,2 and 3) in a freeze-substituted cell; (9) Lysosomes (1,2,3, and 4) in a cell fixed with glutaraldehyde and osmium tetroxide.

cisterna does not belong to the Golgi body but to the trans Golgi network. This network expands from the trans-region of the Golgi body deeply into the cytoplasm. Flat vesicular bodies, seen near the trans-face of Golgi bodies, are often not actual vesicles but portions of the trans Golgi network ('nw', Fig. 6). Between the plasma membrane and the chloroplast, many vesicular bodies are distributed. They are not spherical but are elongated and frequently dumbbell-shaped and contain electron-dense granules (Fig. 5, arrows). They are probably portions of an expanded trans Golgi network. [See Griffiths and Simons (1987) for the function of trans Golgi network].

6. Lysosomes

Lysosomes are the organelles that contribute to the decomposition of small portions of cytoplasm by taking them up and digesting them via the action of endogenous enzymes (De Duve 1963). Golgi bodies produce vesicles that enlarge and become the vesicles of mature lysosomes. In plant cells, the lysosomes have not been as extensively studied as those in animal cells. Functional and morphological similarities between lysosomes and vacuoles have been demonstrated in yeast cells (Matile 1975; Holtzman 1975).

Figure 8 shows 4 vesicles, among which the vesicle labeled with the number 1 contains electron-dense substances and a spherical inclusion. Its appearance is typical of lysosomes. Other vesicles contain small inclusions and the vesicle labeled 3 closely resembles the small vacuolar vesicles. Lysosomes with inclusions are also shown in Figure 1 (labeled with the number 1). The process of transition from lysosomes to large vacuoles can be seen in Figures 2 and 9. The vesicle labeled 1 in Figure 9 is somewhat electron-dense and may be assumed to be a young lysosome. Vesicles labeled 2 and 3 contain membranous or myelin structures which may be derived from the incomplete decomposition of membranes of organelles. These inclusions will be taken up finally by the large vacuoles.

7. Possible taxonomic position of *Cyanidium caldarium*, as deduced from its ultrastructure

As described above, chloroplasts in *C. caldarium* are enveloped with double membranes and contain single running unstacked thylakoids. This characteristic excludes the alga from the following algal divisions: The Chlorophyta which possess a grana structure; the Euglenophyta, the Phaeophyta, the Pyrrophyta, the Eustigmatophyta, the Chrysophyta and the Xanthophyta, all those have membranes composed of 3 stacked thylakoids; the Prasinophyta with scales, and the Haptophyta with haptonema. The distribution of phycobilisomes on the outer surfaces of thylakoids excludes *C. caldarium* from the Cryptophyta members which possess phycobilisomes within their thylakoids. The presence of a nucleus indicates that *Cyanidium* cannot be a cyanobacterium. The phylum that remains is the Rhodophyta. Double membrane of chloroplast envelope, unstacked thylakoids, and the occurrence of girdle thylakoids are the characteristics that are commonly found in cells of Rhodophyta (see chapter of Broadwater and Scott on the ultrastructure of red algae, this volume). Therefore, *C. caldarium* should be classified into the Rhodophyta.

Does *C. caldarium* belong to the group of primitive Rhodophyta? There is no evidence to suggest that its chloroplasts are primitive in terms of ultrastructure. The chloroplasts of this alga do not contain pyrenoids. Pyrenoids are generally included in cells of primitive members of the Rhodophyta and are not included in cells of the advanced members of the Rhodophyta (Hara and Chihara 1974). The construction of Golgi bodies and their accompanying structures in *C. caldarium* also exclude the possibility that they are primitive in terms of structure. They are quite similar to those seen in green algae and in higher plants. Lysosomes and vacuoles in this alga are not very different in structure from those in other plants. In conclusion, no evidence of a primitive nature can be found in the ultrastructure of cytoplasmic organelles of *Cyanidium caldarium*.

References

De Duve, C. (1963). In: *Ciba foundation Symposium on Lysosomes (eds. de Renck, A.V.S. A. and Cameron, M.P.) Little Brown, Boston.* pp 1–28.

De Luca, P., Gambardella, R. and Merola, A. (1979). *Bot. Gaz.* 140: 418–427.

Dodge, J.D. (1973). *The Structure of Algal Cells. Academic Press, London, New York.*

Dodge, J.D. and Crawford, R. M. (1971). *Bot. J. Linnean Soc.* 64: 104–115.

Ford, T. W. (1984). *Ann. Bot.* 53: 285–294.

Gantt, E. (1981). *Ann. Rev. Plant Physiol.* 32: 327–347.

Gibbs, S.P. (1962). *J. Ultrastr. Res.* 7: 418–135.

Gibbs, S.P. (1970). *Ann. New York Acad. Sci.* 175: 454–473.

Griffiths, G. and Simons, K. (1986). *Science* 234: 438–443.

Hara, Y. and Chihara, M.(1974). *Sci. Reports Tokyo Kyoiku Daigaku Ser. B.* 15: 209–236.

Hatano, K. and Ueda, K. (1988). *European J. Cell Biol.* 47: 193–197.

Hibberd, D. J. and Leedale, G. F. (1972). *Annals Bot.* 36: 49–71.

Holtzman, E. (1975). *Lysosomes: A survey. Cell Biology monographs, vol. 3. Springer-Verlag, Wien, New York.*

Howard, R.J. and Aist, J.R. (1978). *J. Ultrastr. Res.* 66: 224–234.

Manton, I. (1966). *J. Cell Sci.* 1: 429–438.

Matile, Ph. (1975). *The Lytic Compartment of Plant Cells. Cell Biol. Monographs vol. 1. Springer-Verlag, Wien, New York.*

Mattox, K.R. and Stewart, K.D. (1984). In: *The Systematics of the Green Algae (eds. Irvine, D. and John, D.) Acad. Press, New York.* pp. 29–72.

Mercer, F.V., Bogorad, L. and Mullens, R. (1962). *J. Cell Biol.* 13: 393–403.

Mita, T., Kanbe, T., Tanaka, T. and Kuroiwa, T. (1986). *Protoplasma* 130: 211–213.

Mita, T. and Kuroiwa, T. (1988). *Protoplasma* Suppl. 1: 133–152.

Nagashima, H. and Fukuda, I. (1981). *Jap. J. Phycology* 29: 237–242.

Pellegrini, M. (1980). *J. Cell Sci.* 46: 313–340.

Pickett-Heaps, J.D. (1972). *Cytobios* 5: 59–77.

Rosen, W.G. and Siegesmund, K.A. (1961). *J. Biophys. Biochem. Cytol.* 9: 910–914.

Seckbach, J. (1972). *Microbios* 5: 133–142.

Seckbach, J. (1987). *Ann. New York Acad. Sci.* 503: 424–437.

Seckbach, J. (1991). *J. Phycol.* 27: 794–796.

Seckbach, J. (1992). In: *Algal Symbioses (ed. Reisser, W) Biopress, UK.* pp. 399–426.

Seckbach, J. and Fredrick, J. F. (1980). *Microbios* 29: 135–147.

Seckbach, J., González, E., Wainwright, I. M. and Gross, W. (1992). *Nova Hedwig.* 55: 99–109.

Staehelin, L.A., Giddings, T.H., Kiss, J.Z. and Sack, F.D. (1990). *Protoplasma* 157: 75–91.

Tôyama, S. (1980). *Cytologia* 45: 779–790.

Ueda, K. (1961). *Cytologia* 26: 344–358.

Ueda, K. (1966). *Cytologia* 31: 461–472.

Ueda, K. and Yokochi, J. (1981). *Bot. Magz. Tokyo* 94: 159–164.

Ueda, K. and Noguchi, T. (1986). *J. Cell Sci.* 82: 217–222.

Ueda, K. and Chida, Y. (1987). *Brit. Phycol. J.* 22: 61–95.

Chapter 19

Cyanidium caldarium as a model cell for studying division of chloroplasts

TSUNEYOSHI KUROIWA, HARUKO KUROIWA, TAKASHI MITA and NIJI OHTA
Division of Developmental Biology, Department of Biology, Faculty of Science, University of Tokyo, Hongo 113, Japan

Abbreviations: cp, chloroplast; DAPI, 4′-6-diamidino-2-phenylindole; cp, chloroplast; ISC, initiation of synchronous culture; mit, mitochondria; SEG, semielectron dense granule; PD, plastid dividing; pp, proplastid; pt, plastid; VIMPICS, video-intensified photon counting system.

Outlines:

1. Introduction
2. The plastid nuclei of *C. caldarium* as simplest pt-nucleus among various plants
3. Organization of the plastid nucleus
4. Division of plastid
   4.1 Division of the plastid nucleus
   4.2 Plastidkinesis by the plastid-dividing ring
5. Summary
References

1. Introduction

During the past 15 years, there have been remarkable advances in molecular biology of the organization of the genes related to the chloroplast genome. Higher plant's chloroplast (cp-) DNA can be isolated as a covalently, closed circular molecule with a molecular mass, depending upon the species, of 85–95 x 10^3 kD (Kolodner and Tewari 1975). Denaturation mapping (Kolodner and Tewari 1975) and restriction endonuclease analysis (Bedbrook and Bogorad 1976) have shown that the majority of the circular molecules in the chloroplasts of a given species are identical in sequence. Then the physical maps of the cp-DNA from various plants such as *Zea may* (Bedbrook *et al.* 1977; Palmer 1985) and *Chlamydomonas reinhardtii* (Rochaix 1978) have been reported. Shinozaki *et al.* (1986) sequenced the entire cp-genome from the chloroplasts of *Nicotiana tabacum* and Ohyama *et al.* (1986) did the same for *Marchantia polymorpha*: the cp-genome of *N. tabacum* and *M. polymorpha* contain 155,844 base pairs and 121,024 base pairs, respectively. These DNAs contain up to 137 genes, which encode tRNA, rRNA (16S, 23S, 4.5S and 5S), and numerous proteins. It has been established that each the chloroplast contains a specific genome that is essential for the semi-autonomy of the organelle.

However, the following basic questions remain to be answered. How are the pt-DNAs organized into organelles? How are pt-chromosomes separated into daughter organelles? How does plastidkinesis occur? How are pt-DNAs transmitted to a cell's descendants? How many copies of pt-DNA does each plastid contain? What interactions are there among the cell-nuclear, mit- and pt-genome?

A DNA-binding fluorochrome, DAPI, which emits stronger fluorescence than do conventional fluorochromes, was synthesized by Dann *et al.* (1971) and applied to observations of organelle DNA (Williamson and Fenell 1975; James and Jope 1978; Coleman 1978; Kuroiwa *et al.* 1981). The use of DAPI combined with an epifluorescence microscopy, which was improved to generate strong fluorescence, makes it possible to visualize extremely low levels of DNA in mitochondria and plastids, and to analyze their behavior. The results of such analyses confirm strongly that cp- and mit-genome are not naked but are organized, with proteins, to form organelle nuclei (Kuroiwa 1982, 1991). Therefore, all of organelle DNA could be observed under DAPI-epifluorescence microscope.

J. Seckbach (ed.), Evolutionary Pathways and Enigmatic Algae: Cyanidium caldarium *(Rhodophyta) and Related Cells,* 239–253, 1994.

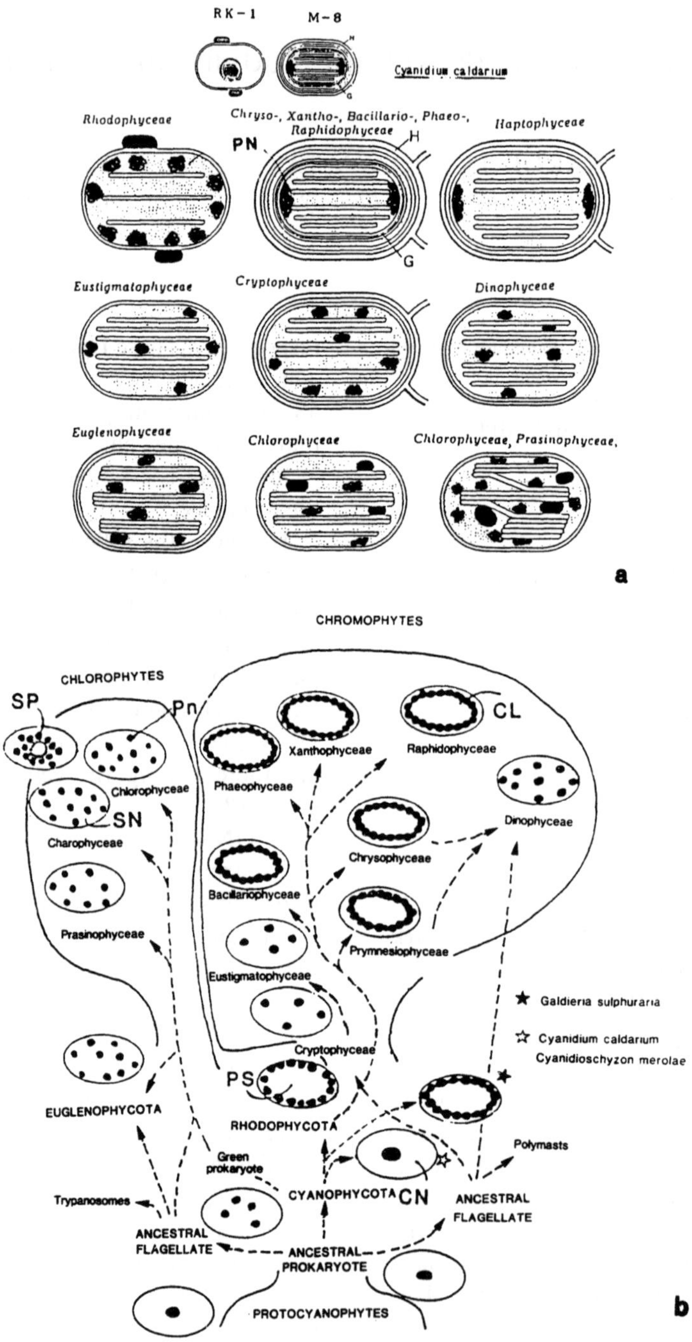

Fig. 1. Systematic representation (a) of the localization of pt-nuclei on the membrane systems of matured chloroplasts in various plants as published by Ettl (1980) and chloroplasts with pt-nuclei of the SN-type, CN-type, CL-type, PS-type or SP-type, on a generalized polygenic tree (b) showing hypothetical relationships among the classes of algae, as published by Dodge (1979). The differences in cp-nuclear patterns are explained by the distribution patterns of the replicated cp-genome in mature chloroplasts which are distributed randomly throughout the chloroplast (SN-type); gathered around the pyrenoid in the chloroplast (SP-type); distributed along the periphery of the chloroplasts (PS-type); fused to form spherules in the central area of the chloroplast (CN-type); or fused to form a circle along the periphery of the chloroplast (CL-type). H, lamellae; G, girdle lamellae; Pn, plastid nucleus. The CN-type can be found through the life cycle of the lower eukaryote, *C. caldarium* RK-1. The CL-type are observed in the red alga, *C. caldarium* M-8, and almost all brown algae. The SN-type is observed in the green algae, mosses, ferns and higher plants. It is very interesting that the CN-

2. The plastid nuclei of *Cyanidium caldarium* as simplest pt-nucleus among various plants

Ris and Plaut (1962) described DNA-like fibers within the chloroplasts of *C. reinhardtii*. These fine fibrils appeared in an electron-transparent area (nucleoid) of the chloroplasts after conventional fixation. Such DNA-containing regions have also been reported in the chloroplasts of many other plants (Kislev *et al.* 1965; Gunning 1965; Werz 1966; Yokomura 1967). However, even within a single species, such DNA-like fibers containing area is not always visible in all chloroplasts. The reason is that a very small amount of DNA is embedded in the semi-electron dense pt-matrix under standard physiological condition treated by conventional fixation techniques. Even when DNA-like fibers are observed in electron-transparent areas, the amounts of DNA fibers are very much lower than anticipated. Probably, some DNA-like fibers are embedded in somewhat electron-dense matrix around the electron-transparent regions. Therefore, it seems that the DNA-containing regions in the electron-transparent area (ETA) of plastids which are conventionally called 'plastid nucleoids', they do not contain the entire DNA of the genome. By contrast, after staining with DAPI, the fluorescence spots and intact isolated pt-DNA regions show all pt-chromosomes. Therefore, we consider these compact isolated structures found *in situ* as 'pt-nuclei' in view of their structural and functional integrity (Kuroiwa *et al.* 1981; Kuroiwa 1982, 1991; Nemoto *et al.* 1988). This was followed the analogy to 'bacterial nuclei' (Robinow 1956) and 'mitochondrial nuclei' (Kuroiwa *et al.* 1976; Kuroiwa 1982). Based on this definition, the nuclei are called 'cell nuclei'. Furthermore, DAPI epifluorescence microscopy and an apparatus that combines DAPI epifluorescence microscopy with conventional microscopic fluorimetry, or VIMPCS, made it very easy to observe the distribution or behavior of cp-DNA and to estimate extremely low levels of DNA per plastid.

Kuroiwa *et al.* (1981) examined the distribution and behavior of pt-nuclei during the development of chloroplasts and the cycle of plastid division in many plants by staining with DAPI. The number, size, shape and distribution of pt-nuclei were found to change markedly during chloroplast development and during the division of plastids, and these parameters differed among various plants. The patterns of distribution of pt-nuclei in mature chloroplasts were found to be unique to plant groups although they changed during the division cycle and the development of plastids (Kuroiwa *et al.* 1981; Coleman 1985). Most eukaryotic plants can be classified into five types according to differences in the shape, size and distribution of the cp-nuclei in their mature chloroplasts (Fig. 1a; Kuroiwa *et al.* 1981). The first, the SN-type (scattered pt-nuclei), is characterized by chloroplasts with small, uniformly dispersed cp-nuclei in the matrix between thylakoid membranes and/or grana. The land plants and algal groups, e.g., Chlorophyceae, Prasinophyceae, Charophyceae, Euglenophyceae, Cryptophyceae, Eustigmatophyceae and Dinophyceae, are of the SN-type. The second, the CN-type (central located pt-nuclei), is characterized by chloroplasts with one or a few cp-nuclei located in the central area surrounded by lamellae.

Cyanidium caldarium RK-1, which was reported to be a red alga, is a typical example of this type (Nagashima *et al.* 1984). The third type, the CL-type (circular pt-nucleus), has chloroplasts with a large, ring-shaped cp-nucleus inside the girdle lamellae. Such circular pt-nucleus isolated from the brown alga *Ectocarpus indicus* appears to be a chain of small spherical particles which may correspond to the small cp-nuclei of the SN-type (Kuroiwa and Suzuki 1981). Algae such as the Chrysophyceae, Xanthophyceae, Bacillariophyceae, Phaeophyceae, Rhaphidophyceae and Haptophyceae are of the CL-type. The shape of the circle changes with the shape of the chloroplast in different species. In some diatoms and brown algae, the chloroplasts are disc-shaped and the cp-nucleus is a ring; in other diatoms and in *C. caldarium* M-8, which has irregularly shaped chloroplasts, the pt-nucleus forms an irregular circle that lies along the periphery of the chloroplast

type, the CL-type and the SN-type appear in that order not only during growth of leaves in plants but also in that order during evolution of plants. The distribution pattern of the nucleus in prokaryote is very similar to that of the CN-type cp-nucleus in *C. caldarium* RK-1, but is not similar to that of the CL type cp-nucleus in *C. caldarium* M-8.

(Nagashima *et al.* 1986; Kuroiwa *et al.* 1989). The fourth type, the PS-type (peripherally scattered pt-nuclei), and the fifth type, the SP-type (spread pyrenoid pt-nucleus), are modifications of the CL-type and the SN-type, respectively. The PS-type is characterized by chloroplasts with pt-nuclei scattered along their peripheries beneath an inner limiting membrane. The rhodophycean algae, such as *Gelidium amansii* and *Symphyocladia latiuscula* are of this type. The last SP-type has numerous small pt-nuclei which form a shell around a pyrenoid in the chloroplast. The green alga *Bryopsis pulumos* is of this type.

It is very interesting in terms of systematic biology that the CN-type, which can be found through the life cycle of the lower eukaryote, *C. caldarium* RK-1, was also observed in prokaryotes and an undifferentiated proplastids in algae and higher plants (Fig. 1b; Kuroiwa *et al.* 1981). The CL-type, to which chloroplasts in the red alga, *C. caldarium* M-8, and almost all brown algae belong, can be observed in etioplasts before the development of chloroplasts in monocotyledons (Sellden and Leech 1981). The SN-type, to which chloroplasts in the green algae, mosses, ferns and higher plants belong, is observed in mature chloroplasts. It is very interesting that the CN-type, the CL-type and the SN-type appear in that order not only during growth of leaves in plants but also in that order during evolution of plants (Fig. 1b). Such diversity in the distribution, number, size and shape of pt-nuclei in various plants must depend basically on the development and evolution of the thylakoid membrane during the evolution of membrane systems and the development of chloroplasts (Kuroiwa 1982; Rose 1988). The distribution pattern of the nucleus in cyanobacteria is very similar to that of the CN-type cp-nucleus in *C. caldarium* RK-1, *C. caldarium* Form A and *Cyanidioschyzon merolae*, but is not similar to that of the CL type cp-nucleus in *C. caldarium* M-8 (Suzuki *et al.* 1992). The DNA contents per cell nucleus in *C. merolae*, *C. caldarium* RK-1, *C. caldarium* Form A and *C. caldarium* M-8 are 0.8 x 10^4 kbp, 1.3 x 10^4 kbp, 1.8 x 10^4 and 2.5 x 10^4, respectively (Ohta *et al.* 1992 and Suzuki *et al.* 1992). Therefore, *C. merolae* and *C. caldarium* RK-1 seem to be a prototype of *C. caldarium* M-8 (Fig. 2b).

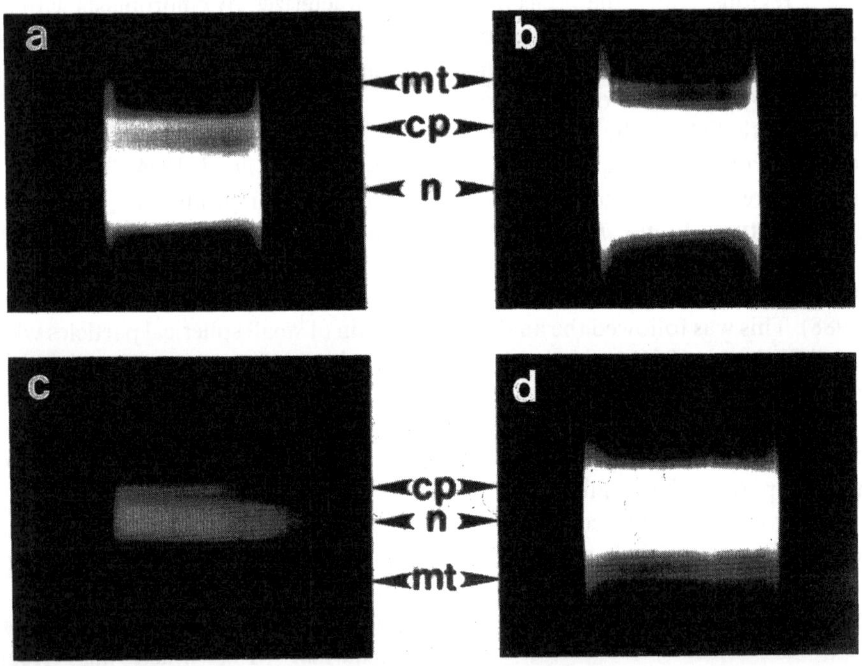

Fig. 2. Profiles after centrifugation on Hoechst 33258-CsCl density gradients of DNAs extracted from *C. caldarium* RK-1 (a, b) and *C. caldarium* M-8 (c, d). Pairs of photographs (a, b) and (c, d) are in the same fields. The buoyant densities of the cell nuclear DNA of RK-1, cp-DNA of RK-1, cell-nuclear DNA of M-8 and cp-DNA of M-8, are 1.715, 1.710, 1.698 and 1.688, respectively. mt = mitochondrial DNA; cp = chloroplast DNA ; n = cell-nuclear DNA.

Ohta *et al.* (1992) have shown a corroborate evidence about the difference of two strains that the GC contents of cell nuclear DNA and cp-DNA of *C. caldarium* RK-1 were 55.5% and 51.5%, respectively while those of *C. caldarium* M-8 were 39.3% and 29.0%, respectively (Fig. 2; Ohta *et al.* 1992). In addition, Suzuki *et al.* (1992) have isolated cell-nuclear DNA, cp-DNA and mit-DNA from *C. merolae*. The profiles of CsCl density gradient of these DNAs are not similar to those of *C. caldarium* M-8 but similar to those of *C. caldarium* RK-1 except the GC content in cp-DNA of *C. merolae* is slightly higher than that of *C. caldarium* RK-1. The base sequences of specific genes such as *rbcL*, *rbcS* and *trnK* in the cp-DNA in *C. caldarium* RK-1 (Ohta *et al.* 1992) are different from those isolated from *C. caldarium* M-8 and *C. caldarium* (Allen) (Valentin and Zetsche 1990). Recently, Seckbach (1991) has reconfirmed that *C. caldarium* M-8 (forma B) or Allen strain, is in fact the more advanced cyanidiophyte *Galdieria sulphuraria* (Merola *et al.* 1981; Nagashima and Fukuda 1981).

3. Organization of the plastid nucleus

As described above, pt-genome is located within a specific area of the plastids. It has been pointed out that the packing mode, that is, the dispersion and condensation of the cp-DNA, is intimately related to the photosynthetic oxygen-evolving activity of the cells in *Chlamydomonas* (Nakamura *et al.* 1986). Therefore, the three-dimensional structure of pt-nuclei may be important for understanding the function of pt-DNA during the division and differentiation of plastids. To elucidate the organizations of pt-nuclei, the intact pt-nuclei must be isolated from the plastids of various plants. Several groups have tried to isolate pt-nuclei from chloroplasts or chromoplasts (Kuroiwa and Suzuki 1981; Briat *et al.* 1982; Reiss and Link 1985; Hansmann *et al.* 1985). It is, however, difficult to purify the isolated pt-nuclei because thylakoid membrane systems in chloroplasts and in chromoplasts are highly developed. By contrast, proplastids, similarly to mitochondria, have very poorly developed membrane systems and they contain the smallest numbers of copies of pt-genome. Thus, they are useful for the isolation of intact pt-nuclei. Nemoto *et al.* (1988) developed a method for isolating morphologically intact proplastids and proplastid nuclei in large quantities from protoplasts (pp) of *N. tabacum* by a method that involves disruption of cells by forcing them through a layer of nylon mesh. The isolated pp-nuclei contain four polypeptides (69 kD, 31kD, 30 kD and 14 kD) which are bound to the pp-DNA. After treatment with proteinase K or deproteinization, a number of DNA fibrilar loops are released from the cp-nuclei but the circular DNA was rarely observed. Therefore, it appears that one to several copies of the pt-DNA molecule, (approximately 53 μm in length) are packed into a pt-nucleus of approximately 0.5 μm in diameter. From these results, it is concluded that the pp-DNAs are not naked *in situ* but are organized by interactions with some basic proteins to form compact structures called the 'pt-nuclei'. Fukuda and Nagashima (1987) first showed with VIMPCS that the cell nuclei in *C. caldarium* RK-1 and of *C. caldarium* M-8 emitted 1.93×10^6 and 1.85×10^6 photons, respectively and then reported that their strains emitted 1.93×10^6 and 3.79×10^6 photons, respectively (Table 1). The results also indicated that the amounts of DNA per cell nucleus and per chloroplast in *C. caldarium* RK-1 were approximately 1/4 times and 1/10 times, respectively, of the value (1.5×10^7bp) of the cell nucleus in *Saccharomyces cerevisiae* (Table 1). By contrast, Ohta *et al.* (1992) have shown with biochemical techniques that the percentages of DNA in the mitochondrial, chloroplast and the cell nuclear bands (of *C. caldarium* RK-1) were approximately 2%, 10% and 88%, respectively. The results do not conflict with the data using the fluorescent fluorimetry that the cell nucleus, chloroplast and mitochondrion of *C. caldarium* RK-1 during G1 phase contained DNA molecules of about 1.3×10^7 bp, 1.7×10^6 bp and 0.5×10^6 bp, respectively. Since the molecular size of chloroplast DNA in *C. caldarium* RK-1 is estimated approximately 1.5×10^5 bp by restriction enzymes (Ohta *et al.* 1994), the mt-DNA content per pt-nucleus is estimated to be approximately 11 molecules per cp-nucleus. These molecules seem to be organized to form the cp-nucleus by special proteins. The cp-nuclei are enveloped by thylakoid membranes.

Table 1. DNA photo-count (photons) and DNA content (nucleotides) per cell nucleus or chloroplast of *Cyanidium caldarium* RK-1 and *C. caldarium* M-8 and *Saccharomyces cerevisiae*

Experiments	*C. caldarium* RK-1:		*C. caldarium* M-8:	*S. cerevisiae*
	Cell nuclei:	Chloroplasts	Cell nuclei	Cell nuclei
1	1.93×10^6	-	1.85×10^6	-
2	1.93×10^6	7.28×10^5	$^*3.79 \times 10^6$	7.58×10^6
3	1.4×10^6	1.83×10^5	-	1.62×10^6
	(1.3×10^7bp)	(1.7×10^6bp)		(1.5×10^7bp)

1, Fukuda and Nagashima (1987); 2, Fukuda *et al.* 1990, * Seckbach *et al.* 1990; 3, Ohta *et al.* (1992).

4. Division of plastid

The concept of chloroplast nuclei has changed our ideas about the division of chloroplasts themselves. It seems clear that the process of chloroplast division must be composed of two main events: division of the cp-nucleus and plastidkinesis (division of the other components of the plastid). The latter term has been adopted as an appropriate analogue of cytokinesis.

4.1 Division of the plastid nucleus

During the mitotic cycle, the cell nucleus at the G1 phase enters into the S phase and duplicates its DNA. In general, when the cell volume becomes approximately double during metaphase, the cell divides into daughter cells with an equal division of chromosomes. Then the mitotic cycle is repeated with a doubling of both DNA and cell volume during the next cell cycle. Similar events may occur in the case of plastids and one might assume that a round of replication of cp-DNA occurs during each cycle of chloroplast division in higher plants (Possingham and Rose 1976; Tymms *et al.* 1983; Rose 1988). Since plastid divisions of the CL- and SN-types have been observed in multiplastidic cells of plants and are often not synchronized, it is difficult to study these division cycles in detail. By contrast, the typical CN-type of division can be observed in proplastids in the young thalli of algae, in dark-grown leaves, and in chloroplasts of *C. caldarium* RK-1. Especially, since *C. caldarium* RK-1 is composed of one cell nucleus, one mitochondrion and one chloroplast (Nagashima and Fukuda 1981) which contain DNA molecules of approximate 13 Mega base pairs, 0.43 Mega base pairs and 1.7 Mega base pairs (Ohta *et al.* 1992), it is very useful for analysis of plastid division cycles. The basic life cycle of *C. caldarium* RK-1 is shown in Figure 7. If we start with an examination of young cells, we see that each cell contains a cell nucleus of about 0.1 μm^3 in volume, a mitochondrion of about 0.06 μm^3 in volume and a chloroplast of about 1 μm^3 in volume. Growth of mother cells takes place for up to 50 h after the initiation of a synchronized culture (ISC) and these growing cells contain mostly four-endospore cells after the second endospore divisions. The divisions of the chloroplast, cell nucleus, mitochondrion and the cell itself occur in that order. In the case of *C. caldarium* M-8, the chloroplast divides just after the nuclear division. The cp-nucleus in *C. caldarium* RK-1 is located in the central area of the chloroplast; it increases in volume with the growth of the chloroplast and it divides just before chloroplast division. The levels of cp-DNA in the chloroplast increase soon after ISC and reach 4 times the initial value before the first division of the chloroplast. The amount of DNA in the chloroplast decreases stepwise after each endospore division, while the DNA in the cell nucleus is duplicated during each cycle of endospore division. The duration of the period of cp-DNA synthesis appears to be about 8 h and the DNA synthesis occurs between 6 h and 22 h after ISC.

On the basis of the observations and previous data, we compared the division cycle of the cell nucleus, with two distinguishing features of plastid division: one is an endoduplication of DNA which is greater than twofold increase in volume during a single plastid division cycle; the other is the presence of a division cycle without the synthesis of cp-DNA or increase in volume. Stepwise reduction in the volume and DNA content of plastids by division has been observed in the chloroplast of *C. caldarium* M-8 (Kuroiwa *et al.* 1989) and in the plastids of spermatocytes, during spermatogenesis, in the green alga *Bryopsis maxima* (Kuroiwa and Hori 1986), in *Chara australis* (Sun *et al.* 1988) and in the fern *Pteris vittata* (Kuroiwa *et al.* 1988), However, the molecular mechanism for separation of pt-chromosomes is unclear. Recently Rose (1988) summarized the literature as it relates to the role of plastid membranes in the segregation of pt-chromosomes and emphasized the role of the association between DNA and membranes in the segregation of pt-chromosomes. Pt- chromosomes are attached to the thylakoid membrane and, thus, can be separated equally into daughter plastids by an elongation of the binding sites that is accompanied by growth of the thylakoid membrane system. Considering various data, Rose (1988) proposed an interesting model in which segregation of pt-chromosome is related to thylakoid membrane systems in dividing chloroplasts, etioplast and proplastids with various types of cp-nucleus, e.g., the SN-and PS-types. However, there is no direct evidence showing such a specific association as is surmised to be essential for the segregation. Nemoto *et al.* (1990) recently obtained evidence that specific sites on the pt-chromosomes in two species of *N. tabacum* were bound to a complex that consists of thylakoid membrane and proteins. They showed, by various experiments, that at least four regions (the region between IRF 168 and *rpo B* in LSC; the region between *rps 16* and *psb A* in LSC; the regions between *rps 7* and *23 S* in inverted repeats) in the circular's genome are bound to membranes. The region containing *psa A* is stronger in terms of binding capacity than the other regions. However, when intact pp-nuclei isolated from tobacco cultured cells were treated with restriction enzymes, their fragment patterns were similar to those of purified DNA. In the chloroplasts of *C. caldarium* RK-1, there is, as yet, no direct evidence at the molecular level for an association between a specific part of the pt-chromosome and a membrane system that is a prerequisite for the separation of pt-chromosomes. If there is such an association, the weak binding at one or all of three binding regions may be related to the separation of pt-chromosomes.

4.2 Plastidkinesis by the plastid-dividing ring

We have been very interested to elucidate the structural basis of the mechanisms of organellekinesis. Since plastids are considerably larger than mitochondria and contain more highly developed membrane systems, a bulkier apparatus may be required for plastidkinesis than is required for mitochondriokinesis. It has proved advantageous for analyses of the apparatus involved in chloroplast division to use synchronous cultures of the monoplastidic cells of *C. caldarium* RK-1. Synchronous cultures of *C. caldarium* RK-1 can be initiated from young cells. The number of cells increases stepwise after ISC and finally reaches a value of about four times the initial number. Growth of mother cells takes place up to 50 h after ISC and these growing cells are mostly four-endospore cells after the second endospore divisions. In control preparations (Fig. 3), when the mother cells have been fixed at 36 h after ISC and excited with UV after staining with DAPI, four spherical cell-nuclei and four irregularly shaped cp-nuclei, emitting blue-white fluorescence, can be seen. When the same field of such cells has been excited with green light, four chloroplasts, emitting red auto-fluorescence, can clearly be seen in the areas where the ct-nuclei are located. When cells have been exposed throughout two sequential endospore divisions to cremart, an inhibitor of the assembly of tubulin, each mother cell can be seen to contain one cell-nucleus and four chloroplasts (Fig. 3). By contrast, when the cells have been exposed throughout two sequential endospore divisions to cytochalasin B, an inhibitor of the polymerization of actin into filaments, each mother cell can be seen to contain four cell nuclei and one large chloroplast (Fig. 3). At that time, the cytoplasmic membrane is not observed between the cell nuclei (Figs. 3 and 4). The results suggest that microtubules are not involved in the division of the chloroplast but are involved in the division of the cell nucleus. On the other hand, polymerized actin filaments are not involved in division of the cell nucleus but play an important role in the division of chloroplasts and in cytokinesis.

When cells of *C. caldarium* RK-1 have been stained with rhodamin-conjugated phalloidin, a fluorescent dye that is specific for actin, a ring that

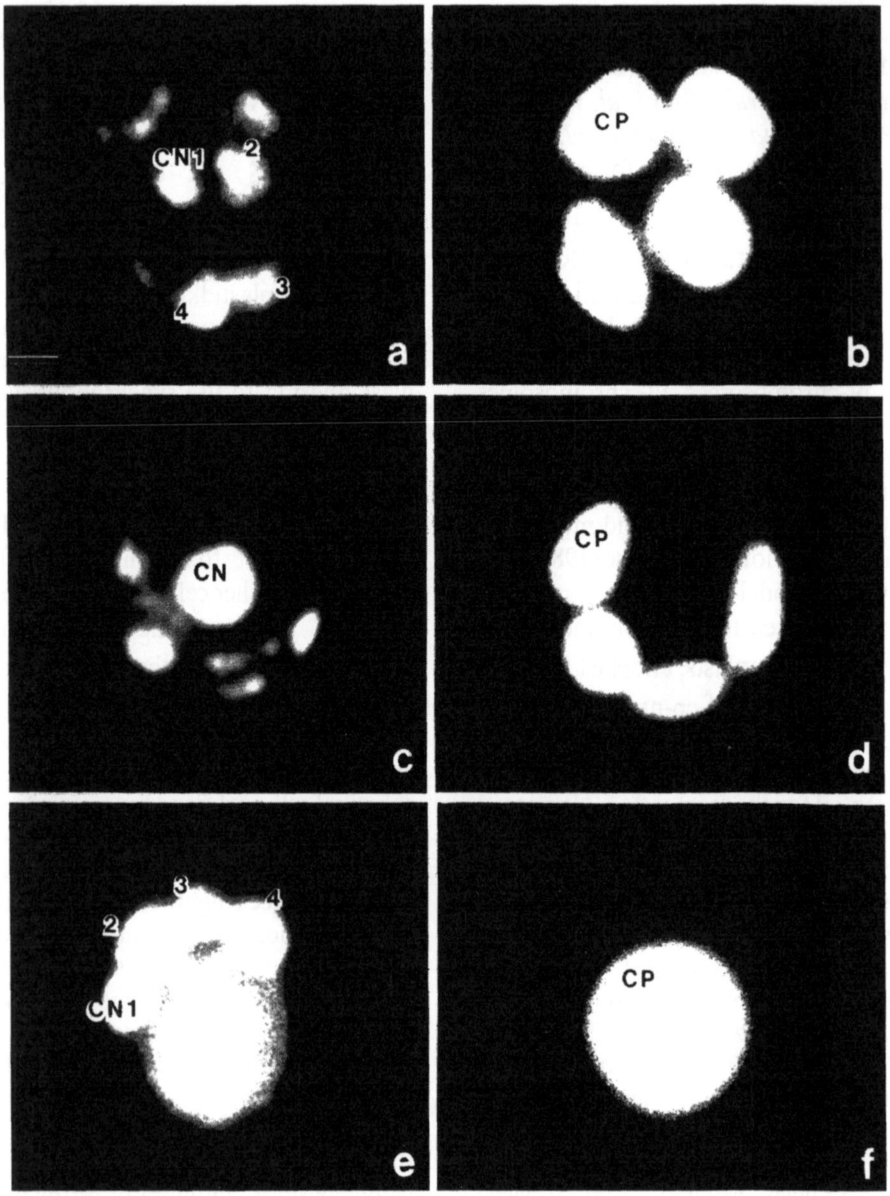

Fig. 3. Epifluorescence photomicrographs illustrating cell nuclei (CN1–4) and the chloroplast (CP) in mother cells *C. caldarium* RK-1 fixed immediately after incubation for 30 h without any inhibitor (a, b); and after incubation for 30 h with cremart at 10 μg/ml (c,d). Fluorescence photomicrographs with UV light (a, c, e) and green light (b, d, f) were taken in the same field for each treatment. In the control (a,b), one mother cell is composed of four discrete endospore cells, each of which contains a cell nucleus and a chloroplast. When cells have been treated with cremart through two sequential endospore divisions, the mother cell contains one cell nucleus and four chloroplasts (c, d). By contrast, when the cells have been treated with cytochalasin B, the mother cell contains four cell nuclei and one large chloroplast (e, f). Scale bar represents 1 μm. (Mita and Kuroiwa 1988; reporduced by permission of Springer-Verlag).

emits orange-colored fluorescence appears faintly in the equatorial region of the dividing chloroplast (Mita and Kuroiwa 1988). However, such a result does not entirely exclude the possibility that the ring may correspond to the cytoplasmic contractile ring involved in cytokinesis.

In synchronized cells of *C. caldarium* RK-1, it is possible to observe plastidkinetic events and fine structures in detail, from the formation of the PD ring during the early stages to the disappearance of

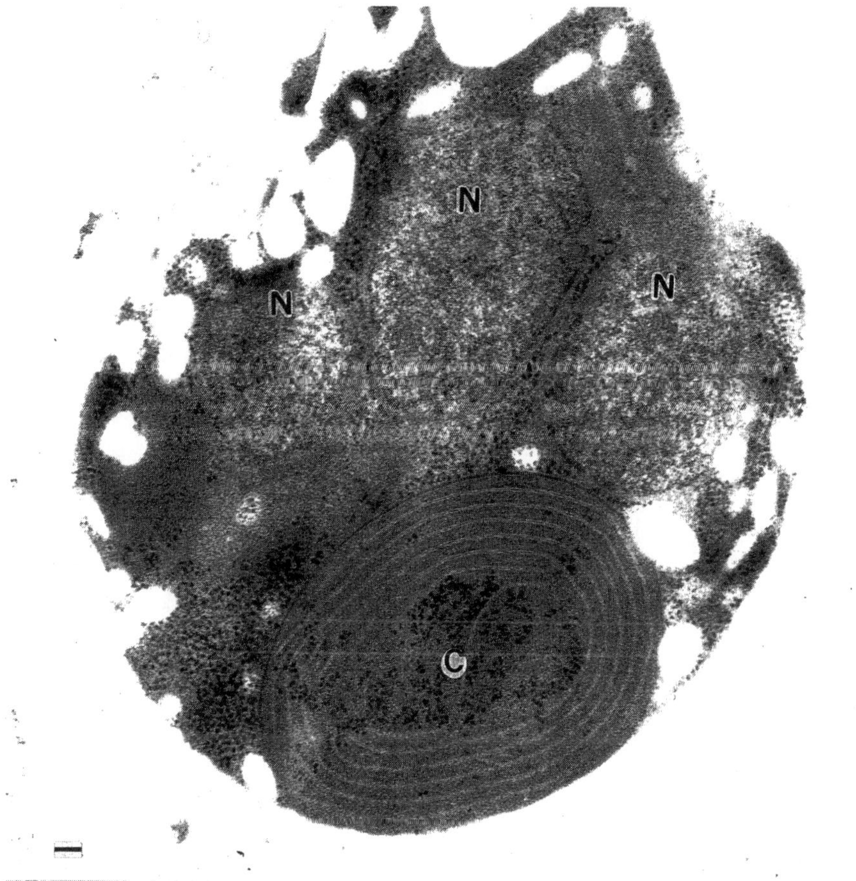

Fig. 4. An electron micrograph of the chloroplast (C) and three cell nuclei (N) in the cell after treatment with cytochalasin B at the concentration of 20 μg/ml. The division of cell nuclei is not inhibited but the division of the chloroplast and a cytokinesis are completely inhibited. Scale bar represents 0.1 μm.

the PD ring during the late stage, as summarized in Fig. 7. When the cell, the cell nucleus, and the chloroplast increase in volume about 3-fold, 2-fold and 3.5 fold, respectively, the shape of the chloroplast changes from a spherule to a football-like structure and concentric. Circular thylakoid membranes in the chloroplast begin to separate into two parts. At that time, many somewhat electron-dense granules (SEG), each 40–90 nm in diameter, and electron-dense deposits appear in the cytoplasm close to the outer envelope membrane and begin to be distributed at the equatorial region of chloroplast (Mita and Kuroiwa 1988). The PD ring is made up of SEG. A portion of the PD ring can be seen as a bar, about 60 nm in width, at the edge of the PD ring. The bar consists of fine filaments, each about 5 nm in diameter, which are aligned parallel to the longitudinal direction of the bar. The arrangement indicates that the PD ring is a bundle of fine filaments. By the time the PD ring starts to contract, the small SEGs have completely disappeared (Figs. 5 and 7). When sequential thin sections are cut through the constricted isthmus of a dividing chloroplast of the alga during the middle phase of division, it appears that the electron-dense deposits at the bridge between daughter chloroplasts are distributed as a close ring or belt-like structure, 60 nm wide and 50 nm thick, lying in close apposition to the outside of the outer envelope of the chloroplast. At the final stage, the width of the deposits that make up the constricted PD ring appears to be somewhat greater than that of the deposits at the early and middle stages of chloroplast division (Figs. 5 and 7). However, the width of the PD ring does not deviate very much from the cited value of 60 nm, regardless of the stage of division or the steepness of the walls of the furrow between the daughter chloroplasts, nor does it vary much among chloroplasts of vastly different volumes. At higher magnification, the cross

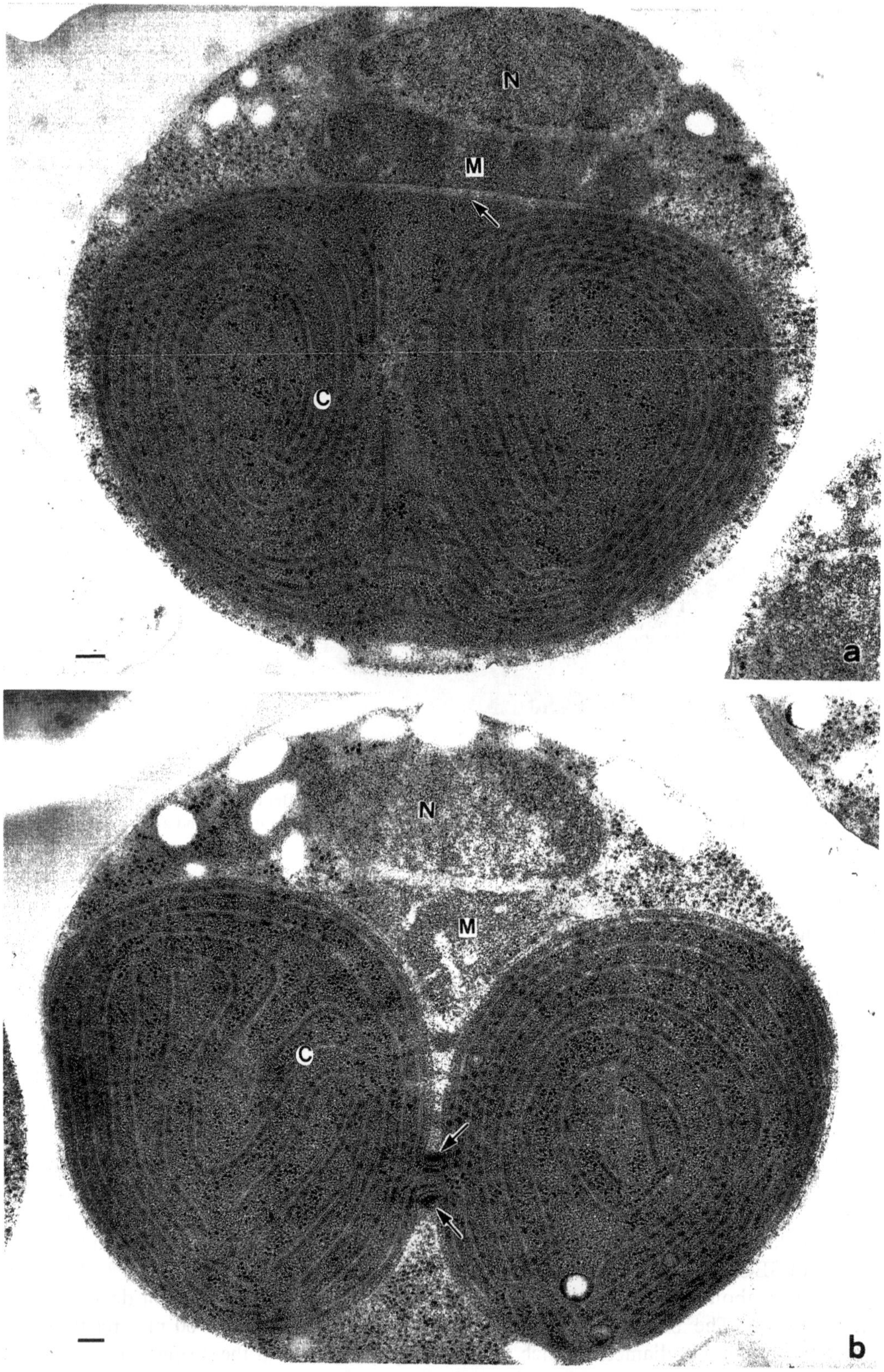
N
M
C
a
N
M
C
b

sections of the PD ring clearly reveal that the PD ring is located on the cytoplasmic side of the outer envelope. Inside the inner envelope, some electron-dense deposits can also be seen but their width and thickness do not change from the early stage of chloroplast division until it is complete.
Similar events can be seen in the second division of the chloroplasts (Figs. 6 and 7). The PD ring, when visualized in sections cut in a direction parallel to the plane of division, is clearly made up of a circular belt that appears to be composed of tightly packed, fine filaments. When the chloroplast has ceased dividing, a centriole-like plaque with microtubules develops outside the cell nucleus and electron-dense deposits, which consist of actin-like filaments and are related to cytokinesis, appear beneath the cell membrane.

Mita *et al.* (1986) and Mita and Kuroiwa (1988) proposed that the main components of the PD ring are actin-like filaments, on the basis of the following pieces of evidence. (i) Cytochlasin B, inhibits the division of chloroplasts without inhibiting division of the cell nucleus. (ii) The ring-like structure around the chloroplasts can be stained with rhodamin-conjugated phalloidin. (iii) Fine filaments, which are observed in the PD-ring, are very similar in diameter to actin filaments. However, such a proposal is not supported by the results of an immunogold staining experiments with actin-specific antibodies: a few gold particles, which showed the localization of actin, were found on the SEGs and none were found on the PD ring (Mita and Kuroiwa 1988).

A cytoplasmic PD ring, like that in *C. caldarium*, was also observed around the plastids of the green alga *Trebouxia* (Senda and Ueda, 1991), in *Pyraminomonas virginica* (Hori, personal communication), and in the moss *Funaria hygrometrica* (Tewinkel and Volkmann 1987) during the division of plastids, but no inner matrix ring was observed in any of these cases. In the green alga *Pyraminomonas virginica*, the division of the chloroplast with a large pyrenoid occurs after the cell-nucleus division and immediately before cytokinesis. The PD ring appears to generate the contractile force that is involved in the division of the chloroplast with a pyrenoid. Using a serial thin-sectioning technique, Hashimoto (1986) observed the presence of an electron-dense 'double' ring structure (PD ring doublet) around the constricted isthmus of dumbbell-shaped plastids of *Avena sativum*. The inner and outer rings of the doublet were reported to coat the inside (stromal side) of the inner envelope membrane and the outside (cytoplasmic side), respectively.

A similar PD ring doublet was reported in tobacco cultured cells (Kuroiwa 1989) and in root tips cells of spinach, bean, tobacco and wheat by Oross and Possingham (1989). Hori (1990, personal communication) observed a PD ring in plastids of the gymnosperm *Ginkgo biloba*. It is likely that the cytoplasmic PD ring is to be found in plastids of red, brown and green algae, in mosses and ferns, and in gymnosperm while the clear PD ring doublet is to be found in all angiosperms if a careful search is made. The PD ring doublet has also been seen in a mutant deficient of the monocot that is deficient in plastid ribosomes (Hashimoto and Possingham 1989). The observation does not conflict with the hypothesis that the PD ring consists of an actin-like protein, which must be encoded in the cell-nuclear chromosomes.

5. Summary

Our present understanding of the replication of plastids can be summarized as follows.

1. Most plants can be classified as being one of five types: the SN, CN, CL, PS or SP-type, based on differences in the shape, size and distribution of the cp-nuclei in their mature chloroplasts. The differences in patterns of chloroplast nuclei can be explained by the pattern of distribution of the replicated chloroplast genome in mature chloroplasts. They may be distributed randomly

Fig. 5. Electron micrographs of a cell nucleus (N), a chloroplast (C) and a mitochondrion (M), in sections cut perpendicular to the plane of division at the equatorial region (a) and the constricted isthmus (b) of the dividing chloroplast, in cells fixed 24 h (a) and 32 h (b) after the initiation of synchronous culture. When the chloroplast is progressively pinched, the PD ring becomes more-electron dense and is seen to increase in thickness (arrows in a and b). The PD rings can be seen clearly to be located on the cytoplasmic side of the outer envelope membrane. Scale bars represent 0.1 μm. (From Mita and Kuroiwa 1988; reproduced by permission of Springer Verlag.)

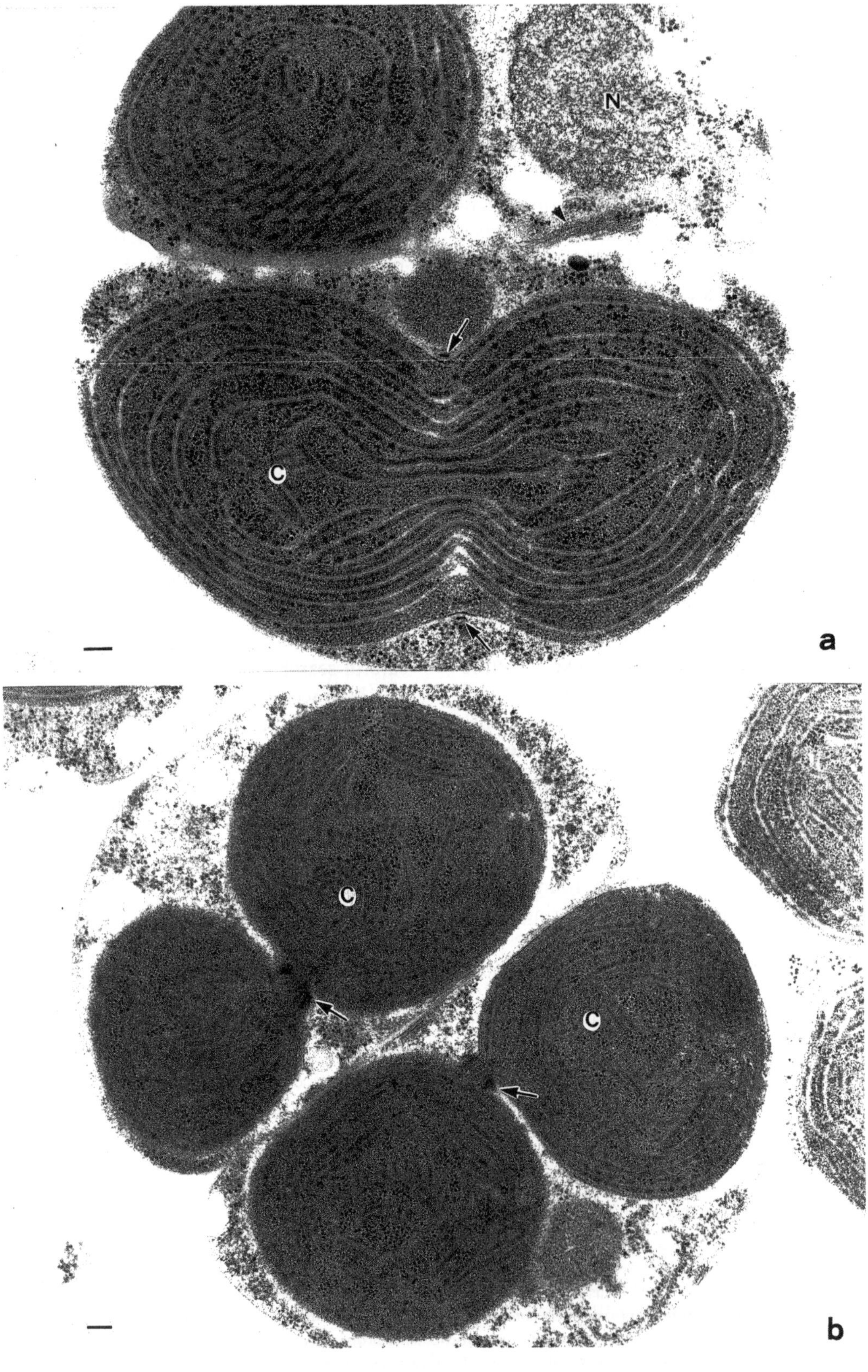
N
C
a
C
C
b

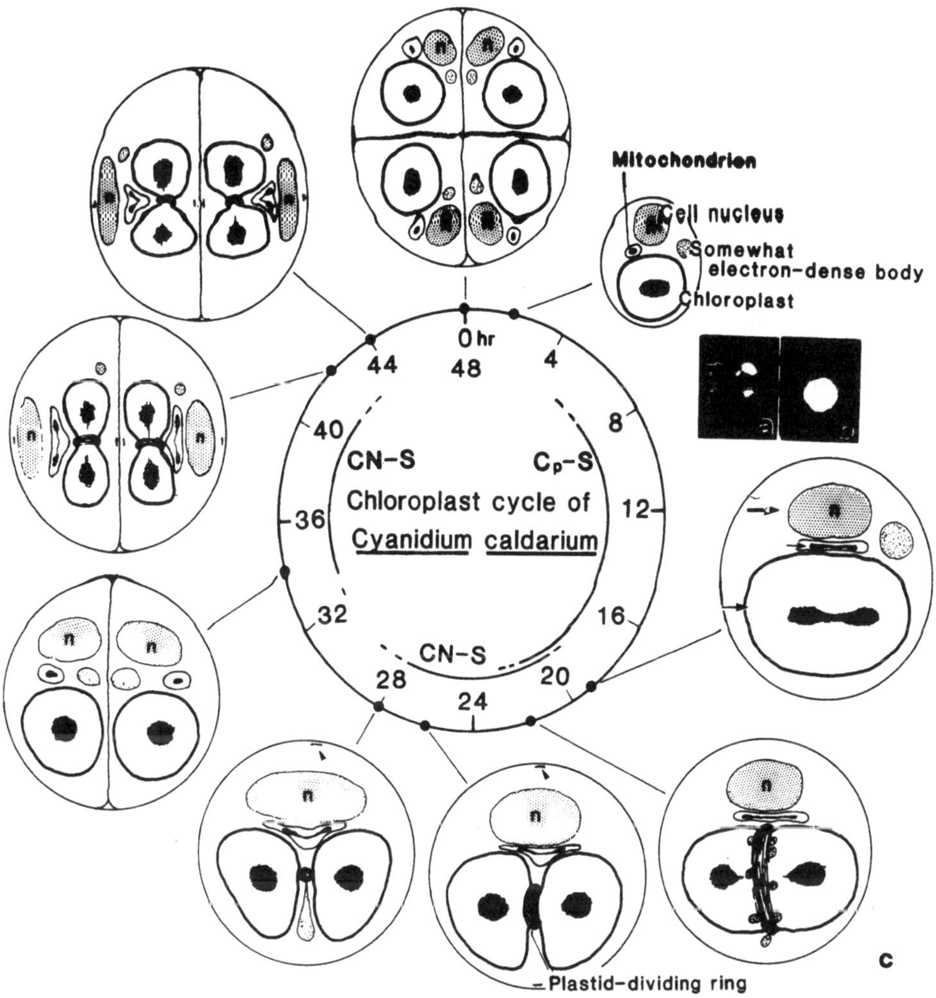

Fig. 7. Epifluorescence photomicrographs of a cell nucleus (large arrows in a, c), a mit-nucleus (small arrows in a, c) and a pt-nucleus (middle arrows in a, c) and a chloroplast (b) in a cell of the red alga *C. caldarium* RK-1 cell after staining with DAPI. a and b are the same field. (c): Diagram of the life cycle of the alga depicting a possible sequence of events based on the observations of cells fixed at various times (hours) after ISC. The divisions of the chloroplast, the mitochondrion and the cell nucleus occur in that order. The process of chloroplast division can be divided into the following four stages, on the basis of morphological and temporal differences, as follows. The chloroplast growth stage: the small, spherical chloroplast increases in volume and becomes a football-like structure. The stage at which formation of the plastid-dividing ring occurs: the somewhat electron-dense body (SEB) is fragmented into many small somewhat electron-dense granules (SEGs), which are aligned along the equatorial region of the chloroplasts and fine filaments are formed from the SEGs in the equatorial region of the chloroplasts. The fine filaments of the plastid-dividing ring align themselves according to the longest axis of their overall domain. Constriction stage: after chloroplast division, the remnants of the PD ring are converted into SEGs. The initial signals of cytokinesis appear beneath the cell membrane (arrowheads). Similar events occur during the second cycle of chloroplast division. The divisions of organelle nuclei occur just before organellkinesis. Cp-S and CN-S indicate the chloroplast DNA-synthetic stage and the cell-nuclear DNA synthetic stage, respectively. (Mita and Kuroiwa 1988; reproduced by permission of Springer-Verlag).

Fig. 6. Electron micrographs of the PD rings of chloroplasts (C) sectioned in directions perpendicular planes of the second division of the chloroplasts in two-endospore cells fixed 40 h (a) and 44 h (b) after the initiation of synchronous culture. The deposits of the PD ring appear on the constricted isthmus of dividing chloroplasts during the second chloroplast divisions (arrows). The width (30 nm) and thickness (15 nm) of the PD rings are about half as large as those seen at the first chloroplast division. However, as the division of chloroplasts progresses, the PD ring increases in both electron density and thickness (arrow in a, b). An arrowhead shows actin-like filaments. Bars represent 0.1 μm.

throughout the chloroplast (SN-type); gathered around the pyrenoid in the chloroplast (SP-type); distributed along the periphery of the chloroplast (PS-type); fused to form spherules in the central area of the chloroplast (CN-type) or fused to form a circle along the periphery of the chloroplast (CL-type). The chloroplast of *C. caldarium* RK-1 is belonged to the most prototype, CN type.

2. The chloroplasts in *C. caldarium* RK-1 contain 11 copies of their own unique circular genome of 1.5 x 10^5 base pairs. The genome is organized to form a proplastid nucleus approximately 0.5 μm in diameter in the central area of a proplastid by interactions with specific proteins.
3. The concept of organelle nuclei, as described above, has changed our ideas about the division of organelles. Thus, the process of organelle division must be composed of two main events: division of the organelle nucleus and organellekinesis (analogue of cytokinesis). The pt-nuclear division occurs after endoduplication of pt-DNA or in the absence of prior DNA synthesis. The weak association between pt-DNA and membrane systems which has not been clearly proven in proplastids. The strong binding between specific sites of pt-DNA and membrane systems is observed in chloroplasts and may be related to the dispersion of pt-chromosomes and the activation of photogenes after greening. The plastidkinesis is mediated by the PD ring in lower eukaryotes, in moss and in gymnosperm which is located on the cytoplasm outside plastids. In angiosperm, the plastid dividing ring is a doublet which is composed of an outer ring and inner ring. The PD ring is composed of a bundle of actin-like filaments. Plastidkinesis occurs by contraction of this bundle.

Acknowledgements

We wish to thank Dr. S. Kawano for his valuable advice and encouragement. This work was supported by grants nos. 05269205 and 05354002 from the Japanese Ministry of Education, Science and Culture and a grant pioneering research project in biotechnology from Ministry of Agriculture, Forestry and Fisheries of Japan.

References

Bedbrook, J. R. and Bogorad, L. (1976). *Proc. Natl. Acad. Sci. USA.* 73: 4309–4313.
Bedbrook, J. R., Kolodner, R. and Bogorad, L. (1977). *Cell* 11: 739–749.
Briat, J.-F., Gigot, C., Laulhere, J. P. and Mache, R. (1982). *Plant Physiol.* 69: 1205–1211.
Chida, Y. and Ueda, K. (1991) *Ann. Bot.* 67: 435–442.
Coleman, A. W. (1978). *Exp. Cell Res.* 114: 95–100.
Coleman, A. W. (1985). *J. Phycol.* 21: 1–16.
Dann, O., Bergen, G., Demant, E. and Volz, G. (1971). *Justus Liebigs Ann. Chem.* 749: 68.
Dodge, J. D. (1979). In: *Biochemistry and Physiology of Protozoa (eds. Hutner, S. H. and Levandowsky M.) Academic Press, New York.* pp. 7–57.
Ettle, H. (1980) In: *'Grundris' der allgeneubeb Algologie, Gustav, Fischer Verlag, Stuttgart.*
Fukuda, I. and Nagashima, H. (1987). *Ann. New York Acad. Sci.* 503: 575–577.
Fukuda, I., Nagashima, H. and Seckbach, J. (1989). In: *Endocytobiology 4 (eds. Nardon, P. et al.) iNRA, Paris.* pp. 585–588.
Gunning, B. E. S. (1965). *J. Cell Biol.* 24: 79–85.
Hansmann, P., Falk, H., Ronai, K. and Sitte, P. (1985). *Planta* 164: 459–472.
Hashimoto, H. (1986). *Protoplasma* 135: 166–172.
Hashimoto, H. and Possingham, J. V. (1989). *Plant Physiol.* 89: 1178–1183.
James, T. W. and Jope, C. (1978). *J. Cell Biol.* 79: 623–630.
Kislev, N., Swift, H., and Bogorad, L. (1965). *J. Cell Biol.* 25: 327–344.
Kolodner, R. and Tewari, K. K. (1975). *Biochim. Biophys. Acta* 402: 372–390.
Kuroiwa, H., Sugai, M. and Kuroiwa, T. (1988). *Protoplasma* 146: 89–100.
Kuroiwa, T. (1982).*Int Rev Cytol.* 75: 1–59.
Kuroiwa, T. (1989). *Bot. Mag. Tokyo* 102: 231–329.
Kuroiwa, T. (1991). *Int. Rev Cytol.* 128: 1–62.
Kuroiwa, T. and Hori, T. (1986). *Protoplasma* 133: 85–87.
Kuroiwa, T. and Suzuki, T. (1981). *Exp. Cell Res.* 134: 457–461.
Kuroiwa, T., Kawano, T. and Hizume, M. (1976). *Exp. Cell Res.* 97: 435–445.
Kuroiwa, T., Nagashima, H. and Fukuda, I. (1989). *Protoplasma* 149: 120–129.
Kuroiwa, T.,Suzuki, T., Ogawa, K. and Kawano, S. (1981). *Plant Cell Physiol.* 22: 322–396.
Merola, A., Castaldo, R., De Luca, P., Gambardella, R., Musacchio, A. and Taddei, R. (1981). *Giorn. Bot. Ital.* 115: 189–195.
Mita, T., Kanbe, T., Tanaka, K. and Kuroiwa, T. (1986). *Protoplasma* 130: 211–213.
Mita, T. and Kuroiwa, T. (1988). *Protoplasma suppl.* 1: 133–152.
Nagashima, H., Kuroiwa, I. and Fukuda, I. (1984). *Experientia* 40: 563–564.
Nagashima, H. and Fukuda, I. (1981). *Jpn. J. Phycol.* 29: 237–242.
Nagashima, H., Kuroiwa, T. and Fukuda, I. (1986). *Plant Cell Physiol.* 28, 315–321.

Nakamura, S., Itoh, S. and Kuroiwa, T. (1986). *Plant Cell Physiol.* 27: 775–784.

Nemoto, Y., Kawano, S., Nagata, T. and Kuroiwa, T. (1990). *Plant Cell Physiol.* 31: 768–776.

Nemoto, Y., Kawano, S., Nakamura, S., Mita, T., Nagata, T. and Kuroiwa, T. (1988). *Plant Cell Physiol.* 29: 167–177.

Ohta. N., Kawano, S. Nagashima, H. and Kuroiwa, T. (1992). *Plant Cell Physiol.* 35: 657–661.

Ohta, N., Kawano, S., and Kuroiwa, T. (1994) Curr. Genet. in press.

Ohyama, K., Fukuzawa, H., Kohchi, T., Shirai, H., Sano, T., Sano, S., Umesono, K., Shiki, Y., Takeuchi, M., Chang, Z., Aota, S., Inokuchi, H. and Izeki, H. (1986). *Nature* 322: 72–574.

Oross, J. W. and Possingham, J. V. (1989). *Protoplasma* 150: 131–138.

Palmer (1985). *Ann. Rev. Genet.* 18: 325–354.

Possingham, J. V. and Rose, R. J. (1976). *Proc. R. Soc. London Ser. B.* 193: 295–305.

Ris, H. and Plaut, W. (1962). *J. Cell Biol.* 13: 383–391.

Reiss, T. and Link, G. (1985). *Eur. J. Biochem.* 148: 207–212.

Robinow, C, F. (1956). Bacteriol. Rev. 20: 207–242.

Rochaix, J.-D. (1978). *J. Mol. Biol.* 126: 597–617.

Rose, R. J. (1988). In: *The Division and Segregation of Organelles (eds. S.A. Boffey and D. Lloyd), Cambridge University Press, Cambridge.* pp. 171–195.

Seckbach, J., Nagashima, H. and Fukuda, I. (1990). In: *Endocytobiology 4 (eds. Nardon, P. et al.) INRA, Paris.* pp. 575–578.

Seckbach, J. (1991). *J. Phycol.* 27: 794–796

Sellden, G. and Leech, R. M. (1981) *Plant Physiol.* 68: 731–734.

Shinozaki, K., Ohme, M., Tanaka, M., Wakasugi, T., Hayashida, N., Obokata, J., Y-Shinozaki, K., Ohto, C., Deno, H., Kamogashira, T., Yamada, K., Kusuda, J., Takaiwa, F., Kato, A., Tohdoh, N., Schimada, H. and Sugiura, M. (1986). *EMBO J.* 5: 2043–2049.

Sun, G.-H, Ueda, T. Q. P. and Kuroiwa, T. (1988). *Protoplasma* 144: 185–188.

Suzuki, K., Ohta, N. and T. Kuroiwa, T. (1992). *Protoplasma* 171: 80–84.

Tymms, M. J., Scott, N. S., and Possingham, J. V. (1983). *Plant Physiol.* 71: 785–788.

Valentin , K. and Zetsche, K, (1990). *Mol Gen Genet.* 222: 425–430.

Werz, G. (1966). Planta 68, 256–268.

Williamson, D. H. and Fennel, D. J. (1975). In: *Methods in Cell Physiology (ed. Prescott, D. M.), Academic Press, New York.* Vol. 12 p. 335.

Yokomura, E. (1967). *Cytologia* 32: 361–377.

Chapter 20

Molecular organization and expression of the plastid genome of *Galdieria sulphuraria* (= *Cyanidium caldarium*[1]) and other unicellular red algae

K. ZETSCHE

Institut für Pflanzenphysiologie der Justus-Liebig-Universität Gießen, HeinrichBuff-Ring 58, 35392 Gießen, Germany

Outlines:
1. Introduction
2. Size and coding capacity of the plastid genome of *Galdieria* and other unicellular red algae
3. Sequence of plastid genes
4. Conclusions
   4.1 Taxonomic position of *Galdieria*
   4.2 Phylogenetic implications
References

1. Introduction

The hypothesis of the endosymbiotic origin of plastids is now generally accepted. An overwhelming abundance of cytological, biochemical and especially molecular data are in favor of this hypothesis. Nevertheless several basic questions remain to be settled as for example whether the present different types of plastids (plastids of chlorophytes, rhodophytes, chromophytes, cryptophytes etc.) can be traced back to an unique endosymbiotic event (monophyletic origin) or whether the different plastids are the products of multiple endosymbiotic events (polyphyletic origin) and further what may be the relation between the different types of plastids. For a decision on this problem one has to compare data about the coding capacity, the structure and the sequence of the genomes of the different types of plastids. While many data are available in this respect for the plastids of higher plants and green algae (plastids of chlorophyll a/b plants-chloroplasts) almost nothing was known until a short time ago about the genome of the plastids of red algae (rhodoplasts).

Regarding *Galdieria sulphuraria* there exists an additional problem of the taxonomic placement of this unicellular alga and hence the nature of its photosynthetic organelle. Two opinions are mostly discussed: Either *Galdieria* and the related alga (*Cyanidium*) form part of the Rhodophyta and their photosynthetic organelle is a rhodoplast or the organelle is an endosymbiotic living cyanobacterium — a cyanelle — similar to the photosynthetic organelle of the alga *Cyanophora paradoxa*. For further data on *C. paradoxa*, see chapter on Glaucocystophyta by Schenk in this volume. An analysis of the plastid genome of *Galdieria* should shed light on this problem too.

2. Size and coding capacity of the plastid genome of *Galdieria* and other unicellular red algae

Because provisionally it is impossible to isolate intact plastids of *Galdieria* and other unicellular red algae (e.g., *Porphyridium*), the plastid DNA was recovered from total cell DNA by caesium chloride density gradient centrifugation. The plastid DNA form a weak band above the large band of nuclear DNA (Valentin and Zetsche 1989, 1990). Compare with data by Kuroiwa *et al.* in this volume.

Analysis of this DNA with restriction endonucleases produce several distinct bands of various sizes after electrophoretic separation. The sum of all bands results in a size of the plastid genome in the range of 170–180 kbp which is in good agreement with now estimated plastid genome sizes of other red algae (Li and Cattolico 1987; Shivji 1991).

Because plastid genomes of Chlorophyta have on

[1] *Cyanidium caldarium* Geitler — Allen strain is now classified as *Galdieria sulphuraria* (Merola *et al.* 1981) see Seckbach (1991).

J. Seckbach (ed.), Evolutionary Pathways and Enigmatic Algae: Cyanidium caldarium *(Rhodophyta) and Related Cells,* 255–262, 1994.

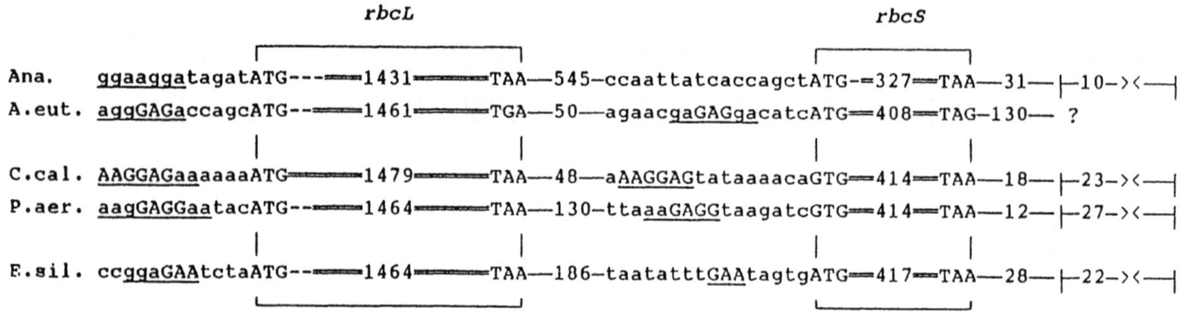

Fig. 1. Organization of Rubisco operons from various sources. Abbreviations are the same as in Table 1. Ana. = *Anabaena* 7120. Putative Shine-Dalgarno sequences are underlined — parts of these sequences present ahead of both genes (rbcL and rbcS) are given in capitals. Numbers indicate the length (in bp) of coding regions, intergenic spacers, inverted repeats and sequences between stop codons and putative transcription terminators (Valentin and Zetsche 1990).

an average sizes of 150 kbp the rhodoplast genome is larger by approximately 20–30 kbp. From this follows that the rhodoplast genome should have a higher coding capacity than the chloroplast genome. This opinion is supported by other facts: (1) Sequenced parts of the *Galdieria* plastid genome show a close package of genes. (2) The inverted repeats of the *Galdieria* plastid genome are rather small and consist with a size of approximately 5.5 kbp (Maid and Zetsche 1992). (3) No introns were detected as yet in sequenced genes.

This assumption was confirmed by the detection of genes on the plastid genome of *Galdieria* and other red algae which are not present on the genomes of the three totally sequenced plastid genomes of green plants *(Nicotiana*, *Oryza*, *Marchantia*).

These additional genes can be grouped into three classes. The first class contains genes which are absent from the plastid genome as well as the nuclear genome of green plants. This class is represented by phycobiliprotein — and phycobiliprotein linker genes. We have identified up to now on the plastid genome of *Galdieria* a gene coding for the β-subunit of allophycocyanin (apcB) and two allophycocyanin linker genes (apcC, apcL 9.5; Zetsche and Valentin 1994). It seems sure that more phycobilisome genes will be identified in the near future.

In the second class, genes are gathered which code for plastidic proteins but are located on the nuclear genome in green plants. In the first place we detected that the gene (rbcS) which code for the small subunit (SSU) of ribulosebisphosphate carboxylase/ oxygenase (Rubisco) is located on the plastid DNA of *Galdieria* and other red algae (Valentin and Zetsche 1989; Valentin and Zetsche 1990; Kostrzewa *et al.* 1990) and form an operon together with the rbcL gene which encode the large subunit (LSU) of Rubisco (Fig. 1). This support our previous work, where we have shown that both subunits of Rubisco are translated from a poly (A-)-messenger RNA (Steinmüller *et al.* 1983).

The location of both Rubisco genes on the plastid DNA is valid also for Chromophytes and Cryptophytes while in all chlorophyll a/b plants so far investigated the rbcS gene is always located on the nuclear genome (Palmer 1985; Douglas and Dunford 1989; Hwang and Tabita 1989; Valentin and Zetsche 1990).

As an another example for this type of genes serve the ATP synthase operon of the plastid genome. Most of the genes which code for the subunits of this complex are organized as a large operon. In higher plants the arrangement of this genes is similar to that of cyanobacteria with the exception that the genes atpG and atpD are located in the nuclear genome (Westhoff *et al.* 1981, 1985; Hermans *et al.* 1988; Miki *et al.* 1988).

In *Galdieria* this operon is organized exactly in the same way as in cyanobacteria, that is with the atpG- and atpD-genes on the plastid genome (Kostrzewa and Zetsche 1993, Fig. 2). A further example for this type of genes is a *Galdieria* gene which is homologous to the *E. coli* heat shock gene groEL (Gatenby and Ellis 1990) and the gene of the Rubisco binding protein of higher plants (Hemmingsen *et al.* 1988). In *Galdieria* this gene is located

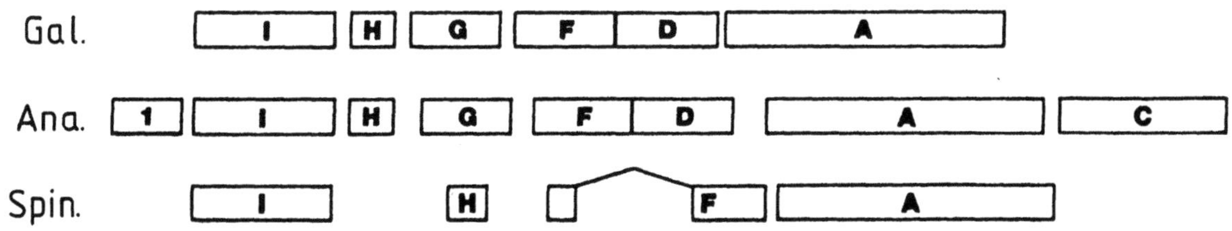

Fig. 2. Gene arrangement of the large plastidic ATP synthase operon of different plants: Gal = *Galdieria sulphuraria*; Ana = *Anabaena* 7120 (Caynobacterium); Spin = *Spinacea oleracea* (Kostrzewa and Zetsche, 1993). The letters specify the individual genes of the operon, which code for the subunits of CF1 and CFo respectively. The *Spinacea* F gene contains a large intron. The F and D genes overlap by four nucleotides in *Galdieria* as well as in *Anabaena.*

on the plastid genome, while in higher plants it is placed in the nuclear DNA. The *Galdieria* groEL homologous gene is expressed and show heat shock responses (Maid *et al.* 1992).

In the third category *Galdieria* plastid genes are gathered not yet identified on the plastid- or on nuclear genome of higher plants; these genes have counterparts on bacterial- and cyano-bacterial genomes. These *Galdieria* genes are homologous to ompR-, secA- and tsf genes of prokaryotes. OmpR genes of bacteria code for a transcription factor which permit the transcription of weak promoters. SecA genes are involved in the secretion of proteins across the bacterial plasma membrane. tsf genes code in *E. coli* for an elongation factor. As yet it is unknown which function these genes may have in the plastids of *Galdieria*.

At least we have detected several open reading frames on the plastid DNA of *Galdieria* which could not be related to known genes or sequences.

Taking all of these facts together, one may conclude that the coding capacity as a whole of the plastid genome of *Galdieria* and of the red algae is distinctly higher than that of higher plants and green algae.

3. Sequence of plastid genes

Up to now we have sequenced thirty plastid genes of *Galdieria* and compared these sequences with corresponding gene sequences of green plants, cyanobacteria, chromophytes and cryptophytes. The results of such comparisons are exemplary documented for the rbcL and rbcS genes.

The tabulation of the sequence homologies between the genes of different plants (Table 1) and the dendrograms constructed from these data (Figs. 3 and 4) clearly show a large evolutionary distance between the rbcL, as well as the rbcS genes of *Galdieria* (and other red algae) and the corresponding genes of green plants and extant cyanobacteria. On the other side a remarkable homology between these genes of rhodophytes and chromophytes including cryptophytes is evident.

Interestingly high homologies exist also between the Rubisco genes of *Galdieria* and the chemolithotrophic bacterium *Alcaligenes eutrophus* (Valentin and Zetsche 1990). Two possible explanations can be envisaged. Firstly the chemolithotrophic bacterium in question originates from a line of cyanobacteria which give also rise to the endosymbiotic ancestor of rhodoplasts or secondly a lateral gene transfer has happened between the bacterium and a cyanobacterium which became the ancestor of rhodoplasts.

Another — but relatively disliked possibility — is the transfer of the Rubisco genes of the bacterium to the rhodoplast of a primitive red alga. Similar differences between green plants and red algae have been obtained with other plastid genes including the 16Sr-DNA sequences (Maid and Zetsche 1991).

4. Conclusions

4.1 Taxonomic position of Galdieria

The taxonomic position of *Galdieria* has been discussed in the past very controversially. (See chapters on the taxonomy by Seckbach and Ott in this volume). While Kremer *et al.* (1978) takes the view that *Cyanidium* (= *Galdieria*) may be an association between an endosymbiotic cyanobacterium (cyanelle) and an eucyte as in the case of

Table 1. Sequence homologies (in %, deletions excluded) between Rubisco proteins from different sources. In the upper section various small subunits (SSU) are compared. In the lower section sequence homologies between different large subunits (LSU) of Rubisco are given

	C.cal.	P.aer.	E.sil.	Cryp.	O.lut.	A.eut.	Cyano.	Chl.*b*
SSU								
C.cal.								
P.aer.	62							
E.sil.	54	70						
Cryp.	62	72	67					
O.lut.	52	66	68	57				
A.eut.	54	59	56	54	56			
Cyano.	38	36	36	39	40	36	64	
Chl.*b*	33	34	28	34	32	39	43	50-78
LSU								
C.cal.								
P.aer.	84							
E.sil.	80	84						
A.eut.	73	70	70					
Cyano.	57	56	56			57		
Chl.*b*	52	50	53			58	80	>82

Abbreviations are: C.cal = *Cyanidium caldarium* = *[Galdieria sulphuraria* (Rhodophyta)]; P.aer. = *Porphyridium aerugineum* (Rhodophyta); Cryp. = *Cryptomonas* (Cryptophyta); O. lut. = *Olisthodiscus luteus* (Chromophyta, Crysophyceae); E.sil. = *Ectocarpus siliculosus* (Chromophyta, Phaeophyceae); A.eut. = *Alcaligenes eutrophus* (chemolithotrophic bacterium); Cyano. = cyanobacteria; Chl.b = chlorophyll a/b plants. Homologies to cyanobacteria and chlorophyll a/b plants are average values.

Cyanophora paradoxa other disagree with this view and regard *Galdieria* as an extraordinary red alga because of its acidophilic and thermophilic features which are unusual among the red algae (Nagashima and Fukuda 1981). Seckbach *et al.* (1981) have presented a compilation of features which point out a closer relation of *Cyanidium* (= *Galdieria*) to red algae rather than to the cyanobacteria. Seckbach *et al.* (1981) also introduce the idea that *Cyanidium* (= *Galdieria*) may be a 'bridge alga' what seems to mean an alga which is a link between cyanobacteria and unicellular (eukaryotic) red algae. The sequence data now available are not compatible with the view that the

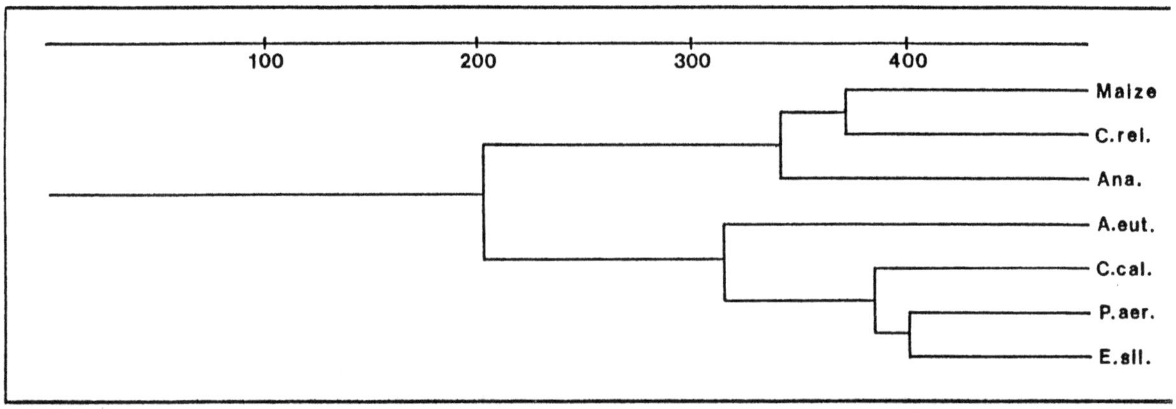

Fig. 3. Dendrogram of different large subunits of Rubisco. The dendrogram was constructed using the CLUSTAL program. Abbreviations are the same as in Table 1. Maze = *Zea mays*; C. rei. = *Chlamydomona reinhardtii.* Numbers give the value of the 'similarity score' which indicates the number of exactly matching residues minus a fixed penalty ('gap penalty') for every gap. The gap penalty was adjusted to 3.

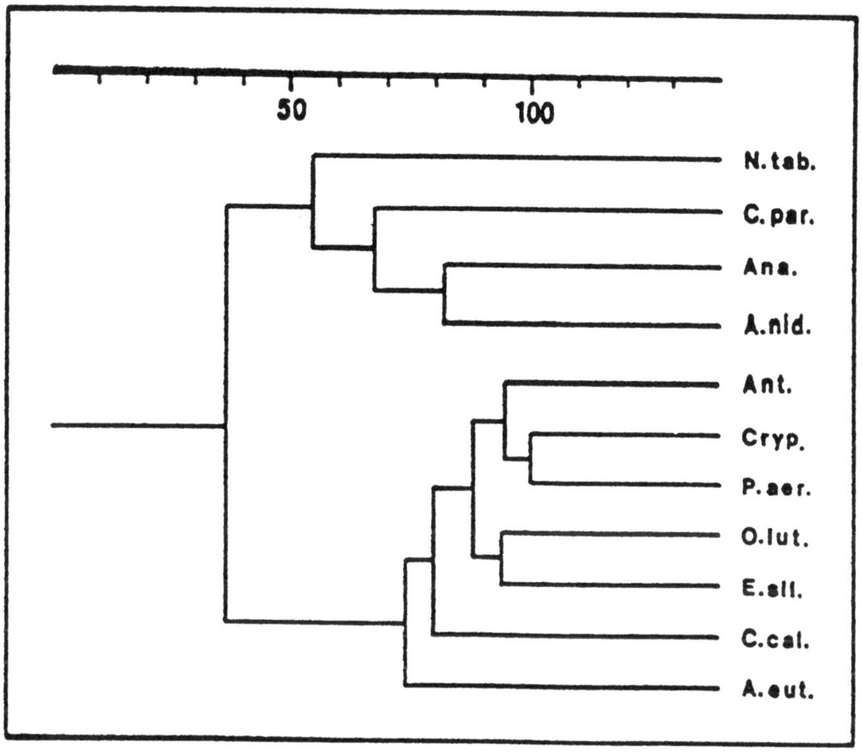

Fig. 4. Dendrogram of different small subunits of Rubisco. Abbreviations are the same as in Table 1 and Fig. 2. N.tab. = *Nicotiana tabacum*, C.rei. = *Chlamydomonas reinhardtii*, C.par. = *Cyanophora paradoxa*, Ant. = *Antithamnion* sp. (Valentin and Zetsche 1990). Numbers indicate the similarity score.

photosynthetic organelle of *Galdiera* (*Cyanidium*) is a cyanelle.

For example the sequence homologies between the Rubisco genes of *Galdieria* on the one side and of cyanobacteria and the cyanelles of *Cyanophora* on the other one are remarkable low, while the homologies between the genes of cyanobacteria, cyanelles and even higher plants are high.

In all dendrograms constructed with the Rubisco genes and other plastid genes, the *Galdieria* genes always cluster with the corresponding genes of unicellular and multicellular red algae. By these data the plastids of *Galdieria* are rhodoplasts and *Galdieria* hence a red alga. The presence of the same additional genes on the plastid genomes of *Galdieria* and other red algae whether uni- or multicellular further strengthen this opinion. But within the group of rhodoplasts the plastids of *Galdieria* hold an outsider position. The rhodoplast genes of the unicellular red alga *Porphyridium* and the multicellular red alga *Antithamnion* are more related than the plastid genes of *Galdieria* with the *Porphyridium* genes. In dendrograms of the Rubisco genes the *Galdieria* genes branch off very early from the main line, earlier than the corresponding genes of Chromophyta and Cryptophyta (see following discussion). This outsider position of *Galdieria* plastids in relation to other rhodoplats can be discussed in differing ways: (1) *Galdieria* could be a very ancient red alga and branch off very early from the main line of red algae evolution. (2) The outsider position is the result of the adaptation of this alga to its hot habitat either because the alga has to synthesize heat stable proteins or because the high temperature raises the rate of mutations. (3) At least one cannot exclude the possibility, that *Galdieria* and its near relatives form their own taxonomic group. Also compare chapter on New Classification by Ott and Seckbach in this volume. Gene sequence data from *Galdieria* relatives within the algal class, such as *Cyanidium caldarium* are needed to make a decision on this problem.

4.2 Phylogenetic implications

The data now available about the coding capacity and the structure and gene sequences of rhodoplast genomes strongly point to a polyphyletic origin of rhodoplasts and chloroplasts. Alternatively, it may present at least a very early separational stage of the evolutionary pathways of these two types of plastids.

In the case of a polyphyletic origin, the immediate ancestors of chloroplasts and of rhodoplasts may be quite different lines of cyanobacteria whereby the line from which the rhodoplasts originate have to be identified.

It is possible to make a relatively clear statement concerning the relationship between the plastids of Rhodophyta, Chromophyta and Cryptophyta. Structural and sequence data of the plastid genomes strongly support the idea of the origin of the chromophytic and cryptophytic plastids from endosymbiotic unicellular red algae. This hypothesis was in particular argued by Gibbs (1981) and Whatley (1981) on the account of cytological data because of the four membranes enclosing the plastids of these algae. Furthermore a vestigial nucleus and 80S ribosomes are located in the space between the two outer and the two inner plastid membranes of cryptophyta (Gillot and Gibbs 1980; Ludwig and Gibbs 1985; Hansmann *et al.* 1985). This hypothesis is now strongly confirmed by sequence data (Valentin and Zetsche 1989; Valentin and Zetsche 1990). The endosymbiont involved must be a *Porphyridium*-like red alga because in several dendrograms of plastid genes, the genes of Chromophyta and Cryptophyta branch off after the plastid genes of *Galdieria* from the main line which leads to the rhodoplasts of more evolved red algae.

References

Douglas, S.E. (1989). *Plant Mol. Biol.* 13: 13–20.
Gatenby, A.A. (1990). *Ann. Rev. Cell Biol.* 6: 125–149.
Gibbs, S. P. (1981). *Ann. New York Acad. Sci.* 361: 193–207.
Gillot, M. A. and Gibbs, S. P. (1980). *J. Phycol.* 16: 558–568.
Hansmann, P., Falk, H. and Sitte, P. (1985). *Z. Naturforsch.* 40c: 933–935.
Hemmingsen, S.M., Woolford, C., van der Vies, S.M., Tilly, K., Dennis, D. T., Georgopoulos, C.P., Hendrix, R.W. and Ellis, R. J. (1988). *Nature* 333: 330–334.
Herman, J., Rother, C., Bickler, J., Steppuhn, J. and Herrmann, R.G. (1988). *Plant Mol. Biol.* 10: 323–330.
Hwang, S. H. and Tabita, F. R. (1989). *Plant Mol. Biol.* 13: 69–79.
Kostrzewa, M., Valentin, K., Maid, U., Radetzky, R. and Zetsche, K. (1990) *Curr. Genet.* 18: 465–469.
Kostrzewa, M. and Zetsche, K. (1993) *Plant Mol. Biol.* 23: 67–76.
Kremer, B. P., Feige, G.B. and Schneider, H.A.W. (1978). *Naturwiss.* 65: 157.
Li, N. and Cattolico, R. A. (1987). *Mol. Gen. Genet.* 209: 343–351.
Ludwig, M. and Gibbs, S.P. (1985). *Protoplasma* 127: 9–20.
Maid, U. and Zetsche, K. (1991). *Plant Mol. Biol.* 16: 537–547.
Maid, U. and Zetsche, K. (1992). *Plant Mol. Biol.* 19: 1001–1010.
Maid, U., Steinmüller, R. and Zetsche, K. (1992), *Curr. Genet.* 21: 521–525.
Merola, A., Castaldo, R., De Luca, P., Gambardella, R., Musacchio, A. and Taddei, R. (1981). *Giorn. Bot. Ital.* 115: 189–195.
Niki, J., Maeda, M., Mukohada, J. and Futai, M. (1988). *FEBS Lett.* 232: 221–226.
Nagashima, H. and Fukuda, I. (1981). *Phytochemistry* 20: 439–442.
Palmer, J.D. (1985). *Ann. Rev. Genet.* 19: 325–354.
Seckbach, J. (1991). *J. Phycol.* 27: 794–796.
Seckbach, J., Hammerman, I. S. and Hananin, J. (1981). *Ann. New York Acad. Sci.* 361: 409–425.
Shivji, M. S. (1991). *Cruu. Genet.* 19: 49–54.
Steinmüller, K., Kaling, M. and Zetsche, K. (1983). *Planta* 159: 308–313.
Valentin, K. and Zetsche, K. (1989). *Curr. Genet.* 16: 203–209.
Valentin, K. and Zetsche, K. (1990). *Mol. Gen. Genet.* 222: 425–430.
Valentin, K. and Zetsche, K. (1990). *Plant Mol. Biol.* 15: 575–584.
Westhoff, P., Nelson, N., Bünemann, H. and Herrmann, R. G. (1981). *Curr. Genet.* 4: 109–120.
Westhoff, P., Alt, J., Nelson, N. and Herrmann, R. G. (1985). *Mol. Gen. Genet.* 199: 290–299.
Whatley, F. R. (1981). *New Phytol.* 87: 223–247.
Zetsche K. and Valentin, K. (1994) Endocytobiosis and Cell Research 10: 107–127.

Note added in proof

a)

In the meantime we have detected and mostly sequenced a lot of further genes not mentioned in this article on the plastid genome of *Galdieria* as well as *Cyanidium caldarium* (strain RK-1). The genes up to now identified in *Galdieria* and Cyanidium are summarized in the following table

Gene	Gene product
Phycobilisomes	
apcA	allophycocyanin α-subunit
apcB	allophycocyanin β subunit
apcE	phycobilisome anchor protein
apcF	phycobilisome linker protein
apcL 9,5	phycobilisome linker protein
Photosystem I	
psaA	apoprotein A1
psaB	apoprotein A2
psaK	8, 4 kD PSI protein
Photosystem II	
psbA	D1 protein
psbD	D2 protein
psbC	CP43 chlorophyll apoprotein
psbN	N protein
psbH	10 kD PSII protein
psbB	CP47 chlorophyll apoprotein
Cytochrom b/f complex	
petB	cytochrom b
petD	subunit 4
petF	ferredoxin
ATPsynthase	
atpA	α-subunit
atpB	β-subunit
atpD	δ-subunit
atpE	ε-subunit
atpF	subunit I
atpG	subunit II
atpH	subunit III
atpI	subunit IV
Rubisco	
rbcL	large subunit (*Cyanidium, Galdieria*)
rbcS	small subunit (*Cyanidium, Galdieria*)
Ribosomes	
16S rDNA	16S ribosomal RNA
23S rDNA	23S ribosomal RNA
5S rDNA	5S ribosomal RNA
rps2	ribosomal protein
rps7	ribosomal protein
rps10	ribosomal protein
rps12	ribosomal protein
rps14	ribosomal protein
rps16	ribosomal protein
rpl21	ribosomal protein

Transcription apparatus

rpoC2	RNA-polymerase subunit
trsB	transcription factor (ompR homologous)

Translation apparatus

infB	initiation factor B
tsf	elongation factor Ts
tufA	elongation factor Tu
trnX	several tRNAs

Biosynthesis of amino acids

glsF	glutamate synthetase (ferredoxin-dependent; *Cyanidium*)

Miscellaneous genes

groEL	chaperonin 60 (*Cyanidium, Galdieria*)
trxA	thioredoxin (*Cyanidium*)
dnaK	heat shock protein 70 (*Cyanidium, Galdieria*)
ccl1	protein involved in cytochrom biogenesis

several orfs

Genes without species name are from Galdieria.

Literature:

Zetsche, K. and Valentin, K. (1994). *Structure, coding capacity and gene sequences of the plastid genome from red alga. Endocytobiosis and Cell Research* 10: 107-127.

b)
Phylogenetic trees constructed for the rbcL-and the rbcS-genes show, that the genes of *Galdieria* and of *Cyanidium* cluster together with the corresponding genes of red algae. But within the cluster of red algae genes, the *Galdieria* genes branch off earlier from the tree than the *Cyanidium* genes. In other words the genes of *Cyanidum* are more related to the corresponding genes of unicellular red alga like *Porphyridium* spec. than those of *Galdieria.*

Chapter 21

The molecular aspects of pigments and photosynthesis in *Cyanidium caldarium*

ROBERT F. TROXLER
Department of Biochemistry, Boston University School of Medicine, 80 East Concord Street, Boston, MA 02118, USA

Outlines:

1. Introduction
   1.1 The organism
2. Photosynthetic pigments
   2.1 Composition
   2.2 Pigment mutants
3. Growth and chloroplast development
   3.1 Culture conditions
   3.2 Chl-a and BSP synthesis in light
4. Pigment biosynthesis
   4.1 The branched porphyrin pathway in plants
   4.2 Biosynthesis of ALA
   4.3 Intermediates between ALA and protoporphyrin IX
   4.4 Synthesis of heme
   4.5 Heme conversion to biliverdin IXα
   4.6 Mechanism of heme oxygenase
   4.7 Biliverdin conversion to PCB
   4.8 Biosynthesis of BSPs
   4.9 Biosynthesis of Chl-a
5. Regulation of pigment synthesis
   5.1 Spectral response curves
   5.2 Photoregulation of gene expression
   5.3 Kinetics of appearance of APC and CPC mRNAs in light
   5.4 Heme regulated genes
6. Photosynthesis
   6.1 Size of photosynthetic units
References

1. Introduction

1.1 The organism

Cyanidium caldarium is a unicellular rhodophyte capable of growing at extremes of temperature and pH (Brock 1978). *C. caldarium* has been isolated in the field at Yellowstone National Park in Wyoming, Lassen National Park and The Geysers in California, Steamboat Springs in Nevada, and from acid, thermal habitats in Italy, Iceland, New Zealand, Japan, Indonesia, Dominica and El Salvador. The requirements for growth, pigmentation and photosynthesis in the natural habitat have resulted in the evolution of remarkable organism with properties ideally suited for investigations on chloroplast biogenesis, biosynthesis of photosynthetic pigments and regulation of chloroplast gene expression (Troxler 1986).

The taxonomic position, and for that matter, the scientific name, of *C. caldarium* has been the subject of considerable interest and speculation. The author has maintained cultures of the original 'Allen strain' of *C. caldarium* (Allen 1959) and of several pigment mutants in his laboratory since 1966. This organism has been used extensively for studies on biosynthesis of δ-aminolevulinate (ALA), heme, phycobilins and phycobiliproteins. Autotrophic cells of the wild-type are spherical, divide by mitosis producing 4 autospores, contain a single cup-shaped chloroplast per cell, appear to have more than one mitochondrion per cell, probably lack vacuoles and are capable of growing heterotrophically, mixotrophically or autotrophically. The Allen strain of *C. caldarium* appears to share features in common with both *C. caldarium* (forma A; RK-1) and *Galdieria sulphuraria* (forma B; M-8) discussed by Seckbach (1991). For further information on these cells see chapters by Seckbach and Ott in this volume.

2. Photosynthetic pigments

2.1 Composition

The principal photosynthetic pigments in the chloroplasts of wild-type cells of *C. caldarium* grown in the light are chlorophyll-a (Chl-a) and two phycobiliproteins (PBPs), allophycocyanin

J. Seckbach (ed.), Evolutionary Pathways and Enigmatic Algae: Cyanidium caldarium *(Rhodophyta) and Related Cells,* 263–282, 1994.

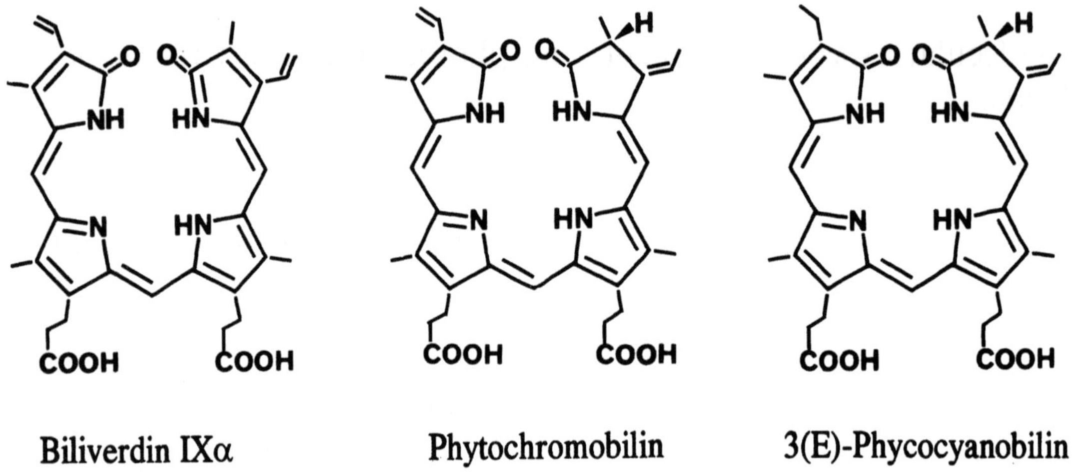

Fig. 1. Structure of biliverdin IXα, phytochrombilin and 3(E)-phycocyanobilin.

(APC) and C-phycocyanin (CPC). Several carotenoids including β-carotene, lutein, zeaxanthin, violoxanthin and cryptoxanthin are also reported to be present (Brock 1978). Like other red algae, *C. caldarium* does not contain Chl-b, Chl-c or Chl-d. Chl-a occurs in thylakoid membranes associated with polypeptides of Photosystem I (PSI) and Photosystem II (PSII). APC and CPC are comprised of α and β subunit polypeptides in the 16–20 kd range. The light absorbing chromophore of APC and CPC is a blue colored bile pigment called phycocyanobilin (PCB). The α and β subunits of APC and the α subunit CPC contain 1 covalently attached residue of PCB and the β subunit of CPC contains two such residues. PBPs occur together with 7–9 linker polypeptides in phycobilisomes (PBS) located on the stroma surfaces of thylakoid membranes. Phycobilisomes are the primary light harvesting antenna for photosynthesis in *C. caldarium* and other red algae (Gantt 1990). APC and CPC are abundant and comprise 25% of the total soluble protein.

PCB is a linear tetrapyrrole structurally related to bile pigments such as biliverdin IXα in man (Fig. 1). It is noteworthy that phytochromobilin, the prosthetic group of phytochrome, is also a bile pigment similar to PCB. Chemically, phytochromobilin contains 2 more electrons and PCB contains 4 more electrons than biliverdin IXα. Both phytochromobilin and PCB are IXα isomers formally arising by removal of the α bridge carbon of heme (see below). PCB possesses an ethylidine side chain at the 3 position on ring I. The ethylidine double bond is the site where PCB becomes attached to the sulfhydryl group of cysteine residues in APC and CPC apoproteins (Brown *et al.* 1979). The chromophore of APC and CPC can be cleaved from apoprotein by refluxing in boiling methanol and the bilin released has been shown to be 3(E)-PCB (Cole *el al.* 1967; Troxler *et al.* 1978).

We have determined the complete amino acid sequences of the α and β subunits of APC and CPC and identified the positions in each polypeptide where PCB is covalently attached (Brown *et al.* 1979; Troxler *et al.* 1981; Offner *et al.* 1981; Offner and Troxler 1983). These sequences are the only case where the complete amino acid sequences of APC and CPC have been experimentally determined in the same species of a red alga (Fig. 2). The α and β subunits of APC contain 160 and 161 amino acid residues and have exact molecular weights (Mrs), based on the sequence, of 18,160 and 18,125, respectively. PCB is attached to Cys 80 of the α subunit and Cys 81 of the β subunit. The α and β subunits of CPC contain 162 and 172 amino acid residues and have exact Mrs (based on the sequence) of 18,393 and 19,572, respectively. PCB is attached to Cys 84 of the α subunit and Cys 82 and Cys 153 of the β subunit. When these sequences are aligned to maximize homology, 19 amino acid residues are identical in all four proteins (Fig. 2). This suggests these residues are functionally important for folding, subunit aggregation and energy transfer in phycobilisomes. The homology

```
        1                                      20                                      40
APC-α   M I V - - T K S I V N A D A E A R Y L S P G E L D R I K S F V L S G Q R R L R I
APC-β   M Q D A I T A V I N T A D V Q G K Y L D S S I I E K L K G Y F Q T G E L R V R A
CPC-α   M K T P I T E A I A A A D N Q G R F L S N T E L Q A V N G R Y Q R A A A S L E A
CPC-β   M L D A F A K V V A Q A D A R G E F L S N T Q L D A L S K M V S E G N K R L D V

                                               60                                      80
APC-α   A Q I L T D N R E R I V K Q A G Q Q L F Q Q R P D I V S P G G N A Y G E E M T -
APC-β   A A T I A A N A A G I I K D A V A K S L L Y S - D I T R P G L D M Y T T R R Y -
CPC-α   A R S L T S N A E R L I N G A A Q A V Y S K F P Y T S Q M P G P Q Y A S S A V G
CPC-β   V N R I T S N A S A I V T N A A R A L F S E Q P Q L I Q P G G T A Y T N R R M -

            PCB                                100                                     120
APC-α   - A T C L R D L D Y Y L R L V T Y G V V A G D I A P I E E I G L V G V K E M Y N
APC-β   - A A C I R D L D Y Y L R Y A T L S M L A G D P S I L D E R V L N G L K E T Y N
CPC-α   K A K C A R D I G Y Y L R M V T Y C L V V G G T G P M D E Y L I A G L E E I N R
CPC-β   - A A C L R D M E I I L R Y V S Y A I I A G D S S I L D D R C L N G L R E T Y Q
            PCB

                                               140                                     160
APC-α   L G T P I S A V A E G I D A M K N V A C S L L S G D D R A - - - - - - - - - - E
APC-β   L G V P I G A T I Q S I Q A M A E V T S S L V G P E A G K - - - - - - - - - - E
CPC-α   F D L S P S W T V E A L N Y I K A N H G - - L S G Q A A N - - - - - - - - - - E
CPC-β   L G V P G A S V A V G I E K M K D S A I A I A N D P S G I T T G D C S A L M A E
                                                                               PCB

                              174
APC-α   A G F Y F D Y K L P A S S
APC-β   M G I Y F D Y I C S G L S
CPC-α   A N T Y I D Y A I N A L S
CPC-β   V G T Y F D R A A T A V Q
```

Fig. 2. Comparison of the amino acid sequences of the α and β subunits of allophycocyanin and phycocyanin from *C. caldarium*. Identical residues in all four proteins are enclosed is boxes. Gaps have been introduced to maximize homology. Residues are numbered as though all four proteins contain 174 amino acids. Note that the second PCB residue in the β subunit of CPC is located in a region not present in the other three proteins. Abbreviations: A, alanine; R, arginine, N, asparagine; D, aspartic acid; C, cysteine; E, glutamic acid; Q, glutamine; G, glycine; H, histidine; I, isoleucine; L, leucine; K, lysine; M, methionine; F, phenylalanine; P, proline; S, serine; T, threonine; W, tryptophan; Y, tyrosine; V, valine; PCB, phycocyanobilin.

between APC and CPC α subunits, and β subunits, ranges from 26–39%. However, when α or β subunits of APC or CPC are aligned with their cyanobacterial counterparts (eg., Frank *et al.* 1978; Sidler *et al.* 1981; DeLange *et al.* 1981) homologies of 80–90% are observed. Consequently, there is a greater degree of sequence similarity between α subunits of APC in red algae and cyanobacteria than between α and β subunits of APC in the same organism. The same is true for the β subunits of APC and for α and β subunits of CPC. This explains why immune serum against APC does not cross-react with CPC, and vice versa, but that immune serum against APC (or CPC) cross-reacts with heterologous APCs (or CPCs) from other species of rhodophytes or cyanobacteria.

2.2 Pigment mutants

Nichols and Bogorad (1960) irradiated wild-type (Allen strain) cells of *C. caldarium* with ultraviolet (UV) light and isolated a series of pigment mutants defective in formation of Chl-a and PBPs. Beale and Chen (1983) described a spontaneous mutant of wild-type *C. caldarium* which forms high levels of Chl-a, APC and CPC in the dark. The phenotypes of these mutants are listed in Table 1. The wild-type and IIIC, GGB and GGBY are unable to make Chl-a or PBPs in the dark. IIID2 makes minute amounts of Chl-a but not PBPs in the dark. CPD is the exception and makes large amounts of Chl-a and PBPs in the dark. Cells cultured in either dark or light produce carotenoids. In light, mutant IIID2 makes more Chl-a and PBPs per cell than does the wild-type, mutant IIIC makes Chl-a (40% less than wild-type) but not PBPs, mutant GGB makes PBPs but not Chl-a and mutant GGBY makes neither Chl-a nor PBPs. The level of pigments in CPD is unaffected by dark or light.

IIID2 and IIIC are able to carry out photosynthesis but GGB and GGBY cannot and thus are obligate heterotrophs. Volk and Bishop (1968) isolated UV mutants of *C. caldarium* and found that a common phenotype was one where APC and CPC were missing. These mutants were similar to

Table 1. Pigments of wild-type and mutant *C. caldarium* cells

Strain	Dark				Light			
	Chl-a	APC	CPC	Carotetenoids	Chl-a	APC	CPC	Carotenoids
Wild-type	–	–	–	+	+	+	+	+
IIID2	–	–	–	+	+	+	+	+
IIIC	–	–	–	+	+	–	–	+
GGB	–	–	–	+	–	+	+	+
GGBY	–	–	–	+	–	–	–	+
CPD	+	+	+	+	+	+	+	+

(a) The symbol – indicates absence and the symbol + indicates the presence of pigment.

mutant IIIC mentioned above. Characterization of one PBP-less mutant, called R-II, revealed that Chl-a was reduced by about 50%. In this strain the Hill reaction was normal, the action spectrum for the quantum yield indicated an efficient photosynthetic mechanism with peak efficiency shifted toward longer wavelengths when compared to the wild-type, and there was no Emerson enhancement effect. This suggests that *C. caldarium* can compensate for the absence of PBPs (which form the principal light-harvesting antenna of PSII) by reducing Chl-a in PSI to equalize energy flow between the two photosystems.

3. Growth and chloroplast development

3.1 Culture conditions

Wild-type cells grown autotrophically in light contain Chl-a and PBPs whereas cells grown heterotrophically in the dark do not make these pigments. Cells inoculated into minimal medium plus 1% glucose (Allen 1959) and incubated with rotary shaking in the dark at 37–42 °C grow to a density of about 2 x 10^8 cells/ml in 5 days. When these pigment-less cells are resuspended in minimal medium minus glucose and incubated in the light, cell division ceases and after a short lag period the cells begin to accumulate Chl-a, APC and CPC. This occurs simultaneously with differentiation of a small proplastid-like organelle into a large cup-shaped chloroplast (Fig. 3). This process is completed within about 24 h. Chloroplast development and synthesis of Chl-a and PBPs is synchronous and occurs in the absence of cell division. Therefore, growth (cell division) and differentiation (chloroplast development) are separated in time and regulatory processes concerned with control of the cell cycle are not superimposed on control of pigment synthesis and expression of chloroplast genes. Further data on the dividing organelles of *C. caldarium* are presented by Kuroiwa *et al.* in this volume.

3.2 Chl-a and PBP synthesis in light

The kinetics of pigment accumulation when dark-grown cells are placed in the light is illustrated in Figure 4. Following a 2–4 h lag period, the duration of which is dependent on cell density and light intensity, synthesis of Chl-a and PBPs begins and continues for about 24 h. Upon continued illumination cells do not resume cell division for a period of about 2 weeks. The factors responsible for controlling the cessation of cell division upon transfer from dark to light are not known. Administration of cycloheximide, chloramphenicol, or rifampicin to cells immediately after transfer from dark to light completely abolishes pigment synthesis. Administration of any one of these inhibitors midway in the greening process prevents further accumulation of Chl-a and PBPs (Fig. 4). The effect of cycloheximide indicates the requirement for protein synthesis on 80S ribosomes in the cytosol and the effect of chloramphenicol indicates the requirement for protein synthesis on 70S ribosomes in the chloroplast for pigment synthesis and chloroplast development to occur. This is not surprising since genes for essential chloroplast proteins are found in both the nuclear and chloroplast genomes.

The ability of rifampicin to abolish greening is interesting. In *Escherichia coli* rifampicin inhibits

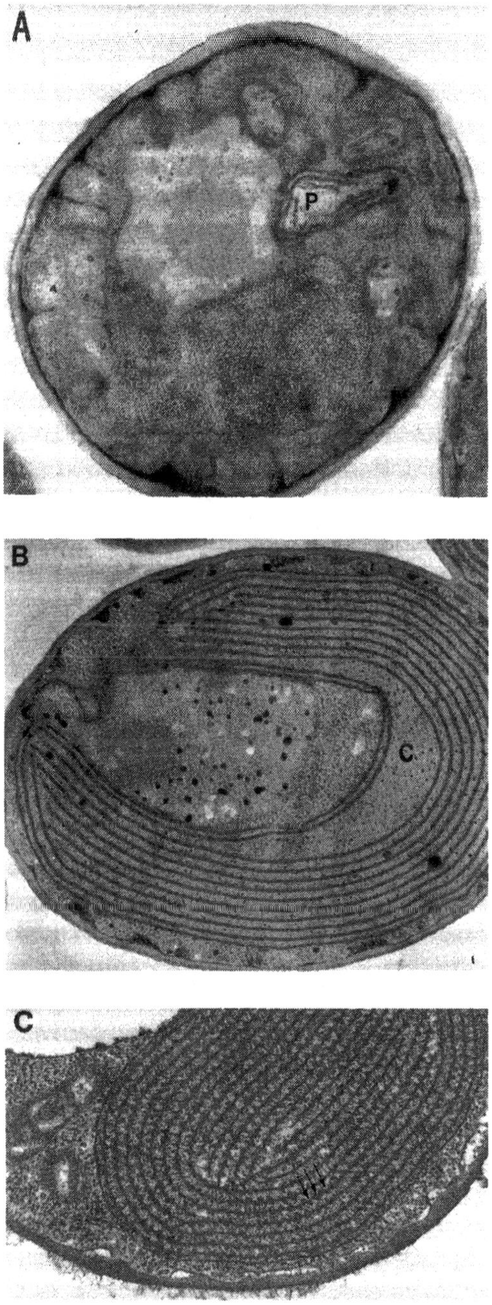

Fig. 3. Ultrastructure of *C. caldarium* cells grown in the dark or in the light. Dark-grown cells contain a small undifferentiated proplastid-like organelle (panel A) and cells grown in light contain a large cup-shaped chloroplast with parallel rows of unstacked thylakoid membranes (panel B). Phycobilisomes (arrows) can be seen on the surfaces of thylakoid membranes (panel C). Abbreviations: P, proplastid; C, chloroplast.

transcription by binding to the β subunit of RNA polymerase and preventing formation of initiation complexes (Jin and Gross 1988).

The similarities between the transcriptional apparatus of bacteria and chloroplasts are well documented. The ability of rifampicin to prevent greening in *C. caldarium* indicates that chloroplast RNA polymerase is rifampicin sensitive and suggests the requirement for transcription of chloroplast genes in order for chloroplast development and synthesis of pigments to occur. It could be argued from this that light regulation of chloroplast gene expression is regulated at least in part at the level of transcription. This is quite different from the situation in higher plants where chloroplast RNA polymerase is not inhibited by rifampicin (Bottomly *et al.* 1971). In addition, expression of chloroplast genes in higher plants is thought to be more under the control of developmental programs and to be regulated primarily by post-transcriptional processes and independently of light (Gruissem 1989).

4. Pigment biosynthesis

4.1 The branched porphyrin pathway in plants

δ-aminolevulinate (ALA) is the rate limiting substrate for tetrapyrrole biosynthesis in living organisms and ALA in plants is synthesized from the carbon skeleton of glutamate through the concerted action of 3 different enzymes (Fig. 5). The porphyrin pathway in plants is branched at protoporphyrin IX with the magnesium branch leading to formation of Chl-a and the iron branch leading to formation of heme. There is no detailed information available on the enzymes or intermediates in the magnesium branch between protoporphyrin IX and Chl-a in *C. caldarium*. However, it is reasonable to assume that Chl-a biosynthesis is similar to that in other algae and higher plants. The first enzyme in the iron branch is ferrocheletase which inserts iron into protoporphyrin IX to form heme. The second step is catalyzed by heme oxygenase which oxidatively cleaves the macrocycle to form biliverdin IXα, carbon monoxide and iron.

Biliverdin IXα is then transformed into PCB by two 2 electron reduction steps involving at least 3 different enzymes. PCB then becomes attached via cysteinyl thioether bonds to APC and CPC apoproteins and this process is thought to be catalyzed by one or more 'thioether synthetases' (Brown *et al.* 1979).

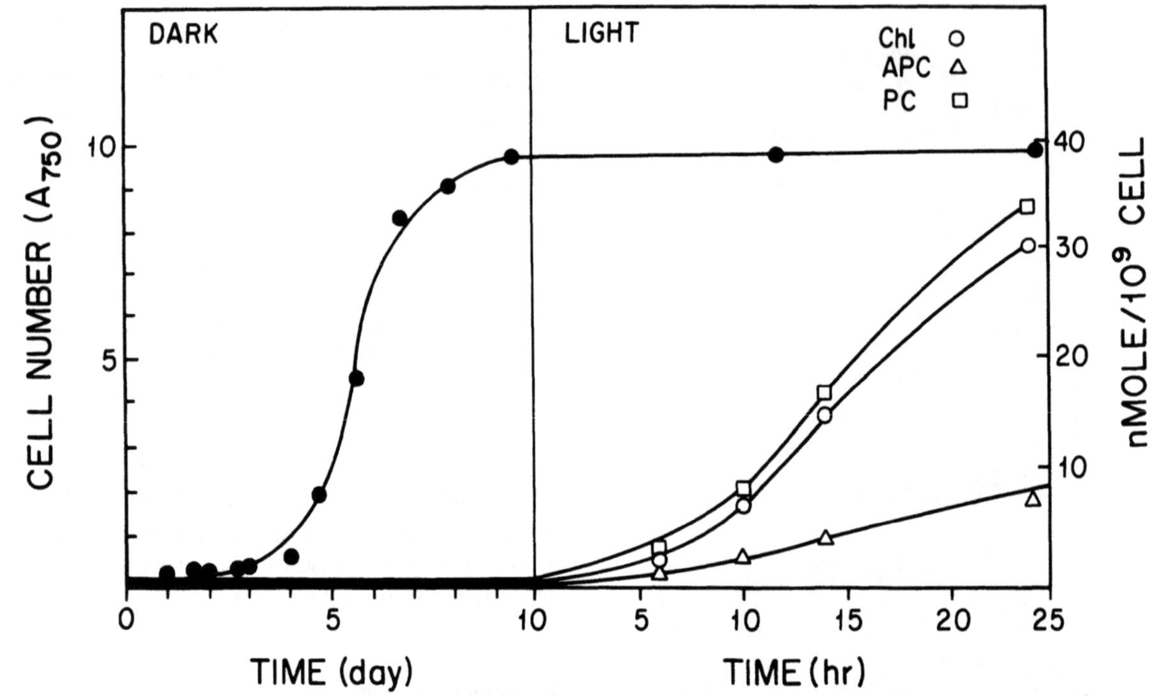

Fig. 4. Synthesis of Chl-a, APC and CPC in the light. Cells were grown from a small inoculum in the dark for 10 days in minimal medium plus glucose at 37 °C and growth (i.e., cell number) was estimated at intervals from scattering at 750 nm. Cells were collected by centrifugation, resuspended in minimal medium minus glucose at 10^8 cells/ml and incubated in the light for 24 h. At intervals, the concentration of Chl-a, CPC and APC was estimated spectrophotometrically in living cells from the absorbance at 675, 650 and 620 nm, respectively, by the opal glass technique (Troxler *et al.* 1989). Addition of rifampicin (10 μg/ml), cycloheximide (1 μg/ml) or chloramphenicol (300 μg/ml) to cell suspensions upon transfer to the light, or after 14h in the light, completely abolished synthesis of pigments (data not shown).

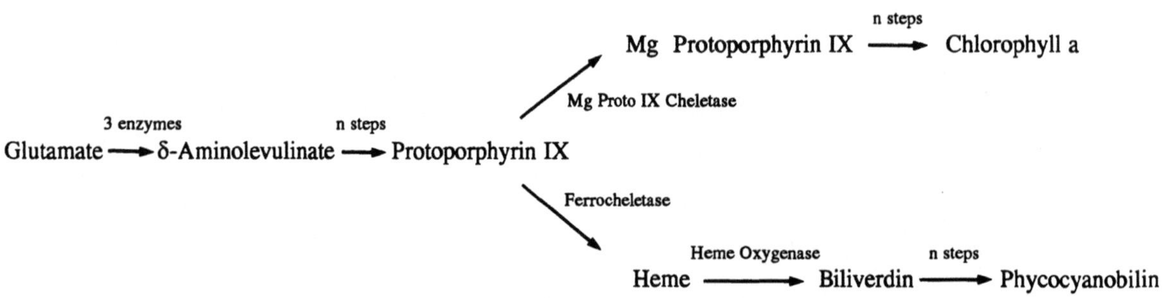

Fig. 5. The branched porphyrin biosynthetic pathway in photosynthetic organisms.

4.2 Biosynthesis of ALA

In animals ALA is made by condensation of glycine and succinyl CoA by the pyridoxal phosphate (PLP) enzyme, ALA synthase (EC 2.3.1.37) (for review, see Jordan 1990). Early attempts to detect ALA synthase activity in plants were unsuccessful and the origin of ALA in plants remained obscure.

Beale (1971) discovered that levulinic acid (LEV), a competitive inhibitor of ALA dehydrase (the second enzyme in the porphyrin pathway) blocked Chl-a synthesis and resulted in accumulation of ALA in *Chlorella*. The use of LEV provided a tool for measuring incorporation of exogenous radiolabled precursors into accumulated ALA in plant tissues. Beale and Castelfranco (1974) found

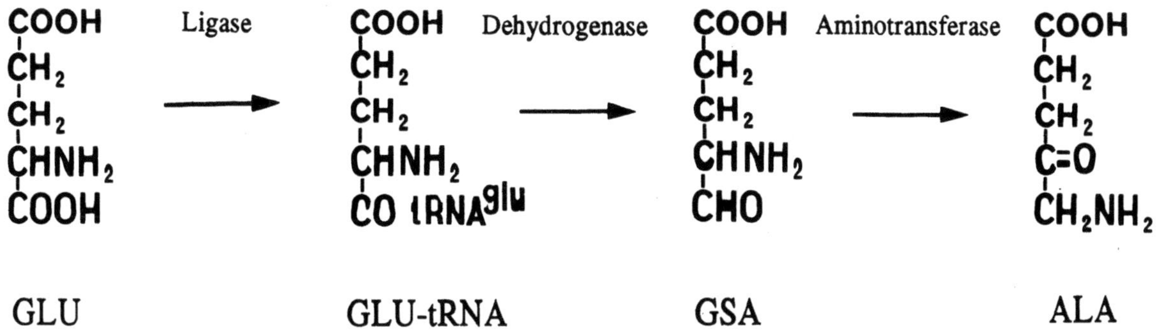

Fig. 6. Scheme illustrating the C-5 pathway for ALA biosynthesis in plants. Abbreviations: GLU, glutamic acid; GSA, glutamate-1-semialdehyde; ALA, δ-aminolevulinic acid.

that greening cucumber cotyledons incubated with LEV incorporated the carbon skeleton of glutamate into accumulated ALA and this cardinal finding lead to discovery of the 'C-5' pathway for ALA synthesis (Fig. 6). Beale and collaborators have since found that the C-5 pathway is widely distributed in living organisms including *E. coli*, some photosynthetic bacteria, cyanobacteria, algae and higher plants (Avissar *et al.* 1989). ALA synthesis from glycine and succinyl CoA by ALA synthase seems to be a more restricted process found in certain photosynthetic bacteria and in animals.

Jurgenson *et al.* (1976) demonstrated the C-5 pathway for ALA synthesis in *C. caldarium*. Dark-grown cells incubated in the light with LEV incorporated radiolabeled glutamate but not glycine into ALA. This was the first instance where the C-5 pathway for ALA synthesis was detected in a red alga. Troxler and Offner (1979) observed that LEV-treated cells of *C. caldarium*, mutant GGBY, made ALA by the C-5 pathway. GGBY is an obligate heterotroph which respires glucose but is unable to make Chl-a and PCB. The accumulated ALA in LEV-treated GGBY was presumably destined for heme in mitochondrial cytochromes. It was concluded that ALA destined for synthesis of all tetrapyrroles is derived via the C-5 pathway and that ALA synthase is not present *C. caldarium*. Weinstein and Beale (1984) confirmed this conclusion by directly measuring incorporation of glutamate into protoheme and heme a (the prosthetic group of cytochrome oxidase).

Three enzymes, a ligase, a dehydrogenase, and an aminotransferase, are required to transform glutamate into ALA in a process requiring ATP, Mg^{++}, NADPH, and tRNAglu (Fig. 6). Kannangara and his colleagues in Copenhagen have pioneered work on isolation and characterization of these enzymes in barley. The first enzyme, glutamyl tRNA ligase (Bruyant and Kannangara 1987) attaches the 5′ end of tRNA Glu to C1 of glutamate to form glutamyl-tRNA in a reaction requiring ATP and Mg++. The second enzyme, glutamyl-tRNA dehydrogenase, reduces the glutamyl moiety to glutamate-1-semialdehyde or its hydrated hemiacetal and releases $tRNA^{glu}$ (Houen *et al.* 1983; Hoober *et al.* 1988). The third enzyme, glutamate-1-semialdehyde aminotransferase, catalyzes a molecular rearrangement leading to formation of ALA in a reaction that does not require an external donor of amino groups (Kannangara *et al.* 1988).

These three enzymes have been partially purified and characterized from *C. caldarium* (Houghton *et al.* 1989) using the same procedures as those developed for barley enzymes. The chromatographic properties and substrate requirements of the *C. caldarium* enzymes were for the most part similar to those in barley. In another study, Houghton *et al.* (1988) preincubated dark-grown *C. caldarium* cells with 5–25 μM gabaculine (3-amino-2,3-dihydrobenzoic acid), an inhibitor of pyridoxal phosphate requiring enzymes (Walsh 1984), and subsequently exposed the cells to light. Chl-a and PBP synthesis were progressively inhibited at increasing concentrations of inhibitor with 50% inhibition at 10 μM. Gabaculine inhibition could be overcome by feeding ALA. It had been established that *in vitro* gabaculine inhibits the aminotransferase which converts glutamate-1-semialdehyde to ALA (Corriveau and Beale 1986; Werck-Reichhart *et al.* 1988; Hoober *et al.* 1988). It seems likely that gabaculine inhibition of Chl-a and PBP synthesis in *C. caldarium* can be ascribed to inhibition of the aminotransferase *in vivo*.

4.3 Intermediates between ALA and protoporphyrin IX

The enzymes in the porphyrin biosynthetic pathway have not been isolated and characterized from *C. caldarium* although there is indirect evidence from ALA-feeding experiments that the enzymes and intermediates are the same as those in other organisms. Troxler and Bogorad (1966) incubated *C. caldarium* with 10 mM ALA in either dark or light and observed that the cells excreted porphobilinogen (PBG), porphyrins and PCB into the suspending medium. The excreted porphyrins were: uroporphyrin III, 7-, 6-, and 5-carboxyl porphyrins, coproporphyrinogen III and protoporphyrin IX. The excreted porphyrins were undoubtedly oxidation products of the corresponding porphyhrinogens which are the actual biosynthetic intermediates in the pathway. In similar experiments, Troxler and Lester (1967) incubated cells with $^{14}C-4$-ALA and measured the specific radioactivity (dpm/μmole) of excreted compounds. They found that the specific radioactivity of PBG was 2 times greater and that the specific radioactivity of excreted porphyrins and PCB was 8 times greater, than the specific radioactivity of added ALA. These data provided direct evidence for synthesis of porphyrins and PCB from ALA in *C. caldarium*.

4.4 Synthesis of heme

Ferrocheletase (EC 4.9.9.1.1) catalyzes insertion of iron into protoporphyrin IX to form heme in the first reaction of the iron branch of the porphyrin pathway (for review, see Dailey 1990) and N-methyl porphyrins are potent inhibitors of this enzyme (De Matteis *et al.* 1980; Tephly *et al.* 1981). Brown *et al.* (1982) preincubated dark-grown *C. caldarium* cells with N-methyl protoporphyrin IX dimethyl ester (NMP) and observed 80% inhibition of PBP synthesis during subsequent incubation in the light. The degree of PBP inhibition in the light increased as the concentration of NMP was increased from 5–20 μM. In similar experiments with *C. caldarium*, Beale and Chen (1983) observed that addition of N-methyl mesoporphyrin IX inhibited synthesis of PBPs in the dark-greening mutant, CPD, of *C. caldarium*. The interpretation of these results is that NMP is taken up by cells, once inside enters the chloroplast and inhibits ferrocheletase, resulting in a reduction in heme synthesis which leads to a deficiency in PCB and failure to make PBPs. The NMP studies are also of interest because Hudson and Smith (1975) had suggested that PCB could just as easily arise from a metalloporphyrin in the magnesium branch of the porphyrin pathway. Inhibition of accumulation of APC and CPC by inhibitors of ferrocheletase provide strong evidence for PCB synthesis from heme.

Brown *et al.* (1984a) measured ferrocheletase in extracts of dark-grown *C. caldarium* and noted that the activity increased substantially after exposure to light. The kinetics of the rise in ferrocheletase activity preceded, but was similar to the kinetics of APC and CPC accumulation in the light. This would be predicted if there was an increased heme demand for making PCB to chromophorylate APC and CPC apoproteins. However, these studies did not address the question of the mechanism underlying the increase in ferrocheletase activity in the light. The gene for ferrocheletase is almost certainly in the nucleus and the increase in ferrocheletase activity could reflect regulation at either the transcriptional or post-transcriptional level. Light could enhance transcription leading to an increase in ferrocheletase mRNA and an increase in enzyme protein. Alternatively, light could positively or negatively regulate the rate of ferrocheletase mRNA turnover, or the efficiency of translation in the cytosol, or the rate of transport into the chloroplast, or possibly chaperonin assisted folding of the enzyme once in the chloroplast.

4.5 Heme conversion to biliverdin IXα

Bile pigments in animals arise from catabolism of heme in senescent erythrocytes and turnover of hepatic hemoproteins. Heme breakdown in animals is catalyzed by microsomal heme oxygenase, a mixed function oxygenase which (a) oxidizes the α-bridge carbon and releases it as carbon monoxide, (b) inserts two oxygen atoms into the pyrrole rings at the positions previously occupied by the α-methene bridge, and (c) releases biliverdin IXα and iron (for review, see Brown and Troxler 1982). Mammalian heme oxygenase requires NADPH as reductant and electrons are transferred from NADPH to heme oxygenase by a second microsomal enzyme, NADPH cytochrome c reductase.

Table 2. Incorporation of 14C-labeled heme into PCB. Packed cells (10 ml) in 100 ml of pH 7.0 culture medium were preincubated with 3.0 mg ^{14}C-heme (23.2 x 10^{10} dpm/mole) for 5 h, resuspended in 1 liter of culture medium, pH 2.0, and incubated in light for 64 h. Cells were disrupted in a French pressure cell, APC and CPC were isolated, PCB was cleaved from apoprotein by methanolysis and purified as the dimethylester (PCBDME). Under these conditions the methanol adduct of PCBDME is produced as well. Both the dimethylester and the methanol adduct of the dimethylester were purified by thin layer chromatography Adsorbosil 5 and assayed for radioactivity in a liquid scintillation spectrometer. Chl-a was extracted from the cell pellet in acetone:water (5:1), purified by thin layer chromatography on Silica gel G, and assayed for radioactivity

Sample	Pigment recovered (nmol)	Radioactivity (dpm)	Specific radioactivity (dpm/mol x 10^{10})	Fraction from added heme (%)
PCBDME	50.9	2322	4.6	20
PCBDME MeOH adduct	100.0	4414	4.4	19
Chl-a	79.1	0	0	0

Troxler *et al.* (1970) provided indirect evidence for heme oxygenase in *C. caldarium* by showing that cells actively making PCB (attached to APC and CPC apoproteins) released one mole of CO for each mole of PCB. This would be expected if heme is an intermediate in biosynthesis of PCB. Cells incubated with ^{14}C-5-ALA synthesized radiolabeled PCB and CO in which the PCB:C0 ratio of specific radioactivy was 7:1 which is exactly what would be predicted if heme oxygenase cleaved heme to biliverdin IXα and the resulting biliverdin was converted to PCB (Troxler 1972). Since *C. caldarium* incubated with ALA excretes tetrapyrroles we reasoned that cells should also take up tetrapyrroles and therefore carried out the direct experiment of incubating cells with radiolabeled heme. Dark-grown cells were allowed to take up ^{14}C-heme in culture medium at pH 8.0 (where heme is soluble) and the medium was then adjusted to pH 2.0 (where heme is insoluble) to trap heme inside and the cells were exposed to light for 24 h. Phycocyanin was isolated, PCB was cleaved from apoprotein, methylated, purified by thin layer chromatography and assayed for radioactivity (Table 2). These data show that exogenous heme fed to cells was converted into PCB within the cells and then attached to APC and CPC apoproteins. No radioactivity from heme was incorporated into Chl-a. This eliminates the possibility that once inside the cell heme was demetallated to protoporphyrin IX and subsequently converted to magnesium protoporphyhrin IX which would then be incorporated into Chl-a. These data essentially exclude the possibility of PCB synthesis from an intermediate in the magnesium branch of the porphyrin pathway.

Beale and Cornejo (1984a) were the first to demonstrate heme oxygenase activity in extracts of *C. caldarium*. Heme oxygenase in the high speed supernatant of broken cells converted mesoheme to mesobiliverdin IXα in a reaction requiring NADPH, molecular oxygen, and a second reductant such as ascorbate. Mesoheme was used as the substrate because product mesobiliverdin IXα is not a substrate for the enzymes which convert biliverdin IXα to PCB. The reaction was inhibited by Sn-protoporphyrin IX, a known inhibitor of mammalian heme oxygenase. Beale and Cornejo (1988) fractionated the soluble heme oxygenase activity into 3 fractions by sequential chromatography on DEAE-cellulose, Reactive Blue 2-Sepharose and 3′,5′-ADP-agarose. Fraction I was retained by DEAE cellulose at low salt concentration and contained a Mr 22,000 ferredoxin-like protein with an Fe/S cluster; fraction II was retained by Blue Sepharose at low salt concentration and contained a Mr 38,000 heme binding protein (heme oxygenase); fraction III was retained by 3′,5′-ADP-agarose are eluted by NADPH and contained a Mr 37,000 ferredoxin-linked cytochrome c reductase. A possible reaction sequence could involve heme binding to heme oxygenase (fraction II), reduction of Fe^{+3} in heme to Fe^{+2} by ascorbate, 0_2 binding to heme iron forming Fe^{+2}-0=0, reduction to Fe^{+2}-0=0- by NADPH with electrons transferred through ferredoxin and NADPH cytochrome c reductase (or vice versa), oxidation and release of the α-methene bridge as carbon monoxide and insertion of oxygen atoms into the pyrrole rings.

In the reconstituted system, fraction I could be replaced by *Porphyra umbilicalis* ferredoxin, frac-

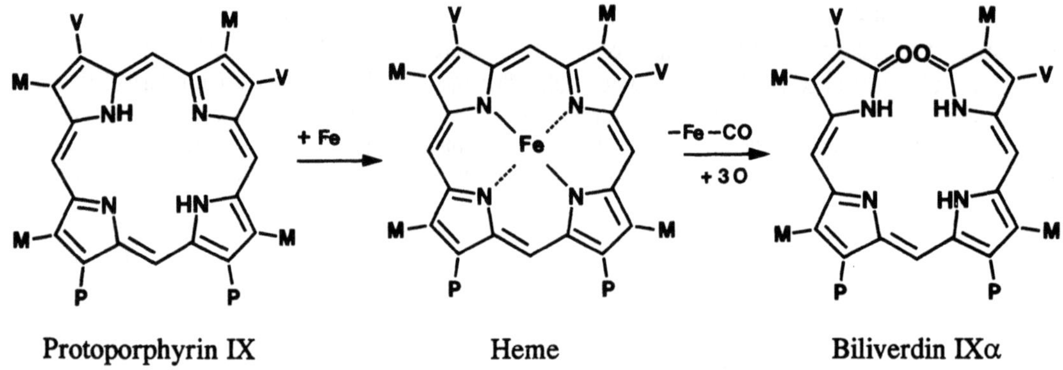

Fig. 7. Heme conversion to biliverdin IXα.

tion II was rate limiting, and none of the fractions could be replaced by bovine spleen heme oxygenase or NADPH cytochrome P^{450} reductase. The components of the algal heme oxygenase system are found in the soluble fraction of the high speed supernatant whereas mammalian heme oxygenase is found in microsomes. This indicates that evolution of the heme oxygenase system in plants and animals has followed divergent paths despite the fact that they catalyze the same reaction, e.g., cleavage of the heme ring.

4.6 Mechanism of heme oxygenase

When the heme macrocycle is cleaved the α-methene bridge carbon is oxidized and removed as CO and two (lactam) oxygen atoms are inserted in its place (Fig. 7). There are four possible mechanisms to account for the origin of the lactam oxygen atoms. In the 'double hydrolytic mechanism' both oxygen atoms originate from water, in the 'hydrolytic mechanism' one oxygen atom originates from water and the other from molecular oxygen, in the 'one molecule mechanism' both oxygen atoms originate from the same oxygen molecule and in the 'two molecule mechanism' both oxygen atoms originate from molecular oxygen but from different oxygen molecules. Troxler *et al.* (1979) investigated the mechanism of heme oxygenase in *C. caldarium* by allowing dark-grown cells to produce APC and CPC in the light under an atmosphere enriched with isotopically labeled molecular oxygen ($^{18,18}0_2$). After pigment synthesis had been completed, CPC and APC were isolated, PCB was cleaved from apoprotein and purified as the dimethyl ester. Incorporaion of $^{18}0$ into the lactam oxygen atoms was measured in the mass spectrometer. The molecular ion, M, of PCB dimethylester occurs at m/e 614 and an increase in peak at M+2 represents incorporation of one atom of $^{18}0$, and an increase in the peak a M+4 represents incorporation of two atoms of $^{18}0$, into the lactam oxygen atoms. Quantitative analyses of the peaks at M, M+2 and M+4 revealed that PCB was made by a two molecule mechanism in which both lactam oxygen atoms come from molecular oxygen but from different oxygen molecules (Table 3). Brown *et al.* (1980) confirmed the mechanism of algal heme oxygenase in similar experiments with *C. caldarium* where somewhat higher ^{18}O incorporations into PCB were achieved. Heme degradation by heme oxygenase in mammals also occurs by a two molecule mechanism (Brown and King 1978). The fact that algal and mammalian heme oxygenase have the identical reaction mechanism with respect to the origin of the lactam oxygen atoms is really quite remarkable. Both enzymes obtain electrons from NADPH but the electron carriers are quite different. Mammalian heme oxygenase is found in microsomes and accepts electrons from NADPH-cytochrome reductase. Cornejo and Beale (1988) fractionated a soluble heme oxygenase activity in *C. caldarium* into three protein components, a ferredoxin-like protein, a ferredoxin-$NADP^+$-reductase and a heme binding protein. The latter probably represents the actual heme oxygenase protein where the heme macrocycle is cleaved. The environment of the heme pocket is probably quite similar in both the algal and mammalian heme oxygenase enzymes in view of the fact

Table 3. Mechanism of the heme oxygenase reaction in *C. caldarium*. Cells were incubated in the light under an atmosphere of isotopically labeled molecular oxygen, APC and CPC were isolated, PCB was cleaved from apoprotein, esterified, and analyzed in the mass spectrometer

			Relative size of peaks[a]				
Gas phase	$^{18,18}O_2$	m/e	Double hydrolytic	Hydrolytic	One molecule	Two molecules	Observed
Control	0	614					99.6
		616					0.4
		618					0
Expt. 1	10.4	614	100	89.4	89.6	80.3	82.5
		616	0	10.4	0	18.6	16.5
		618	0	0	10.4	1.1	1.0
Expt. 2	13.0	614	100	87.0	87.0	75.7	82.5
		616	0	13.0	0	22.6	16.5
		618	0	0	13.0	1.7	1.0

(a) The predicted incorporation of ^{18}O at M+2 and M+4 in the double hydrolytic, the hydrolytic and the one molecule mechanism is self explanatory. Predicted incorporation for the two molecule mechanism in Experiment 1 was calculated as follows: final concentration of $^{18,18}O_2$ in the gas phase of the flask containing algae = 10.4 atom%. The probability of incorporating 2 atoms of ^{16}O (into the lactam oxygen atoms of PCB) is 0.896 x 0.896 x 100 = 80.3%. The probability of incorporating 2 atoms of $^{18,18}O_2$ is 0.104 x 0.104 x 100 = 1.1%. The probability of incorporating 1 atom of ^{16}O and 1 atom of ^{18}O is (0.896 x 0.104) + (0.896 x 0.104) x 100 = 18.6%.

that the lactam oxygen atoms in biliverdin IXα originate by a two molecule mechanism.

This suggests that the mechanism of heme catabolism is an ancient biochemical reaction which evolved before divergence of plants and animals. The linear tetrapyrrole products of the heme oxygenase reaction have followed very different paths of biochemical evolution in that mammalian bile pigments are transient elimination products of hemoglobin catabolism with no known function whereas plant bile pigments carry out essential photoreceptive functions vital to survival of the species.

4.7 Biliverdin conversion to PCB

Formally, PCB is derived from biliverdin IXα by two 2 electron reductions. However, the incorporation of ^{14}C-heme into PCB *in vivo* (Brown *et al.* 1981) or conversion of mesoheme to mesobiliverdin IXα by a crude heme oxygenase preparation in vitro (Beale and Cornejo 1984a, 1988), suggests, but does not prove, that biliverdin IXα is an intermediate in the pathway between heme and PCB. In an attempt to address this point, Beale and Cornejo (1984c) incubated *C. caldarium*, mutant CPD, with ^{14}C-biliverdin and NMP in the dark for 24 h and measured incorporation of radioactivity into PCB cleaved from APC and CPC apoproteins. The ferrocheletase inhibitor, N-methyl protoporphyrin, was employed to minimize endogenous heme synthesis and therefore maximize incorporation of radiolabeled biliverdin IXα into PCB. CPD was used because it is makes PBPs in the dark and potentially phototoxic effects of N-methyl protoporphyrin were excluded. Approximately 50% of the ^{14}C-biliverdin was taken up by cells in the course of the experiments and approximately 30% of this was incorporated into PCB providing good evidence that biliverdin IXα is an intermediate. In a similar experiment, Brown *et al.* (1984b) preincubated dark-grown *C. caldarium* cells with ^{14}C-biliverdin in the dark, subsequently exposed the cells to light for 48 h, and found that approximately 10% of the radiolabeled biliverdin was incorporated into PCB. Brown and Holroyd (1984) showed that mesoheme is not a precursor of PCB in *C. caldarium*. Collectively, these data provide compelling evidence that biliverdin IXα is the first ring-opened intermediate in PCB biosynthesis *in vivo*.

Beale and Cornejo (1984b,c) showed that a crude extract of *C. caldarium* converted biliverdin IX-α to PCB in vitro. Recently, this extract was fractionated into three enzymatic components by ammonium sulfate precipitation and sequential chromatography on ion exchange and affinity columns (Beale and Cornejo 1991a). Two partially purified components were identified and found to

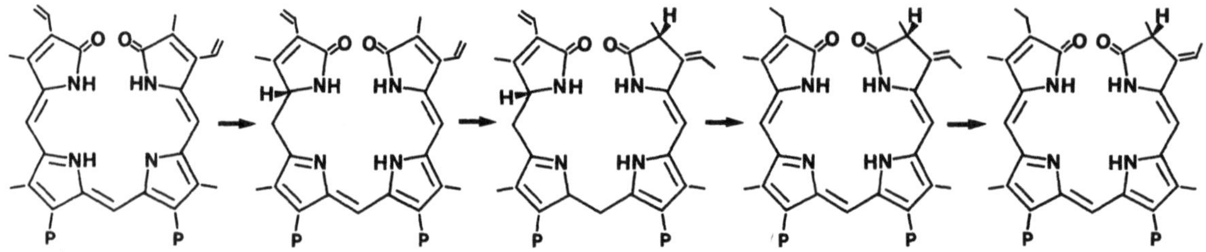

Biliverdin IXα 15,16-Dihydrobiliverdin IXα 3(Z)-Phycoerythrobilin 3(Z)-Phycocyanobilin 3(E)-Phycocyanobilin

Fig. 8. Biosynthetic intermediates between biliverdin IXα and PCB in *C. caldarium* Beale and Cornejo 1991a,b,c).

be identical to ferredoxin and ferredoxin-$NADP^+$ reductase of the heme oxygenase system. The third component of the system was comprised of one or more isomerase enzymes. Biosynthetic intermediates between biliverdin IXα and PCB were isolated from crude or partially reconstituted reaction mixtures and characterized by HPLC retention times by reference to standards, absorption spectra in the visible and NMR spectroscopy (Beale and Cornejo 1991b,c). These studies led to elucidation of the following pathway for biliverdin conversion to PCB (Fig. 8). The first intermediate, 15,16-dihydrobiliverdin IXα is produced from biliverdin IXα by a 2 electron reduction with NADPH as the reductant and electron transfer via ferredoxin and ferredoxin-$NADP^+$-reductase. This reduction takes place at the 15–16 double bond of biliverdin IX-α at the methene bridge carbon connecting rings III and IV. 15,16-Dihydrobiliverdin IXα is the substrate for the second 2 electron reduction which yields 3(Z)-phycoerythrobilin, again with electrons from NADPH transferred via ferredoxin and ferredoxin-NADP-reductase. This reduction occurs at the 2–3 double bond on ring I. 3(Z)-Phycoerythrobilin is then isomerized by the third (isomerase) component to 3(Z)-PCB and this compound is isomerized to 3(E)-PCB. The latter is the compound released from APC or CPC by methanolysis and is presumably the compound added to APC and CPC apoproteins during PBP biosynthesis *in vivo*. An interesting feature of this pathway is that reduction of biliverdin IXα directly at the 2–3 double bond of ring I by the second reduction step above would yield phytochromobilin, the chromophore of phytochrome.

4.8 Biosynthesis of PBPs

Relatively little is known about regulation of PBP biosynthesis in *C. caldarium* or for that matter in other algae. In *C. caldarium* synthesis of APC and CPC is dependent upon light and occurs simultaneously with chloroplast development. Broadly speaking, synthesis of a APC or CPC requires synthesis of PCB via the iron branch of the porphyrin pathway, transcription of the genes for APC and CPC followed by translation of the resulting mRNAS, attachment of PCB to the correct cysteine residues in apoproteins, protein folding to produce native subunits and aggregation to form monomers comprised of one α and one β subunit. Theoretically, one or more of these events could be regulated in some way by light. Once formed monomers aggregate into trimers and then into hexamers, and these associate with linker polypeptides to form phycobilisomes. This process is probably a thermodynamically driven self-assembly process that is not regulated as such although there is no direct evidence for or against this hypothesis.

Troxler and Brown (1970; 1975) measured incorporation of ^{14}C-4-ALA and 4,5-^{3}H-leucine into CPC in *C. caldarium* and found that the ^{14}C was present exclusively in PCB and ^{3}H was present exclusively in apoprotein. This indicated that both PCB and CPC apoprotein were made *de novo* in the light and that there was strict coordinate regulation between PCB biosynthesis and regulation of expression of the genes for PBPs.

Belford *et al.* (1983) fractionated RNA from *C. caldarium* by affinity chromatography on oligo(dT) cellulose into $polyA^+$ and $polyA^-$ fractions and showed that these RNAs stimulated incorporation of radiolabeled methionine incorporation into protein in a reticulocyte cell free system. Proteins with sizes nearly identical to native α and β subunits of APC and PC were immunoprecipitated from the polyA$^-$ (but not the polyA$^+$) RNA fraction using antisera directed against native APC and CPC. Radiosequencing of polypeptides immunoprecipitated from translation reactions with CPC antisera from translation reactions containing different radiolabeled amino acids positively identified these polypeptides as the α and β subunits of CPC. Several meaningful conclusions can be drawn from these results. The occurrence of APC and CPC mRNAs in the polyA$^-$ fraction provided the first evidence to suggest that PBPs may be encoded in the chloroplast genome because chloroplast mRNAs are not polyadenylated. The fact that the Mrs of vitro translated PBP polypeptides were nearly the same as those of the native proteins and that methionine was the amino-terminal amino acid is strong evidence that PBPs in red algae are not made as larger precursor proteins with transit peptides which become cleaved during transport across the chloroplast envelope. Steinmüller and Zetsche (1983) carried out similar kinds of experiments and found that the mRNA for the large and small subunit of ribulose-1,5-bisphosphate carboxylase was present in the polyA$^-$ RNA fraction of *C. caldarium*.

Egelhoff and Grossman (1983) used a somewhat different approach to suggest that the genes for PBPs are encoded in the chloroplast genome and the genes for linker polypeptides are encoded in the nucleus. *C. caldarium* cells were pulse labeled with $^{35}SO_4^{=}$ followed by isolation of phycobilisomes and analysis of the labeling pattern by SDS-PAGE and autoradiography. All phycobilisome polypeptides were labeled in control cells but in cells incubated with cycloheximide APC and CPC apoproteins were labeled but linkers were not. In cells incubated with chloramphenicol linkers were labeled but APC and CPC apoproteins were not. Based on cycloheximide inhibition of protein synthesis on 80S ribosomes in the cytosol and of chloramphenicol inhibition of protein synthesis on 70S ribosomes in the chloroplast, it was concluded that some genes for linkers are nuclear encoded and the genes for PBPs are chloroplast encoded.

Aggregation of PBPs and linker polypeptides and assembly of phycobilisomes is a remarkable process and depends upon complex communication between regulatory processes in different organelles and complex intercellular signal transduction and trafficking of proteins and tetrapyrroles. PCB is synthesized from ALA via heme and becomes attached to PBP apoproteins. Chloroplast genes for APC and CPC apoproteins and nuclear genes for linker polypeptides are expressed simultaneously and in the correct stoichiometry with respect to one another and to PCB. Once this is accomplished PBPs and linker polypeptides self assemble into phycobilisomes and phycobilisomes position themselves over PSII reaction centers on thylakoid membranes.

4.9 Biosynthesis of Chl-a

Chl-a is not normally made in the dark in wild-type cells of *C. caldarium*. The exception is IIID2 which makes minute amounts of Chl-a in the dark and the mutant CPD where the light requirement for pigment synthesis and chloroplast development has been lost. The fact that CPD is a spontaneous mutant indicates the light requirement for chloroplast development is under control of a single regulatory gene. Troxler and Brown (1980) observed that cells of IIID2 incubated with ALA in the dark synthesized large quantities of Chl-a but that these cells were unable to perform photosynthesis. This shows that protochlorophyllide oxido reductase, the enzyme which transforms protochlorophyllide-a to chlorophyllide-a, is not light dependent in *C. caldarium*. This enzyme is light dependent in angiosperms with the result that etiolated plants accumulate only small quantities of protochlorophyllide-a in the dark which is presumed to prevent further synthesis by feedback inhibition on one of the ALA making enzymes and possibly on magnesium protoporphyrin IX cheletase, the first enzyme in the magnesium branch of the porphyrin pathway (for review, see Beale and Weinstein 1990). The fact that ALA-fed *C. caldarium* cells accumulate Chl-a (vs. protochlorophyllide-a) in the dark means that protochlorophyllide oxidoreductase is probably not a control point for regulation of Chl-a synthesis in this organism.

5. Regulation of pigment synthesis

5.1 Spectral response curves

It is reasonable to ask if there are separate and specific photoreceptor for APC and CPC biosynthesis in light or whether there is a general photoreceptor for chloroplast development in which case PBP biosynthesis is one of many processes triggered by a developmental program regulating organelle biogenesis. Nichols and Bogorad (1962) addressed this question by determining the spectral response curve for APC and CPC formation in *C. caldarium*. Mutant GGB was used because this strain does not produce Chl-a which, if present, could act a light absorbing filter and obscure detection of minute quantities of a specific photoreceptor for PBP biosynthesis. Dark-grown cells of GGB were incubated in light provided by interference filters allowing transmission of only narrow ranges of wavelengths of visible light. The effectiveness of different wavelengths of visible light in promoting PBP accumulation was determined in living cells by the increase in absorbance at the *in vivo* absorption maximum of CPC (620 nm). The spectral response curve for CPC displayed peaks at about 420 nm and 550 nm with a 420:550 effectiveness ratio of 10:1. This crude 'action spectrum' was similar to the absorption spectrum of a hemoprotein. It was suggested that the hemoprotein could either be a photoreceptor directing a critical rate limiting step in formation of CPC, or the heme moiety of the hypothetical heme-protein complex could be photooxidized to PCB such that the resulting bile pigment-apoprotein complex could become a subunit of CPC.

Schneider and Bogorad (1979) extended these observations by measuring spectral response curves for both Chl-a and PBPs in IIID2, for Chl-a in IIIC, and for CPC in GGB. The spectral response curve observed for IIID2 displayed peaks of effectiveness at 440 nm, 575 nm, and 645 nm. The spectral response curve for IIIC displayed peaks of effectiveness at 440 nm and 645 nm. The spectral response curve for GGB displayed peaks of effectiveness at 440 nm and 575 nm. These data were interpreted to mean that there are two photoreceptors for pigment synthesis in *C. caldarium*, one absorbing at 440 nm and 645 nm for Chl-a and a second absorbing at 440 nm and 575 nm for PBPs. As yet neither photoreceptor has been isolated and characterized.

5.2 Photoregulation of gene expression

A basic question pertinent to regulation of PBP synthesis in *C. caldarium* is whether cells grown in the dark accumulate APC and CPC apoproteins and which become chromophorylated when PCB synthesis is 'turned on' in the light. We investigated this possibility by performing Northern and Western blotting experiments with RNA and protein, respectively, from cells which had been grown in the dark and or from cells which had been exposed to light for 24 h (Troxler *et al.* 1989). Soluble protein was subjected to SDS-PAGE, blotted onto nitrocellulose and the blots were probed with antisera directed against APC and CPC (Fig. 9A). No immunodetectable APC or CPC apoprotein was found in soluble protein from dark cells. As expected a strong immuno-positive signal for APC and CPC was seen in soluble protein extracted from light cells. Since these antisera can detect as little as 1 ng of protein (Belford *et al.* 1983) it is safe to conclude that PBP apoproteins are either present in minute quantities or absent in the dark. Failure to immunologically detect PBP apoproteins in the dark does not exclude the possibility that mRNAs for APC and CPC are present but are translationally inactive.

Therefore, total RNA from dark and light cells was electrophoresed on agarose denaturing gels, blotted onto nylon membranes, and the blots were hybridized to DNA probes for apcAB and cpcBA (Fig. 9B). A strong hybridization signal was seen for both APC and CPC mRNAs in light but no detectable mRNAs were observed in dark. An oversimplification of this finding is that PBP genes are 'off' in dark and 'on' in light. By definition, this identifies the genes coding for the α and β apoproteins of APC and CPC as positively regulated photogenes. It is not possible to ascertain from these data per se whether light regulation occurs at the transcriptional or post-transcriptional level. Interestingly, cells of *C. caldarium* mutant IIIC (Table 1), which produce Chl-a but not PBPs in the light, do not contain detectable quantities of APC and CPC mRNA or apoproteins (Lin *et al.* 1990).

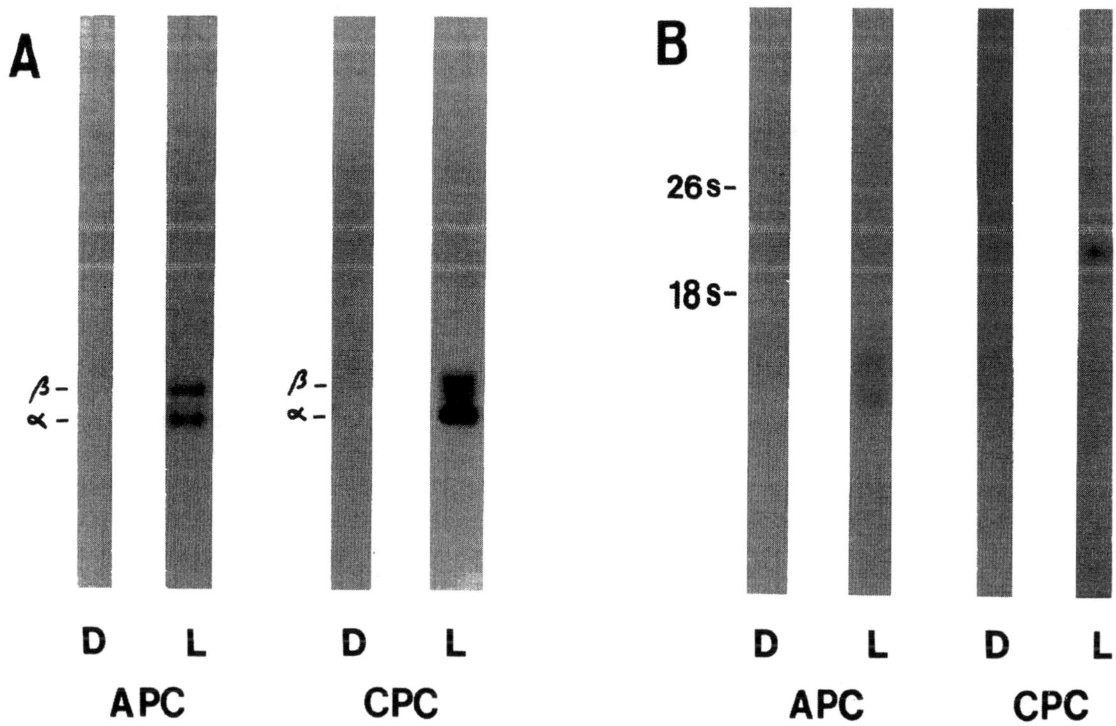

Fig. 9. Effect of light on APC and CPC apoproteins and mRNAs in *C. caldarium*. A. Cells grown in the dark or light were extracted with dimethylformamide to remove Chl-a and carotenoids, suspended in 0.2 M potassium phosphate buffer, pH 7.4, disrupted by passage through a French pressure cell twice at 14,000 p.s.i., centrifuged, and proteins in the pellet were extracted with 8 M urea. Samples (150 μg protein/lane) were subjected to SDS-PAGE on 12.5% gels, electrophoretically transferred to nitrocellulose and the resulting blots probed with immune serum directed against APC and CPC (Belford *et al.* 1983). B. Cells grown in the dark or light were suspended in 6 M guanidine isothiocyanite, disrupted by passage through a French pressure cell once at 14,000 p.s.i., and total RNA was isolated by the method of Chomczynski and Sacchi (1987). RNA was electrophoresed on agarose denaturing gels, blotted onto nylon membrane, and the blots were hybridized with gene specific probes (Troxler *et al.* 1989). The APC probe was the EcoRV subfragment of a 1.5 kb BamH1/EcoRl fragment containing cpcAB from *Cyanophora paradoxa* (Bryant *et al.* 1985). The CPC probe was a 3.0 kb HindIII fragment containing the genes for the α and β subunit polypeptides of CPC from *Agmenellum quadruplicatum* (De Lorimier *et al.* 1984). These probes were kindly provided by Dr. D. A. Bryant, Pennsylvania State University.

5.3 Kinetics of appearance of APC and CPC mRNA in light

When dark-grown cells of *C. caldarium* are placed in the light APC and CPC accumulate over a time span of about 24 h (Fig. 2). APC and CPC apoproteins and mRNAs are not detectable in the dark but are abundant after 24 h in light (Fig. 9). To gain insights into this process, Northern blots of RNA isolated from cells after various times in the light were hybridized with DNA probes for apcAB and cpcBA (Fig. 10). Transcripts became detectable after 1 h and progressively increased in abundance with time in light. After 23 h in light, cells were treated with rifampicin (10 μg/ml)) and incubated for an additional 1–3 h in light. Transcript levels changed little in controls but decreased 30% after 1 h and 95% after 3 h in cells given rifampicin. We have shown that rifampicin blocks greening in *C. caldarium in vivo* (Fig. 2; Troxler *et al.* 1989) and inhibits chloroplast RNA polymerase *in vitro* (R. F. Troxler, unpublished). Thus, the disappearance of PBP transcripts in cells exposed to rifampicin is actually a measure of mRNA turnover because transcription (new mRNA synthesis) does not occur under these conditions. Therefore, rifampicin provides a powerful tool for measuring the relative turnover rates of mRNAs of chloroplast genes in *C. caldarium*. It remains to be determined whether accumulation of apcAB and cpcBA transcripts during the greening process is due to increased transcription, decreased turnover, or both.

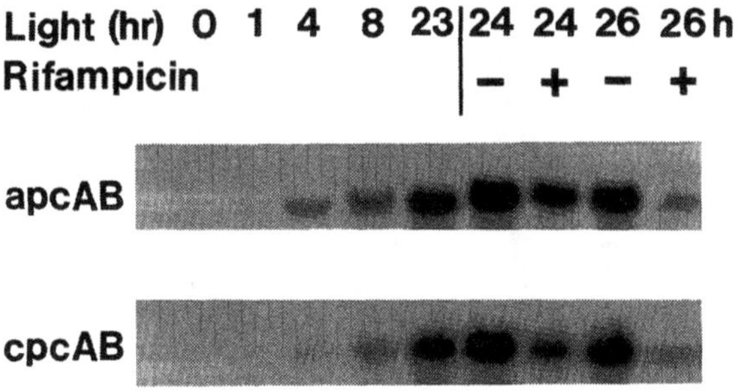

Fig. 10. Appearance of APC and CPC mRNAs in light and effect of rifampicin on PBP message levels in *C. caldarium*. Dark-grown cells were suspended in minimal medium minus glucose at 10^8 cells/ml and incubated in the light. After 23 h cells were treated with rifampicin (10 μg/ml) and incubated for an additional 1–3 h in light. RNA was isolated, electrophoresed, blotted and the blots were hybridized as described in Figure 9. Probes were made by Polymerase Chain Reaction (PCR) from *C. caldarium* genomic DNA with oligonucleotide primers designed on the basis of sequence similarity between the amino acid sequence of APC and CPC from *C. caldarium* and the deduced amino acid sequences of APC from *C. paradoxa* and CPC from *A. quadruplicatum* (De Lorimier et al. 1984; Bryant *et al.* 1985). The 465 bp APC probe was comprised of the region coding for amino acids 12–155 of the β subunit in *C. caldarium* APC. The 428 bp CPC probe was comprised of the region coding for amino acids 12–158 of the β subunit of CPC in *C. caldarium* (Lin 1990).

5.4 Heme regulated genes

The spectral response curves for synthesis of APC and CPC suggested the possible involvement of a hemoprotein as a photoreceptor and/or precursor of PBPs in *C. caldarium*. There are several examples where heme serves as a positive or negative modulator of gene expression in yeast and in mammals (Padmanaban *et al.* 1989). One of the most thoroughly studied cases is that of CYC1 and CYC7, the genes for iso-1-apocytochrome and iso-7-apocytochrome c, in yeast. Guarente and coworkers (Guarente and Mason 1983; Pfeifer *et al.* 1989) discovered a nuclear protein named HAP (heme- activating protein) which when bound to heme promotes transcription of CYC1 and CYC7 by binding to a promoter element called UAS (upstream activating sequences). Heme has also been implicated indirectly as a factor in regulation of Chl-a biosynthesis in higher plants although the mechanism of this process is incompletely understood (for review, see Beale and Weinstein 1990). These findings suggest the possibility that heme could, in some way, regulate expression of the apcAB and cpcBA in *C. caldarium*. Troxler *et al.* (1989) tested this hypothesis by incubating (dark-grown) *C. caldarium* cells with ALA, protoporphyrin IX, heme, biliverdin IXα and PCB in the dark measuring formation of PBP apoproteins on Western blots probed with antisera directed against APC and CPC (Fig. 11). Cells incubated with ALA, protoporphyrin IX and heme synthesized APC and CPC apoproteins in the dark. Cells incubated with biliverdin IXα or PCB in the dark did not. These results strongly implicate heme as a positive modulator of apcAB and cpcBA expression, and it was concluded from these results that heme is the biologically active molecule which substitutes for light and promotes expression of apcAB and cpcAB.

The effect of heme on PBP genes raises the question as to whether expression of other chloroplast genes in *C. caldarium* is also regulated by heme. This was examined in an experiment where dark-grown cells were incubated in the dark with 100 μM heme for 24 h, total RNA was isolated, and compared to total RNA from cells grown in dark or from cells grown in light, on slot blots hybridized with DNA probes for several chloroplast genes (Fig. 12). This experiment showed that heme promoted the appearance of transcripts of apcAB and cpcBA in the dark but had no effect on psaA (gene for a P700 reaction center protein of PSI), psbA (gene for the Q_β quinone binding protein of PSII), and rbcL and rbcS (genes for the large and small subunit of ribulose-1,5-bisphosphatase carboxylase). These results illustrate that there are at least two kinds of photogenes in *C. caldarium*, those regulated by heme and those regulated by an as yet unidentified mechanism. These data show that heme does not serve as a global modulator of the chloroplast genome in this organism.

The mechanism of heme regulation could be complex. By analogy to positive heme regulation of CYC1 and CYC7 in yeast, one could imagine that heme is made in the chloroplast in response to light and binds to a protein factor which is then able to bind to promoter-like elements in the upstream (or downstream) regions of apcAB and cpcBA. The protein factor in question could be a subunit of chloroplast RNA polymerase or it could be a transcription factor encoded in either the chloroplast or nuclear genome. This heme protein complex could

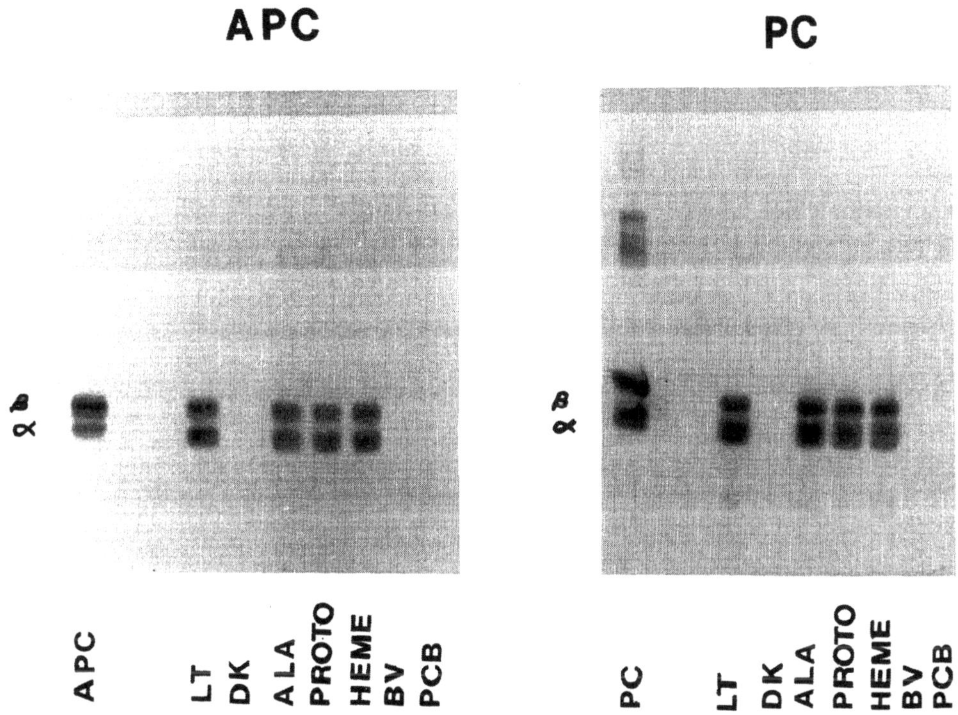

Fig. 11. Effect of ALA and tetrapyrroles on APC and CPC apoprotein in the dark. Cells were grown in the dark, collected by centrifugation, and 1 ml of packed cells was incubated in 10 ml of minimal medium adjusted to pH 8.0 containing 170 μM protoporphyrin IX, heme, biliverdin, or PCB for 2 h in the dark. The cells were centrifuged, resupended in 40 ml minimal medium plus glucose, pH 2.0, and incubated for in the dark for 24 h. For ALA treatment, 1 ml of packed cells was incubated in 40 ml minimal medium plus glucose, pH 2.0, containing 10 mM ALA in the dark for 24 h. Protein isolation, SDS-PAGE and Western blotting were performed as described in the legend to Figure 9.

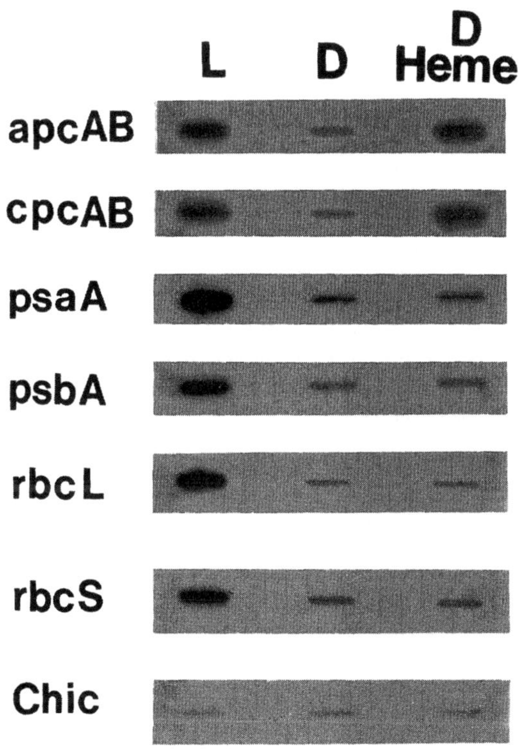

(a) enhance transcription by acting as an general attractant for chloroplast RNA polymerase, (b) displace a repressor protein blocking transcription, (c) effect conformational changes in the chloroplast genome resulting in exposure of promoters and formation of initiation complexes, or (d) interact with other protein factors required to promote tran-

Fig. 12. Effect of heme on other chloroplast genes in *C. caldarium*. RNA was isolated from cells which had been grown in the light, grown in the dark, or which had been incubated in the dark with 170 μM heme for 24 h. Slot blots were prepared with 5 μg RNA/slot and these were hybridized with a series of chloroplast gene probes. Chicken heart RNA (Chic) served as a negative control. The probes for APC and PC were those described in the legend to Figure 10. The psaA (i.e., gene for a P700 reaction center apoprotein) probe was a 2.4 kb BamH1 fragment of the maize chloroplast gene coding for ps1A1. This probe was kindly provided by Dr. L. Bogorad, Harvard University. The psbA, rbcL, rbcS probes contained the complete nucleotide sequences of the coding regions of these genes in *C. caldarium* and were made by Polymerase Chain Reaction from genomic DNA with oligonucleotides designed on the basis of published sequences (Maid *et al.* 1990; Valentin and Zetsche 1990).

scription. Alternatively, by analogy to negative regulation of the lac operon in *E. coli*, a repressor protein could bind to an upstream region of apcAB and cpcBA in the dark and prevent transcription. Binding of heme (normally made in light) to the repressor protein could cause release of the resulting heme-protein complex and enable transcription to proceed. It should be possible to distinguish between these possibilities (eg., positive and negative heme regulation) with tools available today in molecular biology.

6. Photosynthesis

6.1 Size of photosynthetic units

Diner and Mauzerall (1973) described a cell free preparation from *C. caldarium*, mutant IIIC, containing 15 times less Chl-a per photosynthetic unit than normally found in green algae. The preparation was sensitive to 3-(3,4-dichlorophenyl)-1, 1-dimethylurea and displayed an oscillation of 0_2 yield from saturating flashes after a time of dark adaptation. The optical cross section measured by pulsed light saturation experiments suggested that 15 molecules of Chl-a were associated with PSI and PSII, in contrast to about 300 molecules of Chl-a found in most fully greened algae and higher plants. The reduced size of the antenna makes IIIC an attractive subject for studies on the reaction center Chl-a spectrum and kinetics.

Diner and Wollman (1979a) refined these measurements and concluded that the PSI and PSII reaction centers contain about 30–40 Chl-a molecules. However, *Chlorella* and IIIC displayed the same PSII action spectra and low temperature emission spectra and it was suggested that in the PSII center-antenna complex in IIIC represents the core unit found in other algae. Diner and Wollman (1979b) extended these studies by measuring oxygen evolution and fluorescence kinetics during greening of dark-grown cells in light. They found that the size of the PSII antenna was about the same when cells containing 5% of centers was compared to fully greened cells. They also noted that energy transfer between PSII centers occurs early in greening, and that the plastoquinone pool size per PSII did not vary during greening. Diner (1979) measured the distribution of absorbed light energy at 600 nm (absorbed by CPC) and 436 nm (absorbed by Chl-a) by flash yields of 0_2 evolution following illumination at these wavelengths. In the wild-type only half the light absorbed by CPC was transferred to reaction centers of PSII suggesting that only half of the centers were attached to phycobilisomes. Wollman (1979) examined the organization of particles in freeze fractured thylakoid membranes and found that the density of particles was 2 fold greater and the PSII:PSI ratio was 2 times larger in IIIC than in the wild-type. These results indicate that IIIC compensates for missing phycobilisomes by doubling the number of PSII particles containing Chl-a in the PSII antenna. If correct, it would be a very interesting to compare expression of genes coding for PSII polypeptides in IIIC and the wild-type.

Acknowledgements

The author is gratefully acknowledges generous support from The National Science Foundation for his previous research on *C. caldarium* and to National Science Foundation grant – IBN-9119062 for current research support in the preparation of this manuscript.

References

Allen, M. B. (1959). *Arch. Mikobiol.* 32: 270–277.

Avissar, Y. J., Ormerod, J. G. and Beale, S. I. (1989). *Arch. Microbiol.* 151: 513–519.

Beale, S. I. (1971). *Plant Physiol.* 29: 95–102.

Beale, S. I. and Castelfranco, P. A. (1974). *Plant Physiol.* 53: 297–303.

Beale, S. I. and Chen, N. C. (1983). *Plant Physiol.* 71: 263–268.

Beale, S. I. and Cornejo, J. (1984a). *Arch. Biochem. Biophys.* 245: 371–384.

Beale, S. I. and Cornejo, J. (1984b). *Plant Physiol.* 76: 7–15.

Beale, S. I. and Cornejo, J. (1984c). *Arch. Biochem. Biophys.* 227: 279–286.

Beale, S. I. and Cornejo, J. (1988). *J. Biol. Chem.* 263: 11915–11921.

Beale, S. I. and Weinstein, J. D. (1990). In: *Biosynthesis of heme and chlorophylls. (ed. Dailey, H. A.), McGraw-Hill Co., New York.* pp. 287–391.

Beale, S. I. and Cornejo, J. (1991a). *J. Biol. Chem.* 266: 22328–22332.

Beale, S. I. and Cornejo, J. (1991b). *J. Biol. Chem.* 266: 22333–22340.

Beale, S. I. and Cornejo, J. (1991c). *J. Biol. Chem.* 266: 22341–22345.

Belford, H. S., Offner, G. D. and Troxler, R. F. (1983). *J. Biol. Chem.* 258: 4503–4510.
Bottomly, W., Smith, H. J. and Bogorad, L. (1971). *Proc. Nat. Acad. Sci. (US)* 68: 2412–2416.
Brock, T. D. (1978). In: *Thermophilic microorganisms and life at high temperatures, Springer-Verlag, New York.* pp. 255–302.
Brown, A. S., Offner, G. D., Ehrhardt, M. M. and Troxler, R. F. (1979). *J. Biol. Chem.* 254: 7803–7811.
Brown, S. B. and King, R. F. G. J. (1978). *Biochem. J.* 170: 297–311.
Brown, S. B., Holroyd, A. J. and Troxler, R. F. (1980). *Biochem. J.* 190: 445-449.
Brown, S. B., Holroyd, A. J., Troxler, R. F. and Offner, G. D. (1981). *Biochem. J.* 194: 137–147.
Brown, S. B. and Troxler, R. F.(1982). In: *Bilirubin Metabolism. II. (eds. Heirwegh, K. P. M. and Brown, S.B.) CRC Press, Boca Raton, Florida.* pp. 1–38.
Brown, S. B., Holroyd, J. A., Vernon, D. I., Troxler, R. F. and Smith, K. M. (1982). *Biochem. J.* 208: 487–491.
Brown, S. B. and Holroyd, J. A. (1984). *Biochem. J.* 217: 265–572.
Brown, S. B., Holroyd, J. A., Vernon, D. I. and Jones, O. T. G. (1984a). *Biochem. J.* 220: 861–863.
Brown, S. B., Holroyd, J. A. and Vernon, D. I. (1984b). *Biochem. J.* 219: 905–909.
Bruyant, D. A. and Kannangara, C. G. (1987). *Carlsberg Res. Comm.* 52: 99–109.
Bryant, D. A., De Lorimier, R., Lambert, D. H., Dubbs, J. M., Stirewalt, V. L., Stevens, S. E., Porter, R. D., Tam, J. and Jay, E. (1985). *Nat. Acad. Sci. (US)* 82: 3242–3246.
Chomczynski, P. and Sacchi, N. (1987). *Anal. Biochem.* 162: 156–159.
Cole, W. J., Chapman, D. J. and Siegelman, H. W. (1967). *J. Amer. Chem. Soc.* 89: 3643–3645.
Cornejo, J. and Beale, S. I.(1988). *J. Biol. Chem.* 263: 11915–11921.
Corriveau, J. L and Beale, S. (1986). *Plant Science* 45: 9–17.
Dailey, H. A. (1990). In: *Biosynthesis of Heme and Chlorophylls (ed. Dailey, H. A.), McGraw-Hill Publishing Co., New York.* pp. 123–161.
DeLange, R. J., Williams, L. C. and Glazer, A. N. (1981). *J. Biol. Chem.* 256: 9558–9566.
De Lorimier, R., Bryant, D. A., Porter, R. D., Liu, W. Y., Jay, E. and Stevens, S. F. (1984). *Proc. Nat. Acad. Sci. (US)* 81: 7946–7950.
De Matteis, F., Gibbs, A. H. and Smith, A. G. (1980). *Biochem. J.* 189: 645–648.
Diner, B and Mauzerall, D. (1973). *Biochim. Biophys. Acta* 292: 285–290.
Diner, B. (1979). *Plant Physiol.* 63: 30–34.
Diner, B. and Wollman, F. (1979a). *Plant Physiol.* 63: 20–25.
Diner, B. and Wollman, F. (1979b). *Plant Physiol.* 63: 26–29.
Egelhoff, T. and Grossman, A. (1983). *Proc. Nat. Acad. Sci. (US)* 50: 3339–3343.
Frank, G., Sidler, W., Widmer, H. and Zuber, H. (1978). *Hoppe-Seyler's Z. Physiol. Chem.* 359: 1491–1507.
Gantt, E. (1990). In: *Biology of the red algae (ed. Cole, K. M. and Sheath, R. G.), Cambridge University Press, Cambridge, U.K.* pp 203–219.
Gruissem, W. (1989). *Cell* 56: 161–170.
Guarente, L. and Mason, T. (1983). *Cell* 32: 1279–1286.
Houen, G., Gough, S. P. and Kannangara, C. G. (1983). *Carlsberg Res. Comm.* 48: 567–572.
Hoober, J. K., Kahn, A., Ash, D., Gough, S. and Kannangara, C. G. (1988). *Carlsberg Res. Comm.* 53: 11–25.
Houghton, J. D., Turner, L. and Brown, S. B. (1988). *Biochem. J.* 254: 907–910.
Houghton, J. D., Brown, S. B., Gough, S. P. and Kannangara, C. G. (1989). *Carlsberg Res. Commun.* 54: 131–141.
Jin, D. J. and Gross, C. A. (1988). *J. Mol. Biol.* 202: 45–58.
Jurgenson, J. E., Beale, S. I. and Troxler, R. F. (1976). *Biochim. Biophys. Res. Commun.* 69: 149–157.
Kannangara, C. G., Gough, S. P., Bruyant, P., Hoober, J. K., Khan, A. and von Wettstein, D. (1988). *Trends in Biochem. Sci.* 13: 139–143.
Lin, S. (1990). *Ph.D. Thesis, Boston University School of Medicine, Boston, MA.* pp. 1–158.
Lin, S., Offner, G. D. and Troxler, R. F. (1990). *Plant Physiol.* 93: 772–777.
Maid, U., Valentin, K. and Zetsche, K. (1990). *Curr. Genet.* 17: 255–259.
Nichols, K. E. and Bogorad, L. (1960). *Nature* 188: 870–872.
Nichols, K. E. and Bogorad, L. (1962) *Bot. Gaz.* 124: 85–93.
Offner, G. D., Brown-Mason, A. S., Ehrhardt, M. M. and Troxler, R. F. (1981). *J. Biol. Chem.* 256: 12167–12175.
Offner, G. D. and Troxler, R. F. (1983). *J. Biol. Chem.* 2258: 9331–9340.
Padmanaban, G., Venketeswar, V. and Rangarajan, P. N. (1989). *Trends in Biochem. Sci.* 14: 492–496.
Pfeifer, K., Kim, K. S., Kogan, S. and Guarente, L. (1989). *Cell* 56: 291–301.
Schneider, H. A. W. and Bogorad, L. (1979). *Z. für Pflanzenphysiol.* 94: 449–459.
Seckbach, J. (1991). *J. Phycol.* 27: 794–796.
Sidler, W., Gysi, J., Isker, E. and Zuber, H. (1981). *Z. Physiol. Chem.* 362: 611–628.
Steinmüller, K. and Zetsche, K. (1983). *Planta* 159: 308–313.
Tephly, T.R., Coffman, B. L., Ingall, G., AbouZeit-Har, M. S., Goff, H. M., Tabba, H. D. and Smith K. M. (1981). *Arch. Biochem. Biophys.* 212: 120–126.
Troxler, R. F. and Bogorad, L. (1966). *Plant Physiol.* 41: 491–499.
Troxler, R. F. and Lester, R. (1967). *Biochemistry* 6: 3840–3846.
Troxler, R. F., Brown, A. S., Lester, R. and White, P. (1970). *Science* 167: 192–193.
Troxler, R. F. and Brown, A. S. (1970). *Biophys. Acta* 215: 503–511.
Troxler, R. F. (1972). *Biochemistry* 22: 4235–4245.
Troxler, R. F. and Brown, A. S. (1975). *Plant Physiol.* 55: 463–467.
Troxler, R. F., Kelley, P. and Brown, S. B. (1978). *Biochem. J.* 172: 569–576.
Troxler, R. F. and Offner, G. D. (1979). *Biochem. Biophys. Acta* 195: 53–65.
Troxler, R. F., Brown, A. S. and Brown, S.B. (1979). *J. Biol. Chem.* 254:3411–3418.
Troxler, R. F. and Brown, S. B. (1980). *Biol. Bull.* 159: 502.
Troxler, R. F., Ehrhardt, M. M., Brown-Mason, A. S. and Offner, G. D. (1981). *J. Biol. Chem.* 256: 12176–12184.

Troxler, R. F. (1986). In: *Bile Pigments and Jaundice. Molecular, Metabolic and Medical Aspects. (ed. Ostrow, J.D.) Marcel Dekker, Inc., N.Y.* pp. 649–688.

Troxler, R. F., Lin, S. and Offner, G. D. (1989). *J. Biol. Chem.* 264: 29596–20601.

Valentin, K. and Zetsche, K. (1990). *Mol. Gen. Genet.* 222: 425–230.

Volk, S. L. and Bishop, N. L. (1968). *Photochem. Photobiol.* 8: 213–221.

Walsh, C. T. (1984). *Ann. Rev. Biochem.* 53: 493–535.

Weinstein, J. D. and Beale, S. I. (1984). *Plant Physiol.* 74: 146–151.

Werck-Reichhart, D., Jones, O.T.G. and Durst, F. (1988). *Biochem. J.* 249: 473–380.

Wollman, F. (1979). *Plant Physiol.* 63: 375–381.

Chapter 22

A comprehensive bibliography[1] to the world's literature on the algal genus *Cyanidium* Geitler 1933[2] including its ecological cohorts *Galdieria* Merola in Merola *et al.* 1981[2] and *Cyanidioschizon* De Luca, Taddei and Varano 1978[3] with indexing to the major areas of biological interest

FRANKLYN D. OTT and ALETA J. PETRIK-OTT[4]

Introductory statements

Because the majority of the Rhodophycophyta are marine, the presence of about 42 rhodophycean genera (Ott and Sommerfeld 1982) in non-marine habitats has intrigued the senior author of the present paper for a number of years. In order to more fully appreciate the unique biological characteristics of these non-marine rhodophytes, a compilation of the world's literature concerning these algae was initiated in 1962. To the comprehensive bibliography on *Cyanidium* which follows, approximately 950 bibliographical entries have been contributed from this compilation; the completeness of this bibliography testifies to the tedious and countless hours spent in many a dusty library throughout the years.

For each reference, the reader is provided with an abbreviated notation on the general biological disciplines treated in the respective reference. These refer to the following 13 disciplines plus a 'slight mention' category:

Discipline	Abbreviation
1) Biochemistry	(Biochem.)
2) Cytology	(Cytol.)
3) Cultivation	(Cultivat.)
4) Ecology	(Ecol.)
5) Evolution	(Evol.)
6) Floristics	(Florist.)
7) Literature	(Literat.)
8) Morphology	(Morphol.)
9) Photosynthesis	(Photosyn.)
10) Physiology	(Physiol.)
11) Radiobiology	(Radiobiol.)
12) Slight Mention.	(Sl. Ment.)
13) Taxonomy-Systematics	(Tax.-Syst.)
14) Ultrastructure	(Ultrastr.)

At the end of the bibliographical entries there is available to the reader a numerical index to these disciplines; the numbers listed under each category refer to the number of the bibliographical entry. This will permit the reader to concentrate his use of this bibliography to the areas of his interest without having to go through each entry one by one.

While most of the disciplines should be self-evident to the reader, the authors wish to make the following comments regarding their idiosyncrasies in assigning papers to these. The authors found it difficult, if indeed impossible, to separate studies of photosynthetic pigments and their precursors, which are obviously *Biochemical*, from *Photosynthesis* and from *Physiology*. Thus, with the explosion in the number of published papers on the biogenesis of biliproteins, many references have been assigned to all three of these categories. *Cytology* was considered in the classical sense of light microscopy, however, papers dealing with *Ultrastructure* have been assigned also to both disciplines. Not only does *Ecology* deal with the natural phenomena associated with those unique

[1] The literature search for this bibliography was initiated in January, 1962, and terminated on November 26, 1993

[2] Rhodophycophyta, Porphyridiales, Cyanidiaceae Geitler 1933 emend. D.J. Chapman 1974

[3] Rhodophycophyta, Porphyridiales, Porphyridiaceae Kylin [1937] ex Skuja 1939

[4] Present Address: 317 Osborne Avenue, Catonsville, MD 21228, USA

J. Seckbach (ed.), Evolutionary Pathways and Enigmatic Algae: Cyanidium caldarium *(Rhodophyta) and Related Cells,* 283–332, 1994.

habitats of *Cyanidium* and its ecological cohorts, but the authors considered also that papers describing certain conditions of growth in the laboratory as well as those on tolerance and reaction to toxic substances, such as heavy metals, should belong to both *Ecology* and *Physiology*. Investigations on heavy and precious metal concentration and their subsequent removal from the aquatic environment by *Cyanidium* rightly fall to *Ecology* as well as to *Physiology*.

Floristics has been limited, in the main, to the distribution in the natural habitat. However, a close examination of these will give the reader not only the world distribution but also essential ecological information on naturally occurring populations. The category of *Evolution*, including phylogeny, is often closely related to that of *Taxonomy-Systematics* and they usually go hand in hand; papers whose authors make conclusions, directly or indirectly, as to the probable systematic position of the organism are placed in both of these disciplines. *Morphology*, in the strict sense, is restricted to the dynamic, physical stages in the life cycle, but again the authors often found that sharp distinctions between *Morphology* and *Cytology* were impossible to make.

Radiobiology was employed to accommodate studies with radioactive tracers, growth in heavy water, i.e. deuterium, and related substances. *Taxonomy-Systematics*, is basically concerned with the names which have been applied to the organism and how it is related to other similar and dissimilar organisms; this latter systematic endeavor is often considered by some workers distinct from taxonomy. Papers containing statements of possible phylogenetic associations obviously fall to this category. In the search for a 'Natural Arrangement' these two combined categories require considerations on possible *'Evolution'* trends. *Ultrastructure* gives the reader papers on transmission electronmicroscopy, scanning electromicroscopy, and soft x-ray contact microscopy, etc.; these studies are intimately related to *Cytology*.

The category of *Slight Mention* is for those papers which only mention in passing, *Cyanidium*, or one of its ecological cohorts, but essentially give, in the authors' opinion, no new information. While the authors were tempted to omit completely these types of references from this bibliography, these temptations have been overcome by the consideration that it is not the bibliographer's responsibility to make decisions on their worth to the research of others. Thus, they have been included in this bibliography in order that the reader may make their own decision as to their value. This disencumbers the authors from such decisions! However, the reader should note that abstracts of papers presented at various annual meetings of the learned societies and those presented at various annual meetings of the learned societies and those presented at local, state, national, and international colloquia have not been placed in the *Slight Mention* category, but rather these abstracts fall to the other categories as dictated by their contents; abstracts often alert the investigator to significant, new knowledge prior to formal, detailed publications.

Chronological distribution of the bibliography entries

The *c.* 950 references contained in this bibliography span 170 years with the earliest date being 1824 and the latest date being 1993. The chronological distribution of these is given in the following table:

	No. of Ref.:	Interval totals
25 year intervals, 1801-1835:		
1801-1825	1	
1826-1850	8	
1851-1875	4	37
1876-1900	15	
1901-1925	9	
5 year intervals, 1926-1960:		
1926-1930	1	
1931-1935	4	
1936-1940	14	
1941-1945	23	88
1946-1950	8	
1951-1955	11	
1956-1960	27	
One year intervals, in groups of 5, 1961-1993, (November):		
1961	7	
1962	12	
1963	5	45
1964	8	
1965	13	

	No. of Ref.:	Interval totals
1966	16	
1967	18	
1968	19	102
1969	16	
1970	33	
1971	29	
1972	33	
1973	35	164
1974	33	
1975	34	
1976	26	
1977	27	
1978	37	173
1979	46	
1980	37	
1981	45	
1982	23	
1983	30	142
1984	29	
1985	15	
1986	25	
1987	24	
1988	26	136
1989	29	
1990	32	
1991	24	
1992	17	53
1993 (February)	12	
Totals	940	940

Acknowledgements

The preparation of the *Cyanidium* bibliography which follows could not have been realized without the additional cooperation of a number of phycological colleagues and research librarians both in the United States and abroad. While it is impossible here to acknowledge each and every one who had aided the authors in one way or another, the following individuals and respective institutions deserve special thanks for the many kindnesses which they have repeatedly shown us. Dr. Pierre Bourrelly, Sous-Directeur du Laboratoire de Cryptogamie du Muséum National d'Histoire Naturelle, Paris, France; Dr. Thomas D. Burney, Assistant Chief, Rare Book and Special Collections Division, The Library of Congress, Washington, D. C.; Mr. Gennaro J Cacavio, Curatorial Assistant, The Farlow Reference Library of Cryptogamic Botany, Harvard University, Cambridge, Massachusetts; Mrs. Lenore M. Dickinson, Librarian, The Libraries, Gray Herbarium and Arnold Arboretum, Harvard University, Cambridge, Massachusetts; Miss Judith A. Diment, Botany Librarian, British Museum (Natural History), London, England (The United Kingdom); Dr. William N. Doemel, Department of Biology, Wabash College, Crawfordsville, Indiana; Miss Janet Evans, Public Services Librarian, The Academy of Natural Sciences of Philadelphia, Pennsylvania; Miss Jane Fessenden, Librarian, The Library, Marine Biological Laboratory, Woods Hole, Massachusetts; Dr. Alan Fusione, Historical and Preservations Program Specialist, Rare Book Division, National Agriculture Library, Beltsville, Maryland; Herrn Prof. Dr. Lothar Geitler, Botanisches Institut und Botanischer Garten der Universität Wien, Wien (Vienna), Österreich (Austria); Herrn Prof. Dr. J. Gerloff, Botanischer Garten und Botanisches Museum Berlin-Dahlem, Berlin, Bundesrepublik Deutschland (Germany); Mrs. Dorothy T. Hanks, Reference Librarian, History of Medicine Division, Department of Health, Education, and Welfare, Public Health Service, National Institutes of Health, Bethesda, Maryland; Mr. David Jervis, Head of the Reading Room and Admissions Sections, Bibliographical Information Service, Department of Printed Books, The British Library, London, England (The United Kingdom); Dr. Shigeru Kumano, Department of Botany, Faculty of Sciences, Rokkodai, Kobe, Japan; the late Mrs. Carla Lange, Associate Librarian, The Library, Missouri Botanical Garden, Saint Louis, Missouri; Dr. David B. Lellinger, United States National Herbarium, National Museum of Natural History, Smithsonian Institution, Washington, D. C.; Mrs. Lothian Lynas, Head Librarian, The New York Botanical Garden, The Bronx, New York; Dr. Michiyasu Mori, Kora, Shiranuimachi, Utogen, Kumamoto-ken, Japan; Dr. Donald H. Pfister, Curator, The Farlow Reference Library of Cryptogamic Botany, Harvard University, Cambridge, Massachusetts; Mrs. Marge Purk, Library Secretary, The Library, Missouri Botanical Garden, Saint Louis, Missouri; Mr. James Reed, Director, The Library, Missouri Botanical Garden, Saint Louis, Missouri; Dr. Ryozo Seto, Biological Laboratory, Kobe College, Nishinomiya, Japan; Mrs. Ruth F. Schallert, Botanical Librarian, United States National Herbarium, National Museum of Natural History, Smithsonian Institution, Washington, D. C.

As this bibliography was drawing to a close, the following gentlemen kindly sent us, at the last minute, copies of papers which we had been unsuccessful in finding elsewhere: Dr. Michael Wynne, Department of Biology, University of Michigan, Ann Arbor, Michigan; Dr. David J. Chapman, Department of Biology, University of California at Los Angeles, Los Angeles, California; Dr. Richard B. Searles, Department of Botany, Duke University, Durham, North Carolina; Herrn Prof. Dr. E. Wulfgang Becker, Medizinische Universitätklinik, Universität Tübingen, Tübingen, Bundesrepublik Deutschland (Federal Republic of Germany).

Special appreciation and thanks are also due Dr. K. Peter Takayama, Department of Sociology, Memphis State University, Memphis, Tennessee, and his wife, Akiko, who have so kindly provided us with transliterations and translations of titles to various Japanese articles and journals.

The bibliography

1. Aaronson, S. 1970. Molecular evidence for evolution in the algae: a possible affinity between plant cell walls and animal skeletons. pp 531–540 in J.F. Fredrick and R.M. Klein. (Eds). *Phylogenesis and Morphogenesis in the Algae. Ann. N. Y. Acad. Sci.* (New York) 175(2): 413–781. (biochem., evol., physiol.)
2. Aaronson, S. and H. Baker. 1961. Lipid and sterol content of some protozoa. *J. Protozool.* (Utica) 8(3): 274–277.(biochem., physiol.)
3. Abbott, A.G., S.I. Beale, M. Arrese and J.R. Matoon. 1984. Cloned yeast genomic aminolevulinic acid synthase gene does not hybridize with genomic DNA from *Euglena, Cyanidium*, barley, or cucumber. p. 170 in *Program and Abstract of Papers for the Annual Meeting of the American Society of Plant Physiologists. University of California, Davis, August 12–17, 1984. Plant Physiol.* (Lancaster) 74(suppl. 1): (4) + pp 1–207. (biochem.)
4. Adams, B.L. 1970. *Functional Dependence of Fatty Acid Variation in the Thermophilic Alga,* Cyanidium caldarium. Ph.D. Dissertation, Division of Biochemistry, University of Wyoming, Laramie. (1) + VII + 70 pp. (biochem., cultivat., cytol., ecol., photosyn., physiol., ultrastr.)
5. Adams, B.L. and V. McMahon. 1970. Fatty acids in a thermophilic alga. p. 15 in Abstracts of Papers *Presented at the Forty-First Annual Meeting, University of Northern Colorado, Greeley, Colorado, May 1, 2, 1970. J. Colorado-Wyoming Acad. Sci.* (Boulder) 7(1): 5–84. (biochem., physiol.)
6. Adams, B.L. and V. McMahon. 1970. Temperature dependence of fatty acids in the thermophilic alga, *Cyanidium caldarium.* p. 42 in *Program and Abstracts of the Papers for the Annual Meeting of the American Society of Plant Physiologists in Conjunction with A.I.B.S. at Indiana University, Bloomington, August 23–27, 1970. Plant Physiol.* (Lancaster) 46(Suppl.): (2) + pp 1–50. (biochem., physiol.)
7. Adams, B.L., V. McMahon and J. Seckbach. 1971. Fatty acids in the thermophilic alga, *Cyanidium caldarium. Biochem. Biophys. Res. Commun.* (New York) 42(3): 359–365. (biochem., physiol.)
8. Adams, S.M., O.H.W. Kao and D.S. Berns. 1979. Psychrophile *c*-phycocyanin. *Plant Physiol.* (Lancaster) 64: 525–527. (biochem., photosyn., physiol.)
9. Agardh, C.A. 1824. *Systema Algarum, Adumbravit C.A. Agardh.* Typis Berlingianis, Lundae. (2) + XXXVIII pp + 312 pp. (tax.-syst.)
10. Agardh, C.A. 1824 (Reprint, 1965). *Systema Algarum, Adumbravit C.A. Agardh.* Editio Anastatica. A. Asher & Co., Amstelodami. (4) + pp I–IV + Preface by J.Th. Koster (2) + pp V–XXXVIII + 312 pp. (tax.-syst.)
11. Ahlf, W. 1988. Recovery of metals from acid waste water by *Cyanidium caldarium. Appl. Microbiol. Biotechnol.* (Berlin, Heidelberg) 28(4–5): 512–513. (ecol., physiol.)
12. Albertano, P. and G. Pinto. 1988. The action of heavy metals on the growth of three acidophilic algae. *Boll. Soc. Nat. Napoli* (Napoli) 95(1986): 319–328. (biochem., florist., physiol.)
13. Albertano, P., G. Pinto and R. Taddei. 1979. Evaluation of toxical effects of heavy metals to unicellular algae. I-*Delpinoa* (N.S.) (Napoli) 20: 75–86. (biochem., florist., physiol.)
14. Albertano, P., G. Pinto and R. Taddei. 1980. Evaluation of toxic effects of heavy metals on unicellular algae. II-Growth curves with different concentrations of heavy metals. *Delpinoa* (N.S.) (Napoli) 21: 23–34. (biochem., florist., physiol.)
15. Albertano, P. and R. Taddei. 1981. Evaluation of toxic effects of heavy metals on unicellular algae IV. Effects of inoculum size on inhibition type. *Delpinoa* (N.S.) (Napoli) 21–22(1979–1980): 71–77. (ecol., physiol.)
16. Alberte, R.S., T.A. Kursar and R.F. Troxler. 1980. Colorless proteins in phycobilisomes of rhodophycean and cyanobacterial species. *Biol. Bull.* (Woods Hole) 159(2): 498–499. (biochem., photosyn., physiol.)
17. Aliotta, G. and A. Pollio. 1982. Long term effects of copper upon physiological processes and growth of *Chlorella saccharophila* (Kruger) Migula and *Cyanidium caldarium* Geitler. *Giorn. Bot. Ital.* (Firenze) 116(3–4): 123-129. (ecol., physiol.)
18. Aliotta, G. and C. Rigano. 1975. Inactivation by ammonia of nitrate reductase in intact cells of *Cyanidium caldarium.* Inattivazione da ammonio della nitrato riduttasi in cellule intatte di *Cyanidium caldarium. Giorn. Bot. Ital.* (Firenze) 109(3): 160–161. (biochem., physiol.)
19. Allen, C.F., P. Good and R.W. Holton. 1970. Lipid composition of *Cyanidium. Plant Physiol.* (Lancaster) 46(5): 748–751. (biochem., florist., physiol.)

20. Allen, M.B. 1952. The cultivation of Myxophyceae. *Arch. Mikrobiol.* (Berlin, Heidelberg) 17(8): 34–53. (cultivat., ecol., florist., tax.-syst.)
21. Allen, M.B. 1954. Studies on a blue-green *Chlorella*. p. 39 in Huitième Congrès International *de Botanique, Paris, 1954. Comptes-Rendus des Séance et Rapports et Communications déposés lors du Congrès dans la* Section 17 (Phycologie). Pierre André, Imprimeur, Paris. 74 pp + (1). (biochem., sl. ment.)
22. Allen, M.B. 1954. Studies on a blue-green *Chlorella*. pp 41–42 in Huitième Congrès International *de Botanique, Paris, 1954. Rapports et Communications parvenues avant le Congrès à la* Section 17 (Phycologie). Pierre André, Imprimeur, Paris. 192 pp. (biochem., cultivat., cytol., physiol., tax.-syst.)
23. Allen, M.B. 1959. Possible functions of chlorophyll *b*. Studies with green algae that lack chlorophyll *b*. pp 339–342 in R.C. Fuller. (Chmn). *The Photochemical Apparatus. Its Structure and Function. Report of Symposium held June 16–18, 1958. Brookhaven Symposia in Biology* Number 11. BNL [Brookhaven National Laboratory] 512 (C–28) (Biology and Medicine). Biology Department, Brookhaven National Laboratory, Upton. IX + (1) + 336 pp. (biochem., florist., photosyn., physiol., tax.-syst.)
24. Allen, M.B. 1959. Studies with *Cyanidium caldarium*, an anomalously pigmented chlorophyte. *Arch. Mikrobiol.* (Berlin, Heidelberg) 32: 270–277. (biochem., cultivat., ecol., evol., florist., morphol., photosyn., physiol.)
25. Allen, M.B. 1963. *List of Cultures Maintained by the Laboratory of Comparative Biology, Kaiser Foundation Research Institute.* Laboratory of Comparative Biology, Kaiser Foundation Research Institute, Richmond California. 18 pp mimeographed. (cultivat., florist.)
26. Allen, M.B., C.S. French and J.S. Brown. 1960. Native and extractable forms of chlorophyll in various algal groups. pp 33–52 in M.B. Allen. (Ed.). *Comparative Biochemistry of Photoreactive Systems. Symposia on Comparative Biology of the Kaiser Foundation Research Institute.* Vol. I. Academic Press, New York. XII + 437 pp. (biochem., physiol., tax.-syst.)
27. Allen, M.B., T.W. Goodwin and S. Phagpolngarm. 1960. Carotenoid distribution in certain naturally occurring algae and in some artificially induced mutants of *Chlorella pyrenoidosa*. *J. Gen. Microbiol.* (Cambridge) 23: 93–103. (biochem., evol., photosyn., physiol.)
28. Allsopp, A. 1969. Phylogenetic relationships of the procaryota and the origin of the eucaryotic cell. *New Phytol. (Br. Bot. J.)* (London) 68: 591–612. (cytol., evol., morphol., ultrastr.)
29. Andreev, V.P. and I͡u. I. Maslov. 1985. Adaptat͡sii͡a k diuronu pimentnykh struktur *Anabaena variabilis* i *Cyanidium caldarium*. (Adaptation to diuron of *Anabaena variabilis* and *Cyanidium caldarium* pigment structures). *Fizol. Biokhim. Kul 't Rast.* (Kiev) 17(3): 293–297. (biochem., photosyn., physiol., radiobiol.)
30. Andreev, V.P. and I͡u. I. Maslov. 1986. Adaptation to diuron of *Anabaena variabilis* and *Cyanidium caldarium* pigment structure. *Physiol. Biochem. Cultivat. Plants* (USSR, New York) 17(3): (biochem., photosyn., physiol., radiobiol.)
31. Andreev, V.P. and I͡u. I. Maslov. 1989. Izmeneni͡ia v spektrakh flourest͡sent͡sii *Cyanidium caldarium* v khode adaptat͡sii k diuronu. [Fluorescence spectrum of *Cyanidium caldarium* after the adaptation to diuron]. *Vestn. Leningr. Univ.*, Ser. 3, *Biol.* (Leningrad) 1989(2): 96–99. (biochem., photosyn., physiol., radiobiol.)
32. Andrew, P.W., L.J. Rogers, B.G. Haslett and D. Boulter. 1981. Partial structure and properties of the ferredoxin from *Rhodymenia palmata*. *Phytochemistry* (Oxford) 20(4): 579–583. (biochem., evol.)
33. Andrew, P.W., L.J. Rogers, B.G. Haslett and D. Boulter. 1981. Comparative properties of ferredoxins from a marine and freshwater species of *Porphyridium*. *Phytochemistry* (Oxford) 20(6): 1293–1298. (biochem., evol.)
34. Archer, W. 1875. Notes on some collections made from Furnas Lake, Azores, containing algae and a few other organisms. *J. Linn. Soc. London Bot.* (London) 14: 328–340. (florist., tax.-syst.)
35. Ascione, R. 1966. *The Use of Enzymes from Thermophilic Organisms to Investigate the Biological Role of Macromolecular Structure in Soluble Ribonucleic Acid.* A Dissertation Presented to the Faculty of Princeton University in Candidacy for the Degree of Doctor of Philosophy. Recommended for Acceptance by the Department of Biology, Biochemical Science Division. Princeton University, Princeton. (7) + 88 pp. (biochem., cultivat., florist., physiol.)
36. Ascione, R. 1967. The use of enzymes from thermophilic organisms to investigate the biological role of macromolecular structure in soluble ribonucleic acid. *Diss. Abstr. B, Sci. Eng.* (Ann Arbor) 28(4): 1388B–1389B. (biochem., cultivat., physiol.)
37. Ascione, R. and J.R. Fresco. 1964. Heat-stable amino acid activating enzymes from a thermophile, p. 163 in *Abstracts*, 48th Annual Meeting, *Chicago, Illinois, April 12–17, 1964. Fed. Proc., Fed. Am. Soc. Exp. Biol.* (Baltimore) 23(2 part 1): (2) + pp 91–581 + XXXXII. (biochem., physiol.)
38. Ascione, R., W. Southwick and J.R. Fresco. 1966. Laboratory culturing of a thermophilic alga at high temperature. *Science* (Lancaster) 153(3737): 752–755. (biochem., cultivat., ecol., evol., florist., physiol.)
39. Asher, A. and D.F. Spalding. (Eds.). 1982. *Culture Centre of Algae and Protozoa. List of Strains.* [4th Ed.]. Institute of Terrestrial Ecology. Natural Environment Research Council, Cambridge. (102 pp).
-- Asher, A. and D.F. Spalding. (Eds). Please note: For the *first, second and third editions of the Cambridge Collection of Algae and Protozoa*, please see respectively: E.A. George. (Ed.). (1967); E.A. George. (Ed.). (1971); E.A. George. (Ed.). (1976). For the fifth edition please see: A.S. Thompson, J.C. Rhodes and I. Pettman. (Eds). (1988).
40. Baca, B.J. 1978. *Field and Laboratory Studies of Selected Marine Algae of South Texas.* Ph.D. Dissertation, Graduate College, Texas A & M University, College Station. (1) + pp III–XII + 87 pp with 34 figs + 4 tab. (sl. ment., tax.-syst.)
41. Badour, S.S. 1978. Inhibitors used in studies of algal metabolism. pp 479–488 in J.A. Hellebust and J.S. Craigie. (Eds). *Handbook of Phycological Methods. Physiological and Biochemical Methods.* Cambridge University Press, Cambridge, London, New York, Melbourne. XIV + 512 pp. (biochem., sl. ment.)

42. Ba͡ianova, I͡u. I. and I.N. Trubachev. 1981. Comparative evaluation of the vitamin composition of some unicellular algae and higher plants grown under artificial conditions. *Appl. Biochem. Microbiol* (USSR, New York) 17(3): 292-298. (biochem.)
43. Ba͡ianova, I͡u. I. and I.N. Trubachev. 1981. Sravnitel'na͡ia ot͡senka vitaminnogo sostava nekotorykh odnokletochnykh vodorosleĭ i vysshikh rastenii, vyrashchennykh v iskusstvennykh usloviĭ͡akh. (Comparative evaluation of the vitamin composition of the unicellular algae and higher plants grown under artificial conditions). *Prikl. Biokhim. Mikrobiol.* (Moskva) 17(3): 400–407. (biochem.)
44. Bailey, R.W. and L.A. Staehelin. 1968. The chemical composition of isolated cell walls of *Cyanidium caldarium. J. Gen. Microbiol.* (Cambridge) 54(2): 269–276 + pl. I. (biochem., evol., tax.-syst., ultrastr.)
45. Balsamo, F. 1892. Manipolo di alghe napolitane.-Memoria di F. Balsamo. *Boll. Soc. Nat. Napoli* (Ser. I) (Napoli) 6(1): 77–87. (florist., tax.-syst.)
46. Barashkov, V.A., I.N. Trubachev and I.I. Gitel'zon. 1979. Kharkteristike belkov nekotorykh odnokletochykh organizmov kak poten͡tsial'nykh komponentov ekologicheskikh sistem zhizneobespecheniĭ͡a. (Characterization of proteins of certain unicellular organisms as potential components of bioregenerative life support systems). *Kosm. Biol. Aviakosmicheska͡ia* Med. (Moskva) 13(3): 75–80. (biochem.)
47. Barnes, D.S. 1971. Phycocyanins and deuterated proteins. pp 105–148 in S.N. Timascheff and G.D. Fasman. (Eds). *Biological Macromolecules.* Vol. 5. *Subunits in Biological Systems.* Part. A. Marcel Dekker, Inc., New York. (2) + IX + (2) + 408 pp. (biochem., physiol., radiobiol.)
48. Bauld, J. and T.D. Brock. 1974. Algal excretion and bacterial assimilation in hot spring algal mats. *J. Phycol.* (Baltimore, Lawrence) 10(1): 101–106. (ecol., physiol.)
-- Bayanova, Yu. I. Please see: Ba͡ianova, I͡u. I.
49. Beale, S.I. and N.C. Chen. 1982. N-methylmesoporphyrin IX inhibits phycocyanin, but not chlorophyll synthesis in *Cyanidium caldarium.* p. 131 in *Program and Abstracts of Papers for the Annual Meeting of the American Society of Plant Physiologists at University of Illinois, Champaign-Urbana, June 13–17, 1982. Plant Physiol.* (Lancaster) 69(4 Suppl.): + pp 1–163. (biochem., photosyn., physiol.)
50. Beale, S.I. and N.C. Chen. 1983. N-methyl mesoporphyrin IX inhibits phycocyanin, but not chlorophyll synthesis in *Cyanidium caldarium. Plant Physiol.* (Lancaster) 71(2): 263–268. (biochem., photosyn., physiol.)
51. Beale, S.I. and J. Cornejo. 1983. Biosynthesis of phycocyanobilin from exogenous labeled biliverdin in *Cyanidium caldarium. Arch. Biochem. Biophys.* (New York) 227(1): 279–286. (biochem., photosyn., physiol.)
52. Beale, S.I. and J. Cornejo. 1984. Enzymatic heme oxygenase activity in soluble extracts from the unicellular red alga, *Cyanidium caldarium. Arch. Biochem. Biophys.* (New York) 235(2): 371–384. (biochem., photosyn., physiol.)
53. Beale, S.I. and J. Cornejo. 1984. Enzymatic transformation of biliverdin to phycocyanobilin by extracts of the unicellular red alga, *Cyanidium caldarium.* p. 171 in *Program and Abstract of Papers for the Annual Meeting of the American Society of Plant Physiologists at University of California, Davis, August 12–17, 1984. Plant Physiol.* (Lancaster) 74(Suppl. 1): (4) + pp 1–207. (biochem., photosyn., physiol.)
54. Beale, S.I. and J. Cornejo. 1984. Enzymatic transformation of biliverdin to phycocyanobilin by extracts of the unicellular red algae *Cyanidium caldarium. Plant Physiol.* (Lancaster) 76(1): 7–15. (biochem., photosyn., physiol.)
55. Beale, S.I. and J. Cornejo. 1986. Separation and partial characterization of the enzymes of heme oxygenase system from *Cyanidium caldarium.* p. 52 in *Program and Abstracts of Papers for the Annual Meeting of the American Society of Plant Physiologists at Louisiana State University, Baton Rouge, Louisiana, June 8–12, 1986. Plant Physiol.* (Lancaster) 80(4 suppl.): (4) + pp 1–151. (biochem., physiol.)
56. Beale, S.I. and J.D. Weinstein. 1984. Tetrapyrrole precursor biosynthesis in *Euglena* and *Cyanidium.* pp 717–720 in C. Sybesme. (Ed.). *Advances in Photosynthesis Research. Proceedings of the VIth International Congress on Photosynthesis, Brussels, Belgium, August 1–6, 1983.* Vol. IV. Martinus Nijhoff, Dr. W. Junk Publishers (Kluwer Academic Publishers Group), The Hague, Boston, Lancaster. (4) + pp V–XXXI + (1) + 899 pp. (biochem., photosyn., radiobiol.)
57. Beardall, J. and L. Entwisle. 1984. Internal pH of the obligate acidophile *Cyanidium caldarium* Geitler (Rhodophyta?). *Phycologia* (Berkeley, Copenhagen, Vancouver) 23(3): 397–399. (physiol.)
58. Becker, E.W. 1972. *Physiologische Untersuchungen zur Photosynthese von Algen unter extremen Temperaturbedingungen vorgelegt von Wolfgang Becker aus Wriezen/Oderbruch.* Dissertation zur Erlangung des Grades eines Doktors der Naturwissenschaften dem Fachbereich Chemie der Eberhard-Karls-Universität zu Tübingen. (6) + 93 SS + Lebenslauf (1). (biochem., photosyn., physiol.)
59. Becker, E.W. and H. Schmellenkamp. 1972. Temperature dependent ^{14}C incorporation into *Cyanidium caldarium.* pp 2051–2058 in G. Forti, M. Avron and A. Melandri. (Eds). *Photosynthesis, Two Centuries After Its Discovery by Joseph Priestley. Proceedings of the IInd International Congress on Photosynthesis Research, Stresa, June 24–29, 1971.* [Vol. III]. Dr. W. Junk N.V. Publishers, The Hague. XXIII + (1) + pp 1697–2752. (biochem., physiol., radiobiol.)
60. Bedord, C.J. 1975. *The Temperature Induction of Alpha-Linolenic Acid Synthesis in* Cyanidium caldarium. Ph.D. Dissertation, Division of Biochemistry, University of Wyoming, Laramie. (3) + IX + 127 pp. (biochem., physiol.)
61. Bedord, C.J. 1976. The temperature induction of alpha-linolenic acid synthesis in *Cyanidium caldarium. Diss. Abstr. Int. B, Sci. Eng.* (Ann Arbor) 36(12 part 1): 6118–6119. (biochem., physiol.)
62. Bedord, C.J., V. McMahon and B. Adams. 1978. α-linolenic acid biosynthesis in *Cyanidium caldarium. Arch. Biochem. Biophys.* (New York) 185(1): 15–20. (physiol.)
63. Belford, H.S., G.D. Offner and R.F. Troxler. 1983. Phycobiliprotein synthesis in the unicellular rhodophyte, *Cyanidium caldarium*: cell-free translation of the mRNAS for the α and β subunit of polypeptides of phycocyanin. *J. Biol. Chem.* (Baltimore) 258(7):4503–4510. (biochem., photosyn., physiol., radiobiol.)
64. Belly, R.T. and T.D. Brock. 1972. The habitat of *Thermoplasma acidophila.* p. 53 in *Abstracts of the Annual Meeting of the American Society for Microbiology.* 72nd Annual Meet-

ing, *Philadelphia, Pa. 23–28 April 1972*. American Society for Microbiology, Washington, D.C. VIII + 294 pp. (ecol.)

65. Belly, R.T., M.R. Tansey and T.D. Brock. 1973. Algal excretion of ^{14}C-labeled compounds and microbial interactions in *Cyanidium caldarium* mats. *J. Phycol.* (Baltimore, Lawrence) 9(2): 123–127. (biochem., ecol., florist., photosyn., physiol., radiobiol.)
66. Benedikt, E. and H.-P. Köst. 1985. A-α-hydroxymesobiliverdin, a new bile pigment. *Zeitschr. Naturforsch.* Teil C, Biosci. (Wiesbaden, Tübingen) 40(11–12): 755–759. (biochem., photosyn., physiol.)
67. Bennett, A. and L. Bogorad. 1971. Properties of subunits and aggregates of blue-green algal biliproteins. *Biochemistry* (Washington) 10(9): 3625–3634. (biochem., photosyn., physiol.)
68. Bennett, A. and L. Bogorad. 1971(Reprint, 1974). Properties of subunits and aggregates of blue-green algal biliproteins. pp 126–152 in M. Rodriguez-Lopez et al. (Eds). *Blue-Green Alga* I: *Current Research*. MSS Information Corporation, New York. 213 pp. (biochem., photosyn., physiol.)
69. Bergen, J.Y. 1897. Algae in the Solfatara at Pozzuoli, Italy. *Bot. Gaz.* (Crawfordsville, Chicago) 23: 198–199. (ecol., florist.)
70. Berns, D.S. 1967. Immunochemistry of biliproteins. *Plant Physiol.* (Lancaster) 42(11): 1569–1586. (biochem., evol., photosyn., physiol., tax.-syst.)
71. Berns, D.S. 1971. Phycocyanins and deuterated proteins. pp 105–148 in S.N. Timasheff and G.D. Fasman. (Eds). *Biological Macromolecules.* Volume 5. *Subunits in Biological Systems,* Part A. Marcel Dekker, Inc., New York. (2) + IX (2) + 408 pp. (biochem., photosyn., physiol.)
72. Berns, D.S. and B. Berns. 1964. The antigenicity of phycocyanins. *N. Y. State Dep. Health Annu. Rep. Div. Lab. Res.* (Albany) 1964: 80–81. (biochem., evol., photosyn., physiol., tax.-syst.)
73. Berns, D.S., P. Holohan and E. Scott. 1966. Urease activity in blue-green algae. *Science* (Lancaster) 152(3725): 1077-1078. (biochem., evol., photosyn., physiol., tax.-syst.)
74. Berns, D.S., E. Scott and K.T. O'Reilly. 1964. *C*-phycocyanin: minimum molecular weight. *Science* (Lancaster) 145(3636): 1054–1056. (biochem. evol., photosyn., physiol., tax.-syst.)
75. Bisalputra, T. 1974. Plastids, pp 124-160 in W.D.P. Stewart. (Ed.). *Algal Physiology and Biochemistry. Botanical Monographs.* Volume 10. University of California Press, Berkeley and Los Angeles. (Blackwell Scientific Publications, Oxford, London, Edinburgh, Melbourne). XI + (1) + 989 pp. (biochem., cytol., photosyn., physiol.)
76. Bisalputra, T. and A. Baily, 1973. The fine structure of the chloroplast envelope of a red alga, *Bangia fusco-purpurea. Protoplasma (Internat. Zeitschr. Physikal. Chem. Protoplast.*, Leipzig) 76(3/4): 443–454. (morphol., sl. ment., ultrastr.)
77. Bishop, N.I. 1971. Preparation and properties of mutants: *Scenedesmus.* pp 130–143 in A. San Pietro. (Ed.). *Photosynthesis.* Part A. in S.P. Colowick and N.O. Kaplan. (Eds). *Methods in Enzymology.* Volume XXIII. Academic Press, New York, London. XIX + (4) + 743 pp. (physiol., sl. ment.)
78. Bishop, N.I. and H. Senger. 1971. Preparation and photosynthetic properties of synchronous cultures of *Scenedesmus.* pp 53-66 in A. San Pietro. (Ed.). Photosynthesis. Part A. in S.P. Colowick and N.O. Kaplan. (Eds). *Methods in Enzymology.* Volume XXIII. Academic Press, New York, London. XIX + (4) + 743 pp. (cultivat., physiol.)
79. Bjørnland, T. and M. Aquilar-Martinez. 1976. Carotenoids in red algae. *Phytochemistry* (Oxford) 15(2): 291–296. (biochem., evol., tax.-syst.)
80. Bligh, E.G. and W.J. Dyer. 1959. A rapid method of total lipid extraction and purification. *Can. J. Biochem. Physiol.* (Ottawa) 37: 911–917. (biochem., physiol.)
81. Boenzi, D., P. De Luca and R. Taddei. 1977. Fatty acids in *'Cyanidium'. Giorn. Bot. Ital.* (Firenze) 111(3): 129–134. (biochem., evol., physiol.)
82. Bogorad, L. 1962. Chlorophylls. pp 385–408 in R.A. Lewin. (Ed.). *Physiology and Biochemistry of Algae.* Academic Press, New York and London. Frontispiece + XXVII + (1) + 929 pp. (photosyn., sl. ment.)
83. Bogorad, L. 1963. The biogenesis of heme, chlorophylls, and bile pigments. pp 183–231 in P. Bernfeld. (Ed.). *Biogenesis of Natural Compounds.* Pergamon Press, Oxford, London, New York, Paris. XIII + (1) + 930 pp. (biochem., photosyn., physiol.)
84. Bogorad, L. 1965. Studies of phycobiliproteins. *Rec. Chem. Prog.* (Fayette) 26(1): 1–12. (biochem., photosyn., physiol.)
85. Bogorad, L. 1965. Studies of phycobiliproteins. pp 108–119 in D.W. Krogmann and W.H. Powers. (Eds). *Biochemical Dimensions of Photosynthesis.* Wayne State University Press, Detroit. (8) + 124 pp. (biochem., photosyn., physiol.)
86. Bogorad, L. 1975. Phycobiliproteins and complementary chromatic adaptation. *Annu. Rev. Plant Physiol.* (Stanford) 26: 369-401. (biochem., photosyn., physiol.)
87. Bogorad, L. 1976. Chlorophyll biosynthesis, pp 64–148 in T.W. Goodwin. (Ed.), *Chemistry and Biochemistry of Plant Pigments.* Second Edition. Volume I. Academic Press, London, New York, San Francisco. (4) + pp V–XVI + 870 pp. (biochem., photosyn., physiol.)
88. Bogorad, L., F.V. Mercer and R. Mullens. 1963. Studies with *Cyanidium caldarium.* II. The fine structure of pigment-deficient mutants. pp 560–570 in B. Kok and A.T. Jagendorf. (Chmn). *Photosynthetic Mechanisms of Green Plants. Papers Presented at a Symposium by the Committee on Photobiology of the National Academy of Sciences–National Research Council with the Support of the National Science Foundation.14–18 October 1963.* Publication 1145. National Academy of Sciences–National Research Council, Washington, D.C. (4) + IX + (3) + 766 pp. (cytol., photosyn., ultrastr.)

-- Bogorad, L., F.V. Mercer and R. Mullens. Please see: Mercer, F.W., L. Bogorad and R. Mullens (1962), for section 'I' of the above title.

89 Bogorad, L. and R.F. Troxler. 1967. The biogenesis of heme, chlorophylls, and bile pigments. pp 247–313 in P. Bernfeld. (Ed.). *Biogenesis of Natural Compounds. Second Edition.* Pergamon Press, Oxford. XIV + 1209 pp. (biochem., photosyn., physiol.)

90. Bourrelly, P. 1970. *Les Algues d'Eau Douce, Initiation à la Systématique.* Tome III. *Les Algues Bleues et Rouges, Les Eugléniens, Peridiniens et Cryptomonadines.* Éditions N.

Boubée & Cie, Paris. (15) + pp 16–512 + (2) including 134 pls. (cytol., ecol., morphol., tax.-syst.)

91. Bourrelly, P. 1985. *Les Algues d'Eau Douce, Initiation à la Systématique.* Réimpression Revue et Augmentée. Tome III. *Les Algues Blues et Rouges, Les Eugléniens, Peridiniens et Cryptomonadines.* Société Nouvelle des Éditions Boubée, Paris. (15) + pp 16–606 including 134 pls. (cytol., ecol., morphol., tax.-syst.)
92. Brandt, P. and J. Zagon. 1987. Intracellular regime for the assembly of the chlorophyll-protein complex CPI in *Cyanidium caldarium. J. Exp. Bot.* (London) 38(196): 1801–1817. (biochem., evol., photosyn., tax.-syst.)
93. Brandt, P. and J. Zagon. 1988. Intercellular regime for the assembly of the chlorophyll-complex in *Cyanidium caldarium.* p. 15S in C.F. Foote. (Chmn). 16th Annual Meeting *of the American Society for Photobiology. Abstracts of Papers. Colorado Springs, Colorado, 13-17 March 1988. Photochem. Photobiol.* (New York) 47(Suppl.): (4) + pp 1S–99S. (biochem., cult., photosyn., physiol.)
94. Bresciano, E., G. Pinto and M.L. Vuotto. 1978. Indagini preliminari sull'immunosistematica di due specie di alghe termofile ed acidofile dei Campi Flegrei (Napoli). *Delpinoa* (N. S.) (Napoli) 18/19(1976-1977): 17–24. (biochem., ecol., florist., physiol., tax.-syst.)
95. Brierley, J.A. 1966. *Contribution of Chemoautotrophic Bacteria to the Acid Thermal Waters of the Geyser Springs Group in Yellowstone National Park.* Ph.D. Dissertation, Montana State University, Bozeman. VIII + 104 pp. (ecol., physiol.)
96. Brierley, J.A. 1967. Contribution of Chemoautotrophic Bacteria to the acid thermal waters of the Geyser Springs Group in Yellowstone National Park. *Diss. Abstr. B, Sci. Eng.* (Ann Arbor) 27(9): 3201–B. (ecol., physiol.)
97. Brock, T.D. 1967. Life at high temperatures. *Science* (Lancaster) 158(3804): 1012–1019. (ecol., evol., physiol., radiobiol.)
98. Brock, T.D. 1969. Microbial growth under extreme conditions. pp 15–41 + pls 1–2 in P. Meadow and S.J. Pirt. (Eds). *Symposia of the society for general Microbiology* No. XIX. *Microbial Growth.* Cambridge University Press, London. 450 pp. (ecol., florist., physiol.)
99. Brock, T.D. 1970. *Biology of Microorganisms.* Prentice-Hall, Inc., Englewood Cliffs, New Jersey. XII + (2) + 737 pp. (biochem., ecol., photosyn., physiol.)
100. Brock, T.D. 1970. High temperature systems. *Annu. Rev. Ecol. Syst.* (Palo Alto) 1: 191–220. (ecol., florist.)
101. Brock, T.D. 1972. One hundred years of algal research in Yellowstone National Park. pp 393–405 in T.V. Desikachary. (Ed.). *Taxonomy and Biology of Blue-Green Algae. Papers Submitted to the* First International Symposium *on Taxonomy and Biology of Blue-Green Algae Held at Madras, January 8-13, 1970.* University of Madras, Centre for Advanced Study in Botany, Madras. XI + (1) + 591 pp. (ecol., florist., tax.-syst.)
102. Brock, T.D. 1972. The algae of very acid environments. p. 17 in *Program with Abstracts of Papers Scheduled for the Meetings of the Phycological Society of America at the University of Minnesota, Minneapolis, Minnesota, August 27–September 1, 1972. J. Phycol.* (Baltimore, Lawrence) 8(Suppl.): (2) + pp 3–19. (ecol.)
103. Brock, T.D. 1973. Evolutionary and ecological aspects of the cyanophytes. pp 487–500 in N.G. Carr and B.A. Whitton. (Eds). *The Biology of Blue-Green Algae. Botanical Monographs.* Volume 9. Blackwell Scientific Publications, Oxford, London, Edinburgh, Melbourne. (University of California Press, Berkeley, California). Frontispiece + X + 676 pp. (ecol., evol.)
104. Brock, T.D. 1973. Lower pH limit for the existence of blue-green algae: evolutionary and ecological implications. *Science* (Lancaster) 179(4072): 480–483. (ecol., florist., physiol.)
105. Brock, T.D. 1974. *Biology of Microorganisms.* Second Edition. Prentice-Hall, Inc., Englewood Cliffs, New Jersey. XII + 825 pp. (ecol.)
106. Brock, T.D. 1975. Predicting the ecological consequences of thermal pollution from observations on geothermal habitats. pp 599–622 in *Environmental Effects of Cooling Systems at Nuclear Power Plants. Proceedings of a Symposium on the Physical and Biological Effects of the Environment of Cooling Systems and Thermal Discharges at Nuclear Power Stations. Organized by the International Atomic Energy Agency with the Co-operation of the Secretariat of the United Nations Economic Commission for Europe and Held in Oslo, 26–30 August 1974.* Proceedings Series. International Atomic Energy Agency, Vienna (12) + 829 pp + (2). (ecol., florist.)
107. Brock, T.D. 1978. *Thermophilic Microorganisms and Life at High Temperatures.* Springer-Verlag, New York, Heidelberg, Berlin. XI + (1) + 465 pp. (ecol., photosyn., physiol.)
108. Brock, T.D. 1979. *Biology of Microorganisms.* 3rd Edition. Prentice-Hall, Inc., Englewood Cliffs, New Jersey. XIV + 802 pp. (biochem., ecol., photosyn., physiol.)
109. Brock, T.D. and M.L. Brock. 1971. The algae of Waimangu Cauldron (New Zealand): distribution in relation to pH. *J. Phycol.* (Baltimore, Lawrence) 6(4): 371–375. (ecol., florist., physiol.)
110. Brock, T.D. and M.L. Brock. 1971. *Life in the Geyser Basins.* Published by Yellowstone Library and Museum Association in Cooperation with National park Service, U.S. Department of the Interior, [Yellowstone Park?]. (32 pp). (ecol.)
111. Brock, T.D. and M.L. Brock. 1971. Microbiological studies of thermal habitats of the central volcanic region, North Island, New Zealand. *N.Z. J. Mar. Freshwater Res.* (Wellington) 5(2): 233–258. (ecol., florist., physiol.)
112. Brock, T.D. and G.K. Darland. 1970. Limits of microbial existence: temperature and pH. *Science* (Lancaster) 169(3952): 1316–1318. (ecol., florist., physiol.)
113. Brown, A.S. 1979. *Structural Studies on Phycobiliproteins from the Unicellular Rhodophyte* Cyanidium caldarium. Ph.D. Dissertation, Boston University, Boston. 163 pp. (photosyn., ultrastr.)
114. Brown, A.S. 1979. Structural studies on phycobiliproteins from the unicellular rhodophyte *Cyanidium caldarium. Diss. Abstr. Int. B, Sci. Eng.* (Ann Arbor) 40(1): 205. (biochem., photosyn., physiol.)
115. Brown, A.S., J.A. Foster, P.V. Voynow, C. Franzblau and R.F. Troxler. 1975. Allophycocyanin from the filamentous Cyanophyte, *Phormidium luridum. Biochemistry*

(Washington) 14(16): 3581–3588. (biochem., photosyn., physiol.)

116. Brown, A.S., G.D. Offner, M.M. Ehrhardt and R.F. Troxler. 1979. Phycobilin-apoprotein linkages in the α and β subunits of phycocyanin from the unicellular rhodophyte, *Cyanidium caldarium*. Amino acid sequences of ^{35}S-labeled chromopeptides. *J. Biol. Chem.* (Baltimore) 254(16): 7803–7811. (biochem., photosyn., physiol., radiobiol.)
117. Brown, A.S. and R.F. Troxler. 1977. Bilin-apoprotein linkages in rhodophytan phycobiliproteins: the role of cysteine. *FEBS Lett.* (Fed. Eur. Biochem. Soc., Amsterdam) 82(2): 206–210. (biochem., photosyn., physiol.).
118. Brown, A.S. and R.F. Troxler. 1977. Properties and N-terminal sequence of allophycocyanin from the unicellular rhodophyte *Cyanidium caldarium*. *Biochem. J.* (Liverpool) 163(3): 571–581. (biochem., photosyn., physiol.).
119. Brown, S.B. and J.A. Holroyd. 1984. Biosynthesis of the chromophore of phycobiliproteins. A study of mesohaem and mesobiliverdin as possible intermediated and further evidence for an algal haem oxygenase. *Biochem. J.* (Liverpool) 217(1): 265–272. (biochem., photosyn., physiol.)
120. Brown, S.B., J.A. Holroyd and R.F. Troxler. 1980. Mechanism of bile-pigment synthesis in algae. ^{18}O incorporation into phycocyanobilin in the unicellular rhodophyte, *Cyanidium caldarium*. *Biochem. J.* (Liverpool) 190(2): 445–449. (biochem., photosyn., physiol., radiobiol.)
121. Brown, S.B., J.A. Holroyd, R.F. Troxler and G.D. Offner. 1981. Bile pigment synthesis in plants. Incorporation of haem into phycocyanobilin and phycobiliproteins in *Cyanidium caldarium*. *Biochem. J.* (Liverpool) 194(1): 137–147. (biochem., photosyn., physiol.)
122. Brown, S.B., J.A. Holroyd and D.I. Vernon. 1984. Biosynthesis of phycobiliproteins. Incorporation of biliverdin into phycocyanin of the red alga *Cyanidium caldarium*. *Biochem. J.* (Liverpool) 219(3): 905–909. (biochem., photosyn., physiol.)
123. Brown, S.B., J.A. Holroyd and D.I. Vernon. 1989. The biosynthesis of the chromophore of phycocyanin. Pathway of reduction of biliverdin to phycocyanobilin. Biochem. J. (Liverpool) 261(1): 259–264. (biochem., photosyn., physiol.)
124. Brown, S.B., J.A. Holroyd, D.I. Vernon and O.T.G. Jones. 1984. Ferrochelatase activity in the photosynthetic alga *Cyanidium caldarium*. Development of the enzyme during biosynthesis of photosynthetic pigments. *Biochem. J.* (Liverpool) 220(3): 861–863. (biochem., photosyn., physiol.)
125. Brown, S.B., J.A. Holroyd, D.I. Vernon, R.F. Troxler and K.M. Smith. 1982. Effect of N-methylprotoporphyrin IX on the synthesis of photosynthetic pigments in *Cyanidium caldarium*: further evidence for the role of haem in the biosynthesis of plant bilins. *Biochem. J.* (Liverpool) 208(2): 487–491. (biochem., photosyn., physiol.)
126. Brown, S.B., K.M. Smith, G.M.F. Bisset and R.F. Troxler. 1980. Mechanism of photooxidation of bacteriochlorophyll *c* derivatives. A possible model for natural chlorophyll breakdown. *J. Biol. Chem.* (Baltimore) 255(17): 8063–8068. (biochem., photosyn., physiol.)
127. Brown, S.B. and D.I. Vernon. 1988. Photochemotherapy, a newly developing cancer treatment. p. 1000 in *Abstracts of the Association for Radiation Research Winter Meeting, Leeds, 6–8 January, 1988. Internatl. J. Radiat. Biol. Relat. Stud. Phys. Chem. Med.* (London) 53(6): 993–1005. (biochem., physiol., radiobiol.)
128. Brown, T.E. and F.L. Richardson. 1968. The effect of growth environment on the physiology of algae; light intensity. *J. Phycol.* (Baltimore, Lawrence) 4(1): 38–54. (biochem., ecol., photosyn., physiol.)
129. Cammack, R., K.K. Rao, C.P. Bargeron, K.G. Hutson, P.W. Andrew and L.J. Rogers. 1977. Midpoint redox potentials of plant and algal ferredoxins. *Biochem. J.* (Liverpool) 168(2): 205–209. (biochem., sl. ment.)
130. Capasso, L. and G. Pinto. 1983. Evaluation of toxic effects of heavy metals on unicellular algae. V. Influence of extracellular products on toxicity and on type of inhibition. *Giron. Bot. Ital.* (Firenze) 117(3-4): 121–128. (evol., physiol.)
131. Carr, N.G. and I.W. Craig. 1970. The relationship between bacteria, blue-green algae and chloroplasts. pp 119-143 in J.B. Harborne. (Ed.). *Phytochemical Phylogeny. Proceedings of the Phytochemical Society Symposium, Bristol, April 1969.* Academic Press, London, New York. XIII + (1) + 335 pp. (biochem., evol., sl. ment.)
132. Casper, S.J. 1974. *Grundzüge eines natürlichen Systems der Mikroorganismen.* VEB Gustav Fischer Verlag, Jena. 232 SS mit 42 Abbildungen und 32 Tabellen im Text + 2 Tafeln als Beilage. (sl. ment., tax.-syst.)
133. Cassie, V. 1980. Bibliography of the freshwater algae of New Zealand 1849-1980. *N.Z. J. Bot.* (Wellington) 18(3): 433–447. (sl. ment.)
134. Castaldo, R. 1968. Ricerche sull'ultrastruttura del *Cyanidium caldarium* (Tilden) Geitler dei Campi Flegrei (Napoli). *Delpinoa* (N.S.) (Napoli) 8/9(1966–1967): 135–147 + Tav. I–VI. (cytol., morphol., ultrastr.)
135. Castaldo, R. 1970. Ultrastruttura di due forme isolate dalle popolazioni di *Cyanidium caldarium* (Tilden) Geitler. *Delpinoa* (N.S.) (Napoli) 10/11(1968–1969): 91–109 + Tav. I–VI. (cytol., morphol., tax.-syst., ultrastr.)
136. Castaldo, R. and R. Ligrone. 1978. 'Triple' thylakoids in *Cyanidium caldarium* 'forma A' and in the Cyanophyta. A hypothesis on the significance of these structures. *Delpinoa* (N.S.) (Napoli) 18/19(1976–1977): 25–36 + pls I-III. (cytol., morphol., ultrastr.)
137. Castenholz, R.W. 1967. Environmental requirements of thermophilic blue-green algae. pp 55-79 in A.F. Bartsch. (Chmn). *Environmental Requirements of Blue-Green Algae. Proceedings of a Symposium, September 23–24, 1966. Sponsored by Federal Water Pollution Control Administration, Northwest Region, Corvallis, Oregon and University of Washington, Seattle, Washington.* U.S. Department of the Interior, Pacific Northwest Water Laboratory, Corvallis, Oregon. V + 111 pp. (ecol.)
138. Castenholz, R.W. 1969. The thermophilic cyanophytes of Iceland and the upper temperature limit. *J. Phycol.* (Baltimore, Lawrence) 5(4): 360–368. (ecol., florist., sl. ment.)
139. Castenholz, R.W. 1969. Thermophilic blue-green algae and the thermal environment. *Bacteriol. Rev.* (Baltimore) 33(4): 476–504. (ecol.).

140. Castenholz, R.W. 1973. Ecology of blue-green algae in hot springs. pp 379–414 in N.G. Carr and B.A. Whitton. (Eds). *The Biology of Blue-Green Algae. Botanical Monographs.* Volume 9. University of California Press, Berkeley. (Blackwell Scientific Publications, Oxford, London, Edinburgh, Melbourne). Frontispiece + X + 676 pp. (ecol., sl. ment.)
141. Castenholz, R.W. 1976. The effect of sulfide on the bluegreen algae of hot springs. I. New Zealand and Iceland. *J. Phycol.* (Baltimore, Lawrence) 12(1): 54–68. (ecol., sl. ment.)
142. Castenholz, R.W. 1977. The effect of sulfide on the blue-green algae of hot springs II. Yellowstone National Park. *Microb. Ecol.* (New York) 3(2): 79–105. (ecol., sl. ment.)
143. Castenholz, R.W. 1979. Evolution and ecology of thermophilic microorganisms. pp 373-392 in M. Shilo. (Ed.). *Strategies of Microbial Life in Extreme Environments. Report of the Dahlem Workshop on Strategy of Life in Extreme Environments, Berlin, 1978, November 20–24.* Verlag Chemie, Weinheim, New York. (10) + pp0 11–513 + Author Index (1). (ecol., sl. ment.)
144. Castenholz, R.W. and C.E. Wickstrom. 1975. Thermal streams. pp 264–285 in B.A. Whitton. (Ed.). *River Ecology. Studies in Ecology.* Volume 2. University of California Press, Berkeley and Los Angeles; Blackwell Scientific Publications, Oxford. X + 725 pp. (ecol., photosyn., sl. ment.)
145. Cavalier-Smith, T. 1981. Eukaryote kingdoms: seven or nine? *BioSystems* (Amsterdam) 14(3–4): 461–481. (evol., tax.-syst.)
146. Cavalier-Smith, T. 1982. The origins of plastids. *Biol. J. Linn. Soc.* (London) 17(3): 289–306. (evol., tax.-syst.)
147. Cavalier-Smith, T. 1987. Glaucophyceae and the origin of plants. *Evol. Trends Plants* (Warwickshire) 1(2): 75–78. (cytol., evol., tax.-syst.)
148. Chang, Y.-C. and V. McMahon. 1970. Protein production in the alga, *Cyanidium caldarium.* pp 12–13 in *Abstracts of Papers Presented at the Forty-First Annual Meeting, University of Northern Colorado, Greeley, Colorado, May 1, 2, 1970. J. Colorado-Wyoming Acad. Sci.* (Boulder) 7(1): 5–84. (biochem., ecol., physiol.)
149. Chapman, D.J. 1965. *Studies on the Carotenoids of the Flagellate Cryptophyceae, and the Chloroplast Pigments of the Endosymbiotic Algae in* Cyanophora paradoxa *and* Glaucocystis nostochinearum. Ph. D. Dissertation, Department of Marine Biology, University of California, San Diego. XI + 169 pp. (biochem., cytol., morphol., photosyn., physiol., tax.-syst.)
150. Chapman, D.J. 1966. Studies on the carotenoids of the flagellate Cryptophyceae, and the chloroplast pigments of the endosymbiotic algae in *Cyanophora paradoxa* and *Glaucocystis nostochinearum. Diss. Abstr.* (Ann Arbor) 26(10): 5674. (sl. ment., tax.-syst.)
151. Chapman, D.J. 1973. Biliproteins and bile pigments. pp 162-185 in N.G. Carr and B.A. Whitton. (Eds). *The Biology of Blue-Green Algae. Botanical Monographs.* Volume 9. Blackwell Scientific Publications, Oxford, London, Edinburgh, Melbourne. (University of California Press, Berkeley, California). Frontispiece + X + 676 pp. (biochem., evol., photosyn., physiol., sl. ment.)
152. Chapman, D.J. 1974. Taxonomic status of *Cyanidium caldarium*, the Porphyridiales and Goniotrichales. *Nova Hedwigia Zeitschr. Kryptogamenkd.* (Weinheim, Lehre, Braunschweig) 25(3/4): 673–682. (cytol., evol., morphol., tax.-syst.)
153. Chapman, D.J. 1991. Orgin and divergence of protists. pp 477–483 in J.W. Schopf and C. Klein. (Eds). *The Proterozoid Biosphere: a Multidisciplinary Study.* Cambridge University Press, Cambridge, New York, Melbourne. (4) + Frontispiece (1) + (3) + pp VII-XXIV + (6) + pp 7–1348. (biochem., evol., sl. ment.)
154. Chapman, D.J., W.J. Cole and H.W. Siegelman. 1968. Phylogenetic implications of phycocyanobilin and c-phycocyanin. *Am. J. Bot.* (Lancaster) 55(3): 314–316. (biochem., evol., photosyn., physiol.)
155. Chapman, V.J. and D.J. Chapman. 1973. *The Algae.* Second Edition. The Macmillan Press Ltd, London and Basingstoke. XIV + 497 pp. (ecol., evol., morphol., sl. ment.)
156. Charles, D. 1977. Isolation and characterization of DNA from unicellular algae. *Plant Sci. Lett.* (Amsterdam) 8(1): 35–44. (biochem., sl. ment.)
157. Chen, C.-H. and D.S. Berns. 1978. Comparison of the stability of phycocyanins from thermophilic, mesophilic, psychrophilic and halophilic algae. *Biophys. Chem.* (New York) 8(3): 203-213. (biochem., photosyn., physiol.)
158. Chen, C.-H., O.H.W. Kao and D.S. Berns. 1977. Denaturation of phycocyanin by urea and determination of the enthalpy of denaturation by microcalorimetry. *Biophys. Chem.* (New York) 7: 81–86. (biochem., photosyn., physiol.)
159. Christensen, T. 1962. *Alger.* in T.W. Böcher, M. Lange and T. Sørensen. (Redig.). *Botanik.* Bind II. *Systematisk Botanik.* Nr. 2. I Kommission hos Munksgaard, København. (11) + pp 12–178 including 75 figs. (sl. ment., tax.-syst.)
160. Christensen, T. 1966. *Alger.* 2 udgave. in T.W. Böcher, M. Lange and T. Sørensen. (Redig.). *Botanik.* Bind II. *Systematisk Botani.* Nr. 2. 2 udgave. I Kommission hos Munksgaard, København. (11) + pp 12–180 including 76 figs. (sl. ment., tax.-syst.)
161. Christensen, T. 1980. *Algae. A Taxonomic Survey.* Fasc. 1. Aio Tryk as, Odense. (4) + 216 pp + (2) + Preliminary Index (11). (sl. ment., tax.-syst.)
162. Clement-Metral, J.D. 1976. Preparation and some properties of protoplasts from the red alga *Porphyridium cruentum. J. Microsc. Biol. Cell.* (Paris) 26(2/3): 167–172 + 1 col. pl. (biochem., photosyn., physiol., sl. ment.)
163. Cohen-Bazire, G., S. Begin, S. Rimon, A.N. Glazer and D.M. Brown. 1978. Physico-chemical and immunological properties of allophycocyanins. *Arch. Microbiol.* (Berlin) 111(3): 225–238. (biochem., photosyn., physiol.)
164. Collins, F.S., I. Holden and W.A. Setchell. 1901. *Phycotheca Boreali-Americana. A Collection of Dried Specimens of the Algae of North America. Issued by Frank Shipley Collins, Isaac Holden, William Albert Setchell. Assisted by Miss E. Butler, J.E. Duerden, N.L. Gardner, R.E. Gibbs, Mrs G.A. Hall, L.E. Hunt, Miss Jessie Polley, De A. Saunders, Mrs W.C. Simmons and Mrs E. Snyder.* Fascicule XVIII. Malden, Massachusetts. (4) + Specimen Nos 851–900. (florist., morphol.)

165. Collins, F.S., I. Holden and W.A. Setchell. 1905. *Phycotheca Boreali-Americana. A Collection of Dried Specimens of the Algae of North America. Issued by Frank Shipley Collins, Isaac Holden, William Albert Setchell. Assisted by B.M. Davis, N.L. Gardner, Mrs. G.A. Hall, F.D. Lambert, Miss L.J. MacRae, W.J.V. Osterhout.* Fascicule XXV. Malden, Massachusetts. (2) + Specimen Nos 1201–1250. (florist., tax.-syst.)
166. Copeland, J.J. 1936. Yellowstone Thermal Myxophyceae. *Ann. N. Y. Acad. Sci.* (New York) 36(Art. 1): 1–232. (ecol., florist., morphol., tax.-syst.)
167. Cornejo, J. and S.I. Beale. 1988. Algal heme oxygenase from *Cyanidium caldarium.* Partial purification and fractionation into three required protein components. *J. Biol. Chem.* (Baltimore) 263(24): 11915–11921. (biochem., physiol.)
168. Cornejo, J. and S.I. Beale. 1988. Heme oxygenase from *Cyanidium caldarium*: partial purification and fractionation into three required protein components. p. 60 in *Scientific Program (Abstracts of Papers) for the Annual Meeting of the American Society of Plant Physiologists at Bally's Hotel, Reno, Nevada, July 10–14, 1988. Plant Physiol.* (Lancaster) 86(4 Suppl.): (4) + pp 1–164. (biochem., physiol.)
169. Crean, J.G. 1966. *Some Studies of Certain Water-Soluble Hemoproteins of the Alga* Cyanidium caldarium. Ph.D. Dissertation, Department of Botany, University of Chicago, Chicago. VII + 96 pp. (biochem., cultivat., physiol.)
170. Cribb, A.B. 1976. Some algae from Blackdown Tableland. *Queensl. Nat.* (Brisbane) 21(5/6): 132–133. (florist.)
171. Csatorday, K., R. MacColl and D.S. Berns. 1980. Pigment synthesis in *Cyanidium caldarium.* p. 2006 in *Abstracts of Papers. American Society of Biological Chemists, 71st Annual Meeting and Biophysical Society 24 Annual Meeting. New Orleans, Louisiana, 1–5 June, 1980. Fed. Proc., Fed. Am. Soc. Exp. Biol.* (Baltimore) 39(6): 1601–2343. (biochem., photosyn., physiol.)
172. Csatorday, K., R. MacColl and D.S. Berns. 1981. Accumulation of protoporphyrin IX and Zn protoporphyrin IX in *Cyanidium caldarium.* Proc. Natl. Acad. Sci. U.S.A. (Washington) 78(3): 1700–1702. (biochem., photosyn., physiol.)
173. Darnall, D.W., B. Greene and J. Gardea-Torresdey. 1990. Gold binding to algae. pp 487–498 in P.R. Norris and D.P. Kelly. (Eds). *Biohydrometallurgy. Proceedings of the International Symposium on Biohydrometallurgy held at the University of Warwick, July 12–16, 1987.* Science and Technology Letters, Kew, Surrey. (12) + pp 3–578. (ecol., physiol.)
174. Dayhoff, M.O. and R.M. Schwartz. 1980. Prokaryote evolution and the symbiotic origin of eukaryotes. pp 63–83 in W. Schwemmler and H.E.A. Schenk. (Eds). *Endocytobiology. Endosymbiosis and Cell Biology. A Synthesis of Recent Research. Proceedings of the International Colloquium on Endosymbiosis and Cell Research, Tübingen, Germany, April 1980.* Walter de Gruyter, Berlin, New York. (5) + pp VI-XXIV + 1060 pp. (biochem., evol., sl. ment.)
175. De Luca, P., R. Gambardella and A. Merola. 1979. Thermoacidophilic algae of North and Central America. *Bot. Gaz.* (Crawfordsville, Chicago) 140(4): 418–427. (cytol., florist., morphol., tax.-syst., ultrastr.)
176. De Luca, P. and A. Moretti. 1983. Floridosides in *Cyanidium caldarium, Cyanidioschyzon merolae* and *Galdieria sulphuraria* (Rhodophyta, Cyanidiophyceae). *J. Phycol.* (Baltimore, Lawrence) 19(3): 368–369. (biochem., evol., physiol., tax.-syst.)
177. De Luca, P., A. Moretti, A. Musacchio and R. Taddei. 1974. Il primo reperto di *'Cyanidium caldarium'* (Acquasanta, Italia, 1839) e due distinte 'forme' di quest'alga. *Delpinoa* (N.S.) (Napoli) 14/15(1972-1973): 3–11. (florist., morphol., tax.-syst.)
178. De Luca, P., A. Moretti and R. Taddei. 1974. Nuove stazioni di *'Cyanidium caldarium'* nell'Italia meridionale ed in Sicilia. *Delpinoa* (N.S.) (Napoli) 14/15(1972-1973): 49-60. (ecol., florist., tax.-syst.)
179. De Luca, P., A. Moretti and R. Taddei. 1975. Influenza del pH sulle dimensioni cellulari di *Cyanidium caldarium* forma *B. Delpinoa* (N.S.) (Napoli) 14/15(1972–1973): 141–152. (ecol., florist., morphol., physiol., tax.-syst.)
180. De Luca, P., A. Moretti and R. Taddei. 1976-1977. Presenza di *Cyanidioschyzon merolae* De Luca, Taddei, Varano in ambienti acidi extraeuropei (U.S.A. e Indonesia). *Delpinoa* (N.S.) (Napoli) 18/19: 69–76. (florist., tax.-syst.)
181. De Luca, P., A. Musacchio and R. Taddei. 1972. Diverso comportamento in eterotrofia delle due forme di *'Cyanidium caldarium'* dei Campi Flegrei (Napoli). *Delpinoa* (N.S.) (Napoli) 12/13(1970-1971): 19–27. (biochem., cultivat., tax.-syst.)
182. De Luca, P., A. Musacchio and R. Taddei. 1972. Osservazioni sui pigmenti delle due forme de *Cyanidium caldarium* dei Campi Flegrei (Napoli). *Delpinoa* (N.S.) (Napoli) 12/13(1970-1971): 115–123. (biochem., photosyn., physiol., tax.-syst.)
183. De Luca, P., A. Musacchio and A. Taddei. 1981. Acidophilic algae from the fumaroles of Mount Lawu (Java), locus classicus of *Cyanidium caldarium* Geitler. *Giorn. Bot. Ital.* (Firenze) 115: 1–9. (florist., tax.-syst.)
184. De Luca, P. and R. Taddei. 1970. Due alghe delle fumarole acide dei Campi Flegrei (Napoli): *Cyanidium caldarium*? *Delpinoa* (N.S.) (Napoli) 10/11(1968-1969): 79–89. (biochem., cultivat., ecol., florist., morphol., photosyn., physiol., tax.-syst.)
185. De Luca, P. and R. Taddei. 1972. Crescita comparata di due forme di *'Cyanidium caldarium'* dei Campi Flegrei (Napoli) in presenza di diverse fonti de azoto. *Delpinoa* (N.S.) (Napoli) 12/13(1970–1971): 3–8. (biochem., cultivat., ecol., tax.-syst.)
186. De Luca, P. and R. Taddei. 1976. On the necessity of a systematic revision of the thermal acidophilic alga *'Cyanidium caldarium'* (Tilden) Geitler. *Webbia* (*Rac. Scr. Bot.*, Firenze) 30(1): 197–218. (morphol., tax.-syst., ultrastr.)
187. De Luca, P., R. Taddei and L. Varano. 1978. *'Cyanidioschyzon merolae'*: a new alga of thermal acidic environments. *Webbia* (*Rac. Scr. Bot.*, Firenze) 33(1): 37–44. (biochem., florist., morphol., photosyn., ultrastr.)
188. Demoulin, V. 1980. Red algae, neglected key group of protists phylogeny. p. 179 in *Abstracts,* Second International

Congress *of Systematic and Evolutionary Biology.* The University of British Columbia, Vancouver, Canada, July 17-24, 1980. University of British Columbia, Vancouver, Canada. 441 pp + (12). (evol., sl. ment.)

189. De Toni, J.B. 1889. *Syllogue Chlorophycearum Omnium Hucusque Cognitarum* (in 2 Sectio) in J.B. De Toni. *Sylloge Algarum Omnium Hucusque Cognitarum. Digessit Doct. J. Bapt. De-Toni.* Vol. I. *Chlorophyceae.* Sumptibus Auctoris, Typis Seminarii, Patavii. Sectio I. (7) + pp 8–12 + [(1) + pp II-CXXXIX (*Bibliotheca Phycologica seu Catalogus Liborum et Collectinium Exsiccatarum Algas Quascumque Sistentium*)] + (1) + pp 2–592; Section II. (2) + pp 593-1315. (tax.-syst.)
190. Di Martino Rigano, V., C. Di Martino, V. Vona, S. Esposito and C. Rigano. 1989. Nitrogen nutrition and changes in amino acid pools of *Cyanidium caldarium. Phytochemistry* (oxford) 28(11): 2891–2895. (biochem., physiol.)
191. Di Martino Rigano, V., V. Vona, C. Di Martino and C. Rigano. 1986. Effect of darkness and CO_2 starvation on NH_4^+ and NO_3^- assimilation in the unicellular alga Cyanidium caldarium. *Physiol. Plant.* (Copenhagen) 68(1): 34–38. (biochem., cultivat., physiol.)
192. Di Martino Rigano, V., V. Vona, C. Di Martino and C. Rigano. 1987. Regulatory aspects of NH_4^+ utilization in the acidophilic thermophilic unicellular red alga *Cyanidium caldarium. New Phytol.* (*Br. Bot. J.*, London) 102(2): 247–254. (biochem., cultivat., physiol.)
193. Di Martino Rigano, V., V. Vona, A. Fuggi, A. Martello, C. Di Martino and C. Rigano. 1984. Depression of nitrate reductase in the presence of excess ammonium in a unicellular alga growing under conditions of phosphate limitation. *Biochem. Biophys. Res. Commun.* (New York) 119(1): 259–264. (biochem., cultivat., physiol.)
194. Di Martino Rigano, V., V. Vona, A. Fuggi and C. Rigano. 1982. Effect of L-methionine-DL-sulphoxomine, a specific inhibitor of glutamine synthetase on ammonium and nitrate metabolism in the unicellular alga *Cyanidium caldarium. Physiol. Plant.* (Copenhagen) 54(1): 47–51. (biochem., physiol.)
195. Di Martino Rigano, V., V. Vona, L. Manzo and C. Rigano. 1987. NH_4^+ uptake by the unicellular alga *Cyanidium caldarium.* Possible control mechanisms dependent on nitrogen status. *New Phytol.* (*Br. Bot. J.*, London) 107(3): 507–512. (biochem., cultivat., physiol.)
196. Diner, B.A. 1979. Energy transfer from the phycobilisomes to photosystem II reaction centers in wild type *Cyanidium caldarium. Plant Physiol.* (Lancaster) 63(1): 30–34. (biochem., photosyn., physiol.)
197. Diner, A.A. and D. Mauzerall. 1973. Photosynthetic oxygen production and the size of the photosynthetic unit in a cell-free preparation from *Cyanidium caldarium. Biochim. Biophys. Acta B (Bioenergetics)* (New York, Amsterdam) 292(1) B(20): 285–290. (biochem., photosyn., physiol.)
198. Diner, B.A. and F.-A. Wollman. 1979. Biosynthesis of photosystem II reaction centers, antenna and plastoquinone pool in greening cells of *Cyanidium caldarium* mutant III-C. *Plant Physiol.* (Lancaster) 63(1): 26–29. (biochem., photosyn., physiol.)
199. Diner, B.A. and F.-A. Wollman. 1979. Functional comparison of the photosystem II center-antenna complex of a phycocyanin-less mutant of *Cyanidium caldarium* with that of *Chlorella pyrenoidosa. Plant Physiol.* (Lancaster) 63(1): 20–25. (biochem., photosyn., physiol.)
200. Dodge, J.D. 1973. *The Fine Structure of Algal Cells.* Academic Press, A Subsidiary of Harcourt Brace Jovanovich, Publishers, London, New York. (4) + pp V–XII + 261 pp. (cytol., sl. ment., tax.-syst., ultrastr.)
201. Doemel, W.N. 1970. *The Physiological Ecology of* Cyanidium caldarium. Ph.D. Dissertation, Department of Microbiology, Indiana University, Bloomington. X + 222 pp + Vita (1). (biochem., cultivat., cytol., ecol., florist., morphol., photosyn., physiol., radiobiol., tax.-syst.)
202. Doemel, W.N. 1971. The physiological ecology of *Cyanidium caldarium. Diss. Abstr. Int. B, Sci. Eng.* (Ann Arbor) 31(7): 4223B. (biochem., ecol., florist., photosyn., physiol.)
203. Doemel, W.N. and T.D. Brock. 1968. The temperature range of *Cyanidium caldarium* in nature. pp 45–46 in *Bacteriological Proceedings. Abstracts of the* 68th Annual Meeting, *Detroit, Mich., 5–10 May 1968.* American Society of Microbiology, Ann Arbor. (2) + XIII + 222 pp. (biochem., ecol., photosyn., physiol., radiobiol.)
204. Doemel, W.N. and T.D. Brock. 1970. The upper temperature limit of *Cyanidium caldarium.* Arch. Mikrobiol. (Berlin, Heidelberg) 72: 326–332. (evol., florist., physiol., radiobiol.)
205. Doemel, W.N. and T.D. Brock. 1971. The physiological ecology of *Cyanidium caldarium. J. Gen. Microbiol.* (Cambridge) 67(1): 17–32. (biochem., ecol., florist., photosyn., physiol., tax.-syst.)
206. Döhler, G., H. Bürstell and G. Jilg-Winter. 1976. Pigmentzusammensetzung und photosynthetische CO_2-Fixierung von *Cyanidium caldarium* und *Porphyridium aerugineum.* Pigment composition and photosynthetic CO_2 fixation of *Cyanidium caldarium* and *Porphyridium aerugineum. Biochem. Physiol. Pflanz.* (Jena) 170(2): 103–110. (biochem., photosyn., physiol.)
207. Dougherty, E.C. and M.B. Allen. 1960. Is pigmentation a clue to protistan phylogeny? pp 129–144 in M.B. Allen. (Ed.). *Biochemistry of Photoreactive Systems. Symposia on Comparative Biology of the Kaiser Foundation Research Institute.* Vol. I. Academic Press, New York. XII + 437 pp. (evol., photosyn., sl. ment., tax.-syst.)
208. Dring, M.J. 1981. Chromatic adaptation of photosynthesis in benthic marine algae: an examination of its ecological significance using a theoretical model. *Limnol. Oceanogr.* (Baltimore) 26(2): 271–284. (photosyn., sl. ment.)
209. Droop, M.R. 1974. Heterotrophy of carbon. pp 530-559 in W.D.P. Stewart. (Ed.). *Algal Physiology and Biochemistry. Botanical Monographs.* Volume 10. University of California Press, Berkeley and Los Angeles. (Blackwell Scientific Publications, Oxford, London, Edinburgh, Melbourne). XI + (1) + 989 pp. (biochem., cultivat., sl. ment.)
210. Drost-Hansen, W. 1981. Gradient device for the study of temperature effects on biological systems. *J. Wash. Acad. Sci.* (Baltimore) 71(4): 187–201. (cultivat., physiol.)
211. Drouet, F. 1943. New species and transfers in Myx-

ophyceae. *Am. Midl. Nat.* (Notre Dame) 30: 671–674. (tax.-syst.)

212. Drouet, F. and W.A. Daily. 1956. Revision of the coccoid Myxophyceae. *Butler Univ. Bot. Stud.* (Indianapolis) 12: 1–218. (biochem., florist., tax.-syst.)
213. Drouet, F. vand W.A. Daily. 1957. Revision of the coccoid Myxophyceae. Additions and corrections. *Trans. Am. Microsc. Soc.* (Lancaster) 76(2): 219–222. (sl. ment., tax.-syst.)
214. Edwards, M.R. and O.H.W. Kao. 1972. Cell biology of *Cyanidium caldarium*. p. 67a in *Abstracts of Papers Presented at the Twelfth Annual Meeting of the American Society for Cell Biology, The Chase-Park Plaza Hotel, Saint Louis, Missouri, 8–11 November, 1972. J. Cell Biol.* (New York) 55(2 pt. 2): (2) + pp 1a–305a. (biochem., photosyn., physiol., sl. ment.)
215. Edwards, M.R. and J.D. Mainwaring. 1973. Ultrastructural localization of phycocyanin in the acidophilic, thermophilic alga *Cyanidium caldarium*. p. 462 + 1 pl. in C.J. Arceneaux. (Ed.). *Proceedings*, Thirty-First Annual Meeting, *Electron Microscopical Society of America, New Orleans, Louisiana, August 14, 15, 16 en 17, 1973.* Claitor's Publishing Division, Baton Rouge. (22) + 709 pp. (evol., photosyn., ultrastr.)
216. Egelhoff, T. and A. Grossman. 1983. Cytoplasmic and chloroplast synthesis of phycobilisome polypeptides. *Proc. Nat. Acad. Sci. U.S.A.* (Washington) 80(11): 3339–3343. (biochem., photosyn., physiol.)
217. Elenkin, A.A. 1938. *Cinezelenye Vodorosli SSSR. Monografiia Presnovodnykh i Nazemnykh Cyanophyceae. Obnaruzhennykh v Predelakh SSSR. Spetsil'naia* (Sistematicheskaia) *Chast'. Sostavlennaia A. A. Elenkinym, pri Uchastii M. M. Gollerbakha* (Sem. Gloeocapsaceae), *E.K. Kosinskoĭ* (Sem. Scytonemataceae) *i V.* I. *Polianskogo* (Sem. Rivulariaceae). Vyp. I. I. *Chroococceae Geitl.;* II. *Chamaesiphoneae Geitl.*; III. *Hormogoneae (Geitl.) Elenk.: Poriadok Stigonematales Geitl.; Poriadok Mastigocladales Elenk.; Poriadok Diplonematales Elenk.; Poriadok Nostocales (Geitl.) Elenk.; Sem. Nostocaceae Elenk., Sem. Anabaenaceae Elenk., Sem. Aphanizomenonaceae Elenk., Sem. Nodulariaceae Elenk., Sem. Scytonemataceae (Kütz.) Elenk. [Monographia Algarum Cyanophycearum Aqauiculcium et Terrestrium in Finibus USRR Inventarum.* Pars Specialis *(Systematica). Auctore A. A. Elenkin, Particibus M. M. Hollerbach (Fam. Gloeocapsaceae, C. C. Kossinskaja (Fam. Scytonemataceae) et V. I. Poljansky (Fam. Riculariaceae).* Fasc. I. I. *Chroococceae Geitl.;* II. *Chamaesiphoneae Geitl.;* III. *Hormogoneae (Geitl.) Elenk.: Order Stigonematales Geitl.; Ordo Mastigocladales Elenk.; Ordo Diplonematales Elenk.; Ordo Nostocales (Geitl.) Elenk.: Fam. Nosltocaceae Elenk., Fam. Anabaenaçeae Elenk., Fam. Aphanizomenonaceae Elenk., Fam. Nodulariaceae Elenk., Fam. Scytonemataceae (Kütz) Elenk.]* Botanicheskiĭ Institut Akademii Nauk SSSR. Izdatel"stvo Akademii Nauk SSSR, Moskva, Leningrad. (Institutum Botanicum Academiae Scientiarum URSS. Sumptibus Academiae Scientiarum URSS. Mosqua, Leningrad.) (5) + pp 6–984 with 290 figs + 1 folding table. (tax.-syst.)
218. Emoto, Y. 1941. Nihonsan onsen shokubutsu ni tsuite. (Über die Thermalflora von Japan). *Onsen Kagaku (Ber. Jpn. Ges. Balneol.,* Tokyo) 1: 119-138. (ecol., florist., tax.-syst.)
219. Emoto, Y. 1942. Über Flora und Fauna der Thermen. SS C21-C23 in Über des Thermalbad Kusatu. *Mitt. Dtsch. Ges. Nat.-Völk. Ostasiens* (Tokyo, Leipzig) 33(Teil C): C1–C67. (ecol., florist., tax.-syst.)
220. Emoto, Y. and H. Hirose. 1940. Nihonsan onsen shokubutsu no kenkyū (III). Hakone onsengun no saikin rui oyobi sōrui (2). [Studien über die Thermalflora von Japan. (III). Thermale Bakterien und Algen aus den thermalen Quellen von Hakone (2)]. *Shokubutsu Kenkyū Zasshi (J. Jpn. Bot.,* Tokyo) 16: 405–420. (ecol., florist.)
221. Emoto, Y. and H. Hirose. 1940. Nihonsan onsen shokubutsu no kenkyū IV. Adatarasan oyobi Azumasan onsengun no saikinrui to sōrui. [Studien über die Thermalflora von Japan IV. Thermale Bakterien und Algen aus thermalen Quellen der Adatara- und Azuma (eine Teil) Berg]. *Shokubutsu Oyobi Dôbutsu (Bot. Zool. Theor. Appl.,* Tokyo) 8(10): 1577–1586 (ecol., florist.)
222. Emoto, Y. and H. Hirose. 1940. Nihonsan onsen shokubutsu no kenkyū IV. Adatarasan oyobi Azumasan onsengun no saikinrui to sōrui (3). [Studien über die Thermalflora von Japan. IV. Thermale Bakterien und Algen aus thermalen Quellen der Adatara- und Azuma (eine Teil) Berg (3)]. *Shokubutsu Oyobi Dôbutsu (Bot. Zool. Theor., Appl.,* Tokyo) 8(11): 1883–1890. (ecol., florist.)
223. Emoto, Y. and H. Hirose. 1941. Nihonsan onsen shokubutsu no kenkyū X. Nikkō Yumoto onsen no saikin to sōrui. (Studien über die Thermalflora von Japan. X. Thermale Bakterien und Algen aus thermalen Quellen von Yumoto bei Nikko). *Shokubutsugaku Zasshi* (*Bot. Mag.,* Tokyo) 55(658): 477–485. (ecol., florist.)
224. Emoto, Y. and H. Hirose. 1942. Izu hantō onsen no onsen shokubutsu 1. Higashi kaigan onsengun no saikinrui oyobi sōrui. (Thermalflora der heissen Quellen der Halbinsel Izu 1. Thermal Bakterien und Algen aus den Thermen der östlichen Küste). *Onsen Kagaku* (*Ber. Jpn. Ges. Balneol.,* Tokyo) 2(4): 147–161. (There also exists a reprint of this article with the pagination of 1–14). (ecol., florist.)
225. Emoto, Y. and H. Hirose. 1942. Nihonsan onsen shokubutsu no kenkyū XIV. Naruko onsengun no saikinrui to sōrui. (Studies on the thermal-flora of Japan. XIV. Bacteria and algae from the Narugo thermal springs). *Shokubutsugaku Zasshi* (*Bot. Mag.,* Tokyo) 56(661): 25–42. (ecol., florist.)
226. Emoto, Y. and H. Hirose. 1942. Nihonsan onsen shokubutsu no kenkyū XVI. Onikubu onsengun no saikin to sōrui. (Studies on the thermal-flora of Japan. XVI. Bacteria and algae of the Onikobe thermal springs). *Shokubutsugaku Zasshi* (*Bot. Mag.,* Tokyo) 56(663): 120–136. (ecol., florist.)
227. Emoto, Y. and H. Hirose. 1942. Nihonsan onsen shokubutsu no kenkyū XIX. Yahatadaira Yakiyama onsengun saikin to sōrui. (Studien über die Thermalflora von Japan. XIX. Thermale Bakterien und Algen aus thermal Quellen von Hatimantai und Yakeyama). *Shokubut-*

sugaku Zasshi (*Bot. Mag.*, Tokyo) 56(667): 332–343. (ecol., florist.)

228. Emoto, Y. and H. Hirose. 1942. Nihonsan onsen shokubutsu no kenkyū XXI. Tochigiken Shobara onsengun no saikinrui oyobi sōrui (1). [Studien über die Thermalflora von Japan. XXI. Thermale Bakterien und Algen aus den heissen Quellen von Siobara in Totigi-Präfektur. (1)]. *Onsen Kagaku* (*Ber. Jpn. Ges. Balneol.*, Tokyo) 2(2/3): 79–85. (ecol., florist.)
229. Emoto, Y. and H. Hirose. 1942. Nihonsan onsen shokubutsu no kenkyū XXI. Tochigiken Shobara onsengun no saikinrui oyobi sōrui (2). [Studien über die Thermalflora von Japan. XXI. Thermale Bakterien und Algen aus den heissen Quellen von Siobara in Totigi-Präfektur. (2)]. *Onsen Kagaku* (*Ber. Jpn. Ges. Balneol.*, Tokyo) 2(2/3): 86–95. (ecol., florist.)
230. Emoto, Y. and Y. Yoneda. 1940. Nihonsan onsen shokubutsu no kenkyū II. Aso onsengun no saikin rui oyobi sōrui. (Studies on the thermal-flora of Japan. II. Bacteria and algae of the Aso thermal springs). *Seitaigaku Kenkyū* (*Ecol. Rev.*, Sendai) 6(1): 1–16. (ecol., florist.)
231. Emoto, Y. and Y. Yoneda. 1940. Nihonsan onsen shokubutsu no kenkyū VI. Ibusuki onsengun no saikin to sōrui. (Studies on the thermal-flora of Japan. VI. Bacteria and algae of the Ibusuki thermal springs). *Seitaigaku Kenkyū* (*Ecol. Rev.*, Sendai) 6(4): 257–274. (ecol., florist.)
232. Emoto, Y. and Y. Yoneda. 1942. Nihonsan onsen shokubutsu no kenkyū XVII. Toyama kenka sho onsen no saikin to sōrui. (Studies on the thermal-flora of Japan. XVII. Bacteria and algae of hot springs in Toyama Prefecture). *Shokubutsu Bunrui Chiri* (*Acta Phototaxon. Geobot.*, Kyoto) 11(1): 7–26. (ecol., florist.)
233. Enami, I. 1978. Mechanisms of the acido- and thermophily of *Cyanidium caldarium* Geitler V. Acid and heat stabilities of soluble proteins. *Plant Cell Physiol.* (Tokyo) 19(5): 869–876. (biochem., photosyn., physiol.)
234. Enami, I. 1978. Mechanisms of the acido- and thermophily of *Cyanidium caldarium* Geitler VI. Some physicochemical properties of acid and heat stable cytochrome c-553. *Plant Cell Physiol.* (Tokyo) 19(5): 901–905. (biochem., photosyn., physiol.)
235. Enami, I., H. Akutsu and Y. Kyogoku. 1986. Intracellular pH regulation in an acidophilic unicellular alga, *Cyanidium caldarium*: ^{31}p-NMR determination of intracellular pH. *Plant Cell Physiol.* (Tokyo) 27(7): 1351-1359. (biochem., physiol.)
236. Enami, I. and I. Fukuda. 1975. Mechanisms of the acido- and thermo-phily of *Cyanidium caldarium* Geitler I. Effects of temperature, pH and light intensity on the photosynthetic oxygen evolution of intact and treated cells. *Plant Cell Physiol.* (Tokyo) 16(2): 211–220. (biochem., photosyn., physiol.)
237. Enami, I. and I. Fukuda. 1977. Mechanisms of the acido- and thermophily of *Cyanidium caldarium* Geitler III. Loss of these characteristics due to detergent treatment. *Plant Cell Physiol.* (Tokyo) 18(3): 671–680. (biochem., physiol.)
238. Enami, I. and I. Fukuda. 1977. Mechanisms of the acido- and thermophily of *Cyanidium caldarium* Geitler IV. Loss of these characteristics due to enzyme treatment. *Plant Cell Physiol.* (Tokyo) 18(3): 707–710. (biochem., photosyn., physiol.)
239. Enami, I. and M. Kura-Hotta. 1984. Effect of intracellular ATP levels on the light-induced H+ efflux from intact cells of *Cyanidium caldarium. Plant Cell Physiol.* (Tokyo) 25(7): 1107–1113. (biochem., photosyn., physiol.)
240. Enami, I., T. Matsubayasi and I. Fukuda. 1978. A new sensitive polarographic method for determination of hydroquinone in the Hill reaction with special reference to the case of thermal alga *Cyanidium. Photosynthetica* (Praha) 12(1): 83–86. (physiol.)
241. Enami, I., H. Nagashima and I. Fukuda. 1975. Mechanisms of the acido- and thermo-phily of *Cyanidium caldarium* Geitler II. Physiological role of the cell wall. *Plant Cell Physiol.* (Tokyo) 16(2): 221–231. (biochem., physiol., ultrastr.)

-- Engler, A. 1954. *Syllabus der Pflanzenfamilien...* Please see: Schmidt, E.W., Cyanophyta, SS 50–55.

242. Etzler, F.M. and W. Drost-Hansen. 1979. A role for water in biological rate processes. pp 125–164 in W. Drost-Hansen and J.S. Clegg. (Eds). *Cell-Associated Water. Proceedings of a Workshop on Cell-Associated Water Held in Boston, Massachusetts, September, 1976.* Academic Press, New York, San Francisco, London. VIII + 440 pp. (physiol.)
243. Etzler, F.M. and W. Drost-Hansen. 1980. A role for water in growth, metabolism, and intercellular organization. pp 485-497 in M. Blank. (Ed.). *Bioelectrochemistry: Ions, Surfaces, Membranes. Based on a Symposium Sponsored by the Division of Colloid and Surface Chemistry at the 176th Meeting of the American Chemical Society in Miami Beach, Florida, September 11–14, 1978.* Advances in Chemistry Series 188. American Chemical Society, Washington, D.C. XII + (2) + 527 pp. (cultivat., physiol.)
244. Farr, E.R., J.A. Leussink and F.A. Stafleu. 1979. *Index Nominum Genericorum (Plantarum).* Volume I. Ae-Epochnium. Dr. W. Junk b.v. Publishers, The Hague. (*Regnum Vegetible* 100). (6) + pp VII–XXVI + 630 pp. (tax.-syst.)
245. Flaumenhaft, E., S. Bose, H.L. Crespi and J.J. Katz. 1965. Deuterium isotope effects in cytology. *Int. Rev. Cytol.* (New York, London) 18: 313–361. (biochem., physiol., radiobiol.)
246. Fogg, G.E. 1956. The comparative physiology and biochemistry of the blue-green algae. *Bacteriol. Rev.* (Baltimore) 20(1): 148–165. (biochem., physiol., sl. ment.)
247. Fogg, G.E., W.D.P. Stewart, A. Fay and A.E. Walsby. 1973. *The Blue-Green Algae.* Academic Press, London and New York. Frontispiece + VII + (1) + 459 pp. (ecol.)
248. Ford, T.W. 1979. Ribulose 1,5-bisphosphate carboxylase from the thermophilic, acidophilic alga, *Cyanidium caldarium* (Geitler). Purification, characterisation and thermostability of the enzyme. *Biochim. Biophys. Acta* (New York, Amsterdam) 569(2): 239–248. (biochem., physiol.)
249. Ford, T.W. 1984. A comparative ultrastructural study of *Cyanidium caldarium* and the unicellular red alga *Rhodosorus marinus. Ann. Bot.* (London) 53: 285–294 (cytol., evol., tax.-syst., ultrastr.)
250. Ford, T.W. 1986. Thermostability of the photosynthetic

system of the thermoacidophilic alga *Cyanidium caldarium* in continuous culture. *J. Exp. Bot.* (London) 37(184): 1689–1707. (biochem., cultivat., photosyn., physiol.)

251. Ford, T.W. 1988. Effects of temperature on photosynthesis in the thermoacidophilic alga *Cyanidium caldarium.* p. 175 in L.J. Rogers en J.R. Gallon. (Eds). *Biochemistry of the Algae and Cyanobacteria. Proceedings of the Phytochemical Society of Europe,* No. 28. Clarendon Press, Oxford. (9) + pp X–XVI + (3) + pp 4–374. (biochem., physiol.)
252. Fork, D.C. and N. Murata. 1977. The effect of temperature on the physical phase of thylakoid lipids and photosynthesis in algae. p. 22 in *Abstracts of papers scheduled for the combined meetings of the International Seaweed Symposium, The Phycological Society of America, and the International Phycological Society at the University of California, Santa Barbara, California, August 20–27, 1977. J. Phycol.* (Baltimore, Lawrence) 13(Suppl.): (2) + pp 3–78 + author index (2). (biochem., photosyn., physiol., sl. ment.)
253. Fork, D.C. and N. Murata. 1977. The relationship between changes of the physical phase of membrane lipids and photosynthesis in the thermophilic alga *Cyanidium caldarium.* pp 427–436 in S. Miyachi, S. Katoh, Y. Fujita and K. Shibata. (Eds). *Photosynthetic Organelles. Structure and Function.* Special Issue of Plant and Cell Physiology, No. 3. Japanese Society of Plant Physiologists, Center for Academic Publications, Japan. XI + (1) + 460 pp. (biochem., physiol.)
254. Forsberg, H., O. Zetterqvist and L. Engstrom. 1969. Protein-bound phosporylserine in different tissues and organisms. *Biochim. Biophys. Acta* (New York, Amsterdam) 181(1): 171–175. (biochem., physiol.)
255. Forti, A. 1907. *Sylloge Myxophycearum Omnium Hucusque Cognitarum. Digessit Doct. Achilles Forti.* in *J. B. De-toni. Sylloge Algarum Omnium Hucusque Cognitarum. Digessit Doct. J. Bapt. De-Toni.* Vol. V. *Myxophyceae.* Sumptibus Auctoris, Typis Seminarii, Patavii. (11) + pp 2–761. (tax.-syst.)
256. Fredrick, J.F. 1968. Glucosyltransferase isozymes in algae-III. The polyglucoside and enzymes of *Cyanidium caldarium. Phytochemistry* (Oxford) 7: 1573–1576. (biochem., evol., physiol., tax.-syst.)
257. Fredrick, J.F. 1969. Effect of amitrol on biosynthesis of phosphorylases in different algae. *Phytochemistry* (Oxford) 8(4): 725-729. (biochem., evol., physiol., tax.-syst.)
258. Fredrick, J.F. 1970. Introductory remarks: algae are pertinent. pp 415–416 in J.F. Fredrick and R.M. Klein. (Eds). *Phylogenesis and morphogenesis in the algae. Ann. N. Y. Acad. Sci.* (New York) 175(2): 413–781. (ecol.)
259. Fredrick, J.F. 1971. Polyglucan branching isoenzymes of algae. *Physiol. Plant.* (Copenhagen) 24(1): 55–58. (biochem., evol., physiol., tax.-syst.)
260. Fredrick, J.F. 1971. Requirement for pyridoxal in the biosynthesis of 4-glucosyltransferases (phosphorylases) in the algae. *Phytochemistry* (Oxford) 10(5): 1025–1029. (biochem., evol., physiol., tax.-syst.)
261. Fredrick, J.F. 1971. Storage polyglucan-synthesizing isozyme patterns in the Cyanophyceae. *Phytochemistry* (Oxford) 10(2): 395–398. (biochem., evol., physiol., tax.-syst.)
262. Fredrick, J.F. 1972. A non-primer requiring α-1, 4-glucan phosphorylase of *Cyanidium caldarium. Phytochemistry* (Oxford) 11(11): 3259–3262. (biochem., evol., physiol., tax.-syst.)
263. Fredrick, J.F. 1972. Evolutionary significance of storage polyglucoside synthesizing isozymes in blue-green algae. pp 566–570 in T.V. Desikachary. (Ed.). *Taxonomy and Biology of Blue-Green Algae. Papers Submitted to the* First International Symposium *on Taxonomy and Biology of Blue-Green Algae Held at Madras, January 8–13, 1970.* Centre for Advanced Study in Botany, University of Madras, Madras. XI + (1) + 591 pp + 44 pls. (biochem., evol., physiol., tax.-syst.)
264. Fredrick, J.F. 1973. A primordial bifunctional polyglucan-forming enzyme. pp 254–264 in J.F. Fredrick. (Ed. and Conf. Chmn). *Storage polyglucosides. Ann. N. Y. Acad. Sci.* (New York) 210: (3) + pp 4–302. (biochem., evol., physiol., tax.-syst.)
265. Fredrick, J.F. 1973. Differences of the primer-independent phosphorylase isozyme in thermophilic and mesophilic algae. *Plant Cell Physiol.* (Tokyo) 14(3): 443–448. (biochem., evol., physiol., tax.-syst.)
266. Fredrick, J.F. 1973. Electrophoresis of algae glucans. *Plant Cell Physiol.* (Tokyo) 14(6): 1187-1192. (biochem., evol., physiol., tax.-syst.)
267. Fredrick, J.F. 1973. Further studies of primer-independent phosphorylase isozymes in the algae. *Phytochemistry* (Oxford) 12(8): 1933–1936. (biochem., evol., physiol., tax.-syst.)
268. Fredrick, J.F. 1973. Phosphorylase isozyme differences in thermophilic and mesophilic algae. *Isozyme Bull.* (Ithaca) 6: 29. (biochem., evol., physiol., tax.-syst.)
269. Fredrick, J.F. 1975. Effect of environment pH on storage glucan synthesis in three thermophilic algae. *Phytochemistry* (Oxford) 14(9): 1911–1913. (biochem., evol., physiol., tax.-syst.)
270. Fredrick, J.F. 1975. Evolution of isozymes forming storage polyglucans in algae. pp 307–321 in C. Markert. (Ed.). *Isozymes* IV. *Genetics and Evolution. The* Third International Conference *on Isozymes Held at Yale University, New Haven, Connecticut, April 18–20, 1974.* Academic Press, New York, San Francisco, London. XXI + (1) + 965 pp. (biochem., evol., physiol., tax.-syst.)
271. Fredrick, J.F. 1976. *Cyanidium caldarium* as a bridge alga between Cyanophyceae and Rhodophyceae: evidence from immunodiffusion studies. *Plant Cell Physiol.* (Tokyo) 17(2): 317–322. (biochem., evol., physiol., tax.-syst.)
272. Fredrick, J.F. 1977. Glucan and glucosyltransferase isozymes of *Glaucocystis nostochinearum. Phytochemistry* (Oxford) 16: 1489–1490. (biochem., evol., physiol., tax.-syst.)
273. Fredrick, J.F. 1978. Storage glucan branching isozymes of the thermal alga, *Cyanidium caldarium. J. Therm. Biol.* (Oxford, Elmsford) 3(1): 1–4. (biochem., evol., physiol., tax.-syst.)
274. Fredrick, J.F. 1979. Storage glucan biosynthesis in *Cyanidium, Chlorella* and *Prototheca:* evidence for endosymbiosis? *Phytochemistry* (Oxford) 18(11): 1823–1825. (biochem., evol., physiol., tax.-syst.)

275. Fredrick, J.F. 1980. The b. e. and Q types of 1,4-α-D-glucan: 1,4-α-D-glucan-6-glucosyltransferase isozymes in algae. *Phytochemistry* (Oxford) 19(4): 539–542. (biochem., evol., physiol., tax.-syst.)
276. Fredrick, J.F. 1981. The biosynthesis of storage glucans in prokaryotic and eukaryotic algae. pp 426–434 in J.F. Fredrick. (Ed.) *Origins and Evolution of Eukryotic intracellular Organelles. Ann. N. Y. Acad. Sci.* (New York) 361: (8) + p. IX + (1) + 512 pp. (biochem., evol., physiol., tax.-syst.)
277. Fredrick, J.F. 1986. The glucosyltransferase isozymes as markers of evolution from prokaryotic algae to eukaryotic algae. p. 367 in *Abstracts, The Third International Colloquium on Endocytobiology:* 'Endocytobiology III.' *The New York Academy of Science, New York, June 10–12, 1986. Endocytobios. Cell Res.* (Tübingen) 3: 353–369. (biochem., evol., physiol., tax.-syst.)
278. Fredrick, J.F. 1987. Cyanidiophyceae. Transition stages in the evolution of the glucosyltransferase isozymes. pp 438–448 in J.J. Lee and J.F. Fredrick. (Eds). *Endocytobiology. Papers presented at the* Third International Colloquium on Endocytobiology *held June 10–12, 1986, in New York, N. Y and Sponsored by the New York Academy of Science. Ann. N. Y. Acad. Sci.* (New York) 503: (10) + pp XI-XII + 509 pp. (biochem., evol., physiol.)
279. Fredrick, J.F. and J. Seckbach. 1981. Search for primitive life forms. *Explor. J.* (New York) 59(1): 22–25. (biochem., evol., photosyn., physiol., tax.-syst.)
280. Fredrick, J.F. and J. Seckbach. 1983. Glucosyltransferase isozymes of the acidophylic-thermophilic Cyanidiophyceae. *Isozyme Bull.* (Ithaca) 16: 84. (biochem., evol., physiol., tax.-syst.)
281. Fredrick, J.F. and J. Seckbach. 1983. Intraspecies evolution of enzymic mechanisms associated with the synthesis of α-1,4 storage glucans in two thermophilic-acidophilic algae of *Cyanidium* type. *Phytochemistry* (Oxford) 22(5): 1155–1157. (biochem., evol., physiol., tax.-syst.)
282. Fredrick, J.F. and J. Seckbach. 1986. Storage glucan and glucosyltransferase isoenzymes of *Cyandioschyzon merolae:* a primitive eukaryote. *Phytochemistry* (Oxford) 25(2): 363–365. (biochem., evol., physiol., tax.-syst.)
283. Fredrick, J.F. and J. Seckbach. 1991. Eukaryosis in primitave cells: polyglucan synthesis in the cyanidiophyceae (Rhodophyta) as a possible evolutional tracer. p. 43 in *[Abstracts Scheduled to be presented at the] First International Symbiosis Congress, Jerusalem, Israel, November 17–22, 1991.* Congress Organizers: Kenes-Organizers of Congresses and Special Events, Tel Aviv. 70 pp + Index [(2) + 4 pp]. (biochem., evol., tax.-syst.)
284. Friedmann, I. 1964. Progress in the biological exploration of caves and subterranean waters in Israel. *Int. J. Speleol.* (Weinheim) 1(1/2): 29–33. (ecol., florist.)
285. Fritsch, F.E. 1945. The Structure and Reproduction of the Algae. Volume II. *Foreword, Phaeophyceae, Rhodophyceae, Myxophyceae.* At the University Press, Cambridge. Frontispiece + XIV + 939 pp + maps 1–2. (ecol., tax.-syst.)
286. Fuggi, A. 1985. Mechanism of proton-linked nitrate uptake in *Cyanidium caldarium*, an acidophilic non-vacuolated algae. *Biochim. Biophys. Acta* (New York, Amsterdam) 815(3): 392–398. (biochem., physiol.)
287. Fuggi, A. 1989. Competition between nitrate and nitrite as a tool to study the regulation of nitrate metabolism by ammonium in the microalga *Cyanidium caldarium.* Plant Physiol. Biochem. (Paris) 27(4): 563–568. (biochem., photosyn., physiol.)
288. Fuggi, A. 1989. Short-term regulation of nitrate uptake by a 'pump and leak' mechanism in the acidophylic non-vacuolated alga, *Cyanidium caldarium. Biochim. Biophys. Acta* (New York, Amsterdam) 974(2): 141–148. (biochem., physiol.)
289. Fuggi, A., V. Di Martino Rigano, V. Vona and C. Rigano. 1981. Nitrate and ammonium assimilation in algal cell-suspensions and related pH variations in the external medium, monitored by electrodes. *Plant Sci. Lett.* (Amsterdam) 23(2): 129–138. (biochem., physiol.)
290. Fuggi, A., V. Di Martino Rigano, V. Vona and C. Rigano. 1981. Pattern of inhibition of nitrate utilization by ammonium in the acidophilic thermophilic unicellular alga *Cyanidium caldarium. Arch. Microbiol.* (Berlin) 130(5): 349–352. (biochem., physiol.)
291. Fuggi, A., V. Vona, V. Di Martino Rigano, C. Di Martino, A. Martello and C. Rigano. 1984. Evidence for two transport systems for nitrate in the acidophilic thermophilic alga *Cyanidium caldarium. Arch. Microbiol.* (Berlin) 137(4): 281–285. (biochem., physiol.)
292. Fukuda, I. 1958. Physiological studies on a thermophilic blue green alga, *Cyanidium caldarium* Geitler. (Onsensei ransō ideyukogome no seirigakuteki kenkyū). *Shokubutsugaku Zasshi* (*Bot. Mag.,* Tokyo) 71(837): 79–86. (biochem., florist., physiol.)
293. Fukuda, I. 1962. Studies on Hill-Reaction in a thermophilic blue green alga, *Cyanidium caldarium* Geitler. (Onsensei ransō, ideyukogome no hiruhannō ni kansuru kenkyū). *Shokubutsugaku Zasshi* (*Bot. Mag.,* Tokyo) 75(892): 349–355. (biochem., evol., physiol.)
294. Fukuda, I. 1979. *A Possible Literature Survey of a Thermal Alga* cyanidium caldarium *Geitler* (I). Science University of Tokyo, Kagurazaka, Shinjuku-ku, Tokyo, Japan. 6 pp. (literat.)
295. Fukuda, I. 1980. *A Possible Literature Survey on a Thermal Alga* Cyanidium caldarium *Geitler* (II). Science University of Tokyo, Kagurazaka, Shinjuku-ku, Tokyo, Japan. pp 7–10. (literat.)
296. Fukuda, I. 1981. *A Possible Literature Survey on a Thermal Alga Cyanidium caldarium Geitler* (III). Science University of Tokyo, Kagurazaka, Shinjuku-ku, Tokyo, Japan. pp 11–15. (literat.)
297. Fukuda, I. and S. Ishisone. 1965. Ransō *Cyanidium* no kōgōsei ni okeru ESR signals. [Light induced ESR signals from a blue green alga, *Cyanidium*]. *Jpn Electron Opt. News* 44: 4–5. (biochem., photosyn., physiol.)
298. Fukuda, I. and S. Ishisone, 1969. Denshi supin kyōmei kyūshūhō niyoru hikari gosei shoki hannō no kenkyū. [Studies on light induced ESR signals in photosynthesis]. pp 9–13 in K. Shibata. (Ed.). *Mechanism of Energy Conversion in Photosynthesis.* Reports of Cooperative Research from the Ministry of Education. (photosyn., physiol.)
299. Fukuda, I. and H. Nagashima. 1987. Cytochemical determination of DNA content in the nucleus and nucleoids of *Cyanidium* and some microbial cells. pp 575–577 in J.J.

Lee and J.F. Fredrick. (Eds). *Endocytobiology. Papers presented at the* Third International Colloquium *on Endocytobiology held June 10–12, 1986, in New York, N. Y. and Sponsored by the New York Academy of Science. Ann. N. Y. Acad. Sci.* (New York) 503: (10) + pp XI–XII + 509 pp. (biochem., evol., physiol.)

300. Fukuda, I., H. Nagashima and J. Seckbach. 1990. DNA content and the evolution of information in the nucleus and nucleoids of the lowest eukaryote. pp 585–588 in P. Nardon, V. Gianinazzi-Pearson, A.M. Grenier, L. Margulis and D.C. Smith. (Eds). *Endocytobiology IV.* 4th International Colloquium *on Endocytobiology and Symbiosis. 4ème Congrès International sur l'Endocytobiologie et la Symbiose. INSA, Vilerbanne (France), July 4–8, 1989.* Institut National de la Recherche Agronomique, Paris. (7) + pp 8–620. (biochem., cytol., evol., morphol., photosyn., physiol. tax.-syst., ultrastr.)

301. Fukuda, I., H. Nagashima and J. Seckbach. 1991. How *Cyanidioschyzon* envolved. p. 70 in *[Abstracts Scheduled to be Presented at the] First International Synbiosis Congress, Jerusalem, Israel, November 17–22, 1991.* Congress Organizers: Kenes–Organizers of Congresses and Special Events. Tel Aviv. 70 pp + Index [(2) + 4 pp]. (ecol., tax.-syst.)

302. Gabrielson, P.W. and D. Garbary. 1984. Systematics of red algae (Rhodophyta). *CRC (Chemical Rubber Company) Critical Rev. in Plant Sci.* (Boca Raton) 3(4): 325–366. (sl. ment., tax.-syst.)

303. Galdieri, A. 1899. Su di un'alga che cresce intorno alle fumarole della solfatara; nota del dott. Agostino Galdieri. *Rend. Accad. Sci. Fis. Mat. Napoli* (ser. 3a) (Napoli) 5(6–7): 160–164. (cytol., morphol., tax.-syst.)

304. Gambardella, R., A. Moretti and A. Musacchio. 1980. Autoecologia ed ultrastruttura de alghe termoacidofile del nord e centro America. *Delpinoa* (N.S.) (Napoli) 21: 35–45 + Tav. I–III. (florist., physiol., tax.-syst., ultrastr.)

305. Gantt, E. 1975. Phycobilisomes: light-harvesting pigment complexes. *BioScience* (Washington) 25(12): 781–788. (cytol., ultrastr.)

306. Gantt, E. 1981. *Phycobilisomes. Annu. Rev. Plant Physiol.* (Stanford) 32: 327–347. (biochem., photosyn., physiol.)

307. Gantt, E. 1990. Pigmentation and photoacclimation. pp 203–216 in K.M. Cole and R.G. Sheath. (Eds). *Biology of Red Algae.* Cambridge University Press, Cambridge, New York, Port Chester, Melbourne, Sydney. (4) + pp V–IX + (1) + 517 pp. (biochem., photosyn., physiol., sl. ment.)

308. Gantt, E. and S.F. Conti. 1966. Granules associated with the chloroplast lamellae of *Porphyridium cruentum. J. Cell Biol.* (New York) 29(3): 423–434. (cytol., sl. ment., ultrastr.)

309. Gantt, E. and S.F. Conti. 1967. Phycobiliprotein localization in algae. pp 393–405 in *Energy Conversion by the Photosynthetic Apparatus. Report of Symposium Held June 6–9, 1966. Brookhaven Symposia in Biology* Number 19. Brookhaven National Laboratory, Associated Universities, Inc., Biology Department, Upton, New York. IX + Frontispiece + 514 pp. (sl. ment., ultrastr.)

310. Garbary, D.J. and P.W. Gabrielson. 1990. Taxonomy and evolution. pp 477–498 in K.M. Cole and R.G. Sheath. (Eds). *Biology of Red Algae.* Cambridge University Press, Cambridge, New York, Port Chester, Melbourne, Sydney. (4) + pp V–IX + (1) + 517 pp. (evol., sl. ment.)

311. Garbary, D.J., G.I. Hansen and R.F. Scagel. 1980. A revised classification of the Bangiophyceae (Rhodophyta). *Nova Hedwigia, Zeitschr. Kryptogamenkd.* (Weinhein, Lehre, Braunschweig) 33: 145–166. (morphol., tax.-syst.)

312. Gardea-Torresdey, J.L. 1988. *Metal Ion Binding by Algae and its Novel Electroanalytical Applications.* Doctor of Philosophy Dissertation, Department of Chemistry, New Mexico State University, Las Cruces. 262 pp. (ecol., physiol.)

313. Gardea-Torresdey, J.L. 1988. Metal ion binding by algae and its novel electroanalytical applications. *Diss. Abstr. Int. B, Sci. Eng.* (Ann Arbor) 49(8): 3145–3146. (ecol., physiol.)

314. Gardea-Torresdey, J., D. Darnall and J. Wang. 1988. Bioaccumulation and voltammetric behavior of gold at alga-containing carbon paste electrodes. *J. Electroanal. Chem.* (Amsterdam) 252(1): 197–208. (ecol., physiol.)

315. Geitler, L. 1925. Synoptische Darstellung der Cyanophyceen in morphologischer und systematischer Hinsicht. *Beih. Bot. Zentralbl.* (Kassel) 41, Abt. II(3): 163–294 + Taf. XVII–XX. (tax.-syst.)

316. Geitler, L. 1932. *Dr. L. Rabenhorst's Kryptogamen-Flora von Deutschland, Österreich und der Schweiz. Zweite, völlstandig neu bearbeitete Auflage.* Vierzehnter Band. *Die Algen (Herausgegeben von Prof. Dr. R. Kolkwitz). Cyanophyceae von Europa unter Berücksichtigung der andern Kontinento von Dr. Lothar Geitler.* Akademische Verlagsgesellschaft m.b. H., Leizig. VI + 1196 SS. (tax.-syst.)

317. Geitler, L. 1933. Diagnosen neuer Blaualgen von den Sunda-Inseln. *Arch. Hydrobiol.* (Stuttgart) Suppl. Bd XII (*Tropische Binnengewässer* IV): 622-634. (florist., morphol., tax.-syst.)

318. Geitler, L. 1942. *Schizophyta: Klasse Schizophyceae. Bearbeitet von Lothar Geitler.* in *H. Harms und J. Mattfeld Die Natürlichen Pflanzenfamilien nebst ihren Gattungen und wichtigeren Arten, insbesondere den Nutzpflanzen unter Mitwirkung zahlreicher hervorragender Fachgelehrten Begrundet von A. Engler und K. Prantl.* Zweite stark vermehrte und verbesserte Auflage. *Herausgegeben von Adolf Engler. Fortgesetzt von Hermann Harms und Johannes Mattfeld.* Band 1b. *Mit 156 Figuren im Text.* Verlag von Wilhelm Engelmann, Leipzig. (3) + S. IV + 232 SS. (morphol. tax.-syst.)

319. Geitler, L. 1942 (Reprint, 1959). *Schizophyta: Klass Schizophyceae. Bearbeitet von Lother Geitler.* in *H. Harms und J. Mattfeld. Die Näturlichen Pflanzenfamilien nebst ihren Gattungen und wichtigeren Arten, insbesondere den Nutzpflanzen unter Mitwirkung zahlreicher hervorragender Fachgelehrten begrundet von A. Engler und K. Prantl.* Zweite stark vermehrte und verbessert Auflage. *Herausgegeben von Adolf Engler. Fortgesetzt von Hermann Harms und Johannes Mattfeld.* Band 1b. *Mit 156 Figuren im Text.* Duncker & Humblot, Berlin. (3) + S. IV + 232 SS. (morphol., tax.-syst.)

320. Geitler, L. 1959. Die Gattung *Cyanidium. Österr. Bot. Zeitschr.* (Wien) 106: 172–173. (tax.-syst.)

321. Geitler, L. 1960. Schizophyzeen. (2., umgearbeitete Auflage). in W. Zimmermann and O. Ozenda. (Eds). *Handbuch der Pflanzenanatomie.* Band VI, Teil I.

Abteilung: Spezieller Teil. Gebrüder Borntraeger, Berlin-Nikolasse. VII + (1) + 131 pp. (morphol., tax.-syst.)

-- Geitler, L. Please see: W. Titz. (1969) Lothar Geitler. Zum 70. Geburtstag am 18. Mai 1969.

-- Geitler, L. Please also see: M. Hesse (1979) Lothar Geitler. Zum 80. Geburtstagam 18. Mai, 1979.

-- Please also see: A. Weber, M. Kiehn and S. Somtag (1989) Lothar Geitler on the occasion of this 90th birthday on May 18th 1989.

322. Geitler, L. and F. Ruttner. 1935. Die Cyanophyceen der Deutschen Limnologischen Sunda-Expedition, ihre Morphologie, Systematik und Ökologie. Erster Teil. A. Beschreibung der Proben. *Arch. Hydrobiol.* (Stuttgart) Suppl. Bd XIV (*Tropische Binnengewässer* VI): 308–369 mit Taf. XVII–XIX. (florist.)

323. Geitler, L. and F. Ruttner. 1935. Die Cyanophyceen der Deutschen Limnologischen Sunda-Expedition, ihre Morphologie, Systemmatik und Ökologie. Zweiter Teil. B. Systematisch-Morphologischer Tail. *Arch. Hydrobiol.* (Stuttgart) Suppl. Bd XVI (*Tropische Binnengewässer* VI): 371–482 mit Taf. XX–XXI mit 92 Fig. im Text und auf 20 Textbeilagen. (cytol., ecol., morphol., tax.-syst.)

324. Geitler, L. and F. Ruttner. 1936. Die Cyanophyceen der Deutschen Limnologische Sunda-Expedition, ihre Morphologie, Systematik und Ökologie. Dritter Teil. C. Ökologischer Teil. *Arch. Hydrobiol.* (Stuttgart) Suppl. Bd XIV (*Tropische Binnengewässer* VI): 553–715 mit 7 Textbeilagen und 3 Tabellen im Text. (ecol.)

325. George, E.A. (Ed.). 1966. *Culture Collection of Algae and Protozoa. List of Strains.* [1st edition]. University of Cambridge, The Botany School, Cambridge. (2) + 67 pp. (cultivat.)

326. George, E.A. (Ed.). 1971. *Culture Collection of Algae and Protozoa. List of Strains.* [2nd edition]. Culture Centre of Algae and Protozoa, Natural Environment Research Council, Cambridge. (2) + 73 pp. (cultivat.)

327. George, E.A. (Ed.). 1976. *Culture Centre of Algae and Protozoa. List of Strains.* [3rd edition] Institute of Terrestrial Ecology. Natural Environment Research Council, Cambridge. (4) + II + 120 pp. [for the 4th Ed. of this work see Asher and Spalding (1982)]. (cultivat.)

-- George, E.A. (Ed.). Please note: The first, second and third editions of the catalogue for Cambridge culture collection of algae and Protozoa, listed above, give only one isolation of *Cyanidium caldarium*. For the fourth edition please see: A. Asher and F. Spalding. (Eds). (1982) *Culture Centre of Algae and Protozoa, List of Strains* which listed an additional 15 strains of *Cyanidium caldarium*. For the fifth edition, please see: A.S. Thompson, J.C. Rhodes and I. Pettman. (Eds). (1988) *Culture Collection of Alae and Protozoa, Catalogue of Strains* which listed no cultures of *Cyanidium caldarium* but is given in this bibliography to give the reader the most current edition of the holdings of the Cambridge collection.

328. Gerasimenko, L.M., M.A. Pusheva and S.V. Goriunova. 1972. Growth cycle and ultrafine structure of *Cyanidium caldarium. Microbiology* (USSR, New York) 41(2): 282–286. (cultivat., cytol., morphol., photosyn., physiol., ultrastr.)

329. Gerasimenko, L.M., M.A. Pusheva and S.V. Goriunova. 1972. Tsikl razvitiia i ul 'tratonkow stroenie *Cyanidium caldarium.* (The growth cycle and the ultra-fine structure of Cyanidium caldarium.) *Mikrobiologiia* (Moskva) 41(2): 324–326 + ris. 1–2. (cultivat., cytol., morphol., photosyn., physiol., ultrastr.)

330. Gessner, F. 1955. *Hydrobotanik. Die Physiologischen Grundlagen der Pflanzenverbreitung im Wasser.* I. *Energiehaushalt.* VEB Deutscher Verlag der Wissenschaften, Berlin (12) + 517 SS mit 291 Abbildungen im Text und 8 Farbtafeln. (ecol., physiol., sl. ment.)

331. Gessner, F. 1959. *Hydrobotanik. Die Physiologischen Grundlagen der Pflanzenverbreitung im Wasser.* II. *Stoffhaushalt.* VEB Deutscher Verlag der Wissenschaften, Berlin. XIV + 701 SS mit 349 Abbildungen im Text und 8 Farbtafeln. (ecol., physiol., tax.-syst.)

332. Glazer, A.N. 1977. Structure and molecular organization of the photosynthetic accessory pigments of cyanobacteria and red algae. *Mol. Cell. Biochem.* (Den Haag) 18(2/3): 125–140. (biochem., sl. ment.)

333. Glazer, A.N., G.S. Apell, C.S. Hixson, D.A. Bryant, S. Rimon and D.M. Brown. 1976. Biliproteins of cyanobacteria and Rhodophyta: homo'ogous family of photosynthetic accessory pigments. *Proc. Natl. Acad. Sci. U.S.A.* (Washington) 73(2): 428–431. (biochem., evol., photosyn., physiol., sl. ment.)

334. Glazer, A.N., G. Cohen-Bazire and R.Y. Stanier. 1971. Comparative immunology of algal biliproteins. *Proc. Natl. Acad. Sci. U.S.A.* (Washington) 68(12): 3005–3008. (biochem., photosyn., physiol. sl. ment.)

335. Glazer, A.N., G. Cohen-Bazire and R.Y. Stanier. 1971(Reprint, 1974). Comparative immunology of algal biliproteins. pp 154-163 in M. Rodriguez-Lopez *et al. Blue-Green Algae* I: *Current Research.* MSS Information Corporation, New York. (10) + pp 11–213. (biochem., photosyn., physiol., sl. ment.)

336. Goldin, H. 1969, Part I. *On the Interactions of Ribonucleic Acid from* E. coli *and Methylated Bovine Serum Albumin.* Part II. *Studies on the Ribosomes and Ribonucleic Acid from the Thermophilic Algae* Cyanidium caldarium. Ph.D. Dissertation, Division of Biochemistry, University of Wyoming, Laramie. (1) + VII + (Part I) pp 1–39 + (Part II) pp 40–73. (biochem., physiol.)

337. Golf, L.A. and A.W. Coleman. 1990. DNA microspectrofluorometry. pp 43–71 in K.M. Cole and R.G. Sheath. (Eds). Biology of Red Algae. Cambridge University Press, Cambridge, New York, Port Chester, Melbourne, Sydney. (4) + pp V-IX + (1) + 517 pp. (cytol., sl. ment.)

338. Goodwin, T.W. 1960. Algal carotenoids. pp 1-10 in M.B. Allen. (Ed.). *Comparative Biochemistry of Photoreactive Systems. Symposia on Comparative Biology of the Kaiser Foundation Research Institute.* Vol. I. Academic Press, New York. XII + 437 pp. (biochem., evol., physiol., sl. ment., tax.-syst.)

339. Goodwin, T.W. 1966. The carotenoids. pp 121–137 in T. Swain. (Ed.). *Comparative Phytochemistry.* Academic Press, New York, London. XIII + (1) + 360 pp. (biochem., physiol., sl. ment.)

340. Goodwin, T.W. 1974. Carotenoids and biliproteins. pp 176–205 in W.D.P. Stewart. (Ed.). *Algal Physiology and Biochemistry. Botanical Monographs.* Volume 10. Univer-

sity of California Press, Berkeley and Los Angeles. (Blackwell Scientific Publications, Oxford, London, Edinburgh, Melbourne). XI + (1) + 989 pp. (biochem., photosyn., physiol., sl. ment., tax.-syst.)
341. Goodwin, T.W. 1974. Sterols. pp 266–280 in W.D.P. Stewart. (Ed.). *Algal Physiology and Biochemistry. Botanical Monographs.* Volume 10. University of California Press, Berkeley, Los Angeles. (Blackwell Scientific Publications, Oxford, London, Edinburgh, Melbourne). XI + (1) + 989 p. (biochem.)
342. Govindjee and B.Z. Braun. 1974. Light absorption, emission and photosynthesis. pp 346–390 in W.D.P. Steward. (Ed.). *Algal Physiology and Biochemistry. Botanical Monographs.* Volume 10. University of California Press, Berkeley and Los Angeles. (Blackwell Scientific Publications, Oxford, London, Edinburgh, Melbourne). XI + (1) + 989 pp. (biochem., photosyn., physiol.)
343. Govindjee and P.K. Mohanty. 1972. Photochemical aspects of photosynthesis in blue-green algae. pp 171–196 in T.D. Desikachary. (Ed.). *Taxonomy and Biology of Blue-green Algae. Papers submitted to the* First International Symposium *on Taxonomy and Biology of Blue-green Algae held at Madras, January 8–13, 1970.* University of Madras, Centre for Advanced Study in Botany, Madras. (9) + pp 10–591. (biochem., photosyn., physiol.)
344. Granick, S. 1967. The heme and chlorophyll biosynthetic chain. pp 373–410 in T.W. Goodwin. (Ed.). *Proceedings of a NATO Advanced Study Institute held at Aberystuyth, August, 1965.* Academic Press, London. XVIII + 776 pp. (biochem., photosyn., physiol.)
345. Greene, B., R.A. McPherson, D.W. Darnall and J.L. Gardea-Torresdey. 1990. *Recovery of Heavy and Precious Metals from Aqueous Solutions.* Patent Cooperation Treaty, World Intellectual Property Organization, Geneve, 37 pp. (ecol., physiol.)
346. Gribovskai͡a, I.V., O.I. Novoselova, I.A. Gladchenko and T.T. Usti͡ugova. 1980. Mineral'nyĭ sdstav nekotorykh vidov zelenykh i sinezelenykh vodorosleĭ. pp 36–43 in F. I͡a. Sid'ko and V.N. Belianin. (Otvetstvennye Redaktory). *Parametricheskoe Upravlenie Biosintezom Mikrovodorosleĭ.* Akademii͡a Nauk SSSR. Sibirskoe Otdelenie. Institut Fiziki͡a Im. L.B. Kirenskogo. Izpatel'stvo 'Nauka', Sibirskoe Otdelenie, Novosibirsk. (2) + 120 pp. (cultivat., physiol.)
347. Gromov, B.V., I.A. Avilov, L.V. Andreev, F.N. Li͡utikov and V.N. Nikitina. 1979. At͡sidofil'nai͡a vodorosl' *Cyanidium caldarium* iz termal'nykh istochnikov dal'nego vostoka. [Acidophilic alga *Cyanidium caldarium* from far eastern (USSR) thermal springs]. *Biol. Nauki* (Moskva) 1979(10): 78–82. (ecol., florist., physiol.).
348. Gross, W. 1989. Intracellular localization of enzymes of fatty acid-β-oxidation in the alga *Cyanidium caldarium. Plant Physiol.* (Lancaster) 91(4): 1476–1480. (biochem., evol., physiol.)
349. Gross, W. 1990. Membrane bound malate dehydrogenase in mitochondria from the alga *Cyanidium caldarium. Phytochemistry* (Oxford) (29(10): 3081–3085. (biochem., physiol.).
350. Gross, W. and H. Beevers. 1989. Subcellular distribution of enzymes of glycolate metabolism in the alga *Cyanidium caldarium. Plant Physiol.* (Lancaster) 90(3): 799–805. (biochem., evol., physiol.)
351. Grossman, A. and J. Brand 1983. A rapid procedure for the isolation of intact phycobilisomes. pp 116-120 in W.R. Briggs. (Dir.). *Annual Report of the Director, Department of Plant Biology 1892–1983.* Reprinted from the Carnegie Institution of Washington Year Book 82 for the Year July 1, 1982-June 30, 1983, Stanford. (4) + pp 7–129. (biochem., photosyn., physiol.).
352. Grossman, A. and J. Brand. 1983. A rapid procedure for the isolation of intact phycobilisomes. *Carnegie Inst. Washington Yearbook* (Washington) 82(1982-1983): 116-120. (biochem., photosyn., physiol.)
353. Grossman, A., L. Talbott and T. Egelhoff. 1983. Biosynthesis of phycobilisome polypeptides of *Porphyridium aerugineum* and *Cyanophora paradoxa.* pp 112–116 in W.R. Briggs. (Dir.). *Annual Report of the Director, Department of Plant Biology 1982–1983.* Reprinted from Carnegie Institution of Washington Year Book 82 for the Year July 1, 1982-June 30, 1983, Stanford. (4) + pp 7–129. (biochem., photosyn., sl. ment.)
354. Grossman, A., L. Talbott and T. Egelhoff. 1983. Biosynthesis of phycobilisomes polypeptides of *Porphyridium cruentum* and *Cyanophora paradoxa. Carnegie Inst. Washington Year Book* (Washington) 82(1982-1983): 112–116. (biochem., photosyn., physiol., sl. ment.)
355. Guerrero, M.G., J.M. Vega and M. Losada. 1981. The assimilatory nitrate-reducing system and its regulation. *Annu. Rev. Plant Physiol.* (Stanford) 32: 169–204. (biochem., physiol. sl. ment.)
356. Gusev, M.V. 1961. Sinezelenye vodorosli. *Mikrobiologii͡a* (Moskva) 30(6): 1108–1128. (biochem., ecol., morphol., photosyn., physiol., sl. ment., tax.-syst.)
357. Gusev, M.V. 1961. The blue-green algae. *Microbiology* (USSR, New York) 30: 897–911. (biochem., ecol., morphol., photosyn., physiol., sl. ment., tax.-syst.)
358. Halldal, P. and C.S. French. 1958. Algal growth in crossed gradients of light intensity and temperature. *Plant Physiol.* (Lancaster) 33(4): 249–252. (cult., physiol.).
359. Halldal, P., H.S. Huang and C.S. French. 1957. Growth patterns of various algae in crossed gradients of light intensity and temperatures. *Carnegie Inst. Washington Year Book* (Stanford, Calif.) 56: 275–278. (cult., physiol.)
360. Hamana, K. and S. Matsuzaki. 1985. Further study on polyamines in primitive unicellular algae. *J. Biochem.* (Tokyo) 97(2): 1311–1316 (biochem., evol.)
361. Hamana, K., S. Matsuzaki, M. Nittsu, K. Samejima and H. Nagashima. 1990. Polyamines of unicellular thermoacidophilic red alga *Cyanidium caldarium. Phytochemistry* (Oxford) 29(2): 377–380. (biochem., evol., physiol.)
362. Hansgirg, A. 1892. Prodromus Českých Řas Sladkovodních. Díl Druhý Obsahující Řasy Siné (Myxophyceae, Cyanophyceae) s Dodatky k Prvnímy Dílu, Sladkovodními Bakteriemi a Euglenami. (Uveřejněno Podporou České Akademie Císaře Františka Josefa). *Arch. pro Přírodovědecký Výzkum Čech.* (v Praze) VIII, díl, čís. 4(Botanické Oddělení): I–XI + (1) + 1–182. (tax.-syst.)
363. Hansgirg, A. 1892. Prodromus der Algenflora von Bohmen. Zweiter Theil, welcher die blaugrünen Algen

(Myxophyceen, Cyanophyceen), hebst Nachtragen zum ersten Theile und einer systematischen Bearbeitung der in Böhmen verbreiteten saprophytischen Bacterien und Euglenen enthält. Mit dem Opiz-Preise gekrönte Arbeit. Auf Kosten des Opizfondes. *Arch. Naturwissenschaft. Durchforsch. Böhmen* (Prag) VIII, Nro. 4(Botanische Abtheilung): (3) + 4–268. (tax.-syst.)

364. Hara, Y. 1972. An electron microscopic study on the chloroplasts of the Rhodophyta. pp 153–158 in K. Nisizawa. (Ed.). *Proceedings of the* Seventh International Seaweed Symposium, *Sapporo, Japan, August 8–12, 1971. Organized By Science Council of Japan.* A Halsted Press Book, John Wiley & Sons, New York, Toronto. XVI + (1) + 607 pp. (cytol., ultrastr.)
365. Hara, Y. 1975. Yōryokutai bisaikōzō no kenkyū to kōsō shokubutsu no keitōbunrui eno kiyo II. (Studies on the chloroplast ultrastructures and their contributions to the taxonomy and phylogeny of the Rhodophyta II). *Sōrui (Bull. Jpn. Soc. Phycol.,* Sapporo) 23(2): 67–78. (evol., tax.-syst., ultrastr.)
366. Hara, Y and M. Chihara. 1974. Comparative studies on the chloroplast ultrastructure in the Rhodophyta with special reference to their taxonomic significance. *Sci. Rep. Tokyo Kyoiku Daigaku,* Sect. B. (Tokyo) 15(232): 209–235 + pls. I–IV. (cytol., evol., ultrastr.)
367. Harel, E. 1978. Initial steps in chlorophyll synthesis-problems and open questions. pp 33–44 in G. Akoyunoglou and J.H. Argyroudi-Akoyunoglou. (Eds). *Chloroplast Development. Proceedings of the International Symposium on Chloroplast Development Held on the Island of Spetsai, Greece, July 9–15, 1978. Developments in Plant Biology.* Volume 2. Elsevier/North-Holland Biomedical Press, Amsterdam, New York, Oxford. XVI + 888 pp. (biochem., photosyn., physiol.)
368. Hargeaves, J.W. and B.A. Whitton. 1976. Effect of pH on growth of acid stream algae. *Br. Phycol. J.* (London) 11(3): 215–223. (ecol., sl. ment.)
369. Harris, J.U. and D.S. Berns. 1975. Sequences of the N-terminus portions of biliproteins. *J. Mol. Evol.* (Berlin, New York) 5: 153–163. (biochem., evol., photosyn., physiol., tax.-syst.)
370. Harwood, J.L. 1975. Fatty acid biosynthesis. pp 43–93 in T. Galliard and E.I. Mercer. (Eds). *Recent Advances in the Chemistry and Biochemistry of Plant Lipids. Proceedings of a Symposium Arranged by the Phytochemical Society and the Lipid Group of the Biochemical Society, University of East Anglia, Norwich, April, 1974.* Proceedings of the Phytochemical Society *Number* 12. Academic Press, London, New York, San Francisco. XVI + 398 pp. (biochem., physiol.)
371. Hase, T., S. Wakabayashi, K. Wada, H. Matsubara, F. Jüttner, K. Rao, I. Fry and D.O. Hall. 1978. *Cyanidium caldarium* ferredoxin: a red algal type? *FEBS Lett. (Fed. Eur. Biochem. Soc.,* Amsterdam) 96(1): 41–44. (biochem., physiol.)
372. Hesse, M. 1979. Lothar Geitler. Zum 80. Geburtstag am 18. Mai 1979: Wissenschaftliche Veröffentlichungen Lothar Geitlers seit 1968. (Publication numbers 302–344). *Plant Syst. Evol.* (Wien) 131(3-4): 165–168. (literat.)

-- Hetzroni-Leiv, O. Please see: Leiv-Hetzroni, O.

373. Hewitt, E.J. 1974. Aspects of trace element requirements in plants and micro-organisms: the metallo enzymes of nitrate and nitrite reduction. pp 199–245 in D.H. Northcote. (Ed.). *Biochemistry.* Series One. Volume II. *Plant Biochemistry.* MTP International Review of Science. Butterworths, London; University Park Press, Baltimore. (12) + 287 pp. (biochem., physiol.)
374. Hipkin, C.R., P.J. Syrett and B.A. Al-Bassam. 1979. Some characteristics of nitrate reductase in unicellular algae. pp 309–312 in E.J. Hewitt and C.V. Cutting. (Eds). *Nitrogen Assimilation of Plants.* Sixth Long Ashton Symposium 1977. *Proceedings of a Symposium Held at Long Ashton Research Station, University of Bristol, 19–22 September 1977.* Academic Press, London, New York, San Francisco. XX + 708 pp (biochem., physiol., sl. ment.)
375. Hiramatsu, N. 1967. Cyanophyceae from Nagasaki Prefecture. *Shokubutsu Kenkyu Zasshi J. Jpn. Bot.,* Tokyo) 42(1): 33–42. (florist.)
376. Hirose, H. 1947. Tokushu na Ransōrui *Cyanidium* no Yūshokutai ni Tsuite. On a plastid of a special blue-green algae, *Cyanidium caldarium. Kagaku To Kagaku Kyōiku (Sci. Scient. Educ.)* 1(1): 43–45. (cytol., morphol.)
377. Hirose, H. 1950. Ideyukogome *Cyanidium caldarium* (Tilden) Geitler no seitai to bunpu ni tsuite. [Studies on *Cyanidium caldarium* (Tilden) Geitler, with special reference to its ecology and distribution in Japan]. *Shokubutsu Kenkyū Zasshi J. Jpn. Bot.,* Tokyo) 25(8): 179–184. (ecol., florist., physiol.)
378. Hirose, H. 1950. Studies on a thermal alga, *Cyanidium caldarium.* (Onsen shokubutsu *Cyanidium caldarium* ideyukogome ni tsuite). *Shokubutsugaku Zasshi* (*Bot. Mag.,*Tokyo) 63(745/746): 107–111. (biochem., cytol., morphol., tax.-syst.)
379. Hirose, H. 1953. Onsen to sōrui. (Hot springs and algae). *Sōrui (Bull. Jpn. Soc. Phycol.,* Sapporo) 1(1): 10–15. (ecol., florist.)
380. Hirose, H. 1958. Rearrangement of the systematic position of a thermal alga, *Cyanidium caldarium.* (Onsensan ideyukogome no bunruigakuteki chii no saihenkō). *Shokubutsugaku Zasshi* (*Bot. Mag.,* Tokyo) 71(844): 347–352. (biochem., cytol., morphol., tax.-syst.)
381. Hirose, H. 1959. Correction. [Rearrangement of the systematic position of a thermal alga, *Cyanidium caldarium* (Tilden) Geitler]. *Shokubutsugaku Zasshi* (*Bot. Mag.,* Tokyo) 72(855): 348. (tax.-syst.)
382. Hirose, H. 1959. *Sōruigaku Sōsetsu. (General Phycology).* Uchida Rokakuho, Tokyo, Japan. (2) + Frontispiece + (2) + 6 pp + 10 pp + 5 pp + (1) + 2 pp + (2) + 506 pp + (2) + 12 pp + 87 pp + (5). (ecol., morphol., tax.-syst.)
383. Hirose, H. 1966. Noboribetsu onsen no onsen seibutsu, tokuni shokubutsu ni tsuite. (On the thermal plants of Norboribetsu spa). *Onsen Kagaku* (*Ber. Jpn. Ges. Balneol.,* Tokyo) 16(2): 65–70. (ecol., florist.)
384. Hirose, H. 1972. Sōruigaku Sōsetsu. (General Phycology). Uchida Rokakuho, Tokyo, Japan. (2) + Colored Frontispiece + 4 pls + (2) 6 pp + 10 pp + 55 pp + (1) + 2 pp + (2) + 5067 pp + (2) + 12 pp + 87 pp + (5). (ecol., morphol., tax.-syst.)
385. Hofer, I. 1973. *Disc-elektrophoretische Untersuchungen einiger Dehydrogenasen aus Cyanophyceen und eukaryon-*

tischen Algen vorgelegt von Ilse Hofer aus Bludenz. Dissertation zur Erlangung des Grades eines Doktors der Naturwissenschaften dem Fachbereich Chemie der Eberhard-Karls-Universität zur Tübingen. (4) + 68 SS + (1) + Lebenslauf (1). (biochem., physiol., tax.-syst.)

386. Hoffmann, L. 1989. Algae of terrestrial habitats. *Bot. Rev.* (Lancaster) 55(2): 77–105. (ecol.)

387. Holm-Hansen, O. and R.A. Lewin. 1965. Bound ornithine in certain flexibacteria and algae. *Physiol. Plant* (Copenhagen) 18(2): 418–423. (biochem., physiol.)

388. Holm-Hansen, O., R. Prasad and R.A. Lewin. 1965. Occurrence of α, ε-diaminopimelic acid in algae and flexibacteria. *Phycologia* (Berkeley, Copenhagen, Vancouver) 5(1): 1–14. (biochem., evol., physiol., tax.-syst.)

389. Holroyd, J.A., D.I. Vernon and S.B. Brown. 1985. Biliverdin, an intermediate in the biosynthesis of plant pigments. *Biochem. Soc. Trans.* (London) 13(1): 209–210. (biochem., photosyn., physiol.)

390. Holton, R.W. 1962. Isolation, growth, and respiration of a thermophilic blue-green alga. *Am. J. Bot.* (Lancaster) 49(1): 1–6. (ecol., physiol., sl. ment.)

391. Holton, R.W. 1973. Electrophoresis and the taxonomy of algae. pp 297–303 in Y. Schlechter. (Ed.). *Symposium on the use of electorphoresis in the taxonomy of algae and fungi. Bull. Torrey Bot. Club* (Lancaster) 100(5):253–312. (biochem., evol., physiol., tax.-syst.)

392. Hoogenhout, H. and J. Amesz. 1965. Growth rates of photosynthetic microorganisms in laboratory cultures. *Arch. Mikrobiol.* (Berlin, Heidelberg) 50: 10–24. (cult., ecol., physiol., tax.-syst.)

393. Hori, T. and R. Ueda. 1975. The fine structure of algal chloroplasts and algal phylogeny. pp 11–42 in J. Tokida and H. Hirose. (Eds). *Advance of Phycology in Japan.* Dr. W. Junk b.v. Publishers, The Hague. 355 pp. (biochem., ecol., photosyn., physiol.)

394. Houghton, J.D., S.B. Brown, S.P. Gough and C.G. Kannangara. 1989. Biosynthesis of δ-aminolevulinate in *Cyanidium caldarium*: characterization of tRNAglu, ligase, dehydrogenase and glutamate 1-semialdehyde aminotransferase. *Carlsberg Res. Commun.* (Copenhagen) 54(4): 131–144. (biochem., physiol.)

395. Houghton, J.D., J.A. Holroyd and S.B. Brown. 1987. A method for determining phycocyanin and chlorophyll concentration in whole cell suspensions of *Cyanidium caldarium.* pp 507–510 in J. Biggins. (Ed.). *Progress in Photosynthesis Resarch,* Vol. 4. *Proceedings of the* VIIth International Congress *on Photosynthesis, Providence, Rhode Island, U.S.A., August 10–15, 1986.* Martinus Nijhoff Publishers, Dordrecht, Boston, Lancaster. (4) + pp V-XXVII + (1) + 858 pp. (biochem., photosyn., physiol.)

396. Houghton, J.D., L. Turner and S.B. Brown. 1988. The effect of gabaculine on tetrapyrrole biosynthesis and heterotrophic growth in *Cyanidium caldarium. Biochem. J.* (Liverpool) 254(3): 907–910. (biochem., photosyn., physiol.)

397. Ignat'evska͡ia, M.A. and L.A. Mineeva. 1982. Effect of medium acidity on growth and pigment content of *Cyanidium caldarium* in relation to culture conditions. *Sov. Plant Physiol.* (USSR, New York) 29 (3 Part 2): 463–466. (cultivat., ecol., photosyn., physiol.)

398. Ignat'evska͡ia, M. A. and L. A. Mineeva. 1982. Vli͡ianie kislotnosti sredy na rost i pigmentnyĭ sostav *Cyanidium caldarium* v zavisimosti ot sposoba syshchestvovani͡ia. *Fiziol. Rast.* (Moskva) 29(3): 586–590. (cult., ecol., photosyn., physiol.)

-- Ignat'evskaya, M.A. Please see: Ignat'evasta͡ia, M. A.

399. Ikan, R. and J. Seckbach. 1972. Lipids of the thermophilic alga *Cyanidium caldarium. Phytochemistry* (Oxford) 11(3): 1077–1082. (biochem., evol., physiol., tax.-syst.)

400. Inman, O.L. 1940. Studies on the chlorophylls and photosynthesis of thermal algae from Yellowstone National Park, California, and Nevada. *J. Gen. Physiol.* (New York) 23(6): 661–666. (florist., photosyn., tax.-syst.)

401. Ishida, M.R. and T. Kikuchi. 1980. Macromolecular synthesis in algal cells: a review on the kinetics of incorporation *in vivo* of radioisotope-labelled compounds. *Annu. Rep. Res. React. Inst. Kyoto Univ.* (Kyoto) 13: 145–160. (biochem., physiol., radiobiol.)

402. I͡urina, N.P., G.V. Karakashev, M.S. Odint͡sova and N.V. Karapeti͡an. 1989. Influence of norflurazone on the synthesis and composition of the thylakoid polypeptides of the single-cell [unicellular] algae *Cyanidium caldarium. Dokl. Biochem., Akad. Nauk SSSR* (USSR, New York) 306(3): 149–152. (biochem., photosyn., physiol.).

403. I͡urina, N.P., G.B. Karakashev, M.S. Odint͡sova and N.V. Karapeti͡an. 1989. Vli͡ianie norflurazona na sintez i sostav tilokoidnykh polipeptidov odnokletochnoĭ vodorosli *Cyanidium caldarium.* (The effect of norflurazon on the synthesis and composition of thylakoid polypeptides of other unicellular alga *Cyanidium caldarium*). *Dokl. Akda. Nauk SSSR* (Moskva, Leningrad) 306(3): 739–742. (biochem., photosyn., physiol.)

-- Jakob K. Kristjánsson. Please see: Kristjánsson, J.K.

404. Jurgenson, J.E., S.I. Beale and R.F. Troxker. 1976. Biosynthesis of δ-aminolevulinic acid in the unicellular rhodophyte, *Cyanidium caldarium. Biochem. Biophys. Res. Commun.* (New York) 69(1): 149–157. (biochem., photosyn., physiol., radiobiol.)

405. Jüttner, F. 1979. Nor-carotenoids as the major volatile excretion products of *Cyanidium. Zeitschr. Naturforsch.* Teil C, *Biosci.* (Wiesbaden, Tübingen) 34(3/4): 186–191. (biochem., ecol., physiol.)

406. Jüttner, F. 1979. The algal excretion product, geranylacetone: a potent inhibitor of carotene biosynthesis in *Synechococcus. Zeitschr. Naturforsch.* Teil C, *Biosci.* (Wiesbaden, Tübingen) 34(11): 957–960. (biochem., ecol., physiol., sl. ment.)

407. Jüttner, F. 1981. Biologically active compounds released during algal blooms. *Verh. Internat. Ver. Theor. Angew. Limnol.* (Stuttgart) 21: 227–230. (biochem., ecol., physiol.)

408. Jüttner, F., H. Victor and H. Metzner. 1971. Massenanzucht phototropher Organizmen in einer automatischen Kulturanlage. (Mass culture of phototrophic organisms in an Automatic Culture Plant). *Arch. Mikrobiol.* (Berlin, Heidelberg) 77: 275–280. (cultivat., physiol.)

409. Jüttner, F. and K. Wurster. 1979. Einfache Anordnung zur Adsorption von Geruchsstoffen aus Algen an Tenax GC und deren Überführung in Gaschromatographie-Systeme. *J. Chromatogr.* (Amsterdam) 175(1): 178–182. (biochem., physiol.)

410. Kao, O.H.W. and D.S. Berns. 1968. The monomer molecular weight of c-phycocyanin. *Biochem. Biophys. Res. Commun.* (New York) 33(3): 457–462. (biochem., photosyn., physiol.)
411. Kao, O.H.W. and D.S. Berns. 1974. Aggregation properties of c-phycocyanin isolated from thermo-tolerant *Cyanidium caldarium*. p. 1232 in *Abstracts of Papers. American Society of Biological Chemists,* 65th Annual Meeting and Biophysical Society, 18th Annual Meeting *held jointly as the Biochemistry/Biophysical 1974 Meeting. Cosponsored by the Pan-American Association of Biochemical Societies, Minneapolis, Minnesota 2–7, June, 1974. Fed. Proc., Fred. Am. Soc. Exp. Biol.* (Baltimore) 33(5): (4) + pp 1219–1675. (biochem., photosyn., physiol.)
412. Kao, O.H.W. and D.S. Berns. 1977. Similar C-phycocyanins from two strains of thermotolerant cyanophyte *Mastigocladus. Can. J. Microbiol.* (Ottawa) 23(5): 510–517. (biochem., sl. ment.)
413. Kao, O.H.W., D.S. Berns and W.R. Town. 1973. The characterization of c-phycocyanin from an extremely halotolerant blue-green alga, *Coccochloris elabens. Biochem. J.* (Liverpool) 131: 39–50. (biochem., sl. ment.)
414. Kao, O.H.W., M.R. Edwards and D.S. Berns. 1975. Physical-chemical properties of c-phycocyanin isolated from an acidothermophilic eukaryote, *Cyanidium caldarium. Biochem. J.* (Liverpool) 147(1): 63–70. (biochem., photosyn., physiol.)
415. Kao, O.H.W., M.R. Edwards, R. MacColl and D.S. Berns. 1976. Thermophilic phycocyanin. pp 291-305 in H. Zuber. (Ed.). *Enzymes and Proteins from Thermophilic Microorganisms. Structure and Function. Proceedings of the International Symposium, Zürich, July 28 to August 1, 1975.* Birkhäuser Verlag, Basel und Stuttgart. 445 pp. (biochem., photosyn., physiol., ultrastr.)
416. Kaplan, I.R. 1956. Evidence of microbiological activity in some geothermal regions of New Zealand. *N.Z. J. Sci. Technol.* (Wellington) 37: 639–662. (ecol., florist., physiol., sl. ment.)
417. Karakashev, G.V., N.P. I͡urina, N.P. Karapeti͡an and M.S. Odint͡sova. 1989. Uchastie I͡adernogo i Khloroplastnogo genomov v biosinteze bekov tilokoidnykh membran v kletkakh odnokletochnoĭ krasnoi vodorosli *Cyanidium caldarium. Biokhimii͡a (Moskva) 54(10): 1693–1699. (biochem., photosyn., physiol.)*
418. *Karakashev, G.V., N.P. I͡urina, N.V. Karapetian and M.S. Odint͡sova. 1990. Participation of nuclear and chloroplast genomes in the biosynthesis of proteins of the thylakoid membranes in cells of the single-celled red alga Cyanidium caldarium. Biochemistry* (USSR, New York) 54(10, part 2): 1383–1388. (biochem., photosyn., physiol.)
419. Karakashev, G.V., N.P. I͡urina and M.S. Odint͡sova. 1988. Immunokhimicheskiĭ analiz polipeptidov tilakdoidnukh membran t͡sianobaketerii krasnykh volorosleĭ i Vysshikh Rasteniĭ. *Dokl. Akad. Nauk SSSR* (Moskva, Leningrad) 302(1): 228–231) (biochem., physiol.)
420. Karakashev, G.V., N.P. I͡urina and M.S. Odint͡sova. 1989. Immunochemical analysis of polypeptides of the thylakoid membranes of cyanobacteria, red algae, and higher plants. *Dokl. Biochem., Akad. Nauk SSSR* (USSR, New York) 302(1–6): 285–288. (biochem., photosyn., physiol.)
421. Karakashev, G.V., N.P. I͡urina, V.V. Shubin and M.S. Odint͡sova. 1988. Identifikat͡sii͡a komponentova fotosistemy I i t͡sitokhromov tilakoidnykh membran krasnoĭ vodorosli *Cyanidium caldarium. Biokhimii͡a* (Moskva) 53(4): 1569–1577. (biochem., photosyn., physiol.)
422. Karakashev, G.V., N.P. I͡urina, V.V. Shubin and M.S. Odint͡sova. 1989. Identification of the components of photosystem I and cytochromes of the th[y]lakoid membranes of the red alga *Cyanidium caldarium. Biochemistry* (USSR, New York) 54(9, part 2): 1365–1372. (biochem., photosyn., physiol.)
423. Karakashev, G.V., I.M. Stadnichuk, N.P. I͡urina and M.S. Odint͡sova. 1987. Polipeptody tilakoidnykh membran odnokletochnoĭ krasnoĭ vodorosli *Cyanidium caldarium. Biokhimii͡a* (Moskva) 52(9): 1485–1493. (biochem., photosyn., physiol.)
424. Karakashev, G.V., I.M. Stadnichuk, N.P. I͡urina and M.S. Odint͡sova. 1988. Thylakoid membrane polypeptides of the unicellular red alga *Cyanidium caldarium. Biochemistry* (USSR, New York) 52(9, part 1): 1283–1290. (biochem., photosyn., physiol.)
425. Kashino, Y., I. Enami, K. Satoh and S. Katoh. 1990. Immunological cross-reactivity among corresponding proteins of photosystems I and II from widely divergent photosynthetic organisms. *Plant Cell Physiol.* (Tokyo) 31(4): 479–488. (biochem., evol., physiol.)
426. Katz, J.J. and H.L. Crespi. 1966. Deuterated organisms: cultivation and uses. *Science* (Lancaster) 151(3715): 1187–1194. (biochem., photosyn., physiol., radiobiol.)
427. Katz, J.J. and H.L. Crespi. 1969. Nuclear magnetic resonance studies of plant biosynthesis. pp 1–24 in M.K. Seikel and V.C. Runeckles. (Eds). *Recent Advances in Phytochemistry.* Vol. 2. *Proceedings of the* Seventh Annual Symposium *of the Phytochemical Society of North America.* Appleton-Century-Crofts, New York. VIII + 175 pp. (cult., physiol., radiobiol., sl. ment.)
428. Kessler, E. 1974. Hydrogenase, photoreduction and anaerobic growth. pp 456–473 in W.D.P. Stewart. (Ed.). *Algal Physiology and Biochemistry. Botanical Monographs.* Volume 10. University of California Press, Berkeley and Los Angeles. (Blackwell Scientific Publications, Oxford, London, Edinburgh, Melbourne). XI + (1) + 989 pp. (biochem., physiol., sl. ment.)
429. Kies, L. 1980. Morphology and systematic position of some endocyanomes. pp 7–19 in W. Schwemmler and H.E.A. Schenk. (Eds). *Endosymbiosis and Cell Biology. A Synthesis of Recent Research. Proceedings of the International Colloquium on Endosymbiosis and Cell Research, Tübingen, Germany, April 1980.* Walter de Gruyter, Berlin, New York. XXIV + 1060 pp. (biochem., evol., sl. ment., tax.-syst.)
430. Kikuchi, T., M.R. Ishida, T. Matsubara, G. Tsushimoto and N. Mizuma. 1973. Some features of RNA and protein syntheses in thermophilic alga *Cyanidium caldarium*: the effects of antibiotics on *in vivo* incorporation of ^{14}C-uracil and ^{14}C-leucine. *Annu Rep. Res. React. Inst. Kyoto Univ.* (Kyoto) 6: 29–37. (biochem., physiol., radiobiol.)
431. Kimura, Y. 1979. Système et phylogénie du règne végétal. *Rev. Algol.* (N. Sér.) (Paris) 14(4): 285–296 (ecol., evol.)
432. Kirchner, O. 1878. *Kryptogamen-Flora von Schlesien. Im Namen der Schlesischen Gesellschaft für vaterländische*

Cultur herausgegeben Prof. Dr. Ferdinand Cohn. Zweiter Band. Erste Hälfte. *Algen, bearbeitet von Dr. Oskar Kirchner.* J.U. Kern's Verlag (Max Müller), Breslau. IV + Sachliche Abkürzungen (1) + Abkürzungen der Autoren-Namen (1) + 284 SS. (tax.-syst.)

433. Kirk, J.T.O. and R.A.E. Tilney-Bassett. 1967. *The Plastids. Their Chemistry, Structure, Growth and Inheritance.* W.H. Freeman and Company, London and San Francisco. Frontispiece + XVI + 608 pp. (biochem., photosyn., physiol.)

434. Kirk, J.T.O. and R.A.E. Tilney-Bassett. 1978. *The Plastids. Their Chemistry, Structure, Growth and Inheritance.* Revised Second Edition. Elsevier/North-Holland Biomedical Press, Amsterdam, New York, Oxford. XX + 960 pp. (biochem., photosyn., physiol.)

435. Klein, G. 1977. *Versuche zur Proteinbindung und zur Stereochemie der Chromophore in Biliproteiden und Modellverbindungen vorgelegt von Gerhard Klein aus Kaiserslautern.* Inaugural-Dissertation zur Erlangung des Doktorgrades des Fachbereiches Biologie der Ludwit-Maximilians-Universität München, München. (10) + 136 SS + Literaturverzeichnis (8) + Lebenslauf (1). (biochem., photosyn., physiol.)

436. Klein, G. and W. Rüdiger. 1979. Thioether formation of phycocyanobilin: a model reaction of phycocyanin biosynthesis. *Zeitschr. Naturforsch.* Teil C, *Biosci.* (Wiesbaden, Tübingen) 34(3/4): 192–195. (biochem., photosyn., physiol.)

437. Klein, R.M. 1970. Relationships between blue-green and red algae. pp 623–633 in J.F. Fredrick and R.M. Klein. (Eds). *Phylogenesis and morphogenesis in the algae. Ann. N. Y. Acad. Sci.* (New York) 175(2): 413–781. (evol., tax.-syst.)

438. Klein, R.M. and A. Conquist. 1967. A consideration of the evolutionary and taxonomic significance of some biochemical, micromorphological, and physiological characters in the Thallophytes. *Q. Rev. Biol.* (Stony Brook, Baltimore) 42(2): 105–296. (biochem., evol., photosyn., physiol.)

439. Kleinschmidt, M.G. 1969. *Effect of Temperature on Lipid Composition in the Thermophilic Alga,* Cyanidium caldarium. Ph. D. Dissertation, Division of Biochemistry, University of Wyoming, Laramie. (1) + V + 32 pp. (biochem., ecol., physiol.)

440. Kleinschmidt, M.G. and V.A. McMahon. 1968. Effect of temperature on fatty acid composition in *Cyanidium caldarium*. p. S-34 in *Program and Abstracts of Papers to be Presented at the Annual Meetings. University of Massachusetts, Amherst, Massachusetts, August 20–23, 1968. Plant Physiol.* (Lancaster) 43(Suppl.): S-1–S-55. (biochem., physiol.)

441. Kleinschmidt, M.G. and V.A. McMahon. 1970. Effect of growth temperature on the lipid composition of *Cyanidium caldarium* I. *Class separation of lipids. Plant Physiol.* (Lancaster) 46:286–289. (biochem., physiol.)

442. Kleinschmidt, M.G. and V.A. McMahon. 1970. Effect of growth temperature on the lipid composition of *Cyanidium caldarium* II. Clycolipid and phospholipid components. *Plant . Physiol.* (Lancaster) 46: 290–293. (biochem., physiol.)

443. Koller, K.-P. 1979. *Untersuchungen zur Biliproteidkomposition und -Lokalisation in isolierten Phycobilisomen von* Rhodella violacea. Inaugural-Dissertation zur Erlangung des Grades eines Doktors der Naturwissenschaften (Dr. rer. nat.) vorgelegt dem Fachberich Biologie der Philipps-Universität Marburg/Lahn von Dr. Klaus-Peter Koller aus Wanne-Eickel Marburg/Lahn 1979. Druck: Mauersberger, Marburg (2) + 112 SS + Danksagung (1). (biochem., photosyn., physiol., sl. ment.)

444. Köst, H.-P. and E. Benedikt. 1982. Biliverdin IXα, intermediate and end product of tetrapyrrole biosynthesis. *Zeitschr. Naturforsch.* Teil C, *Biosci.* (Wiesbaden, Tübingen) 37(11/12): 1057–1063. (biochem., photosyn., physiol.)

445. Köst-Reyes, E. and H.-P. Köst. 1979. The protein-chromophore bond in b-phycoerythrin from *Porphyridium cruentum:* radiosulfar labeling experiments. *Eur. J. Biochem.* (Berlin) 102(1): 83–91. (biochem., photosyn., physiol., radiobiol.)

446. Kostrzewa, M., K. Valentin, U. Maid, R. Radetsky and K. Zetsche. 1990. Structure of the rubisco operon from the multicellular red alga *Antithamnion* spec. *Curr. Genet.* (Berlin, Heidelberg) 18(5): 465–469. (biochem., evol.)

447. Kremer, B.P. 1978. Blaualgen in einer alge. *Umchau Wiss. Tech.* (Frankfurt am Main) 76(6): 181–182. (biochem., cytol., ecol., evol., photosyn., physiol., radiobiol., tax.-syst.)

448. Kremer, B.P. 1978. Eine ungewöhnliche einzellige Alge: *Cyanidium caldarium. Mikrokosmos* (Stuttgart) 3: 65–67. (cytol., ecol., evol., tax.-syst., ultrastr.)

449. Kremer, B.P. 1980. Carbon metabolism of endosymbiotic algae. pp 89–96 in W. Schwemmler and H.E.A. Schenk. (Eds). *Endocytobiologie. Endosymbiosis and Cell Biology. A Synthetis of Recent Research. Proceedings of the International Colloquium on Endosymbiosis and Cell Research, Tübingen, Germany, April 1980.* Walter de Gruyter, Berlin, New York. (5) + pp VI–XXIV + 1060 pp. (biochem., physiol., sl. ment., tax.-syst.)

450. Kremer, B.P. 1982. *Cyanidium caldarium*: a discussion of biochemical features and taxonomic problems. *Br. Phycol. J.* (London) 17(1): 51–61. (biochem., evol., physiol., radiobiol.)

451. Kremer, B.P. 1983. *Cyanidium caldarium*: rhodophyte, cyanome or transitional species? pp 963–970 in H.E.A. Schenk and W. Schwemmler. (Eds). *Endocytobiology* II. *Intracellular Space as Oligogenetic Ecosystem.* Proceedings Second International Colloquium *on Endocytobiology. Tübingen, Germany, April 10–15, 1983.* Walter de Gruyter, Berlin, New York. XXV + 1071 pp. (biochem., cytol., evol., photosyn., physiol., radiobiol., tax.-syst.)

452. Kremer, B.P. and G.B. Feige. 1979. Accumulation of photoassimilatory products by phycobiliprotein-containing algae with special reference to *Cyanidium caldarium. Zeitschr. Naturforsch.* Teil C, *Biosci.* (Wiesbaden, Tübingen) 34(12): 1209–1214. (biochem., photosyn., physiol., radiobiol., tax.-syst.)

453. Kremer, B.P., G.B. Feige and Hj. A.W. Schneider. 1978. A new proposal for the systematic position of *Cyanidium caldarium. Naturwissenschaften* (Berlin) 65: 157. (biochem., cytol., ecol., radiobiol., tax.-syst.)

454. Krishnamurthy, V. 1980. Patterns of life-history in the Bangiophyceae. pp 185–211 in T.V. Deskachary and V.N. Rao. (Eds). *Taxonomy of Algae. Papers presented at the International Symposium on Taxonomy of Algae held at the Centre of Advanced Study in Botany, University of Madras, December 9–16, 1974.* University of Madras, Madras. (8) + 807 pp + Author Index (4). (sl. ment., tax.-syst.)
455. Kristjánsson, J.K. and G. Á. Alfredsson. 1986. Lífríki hveranna. *Náttúrufraedingurinn* (Reykjavik) 56(2): 49–68. (ecol., florist.)
456. Kumano, S. and H. Hirose. 1977. Kōsō-Kō (Class Rhodophyceae). pp 157-175 in H. Hirose and T. Yamagishi. (Eds). *Nihon Tansuisō Zukan. (Illustrations of the Japanese Freshwater Algae).* Uchida-Rokakuho-Shinsha, Tokyo. 933 pp. (cytol., ecol., morphol., tax.-syst.)
457. Kupelian, R.H. 1964. *The Development and Characteristics of Two Water-Soluble Cytochromes from the Alga,* Cyanidium caldarium. Ph. D. Dissertation, University of Chicago, Chicago. 35 pp. (biochem., physiol.)
458. Kura-Hotta, M. and I. Enami. 1981. Light-induced H^+ efflux from intact cells of *Cyanidium caldarium. Plant Cell Physiol.* (Tokyo) 22(7): 1175–1183. (biochem., photosyn., physiol.)
459. Kura-Hotta, M. and I. Enami. 1984. Respiration-dependent H^+ efflux from intact cells of *Cyanidium caldarium.* Plant Cell Physiol. (Tokyo) 25(7): 1115–1122. (biochem., physiol.)
460. Kuroiwa, T. and T. Mita. 1987. Mechanisms of chloroplast division. p. 199 (No. 4-06a-2) in W. Greuter, B. Zimmer and H.-D. Behnke. (Eds). *Abstracts of the General Lectures, Symposium Papers and Posters presented at the XIV International Botanical Congress, Berlin July 24 to August 1, 1987. Collated and Edited by W. Greuter, B. Zimmer and H.-D. Behnke on behalf of the Organizing Committee of the Congress. Produced with the Assistance of M.I. Hakki, H. Sipman and Staff Members of the Botanical Museum Berlin-Dahlem.* Botanischer Garten und Botanisches Museum, Berlin-Dahlem. (4) + pp 5–480. (cytol., morphol.)
461. Kuroiwa, T., H. Nagashima and I. Fukuda. 1989. Chloroplast division without DNA synthesis during the live cycle of the unicellular alga *Cyanidium caldarium* M-8 as revealed by quantitative fluorescence microscopy. *Protoplasma (Internat. Zeitschr. Physikal. Chem. Protoplast.* Leipzig) 149(2–3): 120–129. (biochem., cytol., morphol., physiol.)
462. Kützing, F.T. 1833. Beitrag zur Kenntniss über die Entstehung und Metamorphose der niedern vegetablischen Organismen, nebst einer systematischen Zusammenstellung der hieher gehörigen niedern Algenformen. *Linnaea* (Berlin) 8: 335–382 + Taf. VI–VIII. (tax.-syst.)
463. Kützing, F.T. 1843. *Phycologia Generalis oder Anatomie, Physiologie und Systemkunde der Tange. Bearbeitet von Friedrich Traugott Kützing. Mit 80 farbig gedruckten Tafeln, gezeichnet und gravirt vom Verfasser.* F.A. Brockhaus, Leipzig. XXXII + 458 SS + Verbesserungen (1) + (1) + 80 Taf. (tax.-syst.)
464. Kützing, F.T. 1845. *Phycologia germanica, d. i. Deutschlands Algaen in bundigen Beschreibungen. Nebst einer Anleitung zum Untersuchen und Bestimmen dieser Gewächse für Anfänger. Bearbeitet von Friedrich Kützing.* Zu finden bei Wilh. Köhne, Nordhausen. VIII + 340 SS + (2) + SS IX–X (Verzeichniss). (tax.-syst.)
465. Kützing, F.T. 1845–1849. *Tabulae Phycologicae oder Abbildungen der Tange. Herausgegeben von Friedrich Traugott Kützing.* I. Band. *Mit 100 Taflen.* Gedruckt auf Kosten des Verfassers (In Commission bei Wilhelm Köhne), Nordhausen. VI + 54 SS + Tab. 1–100. (tax.-syst.)
466. Kützing, F.T. 1845–1849(Reprint, 1977). *Tabulae Phycologicae von F.T. Kützing.* I–V. Bibliotheca Phycologica. Herausgegeben von J. Cramer. Band 32a. J. Cramer, In der A.R. Gantner Verlag Kommanditgesellschaft, Vaduz. (4) + Band I: VI + 54 SS + Tab. 1–100 + Band II: (4) + 37 SS + Tab. 1–100 + Band III: (4) + 28 SS + Tab. 1–100 + Band IV: (4) + XVI + 32 SS + Tab. 1–100 + Band V: (4) + 30 SS + Tab. 1–100. (tax.-syst.)
467. Kützing, F.T. 1849. *Species Algarum. Auctore Friderico Traug. Kutzing.* F.A. Brockhaus, Lipsiae. (2) + VI + 922 pp. (tax.-syst.)
468. Kützing, F.T. 1849(Reprint, 1969). *Species Algarum. Auctore Friderico Traug. Kützing.* Reprint A. Asher & Co., Amsterdam. (4) + VI + 922 pp. (tax.-syst.)
469. Lafraie, M.A. and A. Betz. 1985. Anaerobic fermentation in *Cyanidium caldarium. Planta* (Berlin) 163: 38–41. (biochem., physiol.)
470. Langworthy, T.A., T.D. Brock, W. Castenholz, A.F. Eser, E.J. Johnson, T. Oshima, M. Tsubio, J. Zeikus and H. Zuber. 1979. Life at high temperatures, group report. pp 489–502 in M. Shilo. (Ed.). *Strategies of Microbial Life in Extreme Environments. Report of the Dahlem Workshop on Strategy of Life in Extreme Environments, Berlin, 1978, November 20–24.* Verlag Chemie, Weinheim, New York. (10) + pp 11–513 + Author Index (1). (ecol., sl. ment.)
471. Leclerc, J.-C. 1983. Intérêt de la mesure de la photosynthèse avec une électrode à oxygène en milieu aérien. Application à quelques algues subaériennes et lichens. *Oceanis* (Paris) 9(3): 195–203. (ecol., physiol.)
472. Leclerc, J.-C. 1985. Premières données sur l'activité photosynthétique de quelques algues subaériennes vivant en milieux très peu éclairés. *Can. J. Bot.* (Ottawa) 63(1): 1893–1899. (photosyn., physiol., tax.-syst.)
473. Leclerc, J.-C., A. Couté and P. Dupuy. 1983. Étude des conditions climatiques annuelles de deux stations d'algues subaérienes sciaphiles. *Phycologia* (Berkeley, Copenhagen, Vancouver) 22(4): 445–448. (ecol., florist., physiol.)
474. Leclerc, J.-C., A. Couté and P. Dupuy. 1983. Le climat annuel de deux grottes et d'une église du Potou, ou vivent des colonies pures d'algues sciaphiles. *Cryptogam. Algol.* (Paris) 4(1–2): 1–19. (ecol., florist., physiol.)
475. Lee, R.E. 1974. Chloroplast structure and starch grain production as phylogenetic indicators in the lower Rhodophyceae. *Br. Phycol. J.* (London) 9(3): 291–295. (evol., ultrastr.)
476. Lefort-Tran, M., G. Cohen-Bazire and M. Pouphile. 1973. Les membranes photosynthétiques des algues à biliprotéines observées après Cryodécapage. *J. Ultrastruct. Res.* (New York) 44(3/4): 199–209. (cytol., evol., tax.-syst., ultrastr.)
477. Leiv-Hetzroni, O. 1965. *Atzot Memaarot Haaretz. (Algae*

from Israel Caves). Aavodat Gmar, Musmach Madai Hateva (Final Thesis, Master of Natural Sciences), Haunivesita Haivrit Berushalaim, Ierushalaim (Hebrew University of Jerusalem, Jerusalem). 36 pp + References. (cytol., ecol., florist., morphol.)

478. Lemaux, P.G. and A. Grossman. 1984. Isolation and characterization of a gene for a major light-harvesting polypeptide from *Cyanophora paradoxa*. *Proc. Natl. Acad. Sci. U.S.A.* (Washington) 81(13): 4100–4104. (biochem., evol., photosyn., physiol., tax.-syst.)

479. Levitina, T.P. 1979. Vydelenie i nekotorye svoĭstva khomatine *Euglena gracilis*. [Isolation of some properties of chromatin of *Euglena gracilis*]. *Vestn. Leningr. Univ.*, Ser. 3, *Biol.* (Leningrad) 1979(3): 99–104. (biochem., evol., physiol.)

480. Lewin, R.A. 1961. *Cyanidium caldarium* – a Cryptococcalean? *News Bull., Phycol. Soc. Am.* (Urbana) 14: 6–7. (tax.-syst.)

481. Lewin, R.A. 1968. Biochemistry and physiology of algae: taxnomic and phylogenetic considerations. pp 15–26 in D.F. Jackson. (Ed.). *Algae, Man, and the Environment. Proceedings of an International Symposium Sponsored by Syracuse University and the New York State Science and Technology Foundation, June 18–30, 1967, Syracuse, New York*. Syracuse University Press, Syracuse. VIII + (2) + 554 pp. (biochem., sl. ment.)

482. Lewin, R.A. 1974. *Biochemical taxonomy*. pp 1–39 in W.D.P. Stewart. (Ed.). *Algal Physiology and Biochemistry. Botanical Monographs*. Volume 10. University of California Press, Berkeley and Los Angeles. (Blackwell Scientific Publications, Oxford, London, Edinburgh, Melbourne). XI + (1) + 989 pp. (sl. ment., tax.-syst.)

483. Lewin, R.A. and S.P. Gibbs. 1982. Algae of uncertain taxonomic position: introduction and bibliography. pp 659–662 in J.R. Rosowski and B.C. Parker. (Eds). *Selected Papers in Phycology* II. Phycological Society of America, Inc., Book Division, Lawrence, Kansas. XXI + (1) + 866 pp. (ecol., evol., tax.-syst.)

484. Lichtlé, C. and J.C. Thomas. 1976. Étude ultrastructurale des thylacoïdes des algues à phycobiliproteines, comparaison des résultats obtenus par fixation classique et cryodécapage. *Phycologia* (Berkeley, Copenhagen, Vancouver) 15(3/4): 393–404. (sl. ment., ultrastr.)

485. Lin, S., G.D. Offner and R.F. Troxler. 1990. Studies on *Cyanidium caldarium* phycobiliprotein pigment mutants. *Plant Physiol.* (Lancaster) 93(2): 772–777. (biochem., photosyn.)

486. Lin, S., R.F. Troxler and G.D. Offner. 1988. Mutations affecting phycobiliprotein synthesis in the unicellular rhodophyte, *Cyanidium caldarium*, p. 352a in *Abstracts of Papers Presented at the Joint Meeting of the American Society for Cell Biology and the American Society for Biochemistry and Molecular Biology, 29 January-2 Februari, 1989, San Francisco, California. J. Cell Biol.* (New York) 107(6, part 3): (7) + pp 2a–1007a. (biochem., photosyn., physiol.)

487. Lingens, F. and H.-D. Heilmann. 1966. Zur Biosynthese des Lysins in *Xanthomonas pruni, Streptomyces antibioticus* und *Cyanidium caldarium*. *Hoppe-Seyler's Zeitschr. Physiol. Chem.* (Berlin) 345(4): 249–256. (biochem., evol., radiobiol., tax.-syst.)

488. Lingens, F. and P. Vollprecht. 1964. Zur Biosynthese der Nicotinsäure in Streptomyceten, Algen, Phycomyceten und Hefe. *Hoppe-Seyler's Zeitschr. Physiol. Chem.* (Berlin) 339(1–6): 64–74. (biochem., radiobiol.)

489. Loppes, R., C. Denis and M. Bairiot-Delcoux. 1985. Isolation of *Cyanidium caldarium* and *Porphyridium purpureum* DNA fragments capable of autonomous replication in yeast. *J. Gen. Microbiol.* (Cambridge) 131(7): 1745–1751. (biochem., physiol.)

490. Lüönd, E.B., H.-P. Köst and W. Rüdiger. 1990. Isolation and synthesis of new bile pigments possibly related to the biosynthesis of the phycocyanin chromophore. *Zeitschr. Naturforsch.* Teil C, *Biosci.* (Wiesbaden, Tübingen) 45(11/12): 1099–1110. (biochem., physiol.)

491. MacColl, R. and D.S. Berns. 1979. Evolution of the biliproteins. *TIBS (Trends in Biochemical Sciences,* Amsterdam) 1979: 44–47. (biochem., evol., photosyn., tax.-syst.)

492. Macris, R. 1975. The effects of varied carbon sources on the growth rate and protein content of *Cyanidium caldarium*. *J. Biol.* (Bronx) 26(1): 22–25. (biochem., cultivat., physiol.)

493. Maid, U., K. Valintin and K. Zetsche. 1990. The psbA-gene from a red alga resembles those from Cyanobacteria and Cyanelles. *Curr. Genet.* (Berlin, Heidelberg) 17(3): 255–259. (biochem., physiol.)

494. Maini, P.S. 1957. Nuove osservazioni sull'ecologia de *Entophysalis rivularis* nella solfatara di Pozzuoli. *Ann. Ist. Super. Sci. Lett. 'S. Chiara'* (Napoli) 24[1956-1957 (7)]: 363–369 + 14 figs on 9 pls. (tax.-syst.)

495. Margulis, L. 1968. Evolutionary criteria in thallophytes: a radical alternative. *Science* (Lancaster) 161(3845): 1020–1022. (evol., sl. ment., tax.-syst.)

496. Margulis, L. 1970. *Origin of Eukaryotic Cells. Evidence and Research Implications for a Theory of the Origin and Evolution of Microbial, Plant, and Animal Cells on the Precambrian Earth*. Yale University Press, New Haven and London. XXII + 349 pp. (sl. ment., tax.-syst.)

497. Margulis, L. 1981. *Symbiosis in Cell Evolution. Life and Its Environment on the Early Earth*. W.H. Freeman and Company, San Francisco. XXII + (2) + 419 pp. (evol., tax.-syst.)

498. Marrè, E. 1962. Temperature. pp 541–550 in R.A. Lewin. (Ed.). *Physiology and Biochemistry of Algae*. Academic Press, New York and London. Frontispiece + XXVII + (1) + 929 pp. (physiol., sl. ment.)

499. Marsh, L.E. and D.S. Herson. 1977. Ribosomal RNA of *Cyanidium caldarium*. p. 194 in *Abstracts of the Annual Meeting of the American Society for Microbiology*. 77th Annual Meeting, New Orleans, La. 8–13 May 1977. American Society for Microbiology, Washington, D.C. X + 452 pp. (biochem., evol., physiol., tax.-syst.)

500. Matsumoto, G.I. and H. Nagashima. 1984. Occurrence of 3-hydroxy acids in microalgae and cyanobacteria and their geochemical significance. *Geochim. Cosmochim. Acta* (Oxford) 48(8): 1683–1687. (biochem.)

501. Matsumoto, G.I., M. Shioya and H. Nagashima. 1984. Occurrence of 2-hydroxy acids in microalgae. *Phytochemistry* (Oxford) 23(7): 1421–1423. (biochem., tax.-syst.)

502. Mayer, S.M., S.I. Beale and J.D. Weinstein. 1987. Enzymatic conversion of glutamate to δ-aminolevulinic acid in soluable extracts of *Euglena gracilis. J. Biol. Chem.* (Baltimore) 262(26): 12541–12549. (biochem., physiol., sl. ment.)

503. McMahon, V. 1979. A proposed model for sudden fludity changes. pp 215–218 in L.A. Appleqvist and C. Liljenbert. (Eds). *Advances in the Biochemistry [and] Physiology of Plant Lipids. Proceedings of the Symposium on Recent Advances in the Biology of Plant Lipids held in Goteborg, Sweden, August, 28–30, 1978.* Developments in Plant Biology, Volume 3. Elsevier/North Holland Biomedical Press. Amsterdam, New York, Oxford. (4) + pp V–X + 461 pp. (biochem., physiol.)

504. McMahon, V., Y.O. Chang and B. Adams. 1970. A new source of single cell protein. p. 42 in *Program and Abstracts of the Papers for the Annual Meeting of the American Society of Plant Physiologists in Conjunction with A.I.B.S. at Indiana University, Bloomington, August 23–27, 1970. Plant Physiol.* (Lancaster) 46(Suppl.): (2) + pp 1–50. biochem.

505. Meeks, J.C. 1974. Chlorophylls. pp 161–175 + 1 folded table in W.D.P. Stewart. (Ed.). *Algal Physiology and Biochemistry. Botanical Monographs.* Volume 10. University of California Press, Berkeley and Los Angeles. (Blackwell Scientific Publications, Oxford, London, Edinburgh, Melbourne). XI + (1) + 989 pp. (biochem., photosyn., physiol.)

506. Meer, J.P. van der. 1990. Genetics. pp 103–121 in K.M. Cole and R.G. Sheath. (Eds). *Biology of Red Algae.* Cambridge University Press, Cambridge, New York, Port Chester, Melbourne, Sydney. (4) + pp V–IX + (1) + 517 pp. (biochem., sl. ment.)

-- Melchoir, H. and E. Werdermann. 1954. A. Engler's *Syllabus der Pflanzenfamilien.* Please see: Schmidt, E.W. Cyanophyta. SS 50–55.

507. Melkonian, M. 1984. Taxonomy, systematics and evolution of the algae. *Fortschr. Bot. (Prof. Bot.*, Köln, Leipzig) 46: 248–273. (evol., tax.-syst.)

508. Meller, E. and E. Harel. 1978. The pathway of 50aminolevulinic acid synthesis in *Chlorella vulgaris* and in *Fremyella diplosiphon.* pp 51–57 in G. Akoyunoglou and J.H. Argyroudi-Akoyunoglou. (Eds). *Chloroplast Development. Proceedings of the International Symposium on Chloroplast Development Held on the Island of Spetsai, Greece, July 9–15, 1978. Developments in Plant Biology.* Volume 2. Elsevier/North-Holland Biomedical Press, Amsterdam, New York, Oxford. XVI + 888 pp. (biochem., physiol., sl. ment.)

509. Meneghini, G. 1839. Nuova specie di Alga descritta dal sig. Dott. Giuseppe Meneghini di Padova. *Nuovo Giorn. Lett.* (Pisa) 39: 67–68. (There also exists a single page separate of this article bearing no date, place, publisher, nor pagination). (ecol., florist., tax.-syst.)

510. Meneghini, G. 1842. *Monographia Nostochinearum Italicarum Addito Specimine de Rivulariis. Auctore Professore Iosepho Meneghinio. (Iconibus ab Auctore Depictis).* Ex Officina Regia, Augustae Taurinorum. (2) + 143 pp + Tab. I-XVII. (tax.-syst.)

511. Meneghini, G. 1843. Monographia Nostochinearum Italicarum addito specimine de Rivulariis. Auctore Professore Iosepho Meneghinio. (Iconibus ab auctore depictis). *Mem. R. Accad. Sci. Torino* (Ser. 2) (Torino) 5: 1–143 + Tab. I–XVII. (tax.-syst.)

512. Mercer, F.V., L. Bogorad and R. Mullens. 1962. Studies with *Cyanidium caldarium* I. The fine structure and systematic position of the organism. *J. Cell Biol.* (New York) 13(3): 393-403. [For section 'II' of the above title, see Bogorad, Mercer and Mullens (1963)]. (tax.-syst., ultrastr.)

513. Merola, A., R. Castaldo, P. De Luca, R. Gambardella, A. Musacchio and R. Taddei. 1981. Revision of *Cyanidium caldarium.* Three species of acidophilic algae. *Giorn. Bot. Ital.* (Firenze) 115(4–5): 189–195. (cytol., tax.-syst.)

514. Minami, Y., F. Yamada, T. Hase, H. Matsubara, A. Murakami, Y. Fujita, T. Takao and S. Shimonoishi. 1985. Amino acid sequences of allphycocyanin α- and β-subunits isolated from *Anabaena cylindrica.* Presence of an unknown derivative of aspartic acid in the α-subunit. *FEBS Lett. (Fed. Eur. Biochem. Soc.*, Amsterdam) 191(2): 216–220. (biochem., photosyn., physiol.)

515. Mita, T., T. Kanbe, K. Tanaka and T. Kuroiwa. 1986. A ring structure around the dividing plane of *Cyanidium caldarium* chloroplast. *Protoplasma (Internat. Zeitschr. Physikal. Chem. Protoplast.*, Leipzig) 130(2–3): 211–213. (cytol., morphol., ultrastr.)

516. Mita, T. and T. Kuroiwa. 1989. Division of plastids by a plastid-dividing ring in *Cyanidium caldarium* pp 133-152 in M. Tazawa. (Ed.). *Cell Dynamics.* Vol. I. *Cytoplasmic Streaming, Cell Movement, Contraction and Migration, Cell and Organelle Division, Phototaxis of Cell and Cell Organelle. Protoplasma (Internat. Zeitschr. Physikal. Chem. Protoplast.*, Leipzig) (Suppl. 1): (7) + p. VIII + (2) + pp 3–192 + Index of keywords (1). (biochem., cytol., morphol., physiol., ultrastr.)

517. Mitchell, R. 1974. The evolution of thermophily in hot springs. *Q. Rev. Biol.* (Stony Brook, Baltimore) 49(3): 229–242 (ecol., evol.)

518. Mizutani, H. and E. Wada. 1982. Effect of high atmospheric CO_2 concentration on $\delta\,^{13}C$ of algae. *Origins of Life* (Dordrecht, Boston) 12(4): 377–390. (physiol., radiobiol.)

519. Mohanty, P.B., B.Z. Braun, Govindjee and J.P. Thornber. 1972. Chlorophyll fluorescence characteristics of systeme I chlorophyll *a*-protein complex and system II particles at room and liquid nitrogen temperatures. *Plant Cell Physiol.* (Tokyo) 13(1): 81–91. (biochem., photosyn., physiol.)

520. Mohanty, P. and Govindjee. 1973. Effect of phenazine methosulphate and uncouplers on light induced chlorophyll *a* fluorescence yield changes in intact algal cells. *Photosynthetica* (Praha) 7(2): 146–160. (biochem., photosyn., physiol.)

521. Molisch, H. 1926. *Pflanzenbiologie in Japan auf Grund eigener Beobachtungen.* Verlag von Gustav Fischer, Jena. X + 270 SS mit 84 Abbildungen im Text. (ecol., florist.)

522. Möller, M. and H. Senger. 1972. Photosyntheseleistung synchroner Kulturen von *Cyanidium caldarium. Ber. Dtsch. Bot. Ges.* (Berlin) 85(7–9): 391–400. (biochem., cultivat., photosyn., physiol.)

523. Moretti, A. and R. Nazzaro. 1980. Fatty acids in ther-

moacidophilic algae. *Delpinoa* (N.S.) (Napoli) 21:1–10. (biochem., physiol., tax.-syst.)

524. Morris, I. 1967. *An Introduction to the Algae.* Hutchison University Library, London. 189 pp. (biochem., evol., photosyn., physiol., ultrastr.)

525. Mörschel, E. 1982. Accessory polypeptides in phycobilisomes of red algae and cyanobacteria. *Planta* (Berlin) 154(3): 251–258. (biochem., photosyn., physiol., sl. ment.)

526. Mörschel, E., K.-P. Koller and W. Wehrmeyer. 1980. Biliprotein assembly in the disc-shaped phycobilisomes of *Rhodella violacea:* electron microscopical and biochemical analyses of C-phycocyanin and allophycocyanin aggregates. *Arch. Microciol.* (Berlin) 125(1–2): 43–52. (photosyn., sl. ment.)

527. Moseley, H.N. 1874. Notes on fresh-water algae obtained at the Boiling Springs at Furnas, St. Michael's, Azores, and their neighbourhood. *J. Linn. Soc., Bot.* (London) 14(77): 321–325. (ecol., florist., tax.-syst.)

528. Murata, N. and D.C. Fork. 1975. Temperature dependence of chlorophyll *a* fluorescence in relation to the physical phase of membrane lipids in algae and higher plants. *Plant Physiol.* (Lancaster) 56(6): 791–796. (biochem., photosyn., physiol.)

529. Murata, N. and D.C. Fork. 1977. Temperature dependence of the light-induced spectral shift of carotenoids in *Cyanidium caldarium* and higher plant leaves. Evidence for an effect of the physical phase of chloroplast membrane lipids on the permeability of the membrane to ions. *Biochim. Biophys. Acta* (New York, Amsterdam) 461(3): 365–378. (biochem., photosyn., physiol.)

530. Musacchio, A., G. Pinto, S. Sabato and R. Taddei. 1978. Aloresistenza in diversi ceppi di *Cyanidium caldarium* forma A et forma B. *Delpinoa* (N.S.) (Napoli) 18–19(1976–1977): 37–44. (ecol., florist., physiol., tax.-syst.)

531. Nagashima, H. and Fukuda. 1979. Onsensō ideyukogome no shushu ni yoru yūki eiyo yōkyūsei no sōi ni tsuite. [Organotrophic properties in two strains of a hot spring alga *Cyanidium caldarium*]. p. 176 in *Nihon Shokubutsugakukai dai 44(von jūyon) kai Taikai Kenkyū Happyō Kiroku [Shōwa 54(gojūyonen) nen Jūgatsu 2(futsuka) Nichi 4(yokka) Nichi (Hiroshima)]. (Proceedings of the* 44th Annual Meeting *of the Botanical Society of Japan, Hiroshima).* Nihon Shokubutsugakukai (The Botanical Society of Japan). 269 pp. (biochem., physiol.)

532. Nagashima, H. and I. Fukuda. 1981. A phylogenetic study on two species of unicellular algae in acid hot springs in Japan. p. 288 in *Abstracts of the* XIII International Botanical Congress, *Sydney, Australia, 21–28 August 1981.* Sponsored by the Australian Academy of Science under the Auspices of the International Union of Biological Sciences, (Sydney?). 351 pp. (biochem., physiol., tax.-syst.)

533. Nagashima, H. and I. Fukuda. 1981. Low molecular weight carbohydrates in *Cyanidium caldarium* and some related algae. *Phytochemistry* (Oxford) 20(3): 439–442. (biochem., evol., physiol., tax.-syst.)

534. Nagashima, H. and I. Fukuda. 1981. Morphological properties of *Cyanidium caldarium* and related algae in Japan. (Nihonsan ideyukogome oyobi kinensō no keitai ni tsuite). *Nippon Sōrui Gakki (Jpn. J. Phycol.,* Sapporo, Tokyo) 29(4): 237–242. (cytol., florist., morphol., ultrastr.)

535. Nagashima, H. and I. Fukuda. 1983. Floridosides in unicellular hot spring algae. *Phytochemistry* (Oxford) 22(9): 1949–1951. (biochem., physiol., tax.-syst.)

536. Nagashima, H. and I. Fukuda. 1987. Phylogenetic approach to the primitive eukaryote *Cyanidium* and related algae based upon their chloroplast nucleoids. p. 321 (No. 5–155–6) in W. Greuter, B. Zimmer and H.-D. Behnke. (Eds.). *Abstracts of the General Lectures, Symposium Papers and Posters presented at the* XIV International Botanical Congress, *Berlin July 24 to August 1, 1987. Collated and Edited by W. Greuter, B. Zimmer and H.-D. Behnke on behalf of the Organizing Committee of the Congress. Produced with the Assistance of M.I. Hakki, H. Sipman and Staff Members of the Botanical Museum Berlin-Dahlem.* Botanischer Garten und Botanisches Museum, Berlin-Dahlem. (4) + pp 5–480. (biochem., cytol., physiol., tax.-syst.)

537. Nagashima, H., G. Inoue, M. Iida and I. Fukuda. 1986. Onsenso ideyukogome no shishitsu sosei to keitō. p. 55 in M. Chihara and H. Fukushima. (Chmn). *Nihon Sōrui Gakkai dai 10-kai Taikai Puroguramu Gakkai Kaicho (1986). Shōwa 61-Nen 3 Gatso 30 Nichi (Nichi)-3 Gatsu 31 Nichi (Getsu). Tsukuba Daigaku Seibutsu Noringaku-Beitō A Tō. (The Proceedings of the* 10th Annual Meeting of the Japanese Society of Phycology, *March 30–31, 1986). Nippon Sōrui Gakki (Jpn. J. Phycol.,* Sapporo, Tokyo) 34(1): 43–61. (physiol.)

538. Nagashima, H., T. Kuroiwa and I. Fukuda. 1984. Chloroplast nucleoids in a unicellular hot spring alga *Cyanidium caldarium* and related algae. *Experientia* (Basel) 40(6):563–564. (cytol., morphol., tax.-syst.)

539. Nagashima, H., T. Kuroiwa and I. Fukuda. 1984. *Kyushu Chiho no Sansei Onsen ni Okeru Ideyukogame no Bunpu to Seitai. (Distribution and ecology of* Cyanidium caldarium *of the acid hot springs in the Kyushu area).* pp 103–104 in *Kaikoku Nippon Moruigakukai Dai 8 kai Shunki Taika Program 1984. Kaiki Showa 59 nen 3 Gatsu 31 Nichi (do) 4 Gatsu 1 (Nichi). Kaijo Tokyo Gakugei Daigaku Shin 3 Gokan. [Announcing the 8th Spring Meeting Program of the Japanese Phycological Society. Time: 1984 March 31 (Sat.)–April 1 (Sun.). Place: Tokyo University New Building No. 3.]. Nippon Sōrui Gakki (Jpn. J. Phycol.,* Sapporo, Tokyo) 32(1): 87–108. (evol., physiol.).

540. Nagashima, H., T. Kuroiwa and I. Fukuda. 1987. Chloroplast nucleoids during greening of *Cyanidium caldarium* M-8 type. *Plant Cell Physiol.* (Tokyo) 28(2): 315–321. (biochem., cytol., morphol., photosyn., physiol.)

541. Nagashima, H., G.I. Matsumoto and I. Fukuda. 1986. Hydrocarbons and fatty acids in two strains of the hot spring alga *Cyanidium caldarium. Phytochemistry* (Oxford) 25(10): 2339–2341. (biochem., evol., physiol., tax.-syst.)

542. Nagashima, H., K. Sakkai, M. Kamo and A. Tsugita. 1989. Amino acid sequence of ferredoxin isolated from *Cyanidium caldarium* strain RK-1. *Protein Sequence Data Anal.* (Berlin, Heidelberg) 2(6): 457-460. (biochem., physiol.)

543. Nagashima, H., J. Seckbach and I. Fukuda. 1988. Onsenso shianidioschizon no saibō kōzō. (The cell structure of a hotspring alga *Cyanidioschyzon*). p. 100 in I. Umezake and H. Iwasaki. (Chmn). *Nihon Sōrui Gakkai dai 12-Kai Taikai Puroguramu (1988). Shōwa 63 Nen 3 Gatsu 30 Nichi (Sui);3 Gatsu 31 Nichi (Moku). Mie Gaigaku Seibutsu Shigengakubu Suisangaku-Kōsha. (The Proceedings of the XII Annual Meeting of the Japanese Society of Phycology, March 30–31, 1988). Nippon Sōrui Gakki (Jpn. J. Phycol.,* Sapporo, Tokyo) 36(1): 87–109. (cytol.)

544. Nash, A. 1938. *The Cyanophyceae of the Thermal Regions of Yellowstone National Park. U.S.A., and of Rotorua and Whakarewarewa, New Zealand; with Some Ecological Data.* Ph.D. Dissertation, University of Minnesota, Minneapolis. 215 Unnumbered pages with 108 figures and 18 tables. (ecol., florist., physiol.)

545. Negorō, K. 1942. Keiso *Pinnularia acoricola* Hustedt no Nihon ni okeru sanchi to sono ikusei kankyo. (Fundorte und Lebensbedingungen von *Pinnularia acoricola* Hustedt in Japan). *Igaku To Seibutsugaku (Med. Biol.,* Tokyo) 1(3): 111–114. (ecol., florist., physiol., sl. ment., tax.-syst.)

546. Negorō, K. 1942. Nihon no muisansei suieki ni okeru ranso *Cyanidium caldarium* (Tilden) Geitler no seikatsu jotai ni tsuite. [Über die Lbensbedingungen von *Cyanidium caldarium* (Tilden) Geitler in den mineralogen-azidotrophen Gewässern Japans]. *Rikusui-gaku Zasshi (Jpn. J. Limnol.,* Tokyo) 12(2): 41–49. (ecol., florist., physiol., tax.-syst.)

547. Negorō, K. 1943. Gunmaken Kusatsu Onsen no sōrui shokusei. (Über die Algenvegetation der Thermen von Kusatu, Gunma Präfektur, Japan). *Shokubutsugaku Zasshi (Bot. Mag.,* Tokyo) 57(680): 302–312. (ecol., florist., physiol., tax.-syst.)

548. Negorō, K. 1943. Ryūkikogen shokubutsu no kenkyū. (Study of the plants in solfataric field). *Seibutsugaku No Shinpo (Progress in Biology)* 1: 178. (ecol., florist., physiol., tax.-syst.)

549. Negorō, K. 1944. Untersuchungen über die Vegetation der mineralogen-azidotrophen Gewässer Japans. *Sci. Rep. Tokyo Bunrika Daigaku* Sect. B (Tokyo) 6:101): 231–374 + Taf. I–XII mit Abb. 1–70. (ecol., florist., physiol., tax.-syst.)

550. Negorō, K. 1962. Nihon no mukisansei suieki ni sansuru sōrui to sentairui. (Algae and mosses in the mineralogenous acidotrophic inland-waters of Japan). *Shokubutsu Bunrui Chiri (Acta Phytotaxon. Geobot.,* Kyoto) 20: 314–321. (ecol., florist., physiol., tax.-syst.)

551. Neilson, A.H. and T. Larsson. 1980. The utilization of organic nitrogen for growth of algae: physiological aspects. *Physiol. Plant* (Copenhagen) 48: 542–553. (biochem., physiol.)

552. Neilson, A.H. and L.A. Lewin. 1974. *Phycological reviews* 1. The uptake and utilization of organic carbon by algae: an essay in comparative biochemistry. *Phycologia* (Berkeley, Copenhagen, Vancouver) 13(3): 227–264. (biochem., physiol., sl. ment.)

553. Nichols, K.E. 1962. *Action Spectra Studies of the Formation and Photosynthetic Participation of Phycocyanin in Wild- and Mutant-Type Cells of* Cyanidium caldarium. Ph.D. Dissertation, Department of Botany, University of Chicago, Chicago. IV + 51 pp. (biochem., photosyn., physiol.)

554. Nichols, K.E. and W.W. Bloom. 1970. Report on a carotenoid-mutant of *Cyanidium caldarium*. p. 83 in *Addresses and Contributed Papers. Hanover, Indiana, October 24, 1969. Botany Abstracts. Proc. Indiana Acad. Sci.* (Brookville) 79: 83–84. (photosyn.)

555. Nichols, K.E. and L. Bogorad. 1960. Studies on phycobilin formation with mutants of *Cyanidium caldarium*. *Nature* (London) 188(4753): 870–872. (biochem., photosyn., physiol.)

556. Nichols, K.E. and L. Bogorad. 1962. Action spectra studies of phycocyanin formation in a mutant of *Cyanidium caldarium. Bot. Gaz.* (Crawfordsville, Chicago) 124(2): 85–93. (biochem., photosyn., physiol.).

557. Nichols, K.E., D. Hall and W.W. Bloom. 1986. Evidence of genetic recombination in *Cyanidium caldarium*. *Proc. Indiana Acad. Sci.* (Brookville) 95(1985): 107–109. (biochem., physiol., tax.-syst.)

558. Nichols, K.E. and F.T. Haxo. 1961. Photosynthesis at paired wavelengths in phycocyanin-deficient mutants of *Cyanidium caldarium*. p. 1433 in *National Academy of Sciences. Abstracts of papers presented at the Autumn meeting, 29 October, La Jolla, California, 30 October-1 November 1961, Los Angeles. Science* (Lancaster) 184(3488): 1425–1437. (biochem., photosyn., physiol.)

559. Nishimura, M. 1964. Energy transfer between pigment molecules in a blue-green alga, *Cyanidium*. pp 18–19 in 5th Symposium, *Japanese Society of Plant Physiologists,* Tokyo (biochem., photosyn., physiol.)

560. Nisizawa, K. and T. Ikawa. 1975. Photosynthesis in algae other than *Chlorella*. pp 201–209 in J. Tokida and H. Hirose. (Eds). *Advance of Phycology in Japan. Dr. W. Junk b.v. Publishers, The Hague. 355 pp. (sl. ment., tax.-syst., ultrastr.)*

561. Ó Carra, P. 1970. Algal biliproteins. *Biochem. J.* (Liverpool) 119:2P–3P. (biochem., photosyn., physiol., sl. ment.)

562. Ó Carra, P. and C. Ó hEocha. 1976. Algal biliproteins and phycobilins. pp 328–376 in T.W. Goodwin. (Ed.). *Chemistry and Biochemistry of Plant Pigments.* Second Edition. Volume I. Academic Press, London, New York, San Francisco. XVI + 870 pp (biochem., photosyn., physiol., sl. ment.)

563. Offner, G.D. 1984. *Primary Structure of Red Algal and Cryptophytan Phycobiliproteins.* Ph.D. Dissertation, Boston University, Boston. 236 pp. (biochem., photosyn., physiol.)

564. Offner, G.D. 1984. Primary structure of red algal and cryptophytan phycobiliproteins. *Diss. Abstr. Int. B, Sci. Eng.* (Ann Arbor) 45(4): 1178–B. (biochem., photosyn., physiol.)

565. Offner, G.D., A.S. Brown-Mason, M.M. Ehrhardt and R. Troxler. 1981. Primary structure of phycocyanin from the unicellular rhodophyte *Cyanidium caldarium* I. Complete amino acid sequence of the α subunit. *J. Biol. Chem.* (Baltimore) 256(23): 12167–12175 [For section 'II' of the above title, see Troxler, Ehrhardt, Brown-Mason and Offner (1981)]. (biochem., evol., photosyn., physiol., tax.-syst.)

566. Offner, G.D. and R.F. Troxler. 1982. Complete amino

acid sequence of the allophycocyanin α subunit from the red alga, *Cyanidium caldarium*. *Fed. Proc., Fed. Am. Soc. Expl. Biol.* (Baltimore) 41(4): 1179. (biochem., photosyn., physiol.)

567. Offner, G.D. and R.F. Troxler. 1983. Primary structure of allophycocyanin from the unicellular rhodophyte, *Cyanidium caldarium*: the complete amino acid sequences of the α and β subunits. *J. Biol. Chem.* (Baltimore) 258(16): 9931–9940. (biochem., photosyn., physiol.)

568. Ó hEocha, C. 1962. Phycobilins. pp 421–435 in R.A. Lewin. (Ed.). *Physiology and Biochemistry of Algae*. Academic Press, New York and London. Frontispiece + XXVII + (1) + 929 pp. (biochem., photosyn., physiol., tax.-syst.)

569. Ó hEocha, C. 1966. Biliproteins. pp 406–421 in T.W. Goodwin. (Ed.). *Biochemistry of Chloroplasts*. Vol. I. *Proceedings of a NATO Advanced Study Institute Held at Aberystwyth, August, 1965*. Academic Press, London. XV + 476 pp. (biochem., photosyn., physiol., sl. ment.)

570. Oh-Hama, T. 1990. Evolutionary consideration of two pathways of 5-aminolevulinic acid biosynthesis in organelles and prokaryotes. pp 589–592 in P. Nardon, V. Gianinazzi-Pearson, A.M. Grenier, L. Margulis and D.C. Smith (Eds). *Endocytobiology* IV. *4th International Colloquium on Endocytobiology and Symbioses. 4ème Congrès International sur l'Endocytobiologie et la Symbiose. INSA, Vilerbanne (France), July 4–8, 1989*. Institut National de la Recherche Agronomique, Paris. (7) + pp 8–620. (biochem., physiol., sl. ment.)

571. Ohhata, Y. 1948. 'Mizu no tanjō.' ('Birth of water'). *Rigaku Monogurafu (Science Monographs)* 17. (ecol., florist.)

572. Okada, Y. 1939. Hakkōtasan Sangayu Fukin no Ōnsensan Ransō. (On the Cyanophyceae in the hot springs of Sukayu, Mt. Hakkoda). *Seitaigaku Kenkyū (Ecol. Rev.*, Sendai) 5(4): 259–266. (ecol., florist.)

573. Ono, T.-A. and N. Murata 1977. Temperature dependence of the delayed fluorescence of chlorophyll *a* in blue-green algae. *Biochim. Biophys. Acta* (New York, Amsterdam) 460(2): 220–229. (biochem., photosyn., physiol., sl. ment.)

574. Ott, F.D. 1966. *A Selected Listing of Xenic Algal Cultures*. Systematics-Ecology Program Contribution No. 72. Systematics-Ecology Program, Marine Biological Laboratory, Woods Hole, Massachusetts. II + 45 pp. (cultivat., sl. ment.)

575. Ott, F.D. 1976. Further observations on the freshwater alga, *Flintiella sanguinaria* Ott in Bourrelly 1970. (Rhodophycophyta, Porphyridiales). *Arch. Protistenkd.* (Jena) 118: 34–52 with pls 1–4. (cytol., sl. ment.)

576. Ott, F.D. and M.R. Sommerfeld. 1982. Freshwater Rhodophyceae: introduction and bibliography. pp 671–681 in J.R. Rosowski and B.C. Parker. (Eds). *Selected Papers in Phycology* II. Phycological Society of America, Inc., Book Division, Lawrence, Kansas. XXI + (1) + 866 pp. (literat., sl. ment., tax.-syst.)

577. Papageorgiou, G. 1975. Chlorophyll fluorescence: an intrinsic probe of photosynthesis. pp 319–371 in Govindjee. (Ed.). *Bioenergetics of Photosynthesis*. Academic Press, New York, San Francisco, London. XIV + (2) + 698 pp. (biochem., photosyn.)

578. Pedicino, N. 1873. Poche osservazioni sulla vegetazione presso le terme. *Rend. Accad. Sci. Fis. Mat. Napoli* (Napoli) 12: 60–61. (ecol., florist.)

579. Percival, E.E. and J.R. Turvey. 1973. Polysaccharides of algae. pp 147–165 in A.I. Laskin and H.A. Lechevalier. (Eds). *CRC Handbook of Microbiology*. Volume II. *Microbial Composition*. CRC Press, Cleveland, Ohio. (10) + 1060 pp. (biochem., tax.-syst.)

580. Pilot, T.J. and J.L. Fox. 1984. Cloning and sequencing of the genes encoding the α and β subunits of C-phycocyanin from the cyanobacterium *Agmenellum quadruplicatum*. *Proc. Natl. Acad. Sci. U.S.A.* (Washington) 81(22): 6983–6987. (biochem., sl. ment.)

581. Pinto, G. and R. Taddei. 1975. Nuove stazioni Italiane di '*Cyanidium caldarium*.' *Delpinoa* (N.S.) (Napoli) 14/15(1972–1973): 125–139. (ecol., florist.)

582. Pinto, G. and R. Taddei. 1978. Le alghe delle acque e dei suoli acidi italiani. *Delpinoa* (N.S.) (Napoli) 18/19(1976–1977): 77–106. (ecol., florist.)

583. Pinto, G. and R. Taddei. 1988. Evaluation of toxic effects of heavy metals on unicellular algae. V. Analysis of the inhibition manifesting itself with an increased lag phase. *Boll. Soc. Nat. Napoli* (Napoli) 95(1986): 303–317. (cultivat., physiol.)

584. Pollio, A., M. Della Greca, P. Monaco, G. Pinto and L. Previtera. 1988. Lipid composition of the acidophilic alga *Dunaliella acidophila* (Volvocales, Chlorophyta) I. Nonpolar lipids. *Biochem. Biophys. Acta* (New York, Amsterdam) 963(1): 53–60. (biochem., sl. ment.)

585. Premuzic, E.T., Mou Lin, Hong Li Zhu and A.M. Gremme. 1991. Selectivity in metal uptake by stationary phase microbial populations. *Arch. Environ. Contam. Toxicol.* (New York) 20(2): 234–240. (ecol., physiol.)

586. Profili, P.L. and P.S. Maini. 1950. Ricerche intorno ad una Cianoficea della Solfatara di Pozzuoli. *Ann. Ist. Super. Sci. Lett. 'S. Chiara'* (Napoli) 17[1949–1950 (2)]: 317–327. (ecol., florist., tax.-syst.)

587. Profili, P.L. and P.S. Maini. 1951. Acclimatazione della '*Entophysalis rivularis*' alla zona sulfureo-thermale della Solfatara di Pozzuoli. *Ann. Ist. Super. Sci. Lett. 'S. Chiara'* (Napoli) 18[1950-1951 (3)]: 319–344 + 8 figs on 4 pls. (ecol., physiol.)

588. Provasoli, L. and A.F. Carlucci. 1974. Vitamins and growth regulators. pp 741–787 in W.D.P. Stewart. (Ed.). *Algal Physiology and Biochemistry. Botanical Monographs.* Volume 10. University of California Press, Berkeley and Los Angeles. (Blackwell Scientific Publications, Oxford, London, Edinburgh, Melbourne). XI + (1) + 989 pp. (biochem., sl. ment.)

589. Pueschel, C.M. 1990. Cell structure. pp 7–41 in K.M. Cole and R.G. Sheath. (Eds). *Biology of Red Algae*. Cambridge University Press, Cambridge, New York, Port Chester, Melbourne, Sydney. (4) + pp V–IX + (1) + 517 pp. (cytol., evol., sl. ment., tax.-syst., ultrastr.)

590. Rabenhorst, L. 1863. *Kryptogamen-Flora von Sachsen, der Ober-Lausitz, Thüringen und Nordböhmen, mit Berücksichtigung der benachbarten Länder.* Erste Abtheilung. *Algen im weitesten Sinne, Leber- und Laubmoose Bearbeitet von Dr. L. Rabenhorst. Mit über 200 Illustrationen, sämmtliche Algengattungen bildlich darstellend.*

Verlag von Eduard Kummer, Leipzig. XX + 653 SS. (tax.-syst.)

591. Rabenhorst, L. 1876. *Die Algen Europa's mit Berücksichtigung des ganzen Erdballs. Gesammelt und bearbeitet von Professor Dr. A. Braun. Herausgegeben von Dr. L. Rabenhorst.* Dec. CCXLVI–CCXLVIII. Druck von C. Heinrich, Dresden. (2) + Specimen Nos. 2451–2480. (tax.-syst.)
592. Ragan, M.A. 1981. Chemical Constituents of Sea Weeds. pp 589–626 in C.S. Lobban and M. J. Wynne. (Eds). *The Biology of Sea Weeds.* University of California Press, Berkeley, Los Angeles. (4) + pp V–XI + (3) + pp 3–786. (biochem., sl. ment.)
593. Ragan, M.A. and D.J. Chapman. 1978. *A Biochemical Phylogeny of the Protists.* Academic Press, New York, San Francisco, London. X + (2) + 317 pp. (biochem., evol., physiol.)
594. Raven, J.A. and J. Beardall. 1981. Carbon dioxide as the exogenous inorganic carbon source for *Batrachospermum* and *Lemanea. Br. Phycol. J.* (London) 16(2): 165–175. (ecol., physiol., sl. ment.)
595. Raven, J.A., A.M. Johnston and J.J. MacFarland. 1990. Carbon metabolism. pp 171–202 in K.M. Cole and R.G. Sheath. (Eds). *Biology of Red Algae.* Cambridge University Press, Cambridge, New York, Port Chester, Melbourne, Sydney. (4) + pp V–IX + (1) + 517 pp. (biochem., physiol., sl. ment.)
596. Raven, P.H. 1970. A multiple origin for plastids and mitochondria. Many independent symbiotic events may have been involved in the origin of these cellular organelles. *Science* (Lancaster) 169(3946): 641–646. (cytol., evol., sl. ment.)
597. Raven, P.H. and H. Curtis. 1970. *Biology of Plants.* Worth Publishers, Inc., New York. XI + Frontispiece + 706 pp. (cytol., evol., ultrastr.)
598. Raven, P.H., R.F. Evert and H. Curtis. 1976. *Biology of Plants.* Second Edition. Worth Publishers, Inc., New York. XV + (2) + Frontispiece + 685 pp. (ultrastr.)
599. Raven, P.H., R.F. Evert and H. Curtis. 1981. *Biology of Plants.* Third Edition. Worth Publishers, Inc., New York. XVI + (1) + Frontispiece + 686 pp. (ultrastr.)
600. Reed, R.H. 1983. Taxonomic implications of osmoacclimation in *Cyanidium caldarium* (Tilden) Geitler. *Phycologia* (Berkeley, Copenhagen, Vancouver) 22(4): 351–354. (biochem., ecol., evol., physiol.)
601. Reed, R.H. 1990. Solute accumulation and osmotic adjustment. pp 141–170 in K.M. Cole and R.G. Sheath. (Eds). *Biology of Red Algae.* Cambridge University Press, Cambridge, New York, Port Chester, Melbourne, Sydney. (4) + pp V–IX + (1) + 517 pp. (biochem., physiol., tax.-syst.)
602. Revsbech, N.P. and D.M. Ward. 1983. Oxygen microelectrode that is insensitive to medium composition: Use in an acid microbial mat dominated by *Cyanidium caldarium. Appl. Environ. Microbiol.* (Washington) 45(3): 755–759. (ecol., physiol.)
603. Richter, G. 1961. The lack of diphosphofructose aldolase in two photosynthetic organisms: *Anacystis nidulans* and *Rhodopseudomonas spheroides. Biochim. Biophys. Acta* (New York, Amsterdam) 48(3): 606–608. (biochem., physiol.)
604. Richter, G. 1962. Das Fehlen von Diphosphofructose-Aldolase in Photosynthese-Organismen. SS 133–138 in A. Pirson. (Chmn). *Beiträge zur Physiologie und Morphologie der Algen. Vorträge aus dem Gesamtgebiet der Botanik herausgegeben von der Deutschen Botanischen Gesellschaft.* Neue Folge, Nr. 1. Gustav Fischer Verlag, Stuttgart. (4) + 238 SS. (biochem., sl. ment.)
605. Richter, G., R. Misske, R. Stempka and W. Dirks. 1977. Mode of synthesis of ribosomal RNA in various plant organisms. pp 85–96 in L. Bogorad and J.H. Weil. (Eds). *Nucleic Acids and Protein Synthesis in Plants. Lectures Presented at the NATO Advanced Study Institute on Nucleic Acids and Protein Synthesis in Plants Sponsored by NATO, FEBS, EMBO, at a Colloquium Sponsored by CNRS, Held in Strasbourg, France, July 15–24, 1976. NATO Advanced Study Institutes* Series A: *Life Sciences;* Volume 12. Plenum Press, New York and London. XI + (1) + 417 pp. (biochem., radiobiol.)
606. Rigano, C. 1967. Presenza dell'alga unicellulare *Cyanidium caldarium* (Tilden) Geitler nei terreni fumarolici dei Campi Flegrei e di Ischia (Napoli). *Delpinoa* (N.S.) (Napoli) 6/7(1964-1965): 277–284. (ecol., florist., morphol., physiol.)
607. Rigano, C. 1968. Azione del trattamento termico e del pH sull'attività in 'vitro' di un enzima (nitrato riduttasi) estratto da *Cyanidium caldarium* (ceppo Pozzuoli, alga acidofila e termale. *Delpinoa* (N.S.) (Napoli) 8/9(1966–1967): 75–83. (biochem., physiol.)
608. Rigano, C. 1971. Studies on nitrate reductase from *Cyanidium caldarium. Arch. Mikrobiol.* (Berlin, Heidelberg) 76(3): 265–276. (biochem., physiol.)
609. Rigano, C. 1972. Nitrate reductase from a thermophilic acidophilic alga. pp 2191–2197 in G. Forti, M. Avron and A. Melandri. (Eds). *Photosynthesis, Two Centuries After Its Discovery by Joseph Priestley. Proceedings of the IInd International Congress on Photosynthesis Research, Stresa, June 24–29, 1971.* [Vol. III]. Dr. W. Junk N. V. Publishers, The Hague. XXIII + (1) + pp 1697–2752. (biochem., physiol.)
610. Rigano, C. 1973. Inhibition by reduced pyridine nucleotides of nitrate reductase from *Cyanidium caldarium.* Inibizione de NA(P)H della nitrato riduttasi di *Cyanidium caldarium. Giorn. Bot. Ital.* (Firenze) 107(6): 305. (biochem., physiol.)
611. Rigano, C. 1973. Regulation of nitrate reductase in the unicellular alga '*Cyanidium caldarium*'. pp 113–114 in *Abstracts delle Communicazioni Scientifiche. Società Italiana di Fisiologia Vegetale. Congress Sociale 1972, Modena. 27–29 settembre. Inf. Bot. Ital.* (Firenze) 5(1): 113–114. (biochem., physiol.)
612. Rigano, C. and G. Aliotta. 1975. Electron donors and inhibitors of nitrate reductase from *Cyanidium caldarium. Biochim. Biophys. Acta* (New York, Amsterdam) 384(1): 37–45. (biochem., physiol.)
613. Rigano, C., G. Aliotta and V. Di Martino Rigano. 1975. Observations on enzymes of ammonia assimulation in two different strains of *Cyanidium caldarium. Arch. Microbiol.* (Berlin) 104(3): 297–299. (biochem., evol., physiol.).
614. Rigano, C., G. Aliotta, V. Di Martino Rigano, A. Fuggi and V. Vona 1977. Heterotrophic growth patterns in the

unicellular alga *Cyanidium caldarium*. *Arch. Microbiol.* (Berlin) 113(3): 191–196. (biochem., physiol.)

615. Rigano, C., G. Aliotta and U. Violante. 1974. Presence of high levels of nitrate reductase activity in *Cyanidium caldarium* grown on glutamate as the sole nitrogen source. *Plant Sci. Lett.* (Amsterdam) 2(5): 277–281. (biochem., physiol.)
616. Rigano, C., G. Aliotta and U. Violante. 1974. Reversible inactivation by ammonia of assimilatory nitrate reductase in *Cyanidium caldarium*. *Arch. Microbiol.* (Berlin) 99(1): 81–90. (biochem., physiol.)
617. Rigano, C. and T. Conforti. 1968. Ecologia e distribuzione dell'alga unicellulare *Cyanidium caldarium* (Tilden) Geitler nei Campi Flegrei (Napoli). *Delpinoa* (N.S.) (Napoli) 8/9(1966–1967): 3–11. (ecol., florist., physiol.)
618. Rigano, C., V. Di Martino Rigano and G. Fuggi. 1987. Nitrogen metabolism in thermophilic algae. pp 210–216 in W.R. Ullrich, P.J. Aparicio, P.J. Syrett and F. Castillo. (Eds). *Inorganic Nitrogen Metabolism. Symposium Held at a Meeting of the Federation of European Societies of Plant Physiology, Strausbourg, France, 1984.* Springer-Verlag, Berlin, Heidelberg, New York, London, Paris, Tokyo. (5) + pp VI–XIV + (3) + pp 4–295. (biochem., physiol.)
619. Rigano, C., V. Di Martino Rigano, V. Vona, G. Aliotta and A. Fuggi. 1978. Inibizione della crescita di *Cyanidium caldarium* da metionina sofossimmina e meitonina solfoe. Due inibitori della glutammina sintetasi. *Atti Accad. Naz. Lincei, Cl. Sci. Fis. Mat. Nat. Rend.* (Ser. 8) (Roma) 63[1977(1–2)]: 141–148. (biochem., evol., physiol.)
620. Rigano, C., V. Di Martino Rigano, V. Vona and A. Fuggi. 1978. Effetto della L-metionina-DL-solfossimmina, un inibitore della glutammia sintetasi, sul controllo della riduzione assimilative del nitrato. (Effect of L-methionine-DL-sulphoximine, an inhibitor of glutamine synthetase, on the control of assimilatory nitrate reduction). *Giorn. Bot. Ital.* (Firenze) 113(3): 216. (biochem., physiol.)
621. Rigano, C., V. Di Martino Rigano, V. Vona and A. Fuggi. 1979. Glutamine synthetase activity, ammonia assimilation and control of nitrate reduction in the unicellular red alga *Cyanidium caldarium*. *Arch. Microbiol.* (Berlin) 121(2): 117–120. (biochem., physiol.)
622. Rigano, C., V. Di Martino Rigano, V. Vona and A. Fuggi. 1981. Nitrate reductase and glutamine synthetase activities, nitrate and ammonia assimilation, in the unicellular alga *Cyanidium caldarium*. *Arch. Microbiol.* (Berlin) 129(2): 110–114. (biochem., physiol.)
623. Rigano, C., V. Di Martino Rigano, V. Vona, A. Fuggi and G. Aliotta. 1978. Studies in vivo on the control by ammonia of nitrate reduction to nitrite in the unicellular alga *Cyanidium caldarium*. *Plant Sci. Lett.* (Amsterdam) 13(4): 301–307. (biochem., physiol.)
624. Rigano, C., A. Fuggi, G. Aliotta, V. Di Martino Rigano and V. Vona. 1979. Reduction of nitrate to nitrite *in vivo* at non-physiological pH. Effect of ammonia. pp 307–308 in E.J. Hewitt and C.V. Cutting. (Eds). *Nitrogen Assimilation of Plants.* Sixth Long Ashton Symposium 1977. *Proceedings of a Symposium held at Long Ashton Research Station, University of Bristol, 19–22 September 1977.* Academic Press, London, New York, San Francisco. (4) + pp V–XX + 708 pp. (biochem., physiol.)
625. Rigano, C., A. Fuggi, G. Aliotta, V. Vona and V. Di Martino Rigano. 1978. Riduzione del nitrato a nitrito in un'alga unicellulare. *Atti Accad. Naz. Lincei, Cl. Sci. Fis. Mat. Nat. Rend.* (Ser. 8) (Roma) 63[1977(1–2)]: 149–153. (biochem., evol., physiol.)
626. Rigano, C., A. Fuggi, V. Di Martino Rigano and G. Aliotta. 1976. Studies on utilization of 2-ketoglutarate, glutamate and other amino acids by the unicellular alga *Cyanidium caldarium*. *Arch. Microbiol.* (Berlin) 107(2): 133–138. (biochem., evol., physiol.)
627. Rigano, C. and R. Taddei. 1968. Estrema acido-resistenza dell'alga *Cyanidium caldarium* (Tilden) Geitler vivente alla solfatara di Pozzuoli (Napoli). *Delpinoa* (N.S.) (Napoli) 8/9(1966-1967): 57–63. (ecol., physiol.)
628. Rigano, C. and U. Violante. 1972. A latent nitrate reductase from a thermophilic alga. *Biochem. Biophys. Res. Commun.* (New York) 47(2): 372–379. (biochem., physiol.)
629. Rigano, C. and U. Violante. 1972. Comparative growth of the thermal alga *Cyanidium caldarium* on nitrate and ammonia at different temperatures. *Arch. Mikrobiol.* (Berlin, Heidelberg) 85(1): 13–18. (biochem., physiol.).
630. Rigano, C. and U. Violante. 1972. Effect of heat treatment on the activity in vitro of nitrate reductase from *Cyanidium caldarium*. *Biochim. Biophys. Acta* (New York, Amsterdam) 256(2): 524–532. (biochem., physiol.)
631. Rigano, C. and U. Violante. 1973. Effect of nitrate, ammonia and nitrogen starvation on the regulation of nitrate reductase in *Cyanidium caldarium*. *Arch. Mirkobiol.* (Berlin, Heidelberg) 90(1): 27–33. (biochem., physiol.)
632. Rigano, C., U. Violante and G. Aliotta. 1973. Kinetic aspects of nitrate reductase from *Cyanidium caldarium*. Inhibition by reduced pyridine nucleotides. *Biochim. Biophys. Acta* (New York, Amsterdam) 327(1): 19–23. (biochem., physiol.).
633. Rigano, C., V. Vona, V. Di Martino Rigano and A. Fuggi. 1979. Nitrate reductase and glutamate dehydrogenase of the red alga *Porphyridium aerugineum*. *Plant Sci. Lett.* (Amsterdam) 15(3): 203–209. (biochem., physiol.)
634. Rigano, C., V. Vona, V. Di Martino Rigano and A. Fuggi. 1979. Protezione in vitro da FAD delle forma attiva ed inattiva della nitrato riduttasi. (Protection in vitro by FAD of active and inactive forms of nitrate reductase). pp 216–217 in *Societa Italiana di Fisiologia Vegetale. Congresso Sociale 1978. Camerino. 2–5 Ottobre. Reassunti e Abstracts dei Posters. Giron. Bot. Ital.* (Firenze) 113(3): (2) + pp 3–223. (biochem., physiol.)
635. Rigano, C., V. Vona, V. Di Martino Rigano and A. Fuggi. 1980. Active and inactive nitrate reductase. Effects of mild treatment with denaturing agents of proteins. *Biochim. Biophys. Acta* (New York, Amsterdam) 613(1): 26–33. (biochem., physiol.)
636. Rijgersberg, C.P. and J. Amesz. 1980. Fluorescence and energy transfer in phycobiliprotein-containing algae at low temperature. *Biochim. Biophys. Acta* (New York, Amsterdam) 593(2): 261–271. (biochem., photosyn., physiol.)
637. Rosen, W.G. and K.A. Siegesmund. 1961. Some observations on the fine structure of a thermophilic acidophilic

alga. *J. Biophys. Biochem. Cytol.* (New York) 9(4): 910–914. (ultrastr.)

638. Round, F.E. 1964. The ecology of benthic algae. pp 138–184 in D.F. Jackson. (Ed.). *Algae and Man.* Plenum Press, New York. X + 434 pp. (ecol., sl. ment.)
639. Round, F.E. 1965. The Biology of the Algae. Edward Arnold (Publishers) Ltd., London. VII + (1) + 269 pp. (ecol., sl. ment.)
640. Round, F.E. 1968. *Biologie der Algen.* Eine Einführung. Georg Thieme Verlag, Stuttgart. VIII + 315 SS mit 69 Abbildungen und 8 Tafeln im Text. (ecol., sl. ment.)
641. Round, F.E. 1973. *The Biology of the Algae.* Second Edition. Edward Arnold (Publishers) Limited, London. VII + (1) + 278 pp + pls 1–12. (ecol., sl. ment.)
642. Round, F.E. 1981. *The Ecology of Algae.* Cambridge University Press, Cambridge. (8) + 653 pp. (ecol., physiol.)
643. Rüdiger, W. 1975. Phycobiliproteide. *Ber. Dtsch. Bot. Ges.* (Berlin) 88(1): 125–139. (biochem., photosyn., physiol.)
644. Sahulka, J. 1976. Svláštnosti regulace nitrátreduktázy řas. *Biol. Listy* (Praha) 41(4): 280–282. (physiol., sl. ment.)
645. Sarma, P. and V.J. Chapman. 1975. Additions to the check list of freshwater algae in New Zealand–II. *J. Roy. Soc. N. Z.* (Wellington) 5(3): 289–312. (florist.)
646. Scherer, S., H. Riege and P. Boger. 1988. Light-induced proton efflux of the cyaobacterium *Anabaena variabilis.* pp 121–129 in L.J. Rogers and J.R. Gallon. (Eds). *Biochemistry of the Algae and Cyanobacteria. Proceedings of the Phytochemical Society of Europe,* No. 28. Clarendon Press, Oxford. (9) + pp X–XVI + (3) + pp 4–374. (biochem., physiol., sl. ment.)
647. Schimmelpfeng, L. and C. Thormaelen. 1986. *Method and Apparatus for Decreasing Waste Gas Pollutants by Photosynthesizing Algae.* German Offenlegungschrift. Deutsches Patentamt, Dienstelle, Berlin. 9 pp. (ecol., physiol.)
648. Schlösser, U.G. 1982. Sammlung von Algenkulturen. [Pflanzenphysiologisches Institut der Universität Göttingen (SAG)]. *Ber. Dtsch. Bot. Ges.* (Berlin 95: 181–276. (cultivat., florist.)
649. Schmellenkamp, H. 1971. *Einfluss der Temperatur auf den Kohlenstoffeinbau bei der Alge* Cyanidium caldarium *von Herbert Schmellenkampf aus Stuttgart.* Diplomarbeit vorgelegt dem Diplomprüfungsausschuss für Biochemiker des Fachbereichs Chemie-Biochemie der Eberhard-Karls-Universität Tübingen, Tübingen. (4) + 62 SS + Literaturverzeichnis (4). (biochem., radiobiol.)
650. Schmidt, E.W. 1954. Cyanophyta. Blaualgen, Spaltalgen. SS 50–55 in H. Melchior and E. Werdermann. *A. Engler's Syllabus der Pflanzenfamilien mit besonderer Berücksichtigung der Nutzpflanzen nebst einer Übersicht über die Florenreiche und Florengebiete der Erde.* I. Band. *Allgemeiner Teil. Bakterien bis Gymnospermen.* Zwölfte, völlig neugestaltete Auflage. Gebrüder Borntraeger, Berlin-Nikolasse. (8) + 367 SS mit 141 Abbildungen. (morph., sl. ment., tax.-syst.)
651. Schneegurt, M.A. and S.I. Beale. 1986. Biosynthesis of protoheme and heme-A precursors solely from glutamate in etiolated maize epicotyles. p. 52 in *Program and Abstract of Papers for the Annual Meeting of the American Society of Plant Physiologists at Louisiana State University, Baton Rouge, Louisiana, June 8–12, 1986. Plant Physiol.* (Lancaster) 80(4 suppl.): (4) + pp 1–151. (biochem., photosyn., physiol., radiobiol.)
652. Schneegurt, M.A. and S.I. Beale. 1987. Purification of the RNA required for δ-aminolevulinic acid biosynthesis by affinity chromatography directed againse the glutamate anticodon. p. 50 in *Scientific Program (Abstracts of Papers) for the Annual Meeting of the American Society of Plant Physiologists at Chase Park Plaza Hotel, Saint Louis, Missouri, July 19–23, 1987. Plant Physiol.* (Lancaster) 83(4 Suppl.): (4) + pp 1–74. (biochem., physiol.)
653. Schneegurt, M.A. and S.I. Beale. 1989. Characterization of the RNA required for biosynthesis of δ-aminolevulinic acid from glutamate. *Plant Physiol.* (Lancaster) 86(2): 497–504. (biochem., physiol.)
654. Schneider, Hj. A.W. and L. Bogorad. 1978. On the regulation of phycobiliprotein accumulation in relation to chlorophyll and ALA formation in *Cyanidium caldarium.* pp 823–826 in G. Akoyunoglou and J.H. Argyroudi-Akoyunoglou. (Eds). *Chloroplast Development. Proceedings of the International Symposium on Chloroplast Development Held on the Island of Spetsai, Greece, July 9–15, 1978. Developments in Plant Biology.* Volume 2. Elsevier/North-Holland Biomecial Press, Amsterdam, New York, Oxford. XVI + 888 pp. (biochem., photosyn., physiol.)
655. Schneider, Hj.A.W. and L. Bogorad. 1979. Spectral response curves for the formation of phycobiliproteins, chlorophyll and δ-aminolevulinic acid in *Cyanidium caldarium. Zeitschr. Pflanzenphysiol.* (Stuttgart) 94(5): 449–459. (biochem., photosyn., physiol.)
656. Schneider, Hj. A.W. and W. Liedgens. 1981. An evolutionary tree based on monoclonal antibody-recognized surface features of a plastid enzyme (5 aminolevulinate dehydratase). *Zeitschr. Naturforsch.* (Wiesbaden) 36: 44–50. (biochem., evol., physiol., sl. ment., tax.-syst.)
657. Schnepf, E. and R.M. Brown, Jr. 1971. On relationships between endosymbiosis and the origin of plastids and mitochondria. pp 299–322 in J. Reinert and H. Ursprung. (Eds). Origin and Continuity of Cell Organelles. Springer-Verlag, Berlin, Heidelberg, New York. XII + (2) + 342 pp. (tax.-syst.)
658. Schussnig, B. 1938. *Vergleichende Morphologie der niederen Pflanzen. Eine Einführung in deren Formbildung und Forwechsel auf entwicklungsgeschichtlicher Grundlage.* Erster Teil: *Formbildung.* Verlag von Gebrüder Borntraeger, Berlin. VIII + 382 SS mit 470 Abbildungen im Text. (morphol., tax.-syst.)
659. Schussnig, B. 1960. *Handbuch der Protophytenkunde. Eine vergleichend-mortphologische und biologische Darstellung der niederen Pflanzen für Biologen, Mediziner und Landwirte.* Band II. VEB Gustav Fischer Verlag, Jena. X + 1144 SS mit 880 Abbildungen im Text. (morphol., tax.-syst.)
660. Schuster, A., H.-P. Köst, W. Rüdiger and J. Eder. 1983. Investigations on the apoprotein of phycocyanin from *Cyanidium caldarium. Arch. Microbiol.* (Berlin) 135(1): 30–35. (biochem., photosyn., physiol.)
661. Schuster, A., H.-P. Köst, W. Rüdiger, J.A. Holroyd and S.B. Brown. 1983. Incorporation of haem into phycocyanobilin in levulinic acid treated *Cyanidium caldarium.*

Plant Cell Rep. (Berlin, New York) 2(1): 85–87. (biochem., phytosyn., physiol., radiobiol.)
662. Schwabe, G.H. 1936. Ueber einige Blaualgen aus dem mittleren und südlichen Chile. *Verh. Deutschen Wissenschaftl. Ver. Satiago de Chile* (Valparaiso) (N. F.) 3: 113–146. (ecol., florist., tax.-syst.)
663. Schwabe, G.H. 1942. Thermalokologische Beitrage aus Kusatu. SS C21–C23 in *Über das Thermalbad Kasatu. Mitt. Dtsch. Ges. Nat.-Völk. Ostasiens* (Tokyo, Leipzig) 33(Teil C): C1–C67. (ecol., florist., physiol., tax.-syst.)
664. Schwabe, G.H. 1942. Zur Okologie umraumfremder Quellen. (Ijosui no seitaigaku ni tsuite). *Shokubutsugaku Zasshi (Bot. Mag.,* Tokyo) 56(672): 577–590. (ecol., sl. ment.)
665. Schwabe, G.H. 1944. Umraumfremde Quellen (Umranfrende Quellen (Mit Beiheft, Enthaltend 99 Abbildungen und Plane, 26 Lichtbilden). *Mitt. Dtsch. Ges. Nat.-Völk. Ostasiens* (Schanghai) Supplementband 21: (6) + SS I–VII + (1) + SS 1–239 + [(2) + SS 240–300 (Textbeilagen zu G.H. Schwabe. Umraumfremde Quellen (Shanghai. 1944) (Verlag von Max Nossler & Co., G.m.b.h., Shanghai). ecol., florist., tax.-syst.)
666. Schwabe, G.H. 1947. Blaualgen und Lebensraum, I. Beitrage zur Ökologie und Systematik. *Acta Bot. Taiwanica* (Taipei) 1(1): 3–59 + 1 Doppeltabelle + Abbild. 1–33. (ecol., florist.)
667. Schwabe, G.H. 1960. Über den thermobionten Kosmopoliten *Mastigocladus laminosus* (Cohn. Blaualgen und Lebensraum V. *Schweiz. Zeitschr. Hydrol.* (Basel) 22(2): 759–792. (ecol., sl. ment.)
668. Schwabe, G.H. 1960. Zur autotrophen Vegetation in ariden Boden. Blaualgen und Lebensraum IV. *Österr. Bot. Zeitschr.* (Wien) 107: 281–309. (ecol., sl. ment.)
669. Schwartz, R.M. and M.O. Dayhoff. 1981. The evolution of Blue greens and the origins of chloroplasts. pp 551–558 in Y. Wolman. (Ed.). *Origin of Life. Proceedings of the* Third ASSOL Meeting *and the* Sixth ICOL Meeting, *Jerusalem, June 22–27, 1980.* D. Reidel Publishing Company, Dordrecht, Holland/Boston, U.S.A./London, England. XV + (1) + 613 pp. (biochem., evol., tax.-syst.)
670. Schwertner, H.A., W.I. Butcher and B. Richardson. 1972. Structure and composition of *Oocystis polymorpha* cell walls isolated from the culture medium. *J. Phycol.* (Baltimore, Lawrence) 8(2): 144–146. (biochem., sl. ment.)
671. Seckbach, J. 1969. Life in pure CO_2. I. Growth of algae at room temperature under 1 atmosphere of CO_2. p. 15 in *Abstracts of papers presented at the Annual Sectional Meetings 1969. Plant Physiol.* (Lancaster) 44(Suppl.): (2) + pp 1–44. (cultivat., physiol.)
672. Seckbach, J. 1971. Hayesh Hayim Al Nogah? [Is there life on Venus?]. *Mada, Iton Madai Lakol (Science, Gen. Sci. J.,* Jerusalem) 15(5): 296–303. (biochem., cultivat., ecol., morphol., photosyn., physiol., ultrastr.)
673. Seckbach, J. 1971. Pigmentation of algae under pressure. *Limnol. Oceanogr.* (Baltimore) 16(3): 567–572. (cultivat., photosyn., physiol.)
674. Seckbach, J. 1971. Size, composition and fine structure of *Cyanidium caldarium* cultured under pure CO_2. *Isr. J. Bot.* (Jerusalem) 20(4): 302–310. (biochem., cultivat., cytol., morphol., physiol., ultrastr.)
675. Seckbach, J. 1972. On the fine structure of the acidophylic hot-spring alga *Cyanidium caldarium*: a taxonomic approach. *Microbios* (London) 5(18): 133–142. (biochem., cytol., photosyn., tax.-syst., ultrastr.)
676. Seckbach, J. 1986. Evolution of eukaryotic cells *via* bridge algae: the Cyanidia connection. p. 367 in *Abstracts, The Third International Colloquium on Endocytobiology:* 'Endocytobiology III', *The New York Academy of Sciences, New York, June 10–12, 1986. Endocytobios. Cell Res.* (Tübingen) 3: 367. (evol., florist., tax.-syst.)
677. Seckbach, J. 1987. Evolution of eukaryotic cells *via* bridge algae, the Cyanidia connection. pp 424–437 in J.J. Lee and J.F. Fredrick. (Eds.) *Endocytobiology. Papers presented at the Third International Colloquium on Endocytobiology held June 10–12, 1986, in New York, N. Y. and Sponsored by the New York Academy of Science. Ann. N. Y. Acad. Sci.* (New York) 503: (10) + pp XI–XII + 509 pp. (biochem., evol., florist., morphol., physiol., ultrastr.)
678. Seckbach, J. 1988. Origin of eukaryotic life. The Cyanidia model for algal non-symbiotic evolution. p. 179 in *[Abstracts], The 13th Annual Meeting of the SSOEL-JAPAN, at the Inter-University House of Kansai, July 15–17, 1988. Viva Origino* (Tokyo) 16(3): 178–235. (evol., tax.-syst.)
679. Seckbach, J. 1990. Are the acidothermophilic algae [*Cyanidioschyzon merolae, Cyanidium caldarium, Galdieria sulphuraria* (Cyanidiophyceae)] presenting phylogenetic transitional stages toward Rhodophyta? p. 18 in *Abstracts of Papers Scheduled for the Annual Meeting of the Phycological Society of America at the University of Maryland, College Park, Maryland, 24–28 June, 1990. J. Phycol.* (Baltimore, Lawrence) 26(2 suppl.): (3) + pp 4–22. (evol., tax.-syst.)
680. Seckbach, J., F.A. Baker and P.M. Shugarman. 1970. Algae thrive under pure CO_2. *Nature* (London) 227(5259): 744–745. (cultivat., photosyn., physiol.)
681. Seckbach, J. and J.F. Fredrick. 1980. A primaeval alga bridging the blue-green and the red algae: further biochemical and ultrastructure studies of *Cyanidium caldarium* with special reference to the plastic membranes. *Microbios* (London) 29(117/118): 135–147. (biochem., cytol., evol., tax.-syst., ultrastr.)
682. Seckbach, J. and J.F. Fredrick. 1981. Biphyletic origins of Rhodophyceae and Chlorophyceae: *Cyanidium caldarium* and *Glaucocystis* as bridge algae. p. 333 in *Abstracts of the* XIII International Botanical Congress, Sydney, Australia, 21–28 August 1981. Sponsored by the Australian Academy of Science under the Auspices of the International Union of Biological Sciences, (Sydney?). 351 pp. (biochem., cytol., evol., ultrastr.)
682. Seckbach, J. and J.F. Fredrick. 1981. Biphyletic origins of Rhodophyceae and Chlorophyceae: *Cyanidium caldarium* and *Glaucocystis* as bridge algae. p. 333 in *Abstracts of the* XIII International Botanical Congress, *Sydney, Australia, 21–28 August 1981.* Sponsored by the Australian Academy of Science under the Auspices of the International Union of Biological Sciences, (Sydney?). 351 pp. (biochem., cytol., evol., ultrastr.)
683. Seckbach, J. and J.F. Fredrick. 1981. On the origin of photosynthetic eukaryotic cells: *Cyanidium caldarium* as a

'bridge' alga between prokaryotic cyanobacteria and eukaryotic rhodophytes: evidence from environmental, polysaccharide biochemistry and ultrastructural studies. pp 567–574 in Y. Wolman. (Ed.). *Origin of Life. Proceedings of the* Third ISSOL Meeting *of the* Sixth ICOL Meeting, *Jerusalem, June 22–27, 1980.* D. Reidel Publishing Company, Dordrecht, Holland/Boston, U.S.A./ London, England. IV + (1) + 613 pp. (biochem., cytol., ecol., evol., tax.-syst., ultrastr.)

684. Seckbach, J. and J.F. Fredrick. 1981. Sarid 'Meuban-Hai' Kegesher Evulutzioni Kadum. [On the trail of a living fossil]. *Mada, Iton Madai Lakol (Science, Gen. Sci. J.,* Jerusalem) 25(3): 122–125. (biochem., cytol., evol., ultrastr.)

685. Seckbach, J. and J.F. Fredrick. 1981. Separate evolutionary origins of Rhodophyceae and Chlorophyceae *via Cyanidium caldarium* and *Glaucocystis nostochinearum* as transition or 'bridge' algae. p. 14 in *Proceedings of the* 15th Scientific Meeting *of the Israel Society for Electron Microscopy, Tel Aviv University, May 21st, 1981.* Tel. Aviv. 30 pp. (biochem., cytol., evol.)

686. Seckbach, J. and J.F. Fredrick. 1986. Origin of nucleated cells: a possible non-symbiotic pathway *via* autogenous transitional stages linking the pro-eukaryota through the Prerhodophyta, a new algal taxon positioned between the red and blue-green algae. p. 356 in J.P. Ferris. (Ed.). *Abstract Issue 1986, ISSOL Meeting, Berkeley, California. Origins of life* (Dordrecht, Boston) 16(3/4): (3) + pp IV–XXII + pp 175–542 + Table of Contents to Vol. 6 (2). (biochem., evol., tax.-syst.)

687. Seckbach, J. and J.F. Fredrick. 1987. Hayyim kadmai-im ba-kineret. [Primeval Life in the Galilee Lake]. *Mada, Iton Madai Lakol (Science, Gen. Sci. J.,* Jerusalem) 39(2/3): 134–303. (cytol., florist., ultrastr.)

688. Seckbach, J., J.F. Fredrick and D.J. Garbary. 1983. Auto- or exogenous origin of transitional algae: an appraisal. pp 947–962 in H.E.A. Schenk and W. Schwemmler. (Eds). *Endocytobiology* II. *Intracellular Space as Oligogenetic Ecosystem. Proceedings* Second International Colloquium *on Endocytobiology. Tübingen Germany, April 10–15, 1983.* Walter de Gruyter, Berlin, New York. (5) + pp VI–XV + (3) + pp XVIII–XXV + (4) + pp 4–1071. (biochem., cytol., ecol., evol., tax.-syst., ultrastr.)

689. Seckbach, J. 1991. Systematic problems with *Cyanidium caldarium* and *Galdieria sulphuraria* and their implications for molecular biology studies. *J. Phycol.* (Baltimore, Lawrence) 27(5): 794–796. (evol., tax.-syst.)

690. Seckbach, J., E. Gonzáles, I. Wainwright and W. Gross. 1990. Peroxisomes in the Cyanidiophyceae. p. 18 in *Abstracts of Papers Scheduled for the Annual Meeting of the Phycological Society of America at the University of Maryland, College Park, Maryland, 24–28 June, 1990. J. Phycol.* (Baltimore, Lawrence) 26(2 suppl.): (3) + pp 4–22. (biochem., physiol., tax.-syst.)

691. Seckbach, J., E. Gonzáles, I. Wainwright and W. Gross. 1991. *Cyanidioschyzon morale, Cyanidium caldarium* and *Galdieria sulphuraria* (Cyanophyceae, Rhodophyta): a discussion on microbodies (Peroxisomes) and evolutionary examination. *Nova Hedwigia, Zeitschr. Kryptogammenkd.* (Weinheim, Lehre, Braunschweig) 55(1–2): 99–109. (biochem., evol., physiol., tax.-syst.)

692. Seckbach, J., H. Gross and M.B. Nathan. 1971. Growth and photosynthesis of *Cyanidium caldarium* cultured under pure CO_2. *Isr. J. Bot.* (Jerusalem) 20(2): 84–90. biochem., cultivat., florist., photosyn., physiol.)

693. Seckbach, J., I.S. Hammerman and J. Hanania. 1981. Ultrastructural studies of *Cyanidium caldarium*: contribution to phylogenesis. pp 409–425 in J.F. Fredrick. (Ed.). Origins and Evolution of Eukaryotic Intracellular Organelles. *Ann. N. Y. Acad. Sci.* (New York) 361: (8) + pp IX + 512 pp. (biochem., cytol., ecol., evol., morphol., photosyn., physiol., ultrastr.)

694. Seckbach, J. and R. Ikan. 1972. Sterols and chloroplast structure of *Cyanidium caldarium. Plant Physiol.* (Lancaster) 49(3): 457–459. (biochem., cytol., ecol., evol., physiol., tax.-syst., ultrastr.)

695. Seckbach, J. and I.R. Kaplan. 1973. Growth pattern and $^{13}C/^{12}C$ isotope fractionation of *Cyanidium caldarium* and hot spring algal mats. *Chem. Geol.* (Amsterdam) 12: 161–169. (biochem., cultivat., ecol., evol., florist., physiol., radiobiol.)

696. Seckbach, J. and W.F. Libby. 1969. Life in pure CO_2. 2. Studies with thermophilic algae grcwn at elevated temperatures and under 1 atmosphere of CO_2. pp 15–16 in *Abstract of Papers Presented at the Annual Section Meetings. Plant Physiol.* (Lancaster) 44(Suppl.): (2) + pp 1–44. (cultivat., cytol., ecol., morphol., physiol., ultrastr.)

697. Seckbach, J. and W.F. Libby. 1969. Life on Venus? or algae which grow in pure carbon dioxide in acid media at elevated temperature. p. 195 in XI International Botanical Congress. *Abstracts of the Papers Presented at the* XI International Botanical Congress, *August 24–September 2, 1969 and the International Wood Chemistry Symposium, September 2–4, 1969, Seattle, Washington.* International Botanical Congress, Seattle. (4) + 260 pp. (ecol., morphol., physiol., radiobiol.)

698. Seckbach, J. and W.F. Libby. 1970. Vegetative life on Venus? or investigations with algae which grow under pure CO_2 in hot acid media at elevated pressures. *Space Life Sci.* (Dordrecht) 2: 121–143. (biochem., cultivat., ecol., morphol., photosyn., physiol., ultrastr.)

699. Seckbach, J. and W.F. Libby. 1971. Vegetative life on Venus? or investigations with algae which grow under pure CO_2 in hot acid media and at elevated pressures. pp 62–83 in C. Sagan, T.C. Owen and H.J. Smith. (Eds). *Planetary Atmospheres. International Astronomical Union (Union Astronomique Internationale)* Symposium No. 40 *Held in Marfa, Texas, U.S.A., October 26–31, 1969.* R. Reidel Publishing Company, Dordrecht, Holland; Springer-Verlag New York Inc., New York. XVII + (1) + 408 pp. (biochem., cultivat., cytol., ecol., morphol., photosyn., physiol., ultrastr.)

700. Seckbach, J., H. Nagashima and I. Fukuda. 1990. Autogenous eukaryogenesis: contradiction of completion of endosymbiotic evolution. pp 575–578 in P. Nardon, V. Gianinazzi-Pearson, A.M. Grenier, L. Margulis and D.C. Smith. (Eds). *Endocytobiology* IV. 4th International Colloquium *on Endocytobiology and Symbiosis.* 4ème Congres International *sur l'Endocytobiologie et la Symbiose. INSA, Vilerbanne (France), July 4–8, 1989.* Institut National de

la Recherche Agronomique, Paris. (7) + pp 8–620. (cytol., evol., morphol., tax.-syst., ultrastr.)

701. Seckbach, J., H. Nagshima and I. Fukuda. 1991. Further studies with the thermo-acidophilic algae (Cyanidiophyceae, Rhodophytes): special references to phylogeny, plastid genome, ethylene catalase activity and ultrastructural observations. p. 43 in [Abstracts Scheduled to be presented at the] First International Symbiosis Congress, Jerusalem, Israel, November 17–22, 1991, Program and Abstracts. Congress Organizers: Kenes–Organizers of Congresses and Special Events, Tel Aviv. 70 pp Index [(2) + 4 pp] (biochem., evol., physiol., tax.-syst., ultrastr.)
702. Seckbach, J. and M.B. Nathan. 1971. Method for the separation of algae pigments by thin layer chromatography: CO_2 effect on pigmentation. *Lab. Pract.* (London) 20(12): 933–934. (biochem., ecol., photosyn., physiol.)
703. Seckbach, J., M.B. Nathan and H. Gross. 1970. On algal ubiquity: studies with acidophylic thermophiles adapted to a CO_2 atmosphere at elevated temperatures and pressure. p. 28 in *Programs and Abstracts.* Twenty-third Annual Meeting *of the Society of Protozoologists, Washington, D.C., September 8–11, 1970. Joint Meeting with Second International Congress of Parisitology September 6–12, 1970. J. Protozool.* (Utica) 17(Suppl.): 1–44. (cytol., ecol., physiol., ultrastr.)
704. Seckbach, J., E. Rahamim and J.F. Fredrick. 1987. Non-cytosymbiosis in eukaryotic algal evolution; *Cyanidioschyzon, Cyanidium & Galdieria* as indigenous Prerhodophytes connecting Cyanobacteria and Rhodophyta. p. 321 (No. 5-155-7) in W. Greuter, B. Zimmer and H.-D. Behnke. (Eds). *Abstracts of the General Lectures, Symposium Papers and Posters presented at the* XIV International Botanical Congress, *Berlin July 24 to August 1, 1987. Collated and Edited by W. Greuter, B. Zimmer and H.-D. Behnke on behalf of the Organizing Committee of the Congress. Produced with the Assistance of M.I. Hakki, H. Sipman and Staff Members of the Botanical Museum Berlin-Dahlem.* Botanischer Garten und Botanisches Museum, Berlin-Dahlem. (4) + pp 5–480. (biochem., evol., tax.-syst.)
705. Sli͡akh, I.O., L.A. Mineeva and M.V. Gusev. 1984. Osobennosti ul'stastruktury kletok *Cyanidium caldarium* na raznykh stadi͡iakh rost periodicheskoĭ kul'tury. (The peculiarities of *Cyanidium caldarium* ultrastructure at different stages of growth under periodic culture). *Izv. Akad. Nauk SSSR,* Ser. Biol. (Moskva) 1984(1): 74–81. (cult., cytol., ultrastr.)
706. Sli͡akh, I.O., L.A. Mineeva and M.V. Gusev. 1984. Ultrastructure of *Cyanidium caldarium* cells at various stages of growth of a periodic culture. *Biol. Bull. Acad. Sci. USSR,* Ser. Biol. (USSR, New York 11(1): 55–61. (cultivat., cytol., ultrastr.)
707. Seli͡akh, I.O., N.I. Ramanova, O.I. Baulina, L.A. Mineeva and M.V. Gusev. 1981. Izuchenie membrannykh struktur kletok *Cyanidium caldarium* s primeneniem razlichnykh metodov fikcat͡sii. *Fiziol. Rast.* (Moskva) 28(6): 1210–1215 + ris. 1-2,2,3. (cytol., morphol., ultrastr.)
708. Seli͡akh, I.O., N.I. Romanova, O.I. Baulina, L.A. Mineeva and M.A. Gusev. 1982. Study of the membrane structures of *Cyanidium caldarium* employing different fixation methods. *Sov. Plant Physiol.* (USSR, New York) 28(6, part 2): 880–886. (cytol., morphol., ultrastr.)
-- Selyakh, I.O. Please see: Seli͡akh, I.O.
709. Seng, W.L. 1976. *Biosynthesis and Properties of Phycocyanin.* Master of Arts Thesis, Department of Biochemistry, Boston University, Boston. XI + 54 pp. (biochem., photosyn., physiol.)
710. Senger, H., D. Dornemann, K. Kotzabasis, M. Senge, W. Schmidt and M. Wissenbach. 1987. Chlorophyll RC I, its structure, function and biosynthesis, pp 491-498 in J. Biggins. (Ed.). *Progress in Photosynthesis Research,* Vol. 4. *Proceedings of the* VIIth International Congress on Photosynthesis, Providence, Rhode Island, U.S.A., August 10–15, 1986. Martinus Nijhoff Publishers, Dordrecht, Boston, Lancaster. (4) + pp V–XXVII + (1) + 858 pp. (biochem., physiol.)
711. Senger, H. and G. Strassberger. 1978. Development of photosystems in greening algae. pp 367-378 in G. Akoyunoglou and J.H. Argyroudi-Akoyunoglou. (Eds). *Chloroplast Development. Proceedings of the International Symposium on Chloroplast Development Held on the Island of Spetsai, Greece, July 9–15, 1978. Developments in Plant Biology.* Volume 2. Elsevier/North-Holland Biomedical Press, Amsterdam, New York, Oxford. XVI + 888 pp. (photosyn., sl. ment.)
712. Sent͡sova, O. I͡u., T.R. Kravt͡sova and M.V. Gusev. 1988. Fiziologo-biokhimicheskie osobennosti Atsidotermofil'nykh vodorosleĭ *Cyanidium caldarium* i *Galdieria sulfuraria* i ikh rost pri sovmestnom kul'tivirovanii v fotoavtotrofnykh, miksotrofnykh i geterotrofnykh usloviiakh. *Fiziol. Rast.* (Moskva) 35(3): 516–525. (biochem., cultivat., ecol., evol., florist., physiol., tax.-syst.)
713. Sent͡sova, O. I͡u., T.R. Kravt͡sova and M.V. Gusev. 1988. Physiological and biochemical properties of the acidothermophilic algae *Cyanidium caldarium* and *Galdieria sulfuraria* and their growth in mixed cultures under photoautotrophic, mixotrophic and heterotrophic conditions. *Sov. Plant Physiol.* (USSR, New York) 35(3): 403–411. (biochem., cultivat., ecol., evol., florist., physiol., tax.-syst.)
714. Sent͡sova, O. I͡u., N.I. Romanova, L.A. Mineeva and M.V. Gusev. 1982. Fotosintez obligatno at͡sidofil'noĭ odnakletochnoĭ vodorosli *Cyanidium caldarium*: savisinost' ot pH i formy neorganicheskogo uglerodo. *Dokl. Akad. Nauk SSSR* (Moskva, Leningrad) 266(3): 762–765. (biochem., ecol., photosyn., physiol.)
715. Sent͡sova, O. I͡u., N.I. Romanova, L.A. Mineeva and M.V. Gusev. 1982. Photosynthesis of the obligately acidophilic alga *Cyanidium caldarium*: dependence upon pH and the form of inorganic carbon. *Dokl. Bot. Sci., Akad. Nauk SSSR* (USSR, New York) 265–267: 107–110. (biochem., ecol., photosyn., physiol.)
716. Sheath, R.G. 1984. The biology of freshwater red algae. pp 89–157 in F.R. Round and D.J. Chapman. (Eds). *Progress in Phycological Research.* Volume 3. Biopress Ltd., Bristol. (11) + pp 12–387. (biochem., sl. ment., tax.-syst.)
717. Sheath, R.G., J.A. Hellebust and T. Sawa. 1979. Floridean starch metabolism of *Porphyridium purpureum* (Rhodophyta). I. Changes during ageing of batch culture.

Phycologia (Berkeley, Copenhagen, Vancouver) 18(2): 149–163. (biochem., sl. ment.)

718. Sheath, R.G., J.A. Hellebust and T. Sawa. 1981. Floridean starch metabolism of *Porphyridium purpureum* (Rhodophyta) III. Effects of darkness and metabolic inhibitors. *Phycologia* (Berkeley, Copenhage, Vancouver) (20(1): 22–31. (biochem., physiol., sl. ment.)
719. Sheridan, R.P. 1976. Sun-shade ecotypes of a bluegreen alga in a hot spring. *J. Phycol.* (Baltimore, Lawrence) 12(3): 279–285. (ecol., florist., sl. ment.)
720. Shihira, I. and W.R. Krauss. 1965. *Chlorella. Physiology and Taxonomy of Forty-one Isolates.* University of Maryland, College Park. V + (1) + 97 pp. (tax.-syst.)
721. Silva, P.C. 1962. Classification of algae. pp 827–837 in R.A. Lewin. (Ed). *Physiology and Biochemistry of Algae.* Academic Press, New York and London. Frontispiece + XXVII + (1) + 929 pp. (tax.-syst.)
722. Silva, P.C. 1980. Names of Classes and Families of Living Algae with Special Reference to Their Use in the *Index Nominum Genericorum (Plantarum). Regnum Vegetabile* 103. Bohn, Scheltema & Holkema, Utrecht; Dr. W. Junk b. v., Publishers, The Hague. (6) + 156 pp. (tax.-syst.)
723. Skuja, H. 1970. Alghe cavernicole nelle zone illuminate delle grotte de Castellana. (Murge di Bari). *Le Grotte d'Italia* (Trieste) II(1968–1969) (4): 193-2-2 + Tav. X. (florist.)
724. Slutzky, G. and V. Mcmahon. 1972. Glycolipids of the alga, *Cyanidium caldarium.* p. 43 in *Abstracts. Papers Presented at the* Forty-Eighth Annual Meeting, *Southwestern and Rocky Mountain Division American Association for the Advancement of Science and the* Forty-Third Annual Meeting, *Colorado-Wyoming Academy of Science, April 26–29, 1972, Fort Collins, Colorado, Incorporating the Colorado-Wyoming Academy of Science, Papers Presented at the* Forty-Second Annual Meeting, *April-30 May 1, 1971, Southern Colorado State College, Pueblo, Colorado. J. Colorado-Wyoming Acad. Sci.* (Boulder) 7(2/3): IV + pp 1–117. (biochem., physiol.)
725. Smith. A.J. and L.J. Rogers. 1988. Tetrapyrrole biosynthesis–the C5 pathway, pp 69–96 in L.J. Rogers and J.R. Gallon. (Eds). *Biochemistry of the Algae and Cyanobacteria. Proceedings of the Phytochemical Society of Europe,* No. 28. Clarendon Press, Oxford. (9) + pp X–XVI + (3) + pp 4–374. (biochem., physiol., sl. ment.)
726. Smith, D.W. 1972. *Water Potential and the Ecology of a Soil Alga,* Cyanidium caldarium. Ph.D. Dissertation, Department of Bacteriology, University of Wisconsin, Madison. (1) + VII + 98 pp + (1). (ecol., florist., photosyn., physiol., radiobiol.)
727. Smith, D.W. 1973. Water potential and the ecology of a soil alga, *Cyanidium caldarium. Diss. Abstr. Int. B, Sci. Eng.* (Ann Arbor) 34(1): 329B–330B. (ecol., physiol., radiobiol.)
728. Smith, D.W. and T.D. Brock. 1972. The effect of soil water on microbial activity. p. 8 in *Abstracts of the Annual Meeting of the American Society for Microbiology.* 72nd Annual Meeting, *Philadelphia, P. 23–28 April 1972.* American Society for Microbiology, Washington, D.C. VIII + 294 pp. (photosyn., physiol., radiobiol.)
729. Smith, D.W. and T.D. Brock. 1973. The water relations of the alga *Cyanidium caldarium* in soil. *J. Gen Microbiol.* (Cambridge) 79(2): 219–231. (ecol., florist., photosyn., physiol., radiobiol.)
730. Smith, D.W. and T.D. Brock. 1973. Water potential and the ecology of the alga *Cyanidium caldarium* in soil. p. 48 in *Abstracts of the Annual Meeting of the American Society for Microbiology.* 72rd Annual Meeting, *Miami Beach, Fla. 6–11 May 1973.* American Society for Microbiology, Washington D.C. VIII + 313 pp. (ecol., photosyn., physiol.)
731. Smith, D.W. and T.D. Brock. 1973.Water status and the distribution of *Cyanidium caldarium* in soil. *J. Phycol.* (Baltimore, Lawrence) 9(3): 330–332. (ecol., florist., photosyn., physiol., radiobiol.)
732. Smith, D.W., C.B. Fliermans and T.D. Brock. 1972. Technique for measuring $^{14}CO_2$ uptake by soil microorganisms *in situ. Appl. Microbiol* (Baltimore) 23(3): 595–600. (biochem., ecol., physiol., radiobiol.)
733. Smith, G.M. 1950. *The Fresh-Water Algae of the United States.* Second Edition. McGraw-Hill Book Company, Inc., New York, Toronto, London. VII + (1) + 719 pp. (tax.-syst.)
734. Soeder, C.J., K. Schmidt, J. Komarek and O. Lhotsky. 1981. *Applied Hydrobiology and Wastewater Documentation Code. Spezielle Berichte der Kernforschungsanlage Jülich,* No. 137. Institut für Biotechnologie Jülich, Jülich. (1) + pp 02–41. (tax.-syst.)
735. Soeder, C.J. and E. Stengel. 1979. Physico-chemical factors affecting metabolism and growth rate. pp 714–740 in W.D.P. Stewart. (Ed.). *Algal Physiologigy and Biochemistry. Botanical Monographs.* Volume 10. University of California Press, Berkeley and Los Angeles. (Blackwell Scientific Publications, Oxford, London, Edinburgh, Melbourne). XI + 989 pp. (biochem., cultivat., physiol., sl. ment.)
736. Sonneveld, A., H. Rademaker and L.N.M. Duysens. 1980. Microsecond delayed fluorescence of photosystem II of photosynthesis in various algae: emission spectra and uphill energy transfer. *FEBS Lett. (Fed. Eur. Biochem. Soc.,* Amsterdam) 113(2): 323–327. (photosyn., physiol.)
737. Sonneveld, A., H. Radmaker and L.N.M. Duysens. 1980. Transfer and trapping of excitation energy in photosystem II as studied by chlorophyll *a2* fluorescence quenching by dinitrobenzene and carotenoid triplet: the matrix model. *Biochem. Biophys. Acta* (New York, Amsterdam(593(2): 272–289. (photosyn., physiol.)
738. Staehelin, L.A. 1967. Chloroplast fibrils linking the photosynthetic lamellae. *Nature* (London) 214(5093): 1158. (cytol., ultrastr.)
739. Staehelin, L.A. 1967. Ultrastructural changes of the plasmalemma and the cell wall during the life cycle of *Cyanidium caldarium.* p. 389 in *New Zealand Society for Electron Microscopy,* Second Conference, *Dunedin, May 17–19. 1967. N.Z. Med. J.* (Dunedin) 66(418): 387–393. (cytol., ultrastr.)
740. Staehelin, L.A. 1968. The interpretation of freeze-etched artificial and biological membranes. *J. Ultrastruct. Res.* (New York) 22(3/4): 326–347. (cultivat., cytol., physiol., ultrastr.)
741. Staehelin, L.A. 1968. Ultrastructural changes of the

plasmalemma and the cell wall during the life cycle of *Cyanidium caldarium. Proc. R. Soc. London,* Ser. B (London) 171: 249–259 + pls 7–12. (cytol., ultrastr.)

742. Staehelin, L.A. 1986. Chloroplast structure and supramolecular organization of photosynthetic membranes. pp 1–84 in L.A. Staehelin and C.J. Arntzen. (Eds). *Photosynthesis III.* in A. Pirson and M.H. Zimmermann. (Eds). *Encyclopedia of Plant Physiology,* New Series, Vol. 19. Springer Verlag, Berlin, Heidelberg, New York, Tokyo. (5) + pp VI–XXVIII + (1) + pp 2–802. (cytol., sl. ment., ultrastr.)

-- Stafleu, F.A. Please see: Farr, E.R., J.A. Leussink and F.A. Stafleu. 1979. *Index Nominum Genericorum (Plantarum) ...*

-- Stafleu, F.A.: Please also see: Farr, E.R., J.A. Leussink and G. Zijlstra, 1986. *Index Nominum Genericorum (Plantarum).* Supplementum I

743. Starmach, K. 1977. *Phaeophyta-Brunatnice, Rhodophyta-Krasnorosty. (with Keys for the Identification of Fresh Water Brown- and Red-Algae Mentioned in the Volume).* in K. Starmach and J. Sieminska. (Redak.). *Flora Slodkowodna Polski.* Tome 14. Polska Akademia Nauk, Instytut Botaniki, Panstwowe Wydawnictwo Naukowe, Warszawa, Kraków. (4) + pp 5–444 + (3) + Errata (1). (tax.-syst.)

744. Starr, R.C. 1985. *Additions to the UTEX collection since 1978. (The culture collection of algae at the University of Texas at Austin).* The University of Texas, Austin. (23 pp mimeographed). (cultivat.)

745. Starr, R.C. and J.A. Zeikus. 1987. UTEX–The Culture Collection of Algae at the University of Texas at Austin. *J. Phycol.* (Baltimore, Lawrence) 23(3 Suppl.): 1–47. (cultivat.)

746. Stead, A.D., T.W. Ford, W.J. Myring and D.T. Clarke. 1988. A comparison of soft X-ray contact microscopy with light and electron microscopy for the study of algal cell ultrastructure. *J. Microsc.* (Oxford) 149(3): 207–216. (cytol., morphol., ultrastr.)

747. Steinmuller, K., M. Kaling and M. Zetsche. 1983. In-vitro synthesis of phycobiliproteids and riblose-1, 5-bisphosphate carboxylase by non-poly-adenylated-RNA of *Cyanidium caldarium* and *Porphyridium aerugineum. Planta* (Berlin) 159: 308–313. (biochem., evol., photosyn., physiol., tax.-syst.)

748. Steinmuller, K. and K. Zetsche. 1984. Photo- and metabolite regulation of the synthesis of ribulose bisphosphate carboxylase/oxygenase and the phycobiliproteins in the alga *Cyanidium caldarium. Plant Physiol.* (Lancaster) 76(4): 935/939. biochem., photosyn., physiol.)

749. Stempka, R. and G. Richter. 1978. Maturation of chloroplast and cytoplasmic ribosomal RNAs from single large precursors in two algae species, *Cyanidium* and *Chlorella. Arch. Microbiol.* (Berlin) 119(2): 187–196. (biochem., photosyn., physiol., radiobiol.)

750. Stewart, K.D. and D. Garbary. 1982. Introduction and bibliography on the evolution of algae. pp 641–646 in J. R. Rosowski and B.C. Parker. (Eds). *Selected Papers in Phycology* II. Phycological Society of America, Inc., Book Division, Lawrence, Kansas. XXI + 1) + 866 pp. (evol., tax.-syst.)

751. Stewart, W.D.P. 1974. A note on taxonomy, pp 909–917 in W.D.P. Stewart. (Ed.). *Algal Physiology and Biochemistry. Botanical Monographs.* Volume 10. University of California Press, Berkeley and Los Angeles. (Blackwell Scientific Publications, Oxford, London, Edinburgh, Melbourne). XI + (1) + 989 pp (sl. ment., tax.-syst.)

752. Stewart, W.D.P. and P.J. Bottomley. 1976. Nitrogenase of blue-green algae. pp 257–273 in W.E. Newton and C.J. Nyman. (Eds). *Proceeding of the* 1st International Symposium *on Nitrogen Fixation [June 3–7, 1974, Pullman, Washington].* Volume 1. Washington State University Press, Pullman, Washington. (8) + 311 pp. (biochem., sl. ment.)

753. Stoecker, R.R. 1971. Survival of blue-green algae under primitive atmospheric conditions. *Space Life Sci.* (Dordrecht) 3: 42–45. (photosyn., sl. ment.)

754. Syrett, P.J. 1988. Uptake and utilization of nitrogen compounds. pp 23–39 in L.J. Rogers and J.R. Gallon. (Eds). *Biochemistry of the Algae and Cyanobacteria. Proceedings of the Phytochemical Society of Europe,* No. 28. Clarendon Press, Oxford. (9) + pp X–XVI + (3) + pp 4–374. (physiol., sl. ment.)

755. Taddei, R. 1977. Alghe che vivono in ambienti fortemente acidi nella grotta dell'acqua fredda della Scraio. *Annu. Speleol. Club Alpino Ital. Sez. Napoli* (Napoli) 4(1976–1977): 27–32. (ecol., florist.)

756. Tappan, H. 1980. *The Paleobiology of Plant Protists.* W.H. Freeman and Company, San Francisco. XXI + (3) + 1028 pp. (sl. ment., tax.-syst.)

757. Thompson, A.S., J.C. Rhodes and I. Pettman. (Eds). 1988. *Culture Collection of Algae and Protozoa, Catalogue of Strains, 1988.* [5th edition]. Natural Environment Research Council, Culture Collection of Algae and Protozoa, Freshwater Biological Association, Ambleside, Cumbria. (1) + pp 2–164. (cultivat.)

-- Thompson, A.S., J.C. Rhodes and I. Pettman. (Eds). Please note: For the first, second and third editions of the Cambridge Collection of Algae and Protozoa, please see respectively: E.A. George. (Ed.). (1967); E.A. George (Ed.). (1967); E.A. George. (Ed.). (1976). For the forth edition please see: L.A. Asher and D.F. Spalding. (Eds). (1982)

758. Tilden, J.E. 1896. *American Algae:* Century II. [Minneapolis, Minnesota]. (2) + Specimen Nos 101–200 + Specimen Nos 12b, 12c, 32b, 35b, 47b, 92b on five page leaves. (ecol., florist., tax.-syst.)

759. Tilden, J.E. 1898. *American Algae:* Century III. [Minneapolis, Minnesota]. (2) + Specimen Nos 201–300 + Specimen No. 65b. (ecol., florist., tax.-syst.)

760. Tilden, J.E. 1898. Observations on some West American Thermal algae. *Bot. Gaz.* (Crawfordsville, Chicago) 25(2): 89–105 + pls VIII–X. (ecol., florist., tax.-syst.)

761. Tilden, J.E. 1902. *American Algae:* Century VI. [Minneapolis, Minnesota]. (2) + Specimen Nos 501–600 + Specimen Nos 154b, 498b. (ecol., florist., tax.-syst.)

762. Tilden, J.E. 1910. *Minnesota Algae.* Volume I. *The Myxophyceae of North America and Adjacent Regions Including Central America, Greenland, Bermuda, The West Indies and Hawaii. Report of the Survey,* Botanical Series VIII, Minneapolis. IV + 328 pp + pls 1–20. (ecol., florist., tax.-syst.)

763. Tilden, J.E. 1917. Synopsis of blue-green algae-

Myxophyceae. *Trans. Am. Microsc. Soc.* (Lancaster) 36(4): 179–266. (ecol., sl. ment., tax.-syst.)

764. Titz, W. 1969. Lothar Geitler. Zum 70. Geburtstag am 18. Mai 1969: Wissenschaftliche Veroffentlichungen Lothar Geitler. (Publication numbers 1–301). Österr. Bot. Zeitschr. (Wien) 116(1–5): 6–17. (literat.)

765. Totáro Aloj, E. 1975. Presenza di *Cyanidium caldarium* (?) nella Valle di Ansanto (Avellino). *Boll. Soc. Nat. Napoli* (Napoli) 83(1974): 35–43. (ecol., florist.)

766. Tôyama, S. 1974. Electron miscroscope studies on the morphogenesis of plastids IX. The phylogeny of photosynthetic apparatus with special reference to lamellar structure. *Nippon Shokubutsugaku Shūhō (Jpn. J. Bot.,* Tokyo) 20(7): 413–438. (cytol., evol., florist., ultrastr.)

767. Tôyama, S. 1975. Chloroplast no hōmaku no bisaikōzō to kigen. (The fine structure and origin of chloroplast envelopes). p. 133 in *Nihon Shokubutsugakukai dai 40(yonjū) kai Taikai Kenkyū Happyō Kiroku (Shōwa 50(Gojūnen)nen 11(Jūichigatsu) Gastu 1(tsjuitachi) Nichi 3(mikka) Nichi (Ōsaka)]. [Proceedings of the* 40th Annual Meeting *of the Botanical Society of Japan (Osaka 1975)].* Nihon Shokubutsugakukai (The Botanical Society of Japan). 212 pp. (cytol., ultrastr.)

768. Tôyama, S. 1975. Yōryokutai no kigen to shinka. [The origin and evolution of chloroplasts]. *Iden (Heredity,* Tokyo) 29(4): 33–40. (evol.)

769. Tôyama, S. 1980. Electron microscope studies on the morphogenesis of plastids VIII. Further studies on the fine structure of *Cyanidium caldarium* with special regard to the photosynthetic apparatus. *Cytologia* (Tokyo) 45(4): 779–790. (cytol., evol., florist., ultrastr.)

770. Tôyama, S. 1980. Yōryokutai no shinkateki kigen–kyōsei setsu no tachiba kara. [Evolutional origin of chloroplast–a standpoint of symbiotic theory]. *Kagaku To Seibutsu (Chem. Biol.,* Tokyo) 18(8): 518–526. (evol., ultrastr.)

771. Trench, R.K. 1982. Physiology, biochemistry, and ultrastructure of cyanellae. pp 257–288 in F.E. Round and D.J. Chapman. (Eds). *Progress in Phycological Research.* Vol. I. Elsevier Biomedical Press, Amsterdam, New York, Oxford. (11) + pp 2–383. (biochem., physiol., radiobiol., ultrastr.)

770. Tôyama, S. 1980. Yōryokutai no shinkateki kigen–kyōsei setsu no tachiba kara. [Evolutional origin of chloroplast–a standpoint of symbiotic theory]. *Kagaku To Seibutsu (Chem. Biol.,* Tokyo) 18(8): 518–526. (evol., ultrastr.)

771. Trench, R.K. 1982. Physiology, biochemistry, and ultrastructure of cyanellae. pp 257–288 in F.E. Round and D.J. Chapman. (Eds). *Progress in Phycological Research.* Vol. I. Elsevier Biomedical Press, Amsterdam, New York, Oxford. (11) + pp 2–383. (biochem., physiol., radiobiol., ultrastr.)

772. Troxler, R.F. 1965. *Studies on the Formation of Phycocyanin, Porphyrins, and a Blue Phycobilin in Wild-Type and Mutant Strains of* Cyanidium caldarium. Ph.D. Dissertation, Department of Botany, University of Chicago, Chicago. 44 pp. (biochem., photosyn., physiol.)

773. Troxler, R.F. 1971. An inexpensive technique for measuring carbon monoxide formation in plants. *Plant Physiol.* (Lancaster) 48(3): 376–378. (photosyn., physiol., radiobiol.)

774. Troxler, R.F. 1971. Mechanism of bile pigment formation in plants. p. 46 in *Abstracts of the Papers for the Annual Meeting of the American Society of Plant Physiologists held at the Asilomer Conference Grounds, Pacific Grove, California, August 22–26, 1971. Plant Physiol.* (Lancaster) 47(Suppl.): (2) + pp 1–52. (biochem., photosyn., physiol., radiobiol.)

775. Troxler, R.F. 1972. Synthesis of bile pigments in plants. Formation of carbon monoxide and phycocyanobilin in wild-type and mutant strains of the alga, *Cyanidium caldarium. Biochemistry* (Washington) 11(23): 4235–4242. (biochem., photosyn., physiol., radiobiol.)

776. Troxler, R.F. and L. Bogorad. 1966. Studies on the formation of phycocyanin, porphyrins, and a blue phycobilin by wild-type and mutant strains of *Cyanidium caldarium. Plant Physiol.* (Lancaster) 41(3): 491–499. (biochem., photosyn., physiol.)

777. Troxler, R.F. and L. Bogorad. 1967. Production of bile pigments from δ-aminolevulinic acid by *Cyanidium.* pp 421–426 in T.W. Goodwin. (Ed.). *Biochemistry of Chloroplasts.* Vol. II. *Proceedings of a NATO Advanced Study Institute Held at Aberystwyth, August, 1965.* Academic Press, London and New York. XVIII + 776 pp. (biochem., photosyn., physiol., radiobiol.)

778. Troxler, R.F. and A.S. Brown. 1970. Biosynthesis of phycocyanin *in vivo. Biochim. Biophys. Acta* (New York, Amsterdam) 215(3): 503–511. (biochem., photosyn., physiol., radiobiol.)

779. Troxler, R.F. and A.S. Brown. 1974. Metabolism of L-azetidine-2-carboxylic acid in the alga *Cyanidium caldarium. Biochim. Biophys. Acta* (New York, Amsterdam) 366(3): 341–349. (biochem., photosyn., physiol., radiobiol.)

780. Troxler, R.F. and A.S. Brown. 1975. Metabolism of δ aminolevulinic acid in red and blue-green algae. *Plant Physiol.* (Lancaster) 55(3): 463–467. (biochem., photosyn., physiol., radiobiol.)

781. Troxler, R.F. and A.S. Brown. 1979. Amino acid sequence of the phycocyanin α subunit from the alga, *Cyanidium caldarium.* p. 325 in *Abstracts of Papers. Federation of American Societies for Experimental Biology.* 263 Annual Meeting, *Dallas, Texas, April 1–10, 1979.* Society Grouping I. *April 1–5. American Society of Biological Chemists, American Society for Pharmacology and Experimental Therapeutics, American Institute of Nutrition. Fed. Proc. Fed. Am. Soc. Exp. Biol.* (Baltimore) 38(3): XII + pp 225–876. (biochem., photosyn., physiol., radiobiol.)

782. Troxler, R.F., A.S. Brown and S.B. Brown. 1979. Bile pigment synthesis in plants: mechanism of ^{18}O incorporation into phycocyanobilin in the unicellular rhodophyte, *Cyanidium caldarium.* J. Biol. Chem. (Baltimore) 254(9): 3411–3418. (biochem., photosyn., physiol., radiobiol.)

783. Troxler, R.F., A.S. Brown, M.M. Ehrhardt and G.D. Offner. 1981. Amino acid sequence of the phycocyanin β subunit from the unicellular rhodophyte, *Cyanidium caldarium,* p. 1676 in *American Society of Biological Chemists,* 72nd Annual Meeting, *St. Louis, Missouri, May 31–June 4, 1981. Abstracts of Papers, Indexes. Fed. Proc., Fed. Am. Soc. Exp. Biol.* (Baltimore) 40(6): (4) + pp V–VI + pp 1539–1977. (biochem., physiol., photosyn.)

784. Troxler, R.F., A.S. Brown, J.A. Foster and C. Franzblau. 1974. Properties and partial aminoacid sequences of the α and β subunits of phycocyanin from the alga, *Cyanidium caldarium*. p. 1446 in *Abstracts of Papers. American Society of Biological Chemists*, 65th Annual Meeting *and Biophysical Society*, 18th Annual Meeting *held jointly as the Biochemistry/Biophysical 1974 Meeting. Cosponsored by the Pan-American Association of Biochemical Societies, Minneapolis, Minnesota 2–7, June, 1974. Fed. Proc., Fed. Am. Soc. Exp. Biol.* (Baltimore) 33(5): (4) + pp 1219–1675. (biochem., photosyn., physiol., radiobiol.)
785. Troxler, R.F., A.S. Brown and H. P. Kost. 1978. Quantitative degradation of radiolabeled phycobiliproteins. Chromic acid degradation of c-phycocyanin. *Eur. J. Biochem.* (Berlin) 87(1): 181–189. (biochem., photosyn., physiol., radiobiol.)
786. Troxler, R.F., A.S. Brown, R. Lester and P. White. 1970. Bile pigment formation in plants. *Science* (Lancaster) 167(3915): 192–193. (biochem., photosyn., physiol., radiobiol.)
787. Troxler, R.F. and S.B. Brown. 1977. Mechanism of bile pigment synthesis in plants. p. 8 in *Program and Abstracts of Papers for the Annual Meeting of the American Society of Plant Physiologists and the Canadian Society of Plant Physiologists at the University of Wisconsin, Madison, Wisconsin, August 16 to 20, 1977. Plant Physiol.* (Lancaster) 59(6): (4) + pp 1–134. (biochem., photosyn., physiol., radiobiol.)
788. Troxler, R.F., S.B. Brown and K.M. Smith. 1979. Mechanism of red algal bile pigment formation. *Biol. Bull. (Mar. Biol. Lab.*, Woods Hole) 157(2): 400. (biochem., photosyn., physiol., radiobiol.)
789. Troxler, R.F. and J.M. Dokos. 1973. Formation of carbon monoxide and bile pigments in red and blue-green algae. *Plant Physiol.* (Lancaster) 51(1): 72–75. (biochem., photosyn., physiol., radiobiol., sl. ment.)
790. Troxler, R.F., M.M. Ehrhardt, A.S. Brown-Mason and G.D. Offner. 1981. Primary structure of phycocyanin from the unicellular rhodophyte *Cyanidium caldarium* II. Complete amino acid sequence of the β subunit. *J. Biol. Chem.* (Baltimore) 256(23): 12176–12184. [For section 'I' of the above title, see Offner, Brown-Mason, Ehrhardt and Troxler (1981)]. (biochem., photosyn., physiol.)
791. Troxler, R.F., B. Faris and C. Franzblau. 1973. Effect of azetidine-2-carboxylic acid on chlorophyll and phycocyanin synthesis in the alga *Cyanidium caldarium*. p. 28 in *Program and Abstracts of Papers for the Joint Annual Meeting of the American and Canadian Societies of Plant Physiologists at University of Calgary, Alberta, June 17–21, 1973. Plan Physiol.* (Lancaster) 51(Suppl.): (3) + pp 2–70. (biochem., photosyn., physiol., radiobiol.)
792. Troxler, R.F., J.F. Foster, A.S. Brown and C. Franzblau. 1975. The α and β subunits of *Cyanidium caldarium* phycocyanin: properties and amino acid sequences at the amino terminus. *Biochemistry* (Washington) 14(2): 268–274. (biochem., evol., photosyn., physiol.)
793. Troxler, R.F., L.S. Greenwald and B.A. Zilinskas. 1980. Allophycocyanin from *Nostoc* sp. phycobilisomes. Properties and amino acid sequence at the NH_2 terminus of the α and β subunits of allophycocyanins *I, II, III. J. Biol. Chem.* (Baltimore) 155(19: 9380–9387. (biochem., sl. ment.)
794. Troxler, R.F., P. Kelly and S.B. Brown. 1978. Phycocyanobilin synthesis in the unicellular rhodophyte *Cyanidium caldarium. Biochem. J.* (Liverpool) 172(3): 569–576. (biochem., photosyn., physiol.)
795. Troxler, R.F. and R. Lester. 1967. Bile pigment formation in plants. p. 245 in *Abstracts of Papers Submitted to the Annual National Meeting of the American Federation for Clinical Research. Atlantic City, New Jersey. Sunday, April 30, 1967. Clin. Res.* (Thorofare) 15(2): 195–382. (biochem., photosyn., physiol., radiobiol.)
796. Troxler, R.F. and R. Lester. 1967. Biosynthesis of phycocyanobilin. *Biochemistry* (Washington) 6(12): 3840–3846. (biochem., photosyn., physiol., radiobiol.)
797. Troxler, R.F. and R. Lester. 1968. Composition of phycocyanin: phycocyanobilin. p S-30 in *Program and Abstracts of Papers to be Presented at the Annual Meetings. University of Massachusetts, Amherst, Massachusetts, August 20–23, 1968. Plant Physiol.* (Lancaster) 43(Suppl.): S–1–S–55. (biochem., photosyn., physiol., radiobiol.).
798. Troxler, R.F. and R. Lester. 1968. Formation, chromophore composition, and labeling specificity of *Cyanidium caldarium* phycocyanin. *Plant Physiol.* (Lancaster) 43(10): 1737–1739. (biochem., photosyn., physiol., radiobiol.)
799. Troxler, R.F., S. Lin and G.D. Offner. 1989. Heme regulates expression of phycobiliprotein photogenes in the unicellular rhodophyte, *Cyanidium caldarium. J. Biol. Chem.* (Baltimore) 264(34): 20596–20601. (biochem., photosyn., physiol.)
800. Troxler, R.F. and G.D. Offner. 1978. δ-aminolevulinic acid synthesis in a pigmentless mutant of *Cyanidium caldarium*. p. 84 in *Programs and Abstracts of Papers for the Annual Meeting of the American Society of Plant Physiologists at the Virginia Polytechnic Institute and State University, Blacksburg, Virginia, June 25 to 29, 1978. Plant Physiol.* (Lancaster) 61(4 Suppl.): (4) + pp 1–122. (biochem., photosyn., physiol., radiobiol.)
801. Troxler, R.F. and G.D. Offner. 1979. δ-aminolevulinic acid synthesis in a *Cyanidium caldarium* mutant unable to make chlorophyll *a* and phycobiliproteins. *Arch. Biochem. Biophys.* (New York) 195(1): 53–65. (biochem., photosyn., physiol., radiobiol.)
802. Troxler, R.F. and G.D. Offner. 1981. δ-aminolevulinate synthesis in the alga, *Cyanidium caldarium*. p. 106 in *Program and Abstract of Papers for the Annual Meeting of the American Society of Plant Physiologists at Laval University, Ste.-Foy, Quebec, Canada, June 14–18, 1981. Plant Physiol.* (Lancaster) 67(4 Suppl.): (4) pp 1–169. (biochem., photosyn., physiol., radiobiol.)
803. Trubachev, I.N. and I͡u, I. Ba͡ianova. 1983. Soderzhanie vitaminov v odnok letochnykh organizmakh vozmozhnykh komponentov szho. (Vitamin content in unicellular organisms of possible components of support systems). pp 157–159 *in* L.I. Rubenchik. (Ed.). XI Vsesoiuznoe Rabochee Soveshchanie *po Voprosam Krugovorota Veshchstv v zamknutykh Sistemakh na Osnove Zhiznedeiatel'nosti nizshikh organizmov.* (Kanev, *Sentiabr' 1981 G)* [(Proceedings of the 12th) all Union Conference *on Problems of Metabolism in Closed Systems as Ex-*

emplifiecd by the Activity of Lower Organisms]. Naukova Dumka, Kiev. (2) + pp 3–180. (biochem., physiol.)

804. Trubachev, I.N. and V. Mel"nik. 1978. O soderzhanii nukleinovykh kislot v nekotorykh pishchevykh produktakh i kormakh. (Nucleic acid content in some food products and feeds). *Vopr. Pitan.* (Moskva) 1978(5): 79–80. (biochem.)
805. Ueda, K. and Y. Chida. 1987. Chloroplast structure of *Cyanidium caldarium* shown by freeze-substitution. *Br. Phycol. J.* (London) 22(1): 61–65. (cytol., evol., ultrastr.)
806. Ueda, K. and J. Yokochi. 1981. Structure of *Cyanidium caldarium. Shokubutsugaku Zasshi (Bot. Mag.,* Tokyo) 94(1034): 159–164. (cytol., tax.-syst., ultrastr.)
807. Ulik, T.M. 1975. Adaptation of *Cyanidium caldarium* to variation in light intensity and CO_2 stress. p. 15 in *Program with Abstracts of Papers Scheduled for the Meetings of the Phycological Society of America at Oregon State University, Corvallis, Oregon, August 17–23. J. Phycol.* (Baltimore, Lawrence) 11(Suppl.): (2) + pp 3–24. (biochem., physiol.)
808. Ulik, T.M. 1976. *Adaptation of* Cyanidium caldarium *to Variation in Light Intensity and CO_2 Stress.* Ph.D. Dissertation, University of Montana, Missoula. 162 pp. (biochem., physiol.)
809. Ulik, T.M. 1976. Adaptation of *Cyanidium caldarium* to variation in light intensity and CO_2 stress. *Diss. Abstr. Int. B, Sci. Eng.* (Ann Arbor) (37(5): 2046. (biochem., physiol.)
810. Uptis, V.V., A.F. Nollendorf and D.S. Pakalne. 1980. Die Bedeutung des Nickels für Kultur der Mikroalgen. SS 171–175 in M. Anke, H.-J. Schneider and C. Bruckner. (Hrsg.). 3. *Spurenelement-Symposium Nickel, der Karl-Marx-Universität Leipzig und der Friedrich-Schiller-Universität Jena vom 7. bis 11. Juli 1980 in Jena, Hotel 'Schwarzer Bär' DDR.* Wissenschaftbereich Teirernährumgschemie der Karl-Marx-Universität Leipzig; Urologische Klinik und Poliklinik der Friedrich-Schiller-Universität Jena; VEB Kongress- und Werbedruck Oberlungwitz. (4) + SS V–VIII + 394 SS. (ecol., physiol.)
811. Upitis, V.V., A.F. Nollendorf, D.S. Pakalne and S.A. Trama. 1986. Optimizatsii͡a mineral'nogo pitanii͡a mikrovodorosleĭ. *Latv. PSR Zinat. Akad. Vestis* (Riga) 1986(9): 49–58. (physiol.)
812. Valentin, K. and K. Zesche. 1990. Structure of the rubisco operon from the unicellular red alga *Cyanidium caldarium*: evidence for a polyphyletic origin of the plastids *Molecul. Genet.* (Berlin) 222(2–3): 425–430. (biochem., evol.)
813. Valentin, K. and K. Zetsche. 1989. The genes of both subunits of ribulose-1, 5-biophosphate carboxylase constitute an operon on the plastome of a red alga. *Curr. Genet.* (Berlin, Heidelberg) 16(3): 203–210. (biochem., sl. ment.)
814. Vallentyne, J.R. 1963. Environmental biophysics and microbial ubiquity. pp 342–352 *in* H.E. Whipple. (Ed.). *Life-like Forms in Meterorites and the Problems of the Morphology of Fossile and recent Protobionta. Ann. N. Y. Acad. Sci.* (New York) 108(2): 342–352. (cultivat., ecol., sl. ment.)
815. Venkataraman, G.S. 1969. *The Cultivation of Algae.* Indian Council of Agricultural Research, New Delhi. XVIII + 319 pp. (cult., tax.-syst.)
816. Venkataraman, G.S., S.K. Goyal, B.D. Kaushik and P. Roychoudhury. 1974. *Algae: Form and Function.* Today & Tomorrow's Printers & Publishers, New Delhi. XV + (1) + 562 pp. (ecol., physiol., sl. ment., tax.-syst.)
817. Volk, S.L. and N.I. Bishop. 1968. Photosynthetic efficiency of a phycocyanin-less mutant of *Cyanidium. Photochem. Photobiol.* (New York) 8(3): 213–221. (biochem., photosyn., physiol.)
818. Vouk, V. 1950. *Ein Grundriss zur Balneobiologie der Thermen.* Verlag Birkhauser, Basel. (7) + pp 7–88. (ecol.)
819. Vouk, V. 1954. Die Probleme der Biologie der Thermen im Lichte der neuesten Forschungen. pp 103–113 in H. Osvald and E. Åberg. (Eds). *Proceedings of the* Seventh *International Botanical Congress, Stockholm, July 12–20, 1950.* Almqvist & Wiksell, Stockholm; The Chronica Botanica Co., Waltham, Mass. 899 pp. (ecol., florist.)
820. Walbridge, C.T., Jr. 1963. *Immunochemical Reactions of Various Phycocyanins with Serum Immune to* Cyanidium caldarium *Phycocyanin.* M.S. Thesis, Department of Botany, University of Chicago, Chicago. V + 27 pp. (biochem., evol., photosyn., physiol., tax.-syst.)
821. Wang, J. and T. Martinez. 1990. Uptake of redox couples by algae immobilized on porous cellulosic films. *Electrochim. Acta* (Oxford) 35(9): 1377–1381. (ecol., physiol., ultrastr.)
822. Wang, J., T. Martinez and D. Darnall. 1989. Electrocatalysis at algae modified electrodes. *J. Electroanal. Chem. Interfacial Electorchem.* (Lausanne) 259(1–2): 295–300. (ecol., physiol.)
823. Warden, J.T., Jr., P. Mohanty and J.R. Bolton. 1974. Flash photolysis-electron spin resonance studies of the dynamics of photosystem I: III temperature dependence of the decay of signal I. *Biochem. Biophys. Res. Commun.* (New York) 59(3): 872–878. (photosyn., physiol.)
824. Watanabe, A. 1960. List of algal strains in collection at the Institute of Applied Microbiology., University of Tokyo. *J. Gen. Appl. Microbiol.* (Tokyo) 6(4): 283–292. (cultivat.)
825. Weber, A., M. Kiehn and S. Sontag. 1989. Lothar Geitler on the occasion of his 90th birthday on May 18th 1989: List of publications of Lothar Geitler (since 1979); Subjects and taxa treated in the scientific work of Lothar Geitler. (Publication numers 345–359). *Plant Syst. Evol.* (Wien) 164: 5–15. (literat.)
826. Weed, W.H. 1889. Formation of travertine and siliceous sinter by the vegetation of hot springs. *Ann. Rep. U.S. Geol. Surv.* (Washington) 9: 613–676. (ecol., florist., physiol.)
827. Wegmann, K. and N. Pristavu. 1978. Influence of cycloserine on the photosynthetic ^{14}C incorporation into amino acids in various algae. *Biochem. Physiol. Pflanz.* (Jena) 173(1): 86–90. (biochem., physiol., radiobiol.)
828. Wehrmeyer, W. and H. Schneider. 1975. Elektronenmikroskopische Untersuchungen zur reversiblen Veränderung der Chloroplastenfeinstruktur von *Rhodella violacea* bei Stickstoffmangel. Electron microscopic studies on reversible changes of chloroplast fine structure of *Rhodella violacea* following nitrogen deficiency. *Biochem. Physiol. Pflanz.* (Jena) 168(5/6): 519–532. (sl. ment., ultrastr.)
829. Weinstein, J.D. and S.I. Beale. 1983. Biosynthesis of heme precursors soley from glutamate in the unicellular red alga *Cyanidium caldarium.* p. 19 in *Program and Abstract of*

Papers for the Annual Meeting of the American Society of Plant Physiologists at Colorado State University, Fort Collins, August 7-11, 1983. Plant Physiol. (Lancaster) 72(Suppl. 1): (4) + pp 1-192. (biochem., photosyn., physiol.)

830. Weinstein, J.D. and S.I. Beale. 1984. Biosynthesis of protoheme and heame *a* precursors solely from glutamate in the unicellular red alga *Cyanidium caldarium. Plant Physiol.* (Lancaster) 74(1): 146–151. (biochem., photosyn., physiol.)
831. Weinstein, J.D., S.M. Mayer and S.I. Beale. 1986. Stimulation of δ-aminolevulinic acid formation in algal extracts by heterologous RNA. *Plant Physiol.* (Lancaster) 82(4): 1096–1101. (biochem., photosyn., physiol.)
832. Weinstein, J.D., M.A. Schneegurt and S.I. Beale. 1987. Biosynthetic precursors of δ-aminolevulinic acid in plants and animals. pp 431–434 in J. Biggins. (Ed.). *Progress in Photosynthesis Research,* Vol. 4. *Proceedings of the* VIIth International Congress *on Photosynthesis, Providence, Rhode Island, U.S.A., August 10–15, 1986.* Martinus Nijhoff Publishers, Dordrecht, Boston, Lancaster. (4) + pp V-XXVII + (1) + 858 pp. (biochem., physiol., radiobiol.)
833. West, G.S. 1904. West Indian freshwater algae. *J. Bot., Br. Foreign* (London) 42(502): 281–294 + pl. 464. (florist., tax.-syst.)
834. Whitton, B.A. 1970. Toxicity of heavy metals to freshwater algae: a review. *Phykos* (New Delhi) 9(1/2): 116–125. (ecol., physiol.)
835. Whitton, B.A. 1975. Algae. pp 81–105 in B.A. Whitton. (Ed.). *River Ecology. Studies in Ecology.* Volume 2. University of California Press, Berkeley and Los Angeles; Blackwell Scientific Publications, Oxford. X + 725 pp. (ecol., physiol.)
836. Wille, N. 1925. Vorarbeiten zu einer Monographie der Chroococcaceen. *Nyt Mag. Naturvidensk.* (Kristiania) 62: 169–209 + Taf. VI–XIV. (tax.-syst.)
837. Williams, V.P., P. Freidenreich and A.N. Glazer. 1974. Homology of amino-terminal regions of *c*-phycocyanins from a prokaryote and a eukaryote. *Biochem. Biophys. Res. Commun.* (New York) 59(2): 462–466. (biochem., evol., photosyn., physiol., tax.-syst.)
838. Wittmann, H.G. 1970. A comparison of ribosomes from prokaryotes and eukaryotes. pp 55–76 in H.P. Charles and B.C.J.G. Knight. (Eds). *Organization and Control in Prokaryotic and Eukaryotic Cells.* Twentieth Symposium *of the Society for General Microbiology Held at Imperial College, London, April 1970.* Published for the Society for General Microbiology at the University Press, Cambridge. XI + (1) + 457 pp. (biochem., sl. ment.)
839. Wittrock, V.B. and O. Nordstedt. 1883. *Algae Aquae Dulcis Exsiccatae Praecipue Scandinaviae, quas Adjectis Algis Marinis Chlorophyllaceis et Phycochromaceis Distribuerunt Viet Wittrock et O. Nordstedt. Adjuvantibus F. Cohn, Ch. Flahault, F. Hauck, C.J. Johanson, G. Lagerheim, A. Lofgren, P. Richter.* Fasciculus 12. N:ris 551–600. Central-Tryckeriet, Holmiae. (4) + Specimen Nos 551–600. (florist., tax.-syst.)
840. Wittrock, V.B. and O. Nordstedt. 1883. *Algae Aquae Dulcis Exsiccatae Praecipue Scandinavicae quas Adjectis Algis Marinis Chlorphyllaceis et Phycochromaceis. Distribuerunt Viet Wittrock et Otto Nordstedt. Adjuvantibus F. Cohn, M. Foslie, A. Grunow, C.J. Johanson, G. Lagerheim, A. Lofgren, P. Richter, F. Wolle.* Fasiculus 11. N:ris 501–550. Central-Tryckeriet, Holmiae. (4) + Specimen Nos 501–550. (florist., tax.-syst.)
841. Wittrock, V.B. and O. Nordstedt. 1884. *Algae Aquae Dulcis Exsiccatae Praecipue Scandinavicae quas Adjectis Algis Marinis Chlorophyllaceis et Phycochromaceis. Distribuerunt Viet Wittrock et Otto Nordstedt. Adjuvantibus J. Arechavaleta, P.T. Cleve, W.G. Farlow, Ch. Flahault, M. Foslie, A. Hansgirg, F. Hauck, W. Joshua, G. Lagerheim, P. Richter, F. Wolle.* Fasciculus 14. N:ris 651–700. Excudit Isac Marcus' Boktr.-Aktiebolag, Stockholmae. (4) + Specimen Nos 651–700. (florist., tax.-syst.)
842. Wollman, F.A. 1979. Ultrastructural comparison of *Cyanidium caldarium* wild type and III-C mutant lacking phycobilisomes. *Plant Physiol.* (Lancaster) 63(2): 375–381. (biochem., photosyn., physiol., ultrastr.)
843. Wood, B.J.B. 1974. Fatty acids and saponifiable lipids. pp 236–265 in W.D.P. Stewart. (Ed.). *Algal Physiology and Biochemistry. Botanical Monographs.* Volume 10. University of California Press, Berkeley and Los Angeles. (Blackwell Scientific Publications, Oxford, London, Edinburgh, Melbourne). XI + (1) + 989 pp. (tax.-syst.)
844. Wood. J.M. and F.E. Engle. 1989. *Recovery of Metals from Mine Water by Algae.* Canadian Patent/Brevit Canadien, Patent Office, Ottawa-Hull. 29 pp. (ecol., physiol.)
845. Wood, J.M. and H.K. Wang. 1983. Microbial resistance to heavy metals. *Environ. Sci. Technol.* (Washington) 17(12): 582A–590A (ecol., florist., physiol.)
846. Work, E. and D.L. Dewey. 1953. The distribution of α, ϵ-diaminopimelic acid among various micro-organisms. With a note by R.L.M. Synge on the 'occurrence of diaminopimelic acid in some intestinal micro-organisms from farm animals.' *J. Gen. Microbiol.* (Cambridge) 9: 394–409 + pl. 1. (biochem., sl. ment., tax.-syst.)
847. Yoneda, Y. 1940. Nihonsan onsen shokubutsu no kenkyū VII. Hokkaidō no onsen shokubutsu kenkyū (III). [Studies on the thermal-flora of Japan. VII. Studies on the thermal algae of Hokkaido (3)]. *Shokubutsu Bunrui Chiri (Acta Phytotaxon. Geobot.,* Kyoto) 9(3): 192–202. (ecol., florist.)
848. Yoneda, Y. 1941. Nihonsan onsen shokubutsu no kenkyū IX. Hokkaidō no onsen shokubutsu kenkyū (IV). [Studies on the thermal-flora of Japan. IX. Studies on the thermal algae of Hokkaido (4)]. *Shokubutsu Bunrui Chiri (Acta Phytotaxon. Geobot.,* Kyoto) 10(3): 159–171. (ecol., florist.)
849. Yoneda, Y. 1941. Nihonsan onsen shokubutsu no kenkyū XIII. Hokkaidō no onsen shokbutsu kenkyū (V). [Studies on the thermal-flora of Japan. XIII. Studies on the thermal algae of Hokkaido (V)]. *Shokubutsu Bunrui Chiri (Acta Phytotaxon. Geobot.,* Kyoto) 10(4): 229–253. (ecol., florist.)
850. Yoneda, Y. 1952. A general consideration of the thermal Cyanophyceae of Japan. *Mem. Coll. Agric. Kyoto Univ.* (Kyoto) 62: 1–20. (ecol.).
851. Yoneda, Y. 1952. Nihonsan onsen shokubutsu no kenkyū XXXI. Hokkaidō ni okeru onsen sōrui no kenkyū (VI).

(Konbu onsengun). [Studies on the thermal-flora of Japan XXXI. Studies on the thermal algal of Hokkaido (VI). (Konbu hot springs group)]. *Onsen Kagaku (Ber. Jpn. Ges. Balneol.,* Tokyo) 5(1): 11–15. (ecol., florist.)

-- Yurina, N.P. Please see: I͡urina, N.P.

852. Zagon, J. and P. Brandt. 1985. Coordination between thylakoid differentiation and cell division. *Symbiosis* (Philadelphia, Rehovot) 1(1): 9–18. (biochem., evol., photosyn., physiol.)

853. Zavarzin, G.A. and G.A. Karpov. 1982. Rol" bacterial'nykh faktorov v sovremenom gidrotermal'nom mineraloobrazovanii kal'dery Uzon. (Role of bacterial factors in recent hydrothermal mineral formation at the Uzon caldera). *Dokl. Akad. Nauk SSSR* (Moskva, Leningrad) 264(1): 244–247. (ecol., florist.)

854. Zenvirth, D., M. Voloka and A. Kaplan. 1985. Photosynthesis and inorganic carbon accumulation in the acidophilic alga *Cyanidioschyzon merolae. Plant Physiol.* (Lancaster) 77: 237–239. (biochem., photosyn., physiol.)

Addendum to the bibliography

The following 86 references came to the attention of the authors after the above, numbered listing had been completed. To avoid the renumbering of the above bibliographical entries and the accompanying index which follows, the authors have elected to include these entries in the form of an addendum; this also permits the index numbers of these additional references to fall to their correct numerical sequence in the index.

855. Beal, S.I. and J. Cornejo. 1990. Seperation and characterization of the enzymes catalysing transformation of biliverdin to phycocyanobilin in *Cyanidium caldarium.* p. 62 *in Scientific Program (Abstracts of Papers) for the Annual Meeting of the American Society of Plant Physiologists, Indianapolis Convention Center and the Westin Hotel, Indianapolis, Indiana. Plant Physiol.* (Lancaster) 93(1 suppl.): (4) + 175 pp. (biochem., photosyn., physiol.)

856. Beale, S.I. and J. Cornejo. 1991. Biosynthesis of phycobilins, ferredoxin-mediated reduction of biliverdin catalyzed by extracts of *Cyanidium. J. Biol. Chem.* (Baltimore) 266(33): 22328–22332. (biochem., photosyn., physiol.)

857. Beale, S.I. and J. Cornejo. 1991. Biosynthesis of phycobilins, 3(Z)-phycoerythrobilin and 3(Z)-phycocyanobilin are intermediates in the formation of 3(E)-phycocyanobilin from biliverdin IX. *J. Biol. Chem.* (Baltimore) 266(33): 22333–22340. (biochem., photosyn., physiol.)

858. Beale, S.I. and J. Cornejo. 1991. Biosynthesis of phycobilins. 15,16-dihydrobiliverdin IX is a partly reduced intermidiate in the formation of phycobilins from bilverdin IX. *J. Biol. Chem.* (Baltimore) 266(33): 22341–22345. (biochem., photosyn., physiol.)

859. Brandt, P. 1991. *Evolution der eukaryotischen Zelle.* Georg Theime Verlag, Stuttgart. 157 SS. (biochem., evol., physiol.)

860. Brown, S.B., J.A. Holroyd, D.I. Vernon, Y.D. Shim and K.M. Smith. 1989. The biosyntesis of the chromophore of phycocyanin: pathway of reduction of biliverdin to phycocyanobilin. *Biochem. J.* (Liverpool) 261(1): 359–263. (biochem., photosyn., physiol., radiobiol.)

861. Des Marais, D.J., J. Bauld, A.C. Palmisano, R.S. Summons and D.W. Ward. 1991. The biochemistry of carbon in modern microbial mats. pp 299–308 *in* J.W. Schopf and C. Klein. (Eds.). *The Proterzoic Biosphere: a Multidisciplinary Study.* Cambridge University Press, Cambridge. (4) + Frontispiece (1) + (3) + pp VII–XXIV + (6) + pp 7–1348. (biochem., florist., physiol., radiobiol., sl. ment.)

862. Di Martino Rigano, V., C. Di Martino, V. Vona, S. Esposito and C. Rigano. 1992. Amino acid pools in nutrient limited *Cyanidium caldarium* and responses to resupply. *Phytochemistry* (Oxford) 31(6): 1911–1916. (biochem., physiol.)

863. Di Martino Rigano, V., V. Vona, S. Esposito, C. Di Martino and C. Rigano. 1991. Carbon skeleton sources for ammonium assimilation in N-sufficient and N-limited cells for *Cyanidium caldarium* (Rhodophyta). *J. Phycol.* (Baltimore, Lawrence) 27: 220–223. (biochem., physiol.)

864. Elela, S.A. and R.N. Nazar. 1992. Extended secondary structure as a basis of increased RNA stability in a thermophilic alga *Cyanidium caldarium. Biochim. Biophys. Acta* (New York, Amsterdam) 130(3): 339–342. (biochem., evol., physiol.)

865. Farr, E.R., J.A. Leussink and G. Zijlstra. 1986. *Index Nominum Genericorum (Plantarum).* Supplementum I. Bohn, Scheltema & Holkema, Utrecht, Antwerpen; Dr. W. Junk b. v. Publishers, The Hague. (*Regnum Vegetibile* 113). Cover (2) + (4) + pp VII–XV + (1) + 126 pp. (tax.-syst.)

866. Fliermans, C.B. and T.D. Brock. 1972. Ecology of sulfur-oxidizing bacteria in hot acid soils. *J. Bacterol.* (Baltimore) 111(2): 343–350. (evol., physiol., radiobiol., sl. ment.)

867. Fredrick, J.F. 1992. The biosynthesis of storage branched glucans as markers of protist evolution. p. 69 *in* Endocytobiology V. *Programs & Abstracts. Uji-Kyoto, Japan. June 23–27, 1992. I.S.E. Phillips University-Japan.* Organized by the International Society of Endocytobiology, Uji-Kyoto. Cover with title (2) + (5) + pp 3–121. (biochem., evol., physiol.)

868. Fukuda, I. 1989. Origin of eukaryotic cells from the view point of organel DNA. *Iden (Heredity,* Tokyo) 43(6): 30–34. (biochem., evol.)

869. Fukuda, I. and H. Nagashima. 1992. Seveal new strains of thermal alga *Cyanidioschyzon* as the most prinmitive systems, p. (6) in *Algae as Experimental systems* II. *November 14, 15, Denver, Colorado.* The American Society for Cell Biology, Bethesda. 13 unumbered pages. (evol., sl. ment., tax.-syst.)

870. Fukuda, I. and Nagashima. 1992. Several new strains of thermal alga *Cyanidioschyzon* as the most primitive experimental systems. p. 10 in *On Site Addendum,* ASCB Thiry-second Annual Meeting, *The American Society for Cell*

Biology, Denver, Colorado, November 15–19, 1992. The American Society for Cell Biology, Bethesda. pp 1–5 + (1) + pp 6–13 + (4). (evol., sl. ment., tax.-syst.)

871. Fukuda, I. and H. Nagashima. 1992. Several new strains of thermal alga *Cyanidioschyzon* as the most primitive experimental systems. p. (22) in *Special Poster Session, The American Society for Cell Biology* Thirty-Second Annual Meeting. *Colorado Convention Center, [Denver], Thursday, November 29, 1992, 12:00 Noon-3:00 PM, Halls B/C.* The American Society for Cell Biology, Bethesda, 24 unnumbered pages. (evol., florist., tax.-syst.)
872. Gerasimenkov, L.M. and I͡u. M. Miller. 1985. Gazovyi ogmen termofil'nogo i at͡sidofil'nogo i fototrofa *Cyanidium caldarium* (Gaseous exchange of the extreme thermophilic, adidophilic, phototrophic organism *Cyanidium caldarium*). *Mikrobiologi͡ia* (Moskva) 54(5): 693–698. (florist., physiol.)
873. Gerasimenko, L.M. and I͡u. Miller. 1985. Gas exchange of the thermophilic acidophilic prototrophic organism *Cyanidium caldarium. Microbiology* (USSR, New York) 54(5): 551–555. (florist., physiol.)
874. Gross, W. 1991. Vanadate-sensitive ATPase in the acidophilic red alga *Galdieria sulphuraria.* p. 26 in Fourth International Phycological Congress. *Abstracts of Papers Scheduled for the 1991 Meeting of the* IV International Phycological Congress and the Phycological Society of America *at Duke University, Durham, North Carolina, USA, 4–10 August 1991. The Program will be distributed to Registrants. Duke University, North Carolina, USA. J. Phycol.* (Baltimore, Lawrence) 27(3 Suppl.): 1–91. (biochem., physiol.)
875. Houghton, J.D., A. Livesley and S.B. Brown. 1990. Adaptation to gabaculine inhibition of 5-ALA synthesis in *Cyanidium caldarium.* p. 133 *in* T. Owen, G. Jones and J. Houghton. Botanical and microbiological aspects of porphyrins. pp 131–135 in M.R. Moore. (Ed.). *A Century of Porphyria. Molec. Aspects Med.* (Oxford) 16(1–2): 1–160. (biochem., physiol.)
876. Houghton, J.D., A. Wilks and S.B. Brown. 1990. Haem oxygenase of *Cyanidium caldarium.* pp 133–134 *in* T. Owen, G. Jones and J. Houghton. Botanical and microbiological aspects of porphyrins. pp 131–135 *in* M.R. Moore. (Ed.). *A Century of Porphyria. Molec. Aspects Med.* (Oxford) 11(1–2): 1–160. (biochem., physiol.)
877. I͡urina, N.P., G.V. Karakashev, N.V. Karapeti͡an, M.S. Odint͡sova and A.N. Bach. 1991. Composition and biosynthesis of the thylakoid membrane polypeptides in the red alga *Cyanidium caldarium*: Comparison with the thylakoid polypeptide composition of higher plants and cyanobacteria. *Photosynth. Res.* (Amsterdam) 30(1): 15–33. (biochem., evol., photosyn., physiol., radiobiol.)
878. Jorgensen, B.B. 1991. Chemical analyses. pp 843-953 *in* J.W. Schopf and C. Klein. (Eds). *The Proterozoid Biosphere: a Multicisciplinary Study.* Cambridge University Press, Cambridge, New York, Melbourne. (4) + Frontispiece (1) + (3) + pp VII–XXIV + (6) + pp 7–1348. (biochem., florist., sl. ment.)
879. Kessler, U., U. Maid and K. Zetsche. 1992. An equivalent to bacterial ompR genes is encoded on the plastid genome plastid genome of algae. *Plant Molecul. Biol.* (Dordrecht) 18: 777–780. (biochem., evol., physiol.)
880. Kumano, S. 1977–1981. Rhodophyceae. pp 157–175 *in* H. Hirose. (Representative). *Illustrations of Japanese Algae.* Uchidarokakuho Publishing Co., LTD, Tokyo. (2) + unnumbered plates with 64 figs + (2) + pp 3–933 + (3) and with 255 plates in the text and numerous text figures. (cytol., ecol., florist., morphol., tax.-syst.)
881. Kuroiwa, T. 1989. The nuclei of cellular organelles and the formation of daughter organelles by the 'plastid-dividing ring'. *Bot. Mag.* (Tokyo) 102(1066): 291–329. (biochem., cytol., physiol.)
882. Kuroiwa, T. and T. Mita. 1991. Formation of a plastid-driving ring in *Cyanidium caldarium.* p. 264 *in* The Forty-seventh Annual Meeting *of the Japanese Society of Electron Microscopy. J. Electron Microscopy* (Tokyo) 40(4): 234–300. (cytol., morphol., ultrastr.)
883. Levitt, J. 1972. *Responses of Plants to Environmental Stresses.* Academic Press, New York, Lonond. (4) + pp V–XII + (2) + 697 pp. (biochem., ecol., physiol.)
884. Levitt, J. 1980. *Responses of plants to Environmental stresses.* 2nd Edition. Volume I. *Chilling, Freezing, and High Temperature Stresses.* Academic Press. (A Subsidiary of Harcourt Brace Jovanovich, Publishers), New York, London, Toronto, Sydney, San Francisco. (4) + pp V–XI + (2) + pp 3–479. (biochem., ecol., physiol.)
885. Lüttke, A. 1991. On the origin of chloroplasts and rhodoplasts: protein sequence comparison. *Endocytobios. Cell Res.* (Tübingen) 8: 75–82. (biochem., evol., physiol.)
886. Maid, U., R. Steinmüller and K. Zetsche. 1992. Structure and expression of a plastid-encoded groEL homogous heatshock gene in a thermophilic unicellular red alga. *Cur. Genet.* (Berlin, Heidelberg) 21(6): 521–525. (biochem., evol., physiol.)
887. Maid, U. and K. Zetsche. 1990. Nucleotide sequence of the plastid 16S rRNA gene of the red alga *Cyanidium caldarium. Nucleic Acid Res.* (Oxford) 18(13): 3996. (biochem., physiol.)
888. Maid, U. and K. Zetsche. 1991. Structural features of the plastid ribosomal RNA operons of two red algae: *Antithamnion* sp. and *Cyanidium caldarium. Plant Molecul. Biol.* (Dordrecht) 16(4): 537–546. (biochem., evol., physiol.)
889. Markowicz, Y. and S. Loiseaux-de-Goer. 1991. Plastid genomes of the Rhodophyta and Chromophyta constitute a distinct lineage which differs from that of the Chlorophyta and have a composite phyhlogenetic orgin, perhaps like that of the Euglenophyta. *Curr. Genet.* (Berlin, Heidelberg) 20(5): 427–430. (biochem., evol., physiol., tax.-syst.)
890. Matin, A. 1990. Keeping a neutral cytoplasm; the bioenergetics of obligate acidophiles. pp 307–318 *in* H.P. Durre. (Ed.). *Papers Presented at the Workshop Organized by the Royal Netherlands Academy of Arts and Sciences and the Microbiology Department of the University of Groningen* 'Adaptation of Microorganisms to Extreme Environments' *Haren, The Netherlands, June 1989. FEMS Microbiol. Rev.* (Amsterdam) 75(2–3) [6/2+3)] Special Issue: Cover with title (2) + (2) + pp 103–348. (ecol., sl. ment.)
891. Matsubara, H., T. Hase, S. Wakabayashi and D. Wada. 1980. Structure and evolution of chloroplast- and bacterial-type ferrodoxins. pp 245–266 *in* D.S. Sigman and

M.A.B. Brazier. (Eds). *The Evolution of Protein Structure and Function. A Symposium in Honor of Professor E.L. Smith. UCLA [University of California Los Angeles] Forum in Medical Sciences* No. 21. Academic Press, New York, London, Toronto, Sydney, San Francisco. (2) + Frontispiece (1) + (3) + pp VII–XVII + (1) + 350 pp. (biochem., evol., physiol., tax.-syst.) Elsevier/North-Holland Biomedical Press. Amsterdam, New York, Oxford. (4) + pp V–X + 461 pp. (biochem., physiol., radiobiol.)

892. Nagashima, H., K. Ueki, I. Fukuda and J. Seckbach. 1992. Further investigations of several new strains of thermal alga *Cyanidioschyzon* as one of the most primitive eukaryotes. p. 93 *in* Endocytobiology V. *Programs & Abstracts. Uji-Kyoto, Japan. June 23–27, 1991. I.S.E. Phillips University-Japan.* Organized by the International Society of Endocytobiology, Uki-Kyoto. Cover with title (2) + (5) + pp 3–121. (biochem., cytol., evol., physiol.)

893. Ohta, H., N. Nagashima, S. Kawano and I. Kuroiwa. 1992. Isolation of the chloroplast DNA and the sequence of the trnK gene from *Cyanidium caldarium* strain RK-1. *Plant Cell Physiol.* (Tokyo) 33(5): 657–661. (biochem., evol., photosyn., physiol.)

894. Rothchild, L.N. and R.L. Mancinelli. 1992. Diurnal patterns of carbon fixation and nitrogen cycling in diverse algal mat communities. p. (13) in *Algae as Experimental Systems* II., *November 14, 15, 1992, Denver, Colorado.* The American Society for Cell Biology, Bethesda. 13 Unnumbered pages. (biochem., ecol., photosyn., physiol., radiobiol.)

895. Seckbach, J. 1989. Origin of plant cells: red algal eukaryogenesis via the Cyanidiales. pp 212–213 in Z. Masinovski. (Complier). *Abstracts. The Origin of Life.* The Sixth ISSOL Meeting and the Ninth International Conference, *Prague, Czechoslovakia, July 1989. Host: Czechoslovak Academy of Sciences in Co-operation with the faculty of Science, J.E. Purkyne University Brno.* ISSOL. International Society for the Study of the Origin of Life. Laboratory of Evolutionary Biology, Czechoslovak Academy of Sciences, Prague. (26) + 372 pp.

896. Seckbach, J. 1992. The Cyanidiophyceae and the 'anomalous symbiosis' of *Cyanidium caldarium.* pp 399–426 *in* R. Reisser. (Ed.). *Algae and Symbioses: Plants, Animals, Fungi, Viruses, Interactions Explored.* Biopress Limited, Bristol (2) + pp ii–xii + 746 pp. (biochem., cytol., evol., ecol., morphol., physiol., tax.-syst., ultrastr.)

897. Seckbach, J., I. Ikan, H. Nagashima, and J. Fukuda. 1992. A new phylogenetic proposal for the hot spring algae: a symbiotic origin for *Cyanidioschyzon* (the lowest eukaryotic alga) and autogenous evolution for *Cyanidium* and *Galdieria* (Cyanidiophyceae, Prerhodophyta) as based on several evolutionary markers including recent sterol content. p. 45 *in* Endocytobiology V. *Programs & Abstracts. Uji-Kyoto, Japan. June 23–27, 1992. I.E.S. Phillips University-Japan.* Organized by the International Society of Endocytobiology, Uji-Kyoto. Cover with title (2) + (5) + pp 3–121. (biochem., evol., physiol., tax.-syst.)

898. Seckbach, J., R. Ikan and F.D. Ott. 1992. Minireview and a new taxonomic proposal for the Cyanidiophyceae. *Israel J. Bot.* (Jerusalem) 41(4–6): 319. (biochem., cytol., ecol., evol., physiol., tax.-syst.)

899. Sen͡tsova, O. I͡u. 1991. O raznoobrazii odnokletochnykh at͡sidotermofil'nyka vodorosleiĭ roda *Galdieria* (Rhodophyta, Cyanidiophycedae) [On the diversity of acidothermophilic unicellular algae of genus *Galdieria* (Rhodophyta, Cyanidiophyceae)]. *Bot. Zhurn.* (Moskva, Leningrad) 76(1): 69–79. (cultivat., cytol., morphol., physiol., tax.-syst.)

900. Staehelin, L.A., T.H. Giddings, P. Badami and W.E. Krzymowski. 1978. A comparison of the supramolecule architecture of photosynthetic membranes of blue-green, red, green algae and of higher plants. pp 335–355 *in* D.W. Deamer. (Ed.). *Light transducting membranes: Structure. Function, and Evolution.* Academic Press (A Subsidiary of Harcourt Bruce Jovanovich, Publishers), New York, San Francisco, London. (4) + pp V–XIV + (4) 358 pp. (cytol., sl. ment., ultrastr.)

901. Steinmüller, K. 1983. *Photo- und metabolische Regulation der Synthese von Phycobiliproteiden und der Ribulose-bisphosphat-Carboxylase bei* Cyanidium caldarium *vorgelegt von Klaus Steinmuller aus Giessen.* Inaugural-Dissertation zur Erlangung des Doktorgrades der Naturwissenschaften Fachbereich Biologie der Justus-Liebig-Universität Giessen, Giessen. (8) + 81 SS + lebenslauf (1). (biochem., cytol., evol., photosyn., physiol.)

902. Valentin, K. and K. Zetsche. 1990. Rubisco genes indicate a close phylogenetic relation between the plastids of Chromophyta and Rhodophyta. *Plant Molecul. Biol.* (Dordrecht) 15(4): 575–584. (biochem., evol., physiol.)

903. Ward, D.M., J. Bauld, R.W. Castenholz and B.K. Pierson. 1991. Modern protrophic microbial mats: anoxygenic, intermittenly oxygenic/anoxygenic, thermal, eukaryotic, and terresterial. pp 309–333 *in* J.W. Schopf and C. Klein. (Eds). *The Proterozoic Biosphere: a Multidisciplinary Study.* Cambridge University Press, Cambridge, New York, Melbourne. (4) + Frontispiece (1) + (3) + pp VII–XXIV + (6) + pp 7–1348. (ecol., florist., sl. ment.)

904. Whitton, B.A. 1980. Zinc and plants in rivers and streams. pp 364–400 *in* J.O. Noragu. (Ed.). *Zinc in the Environment.* Part II. *Health Effects.* John Wiley & Sons, New York, Chichester, Brisbane, Toronto. (4) + pp V–XII + (2) + 480 pp. (ecol., physiol., tax.-syst.)

905. Wood, J.M. and F.E. Engle. 1990. *Metal Accumulation and Polysaccharide Production.* United States Patent No. 4,908,317. U. S. Patent Office, Washington, D. C. (biochem., physiol.)

-- Yuina, N.P. Please see: Iurina, N.P.

906. Zetsche, K., M. Kaling and K. Steinmüller. 1983. In vitro synthesis of cyanelle and rhodoplast proteins. pp. 971–978 *in* H.E.A. Schenk and W. Schwemmler. (Eds). *Endocytobiology. Tübingen, Germany, April 10–15.* Walter de Gruyter, Berlin, New York. (5) + pp VI–XXV + (4) + pp 5–1071. (biochem., evol., photosyn., physiol.)

907. Zetsche, K., K. Valentin, U. Maid and M. Kostrzwa. 1992. Structure and sequence of plastic genes of red algae (Rhodophyta)- phylogentic implications. *Endocytobios. Cell Res.* (Tübingen) 8: 227–229. (biochem., evol., tax.-syst.)

908. Arad (Malis), S. 1987. Biochemicals from unicellular algae. *Internat. Indust. Biotech.* (Oxfordshire) June/July: 281–284. (evol., sl. ment., tax.-syst.)

909. Fork, D., N. Murata and N. Sato. 1979. Effect of growth

temperature on the lipid and fatty acid composition and the dependence on temperature of light-induced redox reactions of cytochrome f and of light energy redistribution in the thermophilic blue-green alga *Synechococcus lividus*. *Plant Physiol.* (Lancaster) 653(3): 534–530. (biochem., photosyn., physiol., sl. ment.)

910. Fredrick, J.F. 1992. Biblical hot springs of Israel: search for a missing link. *Explor. J.* (New York) 70(2): 58–59. (ecol., evol., florist.)

911. Fredrick, J.F. 1993. Biosynthesis of branched glucans and the origin of plastids. pp 475–479 *in* S. Sato, M. Ishida and H. Ishikawa. (Eds). *Endocytobiology* V. 5th International Colloquium *on Endocytobiology and Symbiosis. Philips University-Japan, Uji-Kyota, June 23–27, 1992.* Tübingen University Press, Tübingen. (4) + pp V–X + 586 pp. (biochem., evol., physiol.)

912. Gantt, E. 1977. *Yearly Review:* recent contributions in phycobiliproteins and phycobilisomes. *Photochem. Photobiol.* (New York) 26: 685–689. (biochem., photosyn., sl. ment.)

913. Hoffmann, L. and M. Mayele. 1993. Ultrastructure of the plasmalemna of cave *Cyanidium. Phycologia* (Berkeley, Copenhagen, Vancouver) 32(4): 307–309. (ecol., florist., ultrastr.)

914. Kannangara, C.G., S.P. Gough and D. von Wettstein. 1978. The biosynthesis of δ-aminolevulinate and the chlorophyll and its genetic regulation. pp 147–160 *in* G. Akoyunoglou and J.H. Argyroundi-Akoyunoglou. (Eds). *Chloroplast Development.* Proceedings of the International Symposium *on Chloroplast Development Held on the Island of Spetsai, Greece, July 9–15, 1978. Developments in Plant Biology.* Volume 2. Elsevier/North-Holland Biomedical Press, Amsterdam, New York, Oxford. XVI + 800 pp. (biochem., sl. ment.)

915. Lange, W., C. Wilhelm, W. Wehmeyer and E. Mörschel. 1990. The supramolecular structure of photosystem II–phycobilisome-complexes of *Porphyridium cruentum. Bot. Acta* (Stuttgart, New York) 103(3): 250–257. (physiol., sl. ment., ultrastr.)

916. Lee, R.E. 1980. *Phycology.* Cambridge University Press, Cambridge, London, New York, New Rochelle, Melbourne, Sydney. (6) + pp VII–X + (2) + 478 pp. (ecol., evol., sl. ment.)

917. Lee, R.E. 1989. *Phycology.* Second Edition. Cambridge University Press, Cambridge, New York, Port Chester, Melbourne, Sydney. (6) + pp VII–XV + (3) + pp 3–645. (ecol., evol., sl. ment.)

918. Ley, A.C. 1984. Effective absorption cross-section in *Porphyridium cruentum. Plant Physiol.* (Lancaster) 74(2): 451–554. (biochem., photosyn., physiol., sl. ment.)

919. Lips, S.H. and Y.J. Avissar. 1986. Photosynthesis and ultrastructure in microalgae. pp 43–69 *in* A. Richmond. (Ed.). *Handbook of Microalgal Mass Culture.* CRC Press, Inc., Boca Raton (8) + 528 pp. (biochem., photosyn., sl. ment.)

920. Mörschel, E. and K. Mühlethaler. 1993. On the linkage of expoplasmic freeze-fracture particles to phycobilisomes. *Planta* (Berlin) 158(5): 451-457. (photosyn., physiol., sl. ment., ultrastr.)

921. Mörschel, E. and G.H. Schatz. 1987. Correlation of photosystem-II complexes with exoplasmatic freeze-fracture particles of thylakoids of the cyanobacterium *Synechococcus* sp. *Planta* (Berlin) 172(2): 145–154. (photosyn., physiol., sl. ment.)

922. Moretti, A. and A. Musacchio. 1982. Chemotaxonomic notes on *Porphyridium* Nag. (Rhodophyta, Porphyridiales). *G. Bot. Ital.* (Firenze) 116(5–6): 269–274. (biochem., morphol., evol., sl. ment.)

923. Nagashima, H., I. Fukuda and J. Seckbach. 1993. Phylogeny of several thermoacidophilic unicellular algae. p. 217 *in* Abstracts: XV International Botanical Congress, *Yokohama, Japan, Pacifico Yokohama, August 28–September 2, 1993.* Under the Auspices of the International Union of Biological Sciences, 'IUBS', [Yokohama, Tokyo]. (4) + 600 pp. (biochem., ecol., evol., morphol.)

924. Nagashiuma, H., K. Ueki, I. Fukuda and J. Seckbach. 1993. Several new strains of thermal alga *Cyanidioschyzon* as the most primitive Eukaryotes. pp 279–285 *in* S. Sato, M. Ishida and H. Ishikawa. (Eds). *Endocytobiology* V. 5th International Colloquium *on Endocytobiology and Symbiosis. Philips University-Japan, Uji-Kyoto, June 23–27, 1992.* Tübingen University Press, Tübingen (4) + pp V–X + 586 pp. (ecol., evol., morphol., photosyn., physiol., tax.-syst.)

925. Patrone, L.M., S.T. Broadwater and J.L. Scott. 1991. Ultrastructure of vegetative and dividing cells of the unicellular red algae *Rhodella violacea* and *Rhodella maculata. J. Phycology* (Baltimore, Lawrence) 27(6): 742–753. (ultrastr., sl. ment.)

926. Rennis, D.S. and T.W. Ford. 1992. A survey of antigenic differences between phycoerythrins of various red algae (Rhodophyta) species. *Phycologia* (Berkeley, Copenhagen, Vancouver) 31(2): 192–204. (biochem., evol., physiol., tax.-syst.)

927. Rowan, K.S. 1989. *Photosynthetic Pigments in* Algae. Cambridge University Press, Cambridge, New York, New Rochelle, Melbourne, Sydney. (6) + pp VII–XIII + (1) + 334 pp. (biochem., sl. ment., tax.-syst.)

929. Seckbach, J., R. Ikan, F.D. Ott, H. Nagashima and I. Fukuda. 1993. A novel taxonomic proposal, ultrastructure and sterol content of thermophilic algae: *Cyanidium caldarium, Galdieria sulphuraria, Cyanidioschyzon merolae.* p. 216 in *Abstracts:* XV International Botanical Congress, *Yokohama, Japan, Pacifico Yokohama, August 28–September 3, 1993.* Under the Auspices of the International Union of the Biological Sciences, 'IUBS', [Yokohama, Tokyo]. (4) + 600 pp. (biochem., evol., tax.-syst., ultrastr.)

930. Seckbach, J., R. Ikan, H. Nagashima and I. Fukuda. 1993. New Phylogenetic status for acid hot spring algae. pp 241–254 *in* S. Sato, M. Ishida and H. Ishikawa. (Eds). *Endocytobiology* V. 5th International Colloquium *on Endocytobiology and Symbiosis. Philips University-Japan, Uji-Kyoto, June 23–27, 1992.* Tübingen University Press, Tübingen. (4) + pp V–X + 586 pp. (biochem., ecol., cytol., evol., physiol., tax.-syst., ultrastr.)

931. Seckbach, J., R. Ikan and F.D. Ott. 1992. Ha-Cyanidiot – Atzot termofiliot ba'alei hashlakhot evolutziyyoniyot: Skira kelalit ve-hatza'a le-alukaḥ taksonomit ḥadasha. (The Cyanidiaceae-thermophilic algae with evolutionary

ramification. General view and new taxonomic classification). p. 36 in Abstract booklet of the Annual meeting of the Botanical Society of Israel and the Israel Association for Plant Tissue Culture and Molecular Biology. Jerusalem, Israel. 24 February 1993. pp i–iv + (1–5) + 45 pages. See: Israel Journal of Botany *41*(4–6): 319. (tax.-syst., evol., physiol., biochem., ultrastr.)

932. Seckbach, J., R. Ikan, D. Ringelberg and D. White. 1993. Sterols and phylogeny of the acidphilic hot springs algae *Cyanidium caldarium* and *Galdieria sulphuraria*. *Phytochemistry* (Oxford) 34(5): 1345–1349. (biochem., evol., tax.-syst.)
933. South, G.R. and A. Whittick. 1987. *Introduction to Phycology.* Blackwell Scientific Publications, Oxford, London, Edinburgh, Boston, Palo Alto, Melbourne. (6) + pp VII–VIII + 341 pp. (ecol., sl. ment.)
934. Starr, R.C. and J.A. Zeikus. 1993. UTEX-The Culture Collection of Algae at the University of Taxas at Austin. *J. Phycol.* (Baltimore, Lawrence) 29(2 Suppl.): 1–106. (cultivat.)
935. Suzuki, K., K. Iwamota, S. Yokoyama and T. Ikawa. 1991. Glycolate-oxidizing enzymes in algae. *J. Phycol.* (Baltimore, Lawrence) 27(4): 492–498. (biochem., evol., physiol., sl. ment.)
936. Suzuki, K., N. Ohta and T. Kuroiwa. 1992. Isolation of the cell-nuclear, mitochondrial, and chloroplast DNA from the ulta-small eukaryote *Cyanidioschyzon. Protoplasma (Internat. Zeitschr. Physikal. Chem. Protoplast:* Leipzig) 171(1–2): 80–84. (biochem., ecol., evol., morphol., tax.-syst.)
937. Ueda, K. and H. Shibuya. 1993. Molecular phylogeny of *rbcL* and its bearing on the origin of plastids: Bacteria or Cyanophyta? pp 369–376 *in* S. Sato, M. Ishida and H. Ishikawa. (Eds). *Endocytobiology* V. 5th International Colloquium *on Endocytobiology and Symbiosis. Philips University-Japan, Uji-Kyoto, June 23–27, 1992.* Tübingen University Press, Tübingen. (4) + pp V–X + 586 pp. (evol., sl. ment., tax.-syst.)
938. Zilinskas, B.A. and L.S. Greenwald. 1986. Phycobilisome structure and function. *Photosynth. Res.* (Dordrecht) 10(1–2): 7–35. (biochem., photosyn., sl. ment.)
939. Zuber, H. 1986. Comparative biochemistry of light-harvesting systems. pp 238–251 *in* L.A. Staehelin and C.J. Arntzen. (Eds). *Photosynthesis* III. *Photosynthetic Membranes and Light Harvesting Systems* in A. Pirson and M.H. Zimmermann. (Eds). *Encyclopedia of Plant Physiology, New Series,* Vol. 10. Springer Verlag, Berlin, Heidelberg, New York, Tokyo. (5) + pp VI–XXVIII + (1) + pp 2–802. (biochem., photosyn., physiol., sl. ment.)
940. Fredrick, J.F. 1983. Yellowstone: Nature's laboratory. *Explor. J.* (New York) 61(2): 64–67. (ecol., biochem., evol., florist.).

Index for disciplinary fields

Biochemistry

1, 2, 3, 4, 5, 6, 7, 8, 12, 13, 14, 16, 18, 19, 21, 22, 23, 24, 26, 27, 29, 30, 31, 32, 33, 35, 36, 37, 38, 41, 42, 43, 44, 46, 47, 49, 50, 51, 52, 53, 54, 55, 56, 58, 59, 60, 61, 63, 65, 66, 67, 68, 70, 71, 72, 73, 74, 75, 79, 80, 81, 83, 84, 85, 86, 87, 89, 92, 93, 94, 99, 108, 114, 115, 116, 117, 118, 119, 120, 121, 122, 123, 124, 125, 126, 127, 128, 129, 131, 148, 149, 151, 154, 156, 157, 158, 162, 163, 167, 168, 169, 171, 172, 174, 176, 181, 182, 184, 185, 187, 190, 191, 192, 193, 194, 195, 196, 197, 198, 199, 201, 202, 203, 205, 206, 209, 212, 214, 216, 233, 234, 235, 236, 237, 238, 239, 241, 245, 246, 248, 250, 251, 252, 253, 254, 256, 257, 259, 260, 261, 262, 263, 264, 265, 266, 267, 268, 269, 270, 271, 272, 273, 274, 275, 276, 277, 278, 279, 280, 281, 282, 283, 286, 287, 288, 289, 290, 291, 292, 293, 297, 299, 300, 306, 307, 332, 333, 334, 335, 336, 338, 339, 340, 341, 342, 343, 344, 348, 349, 350, 351, 352, 353, 354, 355, 356, 357, 360, 361, 367, 369, 370, 371, 373, 374, 378, 380, 385, 387, 388, 389, 391, 393, 394, 395, 396, 399, 401, 402, 403, 404, 405, 406, 407, 409, 410, 411, 412, 413, 414, 415, 417, 418, 419, 420, 421, 422, 423, 424, 425, 426, 428, 429, 430, 433, 434, 435, 436, 438, 439, 440, 441, 442, 443, 444, 445, 446, 447, 449, 450, 451, 452, 453, 457, 458, 459, 461, 469, 478, 479, 481, 485, 486, 487, 488, 489, 490, 491, 492, 493, 499, 500, 501, 502, 503, 504, 505, 506, 508, 514, 516, 519, 520, 522, 523, 524, 525, 528, 529, 531, 532, 533, 535, 536, 540, 541, 542, 551, 552, 553, 555, 556, 557, 558, 559, 561, 562, 563, 564, 565, 566, 567, 568, 569, 570, 573, 577, 579, 580, 584, 588, 592, 293, 595, 600, 601, 603, 604, 604, 607, 608, 609, 610, 611, 612, 613, 614, 615, 616, 618, 619, 620, 621, 622, 623, 624, 625, 626, 628, 629, 630, 631, 632, 633, 634, 635, 636, 643, 646, 649, 651, 652, 653, 654, 655, 656, 660, 661, 669, 670, 672, 674, 675, 677, 681, 682, 683, 684, 685, 686, 688, 690, 691, 692, 693, 694, 695, 698, 699, 701, 702, 704, 709, 710, 712, 713, 714, 715, 716, 717, 718, 724, 725, 732, 735, 747, 748, 749, 752, 771, 772, 774, 775, 776, 777, 778, 779, 780, 781, 782, 783, 784, 785, 786, 787, 788, 789, 790, 791, 792, 793, 794, 795, 796, 797, 798, 799, 800, 801, 802, 803, 804, 807, 808, 809, 812, 813, 817, 820, 827, 829, 830, 831, 832, 837, 838, 842, 846, 852, 854, 855, 856, 857, 858, 859, 860, 861, 862, 863, 864, 867, 868, 874, 875, 876, 877, 878, 879, 881, 883, 884, 885, 886, 887, 888, 889, 891, 893, 894, 895, 896, 897, 898, 901, 902, 905, 906, 907, 909, 911, 912, 914, 918, 919, 922, 923, 926, 927, 929, 930, 931, 932, 935, 936, 938, 939, 940.

Cultivation

4, 20, 22, 24, 25, 35, 36, 38, 39, 78, 93, 169, 181, 184, 185, 191, 192, 193, 195, 201, 209, 210, 243, 250, 325, 326, 327, 328, 329, 346, 358, 359, 392, 397, 398, 408, 427, 492, 522, 574, 583, 648, 671, 672, 673, 674, 680, 692, 695, 696, 698, 699, 705, 706, 712, 713, 735, 740, 744, 757, 814, 815, 824, 889, 934.

Cytology

4, 22, 28, 75, 88, 90, 91, 134, 135, 136, 147, 149, 152, 175, 200, 201, 249, 300, 303, 305, 308, 323, 328, 329, 337, 364, 366, 376, 378, 380, 447, 448, 451, 453, 456, 460, 461, 476, 477, 513, 515, 516, 534, 536, 538, 540, 543, 575, 589, 596, 597, 674, 675, 681, 682, 683, 684, 685, 687, 688, 693, 694, 696, 699, 700, 703, 705, 706, 707, 708, 738, 739, 740, 741, 742, 746, 766, 767, 769, 805, 806, 880, 881, 882, 892, 895, 896, 898, 899, 900, 901, 910, 930.

Ecology

4, 11, 15, 17, 20, 24, 38, 48, 64, 65, 69, 90, 91, 94, 95, 96, 97, 98, 99, 100, 101, 102, 103, 104, 105, 106, 107, 108, 109, 110, 111, 112, 128, 130, 137, 138, 139, 140, 141, 142, 143, 144, 148, 155, 166, 173, 178, 179, 184, 185, 201, 202, 203, 205, 218, 219, 220, 221, 222, 223, 224, 225, 226, 227, 228, 229, 230, 231, 232, 247, 258, 284, 285, 312, 313, 314, 323, 324, 330, 331, 345, 347, 356, 357, 368, 377, 379, 382, 383, 384, 386, 390, 393, 393, 397, 398, 405, 406, 407, 416, 431, 439, 447, 448, 453, 455, 456, 470, 471, 473, 474, 477, 483, 509, 517, 521, 527, 530, 544, 545, 546, 547, 548, 549, 550, 571, 572, 581, 582, 585, 586, 587, 594, 600, 602, 606, 617, 627, 638, 639, 640, 641, 642, 647, 662, 663, 664, 665, 666, 667, 668, 672, 683, 688, 693, 694, 695, 696, 697, 698, 699, 702, 703, 712, 713, 714, 715, 719, 726, 727, 729, 730, 731, 732, 755, 758, 759, 760, 761, 762, 763, 765, 810, 814, 816, 818, 819, 821, 822, 826, 834, 835, 844, 845, 847, 848, 849, 850, 851, 853, 880, 883, 844, 890, 894, 896, 903, 904, 910, 913, 916, 917, 923, 924, 930, 933, 936, 940.

Evolution

1, 24, 27, 28, 32, 33, 38, 44, 70, 72, 73, 74, 79, 81, 92, 97, 103, 131, 145, 146, 147, 151, 152, 153, 154, 155, 174, 176, 188, 204, 207, 215, 249, 256, 257, 259, 260, 261, 262, 263, 264, 265, 266, 267, 268, 269, 270, 271, 272, 273, 274, 275, 276, 277, 278, 279, 280, 281, 282, 283, 293, 299, 300, 301, 310, 333, 338, 348, 350, 360, 361, 365, 366, 369, 388, 391, 399, 425, 429, 431, 437, 438, 446, 447, 448, 450, 451, 475, 476, 478, 479, 483, 487, 491, 495, 497, 499, 507, 517, 524, 533, 539, 541, 565, 578, 589, 593, 296, 597, 600, 613, 619, 625, 626, 656, 669, 676, 677, 678, 679, 681, 682, 683, 684, 685, 686, 688, 689, 691, 693, 694, 695, 700, 701, 704, 712, 713, 747, 750, 766, 768, 769, 770, 792, 805, 812, 820, 837, 852, 859, 864, 867, 868, 869, 870, 871, 877, 879, 885, 886, 888, 889, 891, 892, 893, 894, 896, 897, 898, 901, 902, 906, 907, 908, 910, 911, 916, 917, 921, 922, 923, 924, 926, 929, 930, 931, 932, 935, 936, 937, 940.

Floristics

12, 13, 14, 19, 20, 23, 24, 25, 34, 35, 38, 45, 65, 69, 94, 98, 100, 101, 104, 106, 109, 111, 112, 138, 164, 165, 166, 170, 175, 177, 178, 179, 180, 183, 184, 187, 201, 202, 204, 205, 212, 218, 219, 220, 221, 222, 223, 224, 225, 226, 227, 228, 229, 230, 231, 232, 284, 292, 304, 317, 322, 347, 375, 377, 379, 383, 400, 416, 455, 473, 474, 477, 509, 521, 527, 530, 534, 544, 545, 546, 547, 548, 549, 550, 571, 572, 578, 581, 582, 586, 606, 617, 645, 648, 662, 663, 665, 666, 676, 677, 687, 692, 695, 712, 713, 719, 723, 726, 729, 731, 755, 758, 759, 760, 761, 762, 765, 766, 769, 819, 826, 833, 839, 840, 841, 845, 847, 848, 849, 851, 853, 861, 871, 872, 873, 878, 880, 903, 910, 913, 940.

Literature

294, 295, 296, 372, 576, 764, 825.

Morphology

24, 28, 76, 90, 91, 134, 135, 136, 149, 152, 155, 164, 166, 175, 177, 179, 184, 186, 187, 201, 300, 303, 311, 317, 318, 319, 321, 323, 328, 329, 356, 357, 376, 378, 380, 382, 384, 456, 460, 461, 477, 515, 516, 534, 538, 540, 606, 650, 658, 659, 672, 674, 677, 693, 696, 697, 698, 699, 700, 707, 708, 746, 880, 882, 892, 896, 897, 899, 922, 923, 924, 936.

Photosynthesis

4, 8, 16, 23, 24, 27, 29, 30, 31, 49, 50, 51, 52, 53, 54, 56, 58, 63, 65, 66, 67, 68, 70, 71, 72, 73, 74, 75, 82, 83, 84, 85, 86, 87, 88, 89, 92, 93, 99, 107, 108, 113, 114, 115, 116, 117, 118, 119, 120, 121, 122, 123, 124, 125, 126, 128, 144, 149, 151, 154, 157, 158, 162, 163, 171, 172, 182, 184, 187, 196, 197, 198, 199, 201, 202, 203, 206, 207, 208, 214, 215, 216, 233, 234, 236, 238, 239, 250, 252, 279, 287, 297, 298, 300, 306, 307, 328, 329, 333, 334, 335, 340, 342, 343, 344, 351, 352, 353, 354, 356, 357, 367, 369, 389, 393, 395, 396, 397, 398, 400, 402, 403, 404, 406, 410, 411, 414, 415, 417, 418, 420, 421, 422, 423, 424, 426, 433, 434, 435, 436, 438, 443, 444, 445, 447, 451, 452, 458, 472, 478, 485, 486, 491, 505, 514, 519, 520, 522, 524, 525, 526, 528, 529, 540, 553, 554, 555, 556, 558, 559, 561, 562, 563, 564, 565, 566, 567, 568, 569, 573, 577, 636, 643, 651, 654, 655, 660, 661, 672, 673, 675, 680, 692, 693, 698, 699, 702, 709, 711, 714, 715, 726, 728, 729, 730, 731, 736, 737, 747, 748, 749, 772, 773, 774, 775, 776, 777, 778, 779, 780, 781, 782, 783, 784, 785, 786, 787, 788, 789,

790, 791, 792, 794, 795, 796, 797, 798, 799, 800, 801, 802, 817, 820, 823, 829, 830, 831, 837, 842, 852, 854, 855, 856, 857, 858, 860, 861, 877, 893, 894, 901, 906, 909, 911, 912, 918, 919, 920, 921, 924, 938, 939.

Physiology

1, 2, 4, 5, 6, 7, 8, 11, 12, 13, 14, 15, 16, 18, 19, 22, 23, 24, 26, 27, 29, 30, 31, 35, 36, 37, 38, 47, 48, 49, 50, 51, 52, 53, 54, 55, 57, 58, 59, 60, 61, 62, 63, 65, 66, 67, 68, 70, 71, 72, 73, 74, 75, 77, 78, 80, 81, 83, 84, 85, 86, 87, 89, 93, 94, 95, 96, 97, 98, 99, 104, 107, 108, 109, 111, 112, 114, 115, 116, 117, 118, 119, 120, 121, 122, 123, 124, 125, 126, 127, 128, 130, 148, 149, 151, 154, 157, 158, 162, 163, 167, 168, 169, 171, 172, 176, 179, 182, 184, 190, 191, 192, 193, 194, 195, 196, 197, 198, 199, 201, 202, 203, 204, 205, 206, 210, 214, 216, 233, 234, 235, 236, 237, 238, 239, 240, 241, 242, 243, 245, 246, 248, 250, 251, 252, 253, 254, 256, 257, 259, 260, 261, 262, 263, 264, 265, 266, 267, 268, 269, 270, 271, 272, 273, 274, 275, 276, 277, 278, 279, 280, 281, 282, 286, 287, 288, 289, 290, 291, 292, 293, 297, 298, 299, 300, 304, 306, 307, 312, 313, 314, 328, 329, 330, 331, 333, 334, 335, 336, 338, 339, 340, 342, 343, 344, 345, 346, 347, 348, 349, 350, 351, 352, 354, 355, 356, 357, 358, 359, 361, 367, 369, 370, 371, 373, 374, 377, 385, 387, 388, 389, 390, 391, 392, 393, 394, 395, 396, 397, 398, 399, 401, 402, 403, 404, 405, 407, 408, 409, 410, 411, 414, 415, 416, 417, 418, 419, 420, 421, 422, 423, 424, 425, 426, 427, 428, 430, 433, 434, 435, 436, 438, 439, 440, 441, 442, 443, 444, 445, 447, 449, 450, 451, 452, 457, 458, 459, 461, 469, 471, 472, 473, 474, 478, 479, 486, 489, 490, 492, 493, 498, 499, 502, 503, 505, 508, 514, 516, 518, 519, 520, 522, 523, 524, 525, 528, 529, 530, 531, 532, 533, 535, 536, 537, 539, 540, 541, 542, 544, 545, 456, 547, 548, 549, 550, 551, 552, 553, 555, 556, 557, 558, 559, 561, 562, 563, 564, 565, 566, 567, 568, 569, 570, 573, 583, 585, 587, 593, 594, 595, 600, 601, 602, 603, 606, 607, 608, 609, 610, 611, 612, 613, 614, 615, 616, 617, 618, 619, 620, 621, 622, 623, 624, 625, 626, 627, 628, 629, 630, 631, 632, 633, 634, 635, 636, 642, 643, 644, 646, 647, 651, 652, 653, 654, 655, 656, 660, 661, 663, 671, 672, 673, 674, 677, 691, 718, 724, 725, 726, 727, 728, 729, 730, 731, 732, 735, 736, 737, 740, 747, 748, 749, 753, 754, 771, 772, 773, 774, 775, 776, 777, 778, 779, 780, 781, 782, 783, 784, 785, 786, 787, 788, 789, 790, 791, 792, 794, 795, 796, 797, 798, 799, 800, 801, 802, 803, 807, 808, 809, 810, 811, 816, 817, 820, 821, 822, 823, 826, 827, 829, 830, 831, 832, 834, 835, 837, 842, 844, 845, 852, 854, 855, 856, 857, 859, 860, 862, 863, 864, 866, 867, 872, 873, 874, 875, 876, 877, 879, 881, 883, 884, 885, 886, 887, 888, 889, 891, 893, 894, 895, 896, 897, 898, 899, 901, 902, 904, 905, 906, 909, 915, 918, 920, 921, 924, 926, 930, 931, 935, 939.

Radiobiology

29, 30, 31, 47, 56, 59, 63, 65, 97, 127, 201, 203, 204, 245, 401, 404, 426, 427, 430, 445, 447, 450, 451, 452, 452, 487, 488, 518, 605, 649, 651, 661, 695, 697, 726, 727, 728, 729, 731, 732, 749, 771, 773, 774, 775, 777, 778, 779, 780, 781, 782, 784, 785, 786, 787, 788, 789, 791, 795, 796, 797, 798, 800, 801, 802, 827, 832, 860, 861, 866, 877, 890, 891.

Slight mention

21, 40, 41, 76, 77, 782, 129, 131, 132, 133, 138, 140, 141, 142, 143, 144, 150, 151, 155, 156, 159, 160, 161, 162, 174, 188, 200, 207, 208, 209, 213, 214, 246, 252, 302, 307, 308, 309, 310, 330, 332, 333, 334, 335, 337, 338, 339, 340, 353, 354, 355, 351, 357, 368, 374, 390, 406, 412, 413, 416, 427, 428, 429, 443, 449, 454, 470, 481, 482, 484, 495, 496, 498, 502, 506, 508, 525, 526, 545, 552, 560, 561, 562, 569, 570, 573, 574, 575, 576, 580, 584, 588, 589, 592, 594, 595, 596, 604, 638, 639, 640, 641, 644, 646, 650, 656, 664, 667, 668, 670, 711, 716, 717, 718, 719, 725, 735, 742, 751, 752, 753, 754, 756, 763, 789, 793, 813, 814, 816, 828, 838, 846, 861, 866, 869, 870, 878, 900, 903, 908, 909, 912, 914, 915, 916, 917, 918, 919, 920, 921, 922, 925, 927, 928, 933, 935, 937, 938, 939.

Taxonomy-systematics

9, 10, 20, 22, 23, 26, 34, 40, 44, 45, 70, 72, 73, 74, 79, 90, 91, 92, 94, 101, 132, 135, 145, 146, 147, 149, 150, 152, 159, 160, 161, 165, 166, 175, 176, 177, 178, 179, 180, 181, 182, 183, 184, 185, 186, 189, 200, 201, 205, 207, 211, 212, 213, 217, 218, 219, 244, 249, 255, 256, 257, 259, 260, 261, 262, 263, 264, 265, 266, 267, 268, 269, 270, 271, 272, 273, 274, 275, 276, 277, 279, 280, 281, 282, 283, 285, 300, 301, 302, 303, 304, 311, 315, 316, 317, 319, 320, 321, 323, 331, 338, 340, 356, 357, 362, 363, 365, 369, 378, 380, 381, 382, 384, 385, 388,

391, 392, 399, 400, 429, 432, 437, 447, 448, 449, 451, 452, 453, 454, 456, 462, 463, 464, 465, 466, 467, 468, 472, 476, 478, 480, 482, 483, 487, 491, 494, 495, 496, 497, 499, 501, 507, 509, 510, 511, 512, 513, 523, 527, 530, 532, 533, 535, 536, 538, 541, 545, 456, 547, 548, 549, 550, 557, 560, 565, 568, 576, 579, 586, 589, 590, 591, 601, 650, 656, 657, 658, 659, 662, 663, 665, 669, 675, 676, 678, 679, 681, 683, 686, 688, 689, 690, 691, 694, 700, 701, 704, 712, 713, 716, 720, 721, 722, 733, 734, 743, 747, 750, 751, 756, 758, 759, 760, 761, 762, 763, 806, 815, 816, 820, 833, 836, 837, 839, 840, 841, 843, 846, 865, 869, 870, 871, 877, 880, 889, 891, 896, 897, 898, 899, 904, 907, 908, 924, 926, 927, 928, 929, 930, 931, 932, 936, 937.

Ultrastructure

4, 28, 44, 76, 88, 113, 134, 135, 136, 175, 186, 187, 200, 215, 241, 249, 300, 304, 305, 308, 309, 328, 329, 364, 365, 366, 415, 448, 475, 476, 484, 512, 515, 516, 524, 534, 560, 589, 597, 598, 599, 637, 672, 674, 675, 677, 681, 683, 683, 684, 687, 688, 693, 694, 696, 698, 699, 700, 701, 703, 705, 706, 707, 708, 718, 739, 740, 741, 742, 746, 766, 767, 769, 770, 771, 805, 806, 821, 828, 842, 882, 895, 896, 900, 913, 915, 920, 925, 929, 930, 931.

Methods for algal investigations

JOSEPH SECKBACH

A. Modified Allen's autotrophic medium (Allen 1959; Seckbach 1971)

General notes

a. Add the following salts and adjust pH in a large volume of water while the nutrient solution is mixed with a magnetic driven stirring bar. In our studies we almost always used the 2x concentration of Allen's autotrophic medium which is well accepted by the cyanidiophytes.
b. Adjust the pH to 2-3 with 1N or 50% H_2SO_4 and 1N NaOH.
Note: One can use nitrate (10 mM KNO_3) for *C. caldarium* instead of ammonium sulphate (while *Galdieria* does not metabolize nitrate).
c. Incubate cultures at 30 to 55°C and aerate with filtered and humidified pure CO_2 or air.
d. For rapid growth in light, aerate the medium with CO_2. For heterotrophic growth in dark, 1% of glucose is added.
e. For routine culture maintenance we supplied pure CO_2 to the screw cap tubes, tightened caps and incubated at 37°C.

1. Assemble the following microelements stock solutions

a. $Na_2MoO_4{\bullet}2H_2O$	0.2 mg/ml
b. $VSO_4{\bullet}H_2O$	0.3 mg/ml
c. $FeCl_3{\bullet}6H_2O$	19.0 mg/ml (or 1% FeNaEDTA)
d. $Na_2B_4O_7{\bullet}10H_2O$	4.4 mg/ml
e. $MnCl_2{\bullet}4H_2O$	1.8 mg/ml
f. $CuCl_2{\bullet}2H_2O$	0.5 mg/ml
g. $ZnSO_4{\bullet}7H_2O$	2.2 mg/ml

1.1 To prepare one liter of medium

a. Dissolve 2.6 g $(NH_4)_2SO_4$ in a large volume of water to prevent precipitation (ppt) once the $CaCl_2$ and $MgCl_2$ have been added.
b. Add 560 mg of KH_2PO_4
c. Add 494 mg of $MgSO_4{\bullet}2H_2O$ (or 241 mg $MgSO_4$)
d. Add 148 mg of $CaCl_2{\bullet}2H_2O$
e. Add 2 ml $FeCl_3$ stock solution (or 3.2 ml FeNaEDTA stock solution)
f. Add 2 ml $MnCl_2$ stock solution
g. Add 2 ml $Na_2B_4O_7$ stock solution
h. Add 0.2 ml $ZnSO_4$ stock solution
i. Add 0.2 ml $CuCl_2$ stock solution
j. Add 0.2 ml Na_2MoO_4 stock solution
k. Add 0.2 ml VSO_4 stock solution
l. Adjust pH and dilute to one liter with distilled water. Sterilize at 15 p.s.i. for 15 min. A white ppt may form but this should dissipate upon cooling.

Note: you may save time by preparing only 3 solution groups of the above microelements. Transfer the above recommended volumes from the following stock solutions: 1. Fe, 2. Mn+B, 3. Zn+Cu+Mo+V.

2. To prepare one liter of solid Allen medium

a. Add 15 g of Bacto Agar to 500 ml of distilled water. Sterilize this solution (15 p.s.i. for 15 min)
b. Adjust the pH of 500 ml Allen 2x medium so that when the medium is diluted with the agar solution, the desired final pH is obtained. Sterilize as above.
c. Cool both solutions in a water bath to ca. 48°C and then add the Allen's 2x to the agar solution and gently mix. This can then be aseptically distributed to tubes for slants or to plates.

Note: Agar is quite acid sensitive and will not gel if heated in the presence of acid nor will it gel if the

J. Seckbach (ed.), Evolutionary Pathways and Enigmatic Algae: Cyanidium caldarium *(Rhodophyta) and Related Cells,* 333–335, 1994.

pH is less than ca. 2.0. We acidified our Allen medium to pH 2.0. After combining this with the agar solution, the result was a final pH of ca. 2.5-3.0.

B. The Carolina Biological Supply Company

See lists in their *Culturing Algae* booklet of 1978 (Burlington, N. Carolina 27215 and Gladstone, Oregon 97027) the following *Cyanidium* medium pH 3.5 (basically the macroelements are similar to the above Seckbach's modified solution).

For each liter of heterotrophic medium required, stock solutions in the amounts indicated are added to 957 ml of glass-distilled water:

20 ml KH_2PO_4	2.80 g/100 ml H_20
20 ml $CaCl_2$	0.74 g/100 ml H_20
20 ml $MgSO_4 \bullet 7H_2O$	2.50 g/100 ml H_20
20 ml $(NH_4)_2SO_4$	13.00 g/100 ml H_20
6 ml trace metals	(see further)
1 g Dextrose	

Adjust pH to 3.5 with 1 N H_2SO_4

Trace metals stock solution

Add 0.750 g of Na_2EDTA to 500 ml of dist. water. After this chelating agent has dissolved, the following salts are added in the amounts indicated:

$FeCl_3 \bullet 6H_2O$	97 mg
$MnCl_2 \bullet 4H_2O$	41 mg
$ZnCl_2$	5 mg
$CoCl_2 \bullet 6H_2O$	2 mg
Na_2MoO_4	4 mg

C. Ascione's supplemented medium (Ascione *et al.* 1966)

1. The heterotrophic nutrient broth contains the following

1.1 Macroelements

KH_2PO_4	0.001M
$CaCl_2$	0.001M
$MgCl_2$	0.001M
$(NH_4)_2SO_4$	0.015M
D(+)-galactose	0.50%
Malt extract	0.05%

Add 0.3 parts per liter trace element mix (see further – Microelements). Use sufficient 10N H_2SO_4 to bring the medium to the desired pH.

1.2 Microelements mix

$Fe(SO_4)_3 \bullet (NH_4)_2SO_4, \bullet 4H_2O$	4.00 g
H_3BO_3	0.50 g
$MnSO_4 \bullet 2H_2O$	0.50 g
$ZnSO_4 \bullet 7H_2O$	0.05 g
$CuSO_4 \bullet 5H_2O$	0.02 g
$(NH_4)MoO_{24} \bullet 4H_2O$	0.01 g
$VSO_4 \bullet 2H_2O$	0.01 g
$CoSO_4$	0.01 g

Distilled water to 300 ml.

The medium is autoclaved after adjusting the pH at 2 atm. absolute pressure for 15 min.

Note: The $(NH_4)_2SO_4$ must be added to a large volume of water to avoid precipitation (once the $CaCl_2$ and $MgCl_2$ have been added).

D. Fukuda's minimum mineral liquid medium (Fukuda 1958)

Some phycologists cultivate their algal cells with supplement of soil extracts, while others grow their cultures on mineral drinking water (personal communication from colleagues). The following Fukuda's medium requires just the macro elements (plus Fe). It is assumed that if the trace elements were not directly supplied to the solution, they have been arrived from the contaminations (impurities) of the major salts.

Composition of Fukuda's *Cyanidium* medium

$Ca(NO_3)_2$	0.5 g
KH_2PO_4	0.2 g
$MgSO_4 \bullet 7H_20$	0.2 g
$FeCl_2$	trace
H_2SO_4 (1N)	10 ml

Distilled water up to one liter, pH 2.4.

E. Determination of unicellular algal growth and chlorophyll content

There are several methods for following growth rates of unicellular algae (see 'cultivation' in the index). We recommend the conventional and simple rapid practice:

a. Volume packed cells (a known volume of suspension is transferred into graduated Hopkins centrifuge tubes and the ppt is measured as ml/l).
b. Determination of the Optical Density of the algal suspension at 580 nm while absorbent units are calibrated against respective standard curves using a known numbers of cells.
c. Direct cell count made with a hemacytometer.
d. Electro-photometric reading are converted into figures in dry or wet weights (mg/ml).
e. Pigment absorption *in vivo* and *in vitro*. For example, extraction of chlorophyll with dimethylformamide and its concentration determined spectrophotometrically (for detail: Volk and Bishop 1968).

F. Electron microscopical techniques

The handling of *Cyanidium* and *Cyanidioschyzon* cells for EM studies is not easy (Seckbach 1992). Most workers have used the $KMnO_4$ fixation method, while a few fixed Cyanidiophytes with aldehyde-OsO_4 $\pm$ 1% sugar or buffered solution (at room temperature). For additional hints see 'Ultrastructure' in the Index.

General index

absorption spectra 100
Acanthocystis turfacea 88
Acaryophrya sp. 88
Achromatium oxaliferum 63
Achromatium sp. 63
acid alga 122
acid *Chlorella* 109, 135
acidophile 114, 118, 140, 196
acidotermophilic 101, 133
acriflavine 9
actin 70, 217, 249
action spectrum 266
aerobacterium 15
aerobically 79
Agmenellum quadruplicatum 207, 277
  A. quadruplicatum 278
ALA 263
  biosynthesis 263
  synthase/dehydrase 268
alanine 91
Alcaligenes eutrophus 256
algae macro 71
algal
  bridge 105
  growth 335
  habitat 115
  investigation 335
  mats 104
alkaline *Cyanidium* 101, 123, 134
alkane 207
Allen medium 106, 335
Allen strain 99, 149, 168
Allium cepa 59
allophycocyanin 27, 99, 135, 178, 234, 256, 264
alpha bridge carbon 264
alpha-methene bridge 270
altruistic 93
amino acids 31, 59, 91, 188, 193, 265, 275
  sequence 7
δ-aminolevulinate, see ALA
aminotransferase 185, 269
ammonia 29
ammonium sulfate 273
amoeba 88
Amphitrema flavum 88
amylopectin 201
Anabaena catenula 54
Anabaena minutissima 56
Anabaena sp. 44, 56, 61
Anabaena variabilis 10, 44, 207
Anacystis nidulans 59, 206
anaerobic 5
anisomycin 14
Anodonta sp. 88
antibiotics 14
antisymbiotic 38
Antithamnion sp. 259
apatite 57
Aphanothece halophytica 46
apochlorotic 140
apoprotein 264
aposymbiotic 87
Arabidopsis thaliana 42
archaebacteria 8, 59, 61, 110
Archaeopsis monococca 26
archidonic acid 29
arginine 55
Arthroptera 63
Ascione's medium 335
Ascomoropha helvetica 88
Ascomycota 62
aspartic acid 33, 55
astaxanthin 80
atmosphere primitive 4
ATP 3, 31, 199, 269
  /ADP 199
  synthase 74, 256
  -ase 13, 42, 61, 82
autogenous 19, 38, 53, 64, 111, 151
autospores 28, 84, 170, 176, 233
autotroph/chemotroph 7
autotrophic 4, 79, 99, 126, 168, 193, 203, 263
  partners 87
Avena sativum 249

azide 82

Bacillariophyceae 12, 202, 241
Bacillus subtilis 8, 10
  B. subtilis 10
bacteria 53, 59, 109, 171, 206
  ammonia-oxidizing 54
  flagellated 61
  green 54
  green sulfur 5
  photosynthetic 5
  purple 54
  purple non-sulfur 6
  purple sulfur 5
  sulfate-reducing 7
  sulfur 54
  nuclei/genome 241, 257
bacteriochlorphyll 6
Bacteriomycota 14
bacterium
  aerobic 13
  spriral 11
Bangia 137
Bangialean 123
bangiales 135, 202
Bangioideae 210
Bangiophyceae 109, 133, 147
Bangiophycidae 118, 135, 154
basal bodies 61
Basidomycota 63
Beggiatoa gigantea 63
Beggiatoa mirabilis 63
bilin 264
biliverdin 263
binary fission 129, 151, 161
biont 47
Blot Northern/Western 277, 276
blotting Northern/Western 276
blue-green algae 53, 83, 99, 125, 133, 175, 188, 196, 201
blue-green bacteria 58
Bostryococcus braunii 133, 207, 209
  B. braunii 134
branching isoenzymes 114, 140
Bridge algae/cell/position 44, 99, 133, 258
  phylogenetic 154
  transitional, algal 133, 222
brown algae, see Phaeophyta
Bryopsis maxima 245
Bryopsis pulumos 242
Buglenophyta 204

C^{13}/C^{12} isotopic fractionation 104
caddis flies 63
Caedibacter 61
calcite 57
calmodulin 225
Calvin cycle 8, 198
Calvin-Benson-cycle 31
campesterol 209
canthaxanthin 80
carbohydrates 90, 110, 114, 138, 154, 183, 199, 201
carbon 34, 57, 126, 193
  dioxide, see CO_2
carbon monoxide 267
carboxisome 27
carboxylation 199
cardiolipin 9
Carolina medium 335
Carophyceae 185
β-carotene 27, 78, 80, 124, 137, 264
carotenoids 27, 80, 99, 135, 265
Castrada viridis 88
catabolism 270
catalase 6, 104, 185, 224
cave *Cyanidium* 101, 165, 175
cavious cyanidiophytes 101
cell
  division 62, 69, 107, 116, 215, 266
  nuclei 241
  wall 11, 27, 64, 69, 133, 176, 186, 215, 231
cellular polarity 118
cellulase 78
centrioles 54, 63, 215, 249
centrosome 74
Cephalodella sp. 88
Ceramiales 202
Chalarodora 26
C. azurea 26
Chamaesiphonales 118
Chaos zoochlorellae 88
chaperonin 270
Chara australis 245
Charophyceae 241
chemosynthesis 4
chemotaxonomy 201
Chlamydomonas 29, 125
Chlamydomonas reinhardii 75, 207, 239, 258
  C. reinhardii 241
Chlophyceae 109
chloramphenicol 14, 31, 83, 266
Chlorella 17, 64, 79, 88, 99, 113, 133, 153, 193, 196, 268
Chlorella caldaria 119, 146
  C. fusca 74
  C. fusca var. *vacuolata* 91
Chlorella pyrenoidosa 207

Chlorella sorokina 94, 193
C. sorokiniana 193
Chlorella variegata 122
Chlorella vulgaris 93, 206
C. vulgaris Beij. 91
Chlorelloidea 84
chloride 197
Chlorococcales 84, 121, 182
Chlorogloea fritschii 7, 15
C. fritschii 15
C. mirabilis Andr. 91
Chlorogonium elongatum 235
Chlorophyceae 87, 121, 153, 185, 202, 241
chlorophycophyta 109, 114, 133, 145
chlorophyll 27, 70, 99, 119, 134, 161, 170, 178, 187, 198, 255, 263
content 335
-a/b/c/d 264
Chlorophyta 46, 136, 157, 167, 204, 237, 255
chloroplast 7, 19, 53, 74, 99, 113, 133, 147, 171, 177, 188, 215, 231, 239, 255, 263
DNA 138, 151, 154, 239
genome 74, 239, 255, 275
nucleoid 107
Cholesterol 208
chondrid 8
chondrillasterol 209
chondriomes 23, 222
chromatin 84, 122, 224
Chromatium buderi 59
chromatophores 9, 19, 25, 121
chromatoplasm 27, 37
chromophore 264, 274
chromophyceae 137
Chromophyta 107, 133, 202, 255
chromoplasts 243
chromosomes 9, 20, 45, 59, 70, 224, 239, 244
Chroococcidiopsis 118, 154, 201
Chroococcidiopsis thermalis 120, 146, 153
C. thermalis 120
C. thermalis var. *nipponica* 120
Chroococcidium 154
Chroococcopsis 108, 135
caldaria 117, 145
Chroococcus 108, 115, 133
C. varius 116
Chroococcus gigantus 63
Chroococcus maximus 63
Chroococcus sp. 117
Chroococcus varius 106, 115, 145, 158
Chrysophyceae 12, 63, 202, 241
Chrysophycophyta 60
Chrysophyta 109, 188, 237
ciliata 88
-alga association 90
cisternae 233
citrate (citric acid) 5, 174
synthase 186
cladistic analysis 60
Climacostomum virens 88
clones 20, 170
Cluster-clone hypothesis 59
clusters 588
CO 269
CO_2 4, 29, 81, 93, 99, 114, 133, 153, 188, 193, 335
assimilation 7
fixation 6
Coccochloris 135
Coccochloris elabens 207
Coccochloris orsiniana 108, 114, 145, 157
C. orsiniana 114, 150, 157, 161
codon 14, 256
Coelenterata 91
coevolution 21
colchicine 59, 82
Coleps
hirtus 88
spetai 88
colonial 25
commensalist 38
commercial gas 81
compartment 21
compartmentalization 10, 139
contractile ring 225
cristae 11, 125, 222
Cryptococcales 124
cryptomonads 138
Cryptomonas 258
Cryptophyceae 12, 109, 124, 136, 241
Cryptophyta 23, 133, 204
cryptophyte 23, 225
cryptoxanthin 27, 80, 264
crystal 83
crystallid 58
cultivation 80
cyanelles 15, 19, 107, 140, 211, 255
Cyanidiaceae 26, 109, 133, 145, 147, 154, 180, 183, 193, 283
Cyanidiales 135, 147
cyanidian 10, 109
bridge 100
Cyanidiophyceae 10, 26, 99, 133, 147, 153, 157, 167, 193, 201
cyanidiophytes 10, 99, 133, 167, 217, 233, 243, 335

Cyanidioschyzon 113, 134, 145, 154, 175, 283
Cyanidioschyzon merolae 10, 26, 64, 100, 154, 161, 183, 201, 219, 242
C. merolae 134, 151, 205, 242
cyanidiospores 99
Cyanidium 133, 145, 153, 175, 183, 193, 231, 255, 263, 283, 335
Cyanidium caldarium 10, 26, 31, 99, 113, 115, 134, 145, 147, 153, 157, 159, 167, 168, 175, 183, 193, 201, 205, 219, 231, 239, 255, 263, 335
C. caldarium Allen strain 149, 243, 263
C. caldarium forma A. 106, 138, 149, 159, 201, 242, 263
C. caldarium forma B. 106, 116, 120, 135, 149, 159, 201, 263
C. caldarium forma *chilensis* 121, 146
C. caldarium KS-1 strain 202
C. caldarium M-8 108, 149, 201, 240, 263
C. caldarium RK-1 108, 149, 201, 242, 263
C. caldarium strain RK-1 138, 149, 154, 233, 240
Cyanidium caldarium f. *grande* 146
Cyanidium caldarium f. *piccola* 146
Cyanidium caldarium forma *rumpens* 146
C. caldarium var. *rumpens* 106, 119
Cyanidium chilense 118, 146, 147, 149, 175
Cyanidium caldarium var. *chilensis* 146
C. caldarium var. *chilense* 175
C. caldarium var. *chilenses* 106
C. caldarium var. *chilensis* 121
Cyanidium caldarium var. *chilensis* f. *rumpens* 145
C. caldarium var. *chilense* f. *rumpens* 175
Cyanidium daedalium 150
C. daedalium 147
Cyanidium maximum 149
C. maximum 147
Cyanidium partitium 149
C. partitium 147
Cyanidium rumpens 149
C. rumpens 147
Cyanidium sulphurarius 151
C. sulphurarius 147
cyanobacteria 3, 7, 25, 37, 53, 59, 71, 83, 99, 113, 133, 153, 184, 194, 201, 206, 237, 242, 255, 265
Cyanocyta korschikoffiana 15
cyanomes 40, 109
Cyanoptyche 26
C. gloeocystis 26
C. korschikoffiana 15
Cyanophora 26, 140, 210
Cyanophora paradoxa 15, 19, 107, 139, 210, 255, 276
C. paradoxa 26, 210, 255, 278
C. tetracyanea 26
Cyanophyceae 140, 153, 343
cyanophycean 113
cyanophycin granules 33, 53
Cyanophycophyta 114, 133, 145
Cyanophyta 14, 126, 133, 157, 167, 222, 201
cyanoplast 19
DNA 34, 153, 239
genome 34
cycle
carboxylic acid 6
cell/life 266, 284
glyoxylate 7, 183
pentose phosphate 5
TCA 5
cycloheximide 14, 31, 83, 266
cylindrical bodies 57
cysteine 262
cytochalasin B 83
cytochrome 6, 138, 269
oxidase/reductase 13, 61, 269, 270
cytokinesis 74, 225, 244
cytoplasm 11, 55, 59, 83, 114, 177, 197, 215
cytosis
endo/exo 11, 21
myco- 21
phago/pino 11, 21
cytoskeleton 215, 225
cytosol 268

Dalyellia viridis 88
DAPI (46-diamidino 2-phenylindol) 10, 239
Dasya baillouviana 217
degeneration phase 38
dehydrogenase 7, 269
delta-aminolevulinate, see ALA
dendrograms 257
Denticula tenuis 180
deoxyribonucleotides 74
Dermocarpa 108, 118, 119, 135
Dermocarpa caldaria 119, 146
D. caldaria 119
Dermocarpa gigantea 63
Dermocarpaceae, 118
desaturase 32
Desulfurococcus mobilis 59

diamopimelic acid 122
diatoms 42, 45, 63, 87
dibiont 45
dicarboxylic acids 31
dictyosome 10, 159, 178, 218
Difflugia oblonga 88
digalactosyldiglyceride 29
digestive vacuoles 90, 233
dimethylformamide 277
dimethylsulfoxide (DMSO) 69
dinitrogen fixation 33, 209
dinoflagellates 62, 87, 137
Dinophyceae 12, 62, 204, 231, 241
direct filiation 64, 69
disaccharides 201
Disematostoma bütschlii 88
Dixoneilla grisea 217
  D. grisea 221, 223
DNA 4, 20, 59, 69, 107, 154, 168, 242, 255
  chloroplast 23, 109, 154, 239
  circular 27, 243
  fiber 17
  hybridization 60
  mitochondrion 151, 243
  nuclear/organelles 134, 151, 239
  organelle 13
  plastid 140, 241, 255
  plastome 23
  probes 276
  /RNA 21
dyssymbiotic 37

echinenon 80
Ectocarpus indicus 241
Ectocarpus siliculosus 256
eikosapentaenoic acid 29
eikosenoic acid 29
electron
  acceptor 4
  microscopical method 335
elongase 32
Embryophyta 34
Emerson enhancement 266
endocyanome 140, 183, 202
endocytobiont 21
endocytobiotic 20, 87
endogenosomes 19
endogenous 19, 94
endomycorrhizae 38
endonuclease 14, 239, 255
endoplasmatic reticulum 74
endoplasmic reticulum 10, 58, 125, 215
endospore division 245
endospores 99, 113, 136, 150, 159, 244
endosymbiont 11, 21, 58, 110, 211, 231
endosymbiotic 37, 53, 107, 140, 151, 255
  association 94, 99, 109, 139
  events 64
  prokaryote 11
Entophysalis 108, 135
  rivularis 119, 135, 146
  rivularis forma *puteolana* 121, 146
enzymes, hydrolytic 11
eocyte 59
Ephydatia fluviatilis 88
epibiotic 87
ER, periplastidial 23
ergosterol 209
Erythrobacter 6
erythrocytes 83, 270
erythromycin 14
Escherichia coli 9, 267
  E. coli 4, 256, 269
ethylene 105
  glycol 195
ethylenediamine-tetraacetate (EDTA) 69
ethylidine 262
etioplast 242
eubacteria 14, 64
eucyte 21, 141, 258
Euglena 10, 138, 233
  gracilis 75
Euglenophyceae 241
Euglenophyta 210, 237
euglenophytes 42, 202
Euglenozoa 41
eukaryogenesis 109
eukaryota 4, 19, 53, 69, 100, 107, 136, 137, 154, 180, 217
eukaryotic 53, 133
  algae 99
  host 110
Euplotes daidaleos 88
Eusinonimi 121
Eustigmatophyta 237
eusymbioses 37, 38
evolution 3, 69, 196, 263, 284
  cell 19
  , chemical 3
  , metabolic 6
  , symbiogenous 19, 151
evolutional
  missing link 46
  process 5
evolutionary
  bridge 100
  clocks 46
  link 139
  tree 107

excretion 92
exogenosomes 19
exospores 118
exsymbiotic 87

Fabales 21
FAD 194
fatty acids 19, 32, 104, 113, 147, 154, 161, 201, 205
fermentation 4
ferredoxin 6, 37, 107, 139, 154, 272
 -NADP$^+$-oxidoredoctase 19, 36, 274
ferrocheletase 267
fibrillar 62
filamentous 61
filaments 53, 217, 249
fixation, EM 54, 234
flagellar apparatus 225
flagellated cells 113, 135
flagellum 11
flavoprotein 6
Flintiella sanguinaria 217
 F. sanguinaria 223
Florideae 210
floridosides 110, 129, 140, 201
food vacuole 90
fossils 69
free living 87
 algae 93
 Chlorella 94
freeze-fracture 180
freeze-substitution 54, 222, 231
Frontonia leucas 88
Frontonia vernalis 88
fructose 91, 140, 201
fucosterol 208
Fukuda's medium 335
fumaroles115
Funaria hygrometrica 249
fungi 64, 83, 170, 196, 206, 242

gabaculine 269
Galdieria 113, 134, 147, 164, 167, 175, 193, 205, 255, 263, 283, 335
Galdieria daedala 146
 G. daedala 100, 130, 145 167
Galdieria maxima 146
 G. maxima 100, 130, 145, 167
Galdieria partita 146
 G. partita 100, 130, 145 167
Galdieria sulphuraria 26, 100, 134, 145, 160, 183, 201, 219, 233, 243, 255
 G. sulphuraria 165, 205
Galdieriaceae, 164, 193
gas vesicles 55

GC content 243
Gelidium amansii 242
gene
 , expression 261
 , package 256
 substitution 42
 transfer 19, 37
genes 4, 13, 107, 140, 154, 239, 255
genetic autonomy 40
genome 4, 21, 59, 255
genome coding 107, 255
genomic hypersystem 38
geochemical indicators 206
Gietleria calcarea 175
 G. floridana 175
Ginkgo biloba 249
girdle 136, 178, 234, 241
Glaucocystis 140, 210
 G. bullosa 26
 G. duplex 26
 G. oocystiforme 26
Glaucocystis Itzigsohn 26
Glaucocystis nostochinearum 19, 139, 184, 231
 G. nostochinearum 26, 210
Glaucocystophyceae 25
Glaucocystophyta 19, 255
Glaucocystopsis africana 26
Glaucophycophyta 15, 133
Glaucophyta 25, 109, 211
Glaucosphaera 25, 101
Glaucosphaera vacuolata 217
 G. vacuolata 26, 217, 223
globulins 127
Gloeocapsa punctata 180
Gloeochaete 26, 140
 G. wittrockiana 26
glucans 105, 114, 140
glucose 5, 31, 91, 126, 136, 160, 169, 201, 265, 335
glucotransferases 114
glutamate 160, 193, 265
glutamine 91
 synthetase 193
glutaraldehyde 222, 231
 -osmium 54, 232
 3-phosphate 5
glycerol galactoside 140, 201
glycine 183, 268
glycogen 122, 140
 granules 53
glycolate 183, 199
 oxidoreductase 185
glycolipids 205
glyoxylate 7, 188

gnorimo- biont/some 21
Golgi 10, 62, 69, 101, 125, 215, 231
Golgi-nucleus association 218
Gracilaria tikvahiae 225
grana 11, 75, 237
granules 53
granuolar area 58
green algae 29, 81, 135, 153, 201, 237, 245, 255
green ciliates 90
greening process 266
Griffithsia pacifica 225

H_2O_2 6, 183
H_2S 4
H_2SO_4 168
Halobacterium 6
Halteria bifurcata 88
Haptocyceae 241
haptonema 237
Haptophyta 204, 237
Heleopara sphagni 88
heme 261
  oxygenase 263
hemoprotein 270
heterologous 87
Heteromastix rotunda 204
Heteromeyenia sp. 88
heterosides 140
heterotrophic 4, 21, 87, 94, 126, 160, 168, 203, 263
hexadecenoic acid 206
hexamers 274
Hill reaction 154, 266
histone gene 84
histones 59, 69
HNO_2 197
HNO_3 197
Holosticha viridis 88
homeostasis 53
homologous 20
hoppers 63
host 13, 21, 87
  genome 37, 42
Hyalosphenia papilio 88
hybridization, Southern 9
hybridize 274
Hydra magnipapillata 88
Hydra viridis 88
hydrocarbons 154, 201
Hydrodictyon 233
hydrogenase 37
hydrothermal vent 63
hydroxy acids 201
hydroxyapatitie 57
hydroxypruvate reductase 183
hyphae 87
hypothesis
  Cluster clone 20
  compartmentation 41
  endogenous 20
  endospore 20
  fusiogenesis 21
  gene duplication segregation 21
  locked cell 20, 53
  neutral mutation random drift 62
  plasmid 20
  transformation 20

immunogold staining 249
inclusion bodies 54
inclusions 54
  cellular 61, 178
  convoluted 58
  crystalloid 53
  intrathylakoidal 53
  lipid 53
  microtubule 57
  polymer-like 53
  spherical 58, 237
insects 63
intermediate status 137
interphase 226
intracellular membrane system 4
intrathylakoidal granules 62
introns 4, 13, 107, 138, 154, 256
invagination 59
iron 73, 265
isofloridosides 129, 140, 202
isozymes 138
Intertaxonic combination 19

Kenetochores 224
kinetoplast 10
Krebs cycle 8, 198

lactam 270
lamella 58, 136
lamella, outermost 136
laminitol 203.
Lecudina tuzetae 63
lichens 87
life 3, 101
ligase 269
light 69, 106, 179, 198
light-harvesting antenna 266
lincomycin 14
Linnaea sp. 88
linoleic acid 32, 206
linolenic acid 31, 129, 205

lipids 29, 32, 107, 201, 234
living fossil 137
lutein 80, 126, 135, 264
Lycopodium clavatum 78
lysosomal compartments 89
lysosome 10, 62, 231

macrobiont 21
Malacophrys sphagni 88
malate dehydrogenase 186
maltose 91, 202
mannitol 201
mannosido glyceride 140
Marchantia 256
Marchantia polymorpha 75, 239
  M. polymorpha 239
marine associations 87
mayflies 63
Mayorella viridis 88
membrane
  chloroplast 125
  convoluted 59
  double (inner/outer) 4, 11, 59, 74, 107, 222, 232
  envelope 21, 231
  monolayer 53
  nuclear 10
  perialgal 21
  peribacteroid 21
  periguest 21
  phagocytotic 90
  phospholipid3
  plasma 3, 29, 57, 220, 231
  plastidal 159
  protective 90
  single 10, 183, 222, 233
  spheres 57
  stacks 57
  symbiosome 21
  system 7
  thylakoid 264
  unit 8, 23
  vacuolic 21
mercurials 195
Mesembryanthemum crystallinum 44
mesophilic 140, 193
mesosomes 8, 57
metabolism 4, 5
  autotrophic 5
  compartmentalizing 12
  fermentative 5
metacyanomes 19
metaphase 244
methane 81
methane-oxidizing bacteria 58
methanol 9, 264
methanolysis 271
methanotrophic organisms 81
methionine 274, 275
  sulfoximine 194
Methylococcus capsulatus 58
methylotrophic bacteria 58
micro-organisms, 5
microalgae 106
microbiont 21
microbodies 62, 136, 183, 215, 233
Microcoleus vaginatus 55
Microcystis 57
microfilaments 57
microsomal enzymes 270
microsomes 272
Microspora amoena var. *thermalis* 115
microtubule organizing centers 62
microtubules 57, 83, 225, 231
mineral medium 120, 136, 168
missing link 141
mitochondria 19, 53, 125, 178, 185, 215, 232, 239
mitochondrial nuclei 101, 241
mitochondriokinesis 245
mitochondrion 7, 69, 101, 133, 147, 154, 159, 177, 233, 263
mitosis 25, 84, 226, 231
mitotic apparatus 69
mitotic cycle 244
mixotrophic 126, 263
Mollusca 88
monadoid 25
Monarthropalpus buxi 63
monobiont 45
monogalactosyl diglyceride 29, 32
monomeres 272
monosaccharides 110, 140, 201
mosquito 63
mosses 242
mRNA 256, 263
mucilage 220
multicellular 217
murein sacculum 29
mutant
  pigment 263
  UV 265
mutants 263
mutations 4, 17
Mycoplasma 8
Mycoplasmomycota 14
myelin 237
Myxophyceae 120

NaCl 203

NAD/NADH 6, 188, 193
NADPH 6, 193, 269
Nannochloris 84
    atomus 64
Nanochloropsis sp. 31
Nanochlorum eucaryotum 64, 69
nanocysts 115
natural
    gas 80
    products 107, 201
    selection 62
needle like crystalline 57
neoxanthin 80
Nephroselmis 64
Neurospora 14
    crassa 42
NH_4 160, 193
Nicotiana 256
Nicotiana tabacum 239, 258
    N. tabacum 239, 258
nitrate 29, 54, 126, 147, 174, 193, 335
nitrate/ammonium 107, 160, 193
    assimilation 33, 193
    reductase 193
nitrite 29, 193
nitrogen 19, 33, 91, 107, 160, 193, 210
    assimilation 33, 193
    deficiency 69, 193
    fixers 58
    metabolism 193
Nitrosolobus multiformis 59
Noctiluca 12
nomenclature 133
non-symbiotic 87
Nostoc 61, 125
    commune 56
    pruniforme 56
nuclear
    division 69, 224
    DNA 36, 74, 107, 138, 243, 257
    encoded genes 42
    envelope 58, 225
    genome 20, 34, 239, 256
    membrane 70, 232
    plaque 63
    pores 58, 224
nuclease 84
nucleic acids 107, 174, 209
nucleocytoplasm 38
nucleoids 107, 147, 241
nucleolus 74, 224, 233
nucleoplasm 59
nucleoplasm-cytoplasm 225
nucleoprotein 84
nucleosome 70, 84, 224
nucleotide 37, 75, 244
    sequence 4
nucleus 20, 34, 53, 58, 69, 101, 118, 133, 154, 177, 185, 215, 232
Nymphon 63

O_2 gas, detoxication, tolerant organism 6
Ochromonas malhamensis 202
oleic acid 32
oligosaccharides 140
Olisthodiscus luteus 60, 258
Oncobyrsa orsiniana 114, 146
Oocystaceae 121
operon 256
Ophrydium versatile 88
organellekinesis 245
organelles 4, 20, 54, 59, 107, 114, 125, 134, 185, 220, 231, 239, 266
    membranous 11
    non-DNA 10, 183
organic acid 5
organism 4, 180, 263
origin
    auto/exogenous 19
    autogenous 23, 151
    endogenous/exogenous 19
    endosymbiogenous 20, 159, 255
    eukaryotic cell 53
    mitochondria 53
    monophyletic 41 225
    nucleus 53
    polyphyletic 41, 255
Oryza 256
Oscillatoria principes 63
Oscillatoria sp. 55, 207
oxaloacetate 6
oxidase 183
β-oxidation 183, 206
oxygen 4, 25, 73, 188

pagocytosis 89
Palmella 109, 114, 135, 157
Palmella orsiniana 145, 158
    P. orsiniana 114
Palmellococcus 117, 135
Palmellococcus thermalis 116, 145
    P. thermalis 117
palmelloid 215
palmitic acid 32, 206
paracrystalline 58
Paramecium 87
Paramecium aurelia 61
Paramecium bursaria 88
paraphyletic 225
Parasinonimi 121

Parasinophyceae 46
parasitic 37
parasitism 17
parasome 38
parasymbiotic 37
pathway
  'C-5' 267
  Embden Meyerhof 5
  Entner-Doudroff 5
pellicula 89
peptidoglucan 27, 29
perialgal vacuole 89
Peridinium sp. 210
perinuclear
  cisterns 20
  dictyosmes 218
peripheral lamellar sheets 57
permeability 4
permease 3
peroxidase 6
peroxisome 10, 62, 101, 183
pH 69, 91, 99, 114, 120, 161, 168, 179, 193, 263, 335
Phaenocora typhlops 88
Phaeophyceae 12, 201
Phaeophyta 188, 201, 207, 237, 241
phagosome 11
Phormidium foveolarum 207
phosphate 32, 193
phosphatidyl
  choline 29, 32
  ethanolamine 9, 33
  glycerol 29, 32
  inositol 32
  serine 32
phosphoenolpyruvate 5
phosphoglycerate 32
phospholipids 205
phosphorus 94
phosphorylases 114
photoassimilation 31
photons 244
photoreceptor 276
photoregulation 263, 276
photorespiration 183
photosynthesis 4, 54, 94, 106, 136, 168, 180, 196, 263, 283
photosynthetic bacteria 269
photosystem I/II 6, 15, 74, 178, 264
Phragmonema sordidum 175
Phragmonemataceae 180
phragmoplast 70
phycobilin 113, 134, 263
phycobiliproteins 31, 136, 178, 222, 256, 263
phycobilisomes 99, 113, 136, 178, 222, 231, 237, 264
phycocyanin 27, 99, 118, 121, 134, 149, 161, 174, 178, 198, 234, 264
phycocyanobilin 264
phycoerythrin 122, 234
phycoerythrobilin 274
phylogenetic 133
  status 133
  trees 60
phylogenic products 7
phylogeny 60, 107, 184, 260
physiological ecology 106, 283
physiology 103, 229, 283
phytochrome 264, 274
phytochromobilin 264
pigment bile 264
pigment, synthesis 266
pigmentation 137, 170, 263
pigments 19, 29, 34, 54, 80, 99, 113, 134, 153, 168, 175, 263
Pisum sativum 44
plankton 85, 170
plasmalemma 20, 77, 94, 178, 196, 216
plasmid 9, 174
plastid 37, 72, 94, 121, 139, 178, 239, 255
  dividing ring 223, 239
  kinesis 239
  nucleus 239
  ribosomes 249
plastidic proteins 256
plastoglobuli 72, 218
plastome 20, 21
plastoquinone 280
plate filament array 57
Platymonas sp. 209
pleated sheet 62
Pleurocapsa 116, 135
Pleurocapsa caldaria 116, 145, 158
  P. caldaria 116
  P. caldarius 118
Pleurochloris cavernosa 180
Pleurococcus 113, 116, 135, 180
Pleurococcus sulphurarius 116, 135, 145, 157, 201
  P. sulphurarius 116, 158, 206
Pluto 108, 135
Pluto caldarius 106, 113, 145
polar fenstrations 74
polar organelles 61, 63
poly-β-hydroxybutyrate granules 57
polyamines 107, 154, 201
polycationic 209
polygenetic trees 41, 240
polyglucans 137, 201

polyglucoside 138
polyhedral bodies 53, 57
polymerase
RNA 264
Chain Reaction 277
polymers, storage 58
polymorphic bodies 57
polyols 201
polypeptides, 243, 264
polyphosphate bodies 53
polyphyletic 215, 262
polysaccharide 122, 140, 201
Polysiphonia 136
polyunsaturated 32, 204
pore complex 64
Porifera 88
poriferasterol 209
Porphyra 137
Porphyra yezoensis 206
Porphyria umbilicalis 271
Porphyridiaceae 108, 113, 122, 145, 151, 283
Porphyridiales 109, 113, 136, 154, 215, 226, 283
Porphyridium 107, 164, 202, 255
Porphyridium aerugineum 31, 60, 136, 196, 206, 216, 256
P. aerugineum 224
Porphyridium cruentum 31, 202
Porphyridium purpureum 188, 206, 216
P. purpureum 211, 216, 223, 224
porphyrin 263
pathway 267
potassium 197, 232
potassium permanganate 222, 232
Prasinophyceae 202, 241
Prasinophyta 237
preplastid 21, 239
Prerhodophyta 10, 106, 134, 154, 201, 255
primeval soup 5
primitive soup 4
Prochloron 60, 70
progenitor 61
Prokaryota 101, 137, 154
prokaryotes 3, 40, 58, 257
prokaryotic 83, 133
prolamellar body 57
promitochondrion 15
prophase 226
Propionibacterium acidi-propionici 7
proplastids 11, 15, 239, 266
Prorodon ovum 88
Prorodon viridis 88
protein 3, 59, 73, 174, 209, 234, 256, 264
import 42
synthesis 14, 83, 196, 266
proteinase 243
protinase 78
protists 63, 75
Protococcus 108, 115, 135
Protococcus botryoides 115, 145
Protococcus botryoides f. *caldarius* 106, 127, 150, 164
P. botryoides f. *caldarius* 115
Protococcus sulphurarius 201
Protococcus vulcanicus 115, 145
protoplasm 234
protoporphyrin 263
Prototheca 110, 140
protozoa 15, 88
Pseudanabaena 59, 256
Pseudocapsa dubia 180
Pseudococcus obscurus 63
Pseudomonas 61
Pseudomonas avenae 61
Psilotricha viridis 88
Pteridinium aquilinum 20
Pteris vittata 245
Pylaiella littoralis 75
Pyraminomonas virginica 249
pyrenoid 28, 89, 115, 136, 149, 177, 216. 231, 249
pyridoxal phosphate 268
Pyrrophyta 210, 237
pyruvate 5
pyruvatekinase 5

R bodies 61
radiobiology 284
rDNA 77, 257
recombination 48
red algae 29, 87, 99, 125, 136, 178, 196, 201, 215, 255, 264
repressor protein 279
reproduction, asexual 113, 137
respiration 4, 196
restiction enzymes 9, 243
reticulocyte 275
Rhaphidophyceae 241
Rhizobiaceae-Fabales symbiosis 38
Rhizobium 21
Rhizosolenia 42
rhodamine-phalloidin 226, 245
Rhodella 223
Rhodella cyanea 217
R. cyanea 223
Rhodella maculata 217
R. maculata 223
Rhodella violacea 217
R. violacea 221

Rhodococcus 108, 135
Rhodococcus caldaria 109, 123, 146
 R. caldarius 123
Rhodophyceae 29, 109, 122, 147, 154, 240, 283
Rhodophycophyta 60, 109, 133, 146, 283
Rhodophyta 10, 101, 154, 215, 237
rhodophytan 34, 133, 202
rhodophytes 107, 123, 134, 136, 161, 164, 167, 183, 222, 263, 283
rhodoplast 20, 107, 113, 136, 255
rhodoplast genome 256
Rhodosorus marinus 216
 R. marinus 223
Rhodospora 136, 223
Rhodosprillum rubrum 7
 R. rubrum 9
Rhopalodia gibba 42
ribosomal proteins 34
ribosomal RNA, see rRNA
ribosomes 14, 61, 75, 83, 220, 249, 260, 266
Richelia intracellularis 42
rifampicin 266
RNA 5, 274
rod-like body 57
Rossettes 58
Rotatoria 88
rRNA 13, 37, 60, 138, 239
 operon 74
 sequence 138
 synthesis 37, 75
RuBCase, see Rubisco
Rubisco 13, 34, 54, 74, 107, 110, 188, 243, 256, 275

Saccharomyces cerevisiae 64, 243
salinity 69, 84
Scytonema julianum 175
segmented rod 58
semiautonomy 20
sequence
 amino acid 18, 154, 264
 DNA 12
 gene 13
 homologous 13, 42
 nucleotide 12, 107
 protein 12
sequences 107, 239, 255, 264
sequencing 114
Serial Endosymbiosis Theory (SET) 21
sexuality 25
sheets 53
Simuliid deptera 63
sitosterol 208
Skujapelta nuda 27
sleeve body 57
sorbitol 32
soy bean 75
species 108, 169
sperm 63
spermatocyte 245
spheroplasts 9
spieces 20
Spinacea oleracea 44, 257
spindle
 apparatus 63, 70
 microtubules 63
 pole 224
spirochaeta 11
Spirolina sp. 44
Spirostomum viridis 88
Spongilla lacustris 88
Spongilla sp. 88
sporopollenin 69
starch 15, 137, 199, 216
starch granules 74
Stentor niger 88
Stentor polymorphus 88
Stentor roeseli 88
sterol 104, 129, 136, 201
stock solution 335
storage
 grains 136
 proteins 34
streptomycin 9, 14
Strobilomonas cyaneus 26
succinyl CoA 268
sucrose 31, 186, 201
sugar 91, 99, 201
Sulfolobus salfataricus 59
sulfur 94
sulfur dioxide 115
sulfurous acid 115
superoxide dismutase 6
symbiogenesis 19, 20
symbiogenous 37
 gene transfer, 42
symbiont 21, 139
 bacterial 10
 cyanobacterial 10
symbiosis 17, 19, 20, 87, 188
symbiotic 107, 164
 association 87, 125, 133, 202
 endo/ecto 87
 view 140
Symphyocladia latiuscula 242
synchronous cultures 245
syncyanosis 48
Synechococcus 43

Synechococcus aeruginosa 63
Synechococcus lividis 114
Synechococcus minutissimus 153
synthetases 32, 114
Syracosphaera carterae 209
system
  electron transfer 8
  electron transport 6
  pigment 6
  transport 196
systematic 133, 284
systematic position 108

taxon 70, 115, 171, 182
taxonomic position 113, 175, 209, 255, 263
taxonomic proposal 108
taxonomy 114, 145, 168, 184, 284
telophase 225
temperature 99, 104, 113, 137, 179
Tetragonidium verrucatum 124
Tetrahymena 138
tetraploid 14
tetrapyrrole 262
Teutophrys trisulca 88
theory
  autogenous 53
  endosymbiotic 5 19, 53
  membrane evolution 5
  replicon 8
  symbiosis 21
  symbiotic 11
thermophilic 108
thermophilic algae 107, 193
thermophily 114
Thermoplasma 110
Thermoplasma acidophilum 59
Thermoproteus tenax 59
Thiothrix sp. 58
thrombocytes 83
thylakoids 11, 20, 53, 72, 113, 125, 136, 178, 222, 231, 237, 267
tobacco 75
transcription 34, 256, 267, 274
transduction 275
transition 59
transitional
  alga/cell 99, 109
  cell/organism 133
  form/status/cell 136, 137, 138
translation 34, 270, 275
transport, active 3
Trebouxia 249
trehalose 201
tri-lamellar bodies 57
tricarboxylic acid cycle 31, 37
trienoic acid 127, 147
trilaminar 77, 78
trimmers 272
tRNA 13, 75, 239, 269
tubular 62
tubular thylakoids 57
tubules 53, 220
tubulin 83, 225, 245
Turbellaria 88
Typhloplana viridata 88

ubiquinone (Q) 5
ultrastructure 53, 102, 113, 136, 215, 231, 267, 283
ultraviolet (UV) light 265
unicellular 25
Unio pictorum 88
ur-eukaryote 25
Uralga 53, 138
urea 195
uricase 183

vacuoles 72, 94, 101, 124, 147, 159, 168, 178, 215, 263
Venus 101
vesicles 57, 183, 215, 234
vesicular crystalloids 57
violaxanthin 80, 264
virus mutant 61
vitamins 91
Vorticella sp. 88

wall bodies 58
wild type 263

xanthin 138
Xanthophyceae 12, 122, 137, 185, 204, 237, 241
xanthophyll 121, 135
xeno-biont/some 21
xylose 91
xylulose 5-phosphate 5

yeast 10, 62, 64, 206, 237, 278

Zea mays 59, 239, 258
zeaxanthin 27, 126, 135, 264
zymolase 78

MIX
Papier aus verantwortungsvollen Quellen
Paper from responsible sources
FSC® C105338

If you have any concerns about our products,
you can contact us on
ProductSafety@springernature.com

In case Publisher is established outside the EU,
the EU authorized representative is:
Springer Nature Customer Service Center GmbH
Europaplatz 3, 69115 Heidelberg, Germany

Printed by Libri Plureos GmbH
in Hamburg, Germany

Ophthalmology Research

Joyce Tombran-Tink, PhD, and Colin J. Barnstable, DPhil
Series Editors

Ocular Angiogenesis: *Diseases, Mechanisms, and Therapeutics,* edited by *Joyce Tombran-Tink,* PhD, *and Colin J. Barnstable,* DPhil, *2006*